**National Center for Construction Education and Research**

# HVAC
*Heating, Ventilating, and Air Conditioning*
# Level Two

Upper Saddle River, New Jersey
Columbus, Ohio

This information is general in nature and intended for training purposes only. Actual performance of activities described in this manual requires compliance with all applicable operating, service, maintenance, and safety procedures under the direction of qualified personnel. References in this manual to patented or proprietary devices do not constitute a recommendation of their use.

---

Copyright © 2001 by the National Center for Construction Education and Research (NCCER), Gainesville, FL 32614-1104 and published by Pearson Education, Inc., Upper Saddle River, New Jersey 07458. All rights reserved. Printed in the United States of America. This publication is protected by Copyright and permission should be obtained from the NCCER prior to any prohibited reproduction, storage in a retrieval system, or transmission in any form or by any means, electronic, mechanical, photocopying, recording, or likewise. For information regarding permission(s), write to: NCCER, Curriculum Revision and Development Department, P.O. Box 141104, Gainesville, FL 32614-1104.

10 9 8 7 6 5 4
ISBN 0-13-060495-X

# Preface

This volume was developed by the National Center for Construction Education and Research (NCCER) in response to the training needs of the construction and maintenance industries. It is one of many in the NCCER's standardized craft training program. The program, covering more than 30 craft areas and including all major construction skills, was developed over a period of years by industry and education specialists. Sixteen of the largest construction and maintenance firms in the United States committed financial and human resources to the teams that wrote the curricula and planned the nationally accredited training process. These materials are industry-proven and consist of competency-based textbooks and instructor's guides.

The NCCER is a non-profit educational entity affiliated with the University of Florida and supported by the following industry and craft associations:

## PARTNERING ASSOCIATIONS

- American Fire Sprinkler Association
- American Society for Training and Development
- American Welding Society
- Associated Builders and Contractors, Inc.
- Associated General Contractors of America
- Association for Career and Technical Education
- Carolinas AGC, Inc.
- Carolinas Electrical Contractors Association
- Citizens Democracy Corps
- Construction Industry Institute
- Construction Users Roundtable
- Design-Build Institute of America
- Merit Contractors Association of Canada
- Metal Building Manufacturers Association
- National Association of Minority Contractors
- National Association of State Supervisors for Trade and Industrial Education
- National Association of Women in Construction
- National Insulation Association
- National Ready Mixed Concrete Association
- National Utility Contractors Association
- National Vocational Technical Honor Society
- North American Crane Bureau
- Painting and Decorating Contractors of America
- Portland Cement Association
- SkillsUSA-VICA
- Steel Erectors Association of America
- Texas Gulf Coast Chapter ABC
- U.S. Army Corps of Engineers
- University of Florida
- Women Construction Owners and Executives, USA

Some of the features of the NCCER's standardized craft training program include the following:

- A proven record of success over many years of use by industry companies.
- National standardization providing portability of learned job skills and educational credits that will be of tremendous value to trainees.
- Recognition: upon successful completion of training with an accredited sponsor, trainees receive an industry-recognized certificate and transcript from the NCCER.
- Compliance with Apprenticeship, Training, Employer, and Labor Services (ATELS) requirements (formerly BAT) for related classroom training (CFR 29:29).
- Well-illustrated, up-to-date, and practical information.

## FEATURES OF THIS BOOK

Capitalizing on a well-received campaign to redesign our textbooks, NCCER is publishing select textbooks in a two-column format. *Heating, Ventilating, and Air Conditioning Level Two* incorporates the design and layout of our full-color books along with special pedagogical features. The features augment the technical material to maintain the trainees' interest and foster a deeper appreciation of the trade.

*Inside Track* provides a head start for those entering the field by presenting tricks of the trade from master HVAC technicians.

*Think About It* uses "What If" questions to help trainees apply theory to real-world experiences and put ideas into action.

We're excited to be able to offer you these improvements and hope they lead to a more rewarding learning experience.

As always, your feedback is welcome! Please let us know how we are doing by visiting NCCER at www.nccer.org or e-mail us at info@nccer.org.

# Acknowledgments

This curriculum was revised as a result of the farsightedness and leadership of the following sponsors:

Black-Haak Heating, Inc.
Central Ohio ABC Chapter
Comfort Systems USA
Encompass Mechanical Services
   Southeast, Inc.
Entek Corporation

Gulfside Mechanical, Inc./
   Comfort Syestems USA
Lee College
Paul Van Zeeland Heating, Inc.
University of Florida Rinker
   School of Building Construction

This curriculum would not exist were it not for the dedication and unselfish energy of those volunteers who served on the Authoring Team. A sincere thanks is extended to:

Robert Haak
Steve McClain
Joe Moravek
Paul Oppenheim
Ricky Sonnier

Thomas J. Swafford
Matthew Todd
Dan Wolfe
Russell Zech

# Contents

**03201-01** Air Distribution Systems . . . . . . . . . . . . . . . . . . . . 1.i

**03202-01** Chimneys, Vents, and Flues . . . . . . . . . . . . . . . . . . 2.i

**03203-01** Maintenance Skills for the Service Technician . . . . 3.i

**03204-01** Alternating Current . . . . . . . . . . . . . . . . . . . . . . . . . 4.i

**03205-01** Basic Electronics . . . . . . . . . . . . . . . . . . . . . . . . . . 5.i

**03206-01** Electric Heating . . . . . . . . . . . . . . . . . . . . . . . . . . . 6.i

**03207-01** Introduction to Control Circuit Troubleshooting . . . . 7.i

**03208-01** Accessories and Optional Equipment . . . . . . . . . . 8.i

**03209-01** Metering Devices . . . . . . . . . . . . . . . . . . . . . . . . . . 9.i

**03210-01** Compressors . . . . . . . . . . . . . . . . . . . . . . . . . . . . 10.i

**03211-01** Heat Pumps . . . . . . . . . . . . . . . . . . . . . . . . . . . . . 11.i

**03212-01** Leak Detection, Evacuation, Recovery, and Charging . . . . . . . . . . . . . . . . . . . . . . . . . . . . 12.i

Module 03201-01

*Air Distribution Systems*

# COURSE MAP

This course map shows all of the modules in the second level of the HVAC curriculum. The suggested training order begins at the bottom and proceeds up. Skill levels increase as you advance on the course map. The local Training Program Sponsor may adjust the training order.

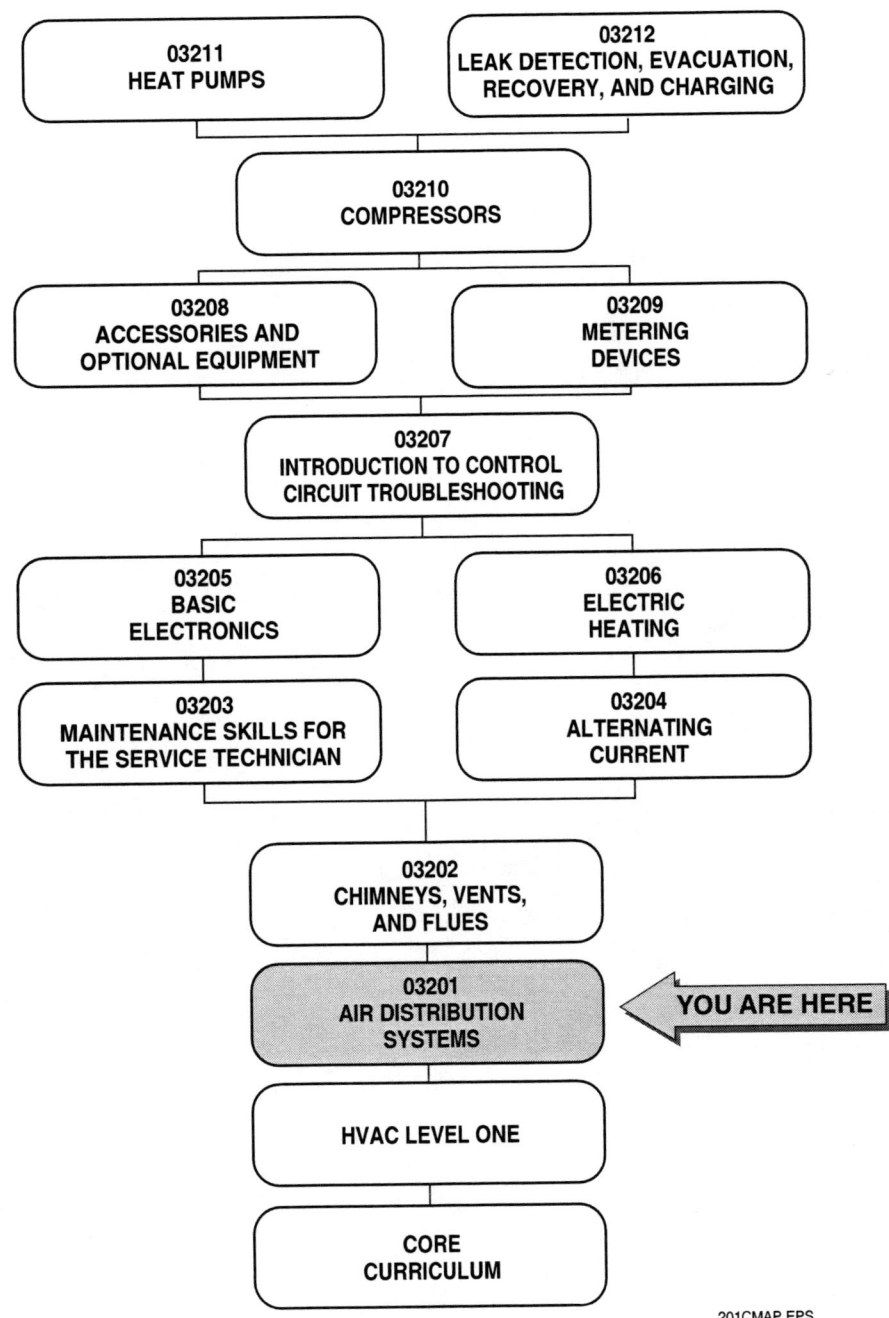

Copyright © 2001 National Center for Construction Education and Research, Gainesville, FL 32614-1104. All rights reserved. No part of this work may be reproduced in any form or by any means, including photocopying, without written permission of the publisher.

# MODULE 03201 CONTENTS

- 1.0.0 INTRODUCTION .................................................. 1.1
- 2.0.0 AIR DISTRIBUTION SYSTEMS ................................. 1.1
  - 2.1.0 Airflow and Pressures in the Distribution System Ductwork .... 1.2
  - 2.2.0 Air Distribution in a Typical Residential System ............ 1.5
- 3.0.0 FANS AND BLOWERS ........................................... 1.6
  - 3.1.0 Belt-Drive and Direct-Drive Blowers and Fans ............... 1.7
  - 3.2.0 Centrifugal Blowers ........................................ 1.7
    - *3.2.1 Forward-Curved Centrifugal Blowers* .................... 1.8
    - *3.2.2 Backward-Inclined Centrifugal Blowers* ................. 1.8
    - *3.2.3 Radial Blowers* ........................................ 1.8
  - 3.3.0 Fans ....................................................... 1.9
    - *3.3.1 Propeller Fans* ......................................... 1.9
    - *3.3.2 Duct Fans* ............................................. 1.10
  - 3.4.0 Fan Laws .................................................. 1.10
  - 3.5.0 Fan Curve Charts .......................................... 1.11
  - 3.6.0 System Air Handler Blowers ................................ 1.12
- 4.0.0 AIR DISTRIBUTION DUCT SYSTEMS ............................. 1.12
  - 4.1.0 Duct Systems Used in Cold Climates ........................ 1.13
    - *4.1.1 Loop Perimeter Duct System* ............................ 1.14
    - *4.1.2 Radial Perimeter Duct System* .......................... 1.14
    - *4.1.3 Extended Plenum Duct System* ........................... 1.14
    - *4.1.4 Reducing Extended Plenum Duct System* .................. 1.15
  - 4.2.0 Duct Systems Used in Warm Climates ........................ 1.15
    - *4.2.1 Overhead Trunk and Overhead Radial Duct Systems* ....... 1.16
    - *4.2.2 Attic Extended Plenum and Attic Radial Duct Systems* ... 1.17
- 5.0.0 DUCT SYSTEM COMPONENTS ..................................... 1.18
  - 5.1.0 Main Trunk and Branch Ducts ............................... 1.19
    - *5.1.1 Galvanized Steel Trunk and Branch Ducts* ............... 1.19
    - *5.1.2 Fiberglass Duct* ....................................... 1.21
    - *5.1.3 Flexible Duct* ......................................... 1.22
  - 5.2.0 Fittings and Transitions .................................. 1.23
  - 5.3.0 Air Diffusers, Registers, and Grilles ..................... 1.24
  - 5.4.0 Dampers ................................................... 1.25
  - 5.5.0 Fire and Smoke Dampers .................................... 1.25
  - 5.6.0 Insulation and Vapor Barriers ............................. 1.27
- 6.0.0 TEMPERATURE AND HUMIDITY MEASUREMENT INSTRUMENTS .......... 1.28
  - 6.1.0 Electronic Thermometers ................................... 1.28
  - 6.2.0 Psychrometers ............................................. 1.29
  - 6.3.0 Hygrometers ............................................... 1.30

## MODULE 03201 CONTENTS (Continued)

**7.0.0 AIR DISTRIBUTION SYSTEM MEASUREMENT INSTRUMENTS** ...1.31
    7.1.0    Manometers .................................1.31
    7.2.0    Differential Pressure Gauge .....................1.31
    7.3.0    Pitot Tubes and Static Pressure Tips ..............1.32

**8.0.0 AIR VELOCITY MEASUREMENT INSTRUMENTS** ...........1.33
**SUMMARY** ..................................................1.34
**REVIEW QUESTIONS** ........................................1.35
**GLOSSARY** .................................................1.37
**APPENDIX** .................................................1.38
**ANSWERS TO REVIEW QUESTIONS** ............................1.39
**REFERENCES & ACKNOWLEDGMENTS** ..........................1.40

## Figures

| | | |
|---|---|---|
| Figure 1 | Basic forced-air distribution system | 1.2 |
| Figure 2 | Pressures in an air distribution system | 1.3 |
| Figure 3 | Causes of friction loss in an air distribution system | 1.3 |
| Figure 4 | Static, total, and velocity pressures | 1.4 |
| Figure 5 | Comparison of atmospheric pressure to inches of water | 1.4 |
| Figure 6 | Typical residential air distribution system | 1.6 |
| Figure 7 | Belt-drive and direct-drive blowers | 1.7 |
| Figure 8 | Centrifugal blowers | 1.7 |
| Figure 9 | Forward-curved centrifugal blower wheel | 1.8 |
| Figure 10 | Backward-inclined centrifugal blower wheels | 1.8 |
| Figure 11 | Radial centrifugal blower wheels | 1.9 |
| Figure 12 | Propeller-type fan | 1.9 |
| Figure 13 | Tube-axial and vane-axial duct fans | 1.10 |
| Figure 14 | Typical fan curve chart | 1.11 |
| Figure 15 | Typical residential air distribution duct system | 1.12 |
| Figure 16 | Room air distribution patterns for a perimeter duct system | 1.13 |
| Figure 17 | Loop perimeter duct system | 1.14 |
| Figure 18 | Radial perimeter duct system | 1.14 |
| Figure 19 | Extended plenum duct system | 1.15 |
| Figure 20 | Reducing extended plenum duct system | 1.15 |

## Figures (Continued)

| | | |
|---|---|---|
| Figure 21 | Room air distribution patterns for high sidewall outlets | 1.16 |
| Figure 22 | Room air distribution patterns for ceiling outlets | 1.16 |
| Figure 23 | Overhead trunk duct system | 1.17 |
| Figure 24 | Overhead radial duct system | 1.17 |
| Figure 25 | Attic extended plenum duct system | 1.17 |
| Figure 26 | Attic radial duct system | 1.18 |
| Figure 27 | Metal ducts | 1.19 |
| Figure 28 | Typical metal duct gauge thickness and aspect ratios | 1.20 |
| Figure 29 | Typical square and rectangular sheet metal duct fasteners | 1.20 |
| Figure 30 | Ductwork vibration and noise control devices | 1.21 |
| Figure 31 | Fiberglass ductboard | 1.21 |
| Figure 32 | Flexible duct | 1.22 |
| Figure 33 | Example of equivalent length | 1.23 |
| Figure 34 | Air registers and diffusers | 1.25 |
| Figure 35 | Typical dampers | 1.26 |
| Figure 36 | Combination fire and smoke damper | 1.26 |
| Figure 37 | Typical duct installation in a ventilated crawl space | 1.27 |
| Figure 38 | Example of R-value calculation using ASHRAE Standard 90-80 | 1.28 |
| Figure 39 | Electronic thermometer | 1.28 |
| Figure 40 | Sling psychrometers | 1.30 |
| Figure 41 | Electronic hygrometers | 1.30 |
| Figure 42 | Manometers | 1.31 |
| Figure 43 | Portable differential pressure gauge | 1.32 |
| Figure 44 | Pitot tube and static pressure tips | 1.32 |
| Figure 45 | Velometers | 1.33 |

# MODULE 03201

# Air Distribution Systems

## Objectives

When you have completed this module, you will be able to do the following:

1. Describe the airflow and pressures in a basic forced-air distribution system.
2. Explain the differences between propeller and centrifugal fans and blowers.
3. Identify the various types of duct systems and explain why and where each type is used.
4. Demonstrate or explain the installation of metal, fiberboard, and flexible duct.
5. Demonstrate or explain the installation of fittings and transitions used in duct systems.
6. Demonstrate or explain the use and installation of diffusers, registers, and grilles used in duct systems.
7. Demonstrate or explain the use and installation of dampers used in duct systems.
8. Demonstrate or explain the use and installation of insulation and vapor barriers used in duct systems.
9. Identify the instruments used to make measurements in air systems and explain the use of each instrument.
10. Make basic temperature, air pressure, and velocity measurements in an air distribution system.

## Prerequisites

Before you begin this module, it is recommended that you successfully complete the following modules: Core Curriculum; HVAC Level One.

## Required Trainee Materials

1. Pencil and Paper
2. Appropriate Personal Protective Equipment

## 1.0.0 ◆ INTRODUCTION

Efficient and proper operation of air conditioning systems requires more than properly operating closed refrigeration and electrical systems. Of equal importance is the delivery of the correct quantity of conditioned air to the occupied space. This requires the use of a properly installed and balanced air distribution system. This module describes the components that form a forced-air air distribution system and the basic methods used to measure air quantity and flow within that system. You will learn how to balance air systems during your Level Three training. In Level Four, you will learn the methods used in air system design.

## 2.0.0 ◆ AIR DISTRIBUTION SYSTEMS

A heating or air conditioning system will perform no better than its air distribution system. All adequate air distribution systems must:

- Supply the right quantity of air to each conditioned space.
- Supply the air in each space so that air motion is adequate but not drafty.
- Condition the air to maintain the proper comfort zones for people, or to maintain the proper conditions needed for a commercial or manufacturing process.

- Provide for the return of air from all conditioned areas to the **air handler**.
- Operate efficiently without excessive power consumption or noise.
- Operate with minimum maintenance.

Most air distribution systems are forced-air systems. The major components that make up a forced-air system are the air handler, air supply system, return air system, and the grilles and registers that allow the circulated air to enter the conditioned space and then return to the conditioning equipment. The air handler is the device that moves the air in a forced-air system. In a split system, it could be a furnace or air handler that contains the blower fan, cooling coil, metering device, air filter, and related housing. The conditioning equipment can be a cooling coil or furnace. *Figure 1* shows the basic components of a forced-air system. The operation of forced-air distribution systems is basically the same for all systems. What changes from system to system is the type of conditioning equipment, size and style of the components, and the installed locations of the components. Additional devices, called *accessories*, may be used in some systems to gain the desired temperature, humidity, air movement, or air cleanliness. This section describes the basics of a forced-air distribution system. You will study the accessories used with forced-air and other systems later on in the HVAC Level Two Module, *Accessories and Optional Equipment*.

### 2.1.0 Airflow and Pressures in the Distribution System Ductwork

Air can be moved by creating a pressure above **atmospheric pressure** (positive pressure) or a pressure below atmospheric pressure (negative pressure). All blowers (or fans) produce both conditions. The air inlet to a blower is below atmospheric pressure, while the exhaust of a blower is above atmospheric pressure.

When a blower is inserted into a duct system, the airflow through the system, except within the blower itself, is the natural flow from a higher to lower pressure area (*Figure 2*). Normal atmospheric pressure exists in the conditioned space. At the return air grille, the air pressure is slightly lower than atmospheric pressure; therefore, air moves into the duct. The pressure decreases to its lowest point at the blower input. Through the action of the blower, the air pressure is increased to its highest level at the blower discharge. From there, the air resumes its normal natural flow from the higher pressure area at the blower discharge to the lower pressure area of the conditioned space.

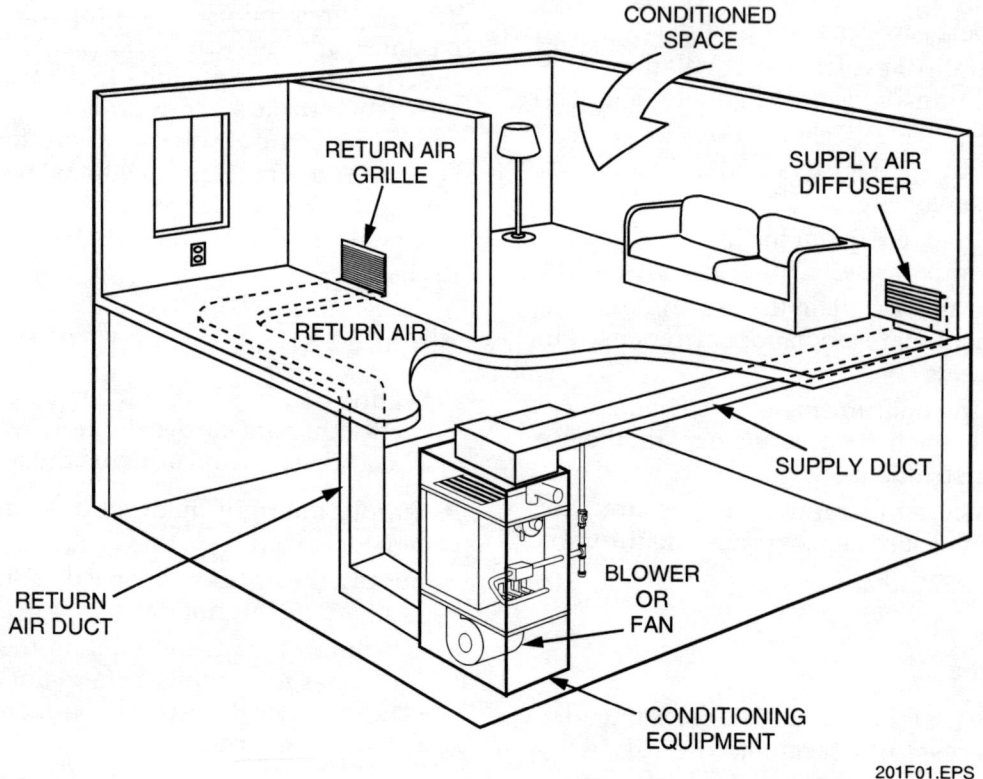

*Figure 1* ◆ Basic forced-air distribution system.

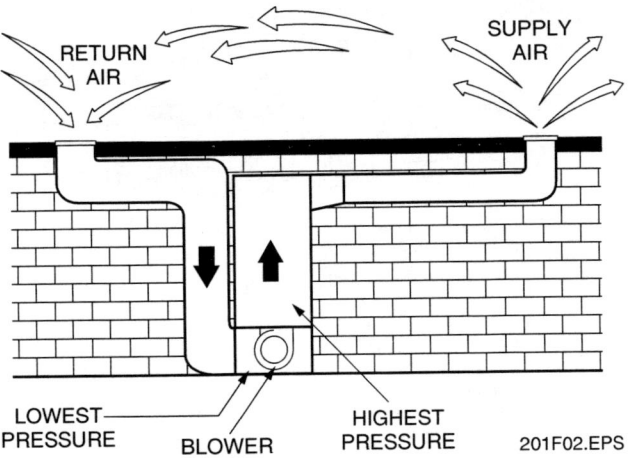

*Figure 2* ◆ Pressures in an air distribution system.

The amount of pressure difference needed to move air through a duct system depends on the **velocity**, the **volume**, the cross-section area of the duct, and the length of the duct. Velocity is how fast the air is moving and is usually measured in feet per minute (fpm). Volume is a measure of the amount of air in cubic feet that flows past a point in one minute. Air velocity and volume can be measured in air distribution systems using anemometers and velometers. You will learn more about these instruments later in this module. Volume in **cubic feet per minute (cfm)** can be calculated by multiplying the velocity of air (in fpm), times the area it is moving through (in square feet) as follows:

cfm = area × velocity

Other variations of the formula are:

Velocity = cfm ÷ area
Area = cfm ÷ velocity

The inside surface of the duct offers resistance to the flow of air. The velocity of airflow within a duct is not uniform. It varies from zero at the duct walls to a maximum in the center of the duct. This variation can be caused by joints or elbows in the duct, screws or fasteners protruding into the airstream, and various duct lining materials.

In addition to the friction loss in the ductwork, the blower must provide the additional pressure needed to overcome the friction or pressure loss caused by duct fittings (*Figure 3*). Fittings such as elbows, takeoffs, or boots that change the direction of airflow, or change its velocity, also add friction and decrease the quantity of air a duct can carry. The quantity and size of the registers and grilles also affects the airflow in a system. Friction losses have a great impact on the size of the blower and ductwork used in a system. These losses are referred to as *static pressure drop*, *static pressure loss*, or *friction loss*.

The three pressures that exist in a duct system are **static pressure (s.p.)**, **velocity pressure**, and **total pressure**.

Static pressure is the pressure which is exerted uniformly in all directions within a duct system. Static pressure is shown as the small arrows in *Figure 4*. In a supply air duct, it is the bursting or exploding pressure that acts on all surfaces of the duct. As shown, static pressure can be applied to a pressure gauge (*manometer*) for measurement via the static pressure openings of a pitot tube or static pressure tip connected to the manometer.

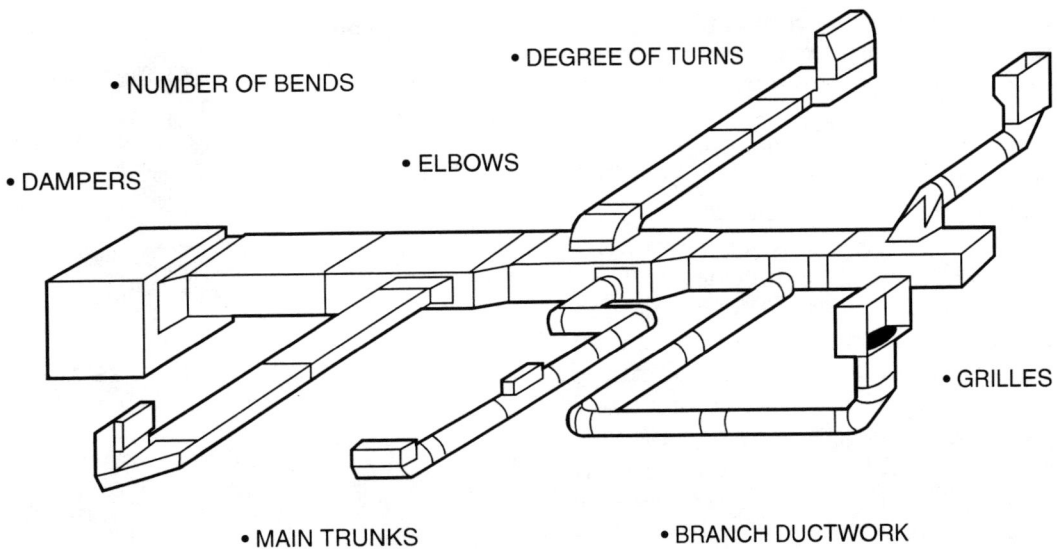

*Figure 3* ◆ Causes of friction loss in an air distribution system.

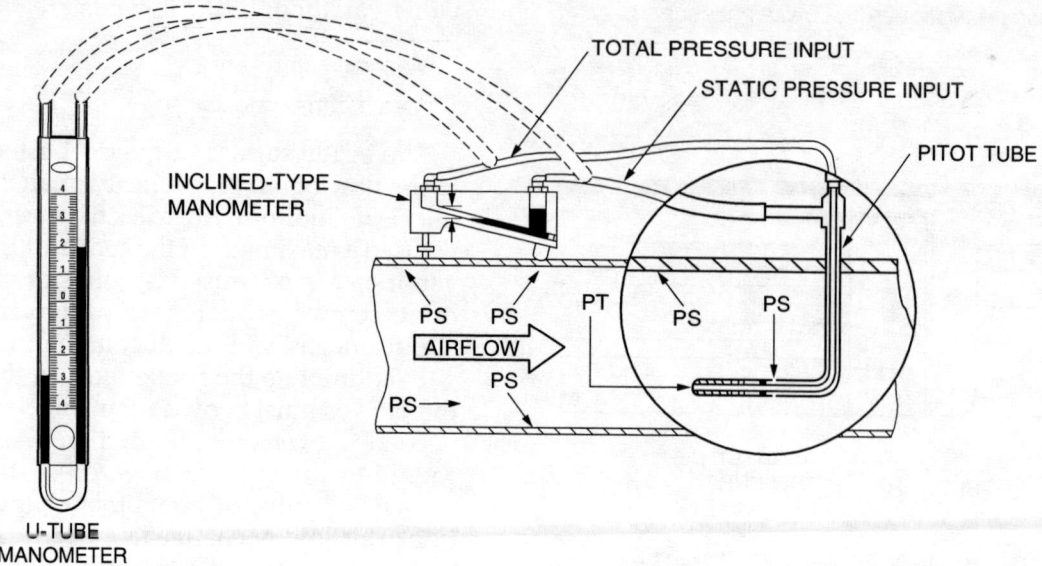

PITOT TUBE SENSES TOTAL AND STATIC PRESSURES. MANOMETER MEASURES VELOCITY PRESSURE (DIFFERENCE BETWEEN TOTAL AND STATIC PRESSURES).

PT = TOTAL PRESSURE
PS = STATIC PRESSURE

- STATIC PRESSURE = TOTAL PRESSURE MINUS VELOCITY PRESSURE (PS = PT − PV)
- TOTAL PRESSURE = STATIC PRESSURE PLUS VELOCITY PRESSURE (PT = PS + PV)
- VELOCITY PRESSURE = TOTAL PRESSURE MINUS STATIC PRESSURE (PV = PT − PS)

*Figure 4* ◆ Static, total, and velocity pressures.

You will learn more about the manometer, pitot tube, and static pressure tips later in this module.

Velocity pressure is the pressure in a duct system caused by the movement of the air. It acts in the direction of airflow only. It is the difference between the total pressure and the static pressure. *Figure 4* shows a manometer connected to give a reading of velocity pressure. As shown, the pitot tube must have its opening pointing in the direction of airflow.

Total pressure is the sum of the static and the velocity pressures in a duct system (*Figure 4*). It is the pressure produced by the fan or blower.

Due to the low pressures inside a duct system, a manometer is used to measure duct static, velocity, and total pressures in inches of water column (in. w.c.). Inches of water column is the height, in inches, to which the pressure will lift a column of water. The atmosphere exerts a pressure of 14.7 psi (14.696 psi) at sea level with 70°F dry air. This atmospheric pressure level of 14.7 psi will support a column of water 33.9', or 406.9" high (*Figure 5*). Therefore, for every one pound per square inch of pressure, a column of water will rise to a height of 27.68" (406.9 ÷ 14.696), or about 2.3'.

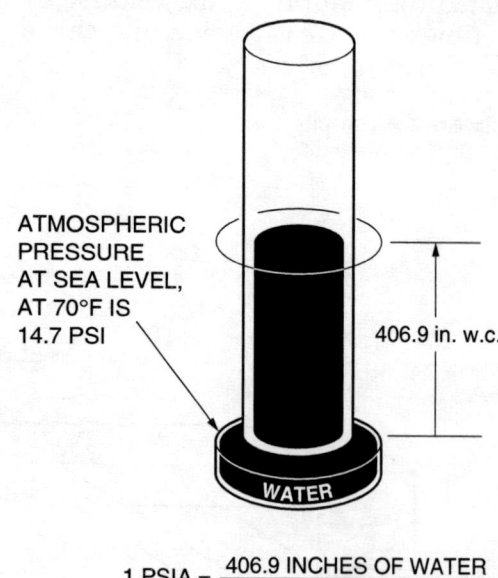

*Figure 5* ◆ Comparison of atmospheric pressure to inches of water.

## Absolute Pressure vs. Gauge Pressure

Every square inch of the Earth's surface at sea level at 70°F has 14.7 pounds of air pressure pushing down on it. We have learned that the atmospheric pressure of 14.7 pounds per square inch (psi) at sea level will support a column of water 406.9" high when measured with a manometer. When measured with a mercury-tube barometer like the ones used by meteorologists, the same atmospheric pressure of 14.7 psi will cause the mercury in the tube to rise to a height of 29.92". The values of 14.7 psi and 29.92 inches of mercury (in. Hg) at sea level, at 70°F, are standards that are used frequently in HVAC work.

The absolute pressure scale is based on the barometer measurements just described. On this scale, pressures are expressed in pounds per square inch (psi) or pounds per square inch absolute (psia) starting from 0 psi. Another scale, called the **gauge pressure** scale, is frequently used to define air pressure levels. Gauge pressure scales use atmospheric pressures as their starting point. Positive pressures above zero (14.7 psi) are expressed in pounds per square inch gauge (psig). Negative pressures (those below 0 psig) are expressed in inches of mercury vacuum (in. Hg vacuum). Gauge pressures can easily be converted to **absolute pressures** by adding 14.7 to the gauge pressure value. Absolute pressures can be converted to gauge pressures by subtracting 14.7. Conversion between absolute and gauge pressure scales is often necessary when making calculations concerned with air pressure relationships.

---

In special low-pressure applications such as blower door testing, pascals (Pa) are used as a unit of measure for air pressure. The normal pressure of the atmosphere at sea level is 101.3 kPa, usually rounded to 100 kPa. To convert psi to kPa, multiply psi by 6.895.

### 2.2.0 Air Distribution in a Typical Residential System

*Figure 6* shows an air distribution system for a fictional single-story house. We will discuss the airflow in this system in detail to demonstrate the concepts and pressure relationships you have learned so far, and some new ones. Generally, more airflow is needed for cooling than for heating. Therefore, the air handler (blower) must be able to supply the volume of air needed for the cooling mode. In this example, assume that 3 tons of cooling are required. As a rule of thumb in HVAC, cooling requires about 400 cfm ±50 of air per ton of cooling. Therefore, the blower in this system must be capable of supplying air at 1,200 cfm (3 tons × 400 cfm) or more. Assume that the blower develops a static pressure of 1.0 in. w.c., and that the supply and return ducts have external static pressures of 0.5 in. w.c. and –0.5 in. w.c., respectively. As shown, the system has 11 air supply outlets, each requiring 100 cfm, and two smaller outlets, each requiring 50 cfm. The return air is taken into the system through two centrally located grilles.

While studying this system, consider the entire house as part of the system. The supply air leaves the supply registers and sweeps the walls of the house. Then, it travels through the conditioned spaces within the house as it flows toward the return air grilles. The air is at room temperature at this time. The duct system begins at the two return-air grilles. Relative to the atmospheric pressure of the rooms, there is a slightly negative pressure at the grilles. As shown, the pressure on the blower side of the return-air grille filters is about –0.03 in. w.c., which is lower than the pressure in the rooms. This results in the higher room pressures pushing the air through the return-air filters and into the return duct. As the air flows down the return duct towards the blower, the pressure continues to decrease as a result of friction losses in the duct. At the inlet to the blower, the air pressure is at its lowest point in the system. For our example, it is at –0.5 in. w.c., which is well below the room pressure (0.03 in. w.c.). The return air is forced through the blower, and at the blower output, is increased to its highest level in the duct system. For our example, it is 0.50 in. w.c., which is well above the room pressure. The difference in static pressure between the input and output of the blower is 1.0 in. w.c.

The air at the blower output is pushed through the furnace heat exchanger and the cooling coil, where it encounters a pressure drop of 0.15 in. w.c. At the input to the supply duct, the air enters at a pressure of 0.35 in. w.c (.50 in. w.c. – 15 in. w.c.). After the air enters the supply duct, it undergoes a slight pressure drop at the tee where the duct is split into two reducing trunks, one to feed each end of the house. It then encounters a discharge

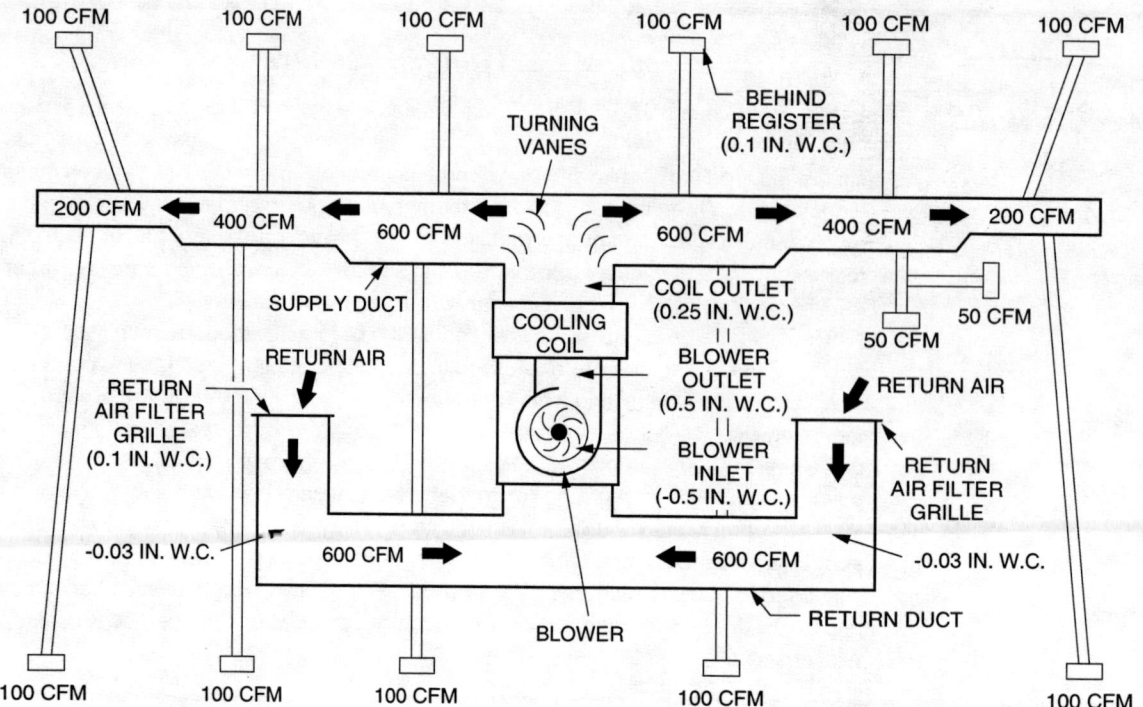

SYSTEM CAPACITY 3 TONS
CFM REQUIREMENT 400 CFM PER TON = 400 x 3 = 1,200 CFM
BLOWER STATIC PRESSURE (1.0 IN. W.C.)
COIL STATIC PRESSURE (0.15 IN. W.C.)
SUPPLY DUCT STATIC PRESSURE (0.35 IN. W.C.)
RETURN DUCT STATIC PRESSURE (-0.5 IN. W.C.)
FILTRATION LOSS (0.2 IN. W.C.)

*Figure 6* ♦ Typical residential air distribution system.

loss into the room through the diffuser of 0.10 in. w.c. for a total supply side blower static pressure of 0.50 in. w.c.

Each first section of the reducing trunk must handle 600 cfm of air. Two branch duct outlets, each with an air capacity of 100 cfm, are supplied from the first trunk section on each side. This reduces the quantity of air supplied to the next sections of the trunk to 400 cfm for each side. These sections each supply 200 cfm of air to the conditioned space. This reduces the quantity of air supplied to the last sections of the trunk to 200 cfm for each side, allowing another reduction in the trunk size for each of these sections. The last section of trunk on each side of the system supplies 200 cfm of air to the remaining outlets on each side. In this example, smaller reducing trunks were used to save the cost of materials. Also, reducing the duct size as air was distributed off the trunk keeps the pressure in the duct system at the desired level all along the duct.

Normally, dampers would be installed in each branch to balance the quantity of air supplied to each room. The system in our example will furnish 100 cfm to each outlet, but if a room does not need that much air, the dampers can be adjusted to reduce the quantity.

## 3.0.0 ♦ FANS AND BLOWERS

The blower or fan provides the pressure difference necessary to force the air into the supply duct system, through the grilles and registers, and into the conditioned space. It must overcome the pressure loss involved in the return of the air as it flows into the return air grilles and through the return ductwork system back to the air handler. In addition, the blower must also overcome the resistance of any other components in the system through which the air passes.

The terms *fan* and *blower* are often used interchangeably, but they usually describe the application. For example, the word *blower* describes applications where the device must work against the resistance of a duct system, like a forced-air system. On the other hand, the word *fan* describes applications where high quantities of air are needed with little resistance to airflow, such as a ventilation fan.

### 3.1.0 Belt-Drive and Direct-Drive Blowers and Fans

Two types of blowers are commonly used in air distribution systems: belt-drive and direct-drive (*Figure 7*). In belt-drive blowers, the blower motor is connected to the blower by a belt and pulley. The blower speed is adjusted mechanically by a change in the pulleys. Belt-drive blowers are commonly used in commercial HVAC products. In direct-drive blowers, the blower wheel is mounted directly on the motor shaft. The blower speed is adjusted electrically by changing the blower motor terminals, or changing the settings of motor speed selection switches on a related motor control board. Most residential equipment uses multi-speed motors with direct-drive. This enables the speed of the motor to be adjusted to match the requirements of the individual heating or cooling air distribution system. It also allows the speed to be changed between heating and cooling seasons.

### 3.2.0 Centrifugal Blowers

Centrifugal blowers (*Figure 8*) are used with forced-air systems because they are designed to work against the resistance of the duct system. They can be used in very large systems that are considered high-pressure systems. High-pressure systems are those that have pressures of 3.5 in. w.c. or greater. In centrifugal blowers, airflow is at right angles (perpendicular) to the shaft on which the wheel is mounted. The wheel is mounted in a scroll-type housing, which is necessary to develop the needed pressures. Centrifugal blowers are identified by the wheel blade position with respect to the direction of rotation. The types of centrifugal blowers include:

- Forward-curved
- Backward-inclined
- Radial

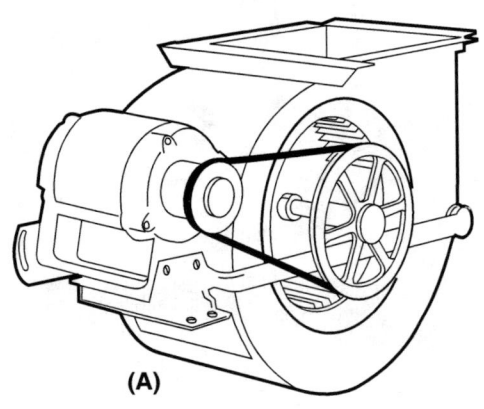

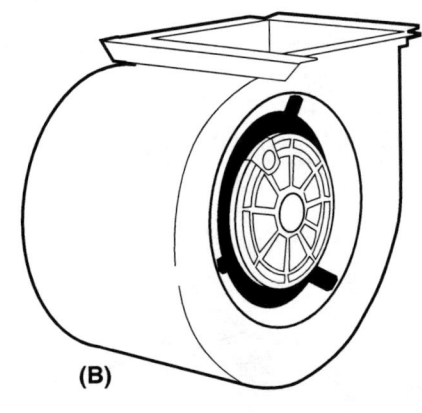

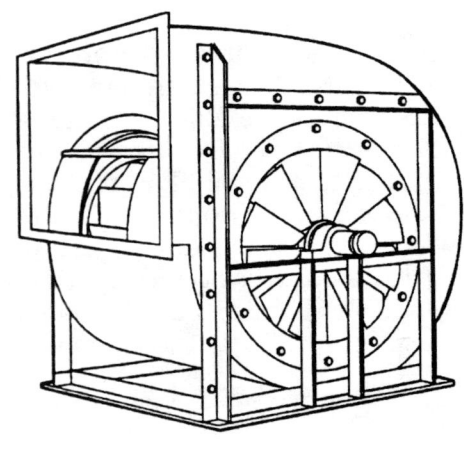

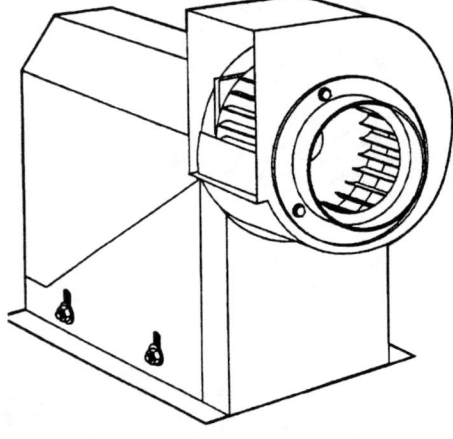

*Figure 7* ◆ (A) Belt-drive blower. (B) Direct-drive blower.

*Figure 8* ◆ Centrifugal blowers.

### 3.2.1 Forward-Curved Centrifugal Blowers

Forward-curved centrifugal blowers are normally used in residential and light commercial heating and air conditioning systems. They are also used in light-duty exhaust systems where maximum air delivery and low noise are required. Typically, these blowers are capable of producing pressures up to 3.0 in. w.c. As shown in *Figure 9*, the tips of the blades in a forward-curved blower are inclined in the direction of rotation.

### 3.2.2 Backward-Inclined Centrifugal Blowers

The blades of the backward-inclined centrifugal blower are inclined away from the direction of rotation. Typically, these blowers are used in commercial and industrial heating and cooling systems which require heavy-duty blower construction and stable air delivery. They are also used extensively as ventilators. Backward-inclined blowers operate at higher efficiencies than forward-curved blowers. They also operate at higher speeds, and therefore tend to be noisier. Typically, these blowers are capable of producing pressures up to 6 in. w.c.

Smaller wheels are usually supplied with flat blades, while larger wheels are supplied with airfoil blades to improve efficiency. Blowers using the airfoil blades generally run more quietly than other types of centrifugal blowers. Also, they do not pulsate within their operating range because air flows through the wheel with less turbulence. *Figure 10* shows a backward-inclined (back-curved) fan wheel and an airfoil fan wheel.

### 3.2.3 Radial Blowers

Radial blower wheels have straight blades which are, to a large extent, self cleaning. This makes radial blowers more suitable for use in air systems that have large amounts of particles or grease in the air. They can also be used in other applications such as pneumatic conveying systems involved with material handling.

The wheels of radial blowers are simple in construction with narrow blades. They can withstand the high speeds needed to operate at higher static pressures (up to 12 in. w.c.). For static pressures above 5.6 in. w.c., the fan rotation speed requires that the wheels and blades be welded. *Figure 11* shows some examples of radial blower wheels.

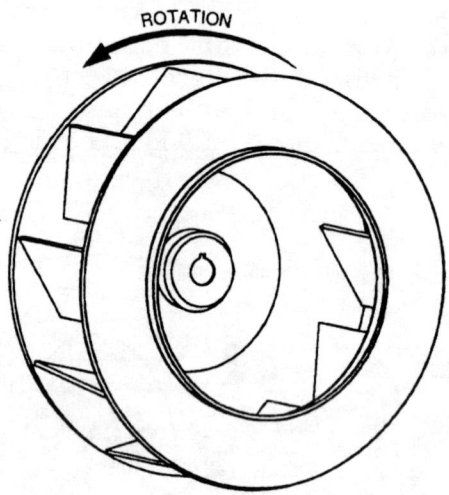

BACKWARD-CURVED FAN WHEEL

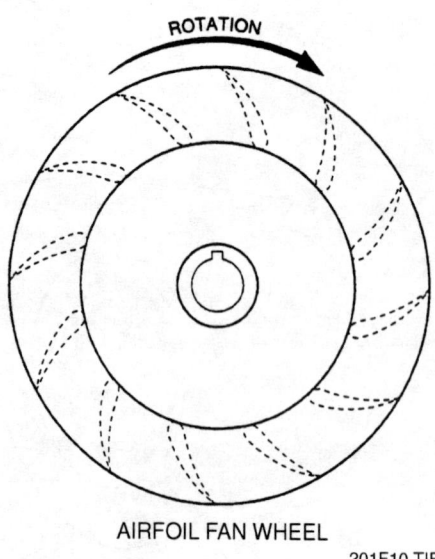

AIRFOIL FAN WHEEL

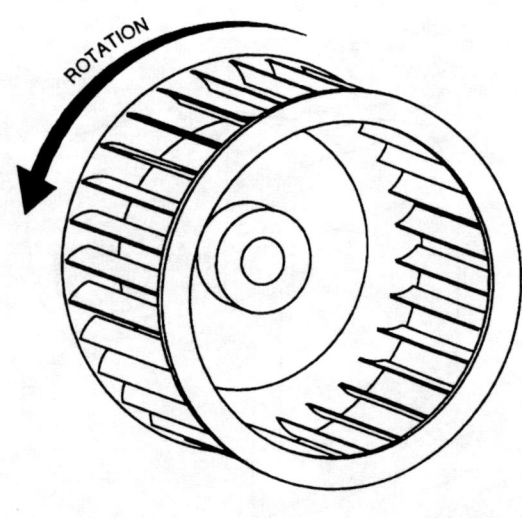

*Figure 9* ◆ Forward-curved centrifugal blower wheel.

*Figure 10* ◆ Backward-inclined centrifugal blower wheels.

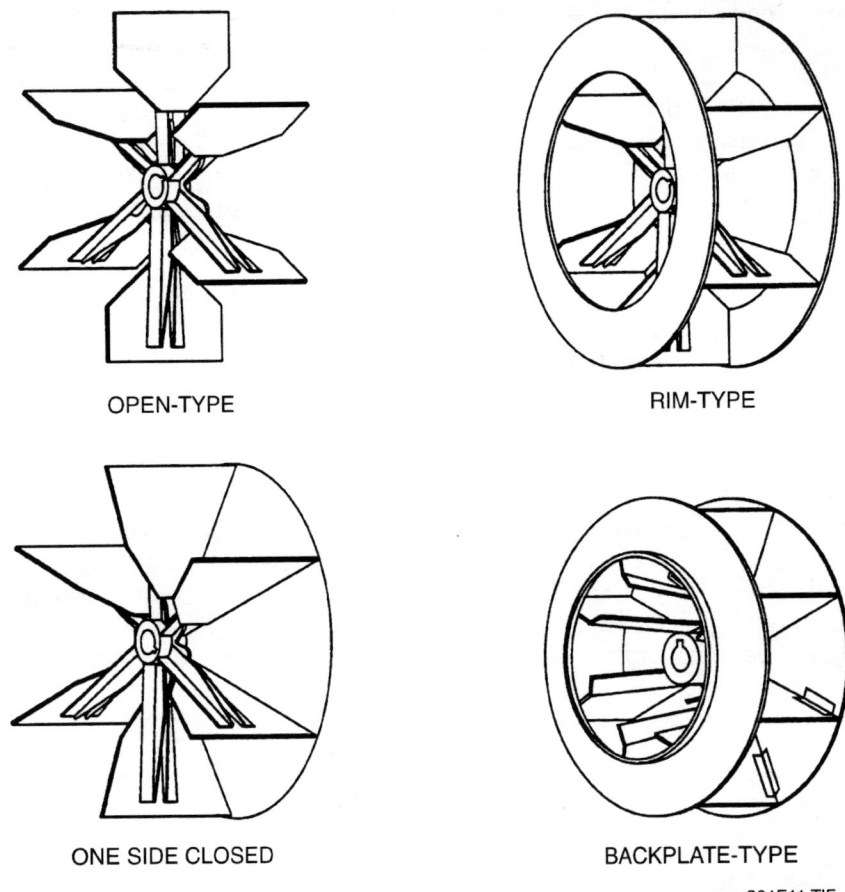

*Figure 11* ◆ Radial centrifugal blower wheels.

## 3.3.0 Fans

Fans are typically used in applications where high quantities of air are needed with little resistance to airflow, such as with air ventilation and exhaust fans. Although there are many variations in fans, most are either propeller fans (axial) or duct fans (tube-axial).

### 3.3.1 Propeller Fans

Propeller or axial fans (*Figure 12*) produce an airflow which is parallel or axial to the shaft on which the propeller is mounted. These fans have good efficiency and near **free air delivery** and are commonly used as condenser fans in HVAC applications. Free air delivery is the condition that exists when there are no effective restrictions to airflow (no static pressure) at the inlet or outlet of an air-moving device (fan). Propeller fans are usually mounted in a **venturi** to cause the air to flow in a straight line from one side of the fan to the other. A venturi is a ring or panel surrounding the blades on a propeller fan. To achieve the best per-

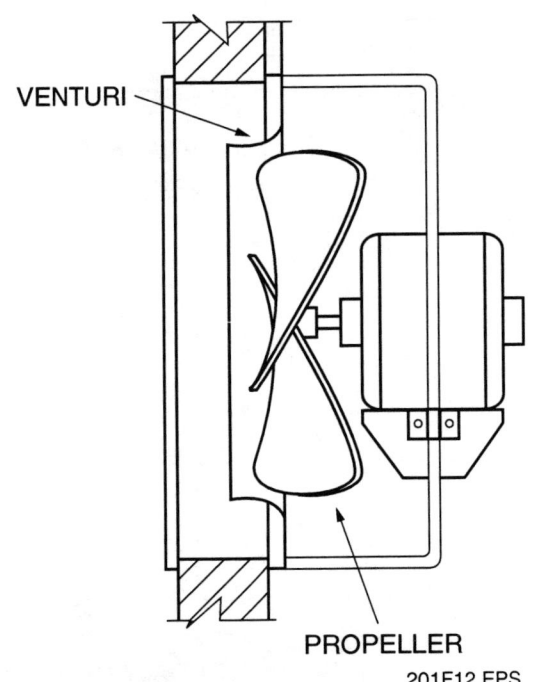

*Figure 12* ◆ Propeller-type fan.

formance from a propeller-type fan, the blade must be properly set in the venturi opening. If the setting is other than that specified by the manufacturer, performance will drop off and the fan might be noisy. Propeller fans make more noise than centrifugal blowers or fans, so they are normally used where noise is not a factor.

### 3.3.2 Duct Fans

In duct fans, airflow is also parallel or axial to the shaft on which the wheel is mounted. However, duct fans have the propeller housed in a cylindrical duct or tube. This design allows duct fans to operate at higher static pressures than propeller fans. Duct fans are commonly used in spray booths and other ducted exhaust systems. A fan is considered ducted if the duct length is more than the distance between the inlet to and the outlet from the fan blades. The two types of ducted fans commonly used are tube-axial fans and vane-axial fans. A tube-axial fan discharges air in a helical or screw-like motion. A vane-axial fan has vanes on the discharge side of the propeller which cause the air to discharge in a straight line. This reduces the amount of turbulence, thereby improving the fan efficiency and pressure capabilities. *Figure 13* shows both tube-axial and vane-axial versions of duct fans.

## 3.4.0 Fan Laws

The performance of all fans and blowers is governed by three rules commonly known as the *Fan Laws*. Cubic feet per minute (cfm), **revolutions per minute (rpm)**, static pressure (s.p.), and horsepower (hp) are all related. For example, when the cfm changes, the rpm, s.p., and hp will also change. The speed at which the shaft of an air-moving device is rotating is measured in rpm. The easiest way to determine the fan rpm is to measure it directly with a tachometer.

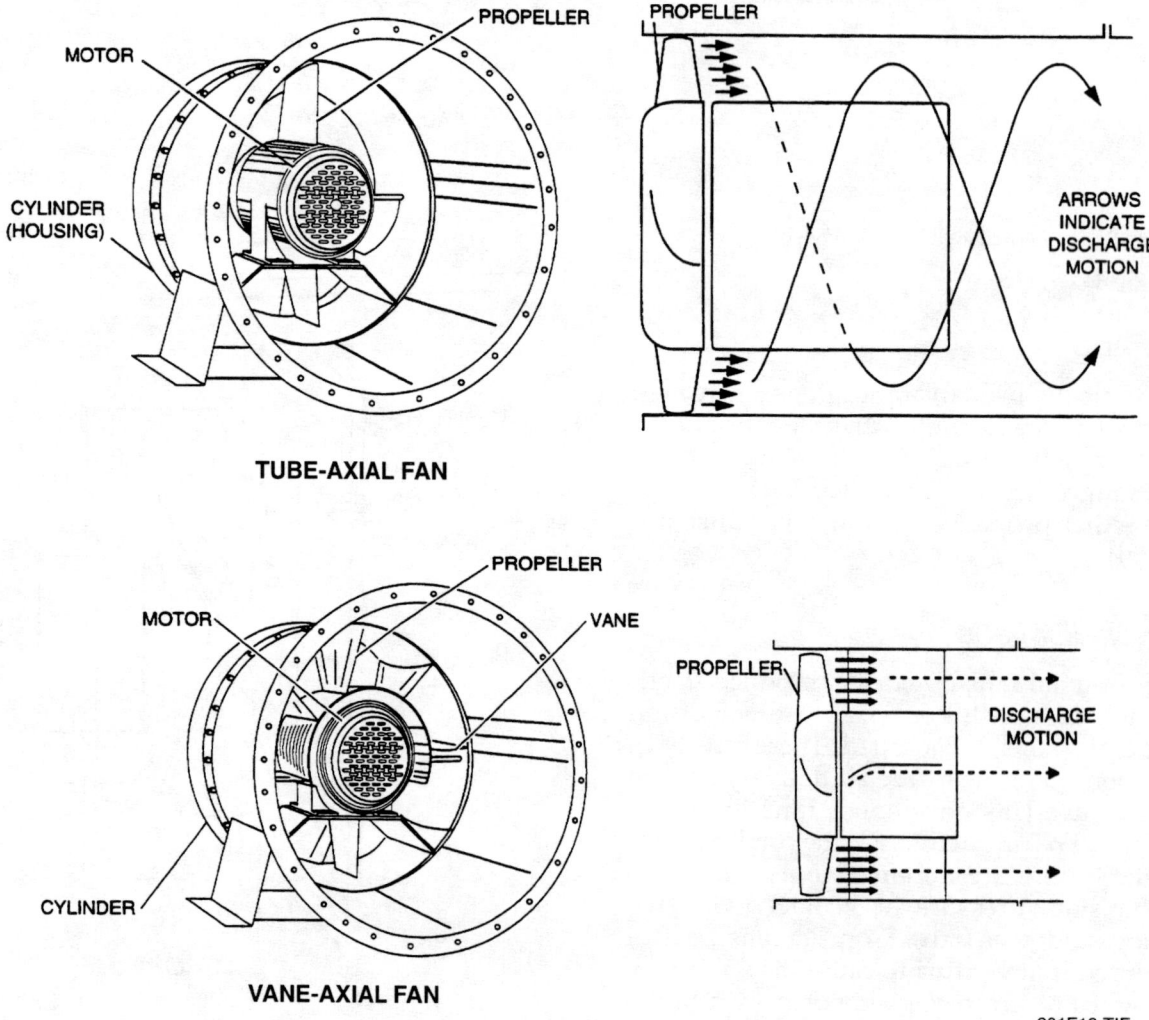

*Figure 13* ◆ Tube-axial and vane-axial duct fans.

Fan Laws 1, 2, and 3 are as follows:

- *Fan Law 1* states that the amount of air delivered by a fan varies directly with the speed of the fan. Stated mathematically:

  New cfm = (new rpm × existing cfm) ÷ existing rpm

  or

  New rpm = (new cfm × existing rpm) ÷ existing cfm

- *Fan Law 2* states that the static pressure (resistance) of a system varies directly with the square of the ratio of the fan speeds. Stated mathematically:

  New s.p. = existing s. p. × (new rpm ÷ existing rpm)$^2$

- *Fan Law 3* states that the horsepower varies directly with the cube of the ratio of the fan speeds. Stated mathematically:

  New hp = existing hp × (new rpm ÷ existing rpm)$^3$

**Study Example**

Assume the existing system conditions are 5,000 cfm, 1,000 rpm, and 0.5 in. w.c., with a fan hp of 0.5. With an increase in airflow to 6,000 cfm, what are the new rpm, s.p., and hp?

- Use Fan Law 1 to calculate the new rpm:

  New rpm = (new cfm × existing rpm) ÷ existing cfm

  New rpm = (6,000 × 1,000) ÷ 5,000

  New rpm = 1,200

- Use Fan Law 2 to calculate the new s.p.:

  New s.p. = existing s.p. × (new rpm ÷ existing rpm)$^2$

  New s.p. = 0.5 × (1,200 ÷ 1,000)$^2$

  New s.p. = 0.72 in. w.c.

- Use Fan Law 3 to calculate the new hp:

  New hp = existing hp × (new rpm ÷ existing rpm)$^3$

  New hp = 0.5 × (1,200 ÷ 1,000)$^3$

  New hp = 0.864

### 3.5.0 Fan Curve Charts

Manufacturer's fan curve charts can also be used to find the relationships that exist for a set of system conditions involving s.p., blower/fan rpm, and cfm. *Figure 14* shows a typical fan curve chart. If you know the values for any two of the three characteristics (s.p., rpm, and cfm) shown on the chart, you can easily find the value for the other characteristic. For example, assume that the s.p. is 1.4 in. w.c. and the blower is running at 900 rpm, as shown on the chart in *Figure 14*. To find the cfm, locate the intersection point of the 1.4 in. w.c. static pressure line and the 900 rpm curve. From this point, drop down vertically to the cfm scale, then read the value of 7,500 cfm. You will study fan curve charts in more detail in HVAC Level Three when you learn how to balance air distribution systems.

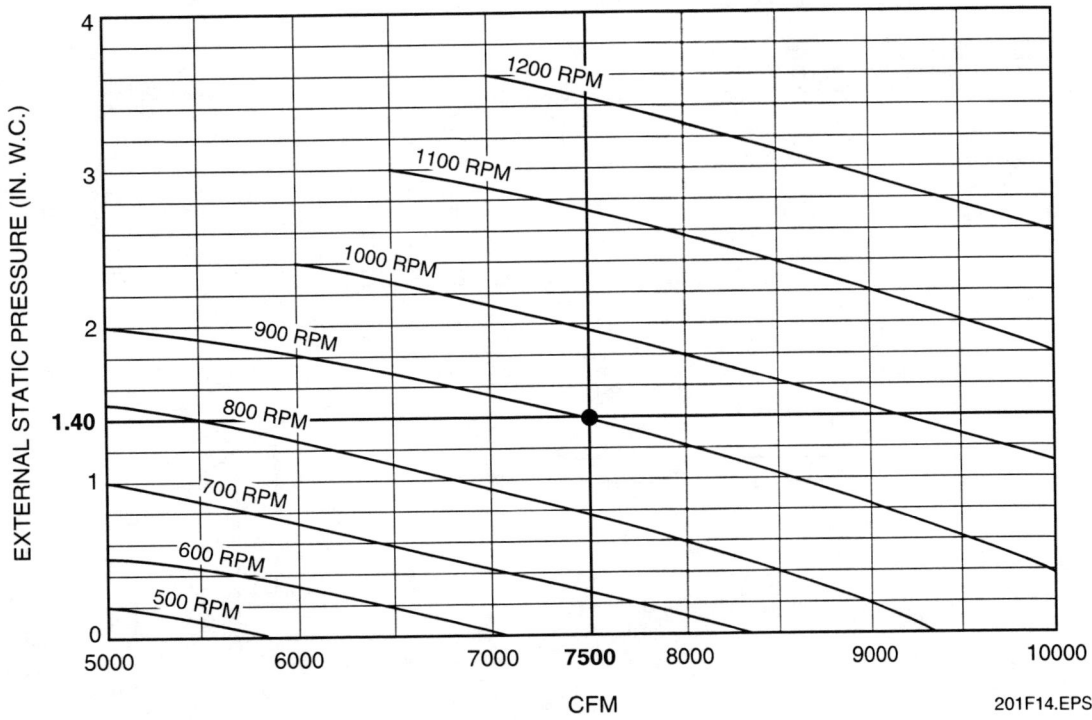

*Figure 14* ◆ Typical fan curve chart.

### 3.6.0 System Air Handler Blowers

The blower used in the system air handler must provide the pressure required to send the air to each conditioned space through the supply ductwork system. It must also overcome the pressure loss involved in pushing the air through the diffusers into each room. Additional pressure loss occurs as the air moves into the return air grilles and through the return ductwork system back to the air handler. The air handler must also overcome the added resistance encountered by the system air as it passes through any other system components. The total pressure loss from the duct system external to the air handler is called the *external static pressure loss*.

Internal resistances of the air handler include pressure loss from the filter, losses in the blower itself, and losses in other components of the air handler. These losses are accounted for by the manufacturer of the air handling unit. Therefore, in the field, you will only be concerned with the external static pressure losses resulting from the system ductwork and its components.

### 4.0.0 ◆ AIR DISTRIBUTION DUCT SYSTEMS

Air distribution systems consist of the supply duct system and the return duct system. The supply duct system receives air from the output of the system air handler, then distributes the air to the terminal units, and through the registers or diffusers into the conditioned space. The return duct system collects and routes air contained in the conditioned space for return to the input of the system air handler. Air distribution systems used in commercial and industrial structures vary depending on the structure and its intended use. Since air distribution systems used for residential applications are more uniform, they will be used as the basis for discussion in the rest of this section. Except for the system layout and size of the parts, the principles of operation and types of parts used in all duct systems are basically the same.

*Figure 15* shows a typical residential distribution duct system which consists of the supply and **return air plenums**, supply and return trunk ducts, branch ducts, supply diffusers, and return grilles. The supply and return trunk ducts attach to their respective plenums. A plenum is a sealed chamber at the inlet or outlet of an air handler.

A good practice is to line the return air plenums with an acoustical duct liner to reduce fan noise, especially if the return grille is close to the furnace. All duct connections to a furnace must extend outside the furnace closet. Return air must not be taken from the furnace room or closet. Also, the height of the return air duct must be high enough to allow the air filter(s) to be removed and replaced. All return air must pass through the air filter after it enters the return air plenum.

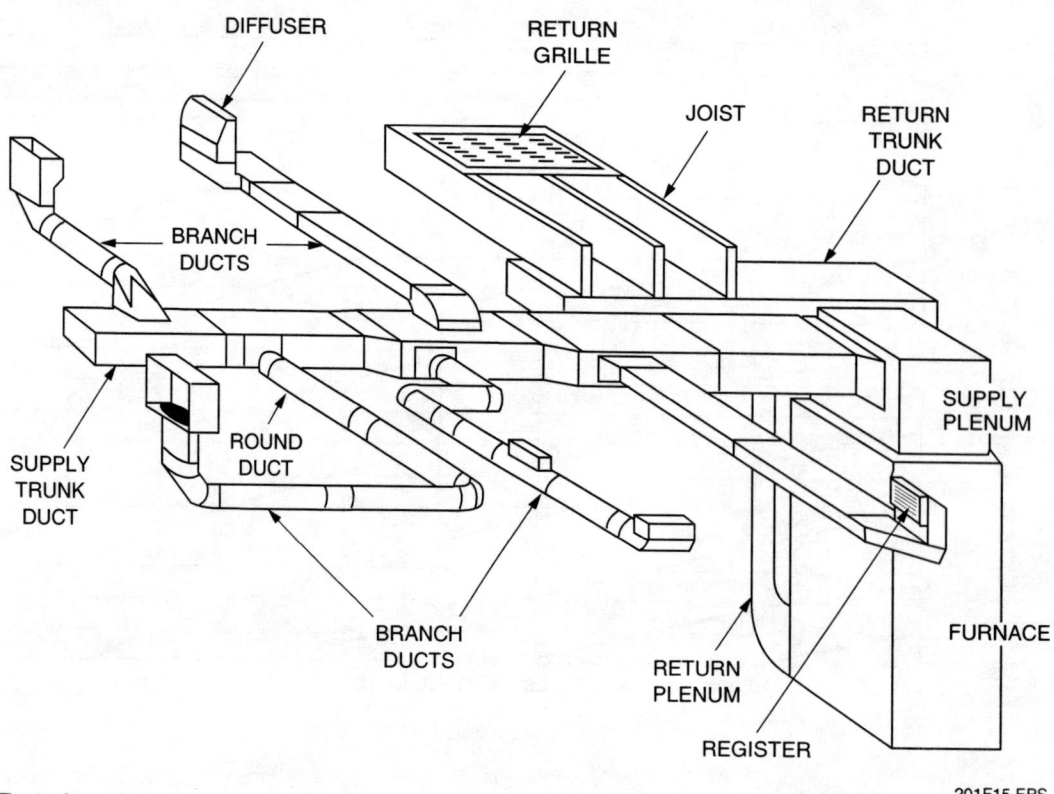

*Figure 15* ◆ Typical residential air distribution duct system.

### Duct System Installation

When installing a duct system, follow these general rules:

- Locate all supply and return registers and grilles, and cut the required openings.
- Pan in all return cavities and carry all returns back to the furnace or air handler location with a return trunk duct.
- Position the furnace or air handler and attach the supply plenum and return air duct.

Work out from the furnace or air handler, attaching lengths of supply duct and individual supply duct runs as you go.

Before getting into the details of duct systems, it is helpful to have a basic understanding of duct system design. The main factors that affect the form of a duct system are climate and building construction.

### 4.1.0 Duct Systems Used in Cold Climates

The type of duct system used in a building is mainly determined by the climate. In cold climates, most buildings use perimeter duct systems. Perimeter systems have floor or baseboard supply diffusers along the perimeter of the building walls. Use of floor or baseboard supply diffusers provides a good trade-off for heating and cooling performance.

In winter, the warm air supplied by the furnace blankets the outside walls and windows. This compensates for the cold downdrafts that tend to develop at the outside walls, windows, and doors. The return air grilles are located on the interior partition walls, at or near the floor. Central returns may be used, or for better performance, individual returns can be installed in each room. Location of the return grilles on the interior walls near floor level helps to remove any cool air from the floor, where it tends to collect or stratify.

*Figure 16* shows the room airflow during the heating and cooling modes of system operation. During the heating mode, the heated air blankets the outside walls and windows. Because it is warmer and lighter than the room air, it spreads across the ceiling and down the inside wall. Room air is drawn (induced) into the flow of warm air and mixes with it. A resulting stratified zone of cool air is formed that tends to collect near the floor. This is mitigated by the use of a low sidewall return.

During the cooling mode, cold supply air travels up the outside wall and windows and strikes the ceiling. Because it is cooler and heavier than the room air, it travels a short distance along the ceiling, then drops back down into the room as

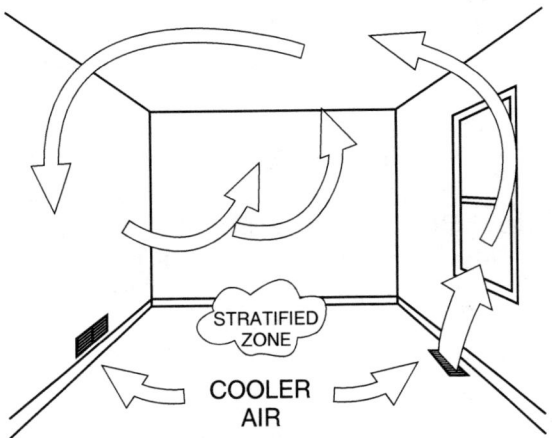

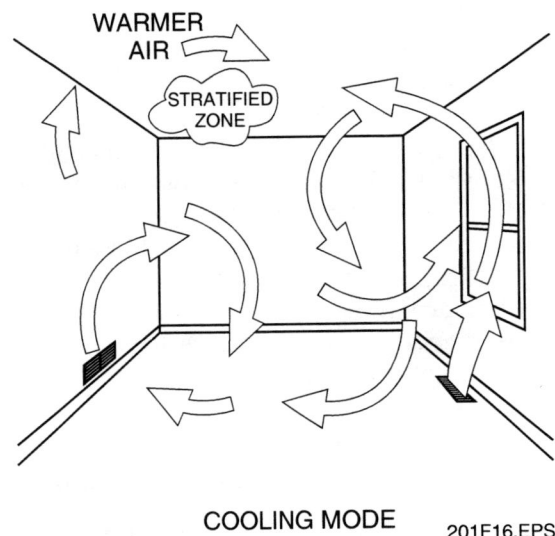

*Figure 16* ◆ Room air distribution patterns for a perimeter duct system.

shown. The cold air induces the room air fairly well, leaving only a small stratified layer of warm air near the ceiling. High sidewall returns would minimize this problem, but would result in a loss of heating performance.

AIR DISTRIBUTION SYSTEMS — TRAINEE MODULE 03201

Perimeter systems can have various layouts. Four common layouts are:

- Loop perimeter
- Radial perimeter
- Extended plenum
- Reducing extended plenum

### 4.1.1 Loop Perimeter Duct System

Loop perimeter duct systems are common in structures built on concrete slabs (*Figure 17*). They are easily used with air handlers that are centrally located. The perimeter loop is a continuous round duct of constant size imbedded in the slab. It runs close to the outer walls with the outlets located next to the wall. The perimeter loop is fed by several branches from the plenum. When the furnace fan is running, there is warm air in the whole loop, which helps to keep the slab at a more even temperature. Heat loss to the outside is minimized by the use of insulation around the slab. The loop has constant pressure around the system and provides the same pressure to all outlets.

### 4.1.2 Radial Perimeter Duct System

In a radial perimeter system (*Figure 18*) each outlet is fed from a central supply plenum through a separate branch duct. This system is most often used in small homes or with additions containing less than 1,200 square feet of space. The ductwork can be installed in a concrete slab or through a crawl space under the floor. It can also be run in an attic. Ductwork running through unconditioned or vented crawl spaces and attics should be insulated. It should also include a proper vapor barrier if used for cooling. This type of system is the most economical. However, it usually has a poor airflow performance because of static pressure losses at the furnace plenum. These losses are due to duct fittings at the plenum that are poor or excessive in number.

### 4.1.3 Extended Plenum Duct System

The extended plenum duct system (*Figure 19*) uses rectangular trunk ducts as the main supply and return ducts. The supply and return trunk ducts are a constant size over the whole length. This is the reason it is called an *extended plenum system*. Separate branch ducts run from the trunk duct to each supply outlet. The extended plenum works best when the air handler is located in the center of the main duct; however, it can be run in one direction. The trunk ducts are normally installed near the center line of the building, and their dimensions are constant over the entire length. The branch ducts are normally round, but can be rectangular. An air volume damper is usually installed in each branch duct near the trunk. This allows the airflow to be balanced with all supply air outlets fully open.

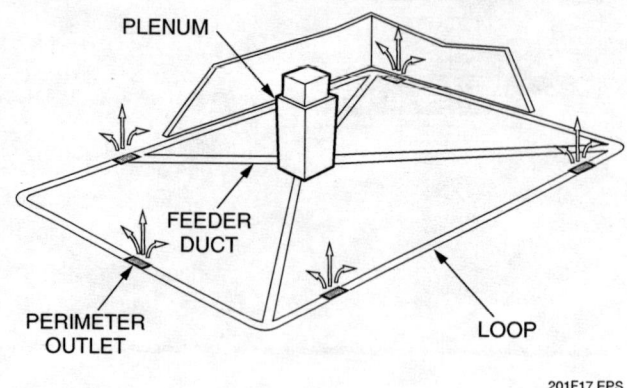

*Figure 17* ♦ Loop perimeter duct system.

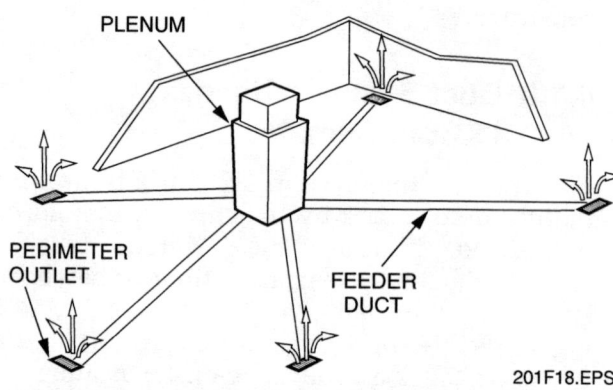

*Figure 18* ♦ Radial perimeter duct system.

### Duct Systems in Concrete Slabs

Duct systems in concrete slabs must be installed to prevent heat loss through the slab into the ground. Rigid foam boards are placed between the ducts and the ground before the concrete is poured. In areas with high water tables, a vapor barrier must be installed to prevent ground water from infiltrating the ducts.

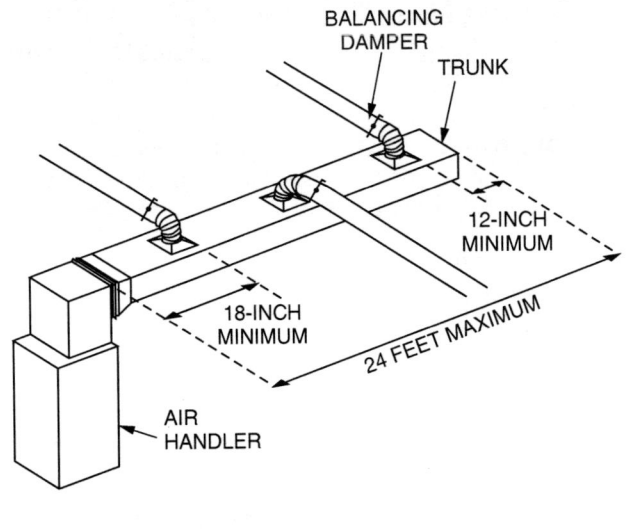

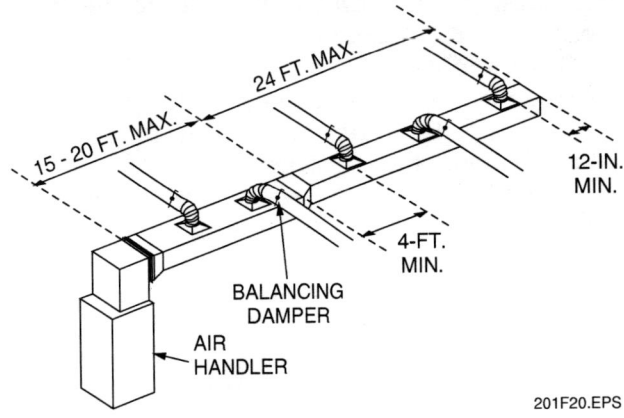

*Figure 20* ◆ Reducing extended plenum duct system.

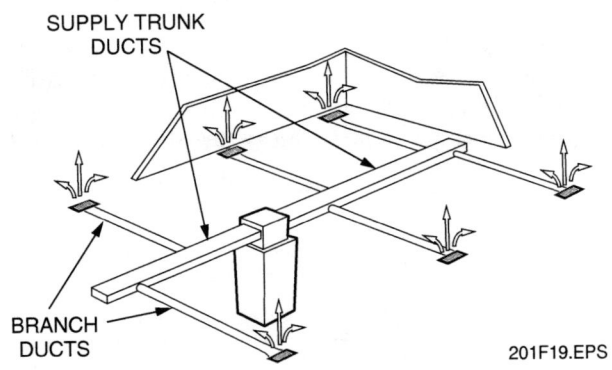

*Figure 19* ◆ Extended plenum duct system.

An extended plenum duct system is recommended for use in many applications. It requires accurate design, but gives the best performance for both the equipment and customer by allowing proper and equal air distribution throughout the system. Some recommended practices for laying out an extended plenum duct system are:

- The supply and return ducts should extend no more than 24' from the air handler.
- The first branch duct should be at least 18" from the beginning of the main duct. This helps to achieve the best balancing of the branch ducts.
- The main trunk should extend at least 12" from the last branch duct.

### 4.1.4 Reducing Extended Plenum Duct System

A reducing extended plenum system is similar to the extended plenum system. *Figure 20* shows an example of a reducing extended plenum duct system. If an extended plenum system is not practical, the reducing extended plenum system is a good option. It works well in larger buildings that require longer duct runs. It is also a better choice for systems where the air handler is installed on one end of the main trunk duct rather than in the middle. When properly designed, the same pressure drop is maintained from one end of the duct system to the other. This allows each branch duct to have about the same pressure pushing the air into its takeoff from the trunk duct. Some recommended practices for laying out a reducing extended plenum duct system are:

- The first main duct section should be no longer than 20'.
- The length of each reducing section should not exceed 24'.
- The first branch duct connection down from a single-taper transition should be at least four feet from the beginning of the transition fitting. This distance allows the air turbulence caused by the fitting to die down before the air is sent into the next branch duct. If the distance is less than four feet, the branch ducts near the transition can be hard to balance and may cause the system to be noisy.
- The trunk duct should extend at least 12" from the last branch duct.

### 4.2.0 Duct Systems Used in Warm Climates

In warm climates, buildings should have duct systems that favor cooling over heating. Perimeter systems like those used in cold climates can work reasonably well in some warm areas. However, their use is normally limited to buildings constructed over a basement or crawl space. Since cold floors and downdrafts from the outside walls, etc., are not normally a problem in warm climates, the air supply outlets do not need to be located at the building perimeter. In warm climates, supply air openings can be mounted high on the interior walls or in the ceiling to intensify cooling. Return openings are usually wall mounted near the baseboards on interior walls.

*Figure 21* shows the room airflow with high sidewall outlets. In the cooling mode, cool air moves across the ceiling and wraps around the far wall. The room air mixes well with the supply air and almost no stratification occurs. In the heating mode, the pulling effect of the low sidewall return air grille tends to draw some of the warm air down from the ceiling and prevents a stratified layer of cold air from building up near the floor.

Ceiling diffusers are one of the best air supply methods used for cooling, but they are the poorest for heating. In the cooling mode, supply air from the diffuser mixes well with the room air (*Figure 22*). Air motion in the room is good with no stagnant areas. In the heating mode, the ceiling diffuser can perform poorly if the return grille is also ceiling mounted. The warm air clings to the ceiling with almost none of it reaching the occupied space in the room. Using a return grille mounted near the floor on an inside wall can help prevent this.

Duct systems used in warm climates can have various layouts. Four typical layouts include:

- Overhead trunk
- Overhead radial
- Attic extended plenum
- Attic radial

### 4.2.1 Overhead Trunk and Overhead Radial Duct Systems

The overhead trunk and overhead radial duct systems are typically used in buildings on concrete slabs. They are often used in buildings with no attic space, or where the entire air system and the

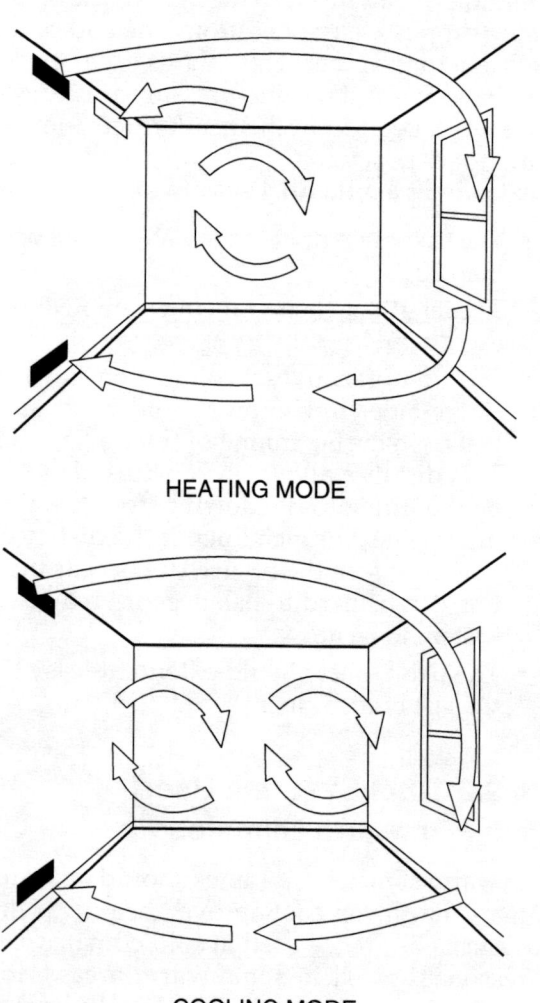

*Figure 21* ◆ Room air distribution patterns for high sidewall outlets.

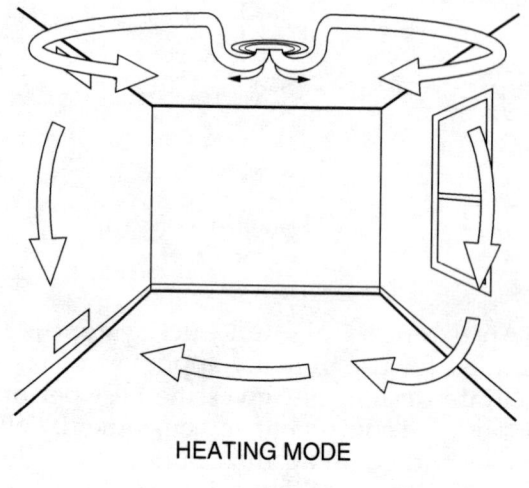

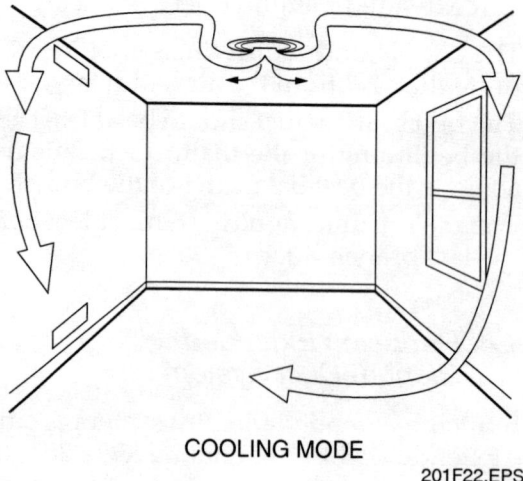

*Figure 22* ◆ Room air distribution patterns for ceiling outlets.

air handler are installed on one floor, such as in apartments. Because the equipment is installed in the conditioned space, duct insulation and vapor barriers are normally not required.

In the overhead trunk duct system (*Figure 23*), the air handler and ductwork are usually placed in a drop ceiling area, such as a hallway or closet. The rectangular trunk ducts run from the side of the supply air plenum straight to each high sidewall outlet located in the rooms. If the trunk duct length extends more than 24' from the air handler, a reducing plenum should be used. Overhead trunk systems most often use a central return.

In the overhead radial duct system (*Figure 24*), separate branch ducts are run from a common supply air plenum to each high sidewall outlet. The branch ducts are round metal or flex duct installed within a soffit or drop ceiling. Overhead radial systems most often use a central return.

### 4.2.2 Attic Extended Plenum and Attic Radial Duct Systems

Attic extended plenum and attic radial duct systems are installed in buildings with attic areas. If a building has an attic, it can be used to house both the air handler and air duct system. This allows the equipment to be installed out of sight. Attic installations eliminate the need for drop ceilings or soffit enclosures in the building. As a result, the amount of living or usable space in the building is increased. The rooms in the building are quieter, because the equipment and its noise have been removed from the rooms. Normally, the air handler is hung from the roof supports with vibration isolators to keep equipment noises from being transmitted through the ceiling.

In the extended plenum duct system (*Figure 25*), the trunk ductwork can be made from insulated fiberglass ductboard or insulated sheet metal with a vapor barrier. If the trunk duct length extends more than 24' from the air handler, a reducing plenum trunk should be used. The branch ducts can be pre-insulated round flex duct, fiberglass ductboard, or round sheet metal covered with an insulating sleeve that has a vapor barrier outside. It should be pointed out that excessive use of flexible duct will result in poor performance and shorter equipment life. Excessive use of flexible duct is considered to be anything more than a 6' section connected at each grille, register, or diffuser. Flexible duct should be used mainly for noise and vibration attenuation. Attic extended plenum systems typically use a central return system. However, because there is full access to the room ceilings, room-by-room returns are also used.

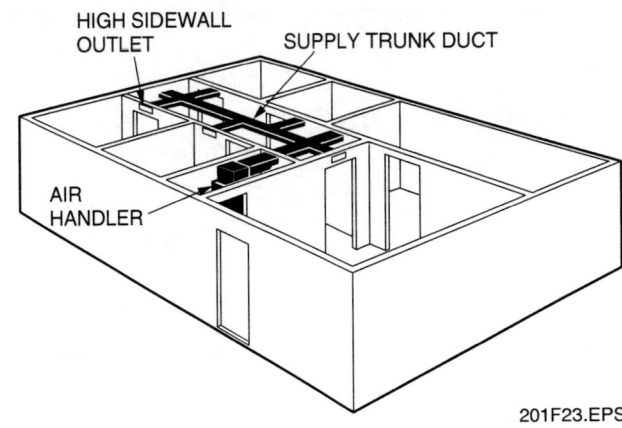

*Figure 23* ◆ Overhead trunk duct system.

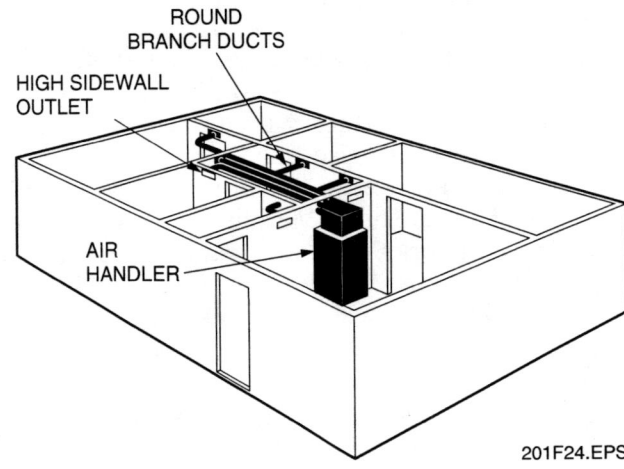

*Figure 24* ◆ Overhead radial duct system.

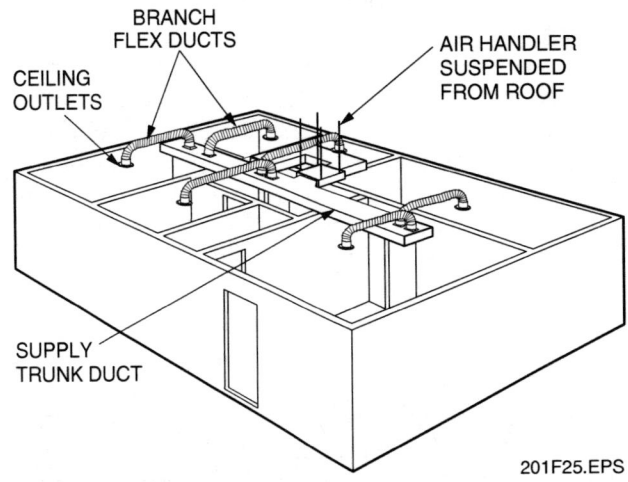

*Figure 25* ◆ Attic extended plenum duct system.

### Installing Overhead Supply Duct

In some locations, a common method of installing an overhead supply duct is to run it concentrically inside the return air plenum. To do this, a chase for the return trunk is framed up around the supply trunk after it and the individual supply ducts are run. Returns are then tied into the framed-in area and the chase is fully enclosed, usually with gypsum drywall.

In operation, return air flows back to the furnace or air handler in the space surrounding the supply duct. To prevent leakage between the supply and return air sides, all supply duct joints should be tight and leak free.

---

The attic radial duct system, also known as a *spider* system, is commonly used with attic-mounted air handlers. This is not a preferred method of system design. Ductwork selection is covered later in this module. Each supply room air outlet is connected to a central supply air plenum on the unit through runout ductwork (*Figure 26*). Flexible, pre-insulated round ductwork is most often used for the runouts, but fiberglass ductboard and round sheet metal ducts with an insulating/vapor barrier sleeve are also used. Depending on the requirements, central or room-by-room return duct systems are used. Manually adjusted volume control dampers should be used in all branch duct runs to facilitate balancing the system airflow. Balancing air volume at the supply air diffusers is poor practice and should be avoided.

### 5.0.0 ◆ DUCT SYSTEM COMPONENTS

Building code requirements pertaining to the installation of air distribution systems are not standard across the nation. Almost all localities have minimum standards or codes that determine the type of materials and methods that must be used. The HVAC technician should become familiar with and follow the local codes that apply to each job.

The selection and size of trunk and branch ducts used in an air distribution system are based on the air volume (cfm) needed to satisfy the heating and/or cooling requirements for the building. For new buildings, this information can usually be found in the design specifications. The method normally used to find the required air volume for an existing building is to survey the structure. Armed with known values for air volume, you can use friction charts, duct sizing charts, and/or duct size calculators to find the correct duct sizes for the job. After the duct sizes are known, the selection of the parts used for the layout is mainly determined by the construction of the building. You will study ductwork selection methods in detail later on in your training.

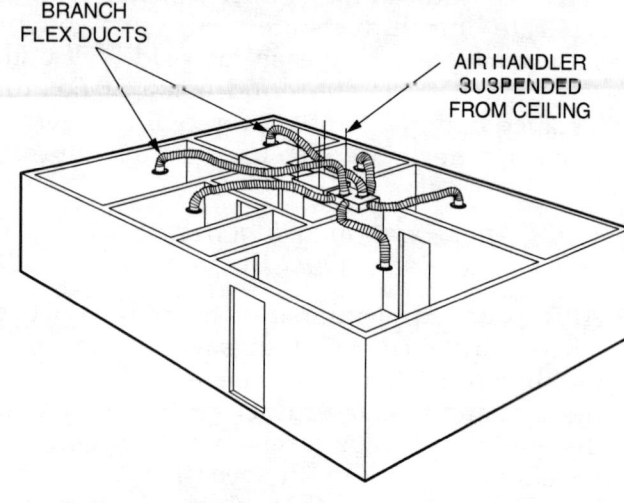

**FLEXIBLE DUCT**

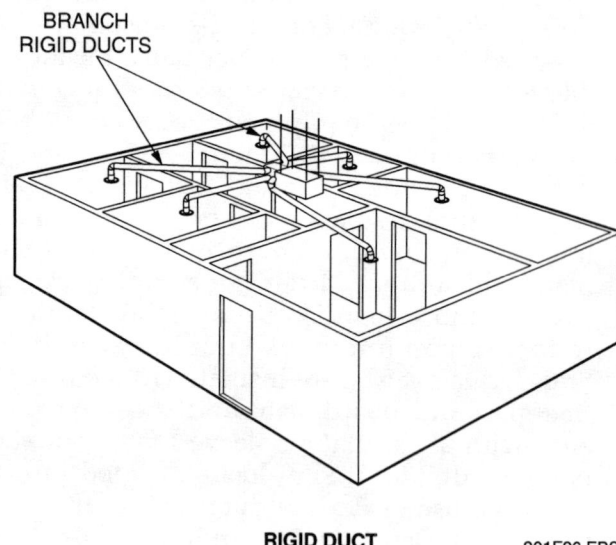

**RIGID DUCT**

201F26.EPS

*Figure 26* ◆ Attic radial duct system.

The tables in the appendix provide typical sizes for residential duct system components based on various airflow requirements. This section describes these components and overviews some

of the application considerations. The basic components of a duct system include the following:
- Main trunk and branch ducts
- Fittings and transitions
- Air diffusers, registers, and grilles
- Dampers
- Insulation and vapor barriers

## 5.1.0 Main Trunk and Branch Ducts

Duct systems can be installed in basements, crawl spaces, attics, and within concrete floors or slabs. Ducts can be made from metal, fiberglass ductboard, ceramic, or plastic materials. Galvanized sheet metal or fiberboard ducts are typically used for heating/cooling air distribution. When installed in a concrete slab, ducts are usually made of metal, plastic, or ceramic. Spiral metal and flexible ducts are also in common use. Where weight is a factor, aluminum duct is sometimes used.

### 5.1.1 Galvanized Steel Trunk and Branch Ducts

Galvanized steel or sheet metal duct can be round, square, or rectangular (*Figure 27*). All three types are often used in the same duct system. Popular sizes of round, square, and rectangular steel ducts, along with an assortment of standard fittings, can be obtained from HVAC supply houses. For large commercial jobs involving customized ductwork, the ducts and fittings are often made separately in a metal shop or fabricated at the job site.

Because sheet metal duct is rigid, the layout must be well planned, and all the pieces cut precisely, or the system will not fit together.

The thickness of galvanized steel and other metal duct is expressed in terms of gauge thickness (*Figure 27*). When a duct is made of 28-gauge sheet metal, this means that the thickness of the duct walls is $1/28"$. Likewise, a sheet metal duct made out of 24-gauge metal has a wall thickness of $1/24"$, etc. Larger ducts are made from thicker metal and are more rigid than smaller ducts. This prevents them from swelling and making popping noises when the system blower starts and stops. Also, lines or ridges, normally called *crossbreaks*, are used on large sheet metal panels or ducts to make them more rigid (*Figure 27*).

The aspect ratio of a duct is often used to classify a duct size and estimate its cost. Aspect ratio is the ratio of the duct's width to its height. For example, if a duct is 18" wide and 6" high, the aspect ratio is $18 \div 6$, or 3 to 1. *Figure 28* shows some typical gauge thicknesses used for rectangular and round metal ducts. It also shows a tabulation of duct aspect ratios.

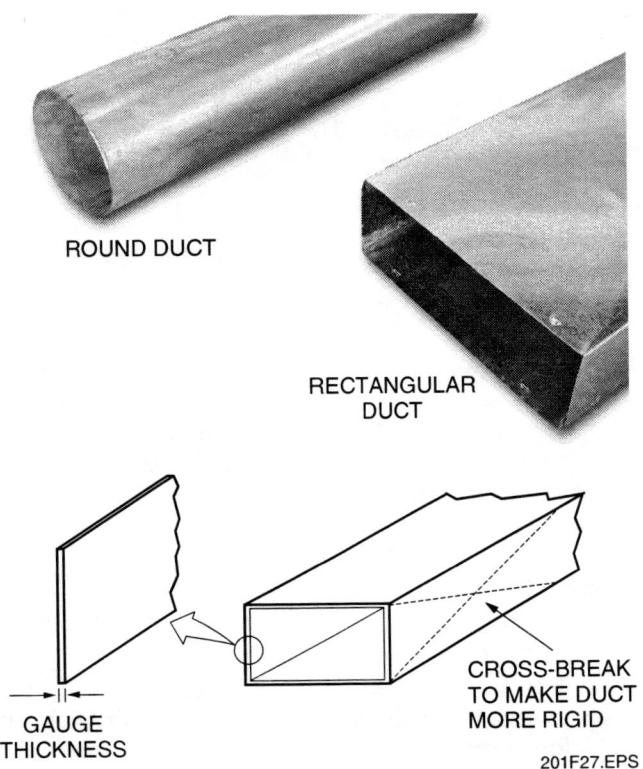

*Figure 27* ◆ Metal ducts.

### RECTANGULAR DUCT

| RECTANGULAR DUCT WIDTH IN INCHES | COMMERCIAL | | RESIDENTIAL |
|---|---|---|---|
| | SHEET METAL GALVANIZED | ALUMINUM | SHEET METAL GALVANIZED |
| UP TO 12 | 26 | .020 | 28 |
| 13 - 23 | 24 | .025 | 26 |
| 24 - 30 | 24 | .025 | 24 |
| 31 - 42 | 22 | .032 | – |
| 43 - 54 | 22 | .032 | – |
| 55 - 60 | 20 | .040 | – |
| 61 - 84 | 20 | .040 | – |
| 85 - 96 | 18 | .050 | – |
| OVER 96 | 18 | .050 | – |

### ROUND DUCT

| ROUND DUCT DIAMETER | COMMERCIAL SHEET STEEL GALVANIZED GAUGE | RESIDENTIAL SHEET STEEL GALVANIZED GAUGE |
|---|---|---|
| UP TO 12 | 26 | 30 |
| 13 - 18 | 24 | 28 |
| 19 - 28 | 22 | – |
| 27 - 36 | 20 | – |
| 35 - 52 | 18 | – |

### ASPECT RATIO

| DUCT CLASS (ASPECT RATIO) | WIDTH IN INCHES | PERIMETER |
|---|---|---|
| 1 | 6 - 8 | 24 - 72 |
| 2 | 12 - 24 | 36 - 72 |
| 3 | 26 - 40 | 70 - 106 |
| 4 | 24 - 88 | 60 - 220 |
| 5 | 48 - 90 | 116 - 216 |
| 6 | 90 - 145 | 210 - 336 |

*Figure 28* ♦ Typical metal duct gauge thickness and aspect ratios.

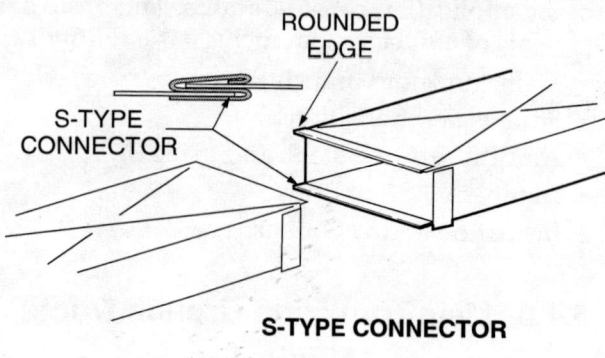

**S-TYPE CONNECTOR**

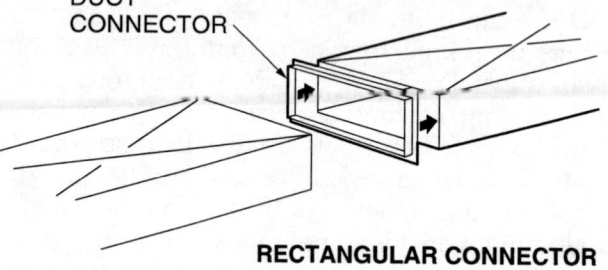

**RECTANGULAR CONNECTOR**

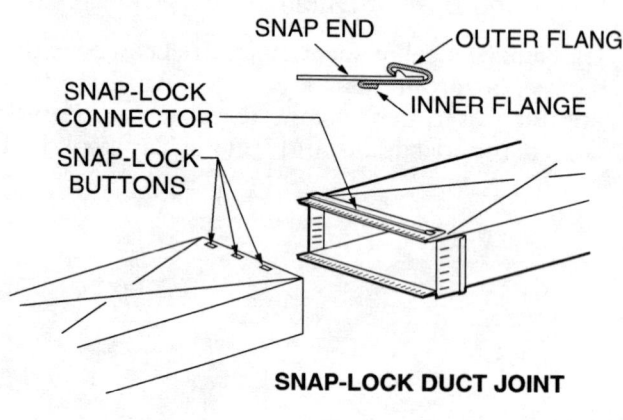

**SNAP-LOCK DUCT JOINT**

*Figure 29* ♦ Typical square and rectangular sheet metal duct fasteners.

Sections of square or rectangular duct are assembled using any one of several fasteners that are available. Typically, S-fasteners and drive clips and/or snap-lock fasteners are used (*Figure 29*). Round duct sections are normally fastened together with self-tapping sheet metal screws. These fasteners make a nearly airtight connection. When further sealing is needed, the joint can be taped using special duct tape or sealed with a flexible duct sealing compound which can be applied with a paint brush or caulking gun. Leaking joints cut down on the amount of air available for delivery to the outlets at the end of long runs.

A ductwork system must be well supported so that it does not move. If it is not properly supported, movement can occur when the fan starts and causes a rush of air through the system. Sheet metal ductwork can also move as a result of expansion and contraction as it heats and cools. This type of movement can be contained by using flexible or fabric joints at different points in the system.

Sheet metal ductwork systems also transmit vibrations from the air handling equipment. Transmission of these vibrations to the duct system can be prevented by using flexible connectors or fabric joints at the main supply and return ductwork connections to the air handler. *Figure 30* shows some typical devices for controlling ductwork noise and vibration.

## 5.1.2 Fiberglass Duct

Fiberglass ductboard can be used instead of metal duct. It has more friction losses than metal duct, but is quieter because the ductboard absorbs blower and air noises better. Fiberglass duct is available in flat sheets for fabrication, or as prefabricated round duct sections. Fiberglass duct is normally 1" thick with an aluminum foil backing. This backing is reinforced with fiber to make it strong. The inside surface of the ductboard is coated with plastic or a similar coating to prevent the erosion of the duct fibers into the supply air. Fiberglass particles released into the air can be harmful to health.

Ductwork is made from sheets of fiberglass ductboard using special machines or knives. These devices cut away the fiberglass to form the edges, and the reinforced foil backing is left intact to support the connections. When two pieces are fastened together, an overlap of foil is left so that one piece can be stapled to the other using special staples (*Figure 31*). The joint is then taped to make it airtight. Round fiberglass duct is also easy to install because it can be cut to size with a knife. Fiberglass ductwork systems must be properly supported or they will sag over long runs. Special hangers designed not to cut the outside cover of the ductboard must be used. The use of fiberglass ductwork is diminishing because of maintenance and indoor air quality issues. The duct walls do not hold up well to duct cleaning procedures and if the duct is damaged, can be difficult to repair. Today, duct cleaning is a widely performed air distribution system maintenance task, and as such needs to be taken into consideration both in design and long-term maintenance of duct systems.

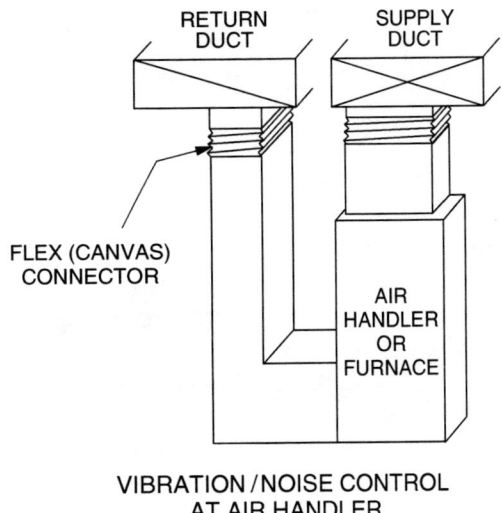

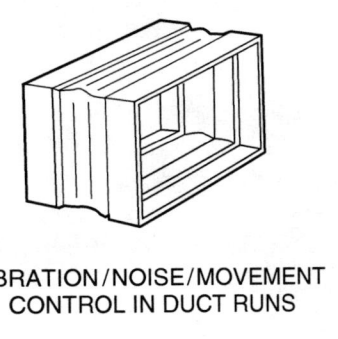

*Figure 30* ♦ Ductwork vibration and noise control devices.

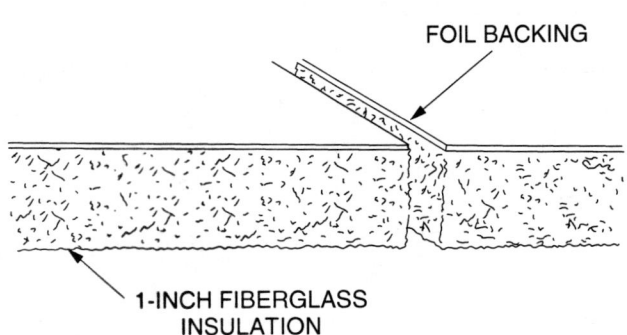

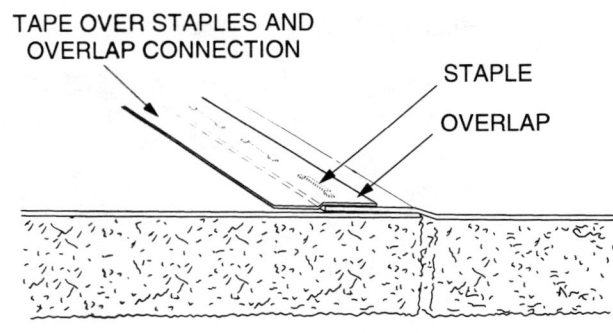

*Figure 31* ♦ Fiberglass ductboard.

### Fiberglass Ductboard

A major disadvantage of using fiberglass ductboard is that it is not as sturdy as sheet metal duct; therefore, it cannot be installed in areas where it might be subject to damage.

AIR DISTRIBUTION SYSTEMS — TRAINEE MODULE 03201    1.21

### 5.1.3 Flexible Duct

Flexible round duct (*Figure 32*) comes in sizes up to 24" in diameter. It is available with a reinforced aluminum foil backing for use in conditioned areas. It also comes wrapped with insulation protected by a vapor barrier made of fiber-reinforced vinyl or foil backing for use in unconditioned areas.

Flexible duct is typically used in spaces where obstructions make the use of rigid duct difficult or impossible. Flexible duct is easy to route around corners and other bends. Duct runs should be kept as short and as straight as possible. Gradual bends should be used, since tight turns can greatly reduce the airflow and may even cause the duct to collapse. If a connection to a ceiling diffuser needs an elbow, it is better to use an insulated metal elbow at the input to the diffuser than to bend flexible duct tightly to form the connection. This is because diffuser performance is disrupted far less by the metal elbow.

Long runs of flexible duct are not recommended unless the friction loss is taken into account. Even when properly installed, most flex ducts cause at least two to four times as much resistance to airflow as the same size diameter sheet metal duct. To avoid sags in the run, flexible duct should be amply supported with 1" wide or wider bands to keep the duct from collapsing and reducing the inside dimension. Some flexible duct comes with built-in eyelet holes for hanging.

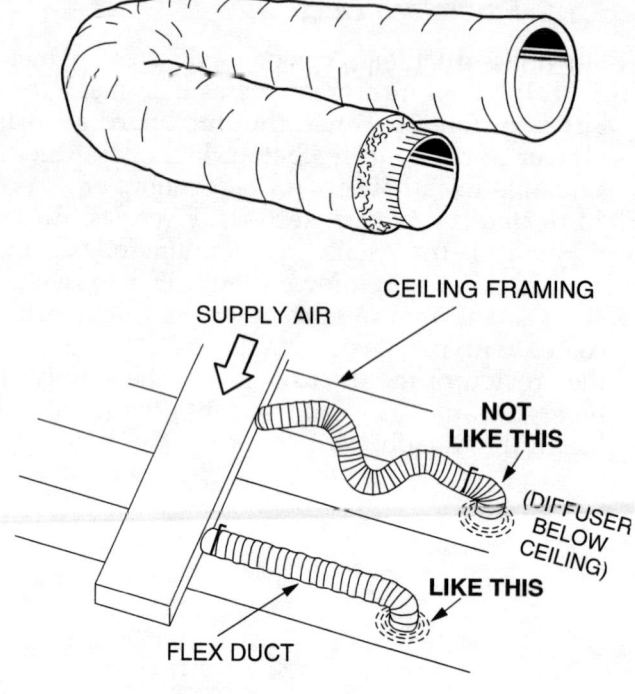

*Figure 32* ◆ Flexible duct.

As mentioned earlier, any duct system that uses excessive flexible duct will result in poor performance and shorter equipment life. Excessive use of flexible duct is considered to be anything more than a 6' section connected at each grille, register, or diffuser. Flexible duct should be used mainly for noise and vibration attenuation.

---

### Flexible Duct

This polyethylene jacketed flexible air duct is an example of a widely used type of flexible duct.

## 5.2.0 Fittings and Transitions

Fittings in ducts, such as elbows, takeoffs, or boots change the direction of airflow or change its velocity. Transitions are typically used to change from one size duct to another. They are also used to change from one duct shape to another.

Air moving in a duct has inertia that makes it want to continue flowing in a straight line. Each fitting in a duct run adds friction and decreases the quantity of air the duct can carry. It takes energy to overcome the resistance (friction) inherent in a fitting. Because of this, the number of fittings used in a duct system, and their types, must be carefully selected in order to minimize the total amount of friction added to the system.

Fittings and/or transitions are rated by equivalent feet of length. This means that a specific fitting produces a pressure drop equal to a certain number feet of straight duct length of the same size. Therefore, adding fittings has the same effect on the pressure loss of a duct as increasing its overall length. This is why duct runs should be made as straight as possible to each room. Also, the use of unnecessary fittings, or ones not best suited for the job, must be avoided. For each standard type of fitting, the pressure drop is known and has been converted to the equivalent feet of duct length. This information is available in a set of charts that show the standard types of fittings and/or transitions and the value for the equivalent feet of length used for each one. These charts are available in **ASHRAE** and **SMACNA** publications.

The total equivalent feet of length for a duct run is calculated by adding all the equivalent lengths for fittings in the run to the actual length of straight duct used. In the example shown in *Figure 33*, an elbow with an equivalent length of 30' is added to a duct with 100' of straight length. The resulting pressure drop is the same as that of a straight duct 130' long.

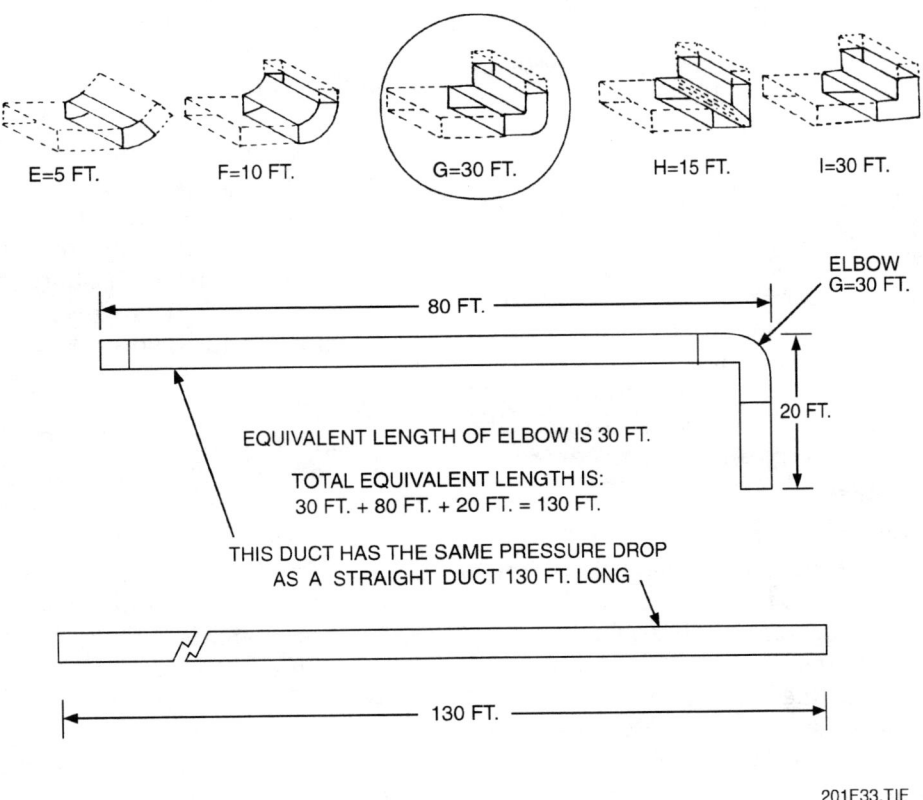

*Figure 33* ◆ Example of equivalent length.

### Equivalent Length

Which elbow would you use if you wanted to shorten the equivalent length of the run in the example shown in *Figure 33* from 130' to 110'?

AIR DISTRIBUTION SYSTEMS — TRAINEE MODULE 03201

### Registers and Grilles

Not all registers and grilles are created equal. Inexpensive grilles and registers can add significant static pressure losses to the duct system, contributing to poor system performance. Always check the register or grille manufacturer's specification sheets to make sure the grilles and registers selected for a job will enhance and not hinder system performance.

### 5.3.0 Air Diffusers, Registers, and Grilles

Air outlets distribute the supply air into the conditioned space. When properly selected, they act to blend the supply air with the room air so that the room is comfortable without excess noise or drafts. The terms *diffuser*, *register*, and *grille* are used to describe different kinds of outlets.

- A diffuser is an outlet that discharges supply air into a room in a widespread fan-shaped pattern.
- A register is an outlet that discharges supply air into a room in a concentrated non-spreading stream. Many have one-way and two-way adjustable air stream deflectors.
- A grille is the louvered covering of an opening created for the passage of air into a room. It controls the distance, height, and spread of airflow, as well as the amount of air delivered to the space. Grilles have many different designs. Some are fixed and can direct air in one direction only. Others are adjustable and can be set to send air in different directions. Grilles with no adjustments are typically used as covers for the return air duct.

*Figure 34* shows several registers and diffusers in common use. As shown, the floor register is relatively long and narrow. It gives excellent performance for both heating and cooling when used in perimeter duct systems. A floor register is usually installed parallel to the room's outside wall at about 6" to 8" away from the wall. Floor registers are fed from below and discharge air upward. Typically, fixed vanes are installed crosswise at various angles to spread the air stream so that it blankets the outside wall and windows with supply air. Floor registers normally have a built-in shutoff damper.

Low sidewall registers are excellent for heating when used in perimeter systems. They also work well for cooling, if designed to discharge air upward. If not so designed, a plastic air scoop accessory may be added to the exterior to redirect the supply air upward for cooling. Low sidewall registers are fed from the back and mounted flush with the wall just above the baseboard trim. Air is discharged outward and slightly upward. Some are made to discharge air in two or three directions. Low sidewall registers normally include a built-in shutoff damper.

High sidewall registers provide poor heating performance in cold climates because they leave a cold layer of air in the lower half of the room, especially with slab floors. However, they are adequate for heating in warm climates. High sidewall registers provide good performance in cooling when used with central returns, and when used with room-by-room returns, the cooling performance is even better. High sidewall registers are mounted flush on the room's inside wall and fed from behind. The top edge is usually mounted 6" to 12" down from the ceiling. They typically discharge air horizontally toward the outside wall. Some are made to discharge air in two or three directions. High sidewall registers normally include a built-in shutoff damper.

Baseboard diffusers are long and narrow. They are mounted on the floor with their back mounted snugly against the wall. Supply air is fed from below and discharged upward close to the wall so as to blanket a wide area of the outside wall and windows. Baseboard diffusers normally have a built-in shutoff damper. They perform well in both heating and cooling when used with perimeter duct systems.

Ceiling diffusers can be round or rectangular in shape. They mount flush to the ceiling and are fed supply air from above. Ceiling diffusers are made that distribute supply air equally in all directions. Other styles are made that distribute air in one, two, three, or four directions. Ceiling diffusers are made both with and without a built-in shutoff damper. They give good performance for cooling when used with a central return, and excellent performance when used with room-by-room returns. Ceiling diffusers give poor performance for heating in cold climates because they leave a cold layer of air in the lower half of the room, especially with slab floors. However, they are adequate for heating in warm climates.

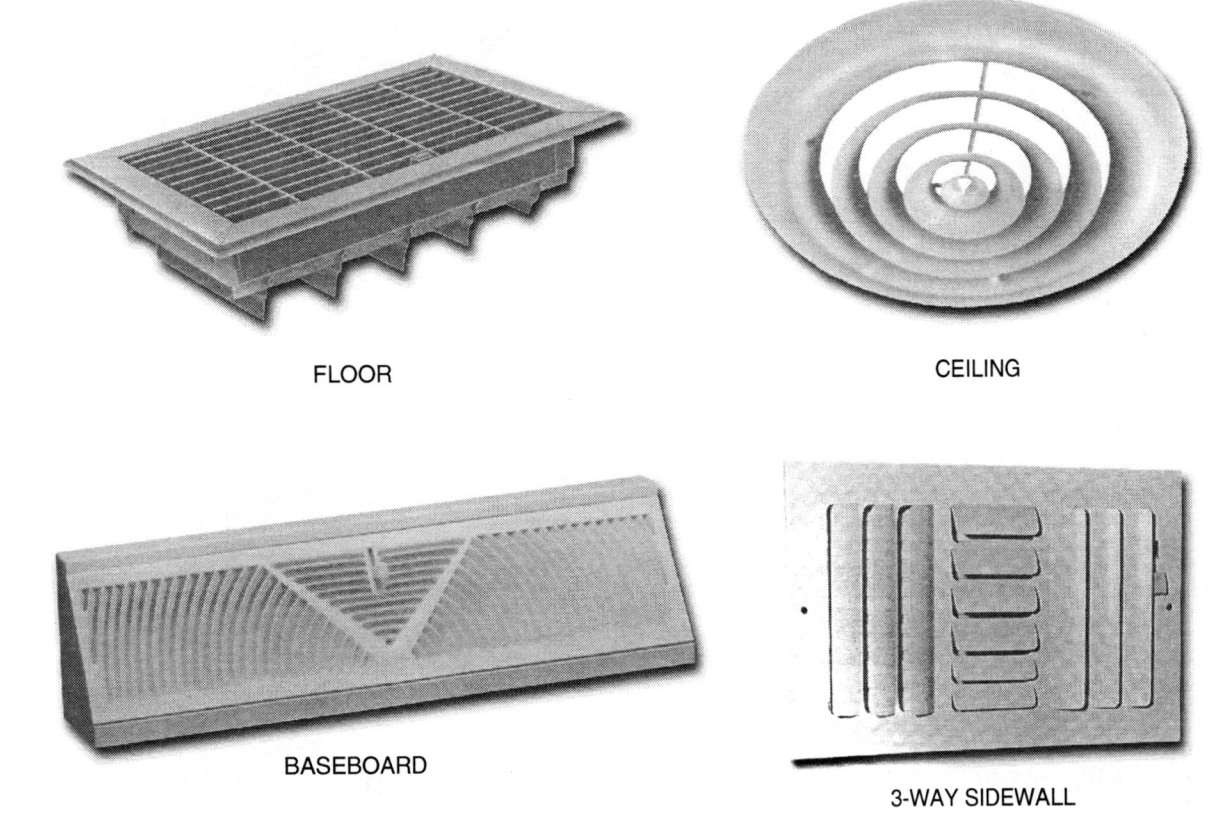

*Figure 34* ◆ Air registers and diffusers.

## 5.4.0 Dampers

Dampers are used to control and balance airflow in duct systems. Without balancing dampers, air distribution systems cannot be properly balanced, causing some rooms to receive too much air while others do not receive enough. Some dampers are made with manual adjustments. Others, used in zoned heating or cooling systems, are automatically controlled. Sometimes dampers are used to mix two airflows, such as with fresh and recirculated air. By code requirements, commercial and industrial buildings normally have automatic fire dampers installed in all the vertical duct runs, since all ducts, especially vertical ones, will spread fumes and flames from a fire.

A damper used to balance a system should be installed in an accessible place in each branch supply duct. The closer the dampers are to the main duct or supply air plenum, the better. They should be tight fitting with minimum leakage. The built-in dampers on supply diffusers and registers should not be used to balance an air system. When partially closed, they disrupt the performance of the diffuser or register and also make it noisy. *Figure 35* shows three types of dampers used in air distribution systems.

## 5.5.0 Fire and Smoke Dampers

Fire dampers are used to maintain the fire-resistance ratings of walls, partitions, and floors that are penetrated by HVAC ducts and to prevent the spread of fire if one should occur. The dampers are normally held open with a fusible link set to melt at 165°F. If a fire occurs, the link will melt and the dampers will close automatically by spring action or gravity.

All fire dampers have a resistance rating of either 1½ or 3 hours. Building partitions with a three-hour fire rating require the use of a three-hour rated damper while partitions with fire ratings of less than three hours would require the use of 1½-hour rated dampers.

In addition to the fire-resistance rating, fire dampers have a static or dynamic air closure rating. Static-rated dampers can only be used in HVAC systems where the HVAC equipment will be automatically shut down in case of fire. In that case, no air would be flowing within the ducts. Dynamic-rated dampers are designed to close even if the HVAC system remains running and air is moving through the ducts.

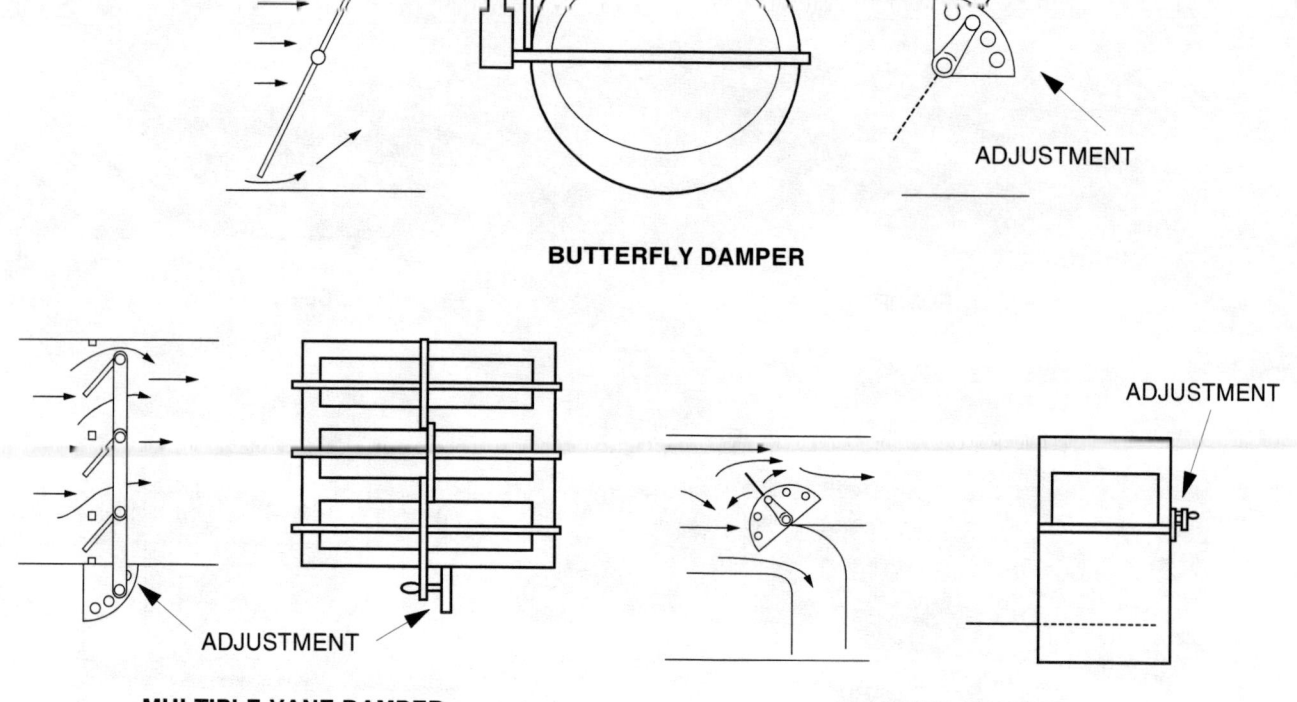

*Figure 35* ◆ Typical dampers.

Smoke can be equally as deadly as fire so controlling the spread of smoke in a building is critical. Smoke dampers in the ducts of HVAC systems perform this critical task. Smoke dampers can be passive in their function where they operate to simply shut off and isolate a section of duct. They also can be part of an engineered smoke control system that allows the building duct system to direct air and/or smoke in such a way as to prevent the spread of fire and smoke or to move the smoke to an area where it poses no problem.

Most smoke dampers are operated electrically or pneumatically and are controlled by a smoke or heat detector, fire alarm, or automated building control system. Smoke dampers are rated for leakage. Class 1 has the lowest leakage rating with classes 2, 3, and 4 having higher leakage ratings. Smoke dampers are rated for temperature and have a velocity and pressure rating which indicates how they will operate against specific airflow and pressure differential conditions within the duct.

Combination fire and smoke dampers (*Figure 36*) are also available. These dampers must conform to the rating agency requirements for both fire and smoke dampers.

*Figure 36* ◆ Combination fire and smoke damper.

## 5.6.0 Insulation and Vapor Barriers

When ductwork passes through an unconditioned space, heat transfer may take place between the air in the duct and the air in the unconditioned space. If the heat exchange adds or removes very much heat from the conditioned air, insulation should be applied to the ductwork. A difference of 15°F from the inside to the outside of the duct is considered the maximum difference allowed before insulation is required. Many installations use preinsulated ductboard for the main supply and return ducts. In this case, the insulation and vapor barrier is the duct itself.

Metal duct can be insulated in two ways: on the outside or on the inside. Insulation inside the duct is installed by the duct manufacturer. It is either glued or fastened to tabs mounted on the inside duct wall. Insulation and a vapor barrier can also be wrapped around the outside of the ductwork after it has been installed. The insulation is usually a foil or vinyl-backed fiberglass. It comes in several thicknesses, with 2" being typical. The backing creates a moisture vapor barrier. If the duct operates below the **dewpoint** temperature of the outside air, use of a vapor barrier is important in order to prevent the moisture in the air outside the ductwork from condensing on the duct. Once installed, all joints must be properly sealed with duct tape. To avoid condensation damage, any punctures, seams, and slits in the vapor barrier must also be sealed.

Ductwork systems with outside insulation have a lower pressure loss, and are therefore more efficient than systems made from ductboard or sheet metal and lined on the inside. Another advantage is that the cost for the metal duct is cheaper because the physical size of the duct can be smaller. A duct with one inch of internal insulation must have a width and height dimension that is two inches larger in order to deliver the same amount of air.

A disadvantage of using duct with outside insulation is that it takes longer to install and there is a greater chance of damaging the insulation during installation. They also tend to be noisier than a lined system.

Special care must be taken when installing duct systems in crawl spaces, attics, and similar unconditioned areas. When an enclosed crawl space itself is properly insulated and protected by an adequate vapor barrier, the ductwork may be installed without any additional insulation or vapor barrier. If a crawl space is ventilated rather than enclosed, or if an existing vapor barrier is questionable, the duct system should be insulated and must include an adequate vapor barrier.

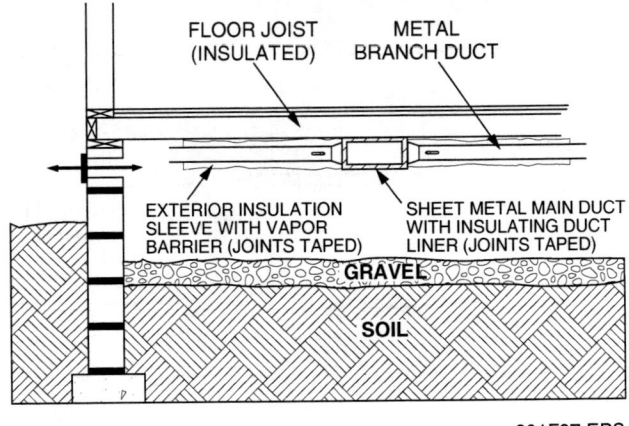

*Figure 37* ◆ Typical duct installation in a ventilated crawl space.

Sheet metal main trunk ductwork with inside insulation and taped joints can be used with no extra vapor sealing. Sheet metal ductwork with outside insulation can also be used, provided it is covered with a vapor barrier.

When branch ductwork is constructed of sheet metal, an external sleeve is normally installed over each branch duct. The insulating sleeve must have a plastic or foil layer outside the insulation to act as a vapor barrier. All vapor barrier joints must be sealed with tape, otherwise the moisture from the air will go through the insulation and condense on the duct, causing the duct to drip and damage the insulation. *Figure 37* shows a typical residential installation for a ventilated crawl space.

In attic installations, the ductwork must be insulated in order to maintain proper cooling/heating in the conditioned rooms of the building. ASHRAE Standard 90-80 specifies the minimum acceptable R-value of insulation that must be used. *Figure 38* shows an example of ASHRAE Standard 90-80 used to calculate the R-value for insulation related to the cooling mode. The R-value must also be calculated for the heating mode in the same way. The amount of insulation actually used for the system is determined by the mode with the greatest need for insulation. Since attic systems are more common in warm climates, it is the cooling mode that usually determines the amount of insulation required.

Uninsulated sheet metal ductwork provides no resistance to heat transmission. This means that the entire resistance for the main supply duct must come from insulation either inside or outside of the duct. Typically, one inch of insulation is equal to an R-value of 4. Branch runs are usually made using pre-insulated, flexible duct, but

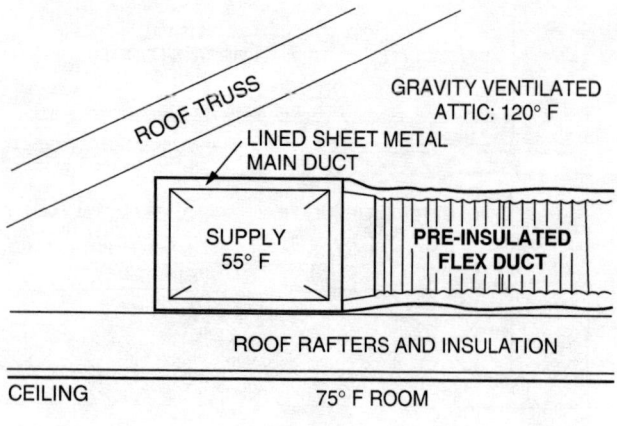

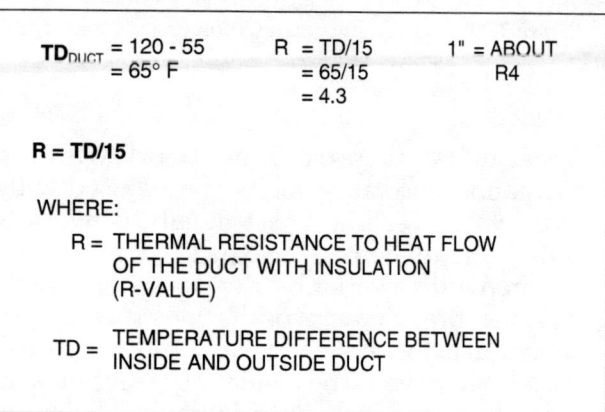

*Figure 38* ◆ Example of R-value calculation using ASHRAE Standard 90-80.

other material combinations can also be used. Even though return duct systems have less difference in air temperature than supply ductwork, they should also be insulated. Any duct system that runs through an unconditioned space and has a temperature difference greater than 15°F should be insulated.

## 6.0.0 ◆ TEMPERATURE AND HUMIDITY MEASUREMENT INSTRUMENTS

Several instruments are used to measure the temperature and humidity of the conditioned air when working on air distribution systems. Three of the most common instruments include:

- Electronic thermometers
- Psychrometers
- Hygrometers

## 6.1.0 Electronic Thermometers

Electronic thermometers (*Figure 39*) are used for measuring **dry-bulb temperatures** and **wet-bulb temperature**s in air distribution systems and other HVAC equipment.

Electronic thermometers display the temperature on either an analog meter or a digital readout. However, digital thermometers are the most commonly used instrument for HVAC field service work. They use either a thermocouple or thermistor-type temperature probe, or both, to sense the heat and generate the temperature reading. Often, several different probes are used with the same instrument to allow measurement of a wide range of temperatures. Many electronic thermometers have two or more probes so that measurements can be made at several locations within the equipment at the same time. Most electronic thermometers of this type can calculate and display the difference in temperature between the different locations being measured.

Many digital multimeters (DMMs) can also be used to measure temperature. This feature requires the use of thermocouple and/or thermistor-temperature probe accessories that convert the DMM into an electronic thermometer. Some DMMs can use a non-contact infrared probe to measure temperature.

Electronic thermometers are precise measuring instruments. Be sure to read and follow the manufacturer's instructions for operating electronic thermometers. Also, be sure to follow the manufacturer's instructions for calibration of the instrument.

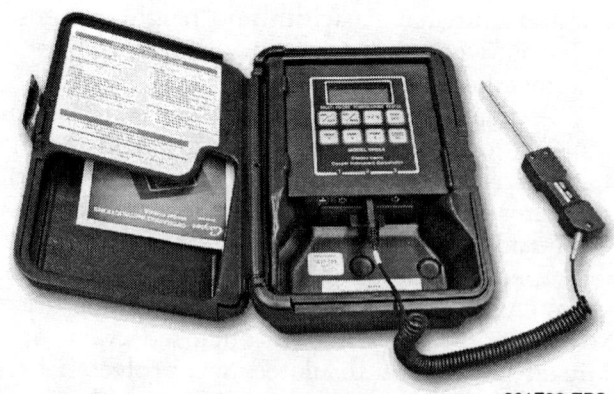

*Figure 39* ◆ Electronic thermometer.

## *Psychrometrics*

Our atmosphere consists of a mixture of air (mostly nitrogen and oxygen) and water vapor. The study of air and its properties is called **psychrometrics**. In 1911, Dr. Willis Carrier presented his Rational Psychrometric Formula to the American Society of Mechanical Engineers. This formula led to the development of the psychrometric chart like the one shown here. It gives a graphical representation of the interrelationships that exist for all the properties of air. A psychrometric chart is typically used by HVAC engineers, designers, and technicians to predict the values for the various properties of air when designing an HVAC system, or before adjusting or modifying an existing HVAC system.

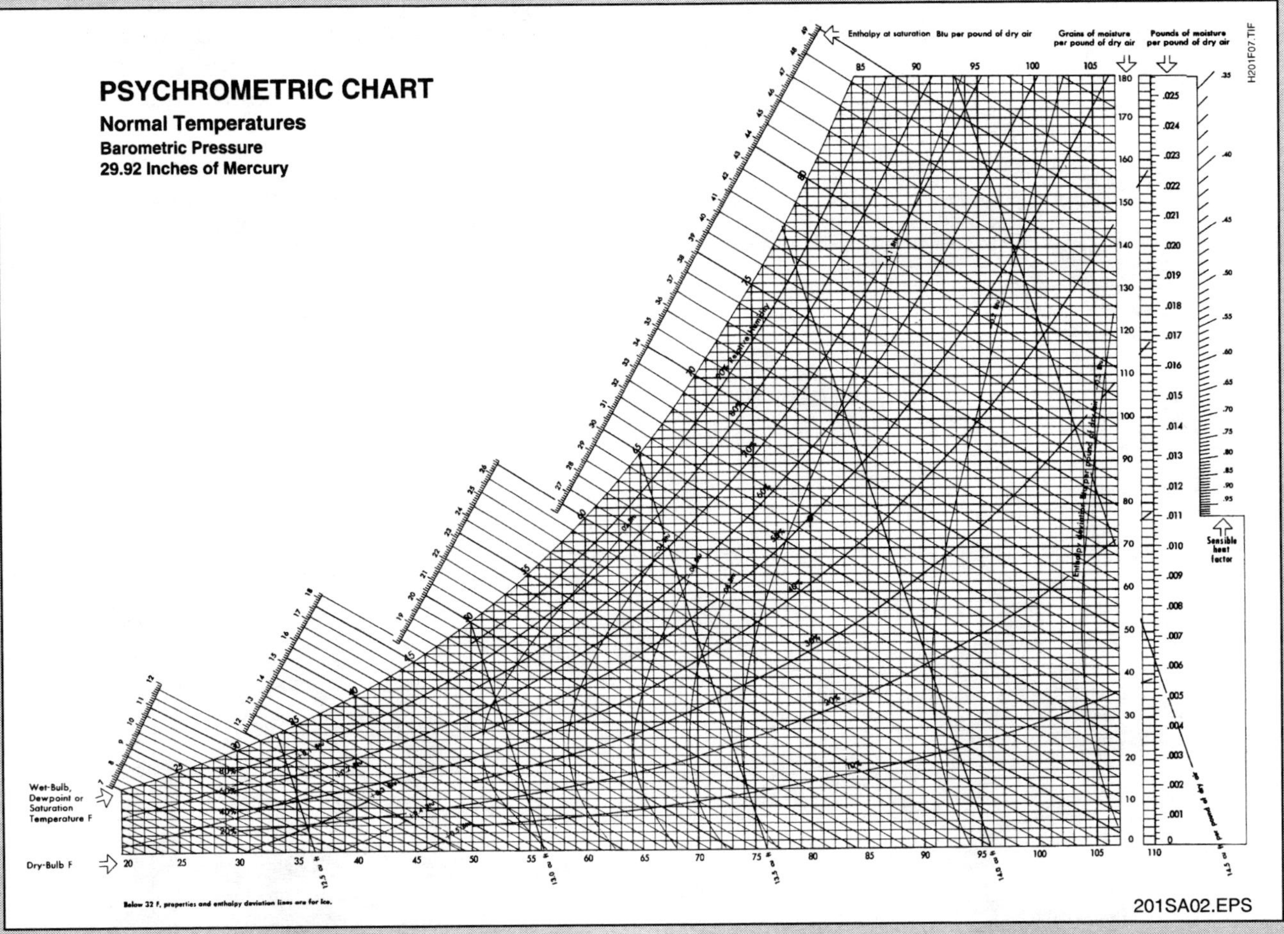

**PSYCHROMETRIC CHART**
Normal Temperatures
Barometric Pressure
29.92 Inches of Mercury

## 6.2.0 Psychrometers

Psychrometers are used to measure temperature. They have two thermometers, one to measure the dry-bulb temperature and the other to measure the wet-bulb temperature. A sling psychrometer (*Figure 40*) is often used to measure temperatures when working on air distribution systems. The sensing bulb of the wet-bulb thermometer is covered with a wick that is saturated with distilled water before taking a reading. To make sure the wet-bulb temperature is accurate, the sling psychrometer must be spun rapidly in the air that is being tested. This is necessary so that evaporation occurs at the wet-bulb thermometer, giving it a lower temperature reading. The measured wet-bulb and dry-bulb temperatures can be used to find the percent of **relative humidity (RH)** in the air. This is done using either a built-in chart on the psychrometer or a separate psychrometric chart. Relative humidity is the ratio of the amount of moisture present in a given sample of air to the amount it can hold at saturation. Relative humidity is expressed as a percentage.

AIR DISTRIBUTION SYSTEMS — TRAINEE MODULE 03201

Squeeze-bulb and battery-operated aspirating psychrometers are also available. These are used in confined spaces where the use of the sling psychrometer is restricted. The squeeze-bulb aspirating psychrometer is operated by rapidly squeezing the bulb to draw air over the thermometers. In the battery-operated psychrometer, a fan draws air over the thermometers.

### 6.3.0 Hygrometers

Hygrometers or relative humidity meters are used to measure and give direct readings of relative humidity. Many varieties of hygrometers are available, including both electronic and dial types.

For field service work, the electronic hygrometer is the most commonly used. Electronic hygrometers (*Figure 41*) have temperature/humidity sensing probes used to measure the conditioned air. The measured relative humidity and/or temperature readings are usually displayed on a digital readout.

Some hygrometers are also capable of measuring and giving a direct reading of the dewpoint. Dewpoint is the temperature at which water vapor in the air becomes saturated and starts to condense into water droplets. Some hygrometers use two or more temperature probes to calculate the differential temperature measured between the probes. Still others are available that can display temperature and relative humidity readings simultaneously. Be sure to follow the manufacturer's instructions for hygrometer usage and calibration.

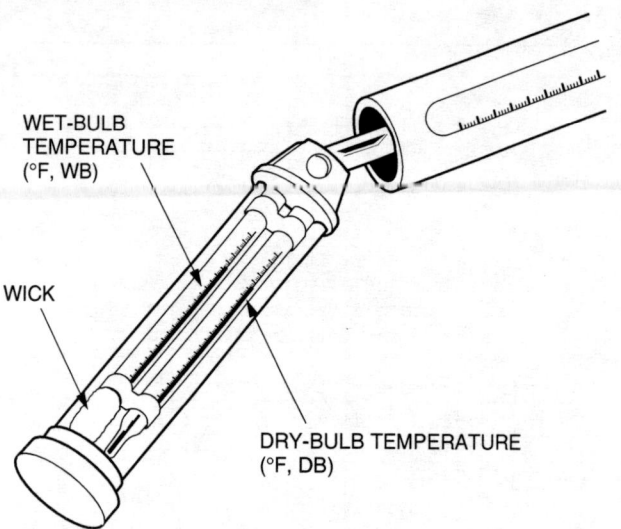

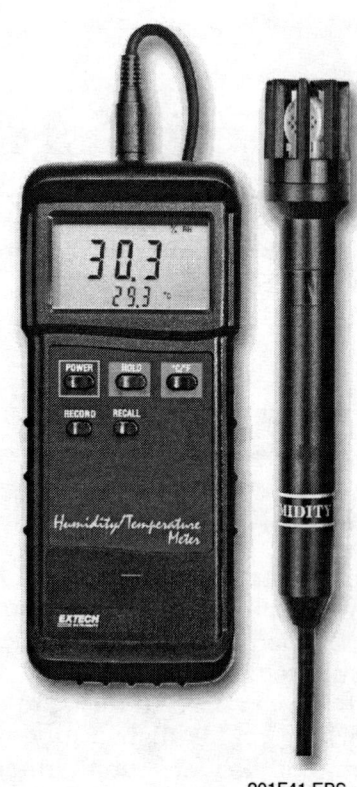

*Figure 40* ◆ Sling psychrometers.

*Figure 41* ◆ Electronic hygrometers.

# 7.0.0 ♦ AIR DISTRIBUTION SYSTEM MEASUREMENT INSTRUMENTS

Several instruments and accessories are used to measure the static, velocity, and total pressures in an air distribution system. Among these are three common devices:

- Manometers
- Differential pressure gauges
- Pitot tubes and static pressure tips

## 7.1.0 Manometers

Manometers are used to measure the low-level static, velocity, and total air pressures found in air distribution duct systems. Manometers used for air distribution servicing are calibrated in inches of water column (in. w.c.). Manometers can use water or oil as the measuring fluid. Popular ones use an oil which has a specific gravity of 0.826 as the measuring fluid. The manufacturer of the gauge specifies the type of oil to be used, so substitution for the specified oil is not recommended. Manometers come in many types, including U-tube, inclined, and combined U-inclined. Electronic manometers are also widely used. Pitot tubes or static pressure tips, described later in this section, are almost always used with manometers when measuring pressures in duct systems. *Figure 42* shows three types of manometers.

Manometers work on the principle that air pressure is indicated by the difference in the level of a column of liquid in the two sides of the instrument. If there is a pressure difference, the column of liquid will move until the liquid level in the low-pressure side is high enough so that its weight and the low air pressure being measured will equal the higher pressure in the other tube.

Individual U-tube and inclined manometers are available in many pressure ranges. Inclined manometers are usually calibrated in the lower pressure ranges and are more sensitive than U-tube manometers. U-inclined manometers combine both the sensitivity of the inclined manometer with the high-range capability of the U-tube manometer in one instrument. Inclined-vertical manometers combine an inclined section for high accuracy and a vertical manometer section for extended range. They also have an additional scale that indicates air velocity in feet per minute (fpm). To get accurate readings with inclined and vertical-inclined manometers, the inclined portion of the scale must be at the exact angle for

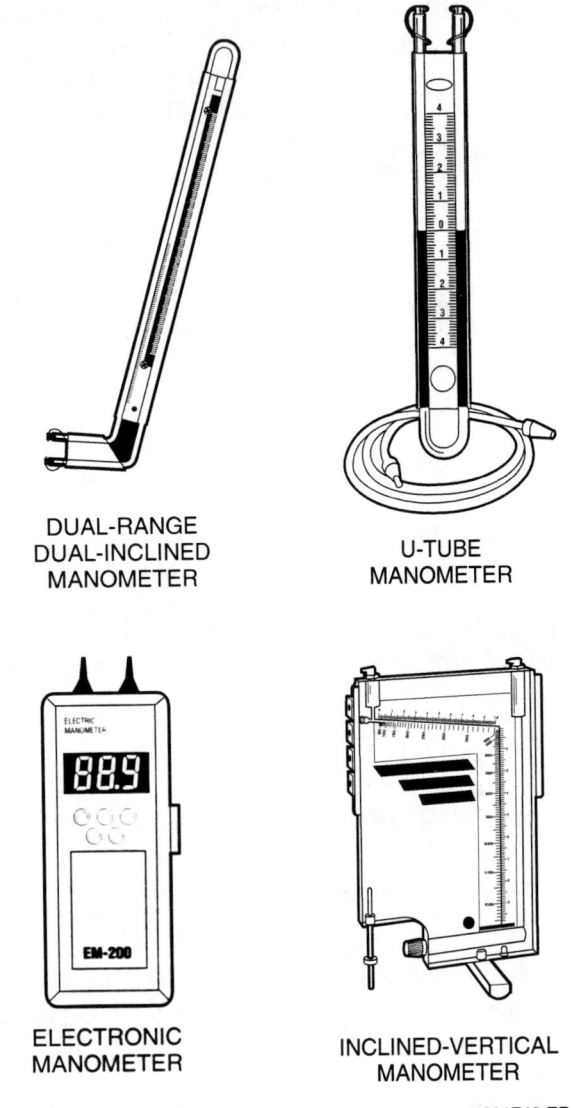

*Figure 42* ♦ Manometers.

which it is designed. A built-in spirit level is used for this purpose. Most also have a screw-type leveling adjustment.

Electronic manometers can typically measure differential pressures of –1 in. wc. to 10 in. w.c. Many can give direct air velocity readings in the range of 300 fpm to 9,990 fpm, eliminating the need for calculations.

## 7.2.0 Differential Pressure Gauge

The differential pressure gauge provides a direct reading of pressure. These gauges are typically used to measure fan and blower pressures, filter resistance, air velocity, and furnace draft. Some

AIR DISTRIBUTION SYSTEMS — TRAINEE MODULE 03201

are capable of measuring just pressure or both pressure and air velocity. Single-scale pressure models are calibrated in either in. w.c. or psi. Dual-scale gauges are normally calibrated for pressure in in. w.c. and for air velocity in fpm. Several models are available covering pressures from 0.0 in. w.c. to 10 in. w.c., and air velocity ranges from 300 fpm to 12,500 fpm. Normally these gauges are installed in the equipment, but portable models are available. Pitot tubes and/or static pressure tips are normally used with portable models to make air pressure and velocity measurements in air distribution system ductwork. *Figure 43* shows a portable differential pressure gauge.

## 7.3.0 Pitot Tubes and Static Pressure Tips

The pitot tube and static pressure tips (*Figure 44*) are probes used with manometers and pressure gauges when making measurements inside the ductwork of an air distribution system. The standard pitot tube used for measurements in ducts 8" and larger has a 5/16" outer tube with eight equally spaced 0.04" diameter holes used to sense static pressure. For measurements in ducts smaller than eight inches, use of pocket size pitot tubes with a 1/8" outer tube and four equally spaced 0.04" diameter holes are recommended.

The pitot tube consists of an impact tube which receives the total pressure input, fastened concentrically inside a larger tube which receives static pressure input from the radial sensing holes around the tip. The air space between the inner and outer tube permits transfer of pressure from the sensing holes to the static pressure connection at the opposite end of the pitot, and then through the connecting tubing to the low- or negative-pressure side of the manometer.

When the total pressure tube is connected to the high-pressure side of the manometer, velocity pressure is indicated directly. To be sure of accurate velocity pressure readings, the pitot tube tip must be pointed directly into the duct air stream. Pitot tubes come in various lengths ranging from 6" to 60", with graduation marks at every inch to show the depth of insertion in the duct.

Static pressure tips, like pitot tubes, are used with manometers and differential pressure gauges to measure static pressure in a duct system. They are typically L-shaped with four radially drilled 0.04" sensing holes.

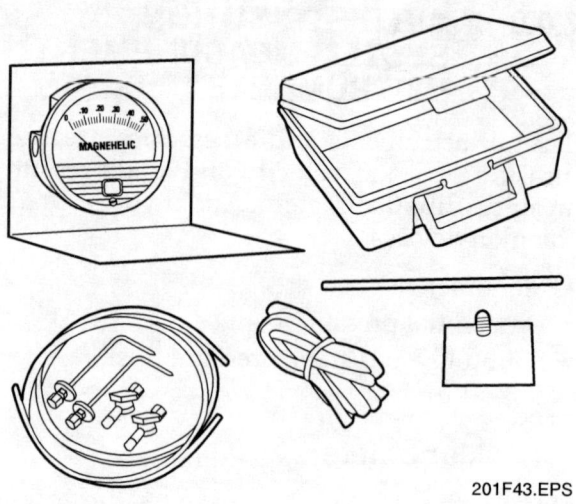

*Figure 43* ◆ Portable differential pressure gauge.

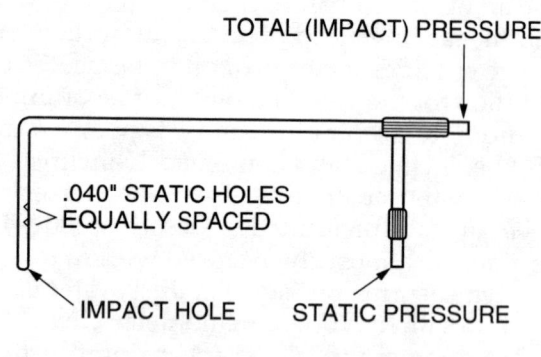

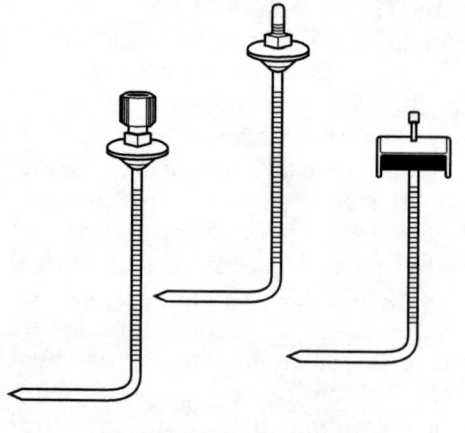

*Figure 44* ◆ Pitot tube and static pressure tips.

## 8.0.0 ◆ AIR VELOCITY MEASUREMENT INSTRUMENTS

Velometers (*Figure 45*) are used to measure the velocity of airflow. Measurement of air velocity is done to check the operation of an air distribution system. It is also done when balancing system airflow.

Most velometers give direct readings of air velocity in fpm. Some can provide direct readings in cfm. Velometers with analog scales and digital readouts are in common use.

Some velometers use a rotating vane (propeller) or balanced swing vane to sense the air movement. When the rotating vane velometer is positioned to make a measurement, the vane rotates at a rate determined by the velocity of the air stream. This rotation is converted into an equivalent velocity reading for display. In the swinging vane velometer, the air stream causes the balanced vane to tilt at different angles in response to the measured air velocity. The position of the vane is converted into an equivalent velocity reading for display.

Another type of velometer, also called a *hot wire anemometer*, gives direct readings of air velocity in fpm. This instrument uses a sensing probe which contains a small resistance heater element. When the probe is held perpendicular to the air stream being measured, the temperature of the heater element changes due to changes in the airflow. This causes its resistance to change, which alters the amount of current flow being applied to the meter circuitry. There, it automatically calculates the air velocity for display on the meter.

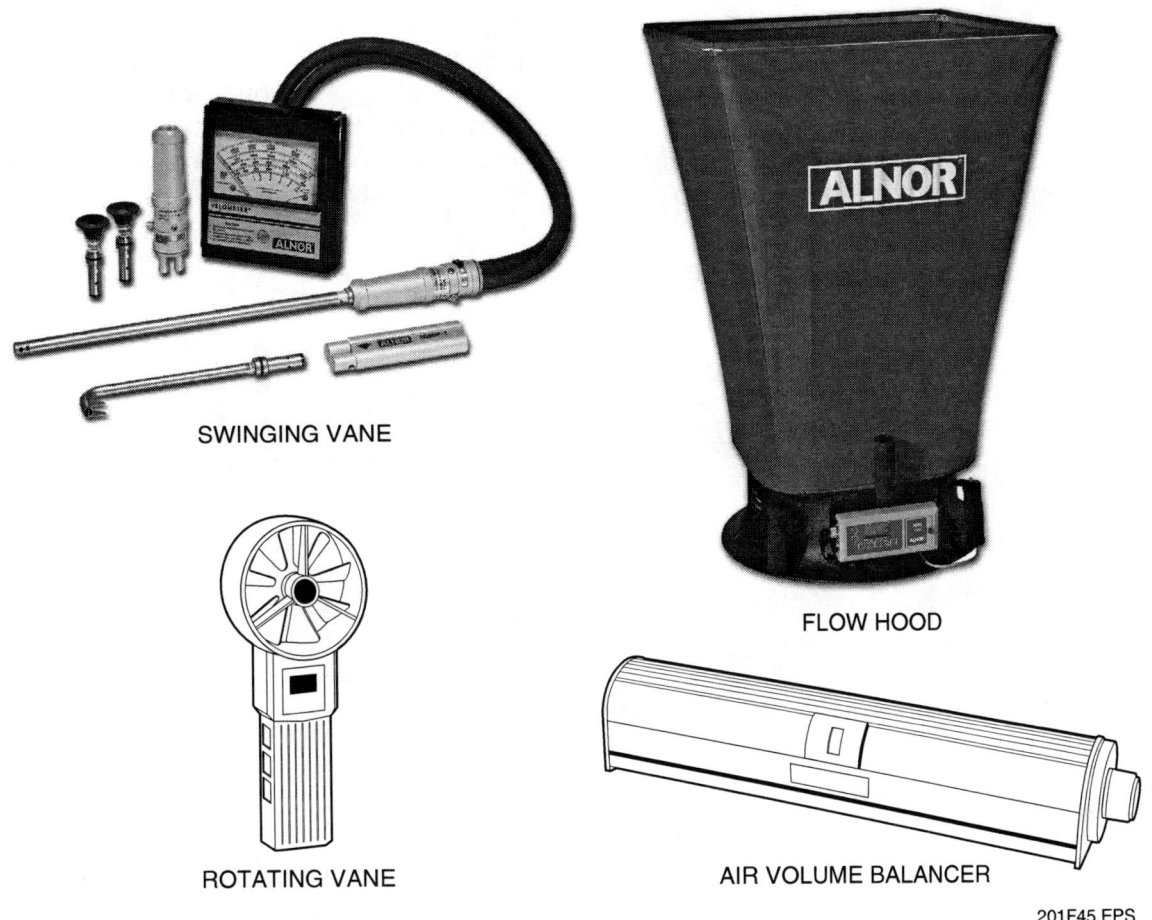

*Figure 45* ◆ Velometers.

Some velometers use probes that have a sensitively balanced vane or a small resistance heater element that, when placed in the air stream, produces a measurement of airflow for display on the velometer meter scale. Depending on the sensing probe or attachment used, velometers can measure air velocities in several ranges within the overall range of 0 to 10,000 fpm. Some electronic velometers that use a microprocessor can automatically average up to 250 individual readings taken across a duct area to provide an average air velocity. Certain velometers also include an optional micro-printer to record the readings.

Special velometers, known as *air volume balancers*, can be used when balancing air distribution systems. This type of velometer is held directly against the diffuser or register to get a direct reading of air velocity. Another type of velometer called a flow hood is frequently used to get direct velocity readings in cfm when measuring the output of large air diffusers in commercial systems.

## Summary

Proper measurement and control of air is necessary so that the human body can feel comfortable in a room environment. It is also critical to many commercial and manufacturing processes. As an HVAC technician, you will need to evaluate the air in a conditioned space, then make knowledgeable decisions pertaining to the service or adjustments that may be needed for the related HVAC equipment. A heating or air conditioning system will perform no better than its air distribution system. All adequate air distribution systems must:

- Supply the right quantity of air to each conditioned space.
- Supply the air in each space so that air motion is adequate but not drafty.
- Condition the air to maintain the proper comfort zones for humans, or the proper conditions needed for a commercial or manufacturing process.
- Provide for the return of air from all conditioned areas to the air handler.
- Operate efficiently without excessive power consumption or noise.
- Operate with minimum maintenance.

The layout and equipment used for air distribution systems is normally determined by the climate in which the system must operate and the construction of the building in which the equipment is installed. Ideally, the system air handler is located in an area that allows the duct length to be as short as possible and the number and types of fittings as few as possible.

## Review Questions

1. Within an air distribution system, the highest pressure level is found at the _____.
   a. conditioned space
   b. input to the return duct
   c. input to the blower
   d. output of the blower

2. The static pressure, velocity pressure, and total pressure measured in a duct system are measured in _____.
   a. psi
   b. inches of mercury
   c. inches of water column
   d. inches of mercury vacuum

3. An air distribution system for a light commercial building is designed to work with a 5-ton cooling system. About how many cfm of airflow should the system blower be capable of supplying?
   a. 1,200 cfm
   b. 1,650 cfm
   c. 2,000 cfm
   d. 2,350 cfm

4. The type of blower or fan most often used in residential heating and air conditioning systems is the _____.
   a. forward-curved centrifugal blower
   b. backward-inclined centrifugal blower
   c. duct fan
   d. radial centrifugal blower

5. An existing air distribution system has an airflow of 6,000 cfm created by a blower operating at a speed of 1,200 rpm. If you wanted to change the airflow to 5,000 cfm, what is the required new blower speed?
   a. 833 rpm
   b. 1,000 rpm
   c. 1,150 rpm
   d. 1,300 rpm

6. External static pressure loss is _____.
   a. the total pressure loss in the air handler and its component parts
   b. the total pressure loss of an air distribution system, excluding the air handler
   c. the total pressure drop across the air handler
   d. None of the above.

7. Perimeter duct systems are used in _____.
   a. warm climates
   b. cold climates
   c. primarily in cold climates, but also homes in warm climates if they are constructed over a basement or crawl space
   d. locations where the outside air temperature difference between the heating and cooling seasons does not exceed 70°F

8. The duct material that has the least friction loss is _____.
   a. sheet metal
   b. fiberglass ductboard
   c. flexible duct
   d. internally insulated sheet metal

9. Each fitting and transition used in a duct systems has _____.
   a. the same friction loss per foot as the same size straight duct
   b. less friction loss per foot as the same size straight duct
   c. a friction loss equal to a predetermined length of the same size straight duct
   d. approximately the same friction loss

10. Dampers built into supply diffusers and registers are used to _____.
    a. add moisture to the air
    b. reduct duct noise
    c. control and balance airflow in a duct system
    d. vent stale air to the outdoors

11. High sidewall registers work best in _____.
    a. cold climates for heating
    b. cold climates for cooling
    c. warm climates for heating
    d. warm climates for cooling

12. A fire damper will close when _____.
    a. the fire has burned for 3 hours
    b. the fusible link melts
    c. the HVAC equipment shuts off
    d. smoke is detected

13. Air duct systems in attics must be insulated with an insulation having an R-value based on the _____.
    a. cooling mode
    b. heating mode
    c. greater value as needed by the cooling or heating mode
    d. insulation manufacturer's recommendations

14. The instrument most likely used to determine relative humidity is a _____.
    a. psychrometer
    b. velometer
    c. pressure gauge
    d. manometer

15. An instrument that can be used to measure pressure losses in duct systems is the _____.
    a. psychrometer
    b. velometer
    c. differential pressure gauge
    d. hygrometer

# GLOSSARY

## Trade Terms Introduced in this Module

*Absolute pressure:* Positive pressure measurements which start at zero (no pressure). Also, gauge pressure plus the pressure of the atmosphere, normally 14.7 psi at sea level at 70°F.

*Air handler:* The device that moves the air across the heat exchanger in a forced-air system. In a split system, it normally contains the blower fan, cooling coil, metering device, air filter, and related housing.

*ASHRAE:* American Society of Heating, Refrigeration, and Air Conditioning Engineers.

*Atmospheric pressure:* The pressure exerted on all things on Earth's surface as the result of the weight of our atmosphere.

*Cubic feet per minute (cfm):* A measure of the amount or volume of air in cubic feet flowing past a point in one minute. Cubic feet per minute can be calculated by multiplying the velocity of air, in feet per minute (fpm), times the area it is moving through, in square feet (cfm = fpm × area).

*Dewpoint:* The temperature at which water vapor in the air becomes saturated and starts to condense into water droplets.

*Dry-bulb temperature:* The temperature measured using a standard thermometer. It represents a measure of the sensible heat of the air or surface being tested.

*Free air delivery:* The condition that exists when there are no effective restrictions to airflow (no static pressure) at the inlet or outlet of an air-moving device.

*Gauge pressure:* The pressure measured on a gauge, expressed as *psig* or *inches water column (in. w.c.)*. Also, pressure measurements that are compared with atmospheric pressure.

*Plenum:* A sealed chamber at the inlet or outlet of an air handler. The duct attaches to the plenum.

*Psychrometrics:* The study of air and its properties.

*Relative humidity (RH):* The ratio of the amount of moisture present in a given sample of air to the amount it can hold at saturation. Relative humidity is expressed as a percentage.

*Revolutions per minute (rpm):* The speed at which the shaft of an air-moving device is rotating.

*SMACNA:* Sheet Metal and Air Conditioning Contractors National Association.

*Static pressure (s.p.):* The pressure exerted uniformly in all directions within a duct system, as measured in in. w.c.

*Total pressure:* The sum of the static pressure and the velocity pressure in an air duct. It is the pressure produced by the fan or blower.

*Velocity:* How fast air is moving. The rate of airflow usually measured in feet per minute.

*Velocity pressure:* The pressure in a duct due to the movement of the air. It is the difference between the total pressure and the static pressure in w.c.

*Venturi:* A ring or panel surrounding the blades on a propeller fan; used to improve fan performance.

*Volume:* The amount of air in cubic feet flowing past a point in one minute (cfm).

*Wet-bulb temperature:* Temperature taken with a thermometer that has a wick wrapped around its sensing bulb that is saturated with clean, distilled water before taking a reading. The reading from a wet-bulb thermometer, through evaporation of the distilled water, takes into account the moisture content of the air. It reflects the total heat content of the air.

# APPENDIX

# *Typical Air Distribution System Duct and Supply Outlet Data*

## SUPPLY OUTLETS

### FLOOR OUTLETS – PERIMETER

| CFM | SIZE (IN.) | APPROX. SPREAD (FT.) | FACE VELOCITY (FPM) | FREE AREA (SQ. IN.) |
|---|---|---|---|---|
| 70  | 2-1/4 x 10 | 9  | 535 | 18.6 |
| 80  | 2-1/4 x 12 | 10 | 565 | 21.1 |
| 100 | 2-1/4 x 14 | 11 | 610 | 23.6 |
| 110 | 4 x 10     | 10 | 500 | 32.4 |
| 135 | 4 x 12     | 13 | 500 | 39.0 |
| 175 | 4 x 14     | 14 | 555 | 45.5 |

### LOW SIDEWALL – PERIMETER

| CFM | SIZE (IN.) | APPROX. SPREAD (FT.) | FACE VELOCITY (FPM) | FREE AREA (SQ. IN.) |
|---|---|---|---|---|
| 80  | 10 x 6 | 13 | 430 | 26.7 |
| 100 | 12 x 6 | 10 | 440 | 32.6 |
| 120 | 14 x 6 | 8  | 450 | 38.4 |

### BASEBOARD

| CFM | SIZE (FT.) | APPROX. SPREAD (FT.) | OUTLET VELOCITY (FPM) | FREE AREA (SQ. IN.) |
|---|---|---|---|---|
| 80 | 2 | 7.5 | 430 | 26.6 |

### HIGH SIDEWALL

| CFM | SIZE (IN.) | HORIZ. THROW (FT.) | FACE VELOCITY (FPM) | FREE AREA (SQ. IN.) |
|---|---|---|---|---|
| 80  | 10 x 4 | 8   | 390 | 29.0 |
| 125 | 10 x 6 | 10  | 415 | 43.3 |
| 150 | 12 x 6 | 10  | 410 | 52.7 |
| 165 | 14 x 6 | 9.5 | 375 | 62.1 |

### ROUND CEILING OUTLETS

| CFM | SIZE (IN.) | HORIZ. THROW (FT.) | OUTLET VELOCITY (FPM) | FREE AREA (SQ. IN.) |
|---|---|---|---|---|
| 45  | 6  | 3    | 500 | 12.2 |
| 105 | 8  | 5    | 580 | 26.1 |
| 185 | 10 | 7    | 580 | 43.8 |
| 285 | 12 | 8.5  | 575 | 65.7 |
| 425 | 14 | 10.5 | 600 | 91.9 |

### SQUARE CEILING OUTLETS

| CFM | SIZE (IN.) | HORIZ. THROW (FT.) | OUTLET VELOCITY (FPM) | FREE AREA (SQ. IN.) |
|---|---|---|---|---|
| 50  | 6 x 6   | 3.5 | 450 | 15.4 |
| 135 | 8 x 8   | 5   | 550 | 35.1 |
| 250 | 10 x 10 | 6   | 620 | 58.1 |
| 325 | 12 x 12 | 7   | 550 | 85.1 |

Tables based on Lima Register Co. Catalog Data, @ .028 inch w.g. drop across outlet.

## EQUIVALENT RECTANGULAR DUCT SIZES

| CFM | Supply | | | Return | | | CFM |
|---|---|---|---|---|---|---|---|
| 200  | 10 x 6 | 8 x 8  |        | 12 x 6 | 10 x 8 |        | 200 |
| 300  | 12 x 6 | 10 x 8 | 10 x 10| 16 x 6 | 12 x 8 | 10 x 10| 300 |
| 400  | 16 x 6 | 12 x 8 | 11 x 10| 20 x 6 | 14 x 8 | 11 x 10| 400 |
| 500  | 18 x 6 | 14 x 8 | 12 x 10| 24 x 6 | 16 x 8 | 12 x 10| 500 |
| 600  | 22 x 6 | 16 x 8 | 12 x 12| 20 x 8 | 16 x 10| 12 x 12| 600 |
| 700  | 18 x 8 | 14 x 10| 13 x 12| 22 x 8 | 18 x 10| 14 x 12| 700 |
| 800  | 20 x 8 | 16 x 10| 14 x 12| 24 x 8 | 19 x 10| 15 x 12| 800 |
| 900  | 22 x 8 | 16 x 10| 16 x 12| 26 x 8 | 20 x 10| 17 x 12| 900 |
| 1000 | 24 x 8 | 18 x 10| 18 x 12| 30 x 8 | 22 x 10| 18 x 12| 1000 |
| 1200 | 28 x 8 | 22 x 10| 20 x 12| 36 x 8 | 26 x 10| 20 x 12| 1200 |
| 1400 | 32 x 8 | 24 x 10|        | 30 x 10| 24 x 12|        | 1400 |
| 1600 | 28 x 10| 22 x 12|        | 34 x 10| 26 x 12|        | 1600 |
| 1800 | 30 x 10| 24 x 12|        | 37 x 10| 30 x 12|        | 1800 |
| 2000 | 32 x 10| 26 x 12|        | 42 x 10| 32 x 12|        | 2000 |

SUPPLY sizes based on friction rate of .08 in. w.g. per 100 ft. equivalent length metal duct.
RETURN sizes based on friction rate of .05 in. w.g. per 100 ft. equivalent length metal duct.

### RETURN AIR GRILLES

| CFM | SIZE (IN.) | FREE AREA (SQ. IN.) |
|---|---|---|
| 100 | 10 x 6 | 36.4  |
| 125 | 12 x 6 | 44.4  |
| 170 | 12 x 8 | 61.0  |
| 145 | 14 x 6 | 52.4  |
| 200 | 14 x 8 | 72.0  |
| 245 | 24 x 6 | 89.6  |
| 335 | 24 x 8 | 122.0 |
| 310 | 30 x 6 | 110.8 |
| 425 | 30 x 8 | 152.0 |

### VERTICAL STACKS

| SUPPLY CFM | STACK SIZE (IN.) | RETURN CFM |
|---|---|---|
| 100 | 3-1/4 x 10 | 75  |
| 125 | 3-1/4 x 12 | 90  |
| 150 | 3-1/4 x 14 | 110 |

2-1/4" stacks = 55% of 3-1/4" stack capacity.

### PANNED JOIST (16 IN. O.C.)

| RETURN CFM | NOMINAL JOIST DEPTH (IN.) | ACTUAL JOIST DEPTH (IN.) |
|---|---|---|
| 260 | 6  | 5-1/2 |
| 375 | 8  | 7-1/2 |
| 525 | 10 | 9-1/2 |

# ANSWER KEY

## Answers to Review Questions

| Answer | Section |
|---|---|
| 1. d | 2.1.0 |
| 2. c | 2.1.0 |
| 3. c | 2.2.0 |
| 4. a | 3.2.1 |
| 5. b | 3.4.0 |
| 6. b | 3.6.0 |
| 7. c | 4.1.0, 4.2.0 |
| 8. a | 5.1.2, 5.1.3, 5.5.0 |
| 9. c | 5.2.0 |
| 10. c | 5.4.0 |
| 11. d | 5.3.0 |
| 12. b | 5.5.0 |
| 13. c | 5.6.0 |
| 14. a | 6.2.0 |
| 15. c | 7.2.0 |

# REFERENCES & ACKNOWLEDGMENTS

## Additional Resources

This module is intended to present thorough resources for task training. The following reference works are suggested for further study. These are optional materials for continued education rather than for task training.

*Modern Refrigeration and Air Conditioning*, 2000. A.D. Althouse, C.H. Turnquist, A.F. Bracciano. Tinley Park, IL: The Goodheart-Willcox Company, Inc.

*Refrigeration & Air Conditioning Technology*, 2000. William C. Whitman, William M. Johnson, John A. Tomczyk. Albany, NY: Delmar Publishers, Inc.

*Residential Air System Design*, 1993. Syracuse, NY: Carrier Corporation.

## Figure Credits

| | |
|---|---|
| Gerald Shannon | 201F27, 201F39 |
| Hart & Cooley, Inc. | 201SA01, 201F34 |
| Nailor Industries, Inc. | 201F36 |
| Carrier, Corporation | 201SA02 |
| Supco, Inc. | 201F40 |
| Extech | 201F41 |
| Alnor® Instrument Company | 201F45 |

# NCCER CRAFT TRAINING USER UPDATES

The NCCER makes every effort to keep these textbooks up-to-date and free of technical errors. We appreciate your help in this process. If you have an idea for improving this textbook, or if you find an error, a typographical mistake, or an inaccuracy in the NCCER's Craft Training textbooks, please write us, using this form or a photocopy. Be sure to include the exact module number, page number, a detailed description, and the correction, if applicable. Your input will be brought to the attention of the Technical Review Committee. Thank you for your assistance.

*Instructors* – If you found that additional materials were necessary in order to teach this module effectively, please let us know so that we may include them in the Equipment and Materials list in the Instructor's Guide.

**Write:** Curriculum Revision and Development Department
National Center for Construction Education and Research
P.O. Box 141104, Gainesville, FL 32614-1104

**Fax:** 352-334-0932

**E-mail:** curriculum@nccer.org

---

Craft _____ Module Name _____

Copyright Date _____ Module Number _____ Page Number(s) _____

Description _____

_____

_____

_____

(Optional) Correction _____

_____

_____

(Optional) Your Name and Address _____

_____

_____

Module 03202-01

# Chimneys, Vents, and Flues

## COURSE MAP

This course map shows all of the modules in the second level of the HVAC curriculum. The suggested training order begins at the bottom and proceeds up. Skill levels increase as you advance on the course map. The local Training Program Sponsor may adjust the training order.

### HVAC LEVEL TWO

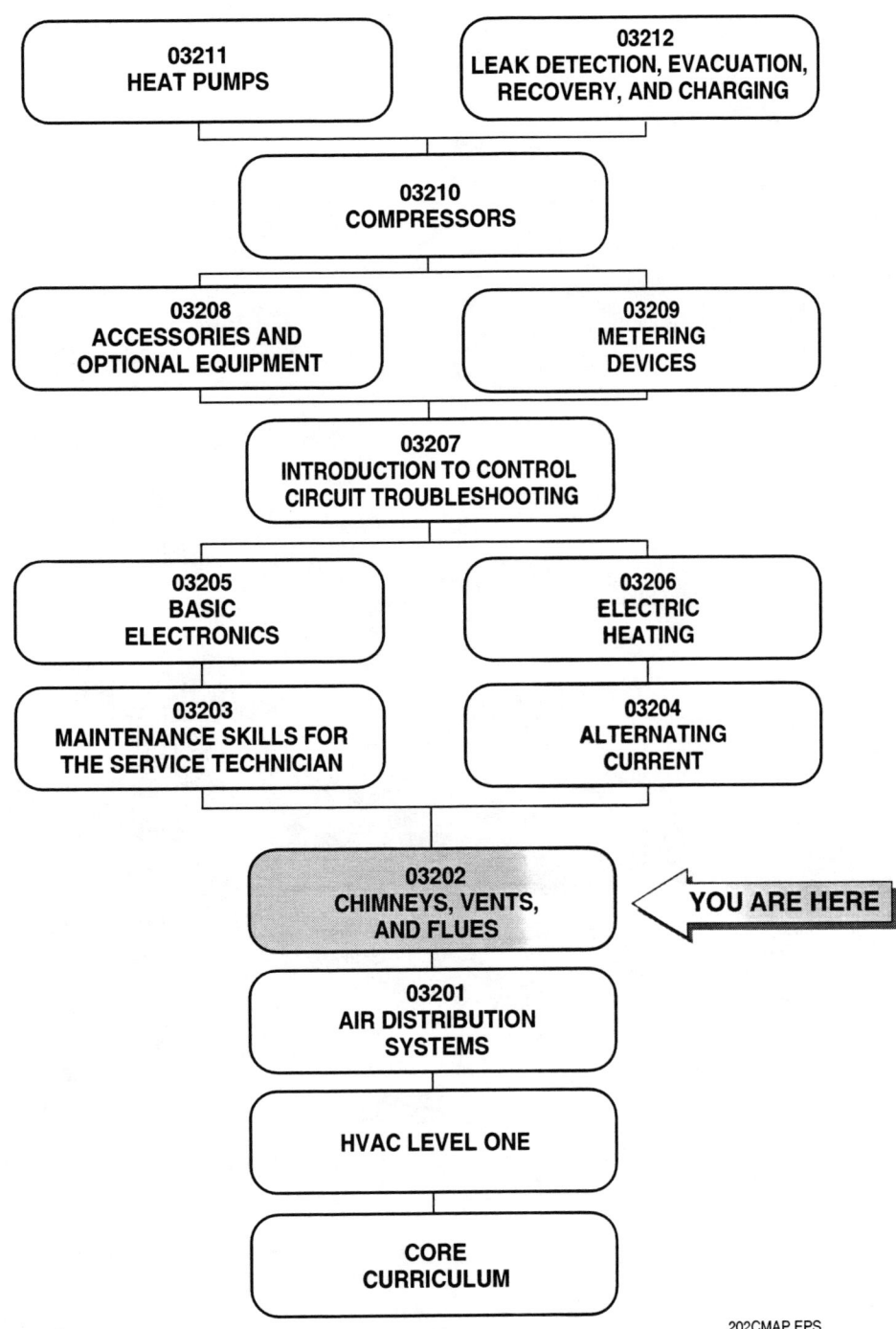

Copyright © 2001 National Center for Construction Education and Research, Gainesville, FL 32614-1104. All rights reserved. No part of this work may be reproduced in any form or by any means, including photocopying, without written permission of the publisher.

# MODULE 03202 CONTENTS

1.0.0 INTRODUCTION .................................................. 2.1
2.0.0 COMBUSTION ................................................... 2.1
    2.1.0 Complete Combustion ..................................... 2.2
    2.2.0 Incomplete Combustion ................................... 2.2
    2.3.0 Combustion Efficiency ................................... 2.2
    2.4.0 Flames .................................................. 2.2
3.0.0 FLUE GASES ................................................... 2.3
4.0.0 FURNACE VENTING .............................................. 2.4
    4.1.0 Requirements ............................................ 2.4
    4.2.0 Clearances .............................................. 2.5
    4.3.0 Air Supply .............................................. 2.5
5.0.0 VENT SYSTEM COMPONENTS ....................................... 2.6
6.0.0 NATURAL-DRAFT FURNACES ....................................... 2.7
7.0.0 INDUCED-DRAFT GAS FURNACES ................................... 2.8
    7.1.0 Furnace Sizing .......................................... 2.8
    7.2.0 Burner Input Adjustment ................................. 2.9
    7.3.0 Temperature Rise Adjustment ............................. 2.9
    7.4.0 Thermostat Heat Anticipator Adjustment .................. 2.9
    7.5.0 Venting Considerations .................................. 2.10
        7.5.1 *General Guidelines for Metal Vents and Vent Connectors* ... 2.10
        7.5.2 *Venting Through a Masonry Chimney* ............. 2.11
8.0.0 CONDENSING GAS FURNACES ...................................... 2.11
9.0.0 DRAFT CONTROLS ............................................... 2.14
    9.1.0 Draft Regulator ......................................... 2.14
    9.2.0 Vent Dampers ............................................ 2.14
    9.3.0 Draft Diverters ......................................... 2.15
SUMMARY ............................................................ 2.16
REVIEW QUESTIONS ................................................... 2.17
GLOSSARY ........................................................... 2.18
ANSWERS TO REVIEW QUESTIONS ........................................ 2.19
REFERENCES & ACKNOWLEDGMENTS ....................................... 2.20

## Figures

Figure 1     Furnace venting ........................................... 2.2
Figure 2     Combustion air ............................................ 2.3
Figure 3     Furnace room venting .................................... 2.5
Figure 4     Components of a factory-built chimney ............ 2.6
Figure 5     Natural-draft furnace .................................... 2.7
Figure 6     Natural drawing action .................................. 2.8
Figure 7     Measuring supply and return air temperature .... 2.9
Figure 8     Thermostat heat anticipator adjustment ........... 2.10
Figure 9     Venting through a metal pipe ........................ 2.10
Figure 10    Flexible chimney liner kit .............................. 2.12
Figure 11    PVC selection chart .................................... 2.13
Figure 12    Terminating PVC vent and combustion air pipes .. 2.14
Figure 13    Concentric termination ................................. 2.14
Figure 14    Draft regulator .......................................... 2.15
Figure 15    Vent damper ............................................. 2.15
Figure 16    Draft diverter ............................................ 2.15

# MODULE 03202

# Chimneys, Vents, and Flues

## Objectives

When you have completed this module, you will be able to do the following:

1. Describe the principles of combustion and explain complete and incomplete combustion.
2. Describe the content of flue gas and explain how it is vented.
3. Identify the components of a furnace vent system.
4. Describe how to select and install a vent system.
5. Perform the adjustments necessary to achieve proper combustion in a gas furnace.
6. Describe the techniques for venting different types of furnaces.
7. Explain the various draft control devices used with natural-draft furnaces.

## Prerequisites

Before you begin this module, it is recommended that you successfully complete the following modules: Core Curriculum; HVAC Level One; HVAC Level Two, Module 03201.

## Required Trainee Materials

1. Pencil and Paper
2. Appropriate Personal Protective Equipment

## 1.0.0 ◆ INTRODUCTION

All fossil-fuel furnaces produce flue gases as a by-product of burning fuel. These gases contain materials that are dangerous. In addition, they contain moisture, soot, and acids that can damage equipment. Flue gases must be vented to the outdoors in order for occupants to avoid their harmful effects (*Figure 1*). The design of the **vent** system depends on the building construction, the type of furnace, and the temperature of the flue gases.

Proper venting is especially important in **induced-draft furnaces**. These are furnaces with an Annual Fuel Utilization Efficiency (AFUE) rating of 78% to 85%. Because of their low flue gas temperatures, these furnaces need to be designed in a way that will prevent the formation of condensation. Moisture can damage the furnace and vent.

**Natural-draft furnaces** are no longer made in large numbers because they cannot meet the minimum AFUE standard of 78% without the use of special accessories. Although you may service them, it is unlikely that you will have to install one.

High-efficiency **condensing furnaces** (AFUE of 90% and higher) are fairly easy to vent. The condensing coil removes much of the moisture before the flue gases reach the vent. The furnace is also equipped to capture and dispose of any condensation that forms.

## 2.0.0 ◆ COMBUSTION

During combustion, oxygen combines with fuel to release stored energy in the form of heat. There are three conditions necessary for combustion to take place:

- First, there must be fuel. The fuel can be gas, such as natural gas; liquid, such as fuel oil; or solid, such as coal. Two elements that all fuels have in common are carbon and hydrogen.

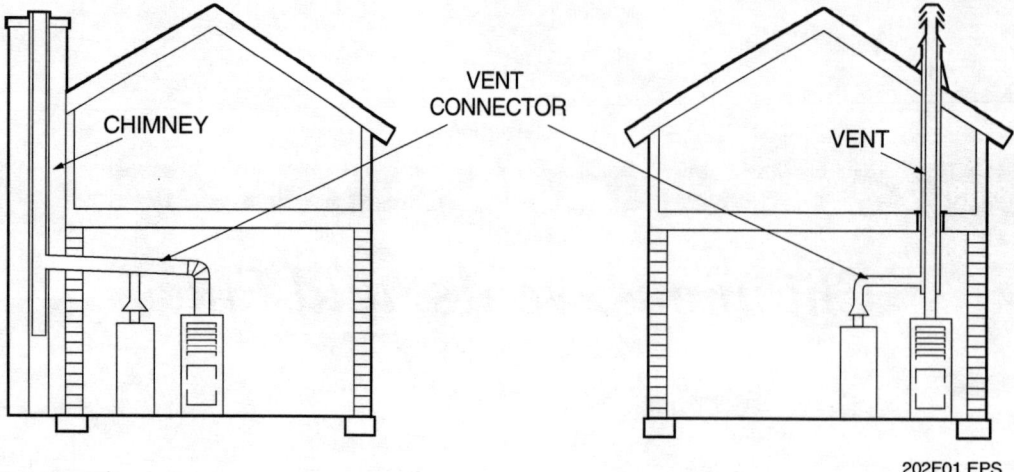

*Figure 1* ♦ Furnace venting.

- Second, fuel must be heated in order to burn or to reach the kindling temperature. A pilot burner or electronic ignition is used to ignite a gas burner, an electric spark is used to ignite fuel oil, and a wood-burning fire is used to ignite coal.
- Third, oxygen must be present for burning to take place.

There are two types of combustion: **complete combustion** and **incomplete combustion**. Incomplete combustion is dangerous; therefore, complete combustion must be obtained in all fuel-burning systems.

## 2.1.0 Complete Combustion

Complete combustion takes place when carbon combines with oxygen to form carbon dioxide. Carbon dioxide is nontoxic and can be exhausted to the outdoors. Hydrogen combines with oxygen to form water vapor, which is also harmless.

## 2.2.0 Incomplete Combustion

Incomplete combustion results from too little oxygen and causes the formation of undesirable products, such as carbon monoxide, pure carbon or soot, and aldeheydes (highly reactive compounds). Both carbon monoxide and aldehydes are toxic. Soot coats the heating surfaces of the furnace and reduces heat transfer.

Enough air must be provided to allow for proper combustion to take place, and to avoid incomplete combustion. In practice, 15% to 30% excess air has been found to provide satisfactory combustion without seriously lowering burner efficiency. Operating burners at a lower percentage of excess air is not practical, because the small improvement in efficiency may not offset the hazards that may be created.

## 2.3.0 Combustion Efficiency

When fuel is burned in a furnace, a certain amount of heat is lost in the hot gases that go out through the vent. This heat loss is necessary to establish a draft in the chimney or vent, but should be minimized to allow the furnace to operate at its peak efficiency. For example, if the amount of heat lost is 20%, the furnace efficiency would be 80%.

Air entering the furnace at room temperature or lower is heated to flue gas temperatures that range from 100°F to 600°F, depending upon the design and adjustments of the furnace. The flue gas temperature in a natural-draft furnace ranges from 350°F to 600°F; from 275°F to 400°F in an induced-draft furnace; and from 100°F to 125°F in a high-efficiency furnace.

The acceptable minimum amount of carbon dioxide should be about 8.5% for natural gas, with no carbon monoxide. For oil, it should be about 10% without any smoke.

## 2.4.0 Flames

The type of flame and the intensity with which it burns affects the efficiency of the heating unit. Pressure-type oil burners burn with a yellow flame. Bunsen-type gas burners burn with a blue flame. The difference is mainly due to the manner in which air is mixed with the fuel.

The color of a gas flame indicates the amount of air being supplied for combustion. A yellow flame is produced when gas is burned by igniting it as it flows out of the open end of a pipe. A blue flame is produced when about 50% of the required air is mixed with the gas prior to ignition. This mix is called **primary air** (*Figure 2*). The balance of air, called **secondary air**, is supplied during combustion to the exterior of the flame. These air mixtures are adjustable. Improper gas

### Furnace Flame Color

When servicing a furnace, inspect the flame to make sure it is the correct color. Any deviation from the proper color indicates a problem. A gas flame should be blue with an orange tip. An oil flame should be solid yellow.

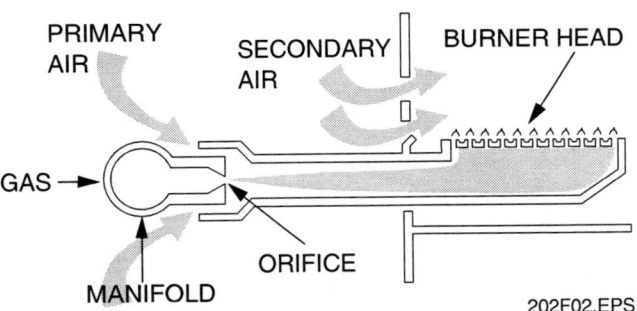

*Figure 2* ◆ Combustion air.

flames are the result of inefficient or incomplete combustion and can be caused by too much primary air, too little secondary air, or by the flame touching a cool surface.

## 3.0.0 ◆ FLUE GASES

Both gas and oil furnaces rely on the combustion of fuel to generate heat. In the process, they also produce wastes in the form of vent gases. The bulk of these waste gases are carbon dioxide, water vapor, excess air, and small amounts of other elements. If incomplete combustion occurs, these gases may also include carbon monoxide, aldehydes, and soot, all of which are potentially dangerous to people. Venting of these gases to the outdoors is an important part of a heating system. Proper gas venting is the removal of all products of combustion, together with excess air and **dilution air**, to the outside of the building. In most furnaces, venting is done through a chimney flue or vertical vent which leads from the furnace area up through the roof. A horizontal metal vent pipe (**vent connector**) is used to connect the furnace to the chimney or a metal flue, which vents to the outdoors. In condensing furnaces, venting may be done with plastic pipe through an outside wall. This is possible because the flue gases from these furnaces are much cooler than those of other furnaces.

There are problems related to the removal of flue gases that must be considered when sizing vents. Often, the true volume of the flue gases or products of combustion is underestimated. Flue gas volume is many times greater than the volume of gas burned, so the inside of the chimney or vent must be large enough to handle the large volume of the various products of combustion. Also, in all types of furnaces, water vapor produced by combustion can be troublesome if allowed to condense into a liquid.

In burning 100 cubic feet of natural gas, a furnace can produce 200 cubic feet of water vapor (about one gallon of water). This water vapor must be prevented from condensing in the vent system. In natural-draft and induced-draft furnaces, the vent temperature stays well above the dewpoint. In condensing furnaces, where the vent temperature is much closer to the dewpoint, the condensing coil removes moisture from the flue gases. In addition, a system to collect and remove condensation is included in the design.

When coal furnaces were widely used, the constant heat from the glowing coals, plus high-temperature flue gases (about 1000°F) helped push the products of combustion up the chimney. Gas furnaces are much different. They usually operate intermittently and produce flue gas temperatures much lower than those of coal. This creates a greater potential for condensation.

For perfect combustion, natural gas is united with oxygen to form one part of carbon dioxide and two parts of water vapor, plus heat. The oxygen needed for combustion comes from air. If 10 cubic feet of air is divided into its elements, it contains about eight cubic feet of nitrogen and other inert gases, and slightly less than two cubic feet of oxygen. Thus, there is roughly 20% oxygen and 80% nitrogen in a given quantity of air. Theoretically then, when one cubic foot of natural gas is burned, its carbon and hydrogen combine with the oxygen present in 10 cubic feet of air to form one cubic foot of carbon dioxide, plus two cubic feet of water vapor, plus heat. The eight cubic feet of nitrogen remain unchanged. Therefore, it takes 10 cubic feet of air to burn one cubic foot of natural gas. If that one cubic foot of natural gas is burned in the presence of less than 10 cubic feet of air, incomplete combustion results.

Incomplete combustion produces carbon monoxide instead of carbon dioxide in the flue gas. Thus, in order to ensure complete combustion, an extra five cubic feet of air are generally supplied for each cubic foot of gas. This extra air is usually termed *excess air*.

## 4.0.0 ♦ FURNACE VENTING

Gas-fired appliances produce flue gases in quantities of about 30 times the volume of gas burned, at temperatures that affect both venting power and moisture condensation. Vents should include the following features:

- Low resistance to flue gas flow
- Small mass to enhance quick warm-up
- Insulating properties to maintain flue gas temperature
- Exact-size availability so that they can be matched to fit specific appliances

Installing gas vents requires the same technical understanding and early-stage planning as the installation of an air system. Nothing should be left to chance. It is necessary to understand the basic principles of vent operation and the factors that interfere with vent action. It is also important to know the rules that apply to proper installation and operation of gas vents.

For example, a 100,000-Btuh gas furnace consumes about 100 cubic feet of gas during each hour of constant operation. Because of the air/gas ratio, about 3,000 cubic feet of air is also consumed during that period. Assume this furnace is in a house with an area of 1,250 square feet and a volume of about 10,000 cubic feet. With an average infiltration rate of one air change per hour, 10,000 cubic feet of outside air will move through the structure every hour. This exceeds the 3,000 cubic feet of air required by the furnace and vent system.

In the past, normal air infiltration was enough to satisfy the furnace needs. However, modern building construction has become tighter. In addition, slab floors have replaced basements, and more dampered exhaust fans have been built into kitchens and bathrooms. Thus, the air leakage rate (air supply to the furnace) has become a critical design factor. Now it is sometimes necessary to deliver outside combustion air to fan-assisted furnaces. Condensing furnaces must use outdoor air for combustion.

### 4.1.0 Requirements

A service technician should know the local codes and regulations that govern vent systems. If local codes or manufacturer's instructions do not cover vent piping, refer to the *National Fuel Gas Code (NFPA 54/ANSI Z223.1)* published by the American Gas Association (AGA) and the National Fire Protection Association (NFPA). All wiring and connections should be made in accordance with the *National Electrical Code* and with any local codes that may apply. Supply gas pipe sizing should be made in accordance with the standards of the AGA.

---

### Gas Furnace Venting

The *National Fuel Gas Code*, which is published jointly by the American Gas Association and the National Fire Protection Association, allows fan-assisted furnaces to use indoor air for combustion in certain circumstances. However, it is important to keep in mind that the environment in which the furnace is installed may change over time. For example:

- Occupants may remodel a basement where the furnace is installed, adding walls and doors that reduce the amount of available open space from which the furnace can draw combustion air.
- Gas appliances such as stoves, water heaters, and clothes dryers might be added, increasing the demand for combustion air.
- Occupants may add insulation and caulk around windows and doors, reducing the amount of infiltration air available.

Many dealers, concerned with the long-term safety of their customers as well as liability issues, require that combustion air be drawn from outside the building, even if existing conditions would allow the use of indoor air. Local governments have also tightened requirements for furnace venting. Be sure to check local codes and your employer's policies before undertaking a gas furnace installation. Also, encourage homeowners to install carbon monoxide detectors.

In general, the vent system should meet the following minimum requirements as defined by the *National Fuel Gas Code*:

- It must not be smaller in diameter than the vent collar on the furnace.
- The combination of the vent and vent connector must not exceed a specified length.
- The installation must not have more than a specified number of elbows.

## 4.2.0 Clearances

Local codes and manufacturers' installation instructions usually specify the minimum distance between the furnace and combustible materials.

**WARNING!**
Flammable materials must not come into contact with the heat exchangers, burners, or any other hot surfaces, such as the flue vent.

Accessibility clearances take precedence over minimum fire protection clearances. Allow at least 24" at the front of the furnace if all parts can be reached from the front. Otherwise, allow 24" on three sides of the furnace if the back must be reached for servicing. When the installation is made in a utility room or closet, the door must be big enough to allow replacement of the appliance. Consult local codes and manufacturer's installation instructions for allowable clearances.

## 4.3.0 Air Supply

Return air plenums should be lined with an acoustical duct liner to reduce fan noise. This is of particular importance when the return air grille is close to the furnace. All duct connections to the furnace must extend outside the furnace closet.

Return air must not be taken from the furnace room or closet. Adequate return air duct height must be provided to allow filters to be removed and replaced. All return air must pass through the filter after it enters the return air plenum. Air required for combustion, draft hood dilution, and ventilation differs somewhat for a furnace in a confined space as opposed to a furnace in an open space. As a general rule, there must be two permanent openings: one within 12" of the ceiling, and one within 12" of the floor. Each opening must have a free area of at least one square inch per 1,000 Btuh of the total input rating of all the gas-fired appliances in the enclosure. (See *Figure 3*.)

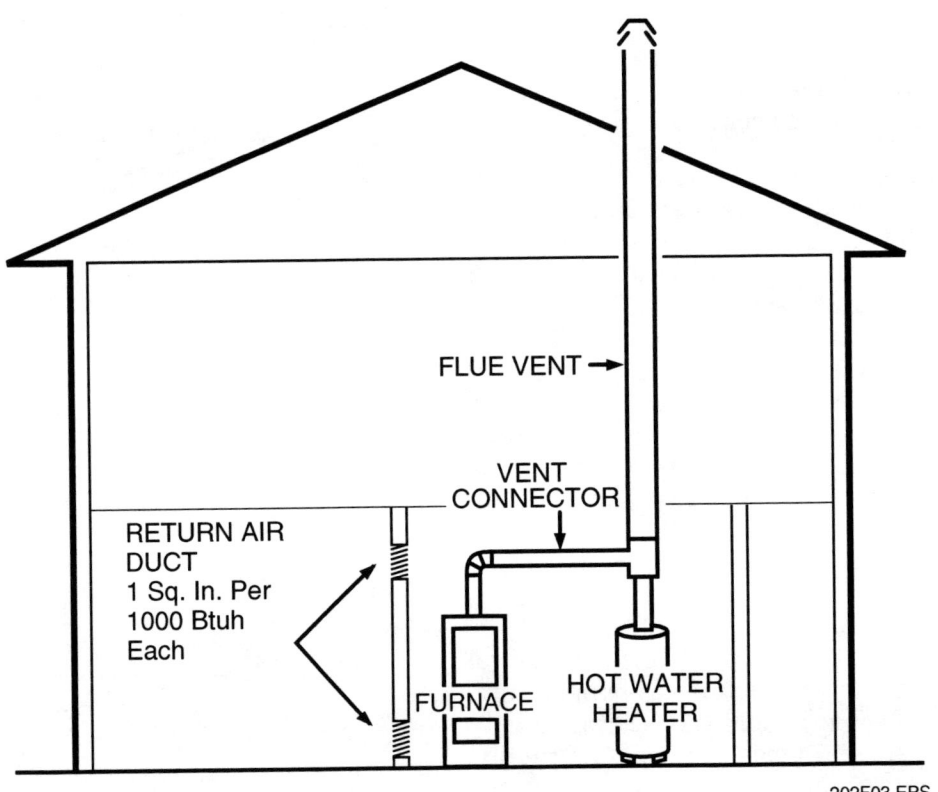

*Figure 3* ◆ Furnace room venting.

## Combustion Air Contamination

Do not install furnaces that use indoor air for combustion near sources of air contamination such as cleaning solvents, aerosol sprays, detergents, bleaches, air fresheners, etc. Some manufacturers will not honor warranties on their heat exchangers unless the combustion air is drawn from outside the building.

### 5.0.0 ♦ VENT SYSTEM COMPONENTS

The type of vent system used is based on the type of furnace and the construction of the building. The *National Fuel Gas Code* identifies vented appliance categories and describes them as follows:

- *Category I* – An appliance that operates with a non-positive vent static pressure and with a vent gas temperature that avoids excessive condensate production in the vent.

- *Category II* – An appliance that operates with a non-positive vent static pressure and with a vent gas temperature that may cause excessive condensate production in the vent.

- *Category III* – An appliance that operates with a positive vent static pressure and with a vent gas temperature that avoids excessive condensate production in the vent.

- *Category IV* – An appliance that operates with a positive vent static pressure and with a vent gas temperature that may cause excessive condensate production in the vent.

This module focuses on Category I (natural-draft and induced-draft furnaces) and Category IV (condensing furnaces). Category II furnaces are natural-draft condensing furnaces. Category III furnaces are sidewall-vented 80% AFUE induced-draft furnaces vented with high-temperature plastic pipe. Neither Category II nor Category III is in common use at this time.

Masonry and factory-built chimneys, along with metal and plastic vents, comprise the basic venting systems for coal, gas, oil, and wood-burning appliances. Factory-built chimneys with an inner wall of stainless steel that are in compliance with Underwriters' Laboratories (UL) Standard No. 959 are suitable for all of these fuels.

*Figure 4* shows an exploded view of a typical factory-built chimney. The flue-gas temperature-rise limit for these applications is set at 1730°F.

There are several types of vent construction approved for use with gas appliances.

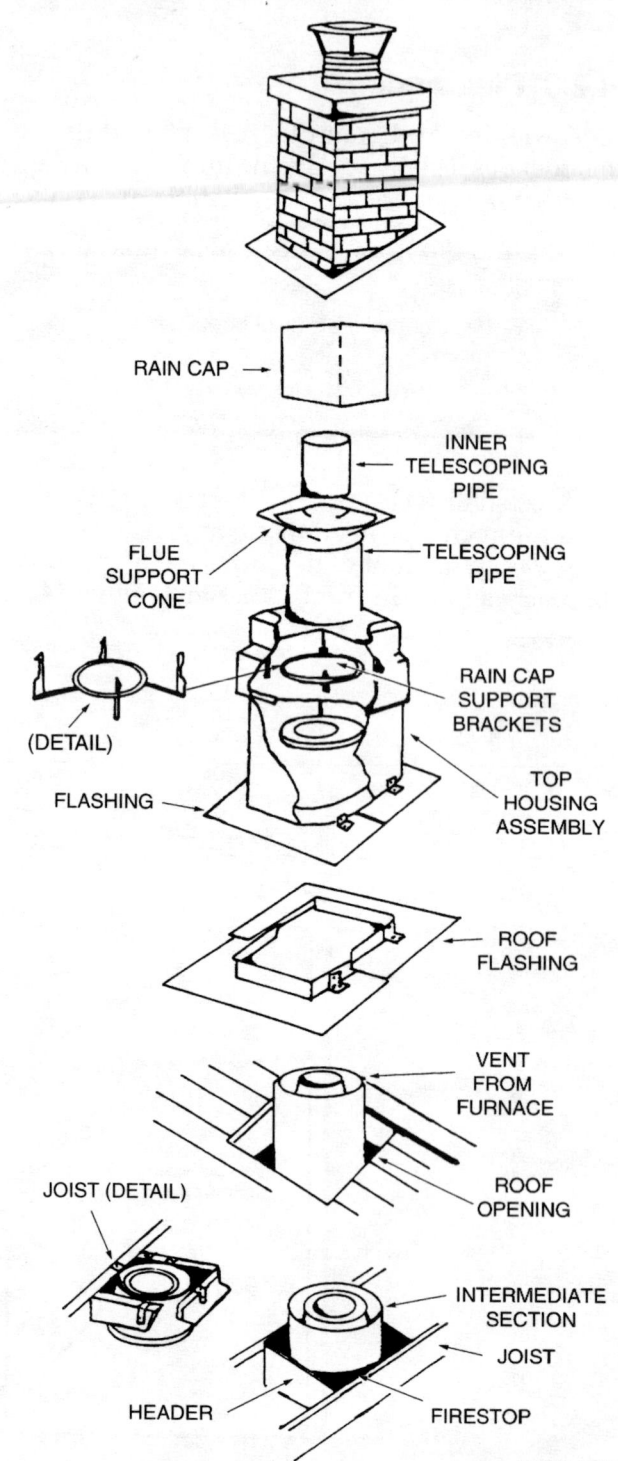

*Figure 4* ♦ Components of a factory-built chimney.

### Chimney-Related Fires

A major cause of chimney-related fires is the failure to maintain the required clearances or air spaces between the chimney and adjacent combustible materials. For this reason, it is essential that a chimney be installed in strict accordance with local codes as well as the manufacturer's instructions.

Type B vents have inner and outer walls made of corrosion-resistant material. They are round and are available in diameters to suit all uses. The double-wall construction of Type B vents helps conserve heat and therefore promotes better draft and reduced condensation. They are often used in new construction because they are cheaper than a lined masonry chimney.

Type B-W vents have the same type of double-wall construction as the B-vent, but are oval. They are designed for venting in-the-wall gas heaters.

Type L vents are also double-wall vents. They are similar to Type B, but are made of materials that are more resistant to heat and corrosion. In general, Type L can be used in any application where Type B is suitable. The reverse is not true, however. Type L vents are also used in venting combination gas/oil appliances, as well as residential incinerators and certain appliances equipped with draft hoods.

High-temperature 3" and 4" diameter plastic pipe is used for venting Category III furnaces. This pipe is made of a special resin that will withstand temperatures up to 480°F. The vent system must be tightly sealed using high-temperature sealants in order to avoid carbon monoxide leakage.

**NOTE**

This high-temperature plastic pipe is no longer used due to safety concerns. Some vents initially built of the high-temperature plastic were later converted to other vent materials.

Schedule 40 PVC pipe is used in venting Category IV (condensing) furnaces. Because these vents are used in positive-pressure applications, they must also be carefully sealed to eliminate vent gas and condensate leakage.

## 6.0.0 ◆ NATURAL-DRAFT FURNACES

Sixteen cubic feet of combustion gases result from burning one cubic foot of natural gas. This does not equal the total gas volume passing up through the vent to the atmosphere. Before leaving the furnace through the vent, the 16 cubic feet of combustion gas is joined by 14 cubic feet of air at the furnace draft hood (*Figure 5*). This additional air is called *dilution air*. Therefore, the total volume in a properly operating vent is 30 cubic feet of vent gas for every cubic foot of natural gas burned.

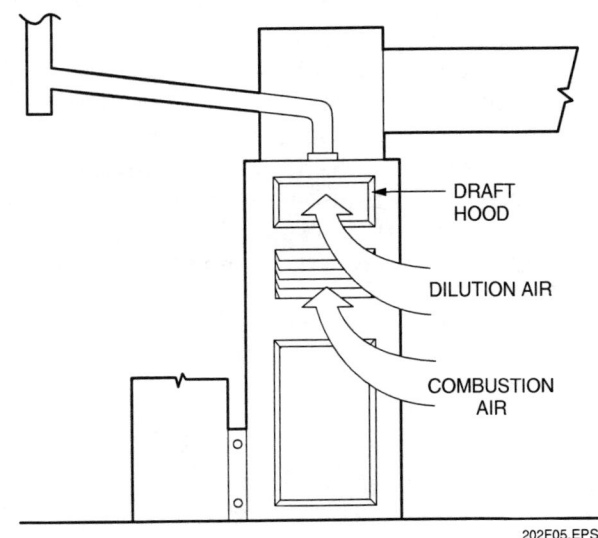

*Figure 5* ◆ Natural-draft furnace.

### Furnace Draft

Based on your knowledge of how a draft is created in a natural-draft furnace, can you think of a simple way to easily tell if a good draft has been established for a furnace or other gas appliance?

In natural-draft furnaces, vent gases are not forced out through the vent pipe and chimney, but are drawn out instead. The chimney does this by producing a suction or drawing action called a *draft*. This principle is shown in *Figure 6*. When no heat is applied to the air or gas, the temperatures in and around the pipe are the same, and no movement occurs. When a fire is lit, however, it heats the air around it. This heated air expands in volume and becomes lighter in weight (less dense). Due to its lighter weight, the warm air rises, creating a draft in the chimney or vent.

When confined within a vertical pipe, the warm air cannot mix with the surrounding air and cool down. When it is within the pipe, it retains its heat and therefore rises at a faster rate. In the process of rising, it drags fresh air in behind to replace it. As this new air is in turn heated, the process is continued and a constant flow of air moves through the pipe as long as heat is applied. The volume of gas the chimney will move, and the amount of draft it will create, depend on two factors: the temperature of the vent gas, and the diameter and height of the chimney.

Poor draft results from a chimney that is too small or too short, or vent gas temperatures that are too low. Increasing the diameter or the height of the chimney will increase the draft. The diameter of the chimney or vent pipe is important because of friction. If the pipe is too small, it will restrict the flow of gases.

Sufficient draft is essential for the proper operation of natural-draft furnaces. Draft allows the products of combustion to be removed safely. It also provides the oxygen used for combustion. Sufficient draft is also necessary with fan-assisted furnaces, but it is less critical because the flue gases receive some push from the inducer fan.

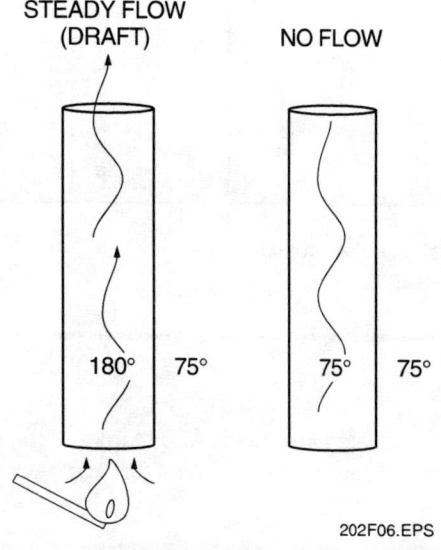

*Figure 6* ◆ Natural drawing action.

## 7.0.0 ◆ INDUCED-DRAFT GAS FURNACES

In induced-draft furnaces, condensation can be a major problem because the flue gases are cooler than those of a natural-draft furnace. As mentioned earlier, natural-draft furnaces have flue gas temperatures of 350°F to 600°F. They are not likely to have condensation problems unless the furnace was greatly oversized or the vent system was improperly designed or installed. On the other hand, the flue gases of induced-draft furnaces run in the range of 275°F to 400°F. Because of that, the moisture in the flue gases condenses more readily.

To avoid condensation that can damage both the vent system and the furnace, there are five key tasks that must be completed when installing an induced-draft furnace:

- The furnace must be properly sized.
- The burners must be firing as close as possible to 100% of their rated input.
- The correct amount of air must be flowing over the heat exchangers in order to maintain the temperature rise at the correct level.
- The thermostat **heat anticipator** must be adjusted correctly.
- The furnace must be properly vented.

### 7.1.0 Furnace Sizing

A furnace must not be severely oversized. However, it is better to slightly oversize than it is to undersize. Generally, the furnace heat output rating should be from 95% to 120% of the heating load. The load is established by an analysis that considers building construction, type and amount of insulation, amount of glass, and other factors. *Manual J*, published by the Air Conditioning Contractors of America (ACCA), provides a method for properly estimating heating and cooling loads.

> **WARNING!**
> If a furnace is oversized, the only effective way to correct the problem is to replace it. Do not attempt to derate the furnace by drilling or changing orifices, or by reducing manifold pressure. It will make the condensation problem worse, and may create a safety hazard. Always follow the installation instructions supplied with the unit.

A correctly sized furnace will have a long operating cycle, particularly when the outdoor temperature approaches the heat loss design temperature. This keeps the furnace and vent warm and prevents condensation. A severely oversized furnace will deliver a large amount of heat in a very short time. During its long off cycle, the furnace and vent will cool, allowing condensation to form when the warm flue gas hits the vent during the next ON cycle.

## 7.2.0 Burner Input Adjustment

The burners must be operated at as close to 100% of their rated input as possible. In most cases, this can be done by adjusting the manifold pressure at the gas valve. This adjustment was covered in the HVAC Level One Module, *Introduction to Heating*.

## 7.3.0 Temperature Rise Adjustment

Temperature rise is the temperature difference between the supply air and the return air. The amount of air flowing over the heat exchangers determines the temperature rise. If there is too much airflow, the heat exchangers will not be able to reach normal operating temperature. Moisture will condense on the heat exchangers and in the vent. If there is too little air, the heat exchangers will become overheated.

The furnace nameplate will specify the correct temperature rise range; 45°F to 75°F is common. The actual rise is determined by drilling holes in the supply and return ducts and measuring the temperatures (*Figure 7*). The difference between the two temperatures must be within the specified range. Ideally, it should be just above the midpoint of the range. If it is not, the blower speed can be changed to compensate. The blower speed is increased for too much rise and decreased for too little rise.

## 7.4.0 Thermostat Heat Anticipator Adjustment

A heat anticipator is a small resistive element that heats up as current flows through the thermostat. It is usually adjustable (*Figure 8*), and the adjustment is required at the time of installation. The heat from the anticipator causes the thermostat to turn off just before the setpoint is reached. This prevents the system from overshooting the desired temperature and allows some of the residual heat from the heat exchangers to be dissipated by the circulating air.

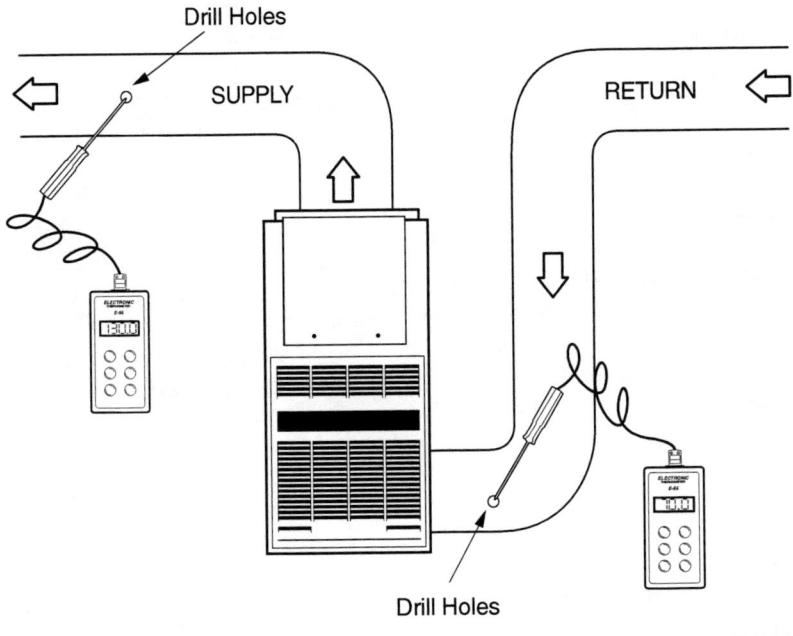

*Figure 7* ◆ Measuring supply and return air temperature.

### High-Altitude Installations

Refer to the manufacturer's instructions for furnaces to be installed at altitudes above 2,000'. The *National Fuel Gas Code* provides guidelines for such installations. Generally, the requirement for high-altitude installations is that the furnace capacity be derated by 4% for each 1,000' above sea level. For example, at 5,000' above sea level, a 100,000-Btu furnace will have an effective capacity of only 80,000 Btus (20% reduction).

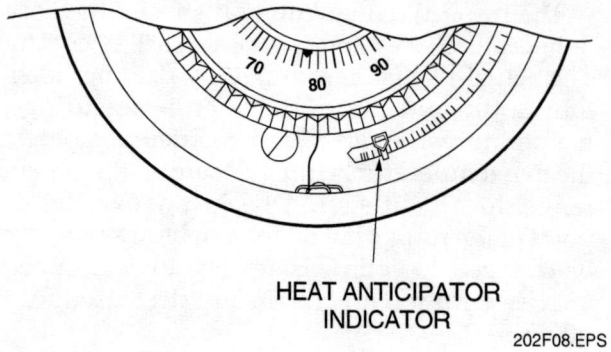

*Figure 8* ◆ Thermostat heat anticipator adjustment.

If the anticipator is set incorrectly, it can cause the furnace to short-cycle and allow moisture to condense. This can occur if the furnace cycles more than six times an hour.

The heat anticipator should be set to the same current that is flowing through the thermostat contacts, which is usually in the range of .15 to 1 amp. Electronic thermostats do not have heat anticipators. If an electronic thermostat is used, set the cycle rate for three cycles per hour or consult the furnace manufacturer's literature for the correct cycle rate.

## 7.5.0 Venting Considerations

Proper furnace sizing and adjustment is meaningless if the furnace is not vented correctly. Vents must be carefully sized. A vent that is too small may not be able to handle the volume of flue gas, while a vent that is too large may not be able to establish a proper draft. This is true of both induced-draft and natural-draft furnaces. Flue gases from gas appliances can cool too quickly in a large vent. This causes condensation. It is important to follow the manufacturer's instructions, along with local and national codes, when selecting and installing furnace vents.

The *National Fuel Gas Code* contains tables and instructions for selecting the diameter and type of metal vents and vent connectors for induced-draft furnaces and other gas appliances. Part of this data is provided in the installation instructions for most furnaces. The following sections summarize some of the important furnace venting requirements specified by the *National Fuel Gas Code*.

### 7.5.1 General Guidelines for Metal Vents and Vent Connectors

Gas appliances may be vented through a metal pipe or lined chimney (*Figure 9*). The vent is the vertical section of the vent system. The vent connector is the horizontal section that connects the appliance(s) to the vent.

Because of the potential for condensation, metal vents must be of double-wall (Type B) construction. Vent connectors for induced-draft furnaces should also be the double-wall type, which heats up faster, thereby limiting the risk of condensation. A single-wall vent connector can sometimes be used with induced-draft furnaces, but only under the following conditions:

- The furnace must be common-vented with another gas appliance, such as a hot water heater. This helps to keep the vent connector from cooling down.
- The length of the vent connector (in feet) cannot exceed 1½ times its diameter (in inches). For example, a 4" diameter vent connector can be no longer than 6'. The selection tables are used to determine the diameter of the vent connector, based on the furnace input (in Btuh) and the height of the vent.

Single-wall pipe has a high heat loss, which allows combustion products to cool rapidly. In general, a single-wall vent connector is suitable only for natural-draft furnaces because their high flue gas temperatures and the use of dilution air prevent condensation. A double-wall vent connector increases the initial cost of the installation, but that is far outweighed by the potential cost of major repairs or furnace replacement due to condensation. The use of a single-wall vent connector also hampers installation flexibility because of the limited length of the vent connector.

Vent connectors should be pitched upward toward the vent at a slope of no less than ¼" per foot and should be as short as possible. Avoid elbows, because they create resistance. Also avoid sharp turns—two 45° connections are better than one 90° connection. If single-wall vent connectors are being used, the use of more than two 90°

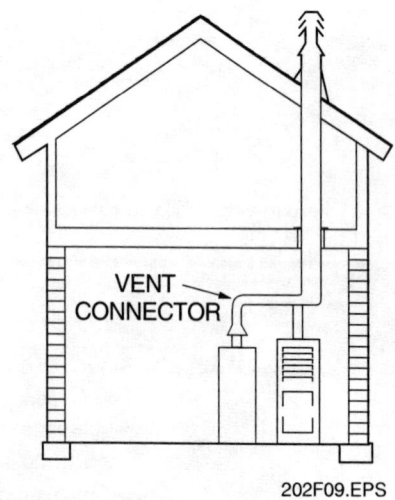

*Figure 9* ◆ Venting through a metal pipe.

elbows requires a 10% reduction in the maximum length of the vent connector for each extra elbow. If the furnace is common-vented with another gas appliance, neither appliance should have a vent damper because dampers can cause condensation. Vent dampers are covered later in this module.

When working with metal vents, the selection tables assume that the vertical run will not be exposed to outdoor air, except above the point where it penetrates the roof. In many locales, exposed metal vents will cause serious condensation problems. The selection tables are not intended for these applications.

### 7.5.2 Venting Through a Masonry Chimney

When a tile-lined masonry chimney is available, it can be used to vent gas appliances. However, a fan-assisted furnace cannot be vented through a masonry chimney unless it is common-vented with another gas appliance or the chimney is suitably lined. In all cases, follow the local codes.

An unlined chimney must not be used in any circumstance. Every chimney must have a liner, because corrosive substances in the flue gases will attack mortar and cause the chimney to deteriorate. This could cause falling mortar and debris to block the vent opening, creating a hazard for occupants. At best, the unlined chimney will be seriously damaged over time. In addition, unlined chimneys are prone to condensation, which can damage the furnace and vent system. Under no circumstances may a furnace or other gas appliance be vented through a chimney that serves a fireplace or other wood- or coal-burning device.

Ideally, a chimney used for furnace venting should run inside the building. If the chimney is exposed below the roofline, it will probably need a metal liner. In some cases, the metal liner will have to be insulated. A chimney with a metal liner is treated the same as an unexposed chimney.

The vent selection tables will identify the chimney size (in square inches) required to match the furnace capacity. An oversized chimney could be dangerous because it will have difficulty in establishing the proper draft, making it hazardous for occupants. Also, it will be more likely to develop condensation. If the opening is too large, a liner *must* be installed. A double-wall metal vent may be used for this purpose; however, it may be difficult to install, especially if there are offsets in the chimney. As an alternative, flexible chimney liners are available (see *Figure 10*). They are expandable; a 30'-long vent might be shipped in a 3'-long box.

Flexible liners are not sized in the same way as solid metal vents. When flexible liners are used, the furnace capacity must be derated by 20% when using the vent selection tables. For example, if a 6" diameter metal vent is suitable for a particular furnace, the same furnace may need an 8" flexible liner.

## 8.0.0 ◆ CONDENSING GAS FURNACES

Because condensing furnaces produce low-temperature flue gases, they can be vented with Schedule 40 PVC pipe. These furnaces also require a lot of combustion air, which may be drawn from outdoors or from inside the building. The diameter of the pipe depends on the furnace input, the length of the pipe run, and the number and type of elbows used. If 45° elbows are used, each one increases the length of the run by an equivalent of 5'. A 90° elbow is equivalent to an increase of 10'. *Figure 11* shows the type of pipe selection chart you might see in the installation instructions for a condensing furnace.

The following are good practices to consider when installing vents and intake piping:

- The vent and intake pipes should always be run together, either through an exterior wall or through the roof. This keeps them in the same pressure zone. They should also be as close together as possible; separations of 3" on a roof and 6" on a sidewall are standard.

- Piping should be sloped back toward the furnace at least ¼" per foot and supported. The slope allows condensate to drain back to the furnace and into the condensate trap for disposal.

- Keep the termination well above expected snow levels. In cold climates, the outdoor portion and any part of the exhaust pipe that runs through an unconditioned space should be insulated.

### Flexible Chimney Liners

Flexible chimney liner kits have made lining chimneys relatively easy. The most difficult part of the job is working on high chimneys or steeply pitched roofs. In those cases, scaffolding or a motorized lift should be used. To install the liner, drop a weighted rope down the chimney. Attach the flexible liner to the rope and have someone pull the liner up or down the chimney. Use the vent cap and all other components of the installation kit to ensure a safe installation.

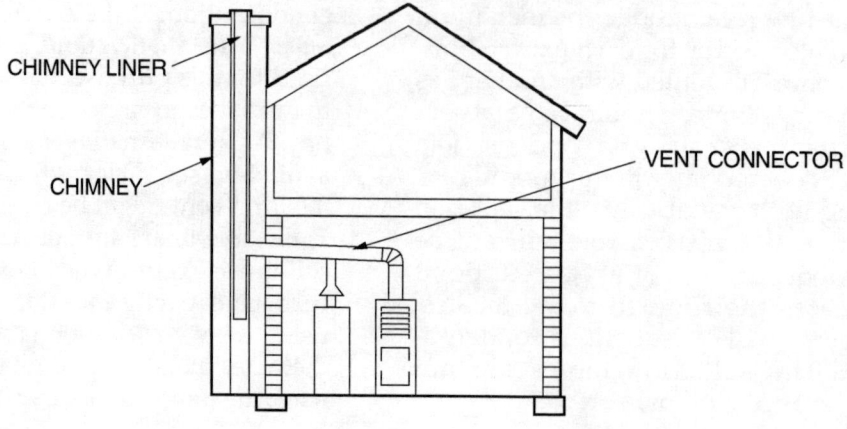

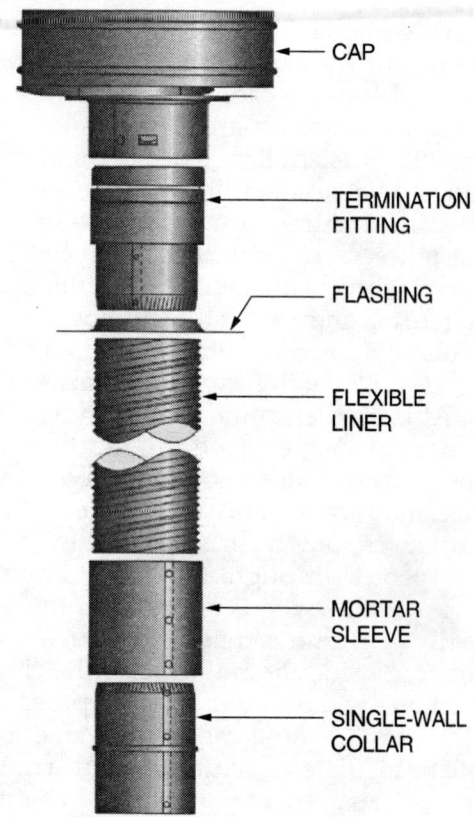

*Figure 10* ◆ Flexible chimney liner kit.

- As shown in *Figure 12*, the combustion air intake should be bent downward to prevent dirt and moisture from entering the system. The exhaust vent must be straight up (roof) or straight out (sidewall). Rooftop terminations are usually best because there is less chance of the pipes being damaged or blocked and less chance of receiving contaminated combustion air.
- When venting through a sidewall, avoid terminating the pipes in a corner, under a deck, or near shrubs or trees, in order to prevent the recirculation of moist vent gases. Also avoid terminating pipes near doors and windows. This can cause flue gases to enter the building and condense on glass.

Special concentric termination devices (*Figure 13*) are available from some manufacturers. These devices allow combustion and exhaust gases to be carried through the same hole in the roof or exterior wall.

**SCHEDULE 40 PVC DIAMETER**

| PIPE LENGTH (FEET) | NUMBER OF 90° ELBOWS | | | | |
|---|---|---|---|---|---|
| | 0 | 2 | 4 | 6 | 8 |
| 5 | 2 | 2 | 2 | 2 | 2 |
| 10 | 2 | 2 | 2 | 2 | 2 |
| 20 | 2 | 2 | 2 | 2 | 2-1/2 |
| 30 | 2 | 2 | 2 | 2-1/2 | 2-1/2 |
| 40 | 2 | 2 | 2-1/2 | 2-1/2 | 2-1/2 |
| 50 | 2 | 2-1/2 | 2-1/2 | 2-1/2 | 2-1/2 |
| 60 | 2-1/2 | 2-1/2 | 2-1/2 | 2-1/2 | 3 |
| 70 | 2-1/2 | 2-1/2 | 2-1/2 | 3 | 3 |
| 80 | 2-1/2 | 2-1/2 | 3 | 3 | 3 |
| 90 | 2-1/2 | 3 | 3 | 3 | 3 |

*Figure 11* ◆ PVC selection chart.

## Condensing Furnaces

Condensing furnaces, such as the one shown here, are usually vented and supplied combustion air through Schedule 40 PVC pipe.

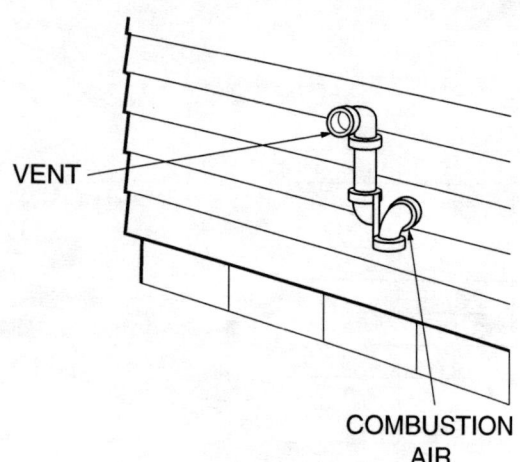

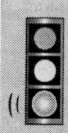

**NOTE**

Always check the prevailing codes and the furnace installation instructions for proper venting requirements before venting a furnace with PVC pipe or through a sidewall. It may not be permitted in all jurisdictions.

## 9.0.0 ◆ DRAFT CONTROLS

Draft controls regulate the amount of air feeding a fire. They are used in the vent systems of natural-draft, gas-fired furnaces as well as oil and solid-fuel furnaces.

### 9.1.0 Draft Regulator

A draft regulator (*Figure 14*) keeps a constant draft over the fire, usually 0.01 to 0.03 inches water column (in. w.c.) on oil-burning furnaces. Too high a draft causes undue loss of heat through the chimney. Too little draft causes incomplete combustion.

A draft regulator consists of a small door in the side of the flue pipe. The door is hinged near the center and controlled by adjustable weights. Basement air is admitted to the flue pipe as required to maintain a proper draft over the fire.

### 9.2.0 Vent Dampers

Some vent dampers (*Figure 15*) are energy-saving devices that can be added to in-service furnaces. They are designed to stay open while the burner is operating in order to vent combustion gases.

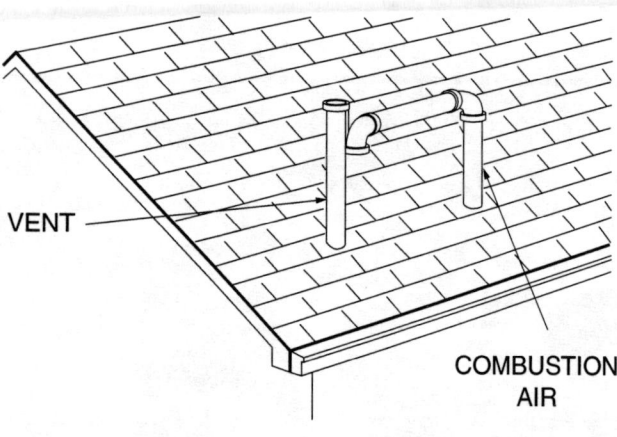

*Figure 12* ◆ Terminating PVC vent and combustion air pipes.

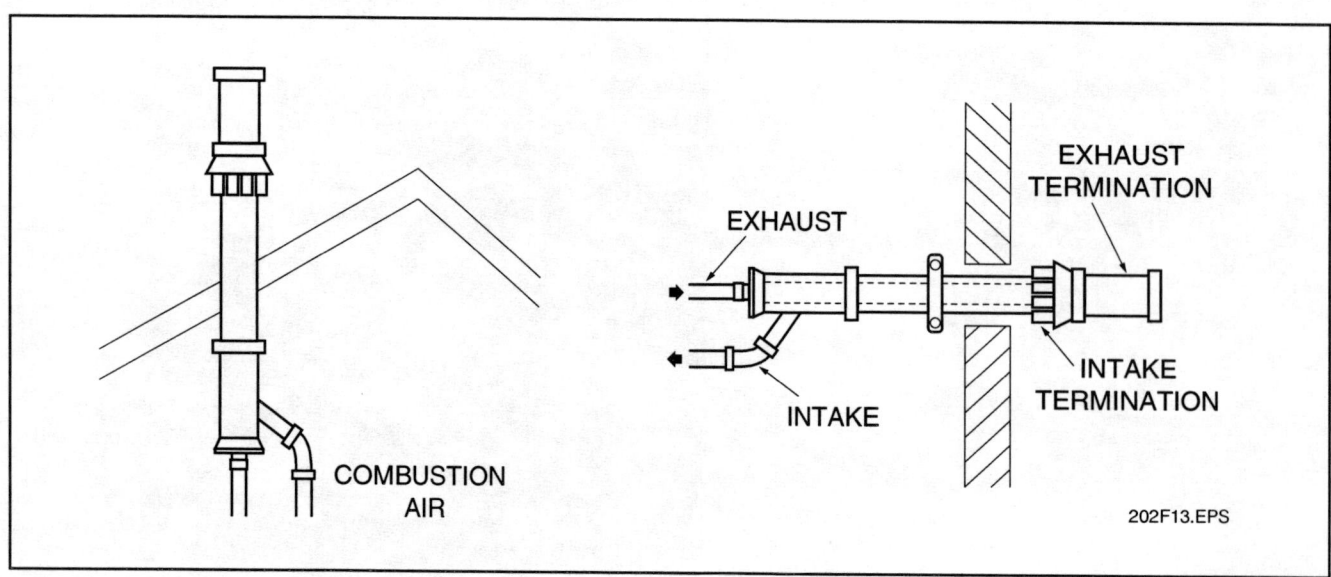

*Figure 13* ◆ Concentric termination.

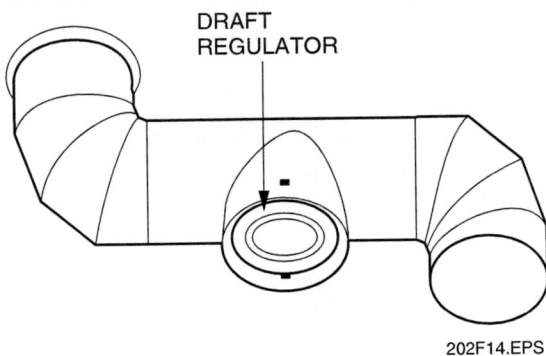

*Figure 14* ◆ Draft regulator.

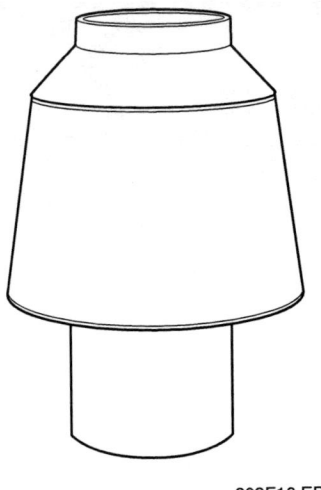

*Figure 16* ◆ Draft diverter.

*Figure 15* ◆ Vent damper.

When the burner shuts off, vent dampers are designed to close and stop the heat from escaping up the flue vent or chimney. They are relatively easy to install, but vent dampers can be both health and fire hazards if they fail to open when the furnace is operating. In addition, some furnace warranties are voided if vent dampers are added. Consult local codes. Some manufacturers are building furnaces with control wiring installed for adding a vent damper. Vent dampers are most effective when combustion air is being drawn from within the house.

## 9.3.0 Draft Diverters

The draft diverter or *draft hood* (*Figure 16*), is designed to provide a balanced draft (slightly negative) over the flame in a gas-fired, natural-draft furnace.

The bottom of the diverter is open to allow air from the furnace room to blend with the products of combustion. The hot vent gases from the furnace normally pass into the draft diverter and then into the vent pipe without any spilling out of the bottom opening. This is because the hot gases tend to stay toward the top of the draft diverter and are removed by the draw from the vent pipe and chimney. The chimney draft is greater than required to remove the vent gases; therefore, additional air from the furnace room is drawn into the bottom opening and passes up the vent along with the gases.

If not enough draft is available to remove the vent gases due to a restriction or a downdraft in the chimney, the draft diverter acts as a relief valve. Since the vent gases are prevented from going up the chimney, they go out the bottom opening in the draft hood. This relief factor prevents combustion from being upset in the furnace. The discharge of gases from the bottom opening in the draft hood or diverter is called *spillage*.

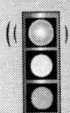 **WARNING!**
When spillage occurs, vent gases are discharged into the structure. The result is that all the water vapor from combustion passes into the conditioned space, causing high humidity. If combustion is complete, relatively harmless carbon dioxide also passes into the space. If combustion is not complete, the vent gases will also contain deadly carbon monoxide. For this reason, the cause of spillage must be found and corrected immediately. Homeowners should be strongly encouraged to install carbon monoxide detectors.

### Power Venting

If venting through a conventional chimney or vent is difficult or impossible, a manufacturer-approved power vent can be installed, such as the one shown here. This device allows the products of combustion to be vented through a side wall. Because power vents are not always permitted, always check local and national codes before installing these devices.

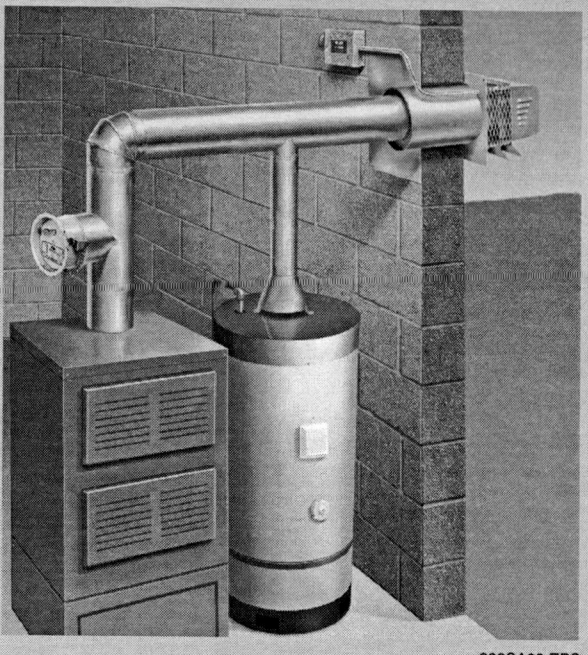

## Summary

There are two important reasons for properly venting a furnace. The first, and most important, is to make sure that harmful flue gases are exhausted from the building and thus do not create a hazard for building occupants. The other reason is to prevent moisture from condensing out of the flue gases and causing serious damage to the vent system and possibly the furnace.

Although proper venting is important with all types of fossil-fuel heating systems, there is a special concern with induced-draft, gas-fired furnaces. The flue gases of these furnaces are cooler than those of natural-draft devices. For that reason, it is very important to select the correct size and type of vent and to install it in accordance with applicable codes and instructions.

## Review Questions

1. If there are indications that incomplete combustion is occurring, the problem is solved by _____.
   a. increasing the flow of fuel
   b. increasing the amount of air
   c. changing the size of the flue vent
   d. adding carbon dioxide

2. The temperature range of the flue gases in an induced-draft furnace is _____.
   a. 100°F to 125°F
   b. 350°F to 600°F
   c. 100°F to 600°F
   d. 275°F to 400°F

3. The horizontal pipe that connects the furnace to the chimney is called the _____.
   a. flex duct
   b. vent connector
   c. vent
   d. flue pipe

4. All _____ furnaces require combustion air to be piped in from outdoors.
   a. natural-draft
   b. induced-draft
   c. condensing
   d. wood-burning

5. Which type of vent does *not* have double-wall construction?
   a. PVC
   b. Type B-W
   c. Type L
   d. Type B

6. If an induced-draft furnace is oversized and is not venting properly, the problem can be solved by changing to a _____.
   a. smaller burner orifice
   b. larger burner orifice
   c. smaller flue vent
   d. different furnace

7. *Temperature rise* is a term that describes the _____.
   a. temperature difference between supply air and return air
   b. natural venting of flue gases
   c. amount of heat in the flue gases
   d. heat remaining in the heat exchanger when the burner is shut off

8. Exterior chimneys may be used to vent gas furnaces as long as they have a _____ liner.
   a. masonry
   b. plastic
   c. suitable
   d. PVC

9. A gas furnace may be vented through an unlined chimney _____.
   a. when it is a natural-draft furnace
   b. when the climate is not too cold
   c. when a double-wall vent connector is used
   d. under no circumstances

10. How many equivalent feet does a 45° elbow add to the length of a PVC vent?
    a. 10
    b. 5
    c. 0
    d. 20

## GLOSSARY

# Trade Terms Introduced in This Module

*Complete combustion:* Burning in which there is enough oxygen to prevent the formation of carbon monoxide.

*Condensing furnace:* A high-efficiency furnace containing a secondary heat exchanger that extracts additional heat from the flue gases.

*Dilution air:* Air added to the flue gases in a natural-draft furnace to aid flue gas removal.

*Heat anticipator:* A resistive heating element in a thermostat that shuts off the furnace before the space temperature reaches the setpoint. It prevents the system from overshooting the desired temperature.

*Incomplete combustion:* Burning in which there is not enough oxygen to prevent the formation of carbon monoxide.

*Induced-draft furnace:* A fan-assisted furnace with an AFUE rating of 78% to 85%.

*Natural-draft furnace:* A furnace that depends on the pressure created by the heat in the flue gases to force them out through the vent system.

*Primary air:* Air that is added to the fuel before it goes to the burner.

*Secondary air:* Air that is added during combustion.

*Vent:* The vertical section of the vent pipe.

*Vent connector:* The horizontal section of the vent system that connects the appliance(s) to the vent pipe or chimney.

# ANSWER KEY

## Answers to Review Questions

| Answer | Section |
|--------|---------|
| 1. b   | 2.2.0   |
| 2. d   | 2.3.0   |
| 3. b   | 3.0.0   |
| 4. c   | 4.0.0   |
| 5. a   | 5.0.0   |
| 6. d   | 7.1.0   |
| 7. a   | 7.3.0   |
| 8. c   | 7.5.2   |
| 9. d   | 7.5.2   |
| 10. b  | 8.0.0   |

# REFERENCES & ACKNOWLEDGMENTS

## Additional Resources

This module is intended to present thorough resources for task training. The following reference works are suggested for further study. These are optional materials for continued education rather than for task training.

*Mid-Efficiency Furnace Installation Awareness,* 1999. Syracuse, NY: Carrier Corporation.

*National Fuel Gas Code (NFPA 54/ANSI Z223.1),* 1999. Quincy, MA: National Fire Protection Association.

*Southern Building Code,* 1999. Birmingham, AL: SBCCI.

## Figure Credits

| | |
|---|---|
| **Hart & Cooley, Inc.** | 202F10 |
| **Gerald Shannon** | 202SA01 |
| **Field Controls L.L.C.** | 202F15, 202SA02 |

# NCCER CRAFT TRAINING USER UPDATES

The NCCER makes every effort to keep these textbooks up-to-date and free of technical errors. We appreciate your help in this process. If you have an idea for improving this textbook, or if you find an error, a typographical mistake, or an inaccuracy in the NCCER's Craft Training textbooks, please write us, using this form or a photocopy. Be sure to include the exact module number, page number, a detailed description, and the correction, if applicable. Your input will be brought to the attention of the Technical Review Committee. Thank you for your assistance.

*Instructors* – If you found that additional materials were necessary in order to teach this module effectively, please let us know so that we may include them in the Equipment and Materials list in the Instructor's Guide.

**Write:** Curriculum Revision and Development Department
National Center for Construction Education and Research
P.O. Box 141104, Gainesville, FL 32614-1104

**Fax:** 352-334-0932

**E-mail:** curriculum@nccer.org

---

Craft _____ Module Name _____

Copyright Date _____ Module Number _____ Page Number(s) _____

Description
_____
_____
_____
_____

(Optional) Correction
_____
_____
_____

(Optional) Your Name and Address
_____
_____
_____

Module 03203-01

*Maintenance Skills for the Service Technician*

# COURSE MAP

This course map shows all of the modules in the second level of the HVAC curriculum. The suggested training order begins at the bottom and proceeds up. Skill levels increase as you advance on the course map. The local Training Program Sponsor may adjust the training order.

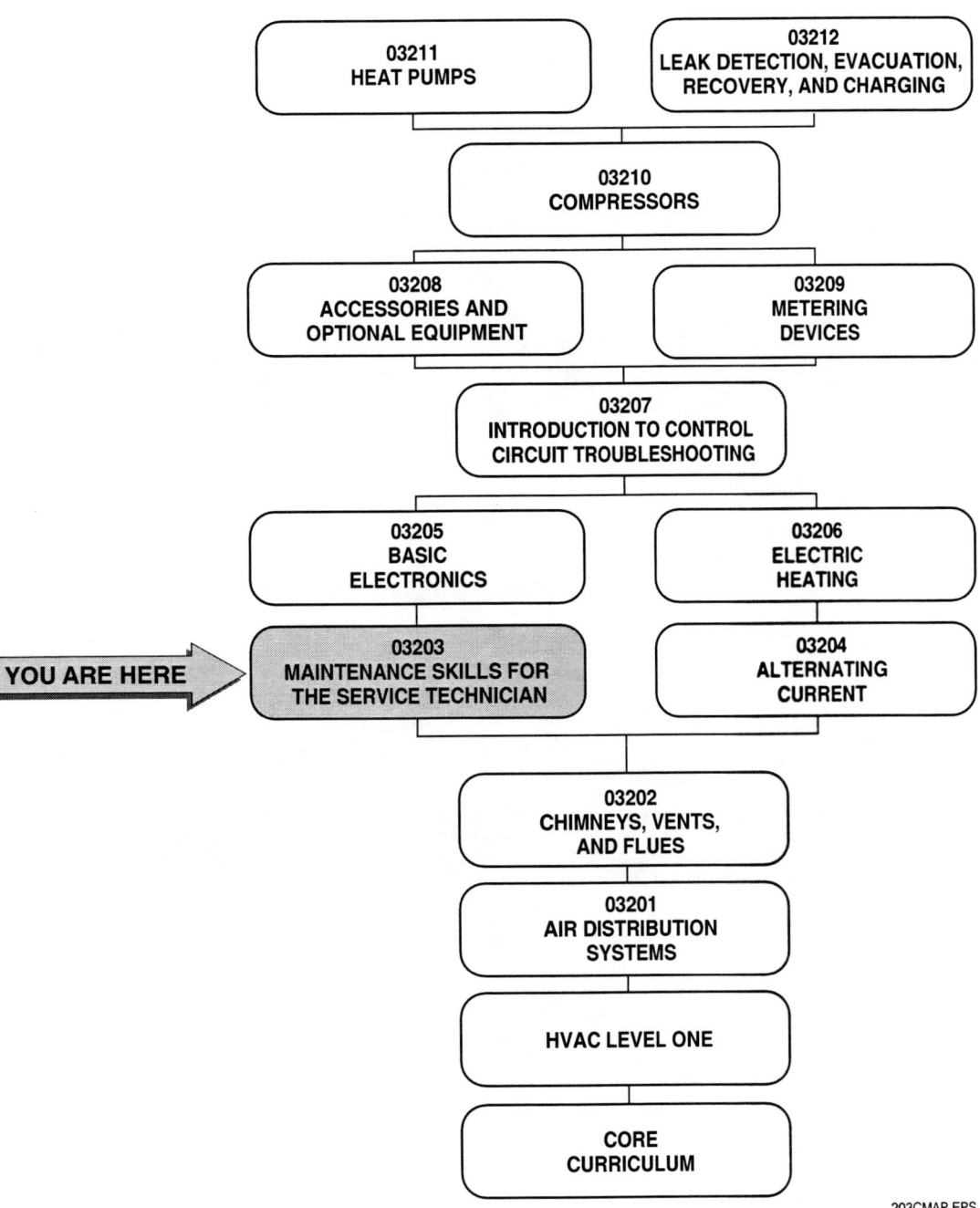

# MODULE 03203 CONTENTS

1.0.0 INTRODUCTION .................................................. 3.1
2.0.0 MECHANICAL FASTENERS ........................................ 3.2
    2.1.0 Thread and Grade Designations ........................ 3.2
    *2.1.1 Thread Designations* .................................. 3.2
    *2.1.2 Fastener Grade Designations* ......................... 3.2
    2.2.0 Threaded Fasteners .................................... 3.2
    *2.2.1 Machine Bolts, Machine Screws, Stud Bolts, and Cap Screws* .................................... 3.4
    *2.2.2 Set Screws* ............................................ 3.5
    *2.2.3 Flat and Lock Washers* ............................... 3.5
    *2.2.4 Nuts* .................................................. 3.5
    *2.2.5 Thread-Forming and Thread-Cutting Screws* ........ 3.7
    *2.2.6 Toggle and Anchor Bolts* ............................. 3.7
    *2.2.7 Thread Repair Inserts* ............................... 3.8
    2.3.0 Non-Threaded Fasteners .............................. 3.8
    *2.3.1 Retainer Rings* ....................................... 3.8
    *2.3.2 Pins* .................................................. 3.8
    *2.3.3 Keys* .................................................. 3.8
    *2.3.4 Rivets* ................................................ 3.9
    2.4.0 Installing Threaded Fasteners ........................ 3.10
    *2.4.1 Torquing Steel Fasteners* ............................ 3.10
    *2.4.2 Flange Tightening Sequences* ....................... 3.12
    *2.4.3 Thread Tapping* ..................................... 3.12
    *2.4.4 Installing Anchor Bolts* .............................. 3.12
3.0.0 GASKETS ...................................................... 3.13
    3.1.0 Gasket Types ......................................... 3.13
    *3.1.1 Flat Gaskets* ......................................... 3.13
    *3.1.2 Ring Gaskets* ........................................ 3.13
    *3.1.3 Spiral-Wound Gaskets* .............................. 3.13
    *3.1.4 Full-Face Gaskets* ................................... 3.13
    *3.1.5 Jacketed Gaskets* ................................... 3.14
    *3.1.6 Envelope Gaskets* ................................... 3.14
    3.2.0 Installing and Removing Gaskets ..................... 3.14

## MODULE 03203 CONTENTS (Continued)

**4.0.0 PACKING** .................................................. 3.15
    4.1.0 Types of Packing ........................................ 3.15
    *4.1.1 Square Braid Packing* ................................... 3.15
    *4.1.2 Braid-Over-Braid Packing* ............................... 3.16
    *4.1.3 Interlocking Braid Packing* ............................. 3.16
    *4.1.4 Twisted Braid Packing* .................................. 3.16
    *4.1.5 Wrapped, Accordion-Fold, and Laminated Packings* ........ 3.16
    *4.1.6 Metal Packing* .......................................... 3.17
    *4.1.7 Graphite Ribbon Packing* ................................ 3.17
    4.2.0 Installing and Removing Packing ......................... 3.17
    *4.2.1 Removing Packing* ....................................... 3.17
    *4.2.2 Installing Compression Packing* ......................... 3.17

**5.0.0 SEALS** .................................................... 3.18
    5.1.0 Nonmechanical Seals ..................................... 3.18
    *5.1.1 O-Rings* ................................................ 3.18
    *5.1.2 Lip and Oil Seals* ...................................... 3.19
    5.2.0 Mechanical Seals ........................................ 3.19
    5.3.0 Installing and Removing Seals ........................... 3.20

**6.0.0 BEARINGS** ................................................. 3.21
    6.1.0 Plain Bearings .......................................... 3.21
    *6.1.1 Sleeve Bearings* ........................................ 3.21
    *6.1.2 Thrust Bearings* ........................................ 3.22
    6.2.0 Anti-Friction Bearings .................................. 3.22
    *6.2.1 Ball Bearings* .......................................... 3.22
    *6.2.2 Roller Bearings* ........................................ 3.22
    6.3.0 Identifying Bearing Failures ............................ 3.23
    6.4.0 Removing and Installing Bearings ........................ 3.23
    *6.4.1 Removing Bearings* ...................................... 3.23
    *6.4.2 Installing Bearings* .................................... 3.24

**7.0.0 LUBRICATION** .............................................. 3.25
    7.1.0 Oils .................................................... 3.26
    *7.1.1 Viscosity* .............................................. 3.26
    *7.1.2 Viscosity Index* ........................................ 3.26
    *7.1.3 Pour Point* ............................................. 3.26
    *7.1.4 Flash Point and Fire Point* ............................. 3.26
    *7.1.5 Oxidation Resistance* ................................... 3.26
    7.2.0 Greases ................................................. 3.26
    7.3.0 Lubrication Equipment ................................... 3.27
    *7.3.1 Lever-Type Grease Gun* .................................. 3.27
    *7.3.2 Lubrication Fittings* ................................... 3.28

## MODULE 03203 CONTENTS (Continued)

**8.0.0 BELTS AND BELT DRIVES** .................................................. 3.28
    8.1.0    *V-Belts* .................................................................. 3.28
    *8.1.1    Fractional Horsepower Belts* ........................... 3.28
    *8.1.2    Standard Multiple Belts* .................................. 3.29
    *8.1.3    Wedge Belts* ................................................. 3.29
    8.2.0    Belt Drive System Maintenance ........................... 3.29

**9.0.0 COUPLINGS AND DIRECT DRIVES** .................................. 3.30
    9.1.0    Coupling Types .................................................. 3.30
    *9.1.1    Rigid Couplings* .............................................. 3.30
    *9.1.2    Flexible Couplings* ......................................... 3.31
    *9.1.3    Soft-Start Couplings* ...................................... 3.31
    9.2.0    General Coupling Removal and Installation Methods ....... 3.31
    *9.2.1    Removing Couplings* ..................................... 3.31
    *9.2.2    Installing Couplings* ....................................... 3.32
    9.3.0    Coupling Alignment ........................................... 3.32
    *9.3.1    Correcting Outer Diameter Alignment, Side View* ......... 3.32
    *9.3.2    Correcting Outer Diameter Alignment, Top View* .......... 3.32
    *9.3.3    Correcting Face Alignment, Side View* ................. 3.33
    *9.3.4    Correcting Face Alignment, Top View* .................. 3.33

**10.0.0 BASIC MAINTENANCE PROCEDURES** ......................... 3.33
    10.1.0    Motor Lubrication ........................................... 3.33
    10.2.0    Air Filter and Screen Maintenance ................. 3.34
    10.3.0    Coil and Condensate System Maintenance ..... 3.36
    10.4.0    Damper Inspection and Cleaning ..................... 3.37

**11.0.0 DOCUMENTATION** ....................................................... 3.37
    11.1.0    The Importance of Make, Model Number, and
                Serial Number .............................................. 3.38
    11.2.0    Service Ticket/Invoice .................................... 3.38
    11.3.0    Commissioning Job Report ............................ 3.38
    11.4.0    Start-Up Report .............................................. 3.39
    11.5.0    Warranty Ticket ............................................. 3.39

**12.0.0 CUSTOMER RELATIONS** ............................................. 3.43
    12.1.0    Why Customer Relations are Important ......... 3.43
    12.2.0    Personal Habits, Behaviors, and Attitudes ..... 3.43
    12.3.0    Customer Relations: Handling Service Calls ... 3.44
    *12.3.1    The Opening* ................................................ 3.44
    *12.3.2    Servicing* ...................................................... 3.44
    *12.3.3    The Closing* .................................................. 3.45

# MODULE 03203 CONTENTS (Continued)

**13.0.0 CUSTOMER COMMUNICATION** .................................................. 3.45
    13.1.0 Keeping Communications Positive ........................ 3.46
    13.2.0 The Positive Approach ........................................... 3.46
    13.3.0 Showing Concern for Customers ........................... 3.46
    13.4.0 Handling Difficult Customers ................................. 3.47
        *13.4.1 Fearful Customers* ............................................. 3.47
        *13.4.2 Opinionated Customers* ..................................... 3.48
        *13.4.3 Argumentative Customers* ................................. 3.48
        *13.4.4 Sloppy Customers* ............................................. 3.49
        *13.4.5 Angry Customers* .............................................. 3.49
        *13.4.6 Critical Customers* ............................................. 3.49
        *13.4.7 Customers With Unresolvable Problems* ............ 3.50
        *13.4.8 Customers Who Request Help With Odd Jobs* .. 3.50
**SUMMARY** ........................................................................................ 3.50
**REVIEW QUESTIONS** ...................................................................... 3.51
**GLOSSARY** ..................................................................................... 3.54
**ANSWERS TO REVIEW QUESTIONS** .............................................. 3.55
**REFERENCES AND ACKNOWLEDGMENTS** ................................... 3.56

## Figures

Figure 1    Thread designations ........................................................ 3.2
Figure 2    Grade markings for steel bolts and screws .................... 3.3
Figure 3    Machine bolts, machine screws, cap screws,
              and stud bolts ................................................................. 3.4
Figure 4    Set screws ....................................................................... 3.5
Figure 5    Flat and lock washers .................................................... 3.5
Figure 6    Nuts ................................................................................. 3.6
Figure 7    Thread-forming and thread-cutting screws ..................... 3.7
Figure 8    Toggle bolts and anchor bolts ........................................ 3.7
Figure 9    Thread repair insert ........................................................ 3.8
Figure 10   Retainer rings ................................................................. 3.8
Figure 11   Pin fasteners .................................................................. 3.8
Figure 12   Keys ................................................................................ 3.9
Figure 13   Rivets .............................................................................. 3.9
Figure 14   Torque specifications ................................................... 3.11
Figure 15   Common fastener tightening sequences ..................... 3.12
Figure 16   Expandable (Redhead™) fastener
              installation process ...................................................... 3.13

## Figures (Continued)

Figure 17　Common gasket types . . . . . . . . . . . . . . . . . . . . . . . . . . . . . . . . . . . 3.14
Figure 18　Types of braid packing . . . . . . . . . . . . . . . . . . . . . . . . . . . . . . . . . . 3.15
Figure 19　Packing . . . . . . . . . . . . . . . . . . . . . . . . . . . . . . . . . . . . . . . . . . . . . . 3.16
Figure 20　Packing puller screwed into packing . . . . . . . . . . . . . . . . . . . . . . . 3.17
Figure 21　Installing packing . . . . . . . . . . . . . . . . . . . . . . . . . . . . . . . . . . . . . . 3.18
Figure 22　Static and dynamic seals using an O-ring . . . . . . . . . . . . . . . . . . . 3.19
Figure 23　Oil seal . . . . . . . . . . . . . . . . . . . . . . . . . . . . . . . . . . . . . . . . . . . . . . 3.20
Figure 24　Typical mechanical seal parts . . . . . . . . . . . . . . . . . . . . . . . . . . . . 3.20
Figure 25　Plain bearings . . . . . . . . . . . . . . . . . . . . . . . . . . . . . . . . . . . . . . . . 3.21
Figure 26　Typical ball bearings . . . . . . . . . . . . . . . . . . . . . . . . . . . . . . . . . . . 3.22
Figure 27　Typical roller bearings . . . . . . . . . . . . . . . . . . . . . . . . . . . . . . . . . . 3.22
Figure 28　Manual puller being used to remove a bearing . . . . . . . . . . . . . . . 3.24
Figure 29　Bearing heater . . . . . . . . . . . . . . . . . . . . . . . . . . . . . . . . . . . . . . . . 3.25
Figure 30　Arbor press . . . . . . . . . . . . . . . . . . . . . . . . . . . . . . . . . . . . . . . . . . 3.25
Figure 31　Bearing clearance measurement . . . . . . . . . . . . . . . . . . . . . . . . . 3.25
Figure 32　Lever-type grease gun and hydraulic
　　　　　　 grease fittings . . . . . . . . . . . . . . . . . . . . . . . . . . . . . . . . . . . . . . . 3.27
Figure 33　V-belt drive . . . . . . . . . . . . . . . . . . . . . . . . . . . . . . . . . . . . . . . . . . . 3.28
Figure 34　Belt-drive misalignment . . . . . . . . . . . . . . . . . . . . . . . . . . . . . . . . . 3.30
Figure 35　Straightedge method for aligning pulleys . . . . . . . . . . . . . . . . . . . . 3.30
Figure 36　Typical couplings . . . . . . . . . . . . . . . . . . . . . . . . . . . . . . . . . . . . . . 3.31
Figure 37　O.D. misalignment, side view . . . . . . . . . . . . . . . . . . . . . . . . . . . . 3.32
Figure 38　O.D. misalignment, top view . . . . . . . . . . . . . . . . . . . . . . . . . . . . . 3.32
Figure 39　Face misalignment, side view . . . . . . . . . . . . . . . . . . . . . . . . . . . . 3.33
Figure 40　Face misalignment, top view . . . . . . . . . . . . . . . . . . . . . . . . . . . . . 3.33
Figure 41　Oiler and lever-type grease gun . . . . . . . . . . . . . . . . . . . . . . . . . . 3.33
Figure 42　Typical mechanical air filter . . . . . . . . . . . . . . . . . . . . . . . . . . . . . . 3.34
Figure 43　Electronic air cleaner . . . . . . . . . . . . . . . . . . . . . . . . . . . . . . . . . . . 3.35
Figure 44　Replacing filter media in a high-efficiency
　　　　　　 mechanical filter . . . . . . . . . . . . . . . . . . . . . . . . . . . . . . . . . . . . . 3.35
Figure 45　Example of a typical repair order/service
　　　　　　 report form . . . . . . . . . . . . . . . . . . . . . . . . . . . . . . . . . . . . . . . . . 3.39
Figure 46　Example of a typical work order form . . . . . . . . . . . . . . . . . . . . . . 3.40
Figure 47　Example of a typical start-up report for a
　　　　　　 heat pump . . . . . . . . . . . . . . . . . . . . . . . . . . . . . . . . . . . . . . . . . . 3.41
Figure 48　Example of a typical start-up report for a furnace . . . . . . . . . . . . . 3.42

## Tables

Table 1　　Lubricant Symptoms Related to Common
　　　　　　Bearing Failures . . . . . . . . . . . . . . . . . . . . . . . . . . . . . . . . . . . . . 3.23

# MODULE 03203

# Maintenance Skills for the Service Technician

## Objectives

When you have completed this module, you will be able to do the following:

1. Identify the types of threaded and non-threaded fasteners and explain their use.
2. Install threaded and non-threaded fasteners.
3. Identify the types of gaskets, packings, and seals and explain their use.
4. Remove and install gaskets, packings, and seals.
5. Identify the types of lubricants and explain their use.
6. Use lubrication equipment to lubricate motor bearings.
7. Identify the types of belt drives and explain their use.
8. Demonstrate and/or explain procedures used to install or adjust a belt drive.
9. Identify the types of couplings and explain their use.
10. Demonstrate and/or explain procedures used to remove, install, and align couplings.
11. Identify the types of bearings and explain their use.
12. Explain causes of bearing failures.
13. Demonstrate and/or explain procedures used to remove and install bearings.
14. Perform basic preventive maintenance inspection and cleaning procedures.
15. List work and personal habits that contribute to good customer relations.
16. Identify steps in the handling of a typical service call that will contribute to good customer relations.
17. Legibly fill out forms used for installation and service calls.

## Prerequisites

Before you begin this module, it is recommended that you successfully complete the following modules: Core Curriculum; HVAC Level One; HVAC Level Two, Modules 03201 and 03202.

## Required Trainee Materials

1. Pencil and Paper
2. Appropriate Personal Protective Equipment

### 1.0.0 ◆ INTRODUCTION

The HVAC technician must be able to properly install and maintain the mechanical components used in HVAC systems. This module describes the mechanical maintenance skills required to install and work with the following components:

- Mechanical fasteners
- Gaskets, packings, and seals
- Lubricants
- Belts and belt drives
- Couplings and direct drives
- Bearings

Mechanical maintenance tasks commonly performed during the scheduled maintenance of most types of HVAC equipment are also covered in this module, including procedures for motor lubrication, air filter and screen maintenance, coil maintenance, and damper maintenance.

In addition to performing the actual installation and maintenance field service tasks, the technician is often required to fill out job reports and checklists used by both the technician and the employer to track the progress of a job and/or record the performance of the equipment. In addition, the technician must also interface with the customer before, during, and after performing the work. For this reason, the last part of this module describes some common types of documentation used in the field, then focuses on guidelines for promoting and maintaining good customer relations.

## 2.0.0 ◆ MECHANICAL FASTENERS

Fasteners are used to install many types of parts and equipment. Common fasteners include bolts, screws, pins, and retainers.

### 2.1.0 Thread and Grade Designations

This section covers thread and grade designations. Fasteners, bolts, and screws are distinguished from one another through the use of standard designations.

#### 2.1.1 Thread Designations

Fastener threads are made to established standards. The most common standard is the unified or *American National Standard*. There are three series or classes of threads defined by the unified standard. These series are based on the number of threads per inch for a fastener of a certain diameter. The three series are:

- *Unified National Coarse (UNC) Thread* – UNC thread is used for bolts, screws, nuts, and other general applications. Fasteners with UNC threads are used for rapid assembly and disassembly of parts where corrosion or slight damage may occur.
- *Unified National Fine (UNF) Thread* – UNF thread is used for bolts, screws, nuts, and other uses where a finer thread than UNC is required.
- *Unified National Extra Fine (UNEF) Thread* – UNEF thread is used for thin-walled tubes, nuts, ferrules, and couplings.

Bolt and screw threads are designated by a standard method (*Figure 1*). The standard designations include:

- Nominal size (diameter)
- Number of threads per inch
- Thread series symbol
- Thread class (Classes 1A, 2A, and 3A are external threads, such as used with a bolt. Classes 1B, 2B, and 3B are internal threads, such as used with a nut.)
- Left-hand thread symbol (Unless shown, the threads are right hand. This symbol is used only for left-hand fasteners.)

Metric screw threads based on the American National Standard are also in common use. Metric

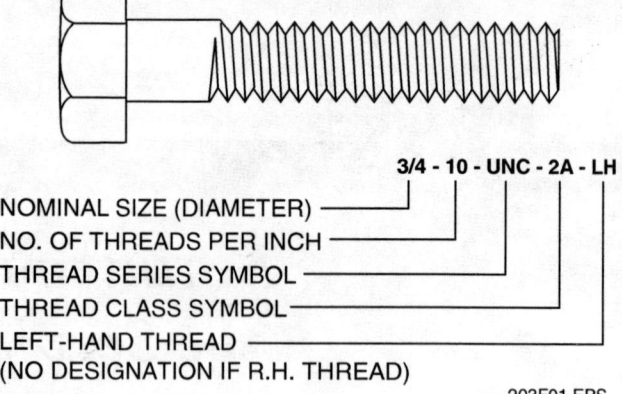

*Figure 1* ◆ Thread designations.

M-profile threaded screws are a coarse-thread series of fasteners used for general fastening purposes. Metric MJ-profile threaded screws are used with aircraft parts and in other high-stress uses requiring extra strength.

#### 2.1.2 Fastener Grade Designations

The strength and quality of a fastener can be determined by special grade markings on the head of the fastener. These markings are standardized by the Society of Automotive Engineers (SAE) and the American Society for Testing of Materials (ASTM). Grade markings are sometimes called *line markings*. *Figure 2* shows the SAE and ASTM markings for steel bolts and screws.

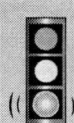 **NOTE**
Always use bolts that have grade markings, since bolts that do not may be inferior.

### 2.2.0 Threaded Fasteners

Threaded fasteners are one of the most common types of fasteners. Many are assembled with nuts and washers. Others are installed in threaded holes. There are many types of threaded fasteners, including:

- Set screws
- Machine bolts, machine screws, stud bolts, and cap screws
- Flat and lock washers
- Nuts
- Thread-forming and thread-cutting screws
- Toggle and anchor bolts
- Inserts

## ASTM AND SAE GRADE MARKINGS FOR STEEL BOLTS & SCREWS

| GRADE MARKING | SPECIFICATION | MATERIAL |
|---|---|---|
| 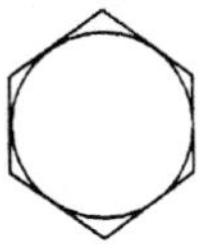 (plain hex) | SAE-GRADE 0 | STEEL |
| (plain hex) | SAE-GRADE 1<br>ASTM-A 307 | LOW CARBON STEEL |
| (plain hex) | SAE-GRADE 2 | LOW CARBON STEEL |
| 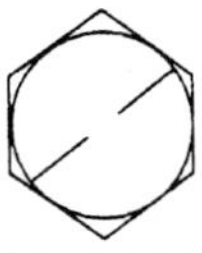 (one line) | SAE-GRADE 3 | MEDIUM CARBON STEEL,<br>COLD WORKED |
|  (A 325) | SAE-GRADE 5 | MEDIUM CARBON STEEL,<br>QUENCHED AND TEMPERED |
| (A 325) | ASTM-A 449 | |
| 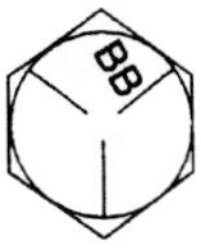 (BB) | ASTM-A 325 | MEDIUM CARBON STEEL,<br>QUENCHED AND TEMPERED |
| 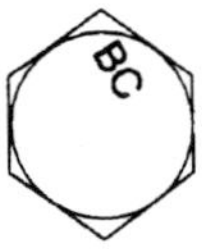 (BC) | ASTM-A 354<br>GRADE BB | LOW ALLOY STEEL,<br>QUENCHED AND TEMPERED |
| 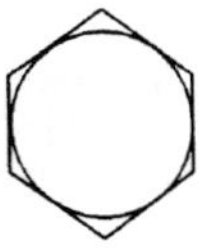 (plain) | ASTM-A 354<br>GRADE BC | LOW ALLOY STEEL,<br>QUENCHED AND TEMPERED |
|  (6 lines) | SAE-GRADE 7 | MEDIUM CARBON ALLOY STEEL,<br>QUENCHED AND TEMPERED<br>ROLL THREADED AFTER<br>HEAT TREATMENT |
| 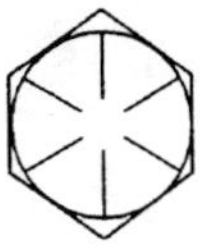 (6 lines) | SAE-GRADE 8 | MEDIUM CARBON ALLOY STEEL,<br>QUENCHED AND TEMPERED |
| (6 lines) | ASTM-A 354<br>GRADE BD | ALLOY STEEL,<br>QUENCHED AND TEMPERED |
|  (A 490) | ASTM-A 490 | ALLOY STEEL,<br>QUENCHED AND TEMPERED |

**ASTM SPECIFICATIONS**
- A 307 - LOW CARBON STEEL EXTERNALLY AND INTERNALLY THREADED STANDARD FASTENERS.
- A 325 - HIGH STRENGTH STEEL BOLTS FOR STRUCTURAL STEEL JOINTS, INCLUDING SUITABLE NUTS AND PLAIN HARDENED WASHERS.
- A 449 - QUENCHED AND TEMPERED STEEL BOLTS AND STUDS.
- A 354 - QUENCHED AND TEMPERED ALLOY STEEL BOLTS AND STUDS WITH SUITABLE NUTS.
- A 490 - HIGH STRENGTH ALLOY STEEL BOLTS FOR STRUCTURAL STEEL JOINTS, INCLUDING SUITABLE NUTS AND PLAIN HARDENED WASHERS.

**SAE SPECIFICATION**
- J 429 - MECHANICAL AND QUALITY REQUIREMENTS FOR THREADED FASTENERS.

*Figure 2* ◆ Grade markings for steel bolts and screws.

### 2.2.1 Machine Bolts, Machine Screws, Stud Bolts, and Cap Screws

Machine bolts (*Figure 3*) are used to assemble parts that do not require close **tolerances**. The tolerance is the amount of variation allowed from a standard. Machine bolts are made with diameters ranging from ¼" to 3" and with lengths from ½" to 30". A machine bolt is tightened and released by turning its mating nut that is usually furnished along with the bolt.

Machine screws are used for general assembly. They have slotted or recessed heads. Machine screws are available in diameters from #6 through #12 (0.060" to 0.2160") and from ¼" to ½", and in lengths from ⅛" to 3". They are also made in metric sizes. A machine screw normally mates with an internally threaded hole into which it is tightened or released. They can also be used with nuts.

Cap screws are generally used on assemblies that need a finished appearance. They pass through a clearance hole in one part of the assembly and are screwed into a threaded hole in the other part. This clamps the parts together when the cap screw is tightened.

Cap screws are made to close tolerances with machined or semifinished bearing surfaces under the head. Cap screws come in coarse and fine threads, and in diameters from ¼" to 2". Lengths from ⅜" to 10" are available. They are also made in metric sizes.

Stud bolts are headless bolts, threaded either along the entire length or on both ends. One end can be screwed into a threaded hole. The part to be clamped is fitted over the other end of the stud, and a nut and washer are screwed on to fasten the two parts together.

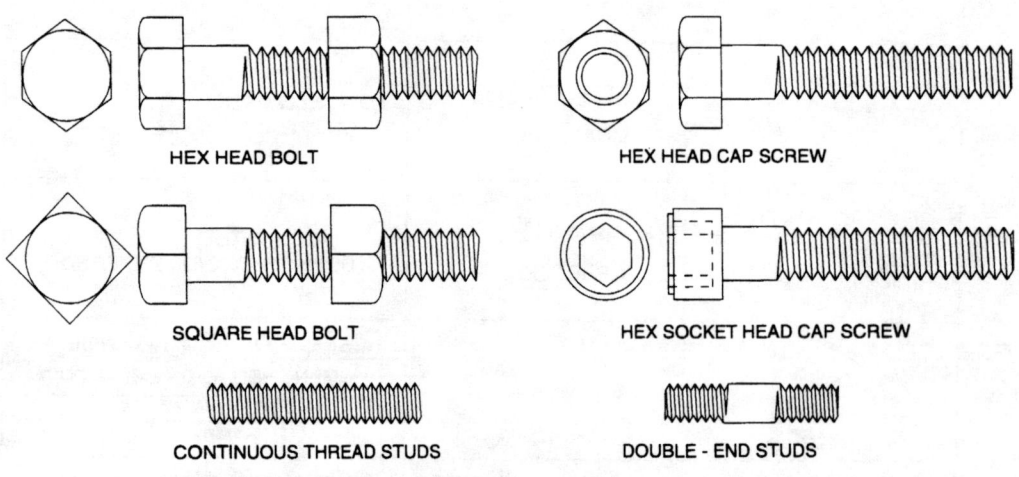

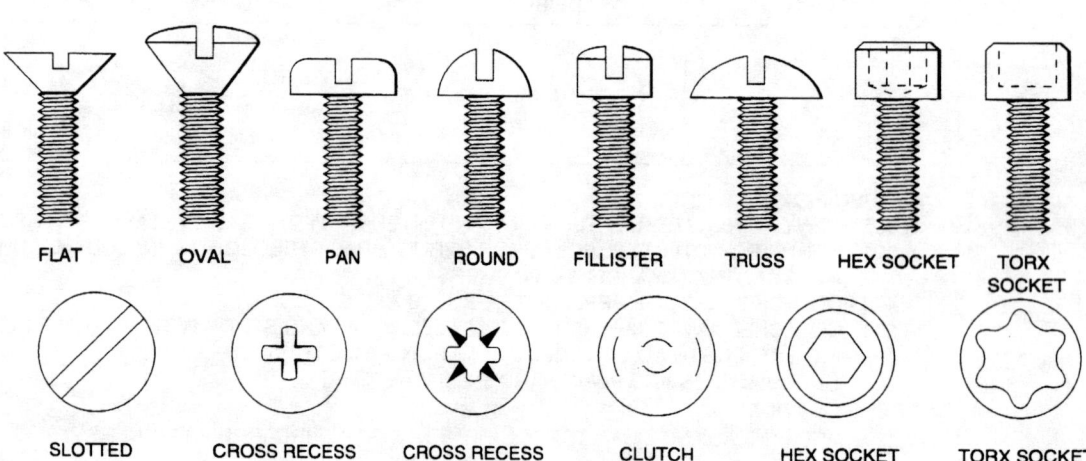

*Figure 3* ◆ Machine bolts, machine screws, cap screws, and stud bolts.

## 2.2.2 Set Screws

Set screws are usually made of heat-treated steel. They are used to fasten pulleys and fan blades on shafts, and to hold collars in place.

Set screws are classified by head styles and point styles. *Figure 4* shows several set screw head and point styles.

## 2.2.3 Flat and Lock Washers

Flat washers (*Figure 5*) provide an enlarged surface used to distribute the load from bolt heads and nuts. Flat washers are made in light, medium, heavy-duty, and extra heavy-duty series. Fender washers have a wide surface area to bridge oversized holes or other wide clearance requirements. Lock washers are used to keep bolts or nuts from working loose. They are placed between the flat washers and the bolts or nuts. Some common types of lock washers include the following:

- *External* – Provides the greatest resistance
- *Internal* – Used on small screws
- *Internal-external* – Used for oversized mounting holes
- *Countersunk* – Used with flat or oval-head screws
- *Split ring* – Commonly used with bolts and cap screws

## 2.2.4 Nuts

Nuts used with most threaded fasteners have hex (hexagonal) or square shapes and are used with bolts having the same shaped head. *Figure 6* shows different types of nuts. Some special-purpose nuts include the following:

- *Acorn nut* – These are used when appearance is important or when there are exposed sharp threads on the fastener.
- *Castellated (or castle) and slotted nuts* – After the nut is tightened, a cotter pin is fitted into one set of slots and through a hole in the bolt. The cotter pin keeps the nut from loosening.
- *Self-locking nut* – This has a nylon insert or is slightly deformed so it cannot work loose. Once a self-locking nut is used and removed, it should be thrown away and replaced with a new one.
- *Wing nut* – These are used where frequent adjustments and service are necessary. Wing nuts allow for loosening and tightening without the use of a wrench.
- *Jam nut* – This is a thin nut used to lock a standard nut in place.

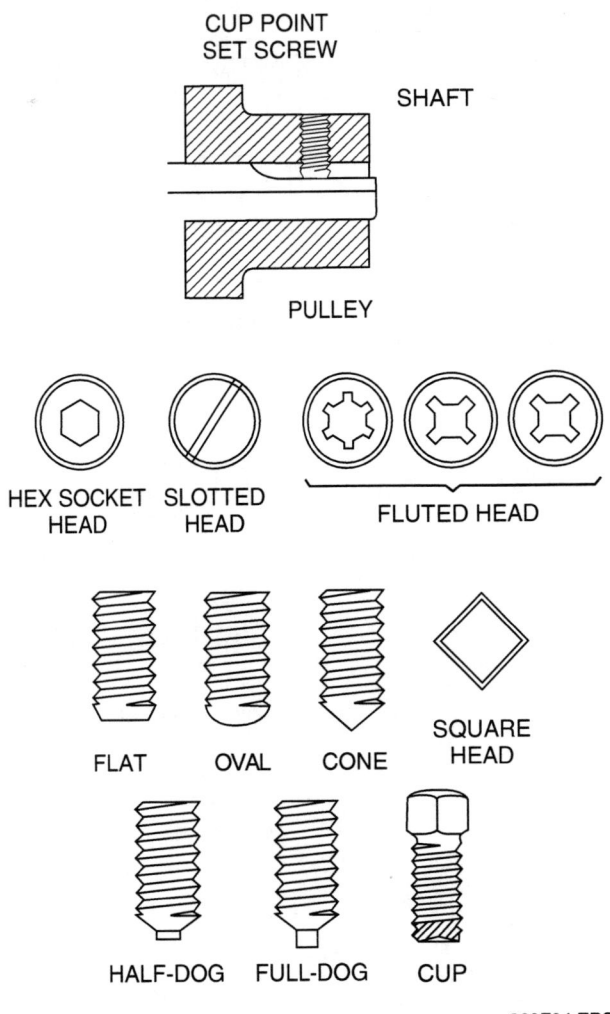

*Figure 4* ◆ Set screws.

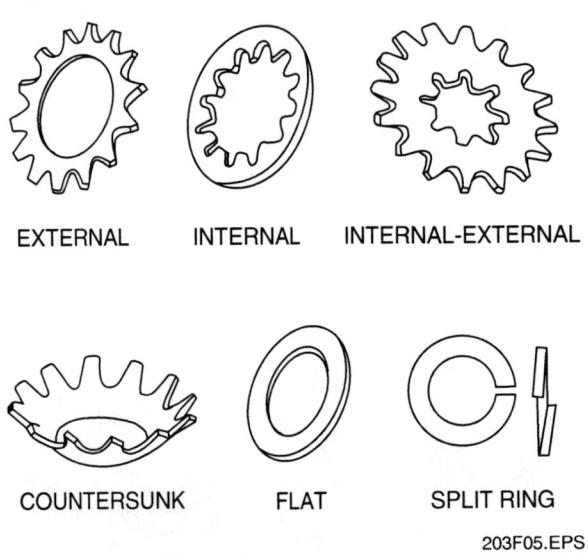

*Figure 5* ◆ Flat and lock washers.

MAINTENANCE SKILLS FOR THE SERVICE TECHNICIAN — TRAINEE MODULE 03203

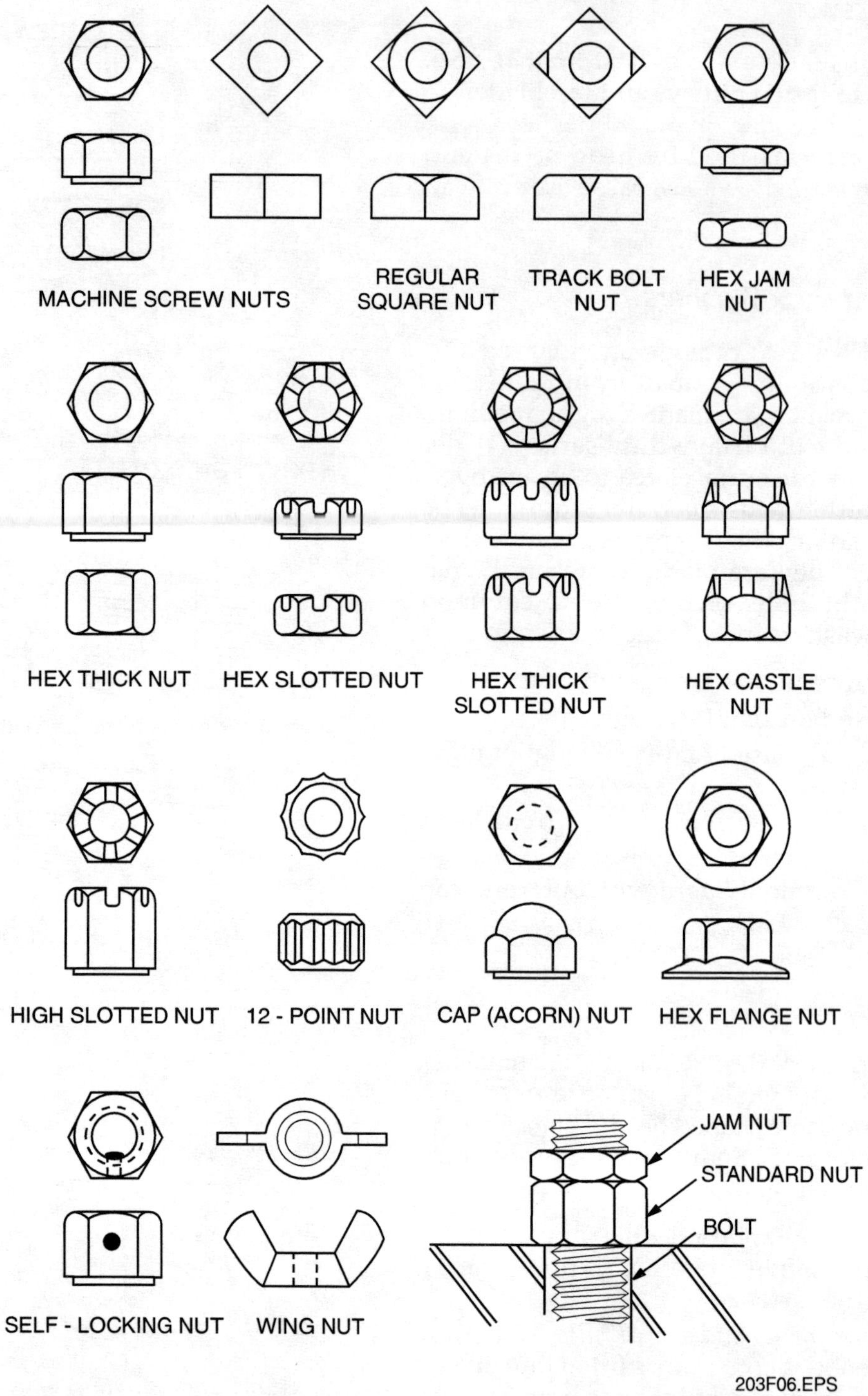

*Figure 6* ◆ Nuts.

## Self-Drilling Sheet Metal Screws

If you are installing a sheet metal duct system, self-drilling sheet metal screws can be real timesavers. Use a cordless drill equipped with a ¼" or ⁵⁄₁₆" magnetized drive socket. Insert the self-drilling screw in the socket and drive the screw in. Be careful not to overtorque the screws as they will strip out.

## 2.2.5 Thread-Forming and Thread-Cutting Screws

Thread-forming screws (*Figure 7*) are used mainly to fasten light-gauge metal parts. They form a thread as they are driven. This eliminates tapping. Some also drill their own hole. This eliminates drilling, punching, and aligning of parts.

Thread-cutting screws cut threads into the metal as they are driven into a pilot hole. They are made of hardened steel and are used to join heavy-gauge sheet metal and nonferrous metal parts.

## 2.2.6 Toggle and Anchor Bolts

Toggle bolts are used to fasten a part to a hollow wall or panel. *Figure 8* shows some common toggle bolts.

Anchor bolts are used to fasten parts and equipment to concrete and masonry. There are several types of non-expansion anchor bolts designed for installation in both wet and hardened concrete and other surfaces. Expansion-type anchor bolts are installed in holes drilled in hardened concrete. When the nut on the anchor bolt is tightened, the bolt base expands to provide the holding force. *Figure 8* shows typical non-expansion and expansion anchor bolts installed in concrete.

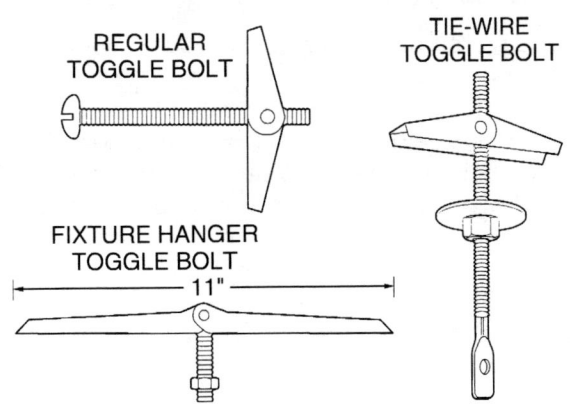

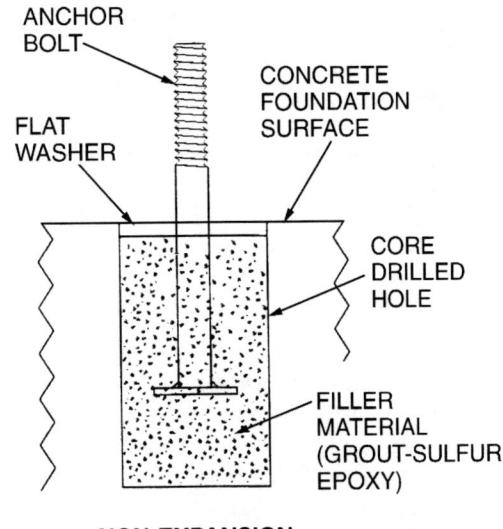

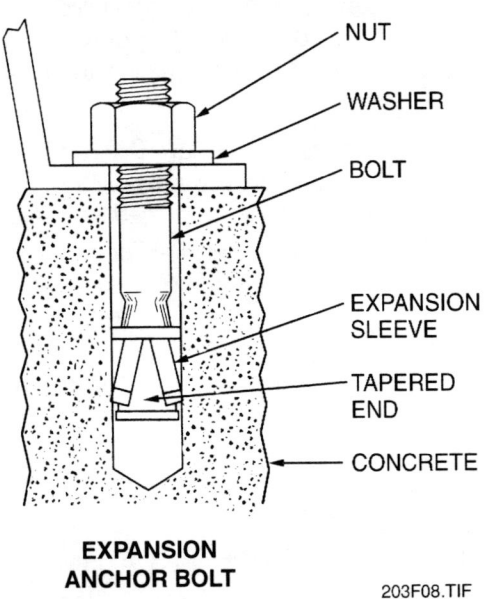

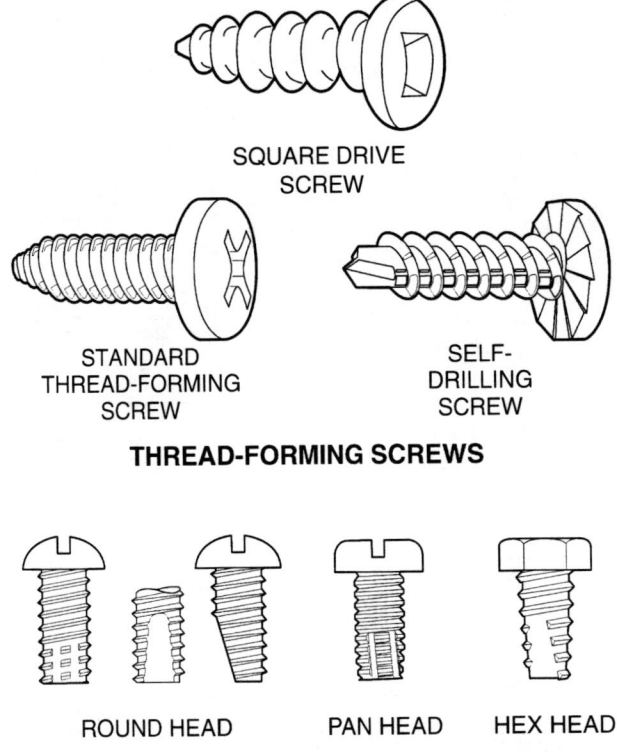

*Figure 7* ◆ Thread-forming and thread-cutting screws.

*Figure 8* ◆ Toggle bolts and anchor bolts.

MAINTENANCE SKILLS FOR THE SERVICE TECHNICIAN — TRAINEE MODULE 03203

### 2.2.7 Thread Repair Inserts

Thread repair inserts, or heli-coils (*Figure 9*), are a special kind of fastener used to provide high-strength threads in soft metals and plastics. They are also used to replace damaged or stripped threads in a tapped hole. Inserts are made in standard sizes and forms including metric sizes. The insert, which is larger in diameter than the tapped hole, is compressed during installation and allowed to spring back, permanently anchoring the insert in the tapped hole.

## 2.3.0 Non-Threaded Fasteners

Non-threaded fasteners have many uses. This section describes the following types of non-threaded fasteners:

- Retainer rings
- Pins
- Keys
- Rivets

### 2.3.1 Retainer Rings

Retainer rings are used for both internal and external fastening. Some retainer rings are seated in grooves. Others are self-locking and do not require a groove. Special pliers are used to remove internal and external rings. *Figure 10* shows typical retainer rings.

### 2.3.2 Pins

Pin fasteners (*Figure 11*) are used to align mating parts, to hold gears and pulleys on shafts, and to secure slotted nuts. Some common pins and their uses are:

- *Dowel pins* – These are fitted into reamed holes to position mating parts. They also support a portion of the load placed on the parts.
- *Taper and spring pins* – These are used to fasten gears, pulleys, and collars to a shaft.
- *Cotter pins* – These are fitted into a hole drilled through a shaft. Cotter pins are used to prevent parts from slipping on or off the shaft, and also to keep slotted nuts from working loose.

### 2.3.3 Keys

Keys are metal parts used to prevent a gear or pulley from rotating on a shaft. One half of the key fits into a keyseat on the shaft. The other half fits into a keyway in the hub of a gear or pulley. *Figure 12* shows some common keys and their trade names.

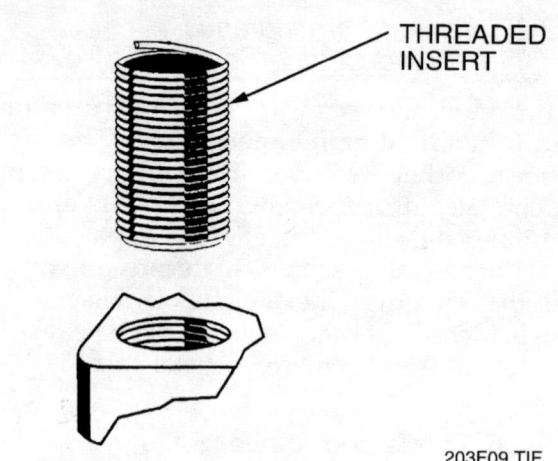

*Figure 9* ◆ Thread repair insert.

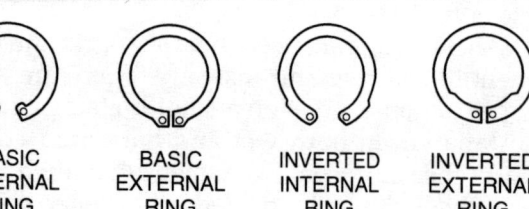
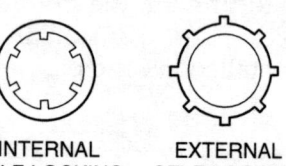

*Figure 10* ◆ Retainer rings.

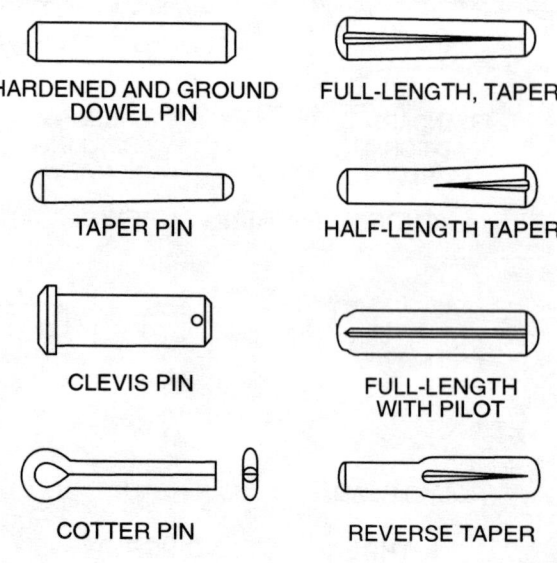

*Figure 11* ◆ Pin fasteners.

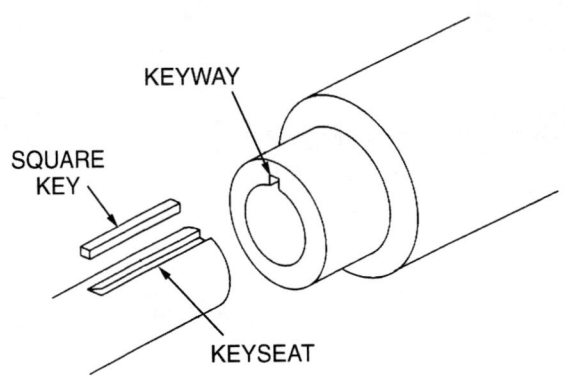

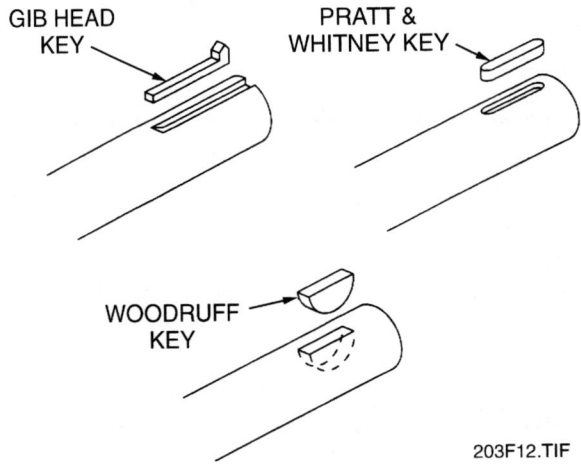

*Figure 12* ◆ Keys.

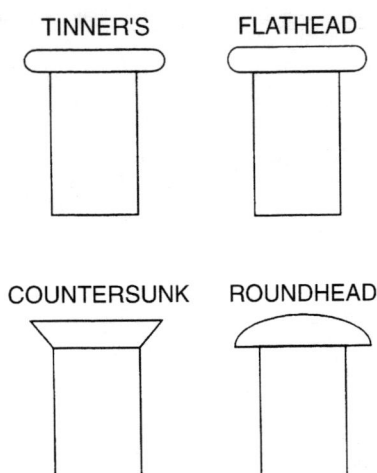

**TINNER'S RIVETS**

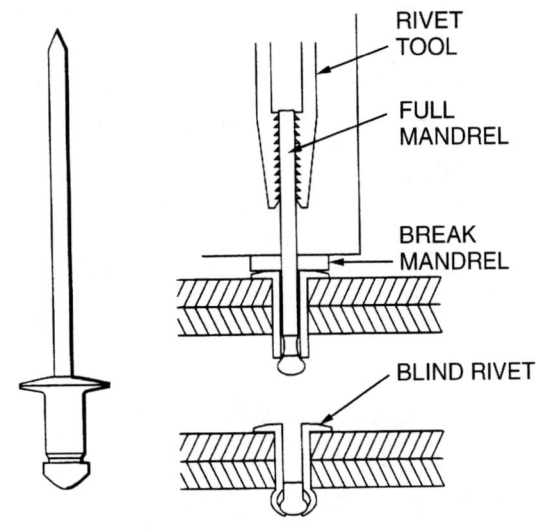

**BLIND RIVET INSTALLATION**

*Figure 13* ◆ Rivets.

## 2.3.4 Rivets

Rivets are used to permanently join two pieces of material. Two common types of rivets are the tinner's rivet and the blind rivet (*Figure 13*). To install a tinner's rivet, the rivet is inserted into a hole that has been drilled or punched. The small end of a tinner's rivet is hammered or set in the form of a head. Hollow tinner's rivets are clinched at the small end with a special tool. Tinner's rivets are sized in ounces or pounds per 1,000 rivets. For example, a 6-ounce rivet means that 1,000 of these rivets will weigh 6 ounces. As the weight increases, so does the diameter and length of the rivet.

Blind rivets are used to fasten sheet metal, fiberglass, and plastics. They are available in various lengths and diameters. Blind rivets are used when the joint can only be reached from one side. To install a blind rivet, the rivet is inserted into a previously drilled or punched hole, then the rivet is set, or popped, using a special tool.

### Removing Rivets

Chisel or grind off the heads of tinner's rivets or steel blind rivets; remove the shaft by striking through with a punch. Drill off blind or countersunk rivet heads using a high-speed electric drill, then punch out the shaft.

### Blind (Pop) Rivet Tool

A manual pop rivet tool like the one shown here can be used to install blind rivets in a shop or at the job site.

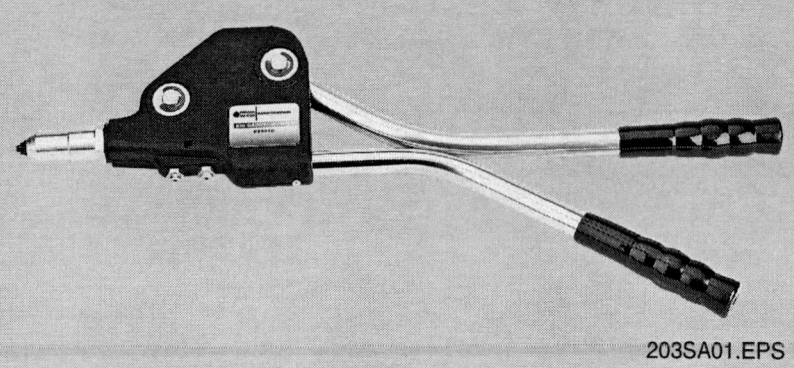

## 2.4.0 Installing Threaded Fasteners

To install threaded fasteners, you must know the type of fastener to use and the correct method of installation. You must also understand the tightening sequence and **torque** specifications. This section describes the guidelines you should follow when installing common fasteners.

### 2.4.1 Torquing Steel Fasteners

When torquing steel fasteners, you must first select the proper type and grade of fastener for the job, then torque (tighten) it to the recommended specifications. *Figure 14* shows typical torque specifications for fasteners. Always follow the manufacturer's instructions and recommended torque values when installing fasteners on equipment. You must use a suitable torque wrench to tighten fasteners to their correct torque specifications. You might want to review the torque wrenches covered in the HVAC Level One Module, *Tools of the Trade*.

Torque is the resistance to a turning or twisting force. The correct size torque wrench for a job is one that will read between 25% and 75% of the scale when the required torque is applied. This allows for adequate capacity and provides satisfactory accuracy. Avoid using an oversized torque wrench because the scale divisions are too coarse, making it difficult to get an accurate reading. Using a wrench that is too small will not allow for extra capacity in the event of seizure or **run-down resistance**. All threaded fasteners must be clean and undamaged in order to get accurate readings.

### Fastening Steel Fasteners

Steel fasteners are commonly fastened using a torque wrench like the one shown here. Whenever possible, apply force to the torque wrench by pulling rather than pushing. This reduces the chance of injury to the fingers or knuckles should the wrench slip. Be sure to grip the handle. Do not use extensions.

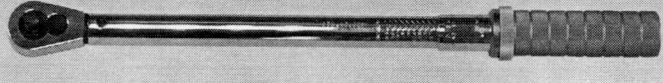

### TORQUE IN FOOT POUNDS

| FASTENER DIAMETER | THREADS PER INCH | MILD STEEL | STAINLESS STEEL 18-8 | ALLOY STEEL |
|---|---|---|---|---|
| 1/4 | 20 | 4 | 6 | 8 |
| 5/16 | 18 | 8 | 11 | 16 |
| 3/8 | 16 | 12 | 18 | 24 |
| 7/16 | 14 | 20 | 32 | 40 |
| 1/2 | 13 | 30 | 43 | 60 |
| 5/8 | 11 | 60 | 92 | 120 |
| 3/4 | 10 | 100 | 128 | 200 |
| 7/8 | 9 | 160 | 180 | 320 |
| 1 | 8 | 245 | 285 | 490 |

### SUGGESTED TORQUE VALUES FOR GRADED STEEL BOLTS

| GRADE | SAE 1 OR 2 | SAE 5 | SAE 6 | SAE 8 |
|---|---|---|---|---|
| TENSILE STRENGTH | 64000 PSI | 105000 PSI | 130000 PSI | 150000 PSI |
| GRADE MARK | (plain) | (3 marks) | (4 marks) | (6 marks) |

| BOLT DIAMETER | THREADS PER INCH | FOOT POUNDS TORQUE | | | |
|---|---|---|---|---|---|
| 1/4 | 20 | 5 | 7 | 10 | 10 |
| 5/16 | 18 | 9 | 14 | 19 | 22 |
| 3/8 | 16 | 15 | 25 | 34 | 37 |
| 7/16 | 14 | 24 | 40 | 55 | 60 |
| 1/2 | 13 | 37 | 60 | 85 | 92 |
| 9/16 | 12 | 53 | 88 | 120 | 132 |
| 5/8 | 11 | 74 | 120 | 169 | 180 |
| 3/4 | 10 | 120 | 200 | 280 | 296 |
| 7/8 | 9 | 190 | 302 | 440 | 473 |
| 1 | 8 | 282 | 466 | 660 | 714 |

*Figure 14* ♦ Torque specifications.

The calibration of torque wrenches must be checked periodically to guarantee accuracy. The following terms, sometimes used in manufacturers' service literature, must be understood when using a torque wrench:

- *Break-away torque* – The torque required to loosen a fastener. This is generally lower than the torque to which it has been tightened. For a given size fastener, there is a direct relationship between tightening torque and breakaway torque. This relationship is determined by actual test. Once known, the tightening torque can be checked by loosening and checking breakaway torque.
- *Set or seizure* – In the last stages of rotation in reaching a final torque, seizing or set of the fastener may occur. When this happens, there is usually a noticeable popping sound and vibration. To break the set, back off and then again apply the tightening torque. Accurate torque settings cannot be made if the fastener is seized.
- *Run-down resistance* – The torque required to overcome the resistance of associated hardware, such as locknuts and lockwashers, when tightening a fastener. To obtain the proper torque value where tight threads on locknuts produce a run-down resistance, add the resistance to the required torque value. Run-down resistance must be measured on the last rotation or as close to the makeup point as possible.

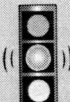

**CAUTION**

Tighten fasteners only small amounts at a time, following the proper tightening sequence for the bolt pattern and fastener type before reaching the final torque. Failure to follow the correct tightening sequence or overtightening can result in damage to the fasteners, any sealing gaskets, or the object being fastened.

### Torque Values
Depending on the type and size of fastener being used, equipment manufacturers may give torque values in inch-pounds or foot-pounds. How do you convert torque values given in inch-pounds to foot-pounds and vice versa?

#### 2.4.2 Flange Tightening Sequences

When tightening bolts on flanges and similar surfaces, the bolts must be tightened to the proper torque and in the proper sequence. Tighten each fastener only a small amount at a time until snug, following the proper tightening sequence for the bolt pattern and fastener type, then tighten to the final torque. This prevents warping or damaging the flange or machine part and also prevents leaks. The numbers in *Figure 15* show the proper tightening sequences for some common bolt patterns.

#### 2.4.3 Thread Tapping

When assembling equipment using a threaded fastener, each fastener should be inspected for damaged threads. If damaged, replace the fastener with a new one. If the threads are dirty or rusty, clean and lightly lubricate them. If the fastener is being installed in a threaded hole, inspect the hole threads for damage. If the threads are damaged, the hole should be retapped. Be sure to use the correct size tap for the fastener being used.

#### 2.4.4 Installing Anchor Bolts

Anchor bolts come in either expandable or non-expandable forms. When installing either type in hardened concrete, make sure the area where the equipment is to be fastened is smooth so that the equipment will have solid footing. Uneven footing might cause the equipment to twist, warp, not tighten properly, or vibrate during operation.

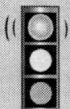

> **WARNING!**
> Drilling in concrete generates noise, dust, and possible flying objects. Always wear safety glasses, ear protectors, and gloves. Ensure that others in the area also wear appropriate personal protective equipment.

Carefully inspect the core drill and bit to make sure they are in good operating condition. If they are not, you could be injured by parts that chip and fly off during operation.

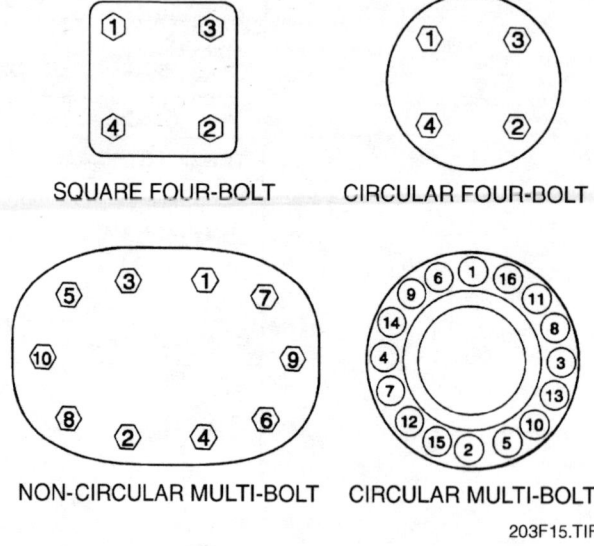

*Figure 15* ◆ Common fastener tightening sequences.

When installing a non-expansion anchor bolt in hardened concrete, it is installed in a drilled hole filled with a filler material (grout). When installing this type of anchor, the drill bit must be slightly larger in diameter than the head of the fastener and the flat washer used with the fastener. The flat washer must be able to fit down inside the hole, just below surface level. Also, the anchor bolt should extend out of the hole far enough for the threads and a little of the unthreaded bolt to be above the surface level. During the drilling process, the drill may be lubricated with water.

After the hole is drilled, the bolt installed, and the hole filled with grout, make sure the bolt is left straight after working the anchor around in the hole and installing the washer. The washer centers the bolt and holds it until the grout hardens. If the grout sets and the bolt is not straight, the bolt will be unusable, and the job will have to be repeated after the bolt is removed. Allow the grout to fully dry before mounting anything to the anchor bolt.

One type of expandable fastener, called a Redhead™, is typical of those in common use (*Figure 16*). This fastener uses a cutting sleeve that first acts as a drill bit and later becomes the fastener

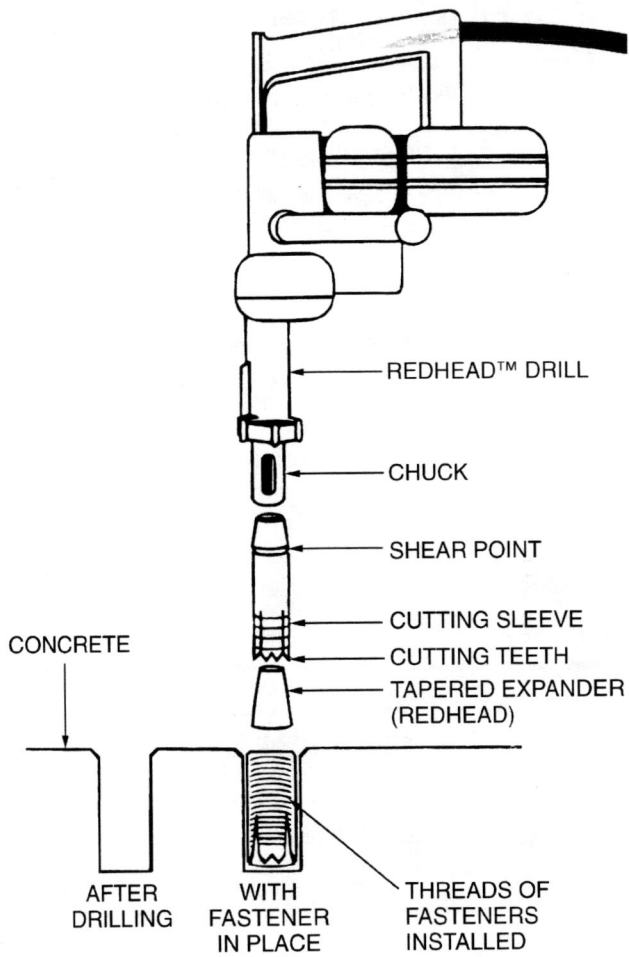

*Figure 16* ◆ Expandable (Redhead™) fastener installation process.

itself. The size of an expandable fastener is based on the size of the bolt that will be used in it. The Redhead™ drill is used with the cutting sleeve to cut the hole and later install the fastener. This tool hammers the cutting sleeve into the hole as it rotates and cuts. Later, the process is repeated, except that the sleeve will have the expander in the tip. As the fastener is hammered down, it hits the bottom, where the tapered expander causes the fastener to expand and lock into the hole.

## 3.0.0 ◆ GASKETS

The basic function of a gasket is to create a seal between two fixed parts, such as joints in pipe systems, flanges in pumps, and compressor cylinder heads. Generally, the thinnest gasket that provides the seal is the most efficient and lasts the longest. The type of gasket used depends on:

- The operating temperature of the system or equipment
- The type of connection being made
- The type of fluid the system or equipment handles
- The system or equipment pressure range

### 3.1.0 Gasket Types

Pre-formed gaskets in standard sizes and patterns, and those used in HVAC equipment and components, are readily available from HVAC equipment suppliers and manufacturers. These factory-made gaskets are cut to the required pattern, with the holes already punched for the specific application. When manufactured gaskets are not available, gaskets must be laid out and cut from the proper material as needed. Depending on the intended use, gaskets are made from natural or man-made materials, or both. *Figure 17* shows some common gasket types, including:

- Flat
- Ring
- Spiral-wound
- Full-face
- Jacketed
- Envelope

### 3.1.1 Flat Gaskets

Flat gaskets are made of various materials and have many uses. They are typically pre-formed and made by equipment manufacturers in patterns for use with specific equipment.

### 3.1.2 Ring Gaskets

Ring gaskets are flat flange gaskets made of various materials for different uses. They are made to fit inside the bolt circle of a flange; therefore, they have no bolt holes. They usually have a pressure range of 150 to 200 psig.

### 3.1.3 Spiral-Wound Gaskets

Spiral-wound gaskets are flat flange gaskets, commonly made of stainless steel with a graphite insert. They are used on high-temperature, high-pressure systems. They are crushable gaskets with high elasticity that allows them to adjust automatically to changes in line pressure, thermal shocks, vibration, and minor flange separation.

### 3.1.4 Full-Face Gaskets

Full-face gaskets are flat flange gaskets. They are similar to ring gaskets but are made to fit the flange and have holes for the flange bolts. They are made of various materials for different uses.

### 3.1.6 Envelope Gaskets

Envelope gaskets are similar in construction to the jacketed gasket. The outer cover is usually Teflon® with various types of fillers. Envelope gaskets have a limited temperature range. Because they have a superior sealing ability, envelope gaskets are frequently used on flanges with imperfections.

## 3.2.0 Installing and Removing Gaskets

The methods used to install and remove gaskets vary depending on the type of gasket material and the type of joint. Before a new gasket can be installed, the old gasket must be removed and the flanges thoroughly cleaned, using solvent if needed. The flanges and old gasket should be inspected for irregularities that might indicate a damaged flange.

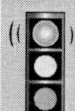

**WARNING!**
Always replace a gasket with one specified by the equipment manufacturer, or an approved equivalent. Substituting the wrong gasket or material may cause equipment failure and possible personal injury.

When installing the new gasket, do not overtighten the bolts because this could damage the gasket, the flange or mating surfaces, or the fasteners. Adhesives, sealers, or lubricants should not be used with gaskets unless specified by the manufacturer. In some cases, adhesives can be used to hold the gasket in place during installation. If used, make sure the adhesive is compatible with the gasket material and will not cause breakdown or contaminate the fluid in the system.

**CAUTION**
Tighten fasteners only small amounts at a time, following the proper tightening sequence for the bolt pattern and fastener type, before reaching the final torque. An incorrect tightening sequence or overtightening can result in damage to the fasteners, any sealing gaskets, or the object being fastened.

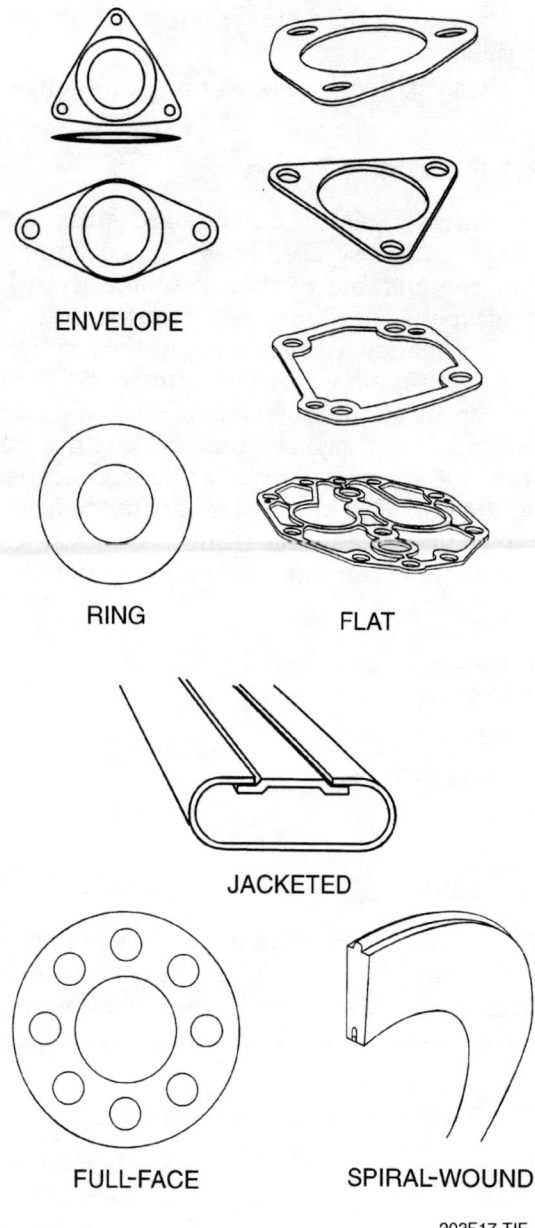

*Figure 17* ◆ Common gasket types.

### 3.1.5 Jacketed Gaskets

Jacketed gaskets have a metal exterior cover with an internal filler. The filler can be of various materials. The outer jacket prevents the fluid in a system from contacting the inner filler material, preventing the filler from contaminating the liquid.

### Removing Gaskets
The use of a commercial liquid gasket remover can make removing old gasket material easier by reducing the amount of scraping and sanding needed, thus preventing possible damage to flange surfaces.

## 4.0.0 ◆ PACKING

Packing is a rope-like material that is impregnated with a lubricant or rubber-like material and is preformed into special shapes. Packing is used to control or prevent leakage in equipment, such as valves and pumps that handle fluids.

## 4.1.0 Types of Packing

Packing is used in almost every type of equipment that has shafts or stems that pass through casings. There are many types of packing to suit the large number of uses. A number of materials are used to make packing. The type of material used depends on the application. Packing can be either compression-type or lip-type. Compression-type packing is mainly used with shafts and valve stem applications. Lip-type packing is often called *automatic* or *hydraulic packing*. It is used to seal against reciprocating motion such as involved with pistons and hydraulic cylinders and plungers. The remainder of this section describes the following types of compression packing:

- Square braid
- Braid-over-braid
- Interlocking braid
- Twisted
- Wrapped, accordion-fold, and laminated
- Metal
- Graphite ribbon

### 4.1.1 Square Braid Packing

Square braid packing (*Figure 18*) is commonly used on reciprocating shafts. It is formed by packing material braided into a square shape. One strand of material is braided over and under other strands that run in the same direction. No strand passes completely through the packing. Because no strand is linked with all of the other strands, the packing is very flexible.

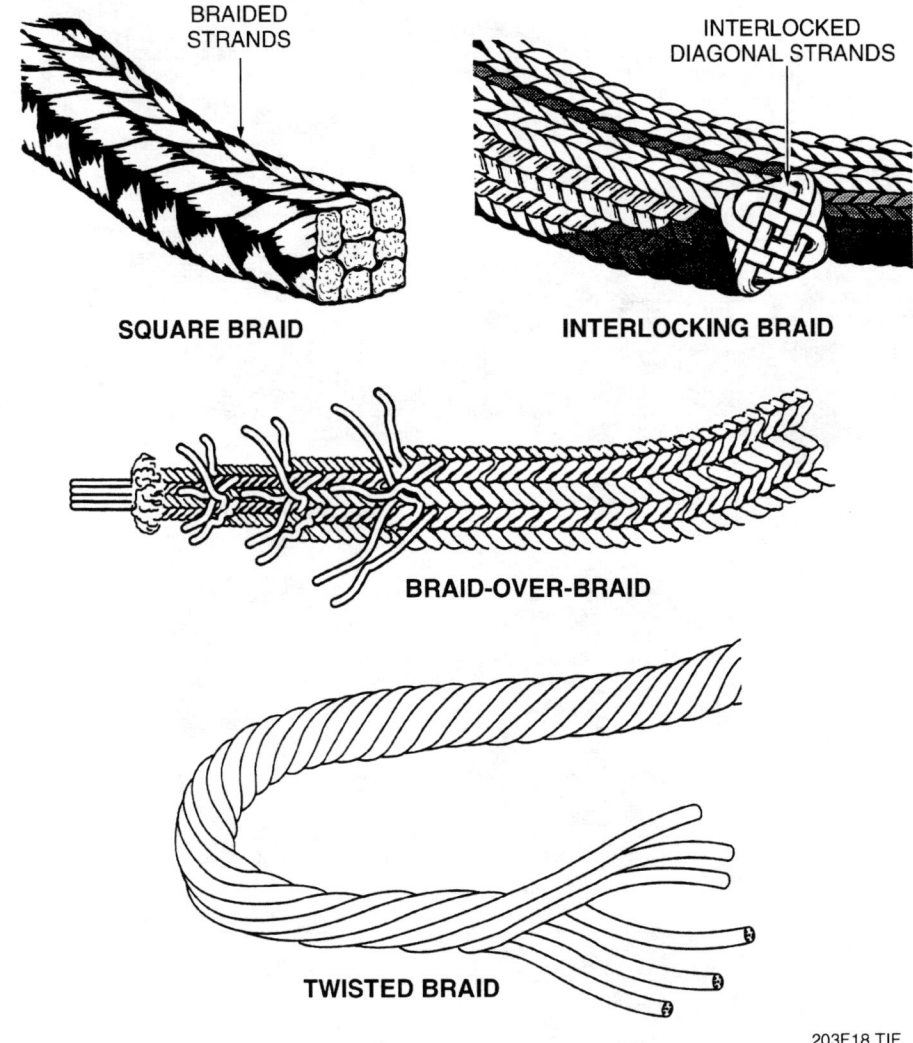

*Figure 18* ◆ Types of braid packing.

### 4.1.2 Braid-Over-Braid Packing

Braid-over-braid packing is made from a series of small braids that have been braided over each other in layers. It is a very dense packing and works well in high-pressure uses.

### 4.1.3 Interlocking Braid Packing

Interlocking braid packing is very dense. It is constructed by passing each strand through the body of the packing at an angle of approximately 45° so that the packing is braided both internally and externally. This type of packing is very strong and resists being deformed or squeezed out of the **packing gland** by high pressures inside a pump. A packing gland is a part used to compress the packing in a **stuffing box**. A stuffing box is the housing that holds the packing in a pump, valve, or piece of equipment.

### 4.1.4 Twisted Braid Packing

Twisted braid packing is made by twisting strands together to form a rope. This type of packing is used where pressure resistance is not a major consideration. Twisted packing can be untwisted and the strands removed and then retwisted to make the packing smaller.

### 4.1.5 Wrapped, Accordion-Fold, and Laminated Packings

Wrapped, accordion-fold, and laminated packings (*Figure 19*) are made from fiber packing material that has been woven into cloth. Wrapped packing is formed by rolling up the material much like rolling up a newspaper. Accordion-fold packing is formed by folding the material back and forth. Laminated packing is formed by laminating the packing material between layers of rubber.

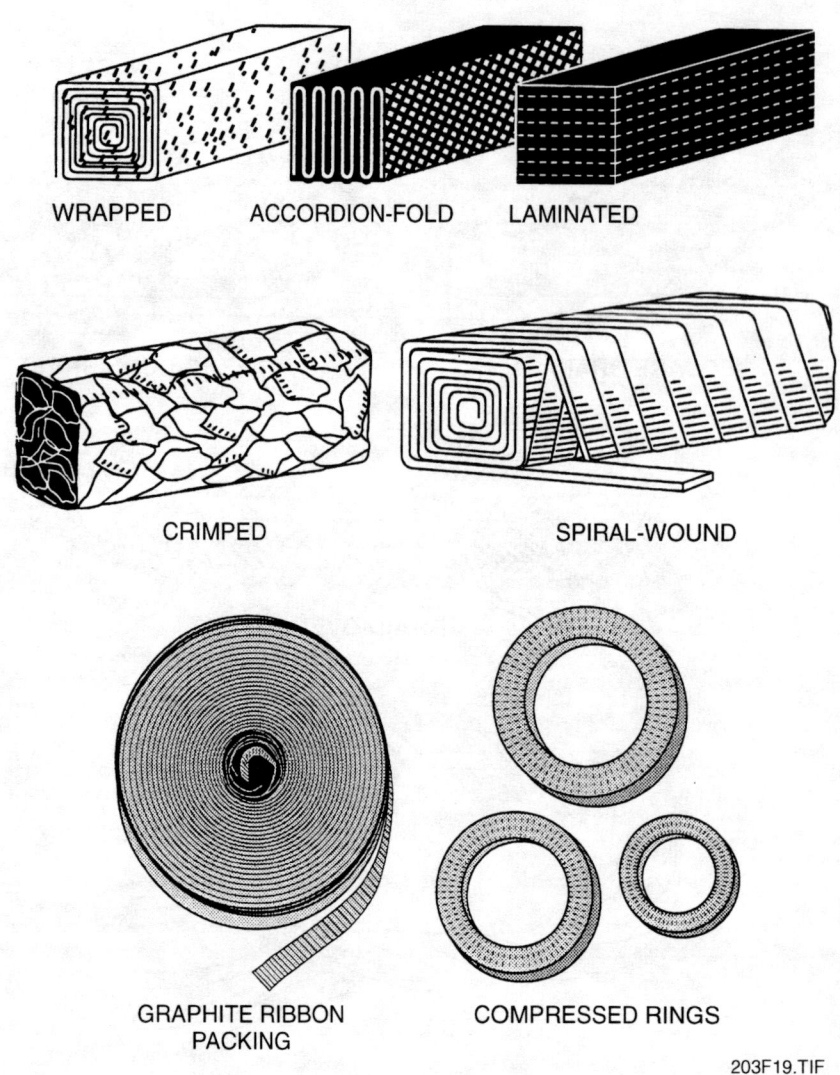

*Figure 19* ◆ Packing.

### 4.1.6 Metal Packing

Metal packing is made of, or reinforced with, various metal foils of lead, steel, or special alloys. It is used in many applications; however, only soft metals are used on rotating shafts because the harder metals will eventually wear down the shaft. Metal packing is usually crimped or spiral-wound.

### 4.1.7 Graphite Ribbon Packing

Graphite ribbon packing usually comes in rolls that can be used to form rings of almost any required thickness. To form packing rings from ribbon packing, the ribbon is wound around the shaft until it will fit into the stuffing box. The packing is then pushed into the box, using a gland follower and split bushing. The gland follower bolts are tightened to compress the packing into a ring. This process is repeated until enough rings are formed to fill the stuffing box.

## 4.2.0 Installing and Removing Packing

This section covers installing and removing packing. Knowing how to properly install and remove packing is critical due to the great impact these actions may have on system performance.

### 4.2.1 Removing Packing

The life of a packing depends on the type of packing used and its application. The most common symptom of packing failure is excessive leakage. When packing fails, it must be replaced.

Before the new packing can be installed, the old packing must be removed. Removal is usually simple, but it is important that the procedure be performed properly. The procedure for removing packing varies slightly from one piece of equipment to another. Always refer to the manufacturer's manual for the specific piece of equipment.

Regardless of the procedure, a packing puller or packing hook is typically used to remove the packing from a shaft installed in a stuffing box. *Figure 20* shows a stuffing box with the gland removed and packing puller screwed into the packing. If the puller is properly screwed into the packing ring, the ring usually comes out in one piece. If the ring breaks during removal, all of the smaller pieces must be removed. To help prevent breakage, two pullers may be used at the same time to pull the packing on both sides.

### 4.2.2 Installing Compression Packing

Packing is a major item of concern in the maintenance of equipment. Selecting the proper type and size of packing and installing it correctly is critical to maintaining optimum equipment operation. Incorrect selection or installation of packing can result in premature failure of the packing or in equipment damage. *Figure 21* shows important points related to installing compression packing. As shown in *Figure 21, View A*, the packing should be wrapped tightly around the shaft. The number of turns should equal the number of packing rings needed for the job. *Figure 21, View B* shows the method used to mark the packing while still wrapped around the shaft in preparation for cutting.

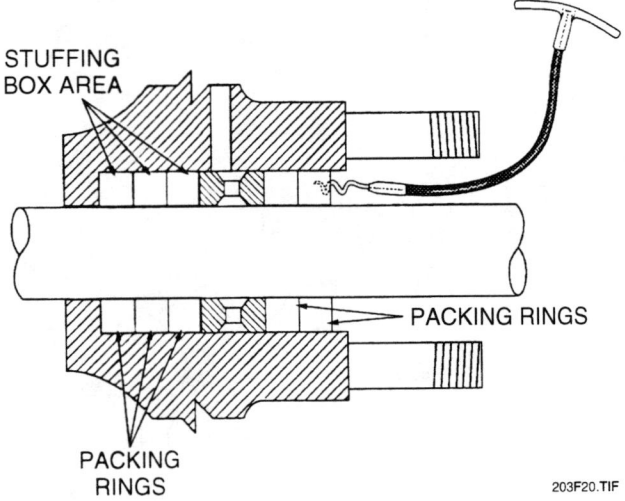

*Figure 20* ♦ Packing puller screwed into packing.

### Using a Packing Puller

When using a packing puller, be careful not to scratch or score the shaft. To prevent damage to the shaft, keep the puller angled away from the shaft at all times. Even minor damage to the shaft will cause accelerated packing wear and failure. If there are any burrs or nicks on the shaft and/or stuffing box, they must be removed before the new packing can be installed. If there are deep scratches or excessive wear, an evaluation must be made as to the need for repairs before new packing is installed.

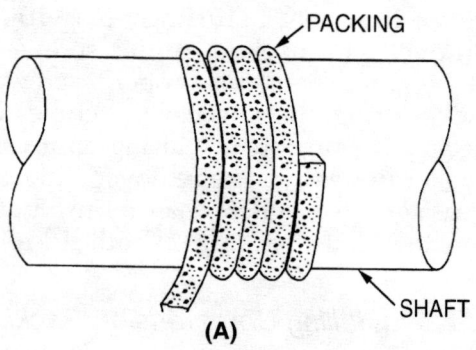

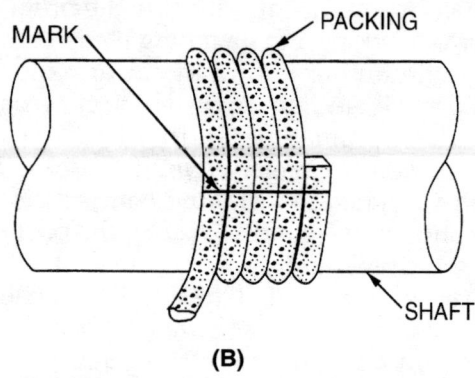

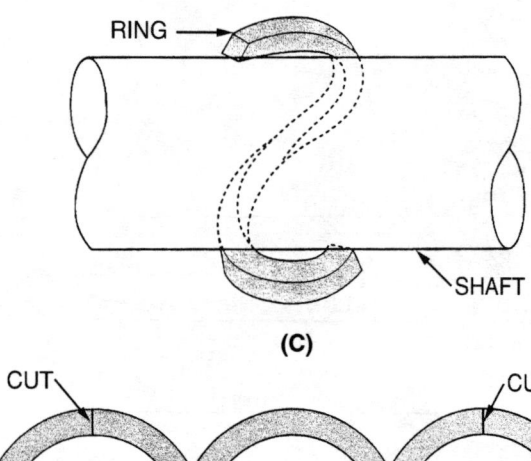

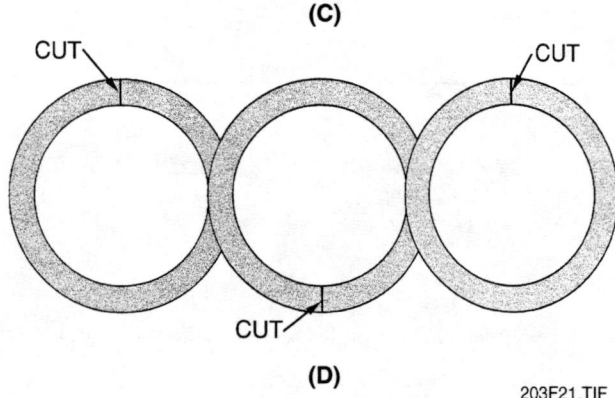

*Figure 21* ◆ Installing packing. (A) Packing on shaft. (B) Mark on packing. (C) Installing packing. (D) Stacking rings.

*Figure 21 (C)* shows the use of an S-twist method for placing the cut packing rings on the shaft. Do not force the rings open because this might break the packing. When installing the rings, install them one at a time, tamping (pressing) each one firmly

**CAUTION**

Do not cut packing on the shaft because the cutting tool may scratch the shaft. Cut the packing on a round steel shaft or pipe of the approximate size of the shaft being packed. Some packing manufacturers recommend a straight cut; others recommend a cut made on an angle (**skive cut**). Check the manufacturer's recommendations before cutting and packing.

into position. The cut in each ring should be rotated 180° from the cut in the ring installed before it. Staggering the cuts as shown in *Figure 21 (D)* provides a better seal than stacking the cuts of the rings in line with one another.

## 5.0.0 ◆ SEALS

Seals are devices used to prevent or control leakage between moving and fixed parts or between two fixed parts. There are two classes of seals: nonmechanical and mechanical.

## 5.1.0 Nonmechanical Seals

Nonmechanical seals are used as both **static seals** and **dynamic seals**. A static seal is one in which there is no movement between the two joining parts or between the seal and the mating part. O-rings are examples of seals often used as a static seal. A dynamic seal is one in which there is movement between two mating parts or between one of the parts and the seal. Lip and oil-type seals are examples of seals used as dynamic seals.

### 5.1.1 O-Rings

O-rings are circular seals used as either static or dynamic seals. They come in a variety of sizes and are made from various materials for use in different applications. When used as a static seal, an O-ring is installed in a slightly wider O-ring groove (*Figure 22*). A static seal needs to be a complete seal, so the O-ring should almost fill the O-ring groove. When used as a dynamic seal, an O-ring is usually placed in a groove or joint that is wider than the diameter of the O-ring. This is because a dynamic seal requires a running (friction) fit.

When an O-ring comes into contact with the areas to be sealed, it is slightly distorted in a motion called *mechanical squeeze*. The pressure caused by mechanical squeeze holds the O-ring in contact with the surfaces to be sealed. This pressure also causes the O-ring to roll and slide to the side of the groove away from the pressure.

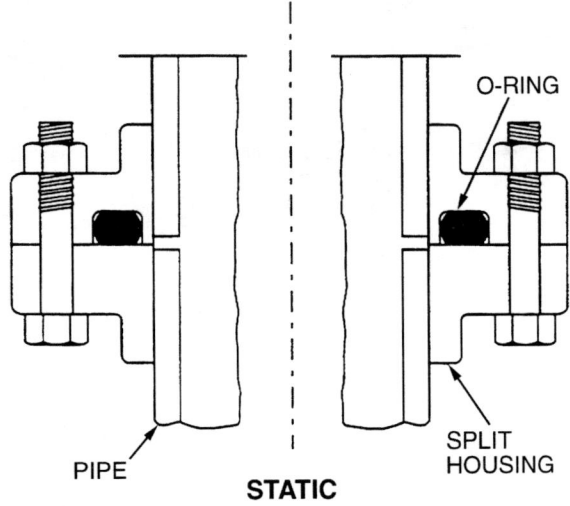

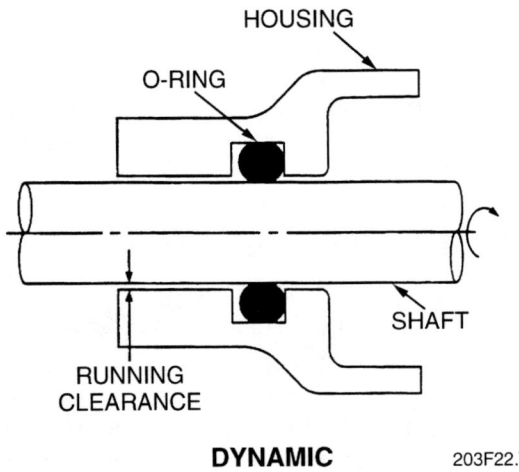

*Figure 22* ◆ Static and dynamic seals using an O-ring.

vacuum behind the lip pushes it against the shaft for a tighter seal. Lip seals are usually installed with the lip facing in to contain lubricant in a housing where the outside conditions are relatively clean. In dirty conditions, the seal is installed with the lip facing out to prevent foreign matter from getting into the housing. In cases where the area being sealed changes from pressure to vacuum conditions, double-lip seals are used to prevent air or dirt from getting in and lubricant from getting out.

The oil seal, or radial lip seal, is a positive-contact seal used on rotating or reciprocating shafts. It is used to keep fluids in or dirt and other foreign matter out. Oil seals are most often used to keep fluids in and are installed with the lip facing in. When an oil seal is used to keep contaminants out, it is installed with the lip facing away from the housing. Double-lip seals that perform both functions are also available. *Figure 23* shows a typical oil seal.

## 5.2.0 Mechanical Seals

When used with rotating shafts, mechanical seals provide a far superior dynamic seal to compression packings and other nonmechanical seals. They are typically used on shafts of open-type compressors and similar devices. Mechanical seals are made in many varieties, but they all basically function in the same way. As shown in *Figure 24*, mechanical seals usually include the following components:

- A set of primary faces is used, one that rotates and one that is stationary (such as the seal ring and insert shown in *Figure 24*).
- A closing mechanism provides the load required to keep the seal faces in contact with each other. These are usually multiple springs, a single spring, or a metal bellows.
- A set of secondary seals is used, such as shaft packing and insert mounting, which can be O-rings, V-rings, wedges, or U-cups.
- The mechanical seal hardware includes gland rings, collars, compression rings, pins, and springs.

To allow for initial mechanical squeeze, general purpose O-rings are made with a cross-section that is 10% larger than the nominal size. The use of O-rings with too much squeeze wears them out quickly, while insufficient squeeze can cause leaks.

### 5.1.2 Lip and Oil Seals

Lip seals are low-pressure, positive-contact seals used with rotating shafts. Pressure on the lip or a

### O-Ring Lubrication

All O-rings require lubrication. Keep in mind, however, that the lubricant and the O-ring material must be compatible. If they are not properly matched, the seal could be damaged. Because new refrigerants and oils are constantly being introduced, it is increasingly important to check the manufacturer's requirements to ensure a match between the O-ring and the lubricant.

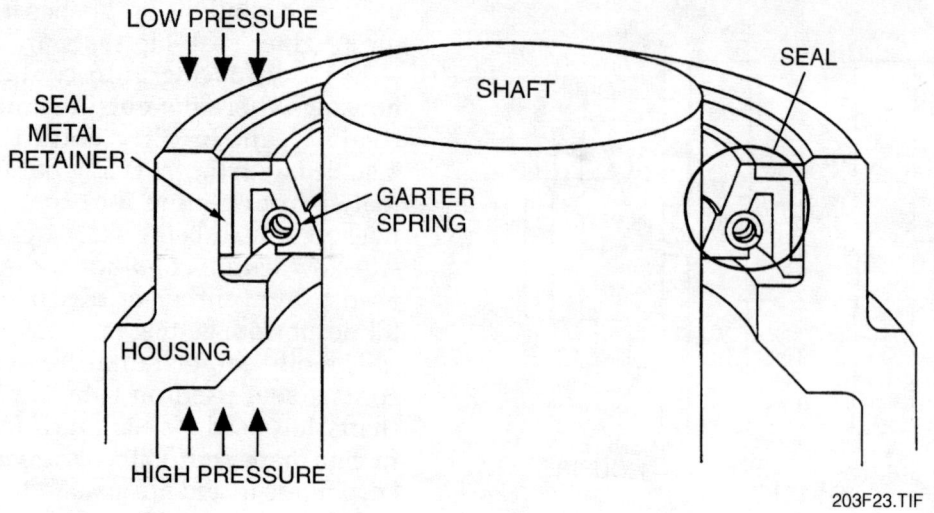

*Figure 23* ◆ Oil seal.

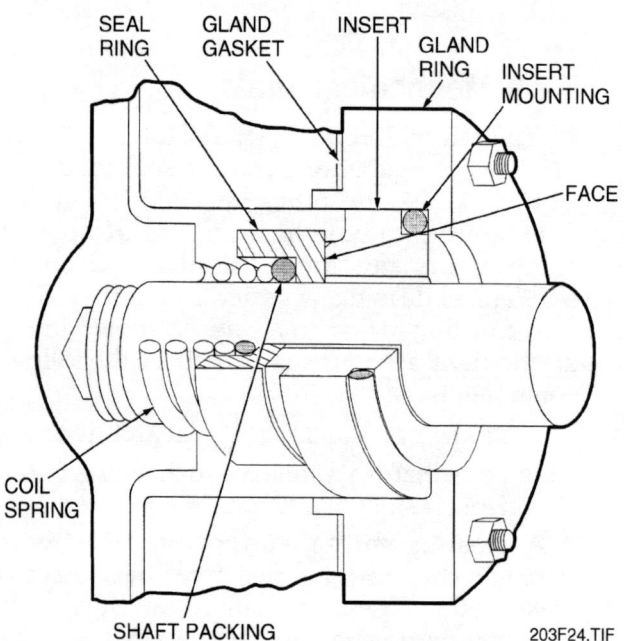

*Figure 24* ◆ Typical mechanical seal parts.

The primary seal is achieved by the compression of two very flat, lapped faces that are held tightly together by springs, metal bellows, or similar devices. Operation of the mechanical seal depends on an extremely thin film of lubricant being applied between the sealed faces. Since no seal is perfect, this lubricant is furnished by virtue of a minuscule level of leakage that occurs at the seal surfaces. Maintenance of this extremely thin film is made possible by machining the seal faces to very high tolerances in respect to flatness and surface finish. To preserve this precise surface finish requires that seal parts be carefully handled and protected.

Mating surfaces should never be placed in contact without lubrication. Each seal application determines the best type of mechanical seal to use.

### 5.3.0 Installing and Removing Seals

The life of a seal depends on the type of seal used and its application. The most common symptom of seal failure is leakage. When a seal fails it must be replaced. Before the new seal can be installed, the old seal must be removed. The removal and installation of an O-ring seal is a relatively simple procedure. The procedures for removing lip, oil, and mechanical seals vary widely from one piece of equipment to another. These seals should be removed by following the instructions given in the manufacturer's service manual for the equipment you are servicing. Inspect the parts of a mechanical seal after they are removed for possible clues as to why the seal failed.

To achieve proper operation of lip, oil, and mechanical seals, it is important that the installation procedures be performed properly, according to the following guidelines:

- The equipment must be properly prepared to receive the new seal. All parts must be clean and free of sharp edges.
- The seal used must be right for the application.
- It is important to install the seal according to the equipment and/or seal manufacturer's instructions, or both. Study the manufacturer's data so that you are familiar with the procedure before attempting to install the seal.
- Lubricate the shaft, O-rings, and seal faces using the lubricant specified by the manufacturer or in the seal catalog.

- Protect precise seal parts from dirt and damage during installation. Do not touch seal faces with your hands. Handle seals by the edges. Inspect seal faces to ensure that there are no nicks or scratches. During assembly, keep the seal centered on the shaft.
- During assembly, protect all O-ring and/or other static seals from damage on sharp edges such as threads, keyways, or the end of the shaft.
- All parts must fit properly without binding.
- Follow the proper bolt-tightening procedure.

## 6.0.0 ◆ BEARINGS

Bearings of one type or another are used in practically every piece of equipment that has rotating parts. Bearings are used to reduce friction between parts in motion. In HVAC equipment, they are commonly used to support shafts in motors, fans, and compressors. Bearings can be divided into two broad classifications: plain and anti-friction.

The following terms are commonly used when describing the operation of bearings:

- *Axial load* – This is an external load that acts lengthwise along a shaft, such as that applied in a direct-drive coupled compressor and motor.
- *Journal* – This is the part of a shaft, axle, or spindle that is supported by and revolves in a bearing.
- *Radial load* – This is the side or radial force applied at right angles to a bearing and shaft, such as that applied from a pulley.
- *Thrust* – This is the force acting lengthwise along the axis of a shaft, either toward it or away from it.

## 6.1.0 Plain Bearings

Plain bearings (*Figure 25*) are simple in construction, operate efficiently, and can support heavy loads. During operation they develop an oil film between the shaft journal and bearing surfaces that overcomes the friction of the sliding motion. There are three general classifications of plain bearings: radial, thrust, and guide bearings. Radial bearings are used to support radial loads. Thrust bearings are used to support axial loads on rotating members. Both radial and thrust bearings are also called *rotational bearings*. Guide bearings are used to guide moving parts in a straight line, such as in machine tools. The most common plain bearings used in HVAC equipment are sleeve and thrust bearings.

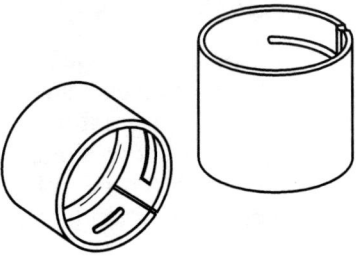

SLEEVE BEARINGS

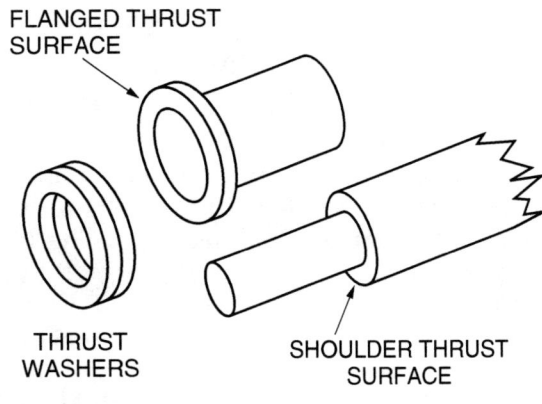

*Figure 25* ◆ Plain bearings.

### 6.1.1 Sleeve Bearings

The sleeve bearing provides the bearing surface for the shaft journal and is usually press-fitted into a supporting member. Sleeve bearings are quiet in operation and have a good radial load capacity, such as needed with a blower motor used in a residential furnace. If made of bronze, sleeve bearings have good resistance to humidity, dirt infiltration, and corrosion.

The simplest and most widely used types of sleeve bearings are cast-bronze and porous bronze. Lubrication in a sleeve bearing is important because the bearing must have an oil film between the shaft and bearing surface. During rotation, the shaft actually floats on this oil film and never touches the bearing surface. Cast-bronze bearings are oil or grease lubricated. This type usually has a reservoir that is filled from the outside by means of an access port. Cast-bronze bearings usually need to be lubricated on a regular basis with the proper type of oil or grease. Porous bearings are impregnated with oil and often have an oil reservoir with a wick to gradually feed the oil to the bearing. Under normal operating conditions, this type of sleeve bearing will operate in a motor for years without the need for lubrication.

MAINTENANCE SKILLS FOR THE SERVICE TECHNICIAN — TRAINEE MODULE 03203

### 6.1.2 Thrust Bearings

Thrust bearings support axial loads and/or restrain lengthwise movement. The simplest type is a thin disc-type thrust bearing called a *thrust washer*. Two or more thrust washers made of low-friction materials are often combined. It is also common to support thrust loads on the end surface of journal bearings. If the area is too small for the applied load, a flange may be provided. A shoulder is usually cut on the mating surface of the shaft.

## 6.2.0 Anti-Friction Bearings

Anti-friction bearings are so named because they operate on the principle of rolling motion, using either balls or rollers between rotating and fixed surfaces. Because of this, friction is reduced to a fraction of that in plain bearings. The rolling action of anti-friction bearings normally makes them noisier than plain bearings. Under certain conditions, anti-friction bearings can rust because they are usually made of steel.

Compared to plain bearings, anti-friction bearings normally require much less maintenance, especially if they are grease packed. Lubricant is used in anti-friction bearings mainly to keep out dirt and moisture. It also helps dissipate the heat that builds up in the bearing, but it does not provide an oil film to reduce bearing friction. Many types are permanently lubricated. There are two general types of anti-friction bearings: ball bearings and roller bearings.

### 6.2.1 Ball Bearings

Ball bearings (*Figure 26*) consist of outer and inner rings, also called *races*, with balls in between. The inner race has a groove around its outside circumference for the balls to roll in. The outer race usually has a similar groove on its inside circumference. A separator or retainer spaces the balls around the track and guides them through the load zone. Ball bearings fall into three classes: radial, thrust, and angular. Angular ball bearings are used in applications that have combined radial and thrust loads, and where precise shaft location is needed. Radial and thrust ball bearings are used in applications with radial loads and axial thrust loads, respectively.

### 6.2.2 Roller Bearings

In general, roller bearings (*Figure 27*) have higher capacities than ball bearings. They are used in heavy-duty, moderate-speed applications. The construction of roller bearings is similar to ball bearings except that rollers are used instead of balls. Also, the outer race is called the *cup*, and the inner race is called a *cone*. The separator or retainer is called the *cage*. Roller bearings fall into three classes:

- *Cylindrical bearing* – These are used for radial loads.
- *Tapered bearing* – These are used for radial loads, axial loads, or both, depending on the design.
- *Spherical bearing* – These are used in applications requiring high load capacity and/or those with a shock load, such as a conveyer or speed reducer.

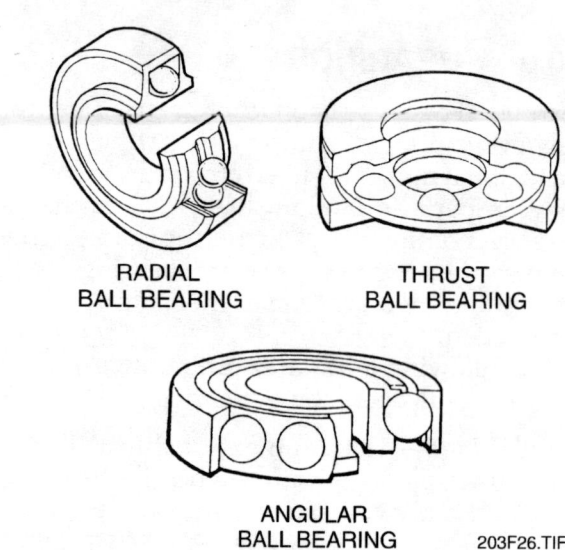

*Figure 26* ◆ Typical ball bearings.

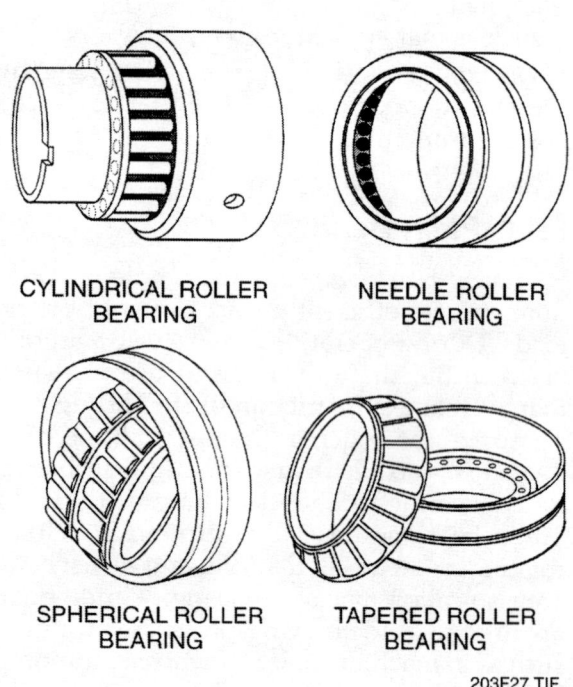

*Figure 27* ◆ Typical roller bearings.

Another bearing considered to be a roller bearing is called a *needle bearing*. Needle bearings do not have cages to separate the rollers. Also, the rollers are longer and more of them are used than in the other roller bearings. This gives the needle bearing more contact area. Variations of this type of bearing allow it to be used for the same applications as other roller bearings. The advantage is that they give more support in less space.

## 6.3.0 Identifying Bearing Failures

Touch and sound are two effective methods used to detect bearing failures. Touching or feeling the bearing housing when a motor or other rotating device is operating can give an indication of a bearing's condition. A bearing is probably in the process of failing or has failed if the bearing housing feels overly hot to the touch.

Listening for the sound of foreign noises coming from a motor or other rotating device will often help you detect a bearing problem.

Another method used to detect a a bearing problem is to place one end of a steel rod on the bearing housing while the other end is held close to the ear. The rod acts as an amplifier, transmitting unusual sounds such as thumping or grinding, which would indicate bearing failure. Special listening devices, such as a stethoscope, can be used for this purpose.

Excessive end play of a motor shaft can also be an indication of bearing wear. Ball-bearing motors should typically have an end play of about $\frac{1}{32}$" to $\frac{1}{16}$". Sleeve-bearing motors may have an end play of up to $\frac{1}{2}$".

Examining the lubricant on a damaged bearing can also be a help in identifying a cause of a bearing failure. *Table 1* summarizes some common problems. Lubricants are discussed in detail later in this module.

## 6.4.0 Removing and Installing Bearings

When bearings wear out, they must be removed and replacements installed. They must also be removed to disassemble a piece of equipment for repair or maintenance.

### 6.4.1 Removing Bearings

The removal of bearings can vary widely from one piece of equipment to another, depending on the type of bearing used and its housing. Bearings should be removed by following the instructions given in the manufacturer's service manual for the equipment. This is important in order to prevent damaging the bearing or the shaft. If the bearing is being removed because it has failed, inspect the bearing after it has been removed for possible clues as to the reason the bearing failed.

The type of bearing removal tool and the method used to remove a bearing generally depend on the size, type, and fit of the bearing to be removed. Bearing pullers, arbor presses, and hydraulic presses can be used. The manual bearing puller is the most common tool for removing bearings in the field. The puller can usually be used to remove the bearing from the shaft while the shaft associated with the bearing is still in the equipment. The manual puller has a bolt that is turned using a wrench to provide the pressure to pull the bearing. *Figure 28* shows a manual bearing puller being used to remove a bearing.

When removing a bearing using a puller, it is important that the procedure be performed properly, according to the following guidelines:

- The puller jaws must apply pressure only to the inner race of the bearing. If pressure is applied to the outer race, the bearing will be damaged and may come apart.

**Table 1** Lubricant Symptoms Related to Common Bearing Failures

| Lubricant | Symptom | Cause |
| --- | --- | --- |
| Oil or grease | Reddish deposit at ball contact | Indicates chafing; look for oscillatory or vibratory motion during operation |
| Oil or grease | Metallic particles accompanied by indications of wear on bearing components | Lubrication failure or external contamination |
| Oil | Dry surface with little evidence of residue, or slight brownish haze on ball or raceway surfaces | Too little lubricant, either initially or because of migration |
| Grease | Grease darkened but still oily | High loads or contaminants |
| Grease | Grease darkened and dry | Extremely high loads or contaminants |
| Grease | Grease surface dry and hard, soft interior | High external heat |

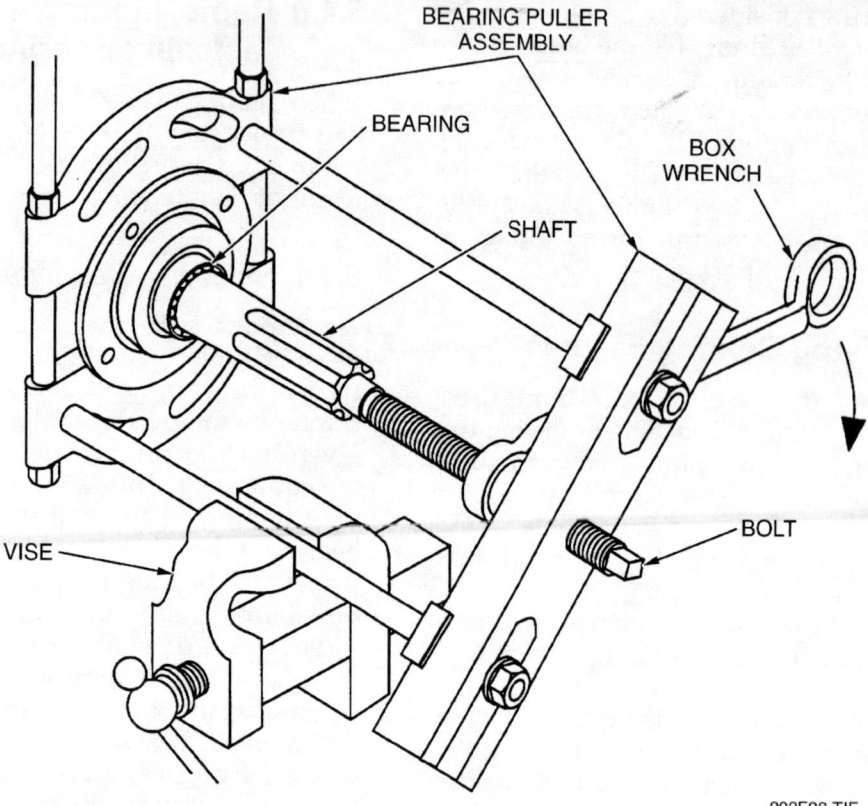

*Figure 28* ♦ Manual puller being used to remove a bearing.

- Be sure that the puller is pulling straight. If it is misaligned, the bearing will become cocked and may damage the shaft.
- Do not let the bearing fall on the floor when it comes off the shaft because it could be damaged and get dirty.

### 6.4.2 Installing Bearings

The tools and methods used to install a bearing generally depend on the size, type, and fit of the bearing to be installed. Three kinds of fits for bearings are the slip fit, press fit, and interference fit. The slip fit is the simplest to install because the bearing fits fairly loosely and can usually be pushed into place by hand. The press fit is much tighter and requires more effort to press the bearing into place. Bearings usually have a slip fit on one ring and a press fit on the other. The ring that rotates is usually press-fitted. In most cases, the inner ring of the bearing rotates. Interference fit bearings must be heated before they are pressed on because the inside diameter is smaller than the shaft. Bearings are usually installed using either a temperature mounting method, press mounting method, or locknut method, depending on the type of bearing. The method used should be the one recommended by the instructions given in the manufacturer's service manual for the equipment that you are servicing.

The temperature mounting method used for interference bearings can usually be performed in the field or shop while the shaft associated with the bearing is still in the equipment. This method uses a bearing heater (*Figure 29*) to heat the bearing to a temperature level specified by the manufacturer. Heating the bearing causes it to expand enough so that it can be slipped onto the shaft. Once on the shaft, it is quickly moved to its proper mounted position and held there to prevent it from moving. When it cools, it shrinks to fit the shaft. When using this method, care must be taken not to overheat the bearing, because overheating can adversely affect the hardness of the bearing steel. The maximum temperature to which a bearing should be heated is 250°F.

 **WARNING!**
Heated bearings can burn your hands. Wear clean, protective gloves and handle the heated bearing in a manner that prevents your skin from coming in contact with the heated metal. To avoid being burned, wait until the bearing has cooled before touching it.

*Figure 29* ◆ Bearing heater.

bearing is at the proper position on the shaft. This position is attained when the amount of clearance initially measured for the bearing is reduced to an amount specified in tables supplied by the bearing manufacturer. When no bearing clearance tables are available, a rule of thumb is to reduce the clearance by about 50%.

## 7.0.0 ◆ LUBRICATION

Any piece of equipment or machinery that has rotating or moving parts produces friction. Friction is the resistance to motion that takes place when two parts move or rub against one another. Lubrication is required whenever friction is a problem and must be controlled. Lubrication is the application of any substance (lubricant) that reduces friction by creating a slippery film between two surfaces.

The press mounting method uses an arbor press (*Figure 30*) to press the shaft into a bearing. This method is typically used when installing bearings where the bearing shaft race is press-fitted and the housing race is slip-fitted, such as with a thrust bearing. In this method, the bearing is placed on the arbor so that the bearing shaft race is well supported. The shaft is then positioned in the bearing bore and pressed into place. The assembled shaft and bearing are then installed in the equipment. When pressing the shaft into the bearing, the shaft must be kept square with the bearing at all times. If the shaft is cocked during the pressing operation, the bearing will gouge the shaft.

The locknut method is typically used when installing tapered-bore bearings, such as a spherical roller bearing. A tapered-bore bearing is mounted on a tapered shaft, or a tapered sleeve. In this method, the bearing is placed on the shaft and positioned in place by tightening a locknut. As the locknut is turned, it causes the bearing to be forced onto the shaft, and the clearance between the races and the rolling elements is reduced. This clearance must be controlled. To do this, the clearance is measured with a feeler gauge before installation and during the tightening process (*Figure 31*). The locknut is turned as needed until the

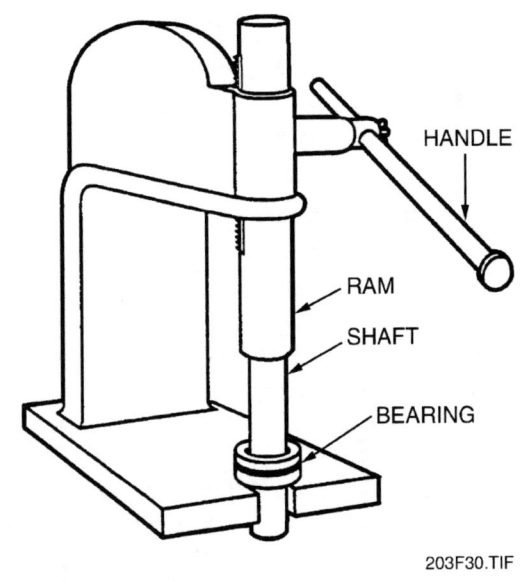

*Figure 30* ◆ Arbor press.

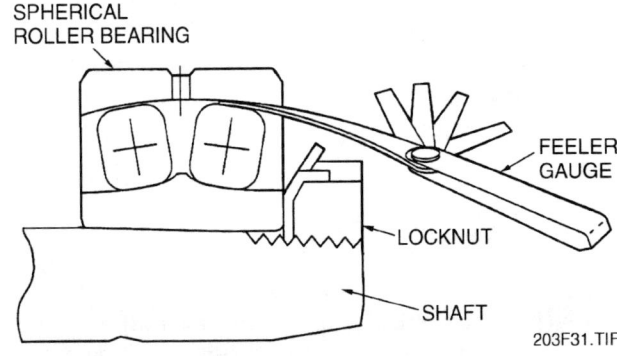

*Figure 31* ◆ Bearing clearance measurement.

Lubricants can be made from animal fats, vegetable oils, mineral oils, or synthetics. However, over 90% of the total lubricants used are made from mineral oils which come from refined crude oil. Lubricants can be in the form of a liquid (oil), semisolid (grease), or solid films. All three are used to lubricate rolling and sliding bearings and gears.

## 7.1.0 Oils

Oils are a broad class of fluid lubricants. Basically, lubricating oils are made from two types of crude oil: naphthenic and paraffinic. Naphthenic oils contain very little wax and are good lubricants for almost any use. Paraffinic oils are very waxy, and lubricants made from them are used mainly in hydraulic equipment and other machinery. All lubricants possess certain properties that govern their suitability for a particular use. These properties include the following:

- **Viscosity**
- **Viscosity index (VI)**
- **Pour point**
- **Flash point**
- **Fire point**
- **Oxidation** resistance

### 7.1.1 Viscosity

Viscosity is the most important property of oil. Viscosity is the thickness of a liquid or the ability of the liquid to flow at a specific temperature. Low-viscosity oils are light oils that flow freely when poured. High-viscosity oils are heavy oils that flow slowly when poured. Medium viscosity oils range somewhere in between.

The viscosity rating of oils can be expressed in two ways. Automotive and gear-lubricating oils are expressed in Society of Automotive Engineers (SAE) ratings. The other viscosity rating system, which applies to industrial lubricants, is the Saybolt Universal Seconds (SSU or SUS). This rating system was developed by the American Society for Testing and Materials (ASTM). Generally, low-viscosity oils are more applicable for use with bearings. Higher-viscosity oils are more applicable for use with gears.

### 7.1.2 Viscosity Index

The viscosity index (VI) is a measure of how viscosity varies with temperature. Not all oils change viscosity at the same rate when subjected to the same temperature changes. If an oil has a low VI, the viscosity changes rapidly with temperature. A high VI represents a lower change in viscosity with temperature changes. Typically, naphthenic oils have a low VI and paraffinic oils have a high VI.

### 7.1.3 Pour Point

The pour point of an oil refers to the lowest temperature at which an oil will flow freely. This must be considered in cold weather applications and refrigeration systems. The lower the pour point of an oil, the more freely it flows at very low temperatures. A high pour point means that the oil stops flowing freely at low temperatures. At the same viscosities, naphthenic oils have lower pour points than paraffinic oils.

### 7.1.4 Flash Point and Fire Point

The flash point is the temperature at which oil gives off ignitable vapors. This temperature is not high enough for the oil to support combustion. When the flash point is reached, the lubricant film between the mating surfaces is destroyed, and damage of the surfaces can occur.

The fire point is the temperature at which oil will burn if ignited. The flash and fire points of an oil are only a consideration when using an oil in high-temperature applications.

### 7.1.5 Oxidation Resistance

Oxidation resistance is related to the service life of an oil. As oils are exposed to air and heat, they take on oxygen from the air. This process is known as oxidation. All lubricants tend to oxidize over time. Oxidation breaks down the chemical structure of oil and eventually destroys its lubricating qualities. Oxidation is a slow process, but if the oil splashes around excessively or is exposed to high temperatures, the process speeds up.

## 7.2.0 Greases

Greases are a solid or semisolid lubricant formed by adding a thickening agent, usually soap, and an additive to oil. An additive (such as a rust inhibitor or antifoam agent) is an extra ingredient used to improve the qualities of the lubricant. Grease produces the same lubricating action as oil, only in a different way. Oil forms a film that keeps surfaces apart to reduce friction and heat. Grease is not a liquid and cannot form a liquid film when it is initially applied. Grease forms the lubricating film when the surfaces begin to move and exert pressure on the grease. When pressure is applied, grease releases some of its oil, allowing the lubrication action to begin. When the motion is stopped and the pressure is reduced, grease tends to solidify again.

Greases are usually rated by their relative hardness or consistency on a scale developed by the National Lubricating Grease Institute (NLGI). The softest greases are rated at 000, with higher numbers indicating harder greases. Most greases fall in the range from 1 to 6. The consistency of grease varies with temperature, and there is generally an increase in the softening of a grease as the temperature increases.

Another common rating method for greases is their **dropping point.** The dropping point is the temperature at which grease is soft enough for a drop of oil to fall away or flow from the bulk of the grease. As a rule of thumb, a temperature that is 50°F less than the dropping point is the maximum operating temperature for the grease.

Normally, NLGI Grade 2 greases are used in most roller-type bearings because they combine good lubricant feeding with resistance to mechanical churning. Frequently, stiffer Grade 3 greases are used for double-sealed bearings and large bearings. Typically, the use of conventional, multipurpose grease is adequate for temperatures between –20°F and +200°F. When choosing a grease, follow the grease, bearing, or equipment manufacturer's recommendations.

## 7.3.0 Lubrication Equipment

Several types of equipment and methods are used to apply lubricants. Some equipment requires manual lubrication; others use mechanical lubricating devices. The lubrication of most HVAC equipment involves the manual method. Manual lubrication equipment is hand-operated and is used to apply oil and grease. The operator must know how to refill and maintain this equipment.

### 7.3.1 Lever-Type Grease Gun

The most common manual lubrication tool used in the field is the lever-type grease gun (*Figure 32*). Lever guns are hand-operated and develop high pumping pressure with little effort. Grease guns are available in low-pressure models, high-pressure models, or both.

Generally, lever guns can be filled with grease in three ways: standard cartridge, bulk fill, and with a grease transfer pump. The standard cartridge is the most common method for HVAC work.

Adapters, such as flexible extension hoses and adjustable swivel fittings, can be used with the lever gun to gain easier access to lubrication fittings in hard to reach or tight places. Also, special grease injectors can be used when lubricating sealed bearings or for other special applications.

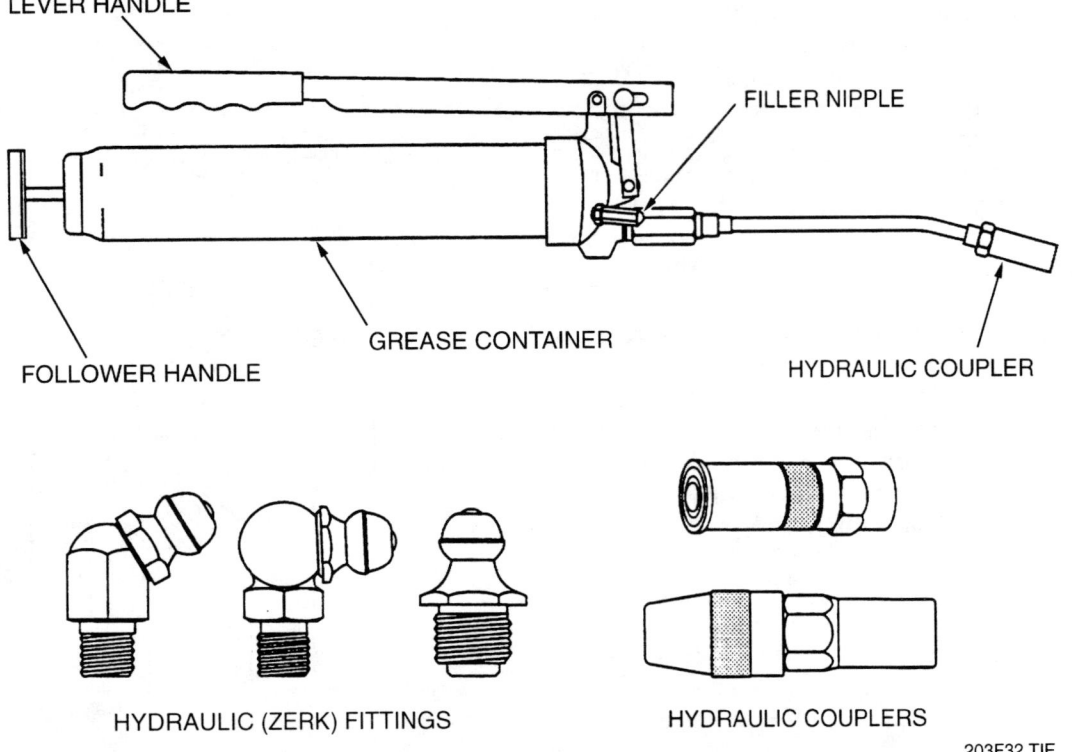

*Figure 32* ♦ Lever-type grease gun and hydraulic grease fittings.

## 7.3.2 Lubrication Fittings

Equipment that is lubricated using grease or other semisolid lubricants normally has fittings that connect with the coupling on the lever gun to form a pressure-tight connection. The most widely used is the hydraulic type, which mates with the jaws in the lever gun coupling. The lubricant pressure seats the coupling. Other types of fittings include button-head and flush types. These fittings are typically found in equipment using large quantities of grease. High pressures are required to pump grease through a fitting and into the grease cavity around a bearing. Lever guns can develop pressures up to 10,000 psi. Overpumping the handle can cause excessive grease pressure that could blow out the bearing seals. Some fittings and grease cavities are equipped with a vent groove or plug that is removed to relieve pressure and allow the old grease to be forced out.

## 8.0.0 ◆ BELTS AND BELT DRIVES

Belt drives are a quiet, smooth, and economical form of power transmission. They are commonly used in commercial HVAC equipment. The drive mechanism transfers a motor's rotating power to the driven device. A belt drive consists of a drive pulley with one or more sheaves, a driven pulley with a matching number of sheaves, and belts to match the sheaves (*Figure 33*). The pulleys or sheaves are available in many shaft sizes, diameters, and types of construction. Multiple-groove pulleys are made for use with equipment that uses two or more belts in the drive. Some air conditioning equipment uses step pulleys for driving the air movement fan. By changing the belt from one groove to another, the speed of the fan can be changed. Special variable-pitch pulleys are also available. These are made with half of the pulley threaded on the hub of the other half. A setscrew locks the variable half in place when it is properly adjusted. By turning the variable half, the V-groove can be widened to let the belt ride lower in the hub. This reduces the speed of the driven fan or other device. Belt drives can be divided into two basic types: V-belts and synchronous belts. HVAC equipment typically uses V-belt drives.

### 8.1.0 V-Belts

V-belts are made of a combination of fabric, cord, and/or metal reinforcement vulcanized with natural rubber compounds. The V-belt has a tapered shape that causes it to wedge firmly into the grooves of the sheave when it is under load. A V-belt works through frictional contact between the sides of the belt and the tapered sheave groove. Most V-belts fall into three general classifications:

- Fractional horsepower belts
- Standard multiple belts
- Wedge belts

### 8.1.1 Fractional Horsepower Belts

Fractional horsepower (FHP) belts are light-duty belts, usually used singularly. The size of FHP belts is indicated by a code marked on the outside of the belt. The first number and letter in the code tell the width of the belt in eighths of an inch. The next three numbers in the code tell the length of the belt in inches, with the last number indicating tenths of an inch. For example, a belt marked 4L300 is ½" wide and 30" long. FHP belts are measured on the outside surface of the belt. FHP belts come in the following standard widths and heights:

2L – ¼" wide × ⅛" high
3L – ⅜" wide × ⁷⁄₃₂" high
4L – ½" wide × ⁵⁄₁₆" high
5L – ⅝" wide × ⅜" high

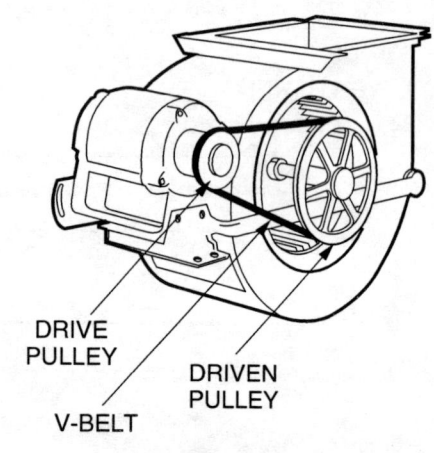

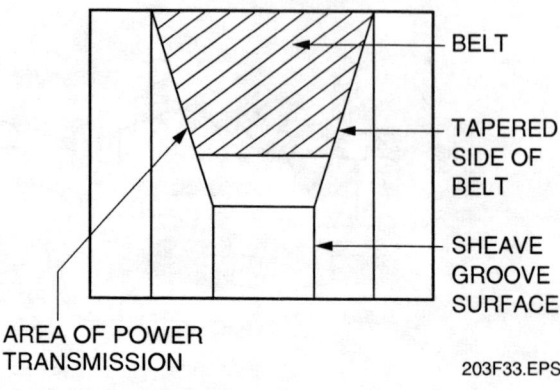

*Figure 33* ◆ V-belt drive.

## 8.1.2 Standard Multiple Belts

Standard multiple belts are used for continuous service. They are used in sets of two or more. The size of standard multiple belts is indicated by a code marked on the belt, with a letter indicating the width and a number indicating the length. For example, a belt marked A42 is ½" wide and 42" long. The length of standard belts is measured on the inside surface of the belt. They are available in various lengths for each width size and in the following standard widths and heights:

A – ½" wide × 11/32" high
B – ⅝" wide × 7/16" high
C – ⅞" wide × 9/16" high
D – 1¼" wide × ½" high
E – 1½" wide × 1" high

## 8.1.3 Wedge Belts

The wedge belt is a multiple belt that has a smaller cross-section per horsepower than the standard multiple V-belt. Also, it is used on smaller diameter sheaves with shorter center distances than the standard belt. Wedge belts are not interchangeable with standard multiple belts and should not be run on sheaves made for standard belts. The size of wedge belts is indicated by a code marked on the outside of the belt. The first number and letter in the code tell the width and cross-section of the belt in eighths of an inch. The next three numbers in the code tell the length of the belt in inches. For example, a belt marked 3V500 has a 3V cross-section (⅜" wide × 5/16" high) and is 50" long. The length of the wedge belt is measured along the pitch line, which runs along the center of the belt thickness. Wedge belts come in the following standard widths and heights:

3V – ⅜" wide × 5/16" high
5V – ⅝" wide × 17/32" high
8V – 1" wide × ⅞" high

## 8.2.0 Belt Drive System Maintenance

Belt drives are intended to give many hours of service in their particular use. Belt-drive failure can often be traced to improper belt tension. All belts should be tightened according to the manufacturer's instructions using a tension gauge. When installing the belts, they should be slipped loosely over the pulleys.

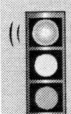

**WARNING!**
Do not try to run the belts on the pulleys. This will place excessive stress on the cords of the belt and can cause it to flop under load and possibly turn over in the sheaves.

Belt-drive failure can also often be traced to pulley misalignment. There are basically three kinds of alignment problems that can occur: angular, parallel, and sheave groove (*Figure 34*).

### Variable-Pitch Pulley

The photo below shows a variable-pitch pulley being adjusted to change the speed of a blower in a belt-drive system.

A simple way to check alignment using a straightedge ruler is shown in *Figure 35*, with arrows indicating the four check points. When the pulleys are properly aligned, the straightedge will be flat on the faces of both pulleys and no light should show at these four points.

## 9.0.0 ◆ COUPLINGS AND DIRECT DRIVES

Couplings are typically used to connect the shaft of a motor (driver) to the shaft of a compressor (driven). This method of coupling is called *direct drive*.

### 9.1.0 Coupling Types

Couplings are made in a variety of types and for many uses. Some couplings allow for slight misalignment and end play between the rotating shafts. Other couplings reduce or absorb vibrations or torque. Three common categories of couplings include:

- Rigid
- Flexible
- Soft-start

#### 9.1.1 Rigid Couplings

Rigid couplings provide a nonflexible connection between the driver and driven shafts. Rigid couplings do not compensate for misalignment and require precise alignment during installation. If a rigid coupling is misaligned and forced together, the drive will be damaged. Even slight misalignment can cause vibration and operating problems. There are three common types of rigid couplings: flanged, sleeve, and ribbed.

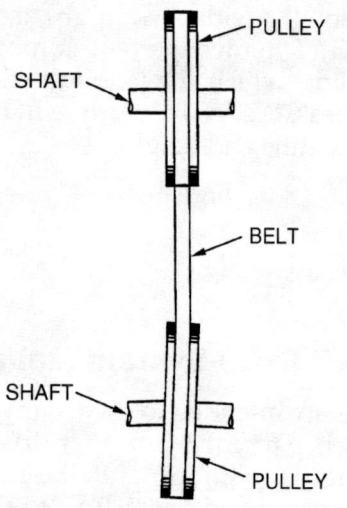
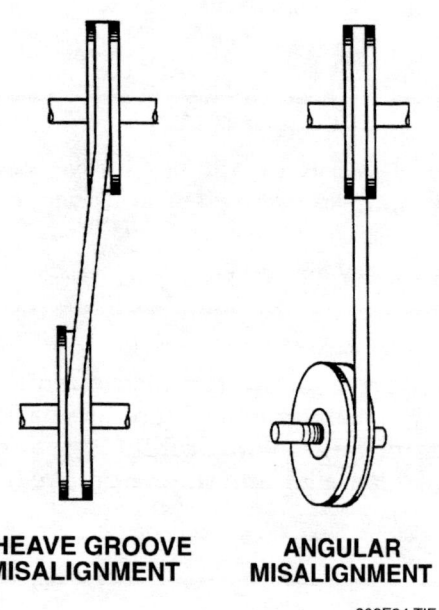

*Figure 34* ◆ Belt-drive misalignment.

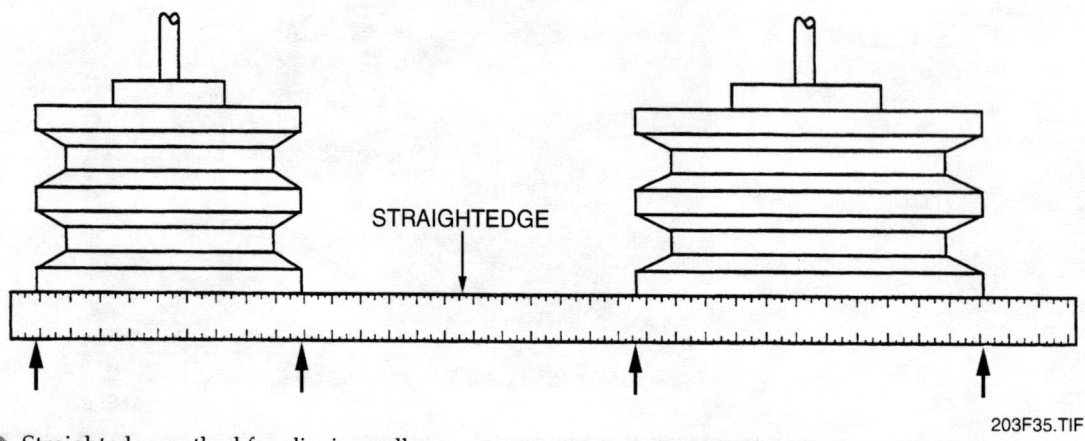

*Figure 35* ◆ Straightedge method for aligning pulleys.

HVAC LEVEL TWO — TRAINEE MODULE 03203

Flanged couplings (*Figure 36*) join the driver and driven shafts using two mating flanges bolted together. Flanged couplings require keys to prevent them from rotating on the shafts.

Sleeve couplings, also called *compression couplings*, are similar to flanged couplings, except that they are taper-bored and have tapered sleeves that fit on the shafts. The wedge principle is used to tighten the coupling on the shafts. As the two halves of the coupling are pulled together over the tapered sleeve by the flange bolts, the coupling halves are tightened on the shafts. Sleeve couplings do not require keys and are normally used on small-diameter shafts. They are not suitable for use with heavy loads.

Ribbed couplings, also called *clamp couplings*, are made in two pieces. They are used when sleeve couplings are difficult to install. One advantage of using ribbed couplings is that they can be installed on shafts that are already in place without moving one of the shafts. They are used when the shafts are the same size and are also used for low-speed drives because of their unbalanced design and weight distribution.

### 9.1.2 Flexible Couplings

Flexible couplings are much more common than rigid couplings because they are usually easier to install and maintain and do not require precise alignment. Although flexible couplings allow for some misalignment, they should be aligned as close as possible, using the same methods employed with rigid couplings. Flexible couplings should not be used when major angular misalignment is known to exist. Deliberate misalignment requires the use of universal joints. Flexible couplings are divided into two categories: mechanical and material. Depending on the application, there are many types of couplings made in both categories.

Mechanical flexible couplings have metal components that may or may not need lubrication. They use the play or clearance in a mechanical device, such as chains or gears, to compensate for misalignment.

Material flexible couplings (*Figure 36*) are made to allow parts of the coupling to flex to compensate for misalignment. These flexing elements can be made of various materials, such as metal, rubber, or plastic. The life of a coupling depends on the life of the flexible material. As the material flexes, it begins to wear. The more the coupling is misaligned, the more the material is flexed and the faster it wears.

### 9.1.3 Soft-Start Couplings

Soft-start couplings (*Figure 36*) are used in applications where smooth, even starts are needed. Soft-start couplings allow the driving motor to pick up speed before the load is engaged, allow the driven device to start slowly and smoothly, and prevent stalls during overload conditions.

## 9.2.0 General Coupling Removal and Installation Methods

This section covers general coupling removal and installation.

### 9.2.1 Removing Couplings

The removal of couplings can vary from one piece of equipment to another, depending on the type of coupling used. The manual coupling puller is the most common field method for removing couplings that are press-fitted on the shafts. This puller

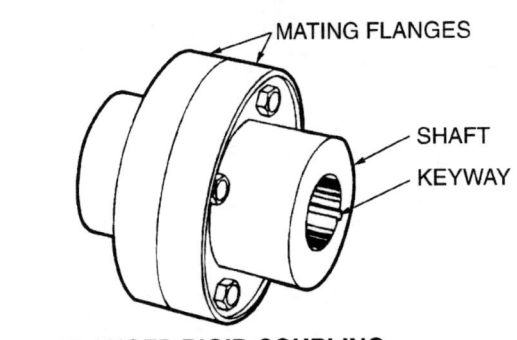

**FLANGED RIGID COUPLING**

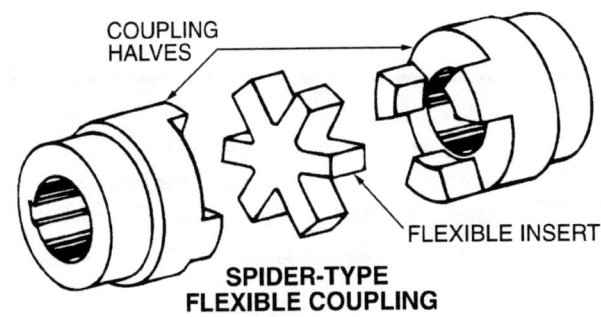

**SPIDER-TYPE FLEXIBLE COUPLING**

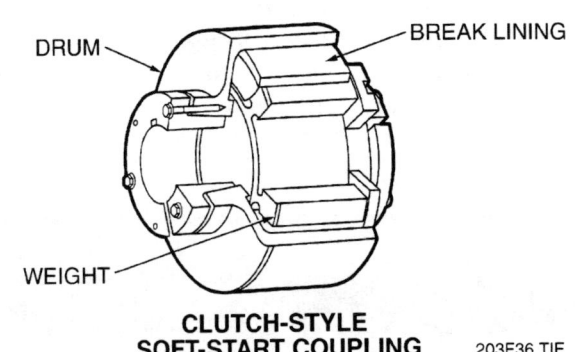

**CLUTCH-STYLE SOFT-START COUPLING**

*Figure 36* ◆ Typical couplings.

is similar to and is used in basically the same way as previously described for the puller used to remove bearings.

### 9.2.2 Installing Couplings

The tools and methods used to install couplings generally depend on the type of coupling to be installed. Many types of couplings are relatively simple to install. Press-fitted couplings are usually installed using either the press-fit mounting method or the interference fit method. The method used should be the one recommended by the manufacturer's service manual for the equipment. The press-fit method of installation uses an arbor press to mount the coupling on the shaft. This requires that the shaft be removed from the equipment. The shaft is securely mounted on the press, then the coupling is positioned on the shaft and pressed into place.

In the interference fit method, the coupling has a bore that is slightly smaller than the shaft diameter. The coupling must be preheated to expand the bore before it can be installed on the shaft. As the coupling cools, it shrinks to grip the shaft securely. This method is basically the same as that previously described for heating and installing a bearing.

## 9.3.0 Coupling Alignment

Rotating shafts are usually aligned with the couplings in place. The measurements are taken from the couplings, and adjustments are made accordingly.

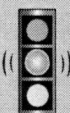
**CAUTION**
Alignment is critical to a few thousandths of an inch. Vibration, coupling wear, and bearing failure may occur if proper alignment is not maintained.

When aligning two coupling halves so that the shafts will be aligned, there are two ways that they must line up. First, the outer diameters, also called the *rims*, of the couplings must be lined up all the way around; then the faces of the couplings must be lined up. This results in four basic ways the coupling must be aligned:

- Outer diameter alignment, side view
- Outer diameter alignment, top view
- Face alignment, side view
- Face alignment, top view

If the couplings are misaligned, the alignment must be corrected to avoid poor equipment operation or damage.

### 9.3.1 Correcting Outer Diameter Alignment, Side View

Outer diameter (O.D.) misalignment, side view is also called *parallel misalignment*. This type of misalignment occurs when one of the coupling halves is not in line with the other when viewed from the side (*Figure 37*). In O.D. misalignment, one coupling is higher than the other, and one of the units must be raised or lowered to align the couplings.

### 9.3.2 Correcting Outer Diameter Alignment, Top View

Outer diameter misalignment, top view (*Figure 38*) occurs when the outer diameters of the couplings are misaligned from side to side as viewed

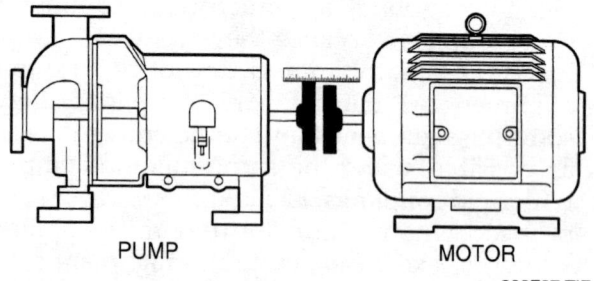
*Figure 37* ♦ O.D. misalignment, side view.

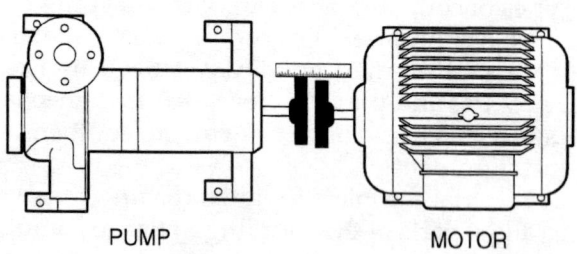
*Figure 38* ♦ O.D. misalignment, top view.

### Using Shims
Shims of varying thickness can be used to correct coupling O.D. misalignment. Insert shims under the motor or motor driven device until alignment is achieved.

from the top. In this type of misalignment, one of the units must be moved to one side or the other to align the couplings.

### 9.3.3 Correcting Face Alignment, Side View

Face misalignment is also called *angular misalignment*. Face misalignment, side view (*Figure 39*) means that the faces of the couplings are not square with one another when viewed from the side. In this type of misalignment, one of the units must be tilted on its base to align the couplings.

### 9.3.4 Correcting Face Alignment, Top View

Face misalignment, top view (*Figure 40*) means that the faces of the couplings are not square with one another when viewed from the top. In this type of misalignment, one of the units must be rotated on its base to align the couplings.

## 10.0.0 ◆ BASIC MAINTENANCE PROCEDURES

This section covers some of the basic maintenance procedures commonly used in the scheduled maintenance of several types of HVAC equipment. These procedures include motor lubrication, system air filter and screen inspection and cleaning, cooling system/heat pump coil inspection and cleaning procedures, and damper inspection and cleaning.

### 10.1.0 Motor Lubrication

To reduce friction and prevent wear, some motor bearings and other rotating and moving components need to be lubricated on a regular basis using a hand oiler (*Figure 41*) or a lever-type grease gun.

The bearings in many motors used in HVAC equipment are permanently lubricated at the fac-

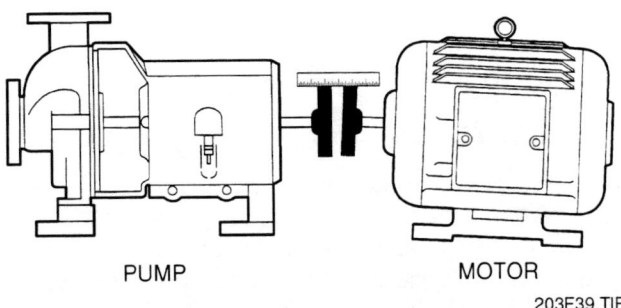

*Figure 39* ◆ Face misalignment, side view.

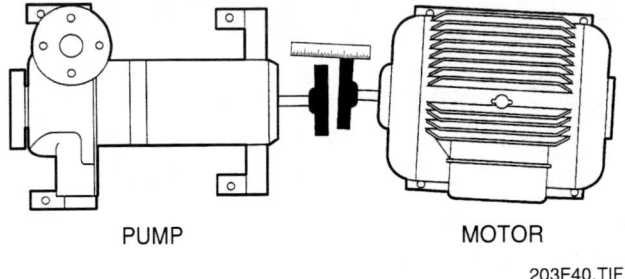

*Figure 40* ◆ Face misalignment, top view.

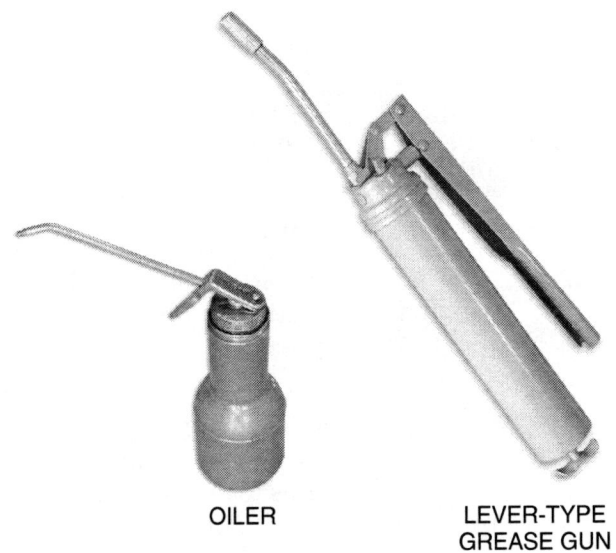

*Figure 41* ◆ Oiler and lever-type grease gun.

tory and require no lubrication in the field. The service literature for the equipment being serviced normally will state when this is the case. Other motors still require periodic lubrication. These motors are normally equipped with oil ports or grease fittings.

When lubricating motor bearings, it is important to follow the lubrication interval and use the type and quantity of lubricant recommended in the equipment manufacturer's service literature.

The most common problem with lubricating motor bearings is over-lubrication. It can cause an increase in operating temperature and a decrease in viscosity. If the lubricant becomes too thin, it cannot carry the load inside the bearing, and the bearing will fail and have to be replaced. Too much grease being forced into a bearing may burst the bearing seals, causing the lubricant to escape from the bearing and allowing contaminants to enter.

To lubricate motors with oil ports, proceed as follows:

*Step 1* Shut off power to the equipment, then lock out and tag the disconnect.

*Step 2* Remove the dust caps from the oil ports on both ends of the motor.

*Step 3* Oil the motor using the type and quantity of oil specified by the manufacturer. Typically, this is about 16 to 25 drops of nondetergent SAE 20 motor oil applied in each oil port. Rotate the shaft while lubricating the bearings to evenly distribute the lubricant.

*Step 4* Wipe up any excess oil, turn on the power, and observe the motor operation.

To lubricate motors or bearings equipped with grease fittings, proceed as follows:

*Step 1* Shut off power to the equipment, then lock out and tag the disconnect.

*Step 2* Wipe away any old or hardened grease from the grease fittings. Remove the relief plug if so equipped.

*Step 3* Using a grease gun, add the grease type and amount specified by the equipment manufacturer. Rotate the shaft while lubricating the bearings to evenly distribute the lubricant. *Do not* over-lubricate.

*Step 4* Wipe up any excess grease, turn on the power, and observe the motor operation.

### 10.2.0 Air Filter and Screen Maintenance

Maintaining the cleanliness of air filters and screens in an HVAC system provides many benefits including maintaining equipment efficiency, improving indoor air quality, and prolonging the life of interior surfaces (such as painted walls).

There are several types of air filters and it is important to use the type specified by the equipment manufacturer. Conventional disposable filters (commonly called *dust stop filters*) are typically constructed of fiberglass, hog hair, polyester, or open-cell foam. Sometimes a coating is applied to the filter media to trap additional particles. This type of filter is very common and is often seen in residential and light commercial HVAC equipment. The main disadvantage of this type of filter is that it only removes the larger particles in the air.

Better filtration can be achieved with an extended surface filter or an electrostatic filter. Extended surface filters have the material folded accordion-style which creates more filtration surface in a given area. Electrostatic filters have an inherent electrostatic charge which polarizes particles in the air stream, causing them to become trapped in the filter. Some of these types of filters can be cleaned and reused.

The best mechanical filter (*Figure 42*) is the high-efficiency mechanical air cleaner. The filter media is several inches thick and folded accordion-style to get a tremendous amount of filtration surface in a small area. This filter type can remove extremely small particles from the air.

The most efficient filter is the electronic air cleaner (*Figure 43*). This device uses high voltage to impart a charge on all particles that enter the device. A plate with an opposite charge then attracts these charged particles before they have a chance to exit the air cleaner. The electronic air cleaner can remove smaller particles than any other type of filter.

Regardless of type, all filters should be cleaned and/or replaced at the intervals recommended by the HVAC equipment manufacturer. Since different filters provide different resistance to air flowing through them, it is important to replace the filter with a similar or identical unit. Failure to do this could affect the operation of the system. If a filter is dirt clogged, the frame is bent, or the filter media is torn, the filter should be replaced. Disposable filters should be replaced and washable filters should be cleaned per the filter manufacturer's instructions.

---

### *Electronic Air Cleaner Maintenance Tip*

Unlike standard mechanical filters, electronic air cleaners become progressively less effective the dirtier they become. Once dirty, they can also generate annoying arcing that sounds like a snap and crackle coming from the filter. If the unit is allowed to operate for long periods of time in this condition, this arcing can create ozone gas ($O_3$), which can cause respiratory irritation in some people. In addition, excessive ozone contamination can initiate and accelerate metal corrosion.

To clean permanent filters and screens, proceed as follows:

*Step 1* Shut off all power to the unit.

*Step 2* Remove the filter and/or screen.

*Step 3* Remove heavy accumulations of dirt by tapping the filter and/or vacuuming it.

*Step 4* Wash the filter with a mild detergent solution until clean.

*Step 5* Dry the filter before re-installing. Do not oil or coat the filter unless directed to do so by the filter manufacturer. If the filter has an airflow arrow, make sure it points in the direction of airflow (usually toward the blower motor).

*Step 6* Replace all panels, turn the power on, and observe the equipment for correct operation.

The media core of high-efficiency mechanical filters can be removed and replaced with a new media core when necessary (*Figure 44*). This should be done in accordance with the manufacturer's instructions, being careful not to tear or damage the filter media during replacement.

*Figure 42* ◆ Typical mechanical air filter.

*Figure 43* ◆ Electronic air cleaner.

*Figure 44* ◆ Replacing filter media in a high-efficiency mechanical filter.

### Sizing an Air Filter

Air must flow through the filter, not around it. If the supply air can bypass the filter, the purpose of the filter is defeated. It is import to use a correctly sized air filter and to install it properly. Also, check gaskets, liners, insulation, slide tracks, filter baffles, blank-off plates, as well as access doors and panels to ensure that all are in place and secure.

### Cleaning Coils

Do not use a high-pressure washer to clean any coil surface. The high-pressure jet of water will flatten the aluminum fin stock on the coil, blocking airflow through the coil. Also, when washing a coil, be sure to prevent water from splashing in any nearby motors or electrical components. A simple plastic bag over a motor can provide a very effective shield. Some units have multi-layer coils. In order to properly clean the coils, carefully separate the layers by 3" to 4" and clean between them.

Because there are many different manufacturers and models of electronic air cleaners, and because of the high voltages involved in their operation, maintenance of these devices, including cleaning, must always be done in strict accordance with the manufacturer's instructions.

## 10.3.0 Coil and Condensate System Maintenance

The condenser and evaporator coils of a cooling system need to be inspected on a yearly basis and cleaned if necessary. Similarly, the indoor and outdoor coils of a heat pump system should be cleaned before each heating and cooling season. Dirt on coil surfaces cuts down on the ability of the coil to transfer heat, causing system inefficiency, poor indoor comfort, and even compressor failure.

To inspect and clean the evaporator/indoor coil, proceed as follows:

*Step 1* Shut off power to the equipment, then lock out and tag the disconnect.

*Step 2* Remove the coil access panel and inspect the coil for dirt accumulation on the coil surface.

*Step 3* Check for debris and/or dirt in the condensate drain pan. Remove any algae by washing the pan with an approved commercial solution.

**NOTE**
Be sure the condensate drain has a trap as called for by the installation instructions (some units have built in traps and do not require a separate trap). The trap prevents condensate from flowing back to the unit.

*Step 4* Use a vacuum cleaner brush attachment to remove light dust or lint. For stubborn dirt, spray the coil with a mild detergent solution. Rinse the coil and condensate pan with clear water. Make sure water runs freely from the condensate pan drain.

*Step 5* If local codes permit, tablets or chemicals can be placed in the condensate drain pan to inhibit the growth of slime or algae.

*Step 6* Straighten any bent coil fins with a fin comb before replacing the coil access panel, then restore power and observe the equipment operation.

To inspect and clean the condenser coil/outdoor coil, proceed as follows:

*Step 1* Shut off power to the unit, then lock out and tag the disconnect.

*Step 2* Gain access to the coil by removing the appropriate panels.

### Coil Cleaners

Many chemical coil cleaners are available to HVAC service technicians. Some of these chemicals are alkaline or acidic and can corrode metal if not thoroughly rinsed off. They also pose a safety hazard if splashed on your skin or in your eyes. If you choose to use these chemicals, first check to see if the equipment manufacturer approves of their use. Follow the application directions for the coil cleaner, and always wear appropriate personal protective equipment.

*Step 3* Vacuum or brush away any light dust or lint. Then spray the coil with a mild detergent solution. Use a low-pressure garden hose to rinse dirt and debris from the coil. Multi-row or multi-section coils may have to be separated to get at hidden dirt.

*Step 4* Replace all access panels, restore power, and observe the equipment operation.

*Step 5* Make sure that plants and shrubs in the vicinity of the unit have not grown to the point that they block or restrict airflow to the condenser/outdoor coil. If they have, recommend that the property owner trim back the plants.

## 10.4.0 Damper Inspection and Cleaning

Commercial HVAC units like economizers and similar equipment are equipped with outdoor air dampers. To inspect and clean these dampers, proceed as follows:

*Step 1* Shut off power to the unit, then lock out and tag the disconnect.

*Step 2* Remove and clean the mesh screen(s) as previously described above under filter and screen maintenance.

*Step 3* Clean the damper blades of any dirt, soot, etc., and make sure none are damaged.

*Step 4* Check any pins, straps, and bushings/bearings for wear, rust, or corrosion.

*Step 5* Check that the seal strips on the damper hood top and sides are not damaged.

*Step 6* With the unit in operation, check to make sure that the damper moves freely, with no binding.

*Step 7* Replace the screen(s).

## 11.0.0 ◆ DOCUMENTATION

One of the most important tasks concerning the installation and maintenance of HVAC equipment involves the completion of various forms and/or reports. Proper completion of forms and/or reports does not end with simply adding the data to the form. The forms used in the HVAC industry are primarily used as communication tools, but they serve many purposes. They must be legible and understandable. In written or oral communications with your customers, remember not to use trade jargon. Customers need to understand what you have done during your service call, how their unit is currently functioning, and what to expect from it in the future. Using trade terms makes this much more difficult. This aspect of your job responsibilities cannot be overemphasized. If you cannot clearly communicate with the customer and your service dispatcher, and leave accurate documentation of what you have done, your success in this industry will always be a struggle.

Forms and reports serve as historical records about the equipment. For this reason, it is important to fill them out completely and accurately. When troubleshooting at some later date, recorded data is useful because it can be compared to current system readings in order to determine areas of possible system degradation. Unfortunately, the popularity of civil and criminal litigation is slipping rapidly into the HVAC industry as well. This trend is forcing the use of the documentation you are generating to be used in court cases. In the event of a lawsuit, these forms or reports are often used in court to prove that specific tasks have been performed and/or that the installed equipment has met design and customer specifications. Forms and paperwork are commonly used to:

- Record the results of system performance tests.
- Verify that required inspection or quality control milestones have been met.
- Verify that installed systems meet design and customer specifications.
- Notify equipment manufacturers of the start of their equipment warranties.
- Record important facts about service calls or other field maintenance activities.
- Communicate to all concerned parties what you have done.

If all of this talk about documentation seems like overkill, you should be reminded that accurate and legible documentation is the key to successfully communicating with all concerned parties about work you and your employer perform. Documentation is not filled out only for the benefit of your service manager or dispatcher. It is done primarily for the benefit of the customer. If improper communication is allowed, then customer dissatisfaction is sure to follow. Without customers willing to pay for your services, there will be no need to worry over any paperwork at all.

Some types of documentation you routinely will be required to fill out include:

- Service ticket/invoice
- Commissioning job report
- Start-up report
- Warranty ticket

### Roof or Campus Maps

In larger commercial projects where multiple units are in service, you may find that the use of a roof or campus map is useful. In these cases, the addition of a mark or unit tag number (such as RTU-I or AC-I) will be required in addition to the standard make, model, and serial number information that you will always provide.

## 11.1.0 The Importance of Make, Model Number, and Serial Number

The three most critical items that should always appear on everything you document are the make, model number, and serial number for the unit being serviced. These three pieces of information can be found on the unit nameplate. The make identifies the name of the manufacturer such as Carrier, Trane, Lennox, York, etc. The model number identifies the specific model of the unit. It describes its style, type, capacity, electrical characteristics, and application. The serial number is the sequential manufacturing number assigned by the manufacturer for the unit. Its use is critical when you need to get replacement parts for the unit. This is because many units are manufactured as a series or group of numbers over the years, and they may even share the same model numbers over time.

Unfortunately, manufacturers often change critical parts in the life of a particular model series, and the serial number is needed to make sure that you get the correct part for use in the unit being serviced. This reduces callbacks and customer dissatisfaction.

## 11.2.0 Service Ticket/Invoice

The service ticket/invoice is the most basic form used by the HVAC technician. Many firms have incorporated the use of computer dispatching and data retrieval. This reinforces the need for accurate and clear communication. The data from the service ticket is then entered into the computer and used for preparing the formal customer invoice or bill. Some systems allow for a computer log-in from a remote site so that the technician can type in the data, then receive an automatic printout of an invoice at the site.

*Figure 45* shows a typical service ticket/invoice form. A service ticket/invoice form is filled out by the HVAC technician for each job. Initially, the form may contain information provided by the shop supervisor/service dispatcher that gives the technician the customer's name, location, and details about the nature of the job or service call.

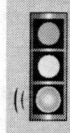

**NOTE**

Many of the newer computer-managed dispatch and service systems do not allow the processing of payroll until all forms and required inputs are correctly entered into the system. This should be an ample incentive for completing accurate paperwork.

Sometimes this information is initially provided on a work order form (*Figure 46*).

At the completion of the service call, the technician fills in all applicable portions of the form to provide a specific description of the work that was done, the quantity and types of materials used, the labor hours expended, and so on. After the service ticket is completed and signed by the technician, it is then given to the customer who also signs the ticket to acknowledge that the service performed was requested and authorized by the customer and that all materials and services have been received. The signed ticket usually serves as a customer billing invoice and a copy is given to the customer.

A typical service ticket is a multi-sheet form consisting of duplicate white, yellow, and pink sheets. Normally, the white sheet is retained by the technician for office use, the yellow sheet is given to the customer, and the pink sheet is retained for the office historical files.

## 11.3.0 Commissioning Job Report

At the completion of most large commercial and industrial HVAC installations, a commissioning process is used to document and verify the performance of the HVAC systems in order to ensure that they operate in conformance with the design intent. Detailed checklists and reports are filled out during the commissioning process to record HVAC equipment readiness, start-up, and performance. At the completion of the process, these checklists and reports are turned over to the building owner or other designated authority.

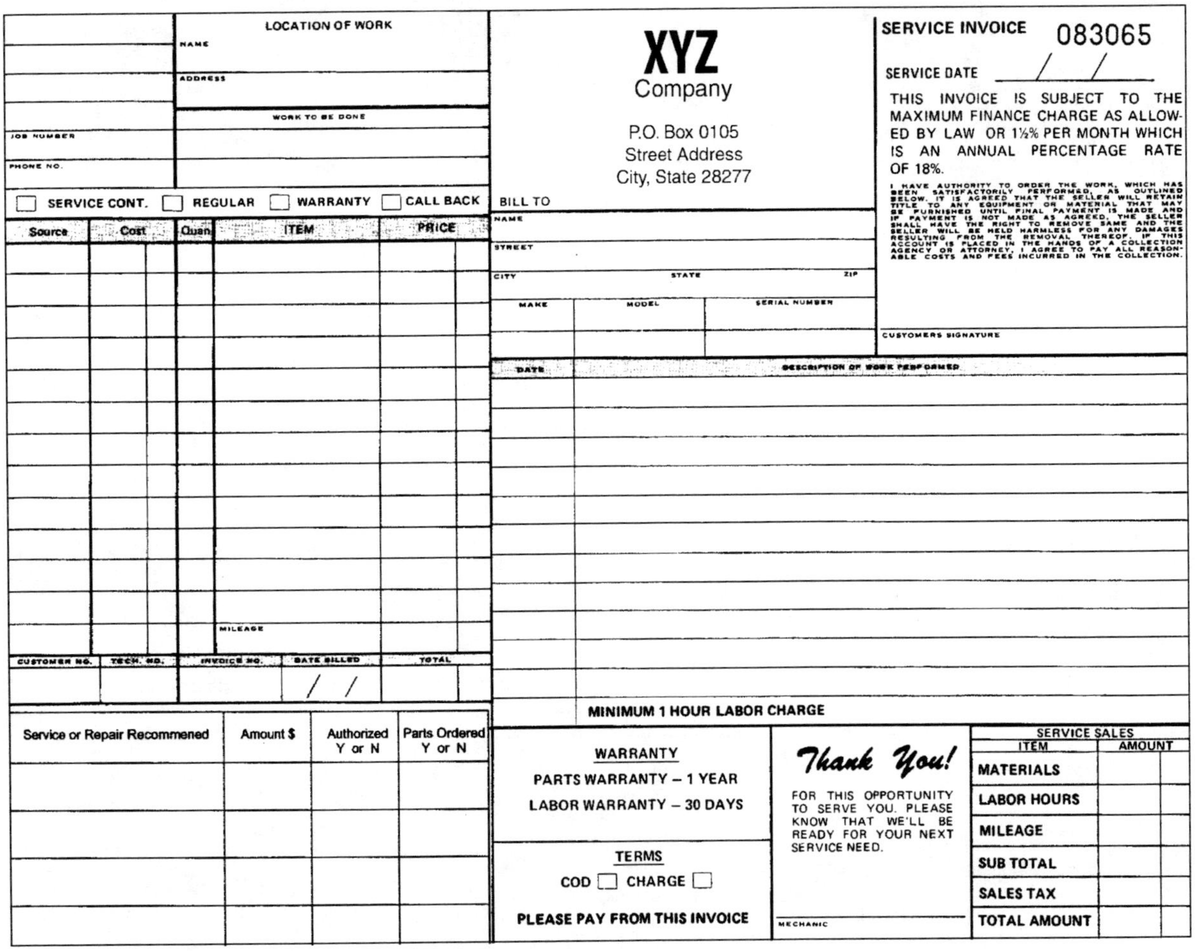

LOCATION OF WORK
NAME
ADDRESS
WORK TO BE DONE
JOB NUMBER
PHONE NO.

XYZ
Company
P.O. Box 0105
Street Address
City, State 28277

SERVICE INVOICE 083065
SERVICE DATE / /
THIS INVOICE IS SUBJECT TO THE MAXIMUM FINANCE CHARGE AS ALLOWED BY LAW OR 1½% PER MONTH WHICH IS AN ANNUAL PERCENTAGE RATE OF 18%.
I HAVE AUTHORITY TO ORDER THE WORK, WHICH HAS BEEN SATISFACTORILY PERFORMED, AS OUTLINED BELOW. IT IS AGREED THAT THE SELLER WILL RETAIN TITLE TO ANY EQUIPMENT OR MATERIAL THAT MAY BE FURNISHED UNTIL FINAL PAYMENT IS MADE AND IF PAYMENT IS NOT MADE AS AGREED, THE SELLER SHALL HAVE THE RIGHT TO REMOVE SAME AND THE SELLER WILL BE HELD HARMLESS FOR ANY DAMAGES RESULTING FROM THE REMOVAL THEREOF. IF THIS ACCOUNT IS PLACED IN THE HANDS OF A COLLECTION AGENCY OR ATTORNEY, I AGREE TO PAY ALL REASONABLE COSTS AND FEES INCURRED IN THE COLLECTION.
CUSTOMERS SIGNATURE

☐ SERVICE CONT. ☐ REGULAR ☐ WARRANTY ☐ CALL BACK

BILL TO
NAME
STREET
CITY STATE ZIP
MAKE MODEL SERIAL NUMBER

| Source | Cost | Quan. | ITEM | PRICE |
|---|---|---|---|---|
| | | | MILEAGE | |

| CUSTOMER NO. | TECH. NO. | INVOICE NO. | DATE BILLED | TOTAL |
|---|---|---|---|---|
| | | | / / | |

| DATE | DESCRIPTION OF WORK PERFORMED |
|---|---|
| | MINIMUM 1 HOUR LABOR CHARGE |

| Service or Repair Recommened | Amount $ | Authorized Y or N | Parts Ordered Y or N |
|---|---|---|---|

WARRANTY
PARTS WARRANTY – 1 YEAR
LABOR WARRANTY – 30 DAYS

TERMS
COD ☐ CHARGE ☐
PLEASE PAY FROM THIS INVOICE

Thank You!
FOR THIS OPPORTUNITY TO SERVE YOU. PLEASE KNOW THAT WE'LL BE READY FOR YOUR NEXT SERVICE NEED.
MECHANIC

| SERVICE SALES | |
|---|---|
| ITEM | AMOUNT |
| MATERIALS | |
| LABOR HOURS | |
| MILEAGE | |
| SUB TOTAL | |
| SALES TAX | |
| TOTAL AMOUNT | |

*Figure 45* ◆ Example of a typical repair order/service report form.

The specific checklists used during a commissioning process are determined by the specific types of installed equipment.

The National Environmental Balancing Bureau (NEBB) has developed a series of forms and check sheets widely used by its members and other NEBB-certified firms for certifying building system commissioning. These forms are grouped into the following categories:

- Administrative forms
- General equipment forms
- Hydronic equipment forms
- Cooling equipment forms
- Air handling equipment forms
- Control system forms

## 11.4.0 Start-Up Report

As the name implies, start-up reports or forms are used to record the specific operating conditions and parameters that exist at the time of initial start-up and operation for all types of HVAC systems and/or individual components of an HVAC system. These reports not only provide a record to verify proper operation at start-up, but they also serve as a record for troubleshooting at some later date where the data can be compared to current system readings in order to determine areas of possible system degradation. *Figures 47* and *48* show examples of start-up reports used for recording the start-up operating conditions for a heat pump and a furnace, respectively.

## 11.5.0 Warranty Ticket

At the completion of a job or HVAC system commissioning process, HVAC equipment warranty forms or tickets should be filled out and given to manufacturers to notify them that their equipment has been put into operation. Copies of all warranties should also be given to the building owner or designated person.

# WORK ORDER

**XYZ** Company
P.O. Box 0105
Street Address
City, State 28277

Date:
Summary:
Reference #:
Tech:
Start Time:

Bill To:

Job Name:

Description of Work

| Material | Labor | Other | Subtotal | Tax 1 | Tax 2 | Total |

All material is guaranteed to be as specified. All work to be completed in a professional manner according to standard practices. Any alteration or deviation from above specifications involving extra costs will be executed only upon written orders and will become an extra charge over and above the estimate. All agreements contingent upon delays beyond our control. Purchaser agrees to pay all costs of collection, including attorney's fees.

*Figure 46* ◆ Example of a typical work order form.

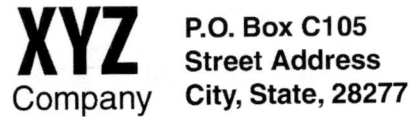

*Figure 47* ◆ Example of a typical start-up report for a heat pump.

**XYZ** Company
P.O. Box 0105
Street Address
City, State 28277

## START-UP REPORT

CUST. NAME _____    DATE _____

ADDRESS _____    TIME _____

CITY _____

PHONE # _____    DATE INSTALLED _____

UNIT MODEL NO. _____    SERIAL NO. _____

## CHECK LIST

_____ GAS LINE PRESSURE         _____ GAS LEAK SEARCH
_____ MANIFOLD PRESSURE         _____ DAMPERS OPEN
_____ FAN ON TEMP               _____ FRESH AIR DAMPER ADJUSTED
_____ TEMPERATURE RISE          _____ REGISTERS ON AND OPEN
_____ TEMP LIMIT OPENS          _____ FLOOR NOISE
_____ TEMP LIMIT CLOSES         _____ VENTING COMPLETE
_____ FAN OFF TEMP              _____ CAULKED GAS LINE & VENTING
_____ CARBON MONOXIDE TEST      _____ STICKERS
_____ FILTER SIZE               _____ DRAIN COMPLETE
_____ BLOWER WHEEL              _____ INSULATION IN BOX SILLS
_____ FURNACE GROUND
_____ LINE VOLTAGE              NOTES: _____
_____ ALL ELECTRICAL            _____
       CONNECTIONS TIGHT        _____

WHICH BLOWER SPEED USED?         START-UP PERFORMED BY:

HEATING _____ COOLING _____      _____

                                 TECHNICIAN

203F48.EPS

*Figure 48* ◆ Example of a typical start-up report for a furnace.

### Customer Technical Knowledge and Skills

When talking with customers, never assume that they do not have technical knowledge or skills. Many customers have a broader technical background and knowledge than some service people.

## 12.0.0 ♦ CUSTOMER RELATIONS

Good customer relations are essential to the success of most businesses, including those in the HVAC industry, where each service technician represents the company and influences what customers think about the business. Keeping these impressions positive requires that all employees work toward generating customer good will. Providing good service is a part of this, but so are good personal habits, good work practices, and customer-pleasing attitudes. Customer good will often results in more business through referrals and repeat servicing of existing customers.

### 12.1.0 Why Customer Relations are Important

Good customer relations consist of the habits, behaviors, and attitudes which guarantee that the customer's first impression of you and your company is a good one; that the customer's needs are clearly understood; and that appropriate customer service is provided to meet these needs.

When you provide this level of effective service to your customers, you can increase your company's share of business. Your ability to relate well to your customers is critical. As a service technician, you're likely to be the only person from your company that the customer ever sees. In the customer's eyes, you are the company.

In this section, personal habits, handling service calls, and handling difficult situations will be explained. Mastering these concepts will not only make your job easier, it will also make it more likely that a customer will call your company the next time HVAC service is needed.

### 12.2.0 Personal Habits, Behaviors, and Attitudes

Appearances do count. First impressions happen only once, in the first sixty seconds. You don't get a second chance to correct a first impression.

Because customers don't have technical skills, they call a servicing company for help with heating and cooling problems. Although they can't judge the servicing technician's professional competence, they can, and do, form an impression of the technician's appearance and attitudes. The customer's first glimpse of a technician can determine that customer's opinion of the worker and the company the technician represents. Examine your first impression. Do you:

- Practice good personal hygiene?
- Get enough sleep and look alert?
- Wear a neat, clean uniform?
- Wear shoe covers or remove shoes when in the home?
- Carry a pencil and pad to take notes?

Once the customer has formed a first impression of you and your company, your on-site work habits will confirm or change that impression.

- Are you on time?
- Is your tool set complete and neatly packed?
- Do you show concern, courtesy, and respect to the customer by listening carefully while the customer identifies the problem?
- Do you tackle the problem promptly and quietly?
- Do you remember to protect the work area? Do you carry rags for cleaning up and dropcloths to protect floors and carpets?

### Personal Habits, Behaviors, and Attitudes

If you were asked to hire a new service technician for your company, what would the worker you choose be like?

- Do you take the time to explain carefully to the customer what's wrong and how the unit will perform satisfactorily after you've repaired it?
- Do you respect the job site by not tracking in dirt and by cleaning up after yourself?
- Do you refrain from smoking on the premises, avoid smelling of smoke, or spitting?
- Do you respect company equipment, tools, and vehicles? Trucks and vehicles displaying the company name are rolling advertisements. Is your vehicle clean and in good repair? Are your driving habits courteous?
- Do you avoid using alcohol and drugs while on the job or around vehicles?
- Do you avoid profanity and horseplay? Customers, especially those with children on the premises, are likely to be offended by inappropriate language.

Every hour of every working day, you are an advertisement for your company. Your appearance and behavior must be consistent with the positive image your company wants you to reflect.

Off the job, you can increase word-of-mouth advertising (free advertising created by satisfied customers who recommend your company to friends) by having a positive attitude about your employers, co-workers, and customers. Positive statements about your job are free advertising for your company.

## 12.3.0 Customer Relations: Handling Service Calls

During each service call, you can do several things to enhance the customer's image of you and your company. The typical service call has three parts:

- *The opening* – Identifying the problem
- *Servicing* – Solving the problem
- *The closing* – Leaving the customer with a positive impression

Remember that your objective in servicing your customer's equipment goes beyond technical competence. Your objective is to win the customer's repeat business.

### 12.3.1 The Opening

As you open the service call, you are already making your first impression. Your personal appearance is good. Your vehicle looks good. You show respect for the customer by appearing promptly and politely at the door. Now what? You should:

- Smile and display confidence and polite respect for the customer.
- Promptly identify yourself. Many customers waiting for a service worker are naturally cautious about admitting a stranger. Identify yourself clearly and show an appropriate ID, if you have one. Give the customer a company business card with your name on it. If your company does not have business cards made with the names of individual service technicians, encourage the company to at least have company business cards available so that you can give one to the customer in case they need to contact the company if they have any follow up questions or concerns.
- Understand your customer's needs. Listen to your customer to learn what the problem is. Ask questions about what the customer has seen, heard, smelled, and felt. Ask when things occurred and how often they happened. Often, the customer can help speed your diagnosis by providing clues to the equipment problem. When you listen attentively, everyone wins.

It's important to ask open-ended questions, the kinds that result in specific, information-gathering answers. You don't want to be led to the equipment without any information from the customer. Ask as many questions as you need to get the exact symptoms. Don't say, "I'm here to fix your furnace. Where is it?" Instead, say, "I came to repair your furnace. What seems to be the problem?"

### 12.3.2 Servicing

While servicing the equipment, you should be adding to your customer's positive impression of you and your company. The customer has an

### Customer Relations
What questions can you ask Mrs. Jones when responding to an air conditioning service call?

### Verify the Unit Works
One of the most important things you can do at the end of a service call is to demonstrate to the customer that the system works. Don't leave them guessing; show them.

equipment problem. You're there to solve it. Build customer confidence by getting right to work to determine the probable cause of the malfunction. Practice good work habits. Avoid general conversation while working. Remember, many service technicians are paid by the hour; that's how the customer pays for their time. The customer must feel that you are filling the time with productive work, not idle chatter.

During this part of the service call, you'll need your professional skills to:

- Analyze the symptoms.
- Isolate the problem.
- Eliminate probable causes that do not apply.
- Isolate probable causes.
- Determine the solution.
- Explain the solution to the customer, including your best estimate of what it will cost.
- If the customer approves, remedy the situation.
- Follow up on the service visit.

Following up is important, but it's often overlooked. If time permits, call the customer a day or two later to ask if your repairs were satisfactory. If you made adjustments, stop by to see if they were acceptable. Often, a simple phone call from you can be the touch that creates a repeat customer. In your company, if the service technicians don't have the time to follow up on service visits, perhaps someone from the office staff can handle these calls.

#### 12.3.3 The Closing

This very important step helps set the tone for your customer's image of you and your company. When you finish the work:

- Neatly pack your tools.

- Return the premises to its original condition. (Replace covers, wipe off dirty fingerprints, clean up dropcloths, etc.)
- Explain to the customer that the problem has been solved. Explain what parts you replaced and offer to show the customer the defective parts.
- Demonstrate that the equipment works.
- Wrap up the call by relating your service to the customer, not the equipment. Don't say: "I fixed the thermostat; your furnace will turn on now." Rather, say: "I repaired your thermostat so you can be warm and comfortable again." A positive closing like this builds confidence in you and your company.

### 13.0.0 ♦ CUSTOMER COMMUNICATION

Having covered the basic customer relations requirements for handling a service call (opening, servicing, and closing), you are ready to examine the customer relations topics that can give your company an extra edge over the competition. These are: keeping communications positive, taking the positive approach, showing concern for customers, and handling difficult customers.

It's important to know the technical requirements of your job. Knowing the steps in completing a typical call is important, too. However, you can gain a little extra leverage by learning how to interact well with customers. This may make all the difference for your company and may determine whether you or your competitor is called the next time. When you practice good public relations throughout the service call, you give the customer confidence in your service and your company and predispose that customer to call you again.

### Closing
After you've replaced the refrigerant charge in Mrs. Jones' air conditioner, what closing statement can you make?

## 13.1.0 Keeping Communications Positive

On every service call, you should strive to keep the communications positive. This means paying close attention to verbal and nonverbal elements. This includes communicating:

- The technical elements of the job
- What the customer needs
- Your problem-solving expertise
- What you've done to make the customer happy

Here are some simple but effective tips:

- Do the best technical job you can, since actions speak louder than words.
- Always treat the customer with courtesy, concern, and respect. Treating others as you'd like to be treated yourself is particularly important, especially in unpleasant situations. Remember, it is always easy to be courteous when things are going well. A truly mature person is courteous even when things aren't going well.
- Treat each service call as if it were an emergency. To the owner, it is. Usually, customers do not call a service technician unless they are without heat in cold weather or without cooling in hot weather. When comfort is at stake, repair becomes a top priority in the customer's mind.
- If your schedule is running late, report this to your supervisor or the customer. If you're running late, the customer must often make other arrangements to fulfill later commitments.

## 13.2.0 The Positive Approach

Here are some important tips on keeping things positive:

- *Smile* – Be genuinely interested in the customer. You may think you were called to repair a furnace. You were really called to make the customer's home comfortable again.
- *Use positive statements with the customer* – Compliment the equipment if you know it is good. Compliment the cleanliness of the furnace area if that contributes to the operation of the equipment. Speak of your employer in positive terms, because that builds customer confidence in your company.
- *Don't criticize or condemn* – Whenever possible, avoid negative statements. Be careful of what you say about the brand of equipment, the design of the system, or the quality of the workmanship, for you may offend the customer. Homeowners do not want to hear that you think they own inferior equipment. Even if it is inferior, fix what is possible or explain respectfully to the customer what must be replaced.

## 13.3.0 Showing Concern for Customers

Another part of good, positive, on-the-job communications is showing concern for your customers. Do this by:

- *Being a good listener* – Listen carefully when the customer explains what went wrong, what was observed, etc. Ask questions as appropriate to gather the information you need to solve the problem. Remember that although the customer may not have the technical vocabulary you do, the customer's own words can supply you with important clues.
- *Talking in terms of the customer's interests* – Always try to look at the problem from the customer's point of view. For example, Mrs. Jones may be upset that you are late. Maybe the emergency you fixed for your last customer caused this. However, to Mrs. Jones, the important thing may be that she's running late in picking up her child from school. If this understandable

### Showing Concern for Customers

What would you do in these situations?

1. Mr. Gray owns a brand of furnace that is no longer being manufactured. You don't think parts are available to fix the burner. What would you say?
2. Mrs. Brown called for a technician to fix her furnace. As you enter the job site, she says, "It's cold in here." What questions would you ask her?
3. Mr. Green says his blower and burner are not operating in sequence and may need to be replaced. You think the problem is a faulty thermostat. How do you build his confidence in your skills without hurting his feelings?
4. Can you suggest some ways of advertising your company that are simple and free?

concern for her child is added to concern over a furnace that isn't heating, Mrs. Jones might be short-tempered. If she speaks to you abruptly, try turning the conversation around by saying, "I know you must be concerned about my being late. I apologize. Please show me your furnace so we can get you back on schedule quickly."

- *Keeping your personal problems to yourself* – Remember, you're on the job to solve the customer's problem.
- *Answering questions honestly, but positively* – If Mr. Brown asks if his equipment is worn out, you may reply, "It needs to be replaced," but you should avoid negative comments such as, "This pile of junk needs to be replaced."
- *Respecting your customer's opinions* – Suggest alternatives that fall short of telling a customer, "You're wrong!" Show the customer his opinion may have value. For example, if Mr. Smith announces that the thermostat is no good and you suspect there is a wiring problem, don't say, "No, you're wrong. It's in the wiring." Instead, say, "I think it's the wiring, but I'll check out the thermostat too."
- *Not socializing while on the call* – Generally, service technicians are paid by the customer by the hour. If you explain pleasantly that you appreciate the offer of a cup of coffee, but that you're sure they'll understand if you decline, the customer will not be offended by your trying to keep the charges to minimum.

Further, you continue to show concern for your customers when you make it easy for them to contact you for repeat business. When possible:

- Leave a business card.
- Put a sticker with your company's phone number on the equipment itself or leave one with the customer for placement near the phone.
- Show the customer some simple maintenance techniques that will prolong equipment life, such as changing filters frequently. Tell the customer why this maintenance is helpful.
- Teach the owner steps for more efficient use of the equipment, if applicable.
- Discuss billing accurately and honestly.
- Perform customer follow-up, if possible.
- If the opportunity arises, discuss additional equipment to improve comfort and efficiency that the customer can purchase from your company. Ask for the customer's permission to notify the sales staff. Follow up by providing this lead to the sales staff.

## 13.4.0 Handling Difficult Customers

When you deal with customers, remember: the customer wants safety, comfort, and convenience. If your habits, behaviors, and attitudes inspire that customer to believe your company will provide for their safety and comfort in a convenient manner, they will return to your company rather than seeking out a competitor next time.

Sometimes, the service technician encounters problems that go beyond failed machinery. For example:

- A customer may be unduly worried about safety.
- A customer may be upset about something totally unrelated to the problem, but is taking out that irritation on the service technician.
- A customer may be angry over equipment performance and your company's bill.

In cases like these, the service technician's actions and attitudes will often determine whether the company loses a customer.

In this section, we will discuss several situations that require more advanced customer relations skills. Decide how you'd handle these customers and compare your ideas with those of your classmates.

Remember, there may be no single right way out of an angry confrontation, but your attitudes and reactions can go a long way to help resolve even the most difficult situations.

### 13.4.1 Fearful Customers

Sometimes customer fears are based on experience, rumors, or misunderstandings. As the technician, you have the technical information that can calm your fearful customer. There is always a reason why your customer has a fear. Find it and work from the facts to ease the customer's mind. If you listen with concern and take customers' fears seriously, they will feel reassured and be more willing to listen to your explanation.

*Exercise Question:*

Mr. Fearful thinks his furnace is leaking gas because he smelled some when the pilot light blew out. You have re-ignited the pilot and are closing the call. What can you say to calm his fear that the house will explode from a leak at the pilot light?

*Exercise Solution:*

Fear is real. Often you can discover the reason for it by asking questions:

- When does this happen? When did it start?
- When it happens, what does it look like?
- What does it sound like?
- What does it smell like?
- Before it happens, are you doing something different from what you used to do?

After you have found the answers, don't just say, "Yes, when the pilot light goes out there is a smell. It is fixed." This has not addressed Mr. Fearful's concerns for the next time it happens. Instead, explain: "When a pilot goes out, there is a smell of gas. A minute quantity escapes but will not ignite. It is also mixed with a definite odor that alerts you to re-light the burner. The small amount of gas will not cause a fire. The vent pipe will exhaust it. Your family is not in any danger." Possibly suggest that a gas sniffer alarm and carbon monoxide sensor be installed to give the customer peace of mind.

Since this explanation not only addresses the technical problem, but also the fear associated with it, the customer will be more satisfied.

### 13.4.2 Opinionated Customers

Opinions are beliefs based not entirely on facts, but also on what seems probable in a person's mind. Everyone has opinions. Everyone also likes to get approval for opinions held. If our opinions are ridiculed, we feel hurt and offended.

It is important that customers do not feel their opinions have been challenged or belittled. With a customer, it's best to turn the conversation to the technical problem at hand (even when you agree with the opinion).

*Exercise Question:*

Mrs. Opinionated has had a bad week, including the failure of her air conditioner. She's trying to draw you into her earlier problems and her opinions about them. You're trying to finish quickly because you have two other service calls before lunch. What do you say?

*Exercise Solution:*

Taking what seems to be a negative approach might offend her. So don't say: "Yeah, we all have problems. My problem is fixing air conditioners. Where is it?" Instead, say: "I've had times like that myself. I can help you with one of your problems, though. Please show me your air conditioner so I can help you solve that problem." This offers sympathy while at the same time diverting the customer to the task at hand.

### 13.4.3 Argumentative Customers

Some customers are just the argumentative type. They want you to challenge their opinions, just to get an argument going. You're sensitive to the customer, and you don't want to seem aloof. You don't want to express negative opinions either. You also know that some topics are really argument-prone, such as politics, religion, children, and in-laws. Here, too, the safest approach is to avoid controversy by diverting the conversation to the task at hand.

*Exercise Question:*

Mr. Argumentative has no airflow from his furnace. You've discovered a broken fan belt. He wants to pull you into a discussion of local politics. He thinks the mayor is a crook. You agree. How would you handle this?

*Exercise Solution:*

The best way to handle this is to state a general fact that will not offend, and to avoid prolonging or expanding the political discussion while you gently turn the conversation to the problem you're there to solve. (Even if you disagreed with the customer, the technique would still be same: shifting attention to the fan belt problem.) Say, "Some people feel that way. But my feeling is that you'll be a lot more comfortable after I replace this fan belt."

### 13.4.4 Sloppy Customers

As you make service calls, you'll see all kinds of housekeeping: some spotless; some cluttered; some downright dirty. Always, it's best to avoid offending your customer. When admitted to the job site, avoid being judgmental in your words and gestures. Since you're a visitor, the owner will often offer an excuse or reason if things are messy. In some homes, the owner's sloppy habits can adversely affect the operation of the equipment.

You can handle such sensitive situations in these ways:

- *Focus on the task at hand* – Don't look like you're judging the job site. Avoid comments like "What a mess!" and raised eyebrows or rolled eyes that say the same thing. Assume that the owner is the unfortunate victim of several small domestic catastrophes that will be remedied soon.
- *Put the owner at ease* – If you must reply, avoid degrading and comparing them unfavorably. Say: "This happens at our home, too."

### Mr. Sloppy

What would you do?

1. Mr. Sloppy takes you into his living room to fix the air conditioner. The musty air smells of cat litter; dusty newspapers fill all the chairs; the cats have knocked two or three flowerpots onto the floor, making a mess. Mr. Sloppy says, "Isn't this a mess?" What do you say?

2. Mr. Sloppy's furnace filter is plugged by two years' worth of dirt. What do you say?

- *Be diplomatic* – If it's cluttered in front of the furnace, move only what's necessary and try to restore things when you're finished. Don't be obvious in moving things. As you clear space, avoid grunts, groans, and comments.
- *Don't sound like an accuser* – Avoid "you" statements. Talk about what's best for the health of the furnace or make up a similar situation that you solved, using your own family as an example.

### 13.4.5 Angry Customers

In the last section, you were advised to avoid "you" statements to a customer when it might suggest a negative judgment; for example, accusing them of bad housekeeping. There is, however, a good use of a "you" statement, and that's when the customer has a right to be angry.

When equipment breaks down twice in three days, or when the customer has tried to meet the service technician two days in a row without success, the customer may feel justified in venting anger at the first available target: the technician who appears at the door. What should you do?

First, let the angry customer vent his or her emotions. Next, acknowledge the customer's feelings. Sometimes, this means saying something sympathetic like, "I'd feel that way, too." Finally, assure the customer that you'll do your best to solve the problem.

*Exercise Question:*

Mrs. Smith's furnace was repaired just last week. It has broken down again, leaving the family without heat for two days. The children have colds. The Smiths have already paid a sizable sum for the repair that did not last. As you walk in the door, Mrs. Smith snarls at you. What do you say?

*Exercise Solution:*

Here is an appropriate answer: "Hello, Mrs. Smith. I hear your furnace isn't working again. I'm sure you're very concerned about your children's comfort. I know just how you feel. My air conditioner quit twice on me last summer during a heat wave. I'll get to the problem as quickly as I can. Let's see what the matter is." You have shown concern for her feelings and worries; you have acknowledged that she is justified in her concern; and you have assured her that you will address the problem quickly and to her satisfaction.

### 13.4.6 Critical Customers

Sometimes you'll encounter a customer who can't be satisfied, no matter what you say or do. Often this happens because the customer is under a lot of stress or has had a bad experience that hasn't been dealt with properly. When this customer explodes, you may be the nearest target. How can you handle these situations with sensitivity and tact?

We've all heard of cases where a parent, angry about a bad day at work, comes home and yells at the kids. This displacement of anger can also happen when a person has an unresolved conflict that is upsetting.

- A man who refused to pay extra for a weekend call now has a mess on his hands because the pipes in his poorly insulated laundry room are frozen.

### Mr. Critical

Mr. Critical is unhappy. The builders who just completed his home left several jobs unfinished. He begins a tirade about builders and service workers in general, including you, in his speech, even though you've only just arrived. How do you handle this difficult customer while fixing his furnace thermostat?

MAINTENANCE SKILLS FOR THE SERVICE TECHNICIAN — TRAINEE MODULE 03203

- A woman having trouble with her sales accounts doesn't really want to be home waiting for a service technician when she should be dealing with her office problems.

The first step in handling these situations is to remember that all of us have bad days. Approach the customer with the attitude that some day you will feel out of sorts. On these days, we all hope those around us will make allowances and be tolerant. Give your customer the same understanding you'd like to have at these times.

If the customer explodes, don't take it personally. After all, you may have just been a convenient target. You may not be responsible for the problem. Review what you have done. If your work was correct, shrug off the comments silently.

If your customer is justifiably upset, let him vent some anger. Empathize with the feelings, explaining that you'd probably feel the same under the same circumstances. Reassure the customer that you'll fix the HVAC technical problem as quickly as you can, so at least part of the day will be a positive experience.

Get the job done as quickly as you can. Get out of the situation as soon as possible without offending the customer. Stick to the facts. Focus on the mechanical problem as you solve it.

### 13.4.7 Customers With Unresolvable Problems

Once in a while, you'll have a customer whose problem you cannot resolve. For example, the wiring may be too outdated to support the air conditioner they want installed, or the existing furnace may be inadequate to heat a newly expanded space. What should you do?

Here is a poor response: "Sorry, sir. Your fuses are too small to handle this air conditioner. I can't do electrical work."

A better response might be: "I found that your electrical service was too small to handle the new air conditioner. We can't do this kind of work, but here is the name of an electrical company we often deal with. Or, you can check the phone book for the names of other electricians. If you prefer, we can subcontract a licensed electrical contractor to do the job, then after the electrical service is fixed, we'll be glad to come back to install the air conditioner."

A typical homeowner might prefer to contact a company that is recommended, rather than searching out several competing companies and taking a chance on the quality of their work. Your positive, helpful attitude can leave a favorable impression, even if the customer already has an electrician. You have identified the problem and proposed a good solution, then offered to complete the job when the customer is ready.

### 13.4.8 Customers Who Request Help With Odd Jobs

Sometimes a customer asks a service technician to perform a task that is outside company regulations. In these situations, explain why you can't help. Then suggest another way to solve the problem.

*Exercise Question:*

While working on her furnace, Mrs. Oddjob asks you to move her freezer. What do you say?

*Exercise Solution:*

A good answer is: "I'm sorry, Mrs. Oddjob, but company regulations won't let me handle appliances we don't service. I know you're anxious to move that freezer. Perhaps a neighbor can help you. It wouldn't be fair to my later customers if I spent too much time at any one place. As you know, you're billed by the hour for my time. I know you want to keep your costs down." This response recognizes that the customer has a problem, but explains why you can't help her solve it.

**Summary**

Mechanical repairs account for a high percentage of HVAC work. Most of these repairs can be traced to improper installation procedures or failure to perform preventative maintenance. The HVAC technician must be able to properly install and maintain the mechanical components used in HVAC systems.

---

### Mr. Can't Help

Mr. Can't Help wants to use his existing furnace to heat the addition he just completed himself. He has asked you to check out the ducts and connect the furnace. You discover that the existing furnace is too small for his new heating needs. Your company does not sell furnaces. What do you say?

The service technician who helps a company acquire and retain happy customers practices good personal and work habits, behaviors, and attitudes. In addition, the technician who learns how to approach problems and work with customers perpetuates the company's positive image. The customer's impression of the technician, who represents the company at the job site, determines if the customer returns to the company for future service calls. Practicing good customer relations increases your value to your company and ensures that you always have customers to serve.

## Review Questions

1. The bolts and screws most likely used on assemblies that must have a finished appearance are called _____.
   a. machine bolts
   b. machine screws
   c. cap screws
   d. stud bolts

2. A jam nut is used _____.
   a. where frequent removal of the nut is required
   b. to lock a standard nut in place
   c. when appearance is important
   d. to lock a split washer in place

3. When installing a ½" Grade 8 steel bolt, it should be tightened to _____.
   a. 37 foot-pounds
   b. 60 foot-pounds
   c. 85 foot-pounds
   d. 92 foot-pounds

4. When installing gaskets at a flange, tighten the _____.
   a. flange bolts only a small amount at a time
   b. bolts one at a time to final torque, before tightening the next bolt in sequence
   c. bolts in the proper sequence
   d. flange bolts only a small amount at a time and in the proper sequence

5. Lip-type packings are used primarily _____.
   a. with shafts
   b. with hydraulic cylinders
   c. with valve stems
   d. as a packing gland

6. O-rings can be used _____.
   a. as a static seal
   b. as a dynamic seal
   c. as both static and dynamic seals
   d. in place of packings

7. A force acting either toward or away from and lengthwise along the axis of a shaft is called _____.
   a. an axial load
   b. thrust
   c. a radial load
   d. a journal load

8. A _____ bearing is a type of anti-friction bearing.
   a. sleeve
   b. ball
   c. thrust
   d. glide

9. When installing a bearing using the temperature method, you should heat the bearing to _____.
   a. the temperature recommended by the bearing manufacturer
   b. a minimum of 275°F
   c. a temperature lower than that recommended by the manufacturer, if the bearing has expanded enough to mount on the shaft
   d. a maximum of 275°F

10. The viscosity of oil refers to the _____.
    a. lowest temperature at which the oil will flow freely
    b. ability of the oil to flow at a specific temperature
    c. temperature at which the oil gives off ignitable vapors
    d. highest temperature at which the oil will flow freely

11. On initial movement, grease lubricates _____.
    a. by forming a liquid film that keeps the surfaces apart to reduce friction and heat
    b. when the oil in the grease flows away from the bulk of the grease
    c. when the surfaces begin to move and exert pressure on the grease
    d. when the surfaces begin to move and friction heats the grease

MAINTENANCE SKILLS FOR THE SERVICE TECHNICIAN — TRAINEE MODULE 03203

12. A V-belt is marked 3L405. Its size is _____.
    a. ¼" wide × ⅛" high × 405" long
    b. ⅜" wide × 7/32" high × 40½" long
    c. ½" wide × 5/16" high × 40½" long
    d. ⅝" wide × ⅜" high × 405" long

13. Flexible couplings _____.
    a. require precise alignment
    b. are generally easier to install than rigid couplings
    c. are used where smooth, even starts are needed
    d. are used where a major angular misalignment exists

14. When looking at the faces of a coupling from a side view, the faces are not square. To achieve correct alignment, one of the units must be _____.
    a. raised or lowered
    b. tilted on its base
    c. moved to one side or the other
    d. rotated on its base

15. Overlubrication of a bearing can cause _____.
    a. contaminants to enter the bearing
    b. a decrease in operating temperature
    c. an increase in viscosity
    d. no problems

16. A document that a technician fills out to indicate that a manufacturer's equipment has been put into service is called a _____.
    a. commissioning job report
    b. start-up job report
    c. warranty ticket
    d. service ticket

17. The customer begins forming a first impression of you and your company _____.
    a. after you've fixed his air conditioner
    b. after you've submitted your bill
    c. as soon as you appear at the door
    d. as soon as you speak

18. When speaking to a customer, always _____.
    a. expect that the customer understands technical jargon
    b. show courtesy, concern, and respect
    c. break the ice by chatting about local politics
    d. tell the customer if you think the unit is low-quality

19. Your customer is worried about furnace safety. You should _____.
    a. tell him to relax
    b. tell him you have everything under control
    c. calm him and work from facts to ease his mind
    d. call your supervisor

20. If you encounter a problem you can't fix because your company doesn't provide the type of service required, you should _____.
    a. tell the customer you fixed part of the problem
    b. call someone you know who can do the job
    c. help the customer find someone else who can do the job
    d. criticize the customer for wasting your time

# GLOSSARY

## Trade Terms Introduced in This Module

*Axial load:* An external load that acts lengthwise along a shaft.

*Break-away torque:* The torque required to loosen a fastener. This is usually lower than the torque to which the fastener has been tightened.

*Dropping point:* The temperature at which grease is soft enough for a drop of oil to fall away or flow from the bulk of the grease.

*Dynamic seal:* A seal made where there is movement between two mating parts, or between one of the parts and the seal.

*Fire point:* The temperature at which oil will burn if ignited.

*Flash point:* The temperature at which oil gives off ignitable vapors.

*Journal:* The part of a shaft, axle, spindle, etc., which is supported by and revolves in a bearing.

*Oxidation:* The process of combining with oxygen. All petroleum products react with oxygen to some degree, and this increases as the temperature increases.

*Packing gland:* A part used to compress the packing in a stuffing box.

*Pour point:* Refers to the lowest temperature at which an oil will flow freely.

*Radial load:* The side or radial force applied at right angles to a bearing and shaft.

*Run-down resistance:* The torque required to overcome the resistance of associated hardware, such as locknuts and lockwashers, when tightening a fastener.

*Set or seizure:* In the last stages of rotation in reaching a final torque, seizing or set of the fastener may occur. This is usually accompanied by a noticeable popping effect.

*Skive cut:* A cut made on an angle.

*Static seal:* A seal made where there is no movement between the two joining parts or between the seal and the mating part.

*Stuffing box:* The housing used to control leaking along a shaft or rod. Typically composed of three parts: the packing chamber (also called the *box*); the packing rings; and the gland follower (also called the *stuffing gland*).

*Thrust:* The force acting lengthwise along the axis of a shaft, either toward it or away from it.

*Tolerance:* The amount of variation allowed from a standard.

*Torque:* The resistance to a turning or twisting force.

*Viscosity:* The thickness of a liquid or its ability to flow at a specific temperature.

*Viscosity index (VI):* A measure of how an oil's viscosity varies with temperature.

# ANSWER KEY

## Answers to Review Questions

| Answer | Section |
|---|---|
| 1. c | 2.2.1 |
| 2. b | 2.2.4 |
| 3. d | 2.4.1 |
| 4. d | 3.2.0 |
| 5. b | 4.1.0 |
| 6. c | 5.1.1 |
| 7. b | 6.0.0 |
| 8. b | 6.1.0; 6.2.0 |
| 9. c | 6.4.2 |
| 10. b | 7.1.1 |
| 11. c | 7.2.0 |
| 12. b | 8.1.1 |
| 13. b | 9.1.2 |
| 14. b | 9.3.3 |
| 15. a | 10.1.0 |
| 16. c | 11.3.0 |
| 17. c | 12.2.0 |
| 18. b | 13.1.0 |
| 19. c | 13.4.1 |
| 20. c | 13.4.7 |

# REFERENCES & ACKNOWLEDGMENTS

## Additional Resources

This module is intended to present thorough resources for task training. The following reference works are suggested for further study. These are optional materials for continued education rather than for task training.

*Modern Refrigeration and Air Conditioning*, 2000. A.D. Althouse, C.H. Turnquist, A.F. Bracciano. Tinley Park, IL: The Goodheart-Willcox Company, Inc.

*Refrigeration & Air Conditioning Technology*, 2000. William C. Whitman, William M. Johnson, John A. Tomczyk. Albany, NY: Delmar Publishers, Inc.

## Figure Credits

| | |
|---|---|
| **Marson** | 203SA01 |
| **Thomas P. Burke** | 203SA02, 203F43, 203F42 |
| **Pruftechnik AG** | 203F29 |
| **Carrier Corporation** | 203SA03, 203F44 |
| **Gerald Shannon** | 203F41 |

# NCCER CRAFT TRAINING USER UPDATES

The NCCER makes every effort to keep these textbooks up-to-date and free of technical errors. We appreciate your help in this process. If you have an idea for improving this textbook, or if you find an error, a typographical mistake, or an inaccuracy in the NCCER's Craft Training textbooks, please write us, using this form or a photocopy. Be sure to include the exact module number, page number, a detailed description, and the correction, if applicable. Your input will be brought to the attention of the Technical Review Committee. Thank you for your assistance.

*Instructors* – If you found that additional materials were necessary in order to teach this module effectively, please let us know so that we may include them in the Equipment and Materials list in the Instructor's Guide.

**Write:** Curriculum Revision and Development Department
National Center for Construction Education and Research
P.O. Box 141104, Gainesville, FL 32614-1104

**Fax:** 352-334-0932

**E-mail:** curriculum@nccer.org

Craft _____ Module Name _____

Copyright Date _____ Module Number _____ Page Number(s) _____

Description
_____
_____
_____
_____

(Optional) Correction
_____
_____
_____

(Optional) Your Name and Address
_____
_____
_____

Module 03204-01

*Alternating Current*

## COURSE MAP

This course map shows all of the modules in the second level of the HVAC curriculum. The suggested training order begins at the bottom and proceeds up. Skill levels increase as you advance on the course map. The local Training Program Sponsor may adjust the training order.

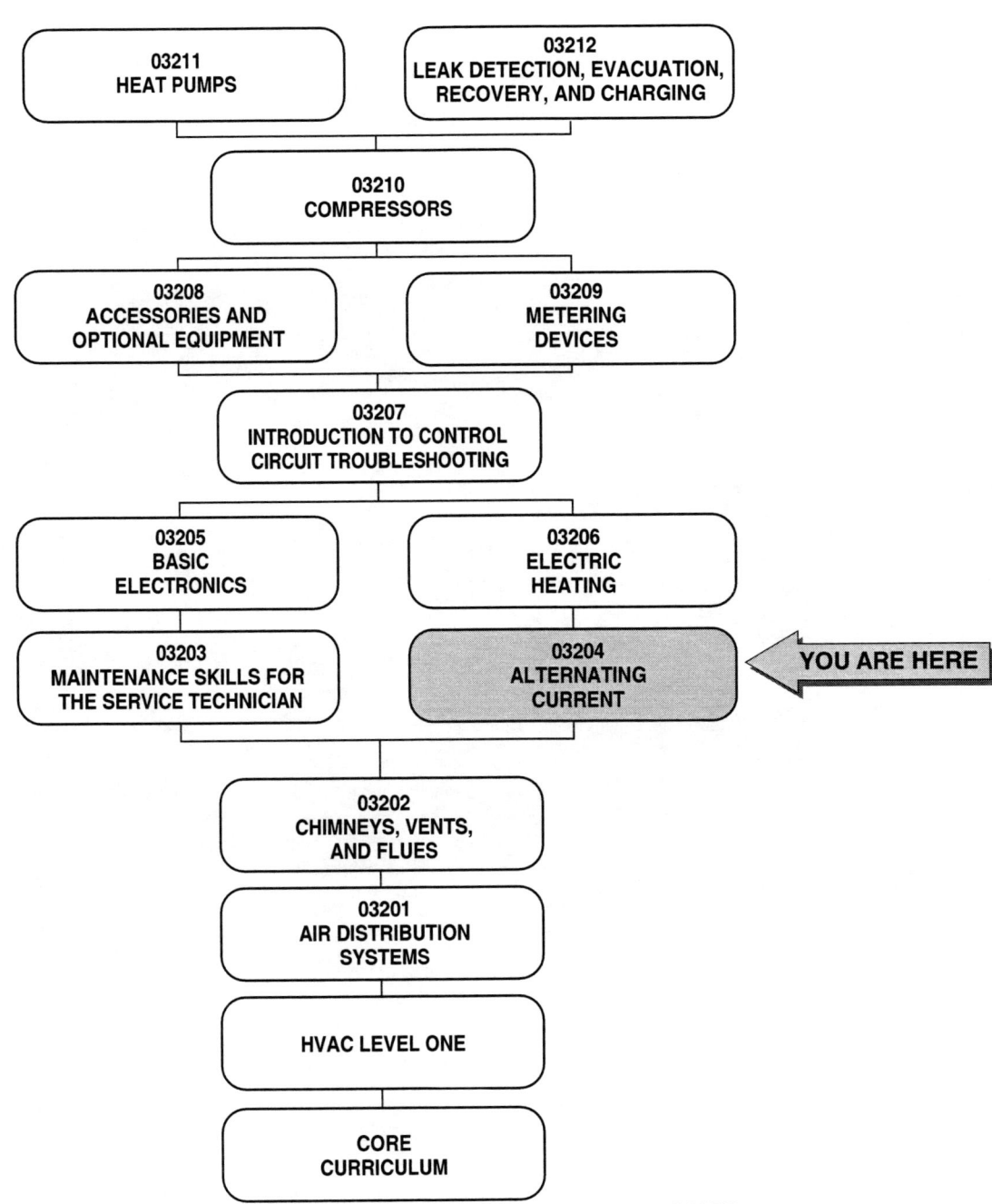

Copyright © 2001 National Center for Construction Education and Research, Gainesville, FL 32614-1104. All rights reserved. No part of this work may be reproduced in any form or by any means, including photocopying, without written permission of the publisher.

# MODULE 03204 CONTENTS

- 1.0.0 INTRODUCTION .................................................. 4.1
- 2.0.0 TRANSFORMERS ................................................. 4.2
  - 2.1.0 Isolation Transformers ................................... 4.3
  - 2.2.0 Autotransformers ......................................... 4.3
  - 2.3.0 Three-Phase Transformers ................................. 4.4
  - 2.4.0 Transformer Selection .................................... 4.5
- 3.0.0 POWER GENERATION ............................................. 4.5
  - 3.1.0 Sine Wave Generation ..................................... 4.6
  - 3.2.0 Frequency ................................................ 4.7
  - 3.3.0 Single-Phase Power ....................................... 4.9
  - 3.4.0 Three-Phase Power ........................................ 4.12
  - 3.5.0 Voltage and Current Imbalance in Three-Phase Systems ..... 4.15
- 4.0.0 USING AC POWER ............................................... 4.15
  - 4.1.0 Resistive Circuits ....................................... 4.15
  - 4.2.0 Inductive Circuits ....................................... 4.16
  - 4.3.0 Capacitors ............................................... 4.16
- 5.0.0 INDUCTION MOTORS ............................................. 4.17
  - 5.1.0 Single-Phase Motors ...................................... 4.18
  - *5.1.1 Split-Phase Motors* ..................................... 4.18
  - *5.1.2 Permanent Split Capacitor Motors* ....................... 4.18
  - *5.1.3 Capacitor Start Motors* ................................. 4.19
  - *5.1.4 Capacitor Start, Capacitor Run Motors* .................. 4.20
  - *5.1.5 Shaded-Pole Motors* ..................................... 4.20
  - *5.1.6 Multi-Speed Motors* ..................................... 4.20
  - 5.2.0 Three-Phase Motors ....................................... 4.21
- 6.0.0 TESTING AC COMPONENTS ........................................ 4.21
  - 6.1.0 Capacitor Analyzer ....................................... 4.21
  - 6.2.0 Wattmeter ................................................ 4.21
  - 6.3.0 Megohmmeter (Megger) ..................................... 4.22
  - *6.3.1 Safety Precautions* ..................................... 4.23
  - 6.4.0 Recording Instruments .................................... 4.24
  - 6.5.0 Checking Inductive Loads ................................. 4.24
  - 6.6.0 Checking Capacitors ...................................... 4.25
  - 6.7.0 Checking Fuses ........................................... 4.26
- 7.0.0 SAFETY ....................................................... 4.27
- 8.0.0 AC VOLTAGE ON CIRCUIT DIAGRAMS ............................... 4.28
- SUMMARY ............................................................ 4.29
- REVIEW QUESTIONS ................................................... 4.30
- GLOSSARY ........................................................... 4.32
- ANSWERS TO REVIEW QUESTIONS ........................................ 4.33
- REFERENCES & ACKNOWLEDGMENTS ....................................... 4.34

## Figures

| | | |
|---|---|---|
| Figure 1 | Typical alternating current sine wave | 4.1 |
| Figure 2 | Basic components of a transformer | 4.2 |
| Figure 3 | Multiple secondary windings | 4.3 |
| Figure 4 | Autotransformer | 4.3 |
| Figure 5 | Three-phase transformers | 4.5 |
| Figure 6 | DC generation | 4.5 |
| Figure 7 | AC generation | 4.6 |
| Figure 8 | Sine wave generation | 4.7 |
| Figure 9 | Sine wave curve | 4.8 |
| Figure 10 | Voltage frequency of 2Hz | 4.9 |
| Figure 11 | AC power distribution | 4.9 |
| Figure 12 | Edison hookup | 4.10 |
| Figure 13 | Service entrance panel | 4.10 |
| Figure 14 | 240V branch circuits | 4.11 |
| Figure 15 | 120V branch circuits | 4.11 |
| Figure 16 | Air conditioner branch circuits | 4.11 |
| Figure 17 | Three-phase voltage | 4.12 |
| Figure 18 | Four-wire closed delta | 4.13 |
| Figure 19 | Four-wire open delta | 4.13 |
| Figure 20 | Four-wire wye | 4.14 |
| Figure 21 | Three-phase compressor circuit | 4.14 |
| Figure 22 | Resistive circuit | 4.15 |
| Figure 23 | Inductive circuit | 4.16 |
| Figure 24 | Capacitor operation | 4.16 |
| Figure 25 | Phase relationships | 4.17 |
| Figure 26 | Basic parts of a motor | 4.18 |
| Figure 27 | AC motor basics | 4.18 |
| Figure 28 | Split-phase AC motor | 4.18 |
| Figure 29 | PSC motor | 4.19 |
| Figure 30 | Capacitor start motor with centrifugal switch | 4.19 |
| Figure 31 | Capacitor start motor with start relay | 4.19 |
| Figure 32 | Capacitor start, capacitor run motor | 4.20 |
| Figure 33 | Shaded-pole motor | 4.20 |
| Figure 34 | Speed taps | 4.20 |
| Figure 35 | Three-phase motors | 4.21 |
| Figure 36 | Capacitor analyzer | 4.21 |
| Figure 37 | Wattmeter schematic | 4.22 |
| Figure 38 | Megger schematic | 4.22 |
| Figure 39 | Digital readout megger | 4.23 |
| Figure 40 | Watt/VAR strip chart recorder | 4.24 |

## Figures (Continued)

Figure 41 Inductive load check ................................. 4.25
Figure 42 Capacitor check ..................................... 4.26
Figure 43 Capacitor discharging tool ........................... 4.26
Figure 44 Fuse checks ......................................... 4.26
Figure 45 High-voltage and low-voltage circuits ................ 4.28
Figure 46 Ladder diagram ...................................... 4.29

## Tables

Table 1 Current Effects on the Human Body ............ 4.27

# MODULE 03204

# Alternating Current

## Objectives

When you have completed this module, you will be able to do the following:

1. Describe the operation of various types of transformers.
2. Explain how alternating current is developed and draw a sine wave.
3. Identify single-phase and three-phase wiring arrangements.
4. Explain how phase shift occurs in inductors and capacitors.
5. Describe the types of capacitors and their applications.
6. Explain the operation of single-phase and three-phase induction motors.
7. Identify the various types of single-phase motors and their applications.
8. Use a wattmeter, megger, capacitor analyzer, and chart recorder.
9. Test inductors and capacitors using an ohmmeter.
10. State and demonstrate the safety precautions that must be followed when working with electrical equipment.

## Prerequisites

Before you begin this module, it is recommended that you successfully complete the following modules: Core Curriculum; HVAC Level One; HVAC Level Two, Modules 03201 and 03202.

## Required Trainee Materials

1. Pencil and Paper
2. Appropriate Personal Protective Equipment

## 1.0.0 ♦ INTRODUCTION

The public utilities that provide us with electricity produce AC voltage. DC voltage, when needed in AC-powered systems, is obtained by converting the available AC to DC. Batteries are also a common source of DC voltage.

Alternating current is defined as current that flows first in one direction, then in the opposite direction. The direction of current flow reverses at established intervals.

The number of cycles that occur each second is known as the **frequency**; the standard frequency for AC systems in the United States is 60 **Hertz (Hz)**, or 60 cycles per second. A single cycle of AC voltage looks like the **sinusoidal (sine) wave** shown in *Figure 1*.

The transformer is the key component in an AC-powered system. Therefore, before we get into a discussion of AC power generation and distribution, we will first discuss transformers.

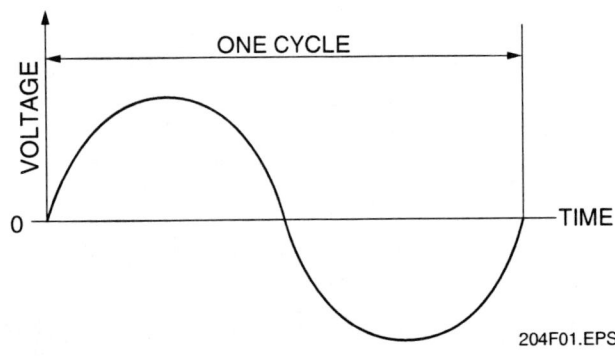

*Figure 1* ♦ Typical alternating current sine wave.

## Current Wars

After inventing the electric lightbulb in 1879, Thomas Edison began work on a system for delivering electricity to homes and businesses. His system relied on direct current (DC)—electric current that always flows in one direction. However, DC transmission over long distances proved impractical. Transmitting direct current at the low voltages useful for lighting or motor operation required the use of thick, expensive copper wire. In fact, the service areas of Edison's DC generating stations were limited to about a square mile and mainly served the downtown areas of large cities. While Edison was pioneering DC distribution, electrical engineers in Europe were experimenting with alternating current (AC), which reverses direction at regular intervals.

The American businessman George Westinghouse saw the value in AC. High-voltage AC power could be distributed over longer distances using thinner, less expensive copper wires. Experts theorized that special devices (transformers) could step the voltage level up and down. Increasing the voltage level would allow it to be distributed across a wider region, while reducing the voltage level would enable the current to be used safely in homes and shops. Westinghouse hired a young electrical engineer named William Stanley, Jr., who developed the first effective transformer and demonstrated the first AC lighting system. Around the same time, the inventor Nikola Tesla filed patents for other devices run by AC. After Westinghouse bought these patents, a full-scale industrial war known as the *current wars* erupted. At stake was whether Edison's direct current or Westinghouse's alternating current would electrify America. Edison claimed that alternating current was extremely dangerous and called for outlawing the high voltages transmitted by AC. Westinghouse countered that transformers safely reduced AC voltages before they entered buildings.

Within ten years, the value of the alternating current system had been convincingly demonstrated. AC proved to be more practical and economical. Eventually, even Thomas Edison was forced to admit he had been wrong, and General Electric, the company he founded, began building and installing high-voltage AC transmission systems.

## 2.0.0 ♦ TRANSFORMERS

As you learned earlier, a transformer generally consists of two or more coils of wire wound around a laminated iron core. One winding is connected to the power source and is called the *primary winding*. A separate winding is placed around the opposite side of the iron core and is called the *secondary winding* (*Figure 2*).

There is no physical electrical connection between the primary and secondary windings; electrical energy is transferred from the primary winding to the secondary winding by **induction**.

Transformers have no moving parts and therefore require very little maintenance. In operation, the primary circuit draws power from the source and the secondary circuit delivers the power to the load device. The power transferred from the primary winding to the secondary winding depends on the current flowing in the primary circuit. The amount of AC voltage induced in the secondary winding is directly related to the number of turns of wire in the secondary side as compared to the primary.

The output voltage of a transformer can be calculated by comparing the number of turns of wire in the primary winding to the number of turns in the secondary winding. If the number of turns in the secondary winding is twice that in the primary winding, its output voltage is twice that of the primary, and it is a step-up transformer. If the secondary winding has half the number of turns of the primary, then its output voltage is half that of the primary voltage, and it is a step-down transformer. This relationship is stated by the formula:

$$\frac{E_p}{E_s} = \frac{N_p}{N_s}$$

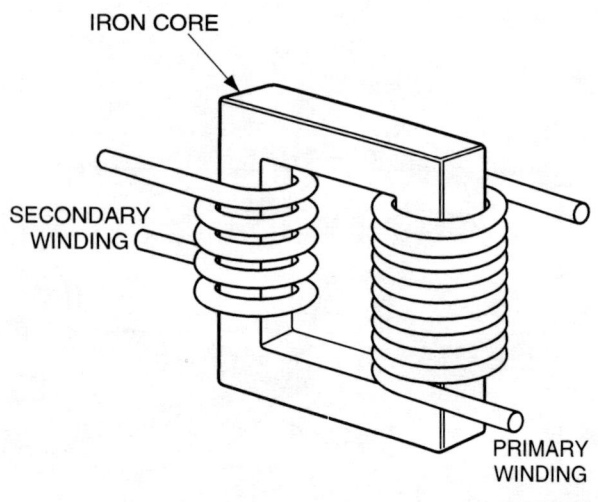

*Figure 2* ♦ Basic components of a transformer.

*Where:*

E = voltage
N = number of turns
P = primary winding
S = secondary winding

The relationship between the primary and secondary windings is called the **turns ratio**. It can be used to calculate unknown values in both step-up and step-down transformers. Some transformers contain more than one secondary winding, as shown in *Figure 3(A)*. They are designed so that more than one voltage can be obtained from the secondary. Others have multi-tap primary windings, allowing them to be used with different levels of input voltage, as shown in *Figure 3(B)*.

The value of current changes inversely with voltage. For example, if the voltage across a step-up transformer is doubled, the current is halved. Sometimes, a replaceable fuse, **fusible link**, or manually resettable external circuit breaker is used to protect the secondary winding from excessive current.

## 2.1.0 Isolation Transformers

Transformers with a one-to-one ratio are designed to deliver an output voltage equal to the input voltage. This special type of transformer is intended either to provide safety to a user or to minimize interference in certain electronic equipment. These transformers are known as **isolation transformers**.

## 2.2.0 Autotransformers

The **autotransformer** is another special type of transformer. It consists of a single, continuous winding that is tapped to provide the necessary step-up or step-down function. When it is used as a step-up transformer, the entire primary winding is a part of the secondary winding (*Figure 4*).

**(A) MULTIPLE-TAP SECONDARY WINDING**

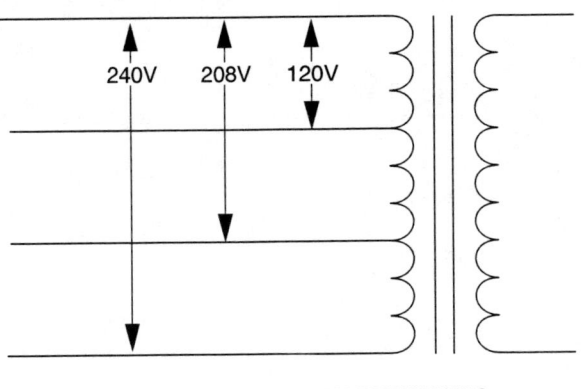

**(B) MULTIPLE-TAP PRIMARY WINDING**

*Figure 3* ◆ Multiple secondary windings.

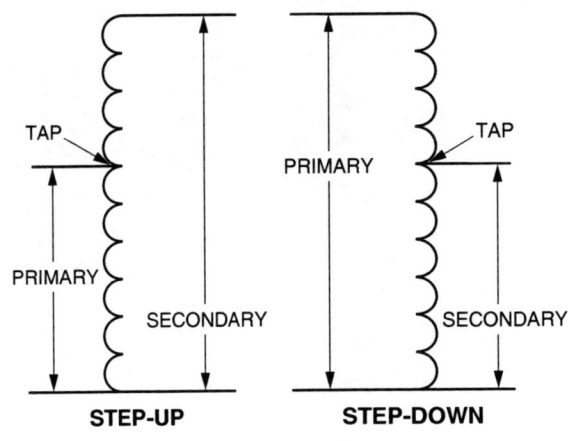

*Figure 4* ◆ Autotransformer.

---

### Step-Down Transformers

Step-down transformers are widely used in HVAC equipment to step down the unit's supply voltage to 24 volts. This low voltage is used in the control circuits of residential and light commercial HVAC equipment.

## Phasing of Transformers

It is sometimes necessary to connect two transformers in parallel to achieve an increased VA rating, in order to carry a specific load. When doing so, it is necessary that the two transformers be identical. Also, the secondary windings of the two transformers must be connected such that their output voltages are in phase; otherwise, the transformers can be damaged or destroyed. Some transformers are color-coded or numbered to indicate phasing. Others are not. A procedure for determining if the transformers are properly phased is described here.

*Step 1*  Connect the primary leads of the two transformers in parallel.

*Step 2*  Connect the secondary lead from one of the transformers to either secondary lead of the second transformer.

*Step 3*  Turn on the power, then use a multimeter to measure the AC voltage between the remaining two open secondary leads.

*Step 4*  If the meter reads zero volts, the connection is correct. Turn off the power and connect the two secondary leads.

*Step 5*  If the meter reads twice the rated voltage, turn off the power, then reverse the secondary leads.

Repeat the test. The meter should now read zero volts, indicating proper phasing.

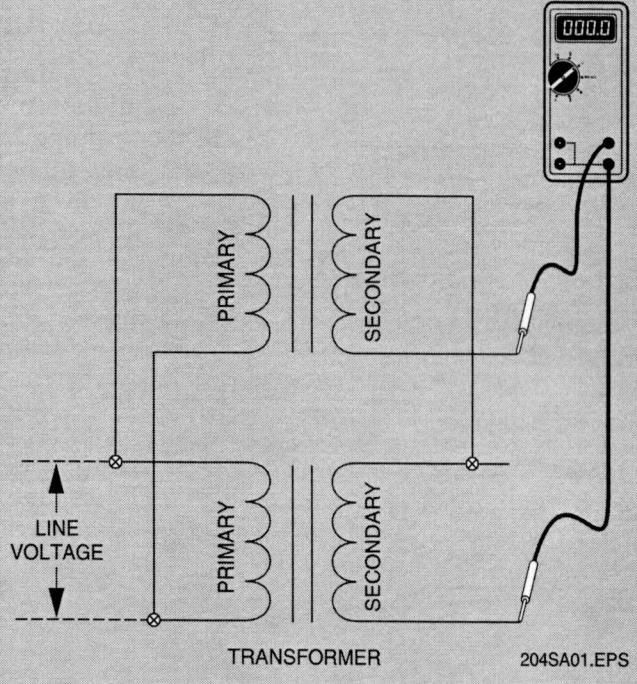

When used as a step-down transformer, the entire secondary winding is part of the primary winding. The major disadvantage of the autotransformer is that it is less safe because the secondary winding is not isolated from the primary winding. Under certain conditions, this situation can cause electrical shock.

### 2.3.0 Three-Phase Transformers

The types of transformers we have discussed previously are designed to work with single-phase voltage. With larger systems, such as those used in commercial buildings and factories, three-phase power is usually required. This requires special transformers.

*Figure 5* shows two types of three-phase transformers. In these transformers, three legs of AC voltage are received on the primary windings, and equal voltage is induced into each of the secondary windings.

Three-phase power distribution arrangements and output voltages are discussed in detail later in this module.

Careful consideration should be given to the selection of transformers that are used to power a low-voltage control system. Inductive devices such as contactors, relays, solenoid valves, and motors require more power on start-up than during steady operation. Thus, the transformer must be able to handle a start-up current surge.

## 3.0.0 ◆ POWER GENERATION

Most electricity is produced by mechanically driven generators. The energy is generated by passing coils of wire through a magnetic field. Generators can be designed to produce DC or AC.

The principles of the direct current, mechanically driven generator are illustrated in *Figure 6*. A simple horseshoe magnet forms a magnetic field as magnetic lines of force travel from one pole to the other. A loop of wire supported on a shaft is rotated within this field. As the loop turns, an electric current is generated and flows to terminals (segments) on the **commutator**. Two stationary contacts, or brushes, touch these commutator segments as the loop of wire rotates. Current generated in the loop is carried by these brushes to an external circuit, where it is used as electrical energy.

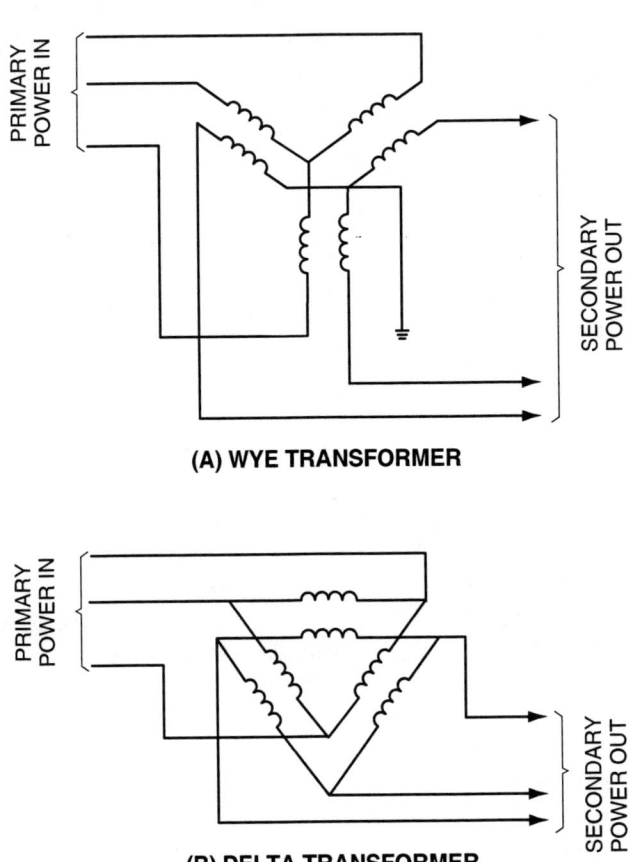

*Figure 5* ◆ Three-phase transformers.

### 2.4.0 Transformer Selection

Transformers are rated or sized according to the amount of power the secondary winding or circuit can handle. This power capability is expressed in VA (voltage times amperage). For example, if the secondary voltage of a transformer is 24V, and the amperage capacity of the circuit is 2A, the VA rating of that particular transformer is 48VA (24V × 2A = 48VA). Thus, the operating capability of that transformer is 48VA, which means that it should not be replaced by a transformer with a smaller VA rating. It can, however, be replaced by one with a larger VA rating.

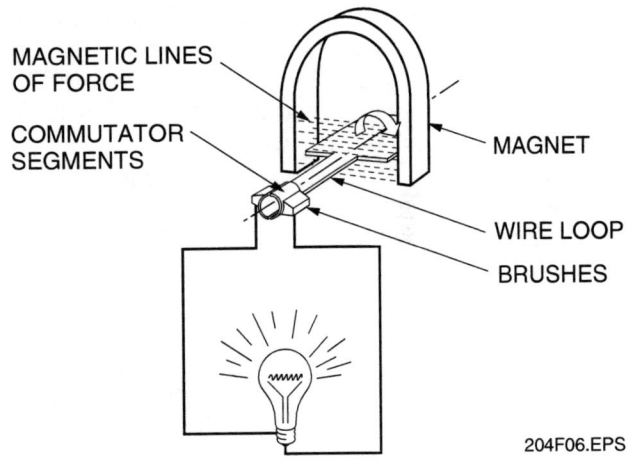

*Figure 6* ◆ DC generation.

---

### Low-Voltage Transformer Load

A furnace is equipped with a 40VA low-voltage transformer. You have determined that the maximum existing load on the transformer is 30VA and this occurs during cooling operation. During heating operation, the load on the transformer is much lower. You wish to add a 24V humidifier that uses a solenoid rated at 12VA. Does the existing 40VA low-voltage transformer need to be replaced with one of a higher VA rating to accommodate the addition of the humidifier?

---

ALTERNATING CURRENT — TRAINEE MODULE 03204

The commutator is in two halves. When the loop of wire rotates one-half turn, each brush makes contact with the other half of the commutator. This is necessary to keep the current flowing in the same direction in the circuit; thus direct current is generated.

An alternating current generator, frequently called an **alternator**, operates on much the same principle as the DC generator (*Figure 7*). The rotation of the conductor through magnetic lines of force generates current in the conductor. Instead of being connected to segments on a commutator, the conductor ends are each connected to slip rings. The slip rings are fastened to the shaft and rotate with it. Stationary brushes are in contact with the slip rings. Current is carried through the brushes to the load.

During one revolution of the conductor, each side passes through the magnetic lines of force, first in one direction and then in the other. Therefore, current flows first in one direction and then the other, thus producing alternating current.

## 3.1.0 Sine Wave Generation

*Figure 8* shows how the sine wave is generated. Voltage is induced as long as the conductor (**armature**) is moving through the magnetic field. The amount of voltage induced depends on the strength of the magnetic field and the angle and speed at which the conductor cuts the lines of force.

In *Figure 8*, the armature has been divided into a dark half and a light half to make our discussion easier. (No load is shown in this illustration.) In part A, the armature loop is moving parallel to the lines of force. The armature is not cutting through any lines of force, so minimum voltage is induced.

As the armature rotates toward the position shown in part B, it cuts more and more lines of force per second, as an increasingly strong voltage is induced. When it reaches the position shown in part B, the maximum voltage is attained because during its rotation, the armature has then cut through the maximum number of lines of force per second.

As the rotation proceeds toward the position shown in part C, the number of lines of force cut per second is reduced. The induced voltage decreases from its maximum value to zero voltage when the armature again reaches this position.

At this point in time (part C), the armature has rotated halfway around. The first half of the curve in *Figure 8* shows the varying strength of the voltage during this half turn. Part A, or the beginning position of the armature, can be labeled 0°. When the armature has reached the C position, it has turned through 180 angular degrees. The voltage generated during that time is one alternation.

As the armature continues moving, it rotates through another 180°, back to its starting point and again one alternation of voltage is generated. Two alternations are equal to 360° of rotation, or one complete rotation.

In this basic generator scheme, with one north pole and one south pole affecting the armature, one complete rotation will always be required in order to generate one pair of alternations.

*Figure 8* also shows the appearance of the AC voltage produced by this kind of generator. The curve shown is a sine wave. One sine wave corresponds to two alternations. With the two-pole generator just discussed, one sine wave is generated by a complete 360° rotation of an armature.

The direction of the voltage induced in this generator (known as the *polarity* of the voltage) is reversed every alternation (180°). This current flows through the load first in one direction, then in the other direction. The top half of the sine wave curve represents flow in one direction. The bottom half of the curve, lying under the horizontal axis, represents the flow of voltage in the opposite direction.

Because the amplitude of the voltage waveform varies between zero and peak, the amount of voltage available to do work—the **effective voltage**—is less than the peak voltage. Effective voltage, also known as **root-mean-square (RMS) voltage**, is .707 times the peak voltage. Therefore, a peak of 169.71 volts produces a working voltage of 120 volts.

$$169.71 \times .707 = 120$$

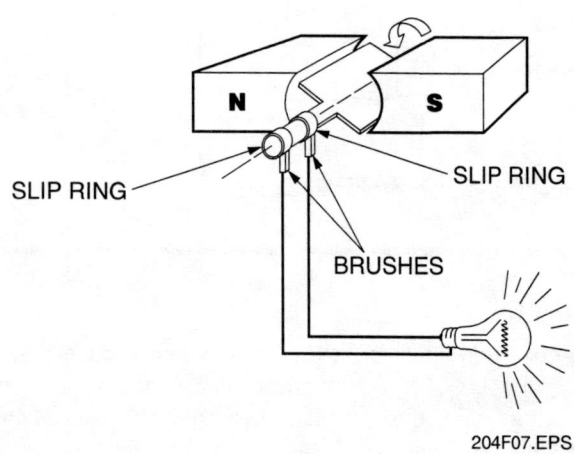

*Figure 7* ◆ AC generation.

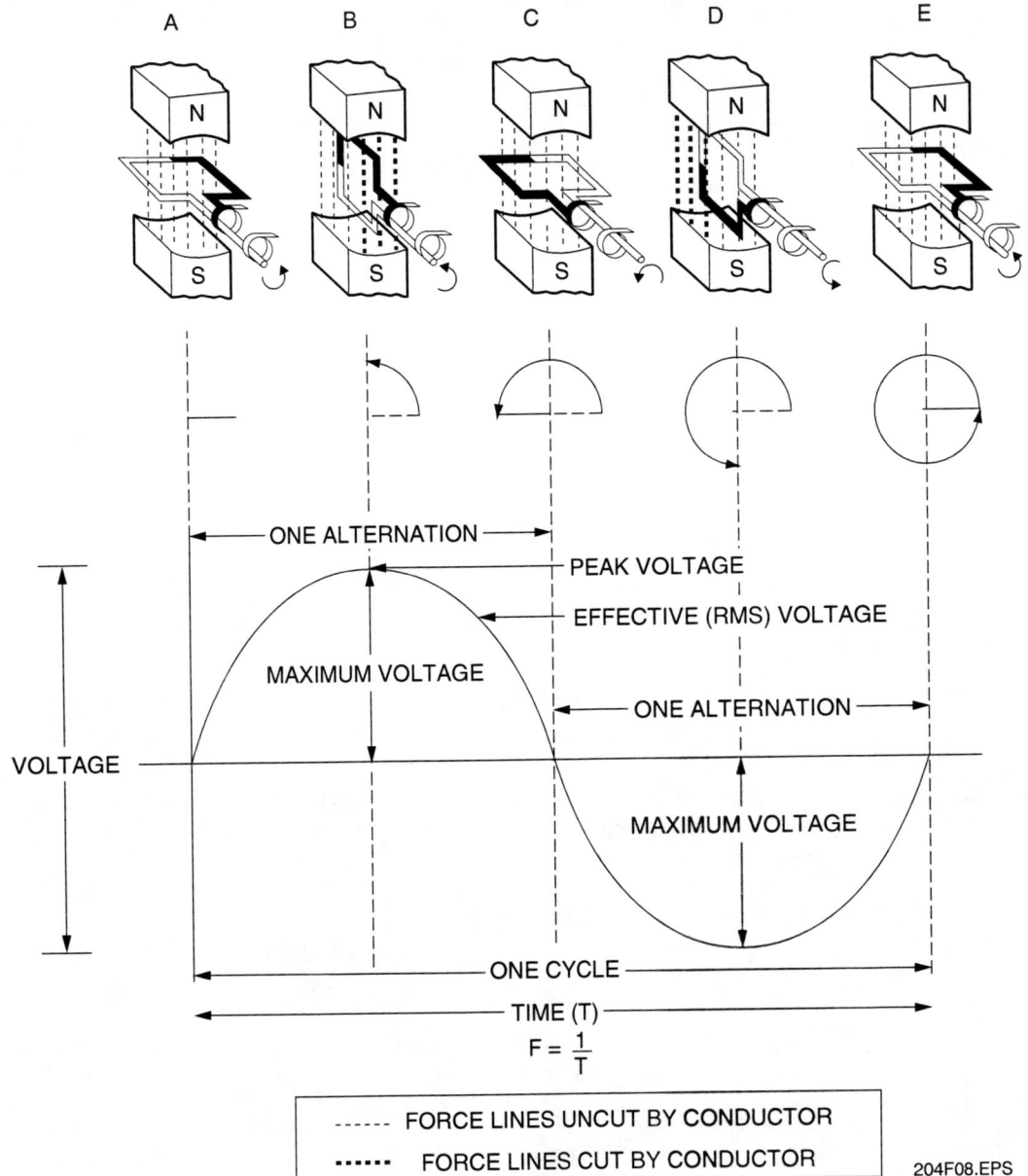

*Figure 8* ◆ Sine wave generation.

If you know the effective voltage and want to determine the peak voltage, multiply the effective voltage by 1.414, which is the square root of 2.

## 3.2.0 Frequency

The speed at which an armature rotates affects the frequency. If the speed of the armature increases, so does the frequency. If a single loop rotates in a two-pole field (produced by one north pole and one south pole), the current flows once in each direction to complete one pair of alternations, or one cycle, as the armature turns through 360° (*Figure 9*). If there is one rotation per second, the frequency will be 1Hz. If the armature makes two complete rotations in one second, the frequency will be 2Hz (*Figure 10*). For three rotations per second, the frequency will be 3Hz, etc.

The more pairs of poles are added to a generator, the higher the frequency will become, given the same armature speed. Generators that operate at low speeds need more pairs of poles than high-speed generators to provide voltage at the same frequency.

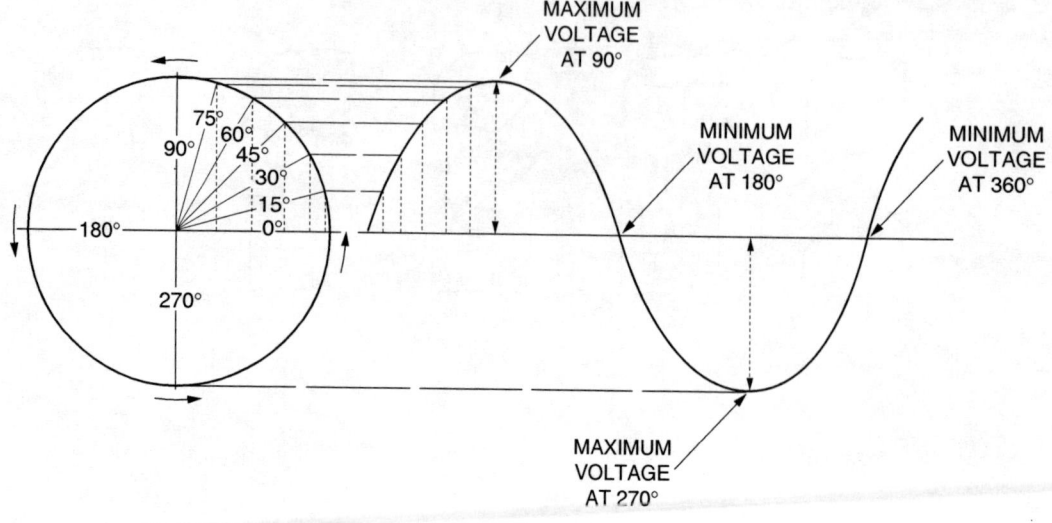

*Figure 9* ◆ Sine wave curve.

## Effective Voltage

In HVAC service work, digital multimeters (DMM) like the one shown here are widely used by technicians to measure effective voltage. Unless stated otherwise, the voltages stamped on the nameplates of HVAC equipment refer to effective voltages.

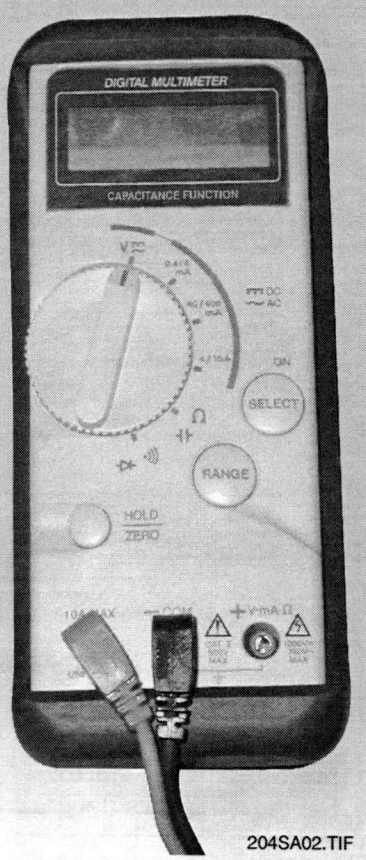

**4.8**          HVAC LEVEL TWO — TRAINEE MODULE 03204

### Frequency

In the United States, commercial power is usually supplied at a frequency of 60Hz. How many cycles occur during each second at a frequency of 60Hz, and what is the time for each cycle?

### 3.3.0 Single-Phase Power

As discussed in your HVAC Level One training, power generated at the power station is transmitted as a very high AC voltage, then stepped down to usable levels using transformers (*Figure 11*). The voltage received at the service entrance panel of a residence is usually about 240 volts.

An Edison hookup (*Figure 12*), is a common wiring arrangement for a power transformer that delivers power to a residence. The neutral line is connected to the building ground, which may be a copper pipe that runs underground and/or a copper rod that has been driven into the ground.

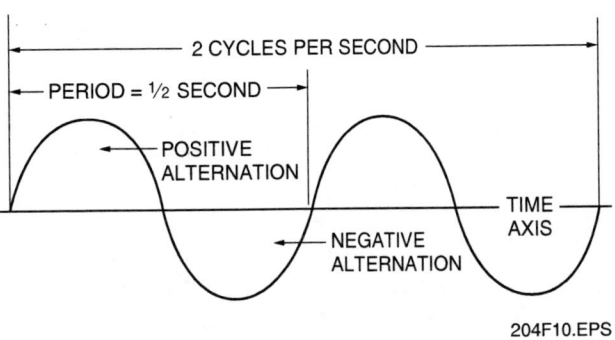

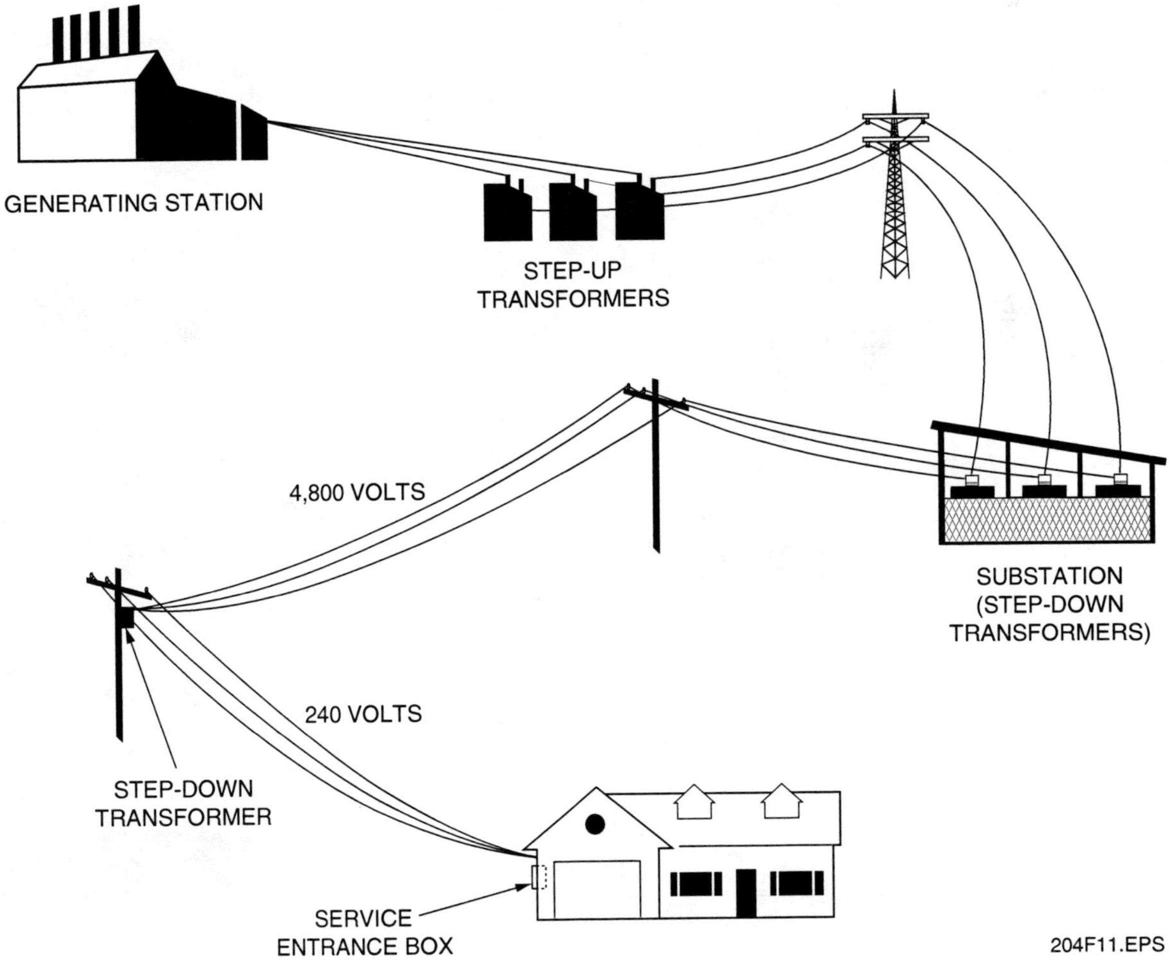

*Figure 10* ◆ Voltage frequency of 2Hz.

*Figure 11* ◆ AC power distribution.

ALTERNATING CURRENT — TRAINEE MODULE 03204

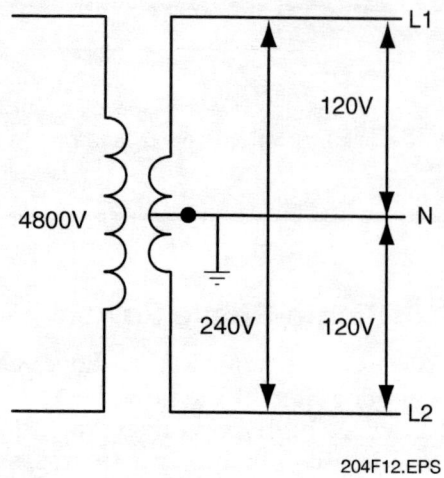

*Figure 12* ◆ Edison hookup.

After being stepped down, the power enters the structure through the service entrance (*Figure 13*). The service entrance is connected to a main fuse box or circuit breaker box. The circuits are broken down into a series of branch circuits within the main entrance box. The 240V branch circuits (*Figure 14*) serve major appliances.

The branch circuits are fused in both legs with fuses rated for the current draw of the appliance.

The wire connecting these circuits to the appliance must be heavy enough to carry the load as determined by the size of the fuse or circuit breaker.

The 120V branch circuits (*Figure 15*) are also broken down into fused connections. Each branch of the 120V circuit is fused or protected in the hot leg only. The 120V branches are taken from each side of the 240V legs in order to balance the loads as closely as possible. Local electrical codes govern the area served by each branch.

The two hot legs (L) are commonly termed L1 and L2. The secondary winding of the transformer is center tapped and this leg is called the *neutral (N)*. If the voltage is measured between L1 and L2, it would be found to be 240, whereas the voltage between either L1 or L2 and N would be 120. The neutral is electrically grounded at the power transformer. Other grounds may exist at the electric meter and the main circuit panel base, depending upon local codes.

*Figure 16* shows how the voltage from a 240V branch circuit might be distributed to an air conditioning compressor and its control circuit.

The *National Electrical Code* requires that a power disconnect be installed on or within sight of the unit. The disconnect has either a manual switch that can be locked out, or a removable fuse plug that disconnects power when pulled out of the unit.

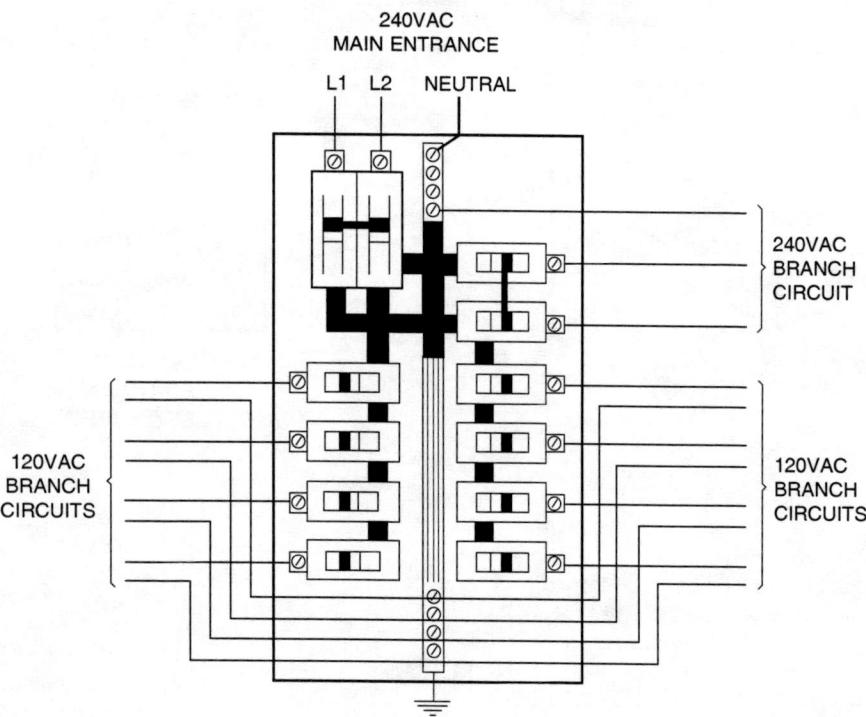

*Figure 13* ◆ Service entrance panel.

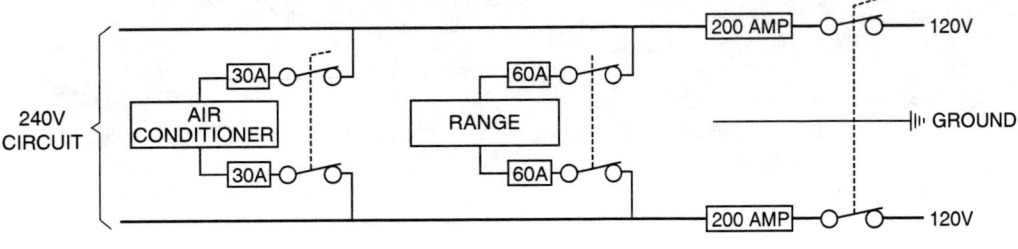

*Figure 14* ◆ 240V branch circuits.

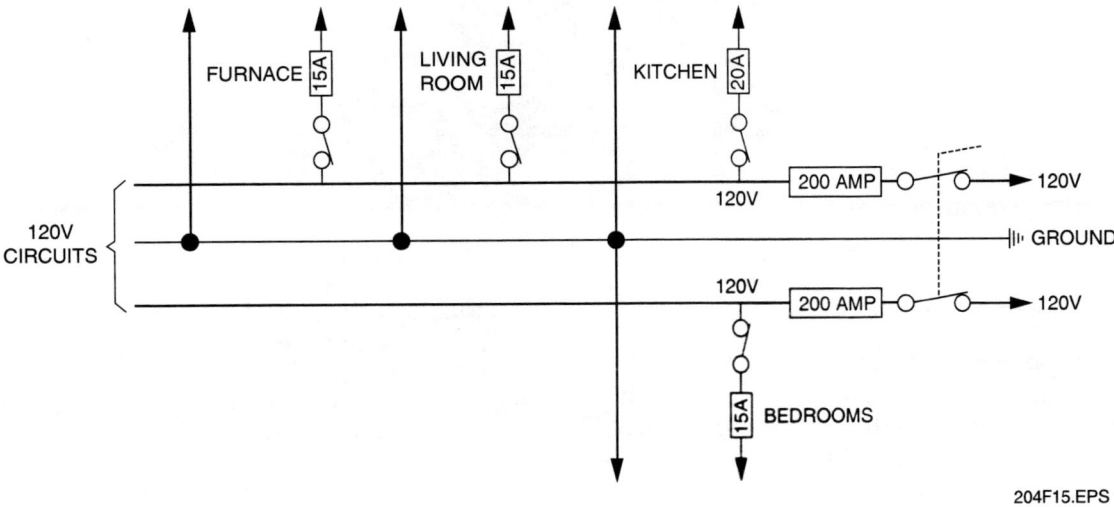

*Figure 15* ◆ 120V branch circuits.

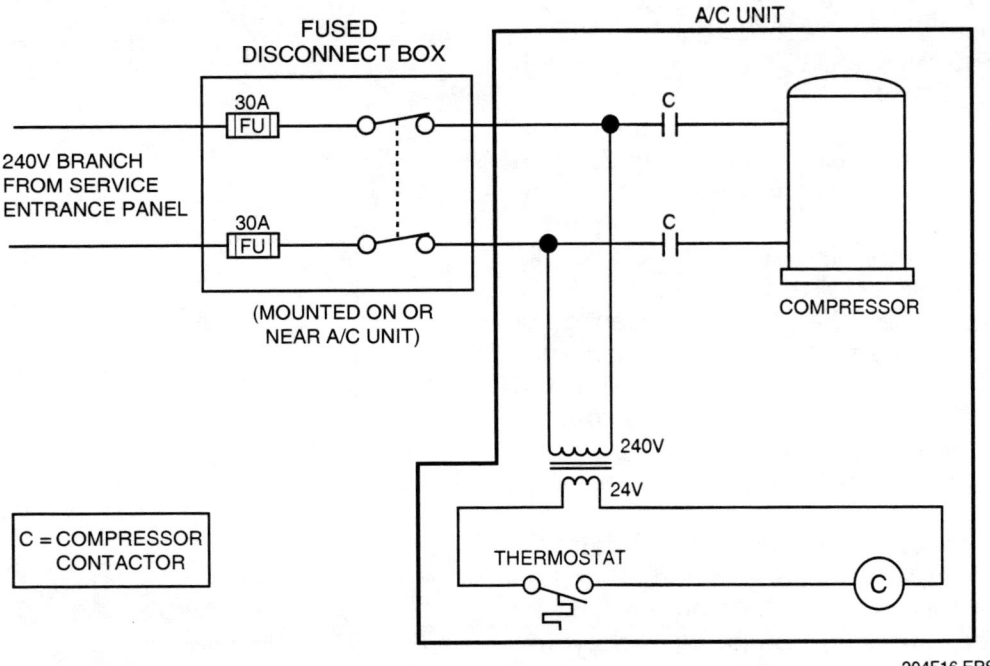

*Figure 16* ◆ Air conditioner branch circuits.

ALTERNATING CURRENT — TRAINEE MODULE 03204

## Disconnect Switches

The *National Electrical Code* requires that disconnect switches used to protect HVAC equipment be installed within sight of the unit and be readily accessible.

### 3.4.0 Three-Phase Power

Single-phase power is adequate to supply residences and small commercial businesses. In commercial and industrial uses, the power demand is greater, especially where large electric motors are used. Motors larger than one horsepower are usually three-phase motors.

The discussion of power generation has focused on a single conductor rotating in a magnetic field. If three rotating conductors are placed 120° apart, three equal voltages are generated. As shown in *Figure 17*, they occur 120° apart in time; in other words, they are 120° out-of-phase with one another.

There are several ways in which three-phase power sources can be connected, depending on the voltage(s) and the amount of current required. The four-wire closed delta arrangement (*Figure 18*) uses three transformers to provide 120V single-phase, 240V three-phase, and 208V single-phase power. The four-wire open delta (*Figure 19*) uses two single-phase transformers to produce 240V three-phase power, and 120V, 240V, and 208V single-phase power. This connection method is not as stable as the closed delta; thus it is more difficult to keep the current in the three legs balanced.

In a three-phase system, a current imbalance in one leg can cause overheating in the other two legs. Three-phase systems therefore need to be checked periodically to make sure they are balanced. If the current is out of balance by more than 10% or the voltage is out of balance by more than 2%, the imbalance must be corrected. Sometimes the problem is at the source and must be corrected by the power company.

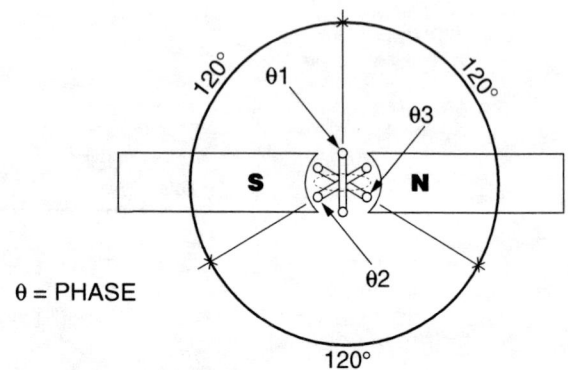

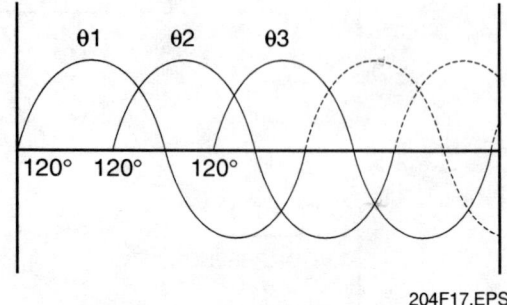

*Figure 17* ♦ Three-phase voltage.

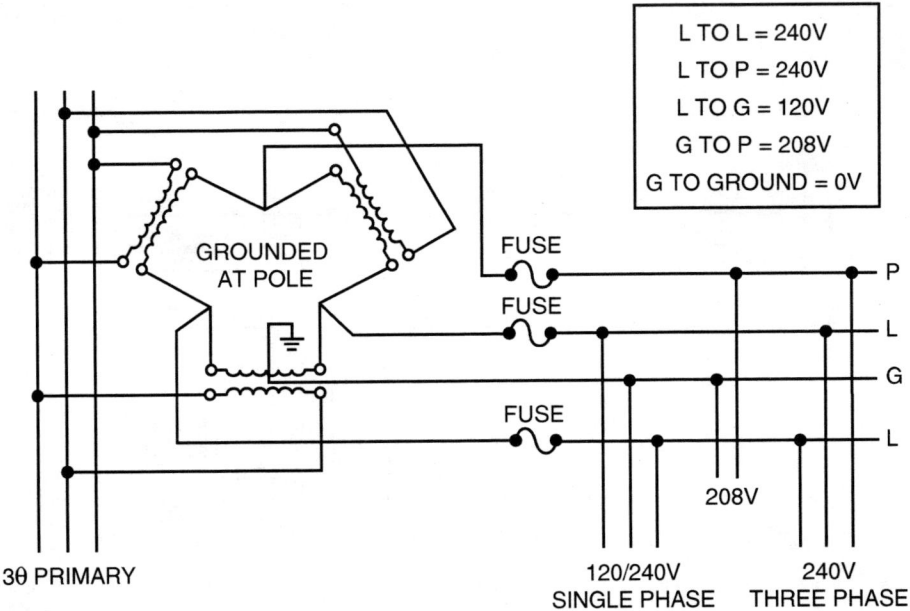

*Figure 18* ◆ Four-wire closed delta.

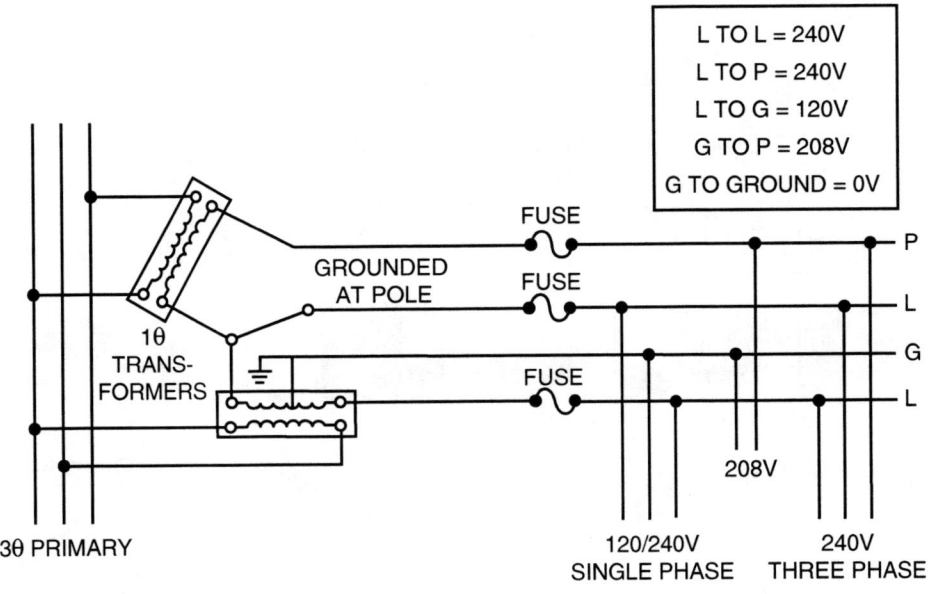

*Figure 19* ◆ Four-wire open delta.

A four-wire wye system (*Figure 20*) may be used to supply power for industrial use where high voltage is required to run large machines. This arrangement produces 480V three-phase power as well as the 277V single-phase power required for fluorescent lighting. *Figure 21* shows a wye-connected source supplying power to a three-phase compressor circuit. In air conditioning systems, single-phase compressor motors are used only up to a cooling capacity of about 60,000 Btuh. Three-phase compressor motors are used for large-capacity systems.

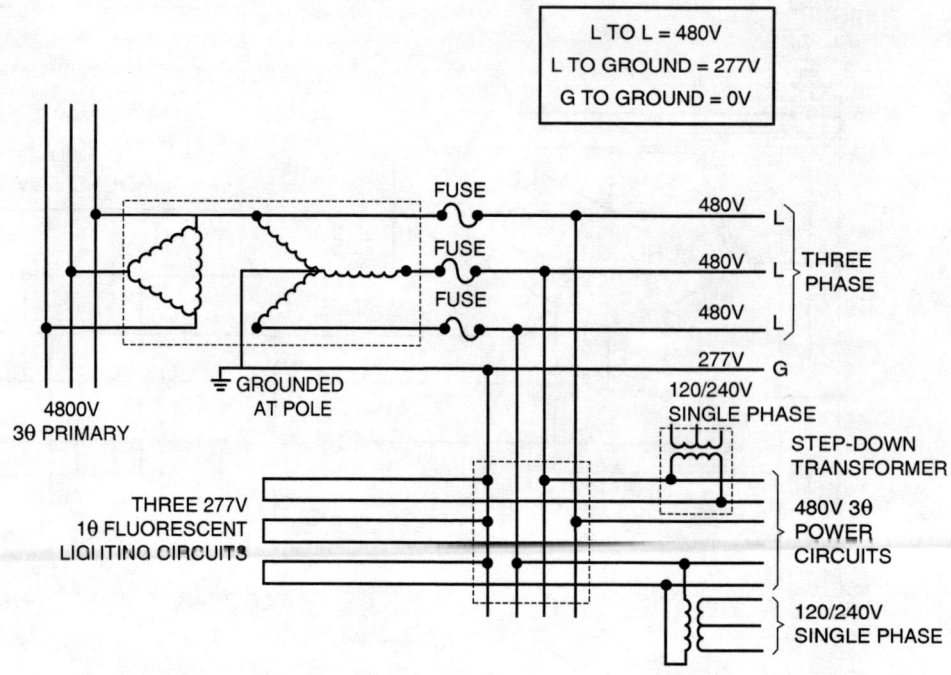

*Figure 20* ◆ Four-wire wye.

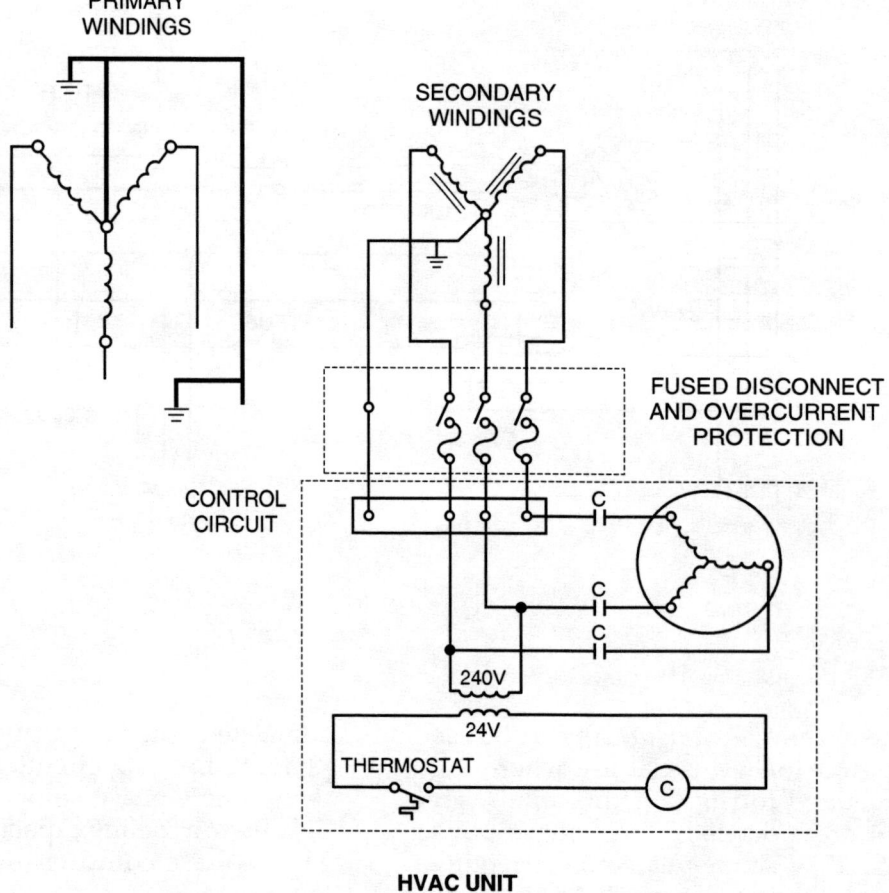

*Figure 21* ◆ Three-phase compressor circuit.

## 3.5.0 Voltage and Current Imbalance in Three-Phase Systems

Voltage imbalance is very important when working with three-phase equipment. A small imbalance in the phase-to-phase voltages can result in a much greater current imbalance. With a current imbalance, the heat generated in motor windings and other inductive loads will be increased. Both current and heat can cause nuisance overload trips and may cause motor/equipment failures. For this reason, the voltage imbalance between any two legs of the voltage applied to a three-phase motor or system should not exceed 2%. If a voltage imbalance of more than 2% exists at the input to the equipment, the problem in the building or utility power distribution system should be corrected before operating the equipment.

Current imbalance between any two legs of a three-phase system should not exceed 10%. A current imbalance may occur without a voltage imbalance. This can happen when an electrical terminal, contact, etc. becomes loose or corroded, causing a high resistance in the leg. Since current follows the path of least resistance, the current in the other two legs will increase, causing more heat to be generated in the devices supplied by those legs.

Procedures for determining the voltage and current imbalance in a three-phase system are covered in detail in the HVAC Level Two Module, *Introduction to Control Circuit Troubleshooting*.

## 4.0.0 ◆ USING AC POWER

One important characteristic that distinguishes AC from DC is that AC acts differently in different kinds of circuits. HVAC components (such as motors) take advantage of this characteristic.

### 4.1.0 Resistive Circuits

Heating elements in an electric furnace are examples of resistive loads. In a resistive circuit, the current waveform occurs in phase with the voltage waveform (*Figure 22*). The amount of power consumed by the load is determined by the formula: P (power) = E (voltage) × I (current).

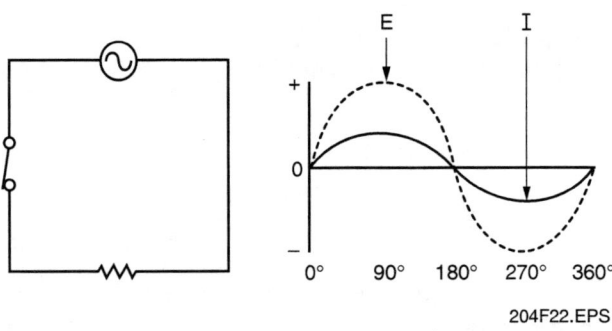

*Figure 22* ◆ Resistive circuit.

---

### Resistive Electric Heating Elements

Resistive electric heating elements like the one shown here are typical of those used in HVAC equipment.

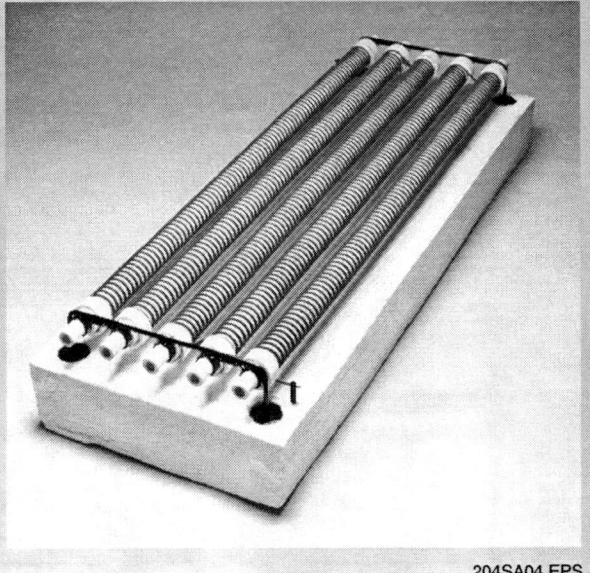

All electrical loads generate some heat. Electric heaters, stove burners, and clothes dryer elements take advantage of this fact. One unit of measure for heat is the British thermal unit (Btu). One watt represents 3.414 Btus. A heater rated at 1,000 watts produces 3,414 Btus. It is important to understand this principle because the heating load of a building is usually stated in Btus, while the capacity of a heater may be stated in watts.

### 4.2.0 Inductive Circuits

When alternating current flows through a coil of wire such as a motor winding, the magnetic field produced by one turn of the wire induces a voltage in adjacent turns. The voltage induced in this manner is opposite in polarity to the applied voltage, and therefore opposes current flow. For that reason, the current through a coil lags the voltage. In a purely inductive circuit, the current waveform will lag the voltage waveform by 90° (*Figure 23*).

### 4.3.0 Capacitors

A **capacitor** (*Figure 24*) is an electrical storage device that charges and discharges as the applied voltage changes. It consists of two metal plates separated by an insulating material known as a **dielectric**. Its capacity, which is measured in **microfarads**, is determined by the size of the plates, the distance between the plates, and the type of dielectric material used.

Current will not flow through a dielectric material. When an AC voltage is applied across the plates of the capacitor, electrons will flow from one plate, through the load, and collect on the other plate, creating a charge with the same polarity as that of the input waveform. When the input waveform changes direction, the capacitor discharges (that is, the charge built up on one plate flows rapidly to the other plate), causing current flow through the load.

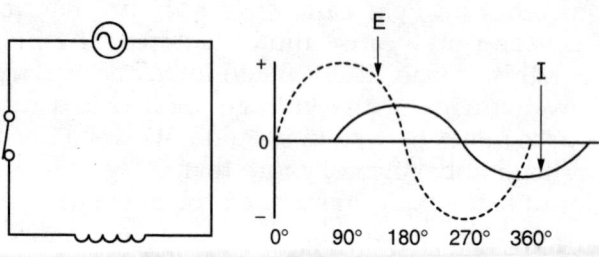

*Figure 23* ◆ Inductive circuit.

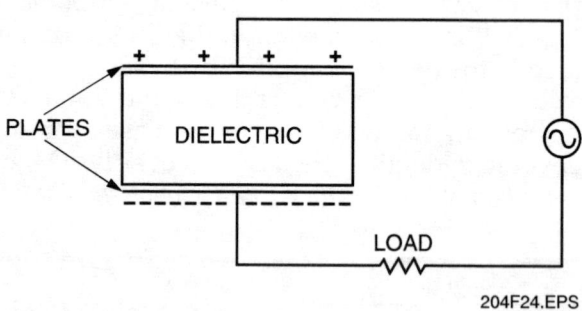

*Figure 24* ◆ Capacitor operation.

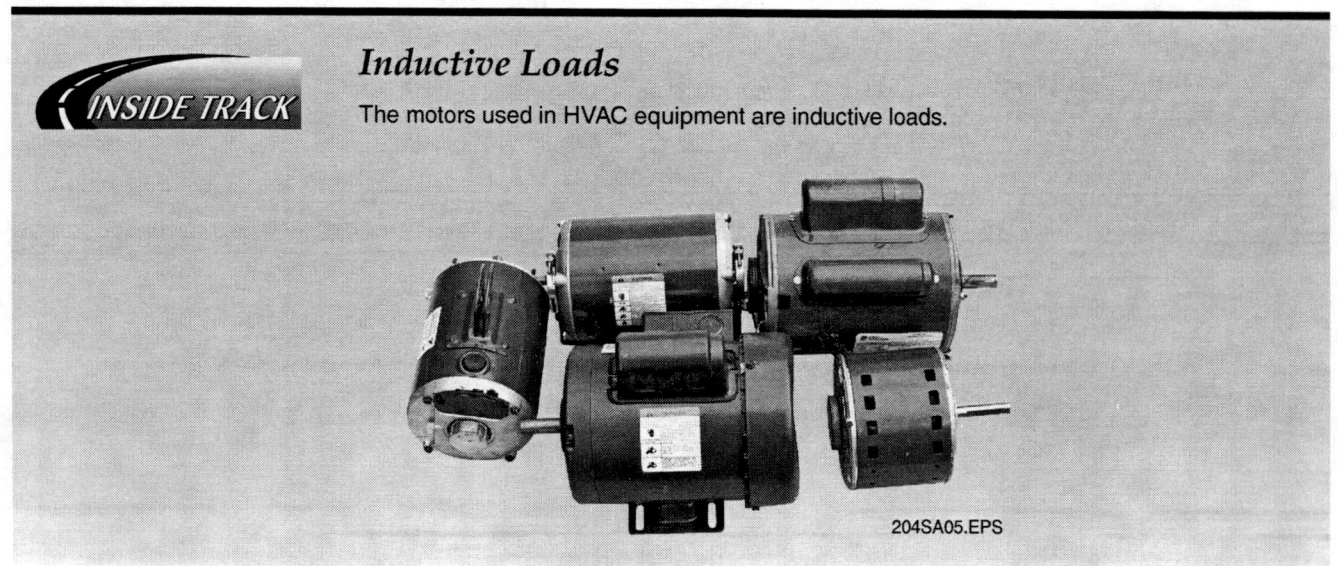

### Inside Track

**Inductive Loads**

The motors used in HVAC equipment are inductive loads.

Because the voltage across the capacitor is created by current flowing from one plate to the other, the voltage across the capacitor lags the current. In a purely capacitive circuit, voltage lags current by 90°. The phase shift is the opposite of that in an inductor, where voltage leads current. *Figure 25* shows the phase relationships in resistive, inductive, and capacitive circuits.

There are two common types of capacitors. In the oil-filled capacitor, paper soaked with an insulating fluid acts as the dielectric. This capacitor has large plates and a large amount of dielectric oil in order to dissipate heat. It can thus remain in the circuit all the time. The electrolytic capacitor is smaller and contains less dielectric material. It will overheat and be damaged if it is left in the circuit.

When electrolytic capacitors are used, they must be switched out of the circuit as soon as the circuit is started.

Capacitors and inductors and the phase shift they provide play an important role in helping to start single-phase **induction motors** and make them operate more efficiently.

## 5.0.0 ◆ INDUCTION MOTORS

AC induction motors are the primary load devices in HVAC equipment. They range in size from fractions of a horsepower to hundreds of horsepower.

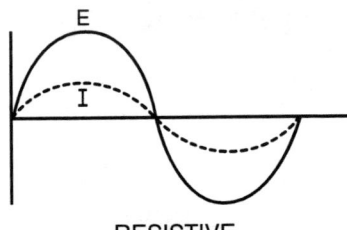
RESISTIVE

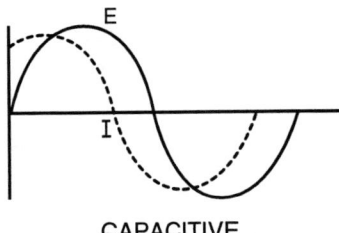
CAPACITIVE

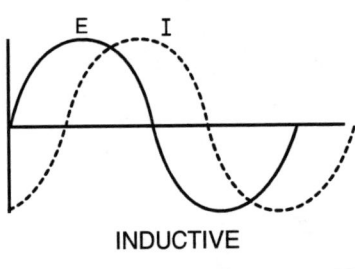
INDUCTIVE

*Figure 25* ◆ Phase relationships.

### Capacitors
Capacitors are widely used with motors in HVAC systems to facilitate motor starting and enhance operation.

They drive compressors, blowers, condenser fans, ventilating fans, induced-draft fans, and humidifiers, among other things. Electromagnetism and induction are the keys to the operation of an AC motor.

## 5.1.0 Single-Phase Motors

The main components of a single-phase motor are the **rotor** and the **stator** (*Figure 26*). The stator is fixed and the rotor turns. The motor accomplishes work by converting electrical energy into mechanical energy that is delivered by attaching a mechanism such as a fan to the rotor shaft.

To operate the motor, voltage is applied to the stator winding (*Figure 27*), which contains many turns of wire. The magnetic field created by the current through the stator winding attracts or repels the rotor, causing it to turn. Because the polarity of the AC voltage applied to the stator is constantly changing, the relationship of the north and south magnetic poles of the stator and rotor constantly changes. This creates a rotating magnetic field that continuously pulls and pushes the rotor and keeps it turning.

The problem with single-phase motors is getting them started. First, it takes extra energy (**torque**) to overcome **inertia** and start the rotor turning. Second, if the rotor stops at a position where the stator field is exerting neither push nor pull, the motor will not restart when power is reapplied. Various ways of dealing with these problems are discussed next.

### 5.1.1 Split-Phase Motors

In the split-phase motor (*Figure 28*), an additional winding (**start winding**) is added to the stator. The start winding contains many turns of very fine wire; therefore the current buildup is slower than that of the **run winding**, which has fewer turns of wire. The phase difference between the two windings creates a torque that starts the rotor turning. A centrifugal switch opens to remove the winding from the circuit once the motor has started.

The rotor of this motor consists of copper bars set in an iron core and connected at the ends by a copper ring. This type of rotor is called a *squirrel cage* rotor because the arrangement of the copper bars resembles a cage. Split-phase motors are used in pumps, oil burners, and other applications requiring ⅓ horsepower or less.

### 5.1.2 Permanent Split Capacitor Motors

In this type of motor (*Figure 29*), the **run capacitor** provides a phase shift between the run and start windings that helps start the motor, then remains in the circuit to improve running efficiency. These motors are often used to drive blowers for applications that require ⅛ to ¾ horsepower.

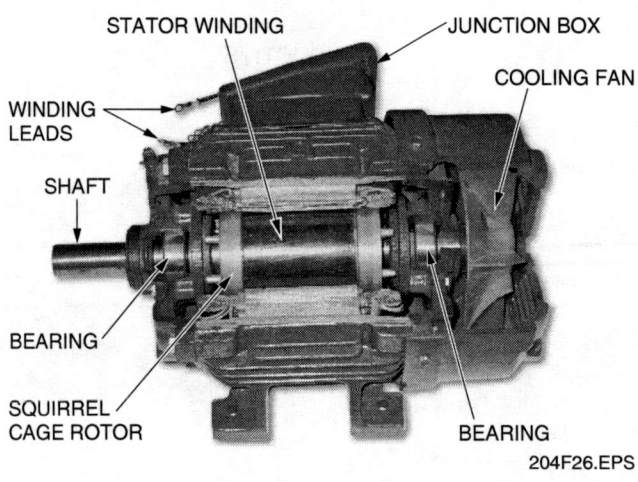

*Figure 26* ◆ Basic parts of a motor.

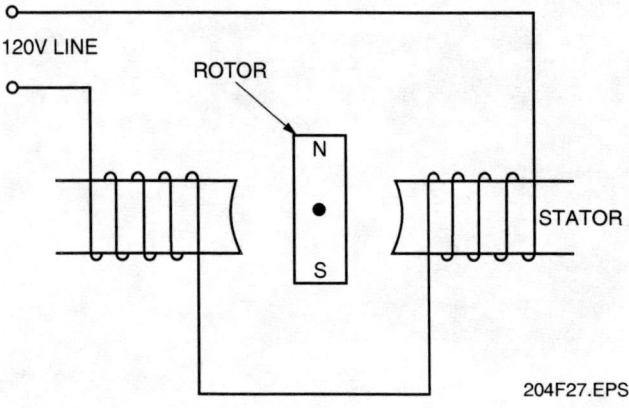

*Figure 27* ◆ AC motor basics.

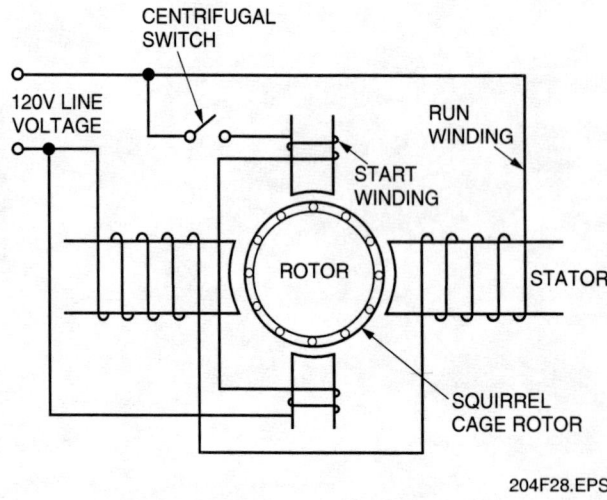

*Figure 28* ◆ Split-phase AC motor.

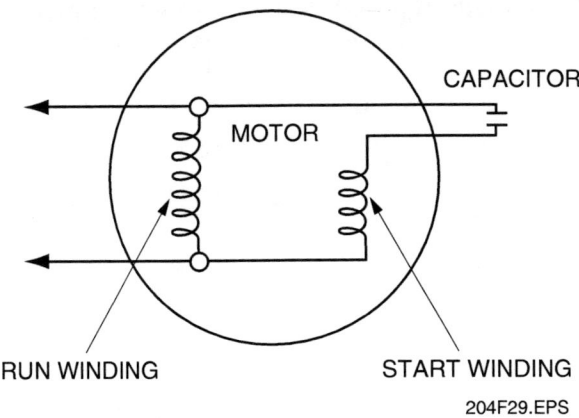

Figure 29 ◆ PSC motor.

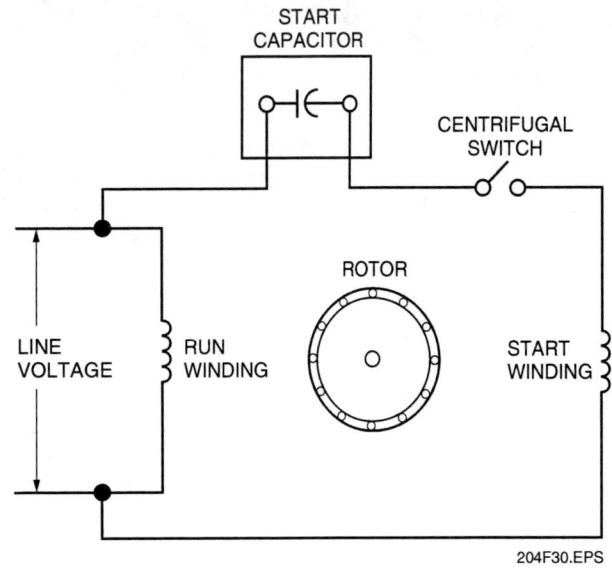

Figure 30 ◆ Capacitor start motor with centrifugal switch.

### 5.1.3 Capacitor Start Motors

Capacitor start motors range in size from fractional horsepower to as high as 10 horsepower. Their high starting torque makes them suitable for powering fans and some refrigeration compressors. The capacitor, which is wired in series with the start winding, provides high starting torque with relatively low starting current. It is an electrolytic capacitor, and must therefore be switched out of the circuit as soon as the motor reaches ⅔ to ¾ of its rated speed. Two methods are commonly used to do this. In the type shown in *Figure 30*, a centrifugal switch is used.

In the more common type shown in *Figure 31*, a potential (voltage) starting relay is used. When operating, all electric motors have some electrical voltage generating capacity resulting from induced voltages generated in the motor stator windings by the motion of the rotating rotor. The voltage generated is known as *back electromotive force (back EMF)*. It is important to know that the back EMF generated in a motor normally has a much higher potential than the line voltage applied to the motor. For example, a motor being driven by a 230V line source can generate back EMF voltages of over 400 volts.

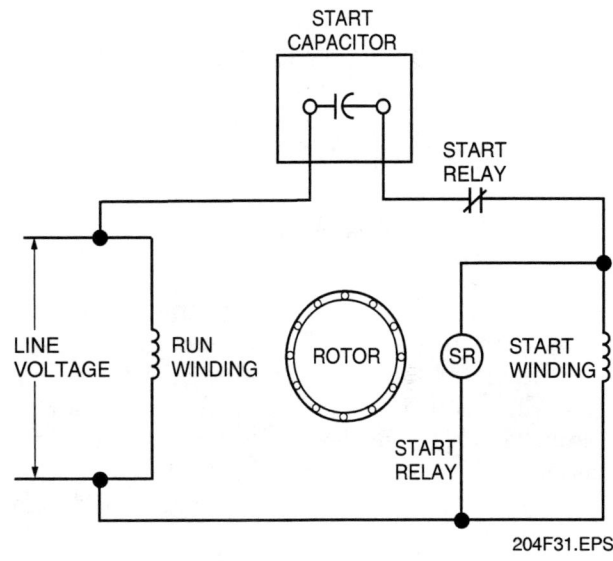

Figure 31 ◆ Capacitor start motor with start relay.

---

### Motor Run and Start Capacitors

The run capacitors used in HVAC motor circuits are of the oil-filled type and remain in the circuit at all times. The start capacitors typically used to start single-phase compressors remain in the circuit for a fraction of a second to help the motor start. A special start relay is commonly used whenever a start capacitor is installed in a system. The start relay functions to quickly remove the start capacitor from the system before it overheats.

It is very important that capacitors and start relays be sized correctly. If you must install these components to help a compressor start, you must use the exact capacitor and start relay specified by the compressor manufacturer.

---

ALTERNATING CURRENT — TRAINEE MODULE 03204

In a capacitor start motor, the back EMF voltage, not line voltage, is used to energize and de-energize the start relay. By design, the start relay will energize at some predetermined back EMF voltage level that is always higher than the line voltage.

When the motor is first turned on, the start relay is de-energized and the start capacitor and start winding are connected in the circuit through the normally closed contacts of the start relay. As the motor picks up speed, the back EMF in the start winding builds up and eventually causes the start relay to energize. This opens the relay contacts and removes the capacitor and start winding from the circuit. As long as the motor runs at or above this speed, the back EMF generated in the start winding remains applied across the parallel-connected start relay coil. This keeps the relay energized and the start capacitor and winding out of the circuit.

### 5.1.4 Capacitor Start, Capacitor Run Motor

The capacitor start, capacitor run motor (*Figure 32*) is used to drive refrigerant compressors. It combines high starting torque with smooth, quiet, efficient operation. Like the capacitor start motor, the start capacitor is switched out of the circuit when the motor comes up to speed. The motor then runs as a permanent split capacitor motor.

### 5.1.5 Shaded-Pole Motor

In the shaded-pole motor (*Figure 33*), the rotor pivots within two pairs of stator windings. A groove in the stator pole separates a small portion of the stator from the remainder. A metal band (shading coil) is placed around the smaller portion. This band causes a slight phase shift, which is enough to provide torque to start the motor. Shaded-pole motors are used to drive small fans and pumps.

### 5.1.6 Multi-Speed Motors

The speed of a motor, which is measured in revolutions per minute (RPM), is determined by the number of stator windings and the frequency of the applied voltage. The maximum speed at which a motor can run is known as its **synchronous speed**. Adding a load to a motor causes some slippage, or inefficiency, in the operation of the motor. Therefore, motors are generally rated at 95% to 97% of their synchronous speed.

The speed of some motors can be changed using taps on the stator winding (*Figure 34*). Instead of using the entire stator coil, a portion is used—the larger the portion, the lower the speed.

This method allows a single motor to be used in different applications. In cases where the same air handler is used for both heating and cooling, the control circuit will automatically select low speed for heating or high speed for cooling.

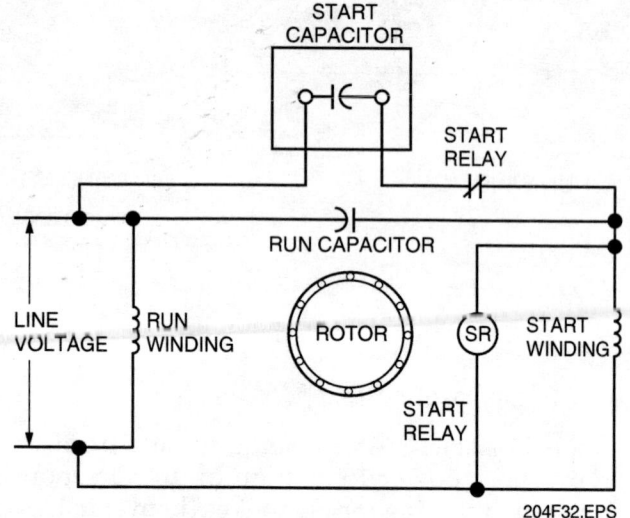

*Figure 32* ◆ Capacitor start, capacitor run motor.

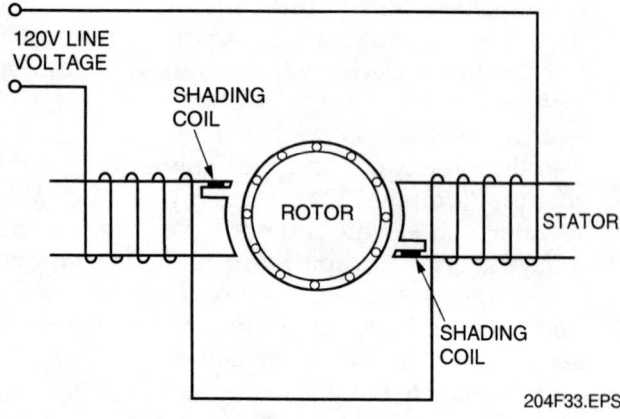

*Figure 33* ◆ Shaded-pole motor.

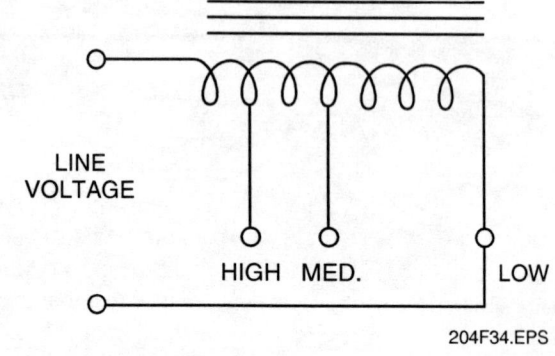

*Figure 34* ◆ Speed taps.

## 5.2.0 Three-Phase Motors

Three-phase motors are used primarily in large commercial systems. They require three-phase voltage, which is not readily available in residential areas. Three-phase motors offer several important advantages over single-phase motors:

- They have a higher starting torque and require no special starting equipment such as capacitors, start relays, or start windings. Because the stator windings are 120° out of phase, at least one of the windings is always applying torque to the rotor.
- The direction of rotation can be easily reversed by reversing any two of the three supply voltage connections.
- They run very smoothly.
- There is less running pulsation because at least one stator winding is always applying torque to the rotor.

Three-phase motor stators are connected in either a delta or wye configuration, depending on the application (*Figure 35*). A delta-connected motor provides more current, and therefore more starting torque, but also consumes more power than the wye-connected motor. The motor's voltage rating is also a factor in selecting the three-phase hookup.

## 6.0.0 ◆ TESTING AC COMPONENTS

Most components of AC circuits can be tested with the standard electrical test meters—the voltmeter, ammeter, and ohmmeter. A multimeter, or VOM, will handle most voltage and resistance readings and a clamp-on ammeter will suffice for current readings. In addition to these instruments, a capacitor analyzer, wattmeter, and **megohmmeter (megger)** will sometimes be needed.

### 6.1.0 Capacitor Analyzer

A capacitor analyzer (*Figure 36*) is a special-purpose instrument used to test capacitors. This analyzer will check capacitors for current leakage, insulation breakdown, capacity, shorts, and opens. An example of its use is to determine the value of a capacitor when you are unable to read the value. This sometimes happens when replacing defective start or run capacitors.

### 6.2.0 Wattmeter

Rather than measure voltage and current and then calculate power, a wattmeter can be connected into a circuit to measure power directly. Not only does a wattmeter simplify power measurements, it also has two other advantages.

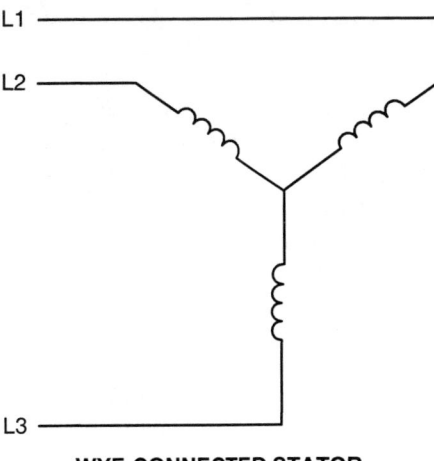

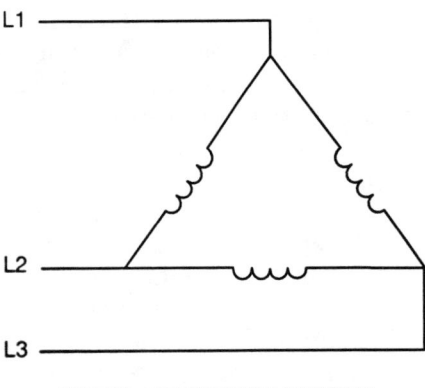

*Figure 35* ◆ Three-phase motors.

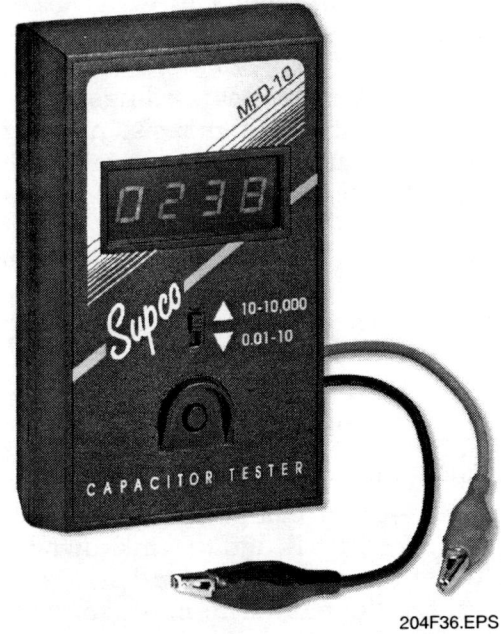

*Figure 36* ◆ Capacitor analyzer.

ALTERNATING CURRENT — TRAINEE MODULE 03204

First, voltage and current in an AC circuit are not always in phase; current sometimes either leads or lags the voltage. When this happens, multiplying the voltage times the current yields apparent power, not true power. However, the wattmeter takes this into account and always indicates true power.

Second, voltmeters and ammeters consume power. The amount consumed depends on the levels of the voltage and current in the circuit, and cannot be accurately predicted. Therefore, very accurate power measurements cannot be made by measuring voltage and current and then calculating power. Some wattmeters compensate for internal power losses so that only the power dissipated in the circuit is measured. If the wattmeter does not compensate for these losses, the power that is dissipated is sometimes marked on the meter or else can be easily determined so that a very accurate measurement can be made. Typically, the accuracy of a wattmeter is within ±1%.

The basic wattmeter consists of two stationary coils connected in a series and one movable coil (*Figure 37*). The movable coil, wound with many turns of fine wire, has a high resistance. The stationary current and voltage coils, wound with a few turns of a larger wire, have a low resistance. The interaction of the magnetic fields around the different coils will cause the movable coil and its pointer to rotate in proportion to the voltage across the load and the current through the load. Thus, the meter indicates E times I, or power.

## 6.3.0 Megohmmeter (Megger)

Normally, an ohmmeter is not used to measure extremely high resistances, such as those involving conductor insulations, insulation between motor or transformer windings, and so on. To adequately test such high resistances, it is necessary to use a much higher potential than is furnished by the battery of an ohmmeter. Test voltages ranging from 50V to 5,000V can be supplied by the megohmmeter, or megger. Meggers measure resistance in megohms (equal to one million ohms). There are three types of meggers: hand, battery, and electric.

The megger has two coils (*Figure 38*). Coil A is in series with resistor $R_2$ across the output of the generator. This coil is wound so that it causes the pointer to move toward the high-resistance end of the scale when the generator is operating. Coil B is in series with $R_1$ and the unknown resistance ($R_X$) to be measured. This coil is wound so that it causes the pointer to move toward the low- or zero-resistance end of the scale when the generator is operating.

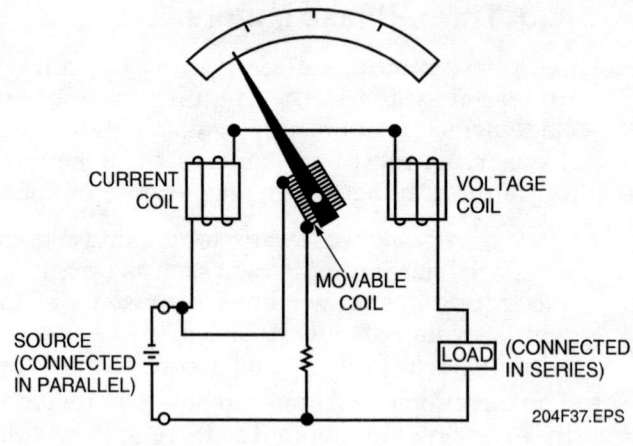

*Figure 37* ◆ Wattmeter schematic.

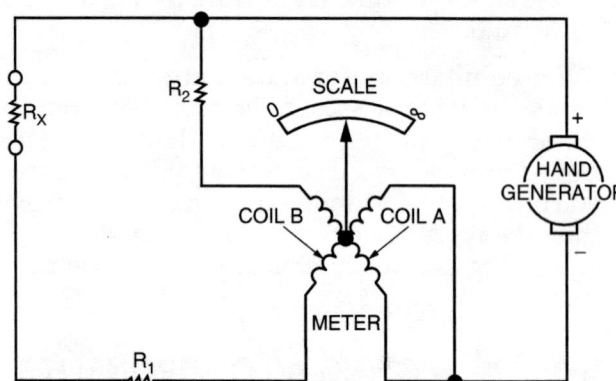

MEGGERS CAN MEASURE RESISTANCE RANGING FROM HUNDREDS TO THOUSANDS OF MEGOHMS.

*Figure 38* ◆ Megger schematic.

When an extremely high resistance appears across the input terminals of the megger, the current through coil A is greater, causing the pointer to deflect toward infinity. Conversely, when a relatively low resistance appears across the input terminals, the current through coil B is greater, and causes the pointer to deflect toward zero.

In this type of megger, hand generators are used to produce the test voltage. To avoid excessive test voltages, most hand meggers are equipped with friction clutches. When the generator is cranked faster than its rated speed, the clutch slips and the generator speed and output voltage are maintained at their rated values.

Newer meggers (*Figure 39*) use the same operational principles. Instead of having a scaled meter movement, these meters give the value of resistance in a digital readout. The digital readout makes reading the measurement much easier and helps eliminate errors.

*Figure 39* ◆ Digital readout megger.

> **WARNING!**
> When a megger is used, the generator voltage is present on the test leads. This voltage could be hazardous to you or the equipment you are testing. *Never touch the test leads while the tester is being used.* Isolate the item you are testing from the circuit before using the megger.

### 6.3.1 Safety Precautions

When you use a megger, you could be injured or cause damage to the equipment that you are working on if the following minimum safety precautions are not observed:

- Use meggers on high-resistance measurements *only* (such as insulation measurements or to check two separate conductors in a cable).

## INSIDE TRACK

### Motor Insulation Tests

Megohmmeters are commonly used to perform motor insulation tests in order to prevent electrical shock and/or motor failure caused by deterioration of the motor winding insulation.

Megohmmeter readings of a motor's insulation resistance should be taken when a motor is first installed, and at least semiannually thereafter. These readings should test the resistance between the individual windings and between the windings and ground. The resistance readings should be recorded and compared in order to detect any deterioration in the insulation resistance indicating the potential for future motor failure. A motor needs to be replaced or repaired if the megohmmeter reading drops below the recommended minimum resistance. A rule of thumb is that the minimum resistance between the motor windings or between the windings and ground should be at least 1,000Ω per volt, based on the operating voltage for the motor. For example, a 230V motor should have a minimum resistance of about 230,000Ω. Note that a motor with good insulation can have a resistance reading many times the minimum acceptable resistance.

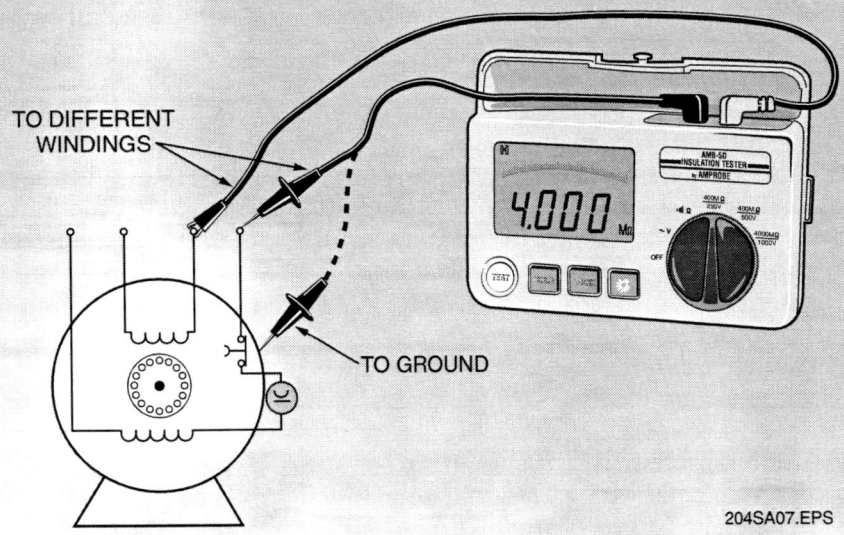

- *Never* touch the test leads while the handle is being cranked.
- De-energize the circuit and verify that it is off before connecting the meter.
- Disconnect the item being checked from other circuitry, if possible, *before* using the meter.

### 6.4.0 Recording Instruments

It is often necessary to know the conditions that exist in an electrical circuit over a period of time to determine such things as peak loads, voltage fluctuations, etc.

It may be neither practical nor economical to assign a worker to watch an indicating instrument and record its readings. An automatic recording instrument can be connected to take continuous readings, and the record can be collected for review and analysis.

The term *recording instrument* describes many instruments that make a permanent record of measured quantities over a period of time. Recording instruments can be divided into three general groups:

- Instruments that record electrical quantities, including potential difference, current, power, resistance, and frequency
- Instruments that record nonelectrical quantities by electrical means (such as a temperature recorder that uses a potentiometer system to record thermocouple output)
- Instruments that record nonelectrical quantities by mechanical means (such as a temperature recorder that uses a bimetallic element to move a pen across an advancing strip of paper)

Recording instruments are basically the same as the indicating meters we have already looked at, but they have recording mechanisms attached to them. They are generally made of the same parts, use the same electrical mechanisms, and are connected in the same way. The only basic difference is the permanent record.

Strip chart recorders are the most widely used recording instruments for electrical measurement. Their name comes from the fact that the record is made on a strip of paper, usually four to six inches wide and perhaps up to 60' long. A watt/VAR strip chart recorder is shown in *Figure 40*.

Strip chart recorders offer several advantages in electrical measurement. The long charts allow the recording to cover a considerable length of time with little attention. Also, strip chart recorders can be operated at a relatively high speed if very detailed records are needed.

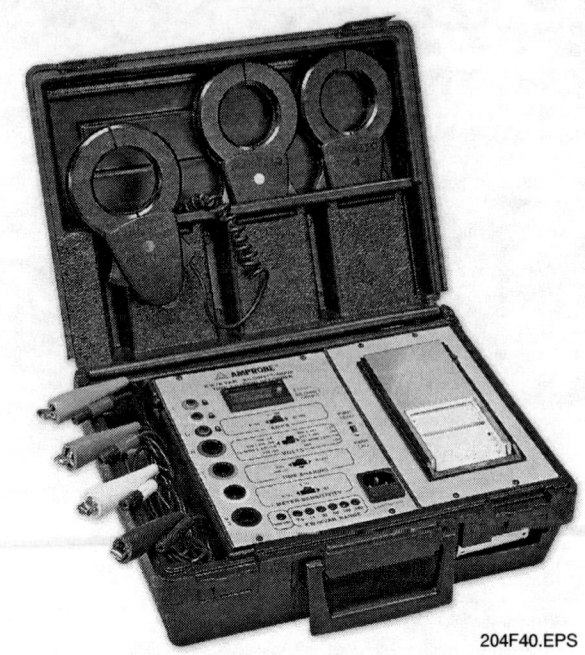

*Figure 40* ♦ Watt/VAR strip chart recorder.

### 6.5.0 Checking Inductive Loads

The most common inductive load is the stator winding of a motor. Stator windings typically have a very low resistance; one or two ohms is fairly common. The resistances will vary from manufacturer to manufacturer and from one type of motor to another. In some fractional-horsepower motors, readings of less than one ohm are possible. The start winding of a single-phase motor will have a higher resistance than the run winding; perhaps three to four times higher. The winding resistance is not provided on the motor nameplate, so the resistance values themselves may not be of much use unless you are familiar with the particular motor. An open winding will be readily apparent (*Figure 41*), as will a severe short. A partial short may only be recognizable if you are familiar with the motor or if there is an unusual difference between the resistances of the start and run windings.

A three-phase motor is easier to check; the resistances of the three windings should be the same.

When resistance and continuity checks are made, one end of the target component or series of components must be disconnected from the circuit. Otherwise, the ohmmeter circuit current might read the resistance of a parallel circuit. In the upper section of *Figure 41*, for example, the meter would read the resistance of the parallel path because it offers much less resistance than the open coil. That is why the target coil is disconnected from the circuit.

## Checking a Motor for Grounded Windings

If you have a branch or equipment circuit breaker connected to HVAC equipment that keeps tripping, you might suspect a grounded motor as the source of the problem. Here's a quick way to check an electric motor for a grounded winding using a multimeter:

***Step 1***    Shut off power to the equipment and discharge any capacitors in the motor circuit.

***Step 2***    Disconnect the motor leads to isolate the motor from the rest of the circuit.

***Step 3***    Place the multimeter range switch to the OHMS × 1,000 setting.

***Step 4***    Place one meter lead on ground (usually the motor case) and the other lead on one of the motor winding leads.

A good motor should have a resistance to ground of at least 1,000Ω per operating voltage of the motor. For example, a 230V motor should have a minimum resistance to ground of 230,000Ωs. A resistance reading higher than that indicates a good motor. Any lower resistance indicates a grounded motor.

### 6.6.0 Checking Capacitors

A simple test for a capacitor can be done with a multimeter set to measure resistance. As shown in *Figure 42*, the capacitor is disconnected from the circuit. Before connecting the multimeter across the capacitor leads, bleed off any capacitor charge using a capacitor discharging tool like the one shown in *Figure 43*. It is a good idea to do this even if the capacitor is equipped with a bleeder resistor, in case the bleeder is defective. Don't use a screwdriver placed directly across both terminals of the capacitor to discharge a capacitor, as this can damage the capacitor.

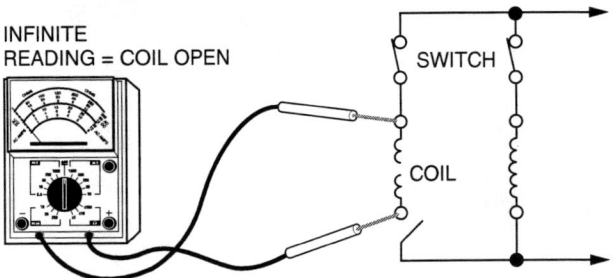

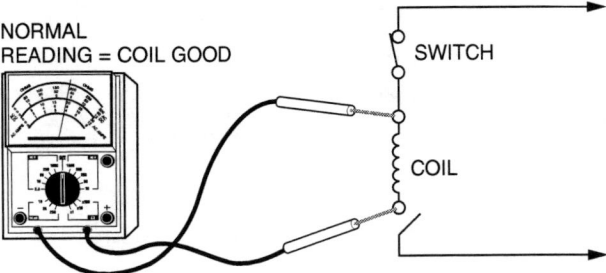

 **WARNING!**
Never place your fingers across the capacitor terminals; the residual charge on the capacitor can be dangerous.

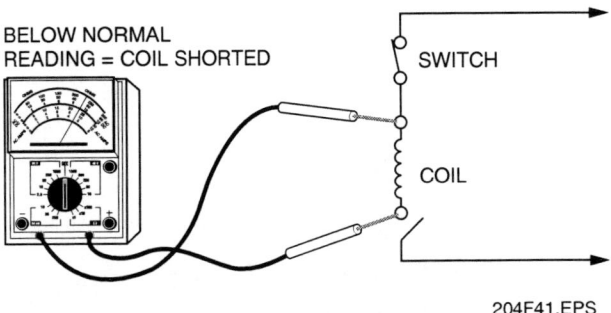

204F41.EPS

*Figure 41* ♦ Inductive load check.

When the multimeter test leads are placed across the capacitor terminals, the meter will immediately register a resistance reading and then return to infinity if the capacitor is good. If there is a short in the capacitor, the meter reading will not return all the way to infinity. If the capacitor is defective, it is important to replace it with one of the same capacity and voltage rating. If an exact replacement is not available, two capacitors with a combined value equal to the original can be connected in series to obtain the correct voltage, or in parallel to obtain the correct capacitance.

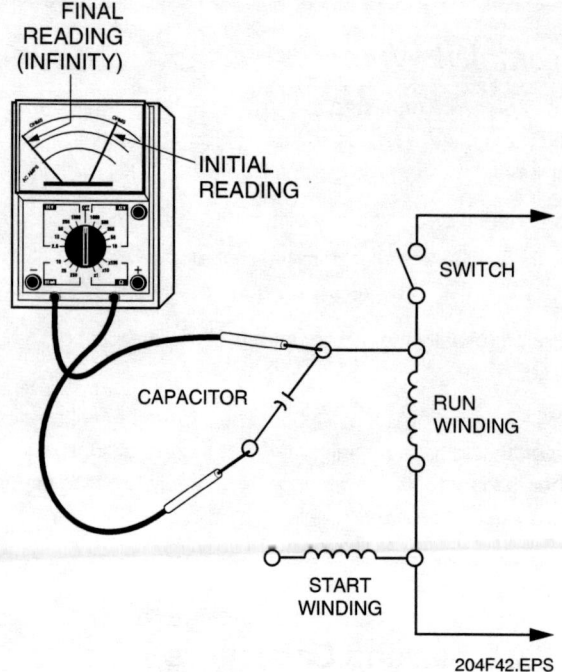

*Figure 42* ♦ Capacitor check.

## 6.7.0 Checking Fuses

The best way to test a fuse is by measuring continuity (*Figure 44*). To check fuses, always open the unit disconnect switch, then remove the fuses using an insulated fuse puller. Test the fuses for continuity using an analog or digital multimeter (VOM/DMM).

If a short exists (zero ohms) across the fuse, it is usually good. If an open exists (infinite resistance) across the fuse, it is blown. A blown fuse is usually caused by some abnormal overload condition, such as a short circuit within the equipment or an overloaded motor.

Replacing a blown fuse without locating and correcting the cause can result in damage to the equipment.

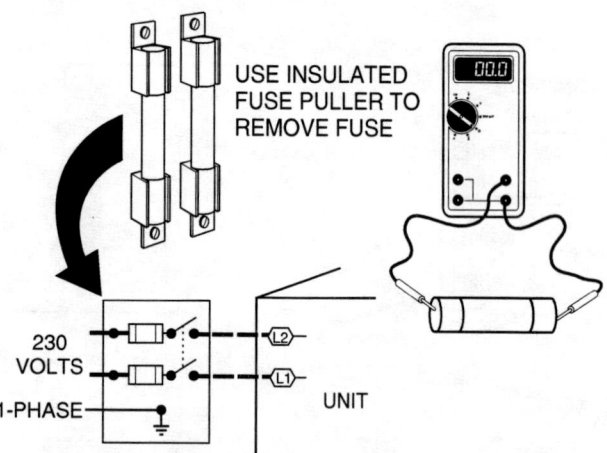

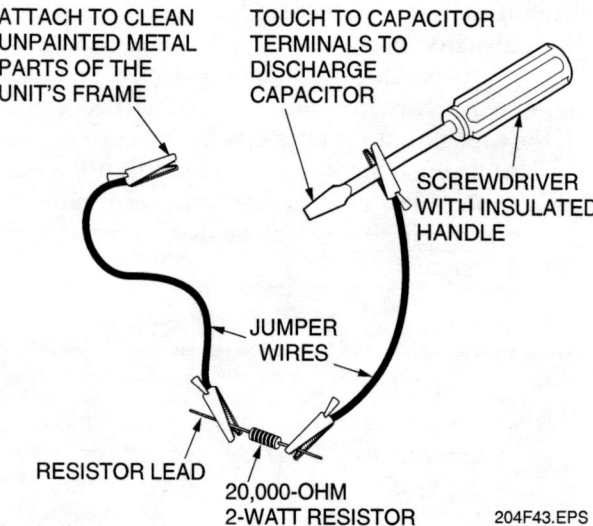

*Figure 43* ♦ Capacitor discharging tool.

### WARNING!

Some older capacitors may contain PCB (polychlorinated biphenyl), which was at one time widely used as a dielectric in capacitors and power distribution transformers. PCB is known to cause cancer in humans. If you have a defective oil-filled capacitor or transformer, do not attempt to pry it open or puncture it. If it is leaking, don't touch the oil or breathe the fumes. Dispose of it in accordance with applicable local or national codes.

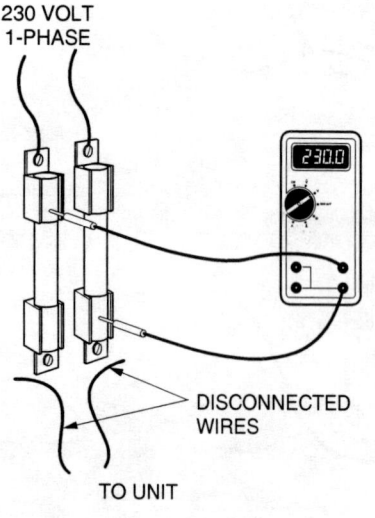

*Figure 44* ♦ Fuse checks.

Fuses can also be tested with the circuit energized. This is done by shutting off the power to the unit, then disconnecting the wires from the load side of each fuse. This eliminates the possibility of current being fed back to the multimeter through a short circuit within the unit. Set the meter to AC voltage on a range that is higher than the highest voltage expected. Turn on the power and place one of the meter leads on the input (line) side of the fuse. Touch the other test lead to the load side of another fuse. If voltage is measured on the load side of the fuse, the fuse is good; if not, the fuse is bad. Repeat this procedure so that all fuses are measured with one test lead on the load side and the other test lead on the line side of a different fuse. This method tests one fuse at a time. If the measurement was performed with both test leads on the load side of the fuses, and the meter showed no reading, you would know that a fuse was blown, but not which one.

## 7.0.0 ◆ SAFETY

Alternating current can be a deadly force if carelessly handled. One tenth of an ampere of alternating current flowing through a vital organ can prove fatal; therefore, safety precautions must be observed whenever working around or with electricity. The amount and duration of the current flow, the parts of the body involved, and the frequency of the current determine the extent of body damage. Damage is greatest when the current flow is through or near nerve centers and vital organs.

People differ in their resistance to electric shock. Consequently, an amount of current that may cause only a painful shock to one person might be fatal to another. *Table 1* presents the effect of 60Hz current flowing through the body from hand-to-hand or hand-to-foot. The table shows that at approximately one milliampere or mA (0.001A), shock is perceptible. At approximately 10mA, the shock would be sufficient to prevent voluntary control of the muscles; at approximately 100mA (0.1A), the shock is fatal if it lasts more than one second.

High-frequency currents (200Hz and above) have a tendency to flow along the surface of the skin, usually causing severe burns. The current may not penetrate the body. In addition to the possibility of burns and death, involuntary movements as a result of electrical shock can cause other types of serious injuries resulting from falls or contact with rotating machinery or hot surfaces.

Two conditions must be present for an electric current to flow through the body and cause electric shock. First, the body or some part of the body must form part of a closed circuit. Second, there must be a voltage somewhere in the closed circuit. To prevent electric shock, you must make certain that your body never forms part of a closed circuit. Your body must also be well insulated from the ground.

Practically all electric shocks are due to human error, rather than equipment failure. Nearly all deaths due to electrical shock are due to the worker's failure to observe safety precautions, failure to repair equipment for electrical defects, or failure to remedy all defects found by tests and inspections.

The following are recommended precautions:

- Never cut off the ground prong of a grounded plug.
- Never touch any electrical wire without ensuring that it is not a live wire.
- Never switch an appliance on or off while standing in or touching a wet surface or area.

**Table 1** Current Effects on the Human Body

| Current Value | Typical Effects |
|---|---|
| Less than 1 milliamp | No sensation. |
| 1 to 20 milliamps | Sensation of shock, possibly painful. May lose some muscular control between 10 and 20 milliamps. |
| 20 to 50 milliamps | Painful shock, severe muscular contractions, breathing difficulties. |
| 50 to 200 milliamps | Up to 100 milliamps, same symptoms as above, only more severe. Between 100 and 200 milliamps ventricular fibrillation may occur. This typically results in almost immediate death unless special medical equipment and treatment are available. |
| Over 200 milliamps | Severe burns and muscular contractions. The chest muscles contract and stop the heart for the duration of the shock, resulting in death. |

- Always turn off power at the main disconnect before working on an electrical circuit or device. Lock and tag the power switch.
- Always unplug an electrical appliance before working on it.
- Replace all worn power cords.
- Unplug cords by pulling on the plug; do not pull the cord.
- Notify the power company or utility whenever a power line is touching the ground.
- If it becomes necessary to work on live electrical wiring, try to use one hand only. If you are shocked while using only one hand, current will probably flow through the hand and arm, then through the feet to the ground. If a shock is conducted through both hands and arms, the electrical path would be through the heart and lungs and would be more likely to be fatal.
- Use protective equipment such as rubber gloves and insulated boots.
- Use tools with dielectric insulation.
- Remove metal jewelry such as rings and watches.

The *National Electrical Code (NEC)*, when used together with the electrical code for your local area, provides the minimum requirements for the installation of electrical systems. Always use the latest edition of the Code as your on-the-job reference. It specifies the minimum provisions necessary for protecting people and property from electrical hazards. In some areas, different editions of the Code may be in use, so be sure to use the edition specified by your employer.

## 8.0.0 ♦ AC VOLTAGE ON CIRCUIT DIAGRAMS

The schematic diagrams you will see in your work will generally be divided into high-voltage sections and low-voltage sections. (See *Figure 45*.) The high-voltage section will contain the line voltage distribution circuits and the primary loads. In an air conditioning system, these would be the compressor motor, fan motors, and resistance heaters. The low-voltage section, which contains the control devices such as the thermostat and control relays, will often operate at 24 volts. The low voltage is obtained by using a control transformer to step down the line voltage. In some large systems using three-phase line voltages of 240V and higher, a control voltage of 120V may be used for some of the control devices.

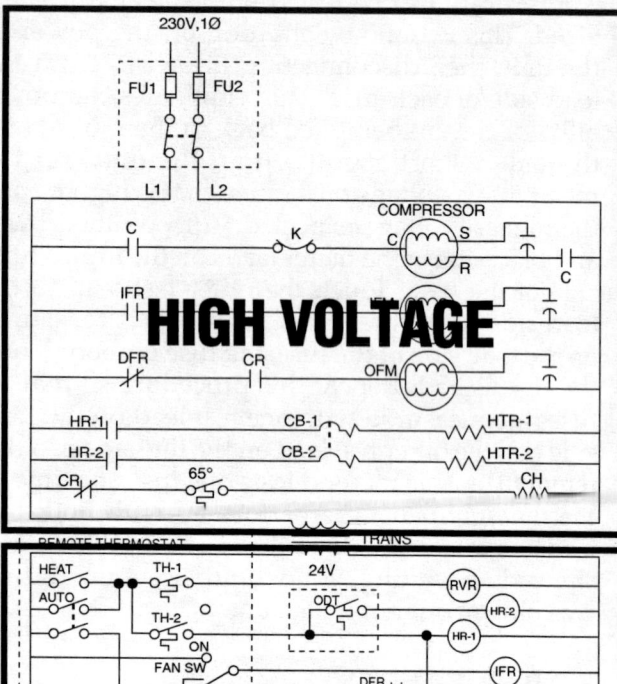

*Figure 45* ♦ High-voltage and low-voltage circuits.

### Ground Fault Circuit Interrupters

To minimize your risk of shock when using power tools, use an extension cord with a built-in ground fault circuit interrupter (GFCI). If there is any current leakage to ground, the GFCI will trip long before a conventional circuit breaker trips and before any potentially harmful levels of current are reached. A GFCI-equipped extension cord can save your life.

As shown in *Figure 46*, single-phase line voltage is often represented on ladder diagrams as two vertical lines labeled L1 and L2, representing the two 240V lines from the secondary of the pole transformer.

## Summary

The HVAC equipment you encounter will be powered by AC voltage. Some of the internal circuits may use DC, which will be obtained by rectifying the AC. Single-phase AC is used to power most homes and small commercial operations. Where more power is needed, three-phase power is available from the local utility.

AC induction motors drive the compressors and fans used in HVAC equipment. There are several types of single-phase motors used in HVAC equipment. They are selected for their starting torque and running characteristics, which are determined by the arrangement of stator windings and the use of capacitors to provide phase shift. Three-phase motors are used where higher torque is required.

In addition to the basic test instruments introduced in the preceding level, troubleshooting of AC circuits may require a capacitor, wattmeter, megger, and sometimes a chart recorder.

The importance of safety in the lab and on the job cannot be overstated. Learn and follow the established practices for the safety of yourself and your co-workers.

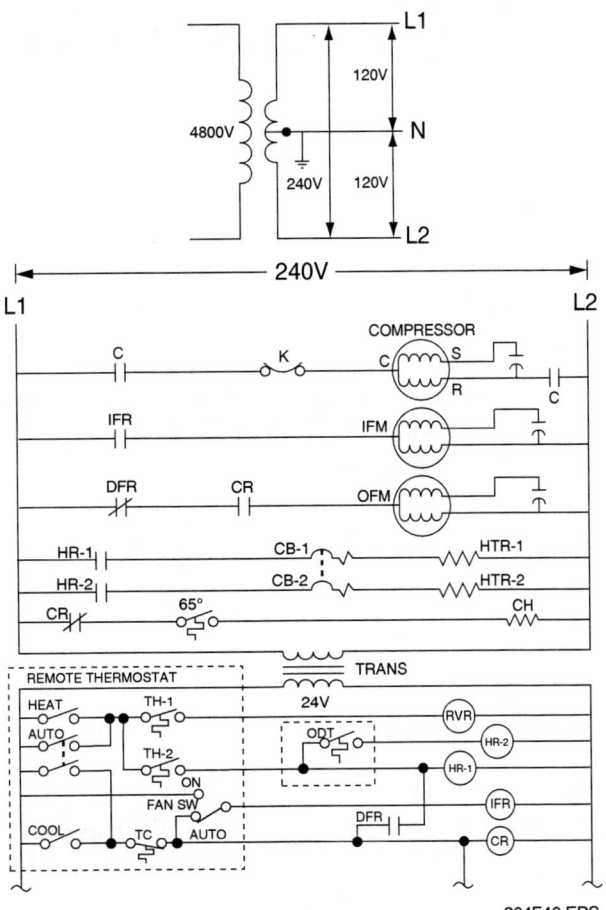

*Figure 46* ◆ Ladder diagram.

## Review Questions

1. An autotransformer is unique because it _____.
   a. has a single winding
   b. is found only on cars
   c. automatically shuts itself off
   d. has more than one secondary winding

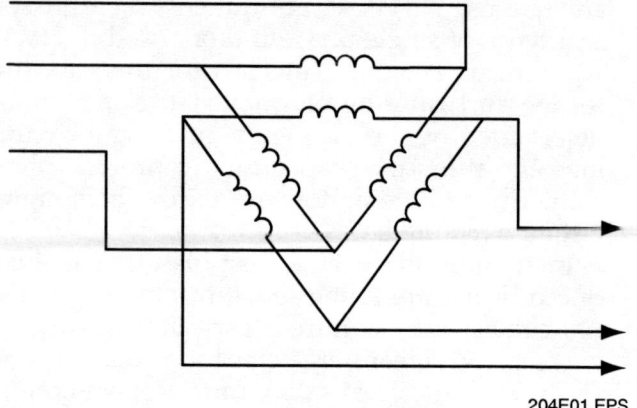

204E01.EPS

2. The schematic diagram above represents a _____.
   a. delta-connected three-phase transformer
   b. wye-connected three-phase transformer
   c. three-phase motor
   d. autotransformer

3. The effective (RMS) voltage of a sine wave with a peak voltage of 200V is _____.
   a. 100
   b. 120.5
   c. 141.4
   d. 200

4. The voltage of a sine wave with a maximum voltage of 10V at the 180° point is _____.
   a. −10V
   b. 0V
   c. 5V
   d. +10V

5. The voltage of a sine wave with a maximum voltage of 10V at the 270° point is _____.
   a. −10V
   b. 0V
   c. 5V
   d. +10V

6. In an inductive circuit _____.
   a. current leads voltage
   b. voltage leads current
   c. voltage and current are in phase
   d. voltage lags current by 90°

7. A run capacitor in a single-phase motor can remain in the circuit after the motor starts because _____.
   a. single-phase motors don't draw much current
   b. it has a paper dielectric and foil plates
   c. single-phase motors don't generate much heat
   d. it has large plates and dielectric oil to dissipate heat

8. The stator winding of a single-phase motor _____ when voltage is applied.
   a. rotates
   b. remains stationary
   c. moves up and down like a piston
   d. remains stationary for a few seconds, then begins rotating

9. The start winding of a split-phase motor _____.
   a. has fewer turns of wire than the run winding
   b. is always left in the circuit after the motor has started
   c. has more turns of wire than the run winding
   d. is 120° out of phase with the run winding

10. You are likely to find shaded-pole motors in _____.
    a. large blower units
    b. compressors
    c. small fans
    d. centrifugal chillers

11. A _____ single-phase motor is most likely to be used to drive a refrigeration compressor.
    a. permanent split capacitor
    b. capacitor start capacitor run
    c. shaded pole
    d. split phase

12. Which of the following applies to three-phase motors?
    a. They must be delta-connected.
    b. They require starting devices.
    c. They generate less starting torque than single-phase motors.
    d. They may be delta-connected or wye-connected.

13. In a single-phase motor, the resistance of the run winding will be _____ that of the start winding.
    a. less than
    b. slightly more than
    c. equal to
    d. exactly twice

14. If you suspect that a capacitor has failed and do not have a capacitor tester, you can test it using a(n) _____.
    a. voltmeter
    b. ammeter
    c. wattmeter
    d. ohmmeter

15. The majority of electrical shocks are caused by _____.
    a. lightning
    b. incorrect use of GFCIs
    c. workers' failure to observe safety precautions
    d. defective test equipment

# GLOSSARY

## Trade Terms Introduced in this Module

*Alternator:* A device that generates alternating current by means of conductors rotated in a magnetic field.

*Armature:* The rotating component of a generator.

*Autotransformer:* A transformer made from a single winding that is tapped to establish the primary and secondary voltage.

*Capacitor:* An electrical storage device containing two metal plates separated by an insulating (dielectric) material.

*Commutator:* The movable contact surface on an electric generator or motor.

*Dielectric:* A material that strongly resists the passage of current.

*Effective voltage:* See *RMS voltage*.

*Frequency:* The number of complete cycles of an alternating current, sound wave, or vibrating object that occur in a period of time.

*Fusible link:* A circuit protective device that melts, opening the circuit, when the current is excessive.

*Hertz (Hz):* The unit of measure for the frequency of alternating current. One Hertz equals one cycle per second.

*Induction:* To generate a current in a conductor by placing it in a moving magnetic field.

*Induction motor:* An AC motor.

*Inertia:* The tendency of a body in motion to remain in motion and a body at rest to remain at rest.

*Isolation transformer:* A transformer with a one-to-one turns ratio. It is used for personnel safety and to prevent electrical interference.

*Megohmmeter (megger):* A test instrument used to test high-resistance circuits.

*Microfarad:* One-millionth of a farad. Used to rate capacitors.

*Root-mean-square (RMS) voltage:* The value of AC voltage that will produce as much power when connected across a load as an equivalent amount of DC voltage. Also known as *effective voltage*.

*Rotor:* The rotating component of an induction motor.

*Run capacitor:* A capacitor that remains in the motor circuit while the motor is running to improve running efficiency.

*Run winding:* The stator winding of a motor that draws current during the entire running cycle of the motor.

*Sinusoidal (sine) wave:* The waveform created by an AC generator.

*Start winding:* The stator winding of a motor that is used to provide starting torque.

*Stator:* The stationary windings of a motor.

*Synchronous speed:* The maximum rated speed of a motor.

*Torque:* The force that must be generated to turn a motor.

*Turns ratio:* The ratio between the number of turns in the primary and secondary windings of a transformer.

# ANSWER KEY

## Answers to Review Questions

| Answer | Section |
|---|---|
| 1. a | 2.2.0 |
| 2. a | 2.3.0; Figure 5 |
| 3. c | 3.1.0 |
| 4. b | 3.2.0 |
| 5. a | 3.2.0 |
| 6. b | 4.2.0 |
| 7. d | 4.3.0 |
| 8. b | 5.1.0 |
| 9. c | 5.1.1 |
| 10. c | 5.1.5 |
| 11. b | 5.1.4 |
| 12. d | 5.2.0 |
| 13. a | 6.5.0 |
| 14. d | 6.6.0 |
| 15. c | 7.0.0 |

# REFERENCES & ACKNOWLEDGMENTS

## *Additional Resources*

This module is intended to present thorough resources for task training. The following reference works are suggested for further study. These are optional materials for continued education rather than for task training.

*General Training—Electricity (GTE)*, 1993. Syracuse, NY: Carrier Corporation.

*HVAC Servicing Procedures*, 1995. Syracuse, NY: Carrier Corporation.

*Modern Refrigeration and Air Conditioning*, 2000. A.D. Althouse, C.H. Turnquist, A.F. Bracciano. Tinley Park, IL: The Goodheart-Willcox Company, Inc.

*Refrigeration & Air Conditioning Technology*, 2000. William C. Whitman, William M. Johnson, John A. Tomczyk. Albany, NY: Delmar Publishers, Inc.

*Pocket Guide to Electrical Installations Under NEC 2002, Volumes I and II*, 2001. Quincy, MA: National Fire Protection Association.

## *Figure Credits*

| | |
|---|---|
| **Gerald Shannon** | 204SA02 |
| **Square D/Schneider Electric** | 204F13 |
| **Carrier Corporation** | 204SA03, 204F45, 204F46 |
| **Kenthal** | 204SA04 |
| **Supco, Inc.** | 204F36 |
| **Amprobe** | 204F39, 204F40 |
| **Walter Johnson** | 204SA05, 204SA06, 204F26 |

# NCCER CRAFT TRAINING USER UPDATES

The NCCER makes every effort to keep these textbooks up-to-date and free of technical errors. We appreciate your help in this process. If you have an idea for improving this textbook, or if you find an error, a typographical mistake, or an inaccuracy in the NCCER's Craft Training textbooks, please write us, using this form or a photocopy. Be sure to include the exact module number, page number, a detailed description, and the correction, if applicable. Your input will be brought to the attention of the Technical Review Committee. Thank you for your assistance.

*Instructors* – If you found that additional materials were necessary in order to teach this module effectively, please let us know so that we may include them in the Equipment and Materials list in the Instructor's Guide.

**Write:** Curriculum Revision and Development Department
National Center for Construction Education and Research
P.O. Box 141104, Gainesville, FL 32614-1104

**Fax:** 352-334-0932

**E-mail:** curriculum@nccer.org

---

Craft _____ Module Name _____

Copyright Date _____ Module Number _____ Page Number(s) _____

Description

_____

_____

_____

_____

(Optional) Correction

_____

_____

_____

(Optional) Your Name and Address

_____

_____

_____

Module 03205-01

# *Basic Electronics*

## COURSE MAP

This course map shows all of the modules in the second level of the HVAC curriculum. The suggested training order begins at the bottom and proceeds up. Skill levels increase as you advance on the course map. The local Training Program Sponsor may adjust the training order.

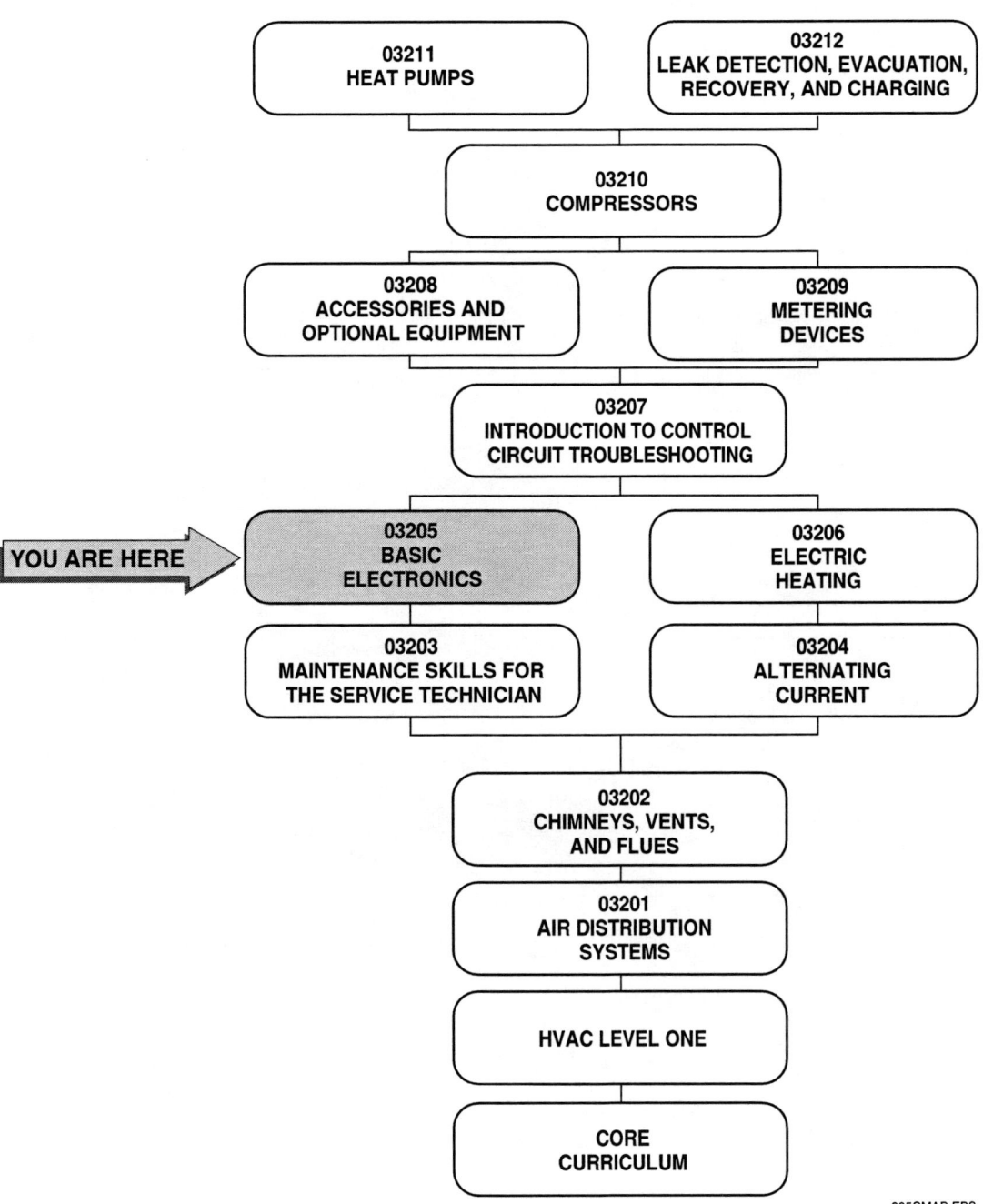

## MODULE 03205 CONTENTS

- **1.0.0 INTRODUCTION** .................................................. 5.1
- **2.0.0 THEORY OF ELECTRONICS** ................................ 5.1
- **3.0.0 SEMICONDUCTOR FUNDAMENTALS** ................. 5.2
  - 3.1.0 Conductors ................................................. 5.3
  - 3.2.0 Insulators .................................................... 5.3
  - 3.3.0 Semiconductors .......................................... 5.3
- **4.0.0 ELECTRONIC COMPONENTS AND CIRCUITS** ... 5.4
  - 4.1.0 Diodes ........................................................ 5.4
  - *4.1.1 Rectifiers* ................................................... 5.5
  - 4.2.0 Light-Emitting Diode ................................... 5.7
  - 4.3.0 Resistors .................................................... 5.9
  - *4.3.1 Resistor Color Codes* ................................ 5.10
  - 4.4.0 Thermistors ................................................ 5.11
  - *4.4.1 Testing a Thermal-Electric Expansion Valve Sensor* ... 5.12
  - *4.4.2 Testing Motor Protection Thermistors* ....... 5.13
  - 4.5.0 Cadmium Sulfide Detector ......................... 5.13
  - 4.6.0 Electronically Commutated Motors ............ 5.14
  - 4.7.0 Variable Frequency Drives ........................ 5.15
- **5.0.0 PRINTED CIRCUIT BOARDS** ........................... 5.15
  - 5.1.0 Integrated Circuit Chips ............................. 5.16
  - 5.2.0 Microprocessors ........................................ 5.17
  - 5.3.0 Diagnostic Capability ................................. 5.17
  - 5.4.0 Electrostatic Discharge Sensitivity ............. 5.17
- **6.0.0 INTRODUCTION TO COMPUTERS** ................. 5.18
  - 6.1.0 Special Terms ........................................... 5.18
  - 6.2.0 Mainframe Computers ............................... 5.19
  - 6.3.0 Personal Computers ................................. 5.19
  - *6.3.1 Monitors* ................................................... 5.21
  - *6.3.2 Connections* ............................................. 5.22
  - *6.3.3 Inside the PC* ........................................... 5.22
  - 6.4.0 Computer Storage Media .......................... 5.24
- **SUMMARY** ................................................................ 5.25
- **REVIEW QUESTIONS** ............................................... 5.26
- **GLOSSARY** .............................................................. 5.27
- **ANSWERS TO REVIEW QUESTIONS** ..................... 5.28
- **REFERENCES** .......................................................... 5.29

**Figures**

| | | |
|---|---|---|
| Figure 1 | Structure of an atom | 5.2 |
| Figure 2 | Electron in orbit around the nucleus | 5.2 |
| Figure 3 | Atom of a conductor | 5.3 |
| Figure 4 | Atom of an insulator | 5.3 |
| Figure 5 | Atom of a semiconductor | 5.4 |
| Figure 6 | Material structure of a diode | 5.4 |
| Figure 7 | Forward and reverse bias | 5.5 |
| Figure 8 | Diode component identification | 5.5 |
| Figure 9 | Testing a diode with an ohmmeter | 5.6 |
| Figure 10 | Half-wave rectifier | 5.6 |
| Figure 11 | Full-wave rectifier | 5.7 |
| Figure 12 | Bridge rectifier | 5.7 |
| Figure 13 | Three-phase rectifier | 5.7 |
| Figure 14 | Electroluminescence in an LED | 5.8 |
| Figure 15 | Schematic symbols for LEDs and photo diodes | 5.8 |
| Figure 16 | Seven-segment display | 5.9 |
| Figure 17 | Common resistors | 5.9 |
| Figure 18 | Symbols used for variable resistors | 5.10 |
| Figure 19 | Resistor color codes | 5.11 |
| Figure 20 | Sample color codes on a fixed resistor | 5.11 |
| Figure 21 | Bridge circuit | 5.12 |
| Figure 22 | TE expansion valve | 5.12 |
| Figure 23 | Overload sensor test | 5.13 |
| Figure 24 | Cadmium sulfide flame detector | 5.13 |
| Figure 25 | Flame detector testing | 5.14 |
| Figure 26 | Simplified ECM circuit | 5.14 |
| Figure 27 | Simplified VFD control circuit | 5.15 |
| Figure 28 | Printed circuit board | 5.16 |
| Figure 29 | An integrated circuit (IC) chip | 5.16 |
| Figure 30 | A microprocessor chip | 5.17 |
| Figure 31 | Mainframe computer | 5.19 |
| Figure 32 | Personal multimedia computer (desktop processor) | 5.20 |
| Figure 33 | Personal multimedia computer (tower configuration processor) | 5.20 |
| Figure 34 | Computer input devices | 5.21 |
| Figure 35 | Rear view of a tower configuration PC | 5.22 |
| Figure 36 | Interior of a tower configuration PC | 5.23 |
| Figure 37 | Data storage devices | 5.24 |

# MODULE 03205

# Basic Electronics

## Objectives

When you have completed this module, you will be able to do the following:

1. Explain the basic theory of electronics and semiconductors.
2. Explain how various semiconductor devices such as diodes, LEDs, and photo diodes work, and how they are used in power and control circuits.
3. Identify different types of resistors and explain how their resistance values can be determined.
4. Describe the operation and function of thermistors and cad cells.
5. Test semiconductor components.
6. Identify the connectors on a personal computer.

## Prerequisites

Before you begin this module, it is recommended that you successfully complete the following modules: Core Curriculum; HVAC Level One; HVAC Level Two, Modules 03201 through 03204.

## Required Trainee Materials

1. Pencil and Paper
2. Appropriate Personal Protective Equipment

## 1.0.0 ♦ INTRODUCTION

The science of **electronics** plays a large role in the control of HVAC systems as it does in many other aspects of our lives. Most of the switching and sensing functions performed by the **electromechanical components** you studied in earlier lessons can now be done with electronic circuits and devices. You will encounter electromechanical devices for some time to come. During your career, however, you can expect to see electronic controls completely replace controls with moving parts.

Electronic circuits have some major advantages. For one thing, they are very small; thousands of circuits can fit on an **integrated circuit** or **chip** no larger than the end of your thumb. For contrast: computers built in the 1950s needed rooms full of equipment; by the 1990s, more processing power than those early models possessed would fit easily in the palm of your hand. **Microminiaturization** is a term that was coined in the computer age; the circuits used in modern computers are so tiny, they can only be seen with a powerful microscope.

Electronic circuits can do a lot more than conventional circuits. They can, for example, process a lot of information about the status of the system and the conditioned space, and use the information to precisely control system operation. This capability results in improved comfort control and operating efficiency. Electronic circuits are also easier to service and less likely to fail than conventional circuits.

## 2.0.0 ♦ THEORY OF ELECTRONICS

If you could view the flow of electrons through a high-powered microscope, at first glance, you might think you were studying astronomy rather than electricity. The atom consists of a central nucleus composed of protons and neutrons, surrounded by orbiting electrons, as shown in *Figure 1*. The nucleus is relatively large when compared with the orbiting electrons, just as our sun is large in comparison to its orbiting planets.

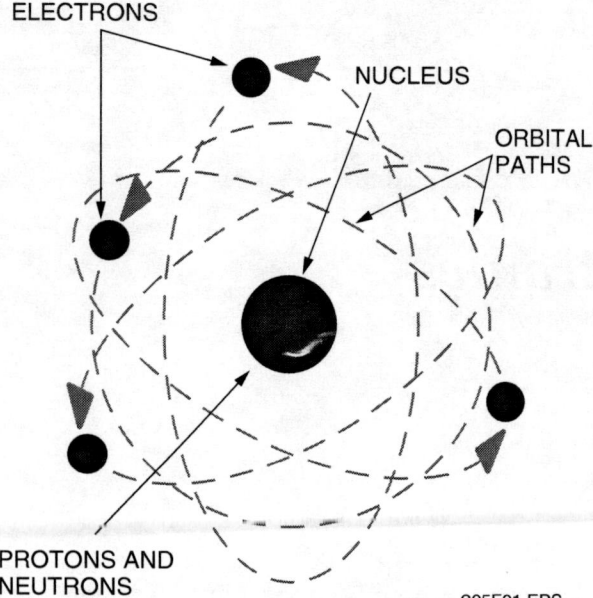

*Figure 1* ◆ Structure of an atom.

In an atom, the orbiting electrons are held in place by the electric force between the electron and the nucleus. It is similar to how Earth's gravity keeps its satellite (the moon) from drifting off into space. The law of electric charges states that opposite charges attract and like charges repel. The positively charged protons in the nucleus, therefore, attract the negatively charged electrons. If this force of attraction were the only one in effect, the electrons would be pulled closer and closer to the nucleus and eventually be absorbed into it. However, this force of attraction is balanced by the **centrifugal force** that results from the motion of the electrons as they orbit around the nucleus (*Figure 2*). The law of centrifugal force states that a spinning object will pull away from its center point. The faster an object spins, the greater the centrifugal force becomes.

Since the protons and electrons of an atom are equal in number, and equal and opposite in charge, they neutralize each other electrically. Thus, each atom is normally electrically neutral— that is, it exhibits neither a positive nor a negative charge. However, under certain conditions, an atom can become unbalanced by losing or gaining electrons. If an atom loses a negatively charged electron, the atom will exhibit a positive charge and is then referred to as a *positive ion*. Similarly, an atom that gains an additional negatively charged electron becomes negatively charged itself and is then called a *negative ion*. In either case, an unbalanced condition is created in the atom, causing the formerly neutral atom to become charged. When one atom is charged and there is an unlike charge in another nearby atom, electrons can flow between the two. This flow of electrons is an electrical current.

## 3.0.0 ◆ SEMICONDUCTOR FUNDAMENTALS

**Semiconductors** are the basis for what is known as *solid-state electronics*. Solid-state electronics is in turn the basis for all modern microminiature electronics such as the tiny integrated circuit and **microprocessor** chips used in computers.

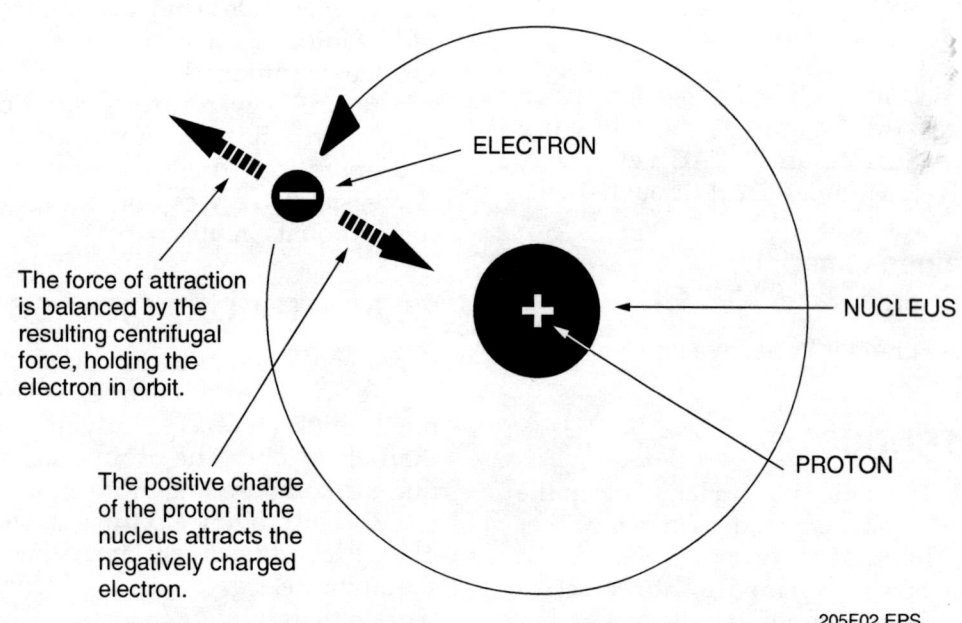

*Figure 2* ◆ Electron in orbit around the nucleus.

The ability to control the amount of conductivity in semiconductors makes them ideal for use in integrated circuits. In order to understand how semiconductors work, it is first necessary to review the principles of conductors and insulators.

## 3.1.0 Conductors

Conductors readily carry electrical current. Good electrical conductors are usually also good heat conductors. Conductors are generally made from materials such as metals that have comparatively large, heavy atoms.

In each atom, there is a specific number of electrons that can be contained in each orbit, or shell. The outer shell of an atom is the valence shell, and the electrons contained in the valence shell are known as **valence electrons**.

Conductors are materials that have only one or two valence electrons in their atoms, as shown in *Figure 3*. These electrons can be easily knocked out of their orbits and are therefore known as **free electrons**. An atom that has only one valence electron makes the best conductor because the electron is loosely held in orbit and is easily released to create current flow.

Gold and silver are excellent conductors, but they are too expensive to use on a large scale. However, in special applications requiring high conductivity, contacts may be plated with gold or silver. You would be most likely to find such conductors in precision devices where small currents are common and a high degree of accuracy is essential.

Copper is the most widely used conductor because it has excellent conductivity, while being much less expensive than precious metals such as gold and silver. Copper is used as the conductor in most types of wire and provides the printed current path on printed circuit boards. Aluminum is also used as a conductor, but it is not as good as copper. Aluminum may be prohibited in some applications such as household wiring because of its tendency to overheat.

## 3.2.0 Insulators

As you already know, insulators are materials that resist (and sometimes totally prevent) the passage of electrical current. Rubber, glass, and some plastics are common insulators. The atoms of insulating materials are characterized by having more than four valence electrons in their atomic structures. *Figure 4* shows the structure of an insulator atom. Note that it has eight valence electrons; this is the maximum number of electrons for the third shell of an atom. Therefore, this atom has no free electrons and will not easily pass electric current.

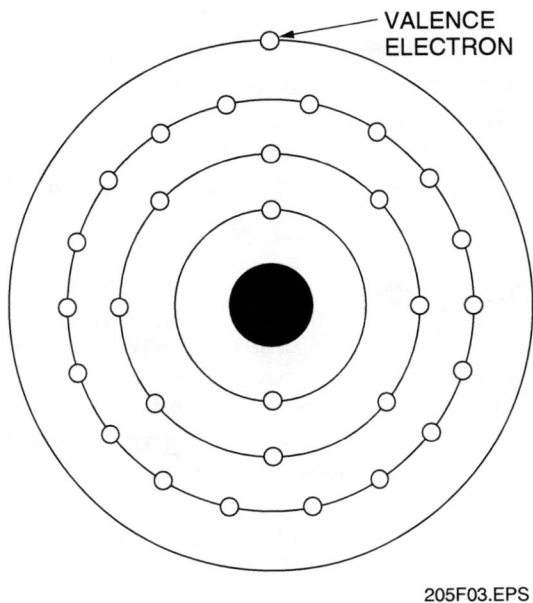

*Figure 3* ◆ Atom of a conductor (copper).

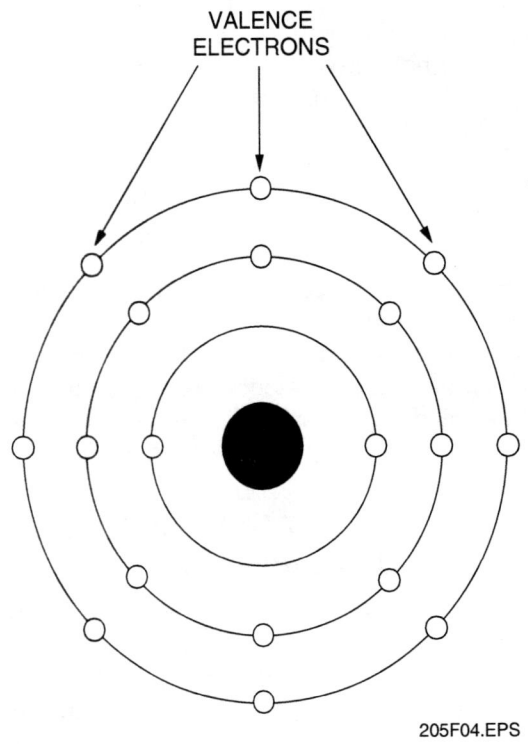

*Figure 4* ◆ Atom of an insulator.

## 3.3.0 Semiconductors

Semiconductors (*Figure 5*) are materials that are neither good conductors nor good insulators. The materials used as semiconductors, such as germanium and silicon, have more free electrons than an insulator, but fewer than a conductor. Silicon is more commonly used because it withstands heat better.

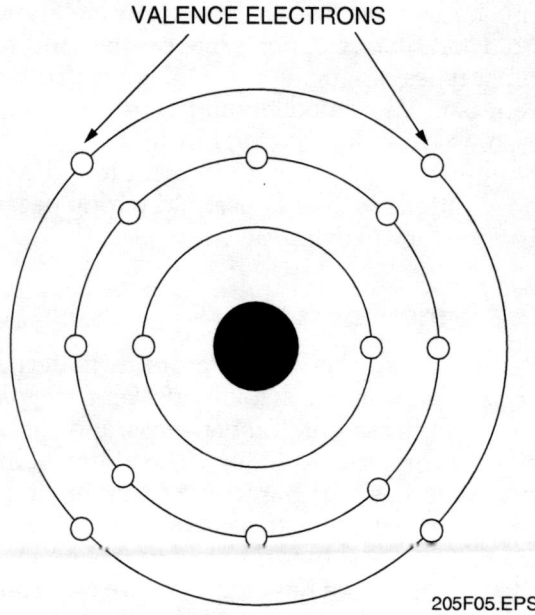

*Figure 5* ◆ Atom of a semiconductor.

*Figure 6* ◆ Material structure of a diode.

The factor that makes semiconductors valuable in electronic circuits is that their conductivity can be readily controlled. Semiconductors can be made to have positive or negative characteristics by adding certain impurities through a process known as *doping*.

When a substance with five valence electrons (e.g., indium or gallium) is added to the semiconductor material, the semiconductor material will no longer be electrically neutral. Instead, it will take on a positive charge and be known as a *P-type material*.

When substances like arsenic or antimony, which have three valence electrons, are added to the semiconductor material, the material takes on a negative charge and is known as an *N-type material*.

## 4.0.0 ◆ ELECTRONIC COMPONENTS AND CIRCUITS

All solid-state (semiconductor) devices are made from a combination of P-type and N-type materials. The type of device formed is determined by how the P-type and N-type materials are connected or joined, the number of layers of material, and the thickness of various layers. For instance, a diode is often called a *PN junction* because it is made by joining a piece of P-type material and a piece of N-type material, as shown in *Figure 6*. The contact surface is the PN junction.

Diodes allow current to flow in one direction, but not in the other. This unidirectional current capability is the distinguishing feature of the diode. The activity occurring at the PN junction is responsible for the unidirectional property of the diode. Diodes are discussed at length in the following section.

### 4.1.0 Diodes

Modern HVAC systems rely heavily on electronic controls, which use low-level DC voltages. Some HVAC systems also use special controls powered by DC motors when very precise control is required. The electricity furnished by the power company is AC; it must be converted to DC to be suitable for most electronic circuits. The process of converting AC to DC is known as **rectification**.

Diodes are used extensively to convert AC to DC. A diode conducts current only when the volt-

### How Much Do I Need to Know About Electronics?

Do you need to be an electronics technician as well as an HVAC technician to work on modern electronically controlled equipment? No, you don't. While it can be helpful to know the internal workings of an electronic device, you only need to know what the device does in the system, and what its inputs and outputs are supposed to be, in order to troubleshoot the system.

age at its **anode** is positive with respect to the voltage at its **cathode** (*Figure 7*). At that time, it is said to be forward biased. When the voltage at the anode is negative with respect to the cathode (reverse bias), current will not flow unless the voltage is so high that it overwhelms the diode. Most circuits using diodes are designed so that the diode will not conduct current unless the anode is positive with respect to the cathode.

There are several ways in which diodes are marked to indicate the cathode and the anode (*Figure 8*). Note that in some cases, diodes are marked with the schematic symbol, or there may be a band at one end to indicate the cathode. Other types of diodes use the shape of the diode housing to indicate the cathode end; that is, the cathode end is either beveled or enlarged to ensure proper identification. When in doubt, the polarity of a diode may be determined with an ohmmeter as shown in *Figure 9*.

Diodes can also be tested with an ohmmeter. The leads of the ohmmeter are placed on the anode and cathode of the diode. The meter selector should be placed at the lowest ohms scale.

The diode will only conduct an electric current in one direction. If the ohmmeter shows a low resistance reading in both directions, the diode is faulty. A good diode will block current flow in one direction and not in the other. Therefore, if the diode indicates flow in both directions or no flow in both directions (leads or meter polarity reversed), it is defective. LEDs can be checked in the same manner as regular diodes.

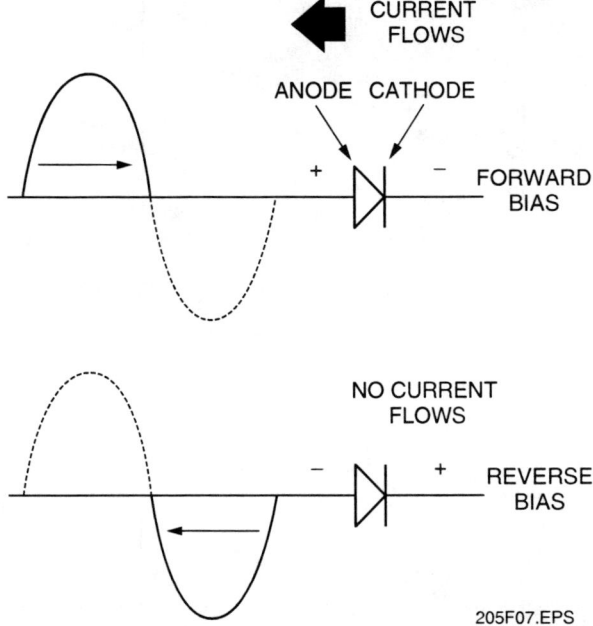

*Figure 7* ◆ Forward and reverse bias.

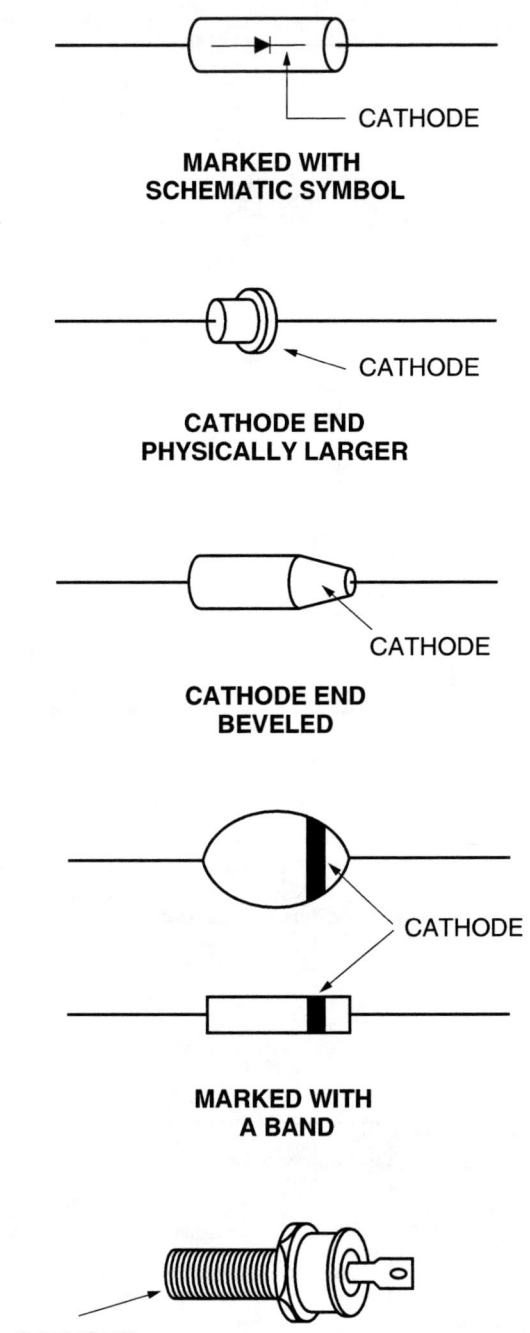

*Figure 8* ◆ Diode component identification.

### 4.1.1 Rectifiers

In a **half-wave rectifier** (*Figure 10*), the single diode conducts current only when the AC applied to its anode is on its positive half-cycle. The result is a pulsating DC voltage. A capacitor connected across the load can filter some of the AC component (ripple), but the voltage is not clean enough to operate electronic circuits.

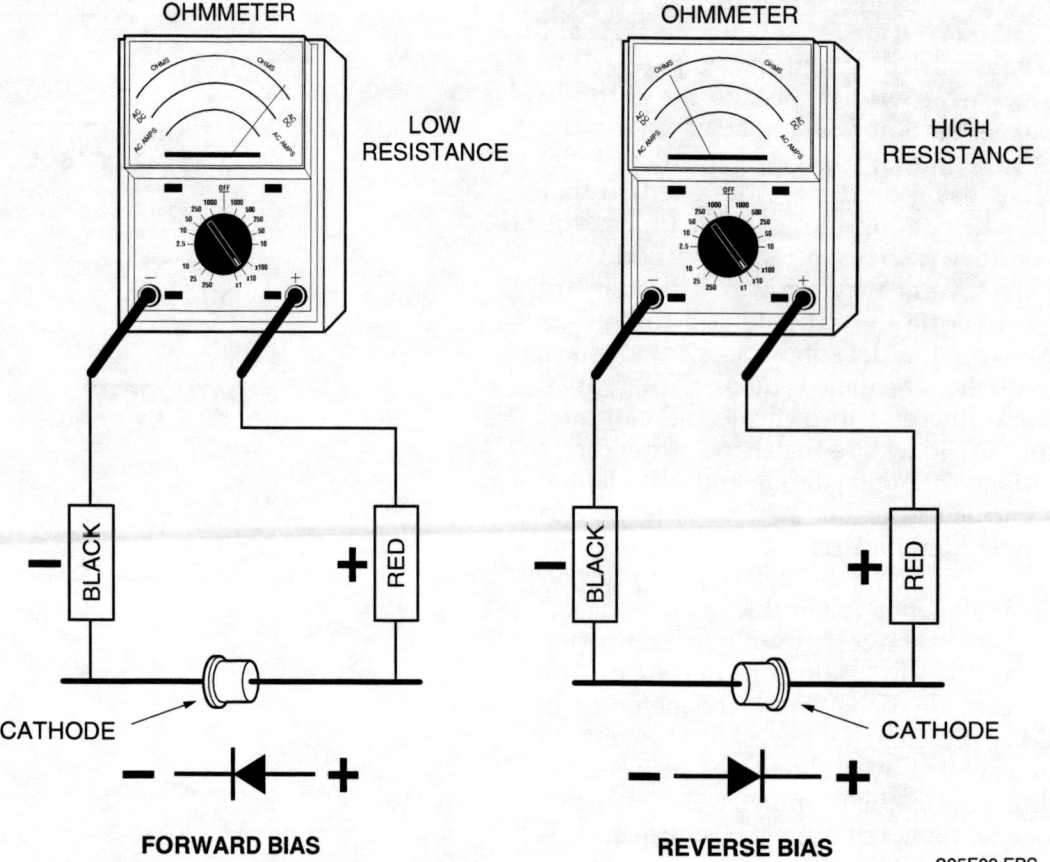

*Figure 9* ◆ Testing a diode with an ohmmeter.

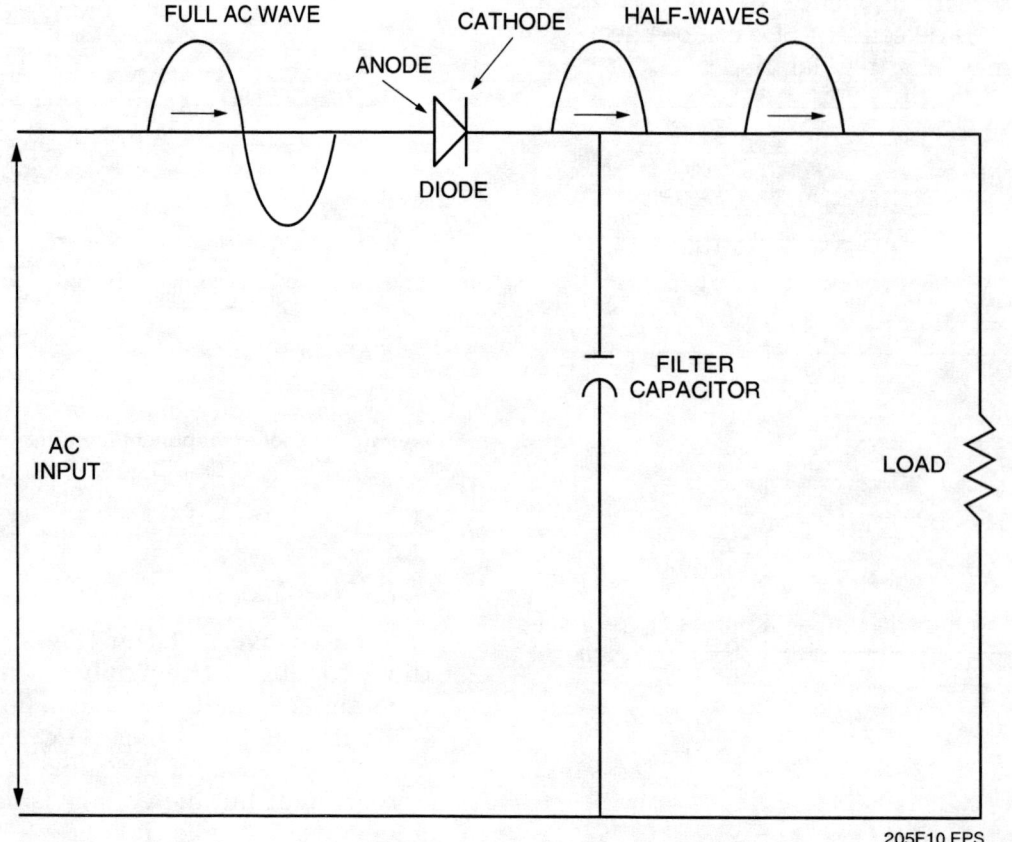

*Figure 10* ◆ Half-wave rectifier.

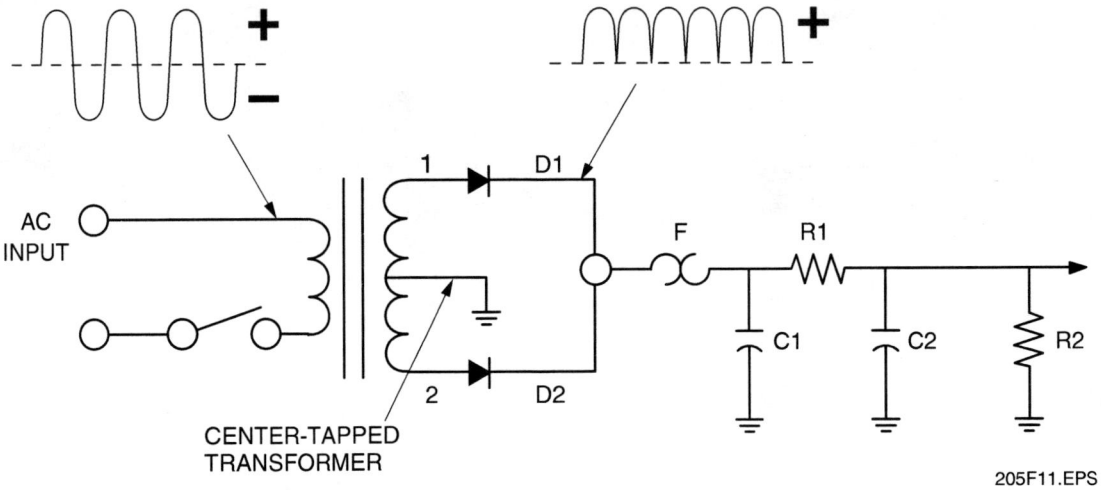

*Figure 11* ♦ Full-wave rectifier.

Figure 11 shows a **full-wave rectifier** with a special center-tapped transformer. In this circuit, one of the diodes conducts on each half-cycle of the AC input. This produces a smoother pulsating DC voltage. Again, a filter capacitor can be used to eliminate almost all of the ripple.

The **bridge rectifier** (*Figure 12*) contains four diodes, two of which conduct on each half-cycle. The bridge rectifier provides a smooth DC output, and is the type most commonly used in electronic circuits.

An advantage of the bridge rectifier is that it does not need a center-tapped transformer. A filter and voltage regulator added to the output of the rectifier provide the precise, stable DC voltage needed for electronic devices.

In a three-phase power system, the three-phase rectifier shown in *Figure 13* is used.

## 4.2.0 Light-Emitting Diode

A **light-emitting diode (LED)** is, as the name implies, a diode that will give off visible light when it is energized. In any forward-biased diode, some energy is given off in the form of photons. In some types of diodes, the number of photons of light energy emitted is sufficient to create a very visible light source.

The process of giving off light by applying an electrical source of energy is called *electroluminescence* (*Figure 14*).

Note in *Figure 15* that the symbol for an LED is similar to that of a conventional diode except that an arrow is pointing away from the diode.

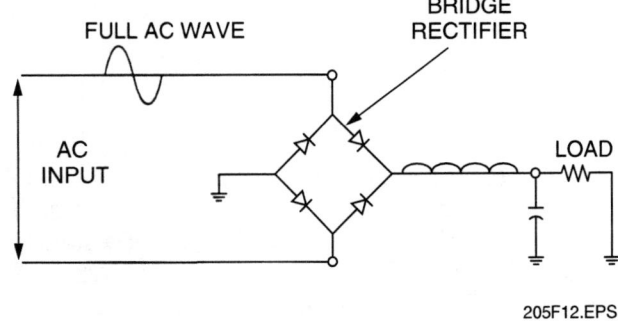

*Figure 12* ♦ Bridge rectifier.

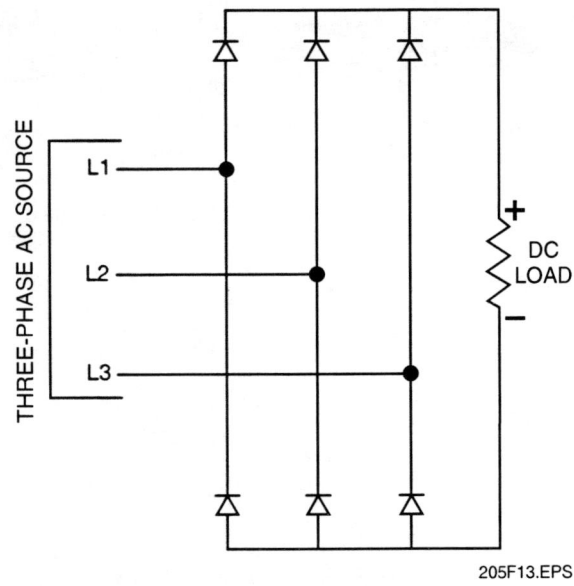

*Figure 13* ♦ Three-phase rectifier.

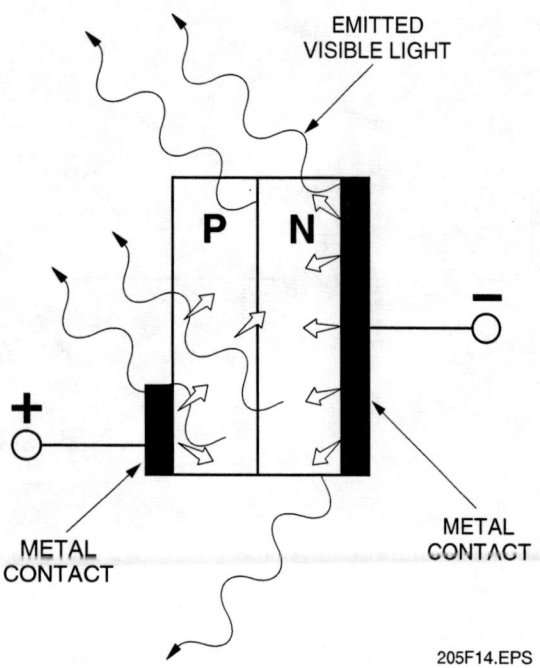

*Figure 14* ◆ Electroluminescence in an LED.

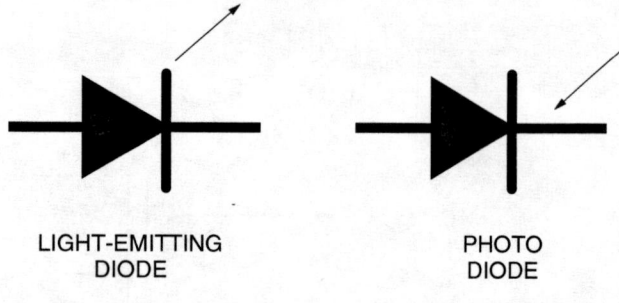

*Figure 15* ◆ Schematic symbols for LEDs and photo diodes.

The **photo diode** is another solid-state device that is turned on by light. The schematic symbol for the photo diode is exactly like that of a standard LED except that the arrow is reversed, as shown in *Figure 15*. The photo diode must have light in order to operate. It acts like a conventional switch. That is, light turns the circuit on, and the absence of light opens the circuit.

The liquid crystal display (LCD) is another method used to display information in electronic systems. The LCD is a segmented display containing conductive material in a semi-liquid state. Different combinations of segments are electrically excited to create the display of a number or letter. *Figure 16* shows how numbers are formed. The number 8 requires all seven segments to be excited, while other numbers use fewer segments. Letters are formed in a similar manner, except that diagonal segments are needed to represent letters such as *X* and *M*.

### LEDs

LEDs are sometimes used to display fault messages on electronic control boards used in heating and air conditioning systems. The LED shown on this PC board flashes a code to indicate one of ten possible faults. If the light remains on all the time, it indicates a failure on the PC board. Otherwise, the codes represent failures that occurred in other parts of the system.

Automatic diagnostic systems like this use sensors to determine if required values are present. The system senses enough points to be able to indicate with high probability where the fault is located. Keep in mind that the fault code indicates the probable fault. The troubleshooter must still confirm the failure and verify the repair.

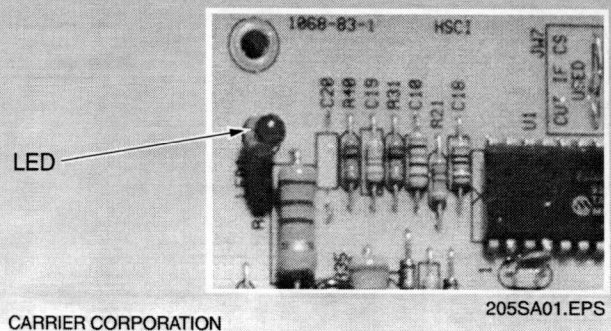

CARRIER CORPORATION

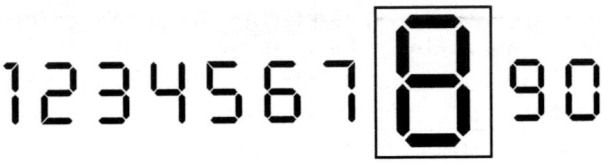

*Figure 16* ♦ Seven-segment display.

When used in a circuit, an LED is generally operated at about 20mA or less. For example, if an LED is to be connected to a 9VDC circuit, a current-limiting resistor must be connected in series with the LED. Ohm's law may be used to calculate the required resistance as follows:

$$R = \frac{E}{I}$$

$$R = \frac{9VDC}{.020A}$$

$$R = 450\Omega$$

Therefore, a 450Ω resistor or a resistor of the closest standard size (without going under 450Ω) should be used to limit the current flow through the LED.

> **NOTE**
> LEDs use the same identifying marks as conventional diodes. Most manufacturers use a flat surface in the LED case near one of the leads. The lead closest to this flat surface is the cathode.

LEDs are used as pilot lights on electronic equipment and as numerical displays. Many programmable HVAC controls use LEDs to indicate when a process is in operation. LEDs are also used in the opto-isolation circuit of solid-state relays for both motor controls and HVAC control systems.

A light-sensing diode (photo diode) can be checked with an ohmmeter by varying the amount of light available to the sensor. To test a photo diode, set the ohmmeter at the lowest scale and connect the leads to the anode and cathode. If the polarity of the leads is correct, the meter reading will fluctuate with the varying light input. If there is no instrument needle deflection, reverse the meter leads and check again. The photo diode should conduct current in only one direction.

### 4.3.0 Resistors

The two most common types of resistors used in electronic circuits are the wire-wound and the carbon composition. A wire-wound resistor consists of a length of nickel wire wound on a ceramic tube and covered with porcelain. Low-resistance connecting wires are provided and the resistance value is usually printed on the side (*Figure 17*). Carbon composition resistors are constructed by molding mixtures of powdered carbon and insulating materials into a cylindrical shape. An outer sheath of insulating material provides mechanical and electrical protection; connecting wires are provided at each end. Carbon composition resistors are smaller and less expensive than the wire-wound type. However, the wire-wound type is the more rugged of the two, and is able to handle more power than the carbon type.

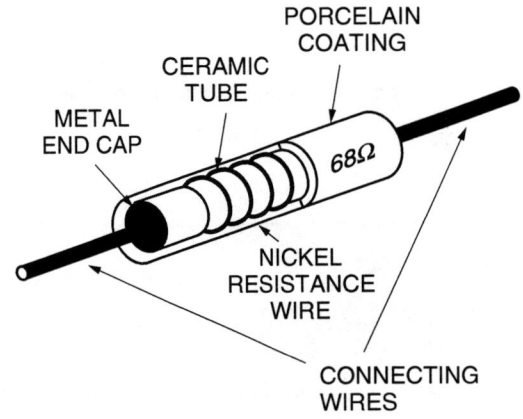

**WIRE-WOUND RESISTOR**

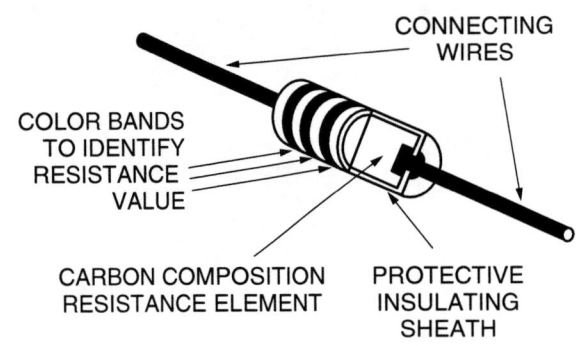

**CARBON COMPOSITION RESISTOR**

*Figure 17* ♦ Common resistors.

While most resistors have standard fixed values, variable or adjustable resistors are also used a great deal in electronics. The two most common symbols for a variable resistor are shown in *Figure 18*.

A variable resistor consists of a coil of closely wound insulated resistance wire formed into a partial circle. The coil has a low-resistance terminal at each end; a third terminal is connected to a movable contact. The movable contact can be set to any point on a connecting track that extends over one (uninsulated) edge of the coil.

Using the adjustable contact, the resistance from either end terminal to the center terminal may be adjusted from zero to the maximum resistance.

### 4.3.1 Resistor Color Codes

Because carbon resistors are physically small (some are less than 1 cm in length), it is not practical to print the resistance value on the side. Instead, a color code in the form of color bands is used to identify the resistance value and tolerance. The color code is illustrated in *Figure 19*.

Starting from one end of the resistor, the first two bands identify the first and second digits of the resistance value, and the third band indicates the number of zeros. An exception to this is when the third band is either silver or gold, which indicates a 0.01 or 0.1 multiplier, respectively.

The fourth band is always either silver or gold. In this position, silver indicates a ±10% tolerance and gold indicates ±5% tolerance.

Where no fourth band is present, the resistor tolerance is ±20%.

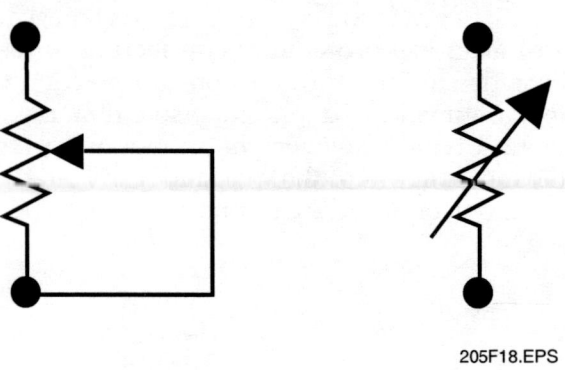

*Figure 18* ◆ Symbols used for variable resistors.

---

## Resistor Applications

A variable resistor is commonly used as a heat anticipator in a room thermostat. The heat anticipator causes the heating thermostat to open just before the room reaches the thermostat set point. This prevents the room temperature from overshooting the set point and causing the room to become too warm.

Fixed resistors are used in thermostats as cooling compensators. A cooling compensator is used to make up for the lag between the call for cooling and the time the system actually begins cooling the space. It turns the thermostat on just before the room temperature reaches the setpoint.

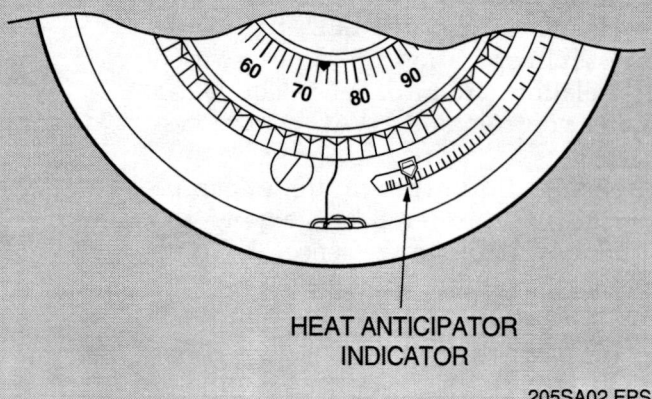

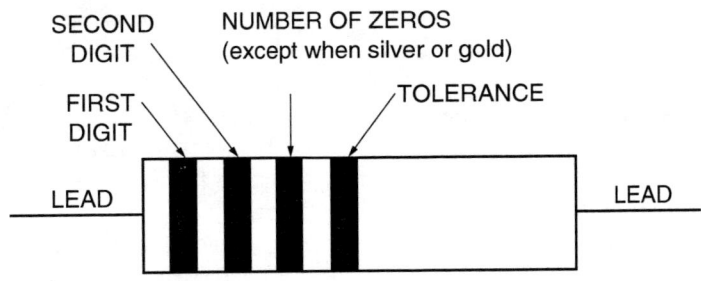

| 0 | BLACK  | 7    | VIOLET           |
|---|--------|------|------------------|
| 1 | BROWN  | 8    | GREY             |
| 2 | RED    | 9    | WHITE            |
| 3 | ORANGE | 0.1  | GOLD             |
| 4 | YELLOW | 0.01 | SILVER           |
| 5 | GREEN  | 5%   | GOLD - TOLERANCE |
| 6 | BLUE   | 10%  | SILVER - TOLERANCE |

*Figure 19* ◆ Resistor color codes.

Let's put this information to practical use by determining the range of values for the carbon resistor in *Figure 20*.

Brown = 1    Black = 0    Red = 2 zeros    Gold = ±5%
First digit of 1 + second digit of 0 + 2 zeros = 1,000Ω

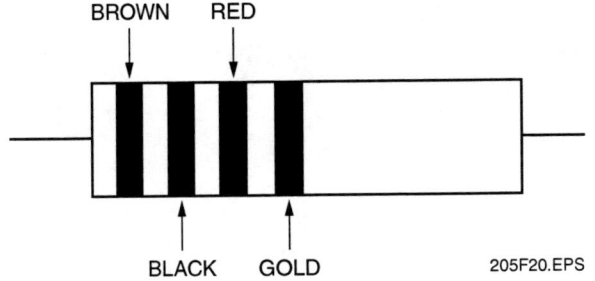

*Figure 20* ◆ Sample color codes on a fixed resistor.

Since this resistor has a value of 1,000Ω ±5%, the resistor can range in value from 950Ω to 1,050Ω.

### 4.4.0 Thermistors

Thermistors are temperature-sensitive semiconductor devices. Their resistance varies in a predictable way with variations in temperature. This allows them to be used in a variety of HVAC control applications. Thermistors have either a positive or a negative coefficient of resistance. If the resistance increases as temperature rises, it has a positive coefficient of resistance. If resistance increases as the temperature drops, it has a negative coefficient of resistance. Thermistors come in different sizes and shapes to suit a variety of applications.

---

### Replacing Resistors

An HVAC technician normally would not have to replace a failed resistor on an electronic device, but it can happen. Here are some rules to follow if you ever have to replace a failed resistor.
- Replace with the same resistance value.
- Replace with an equal or lower tolerance value. For example, a 10% tolerance could be replaced with a value of 5%, but not 20%.
- Replace with the same power rating or higher. Resistors of the same resistance but with a higher power rating will be larger in size.

Thermistors are used to sense temperature changes. They are also used as motor protective devices. Some typical applications of thermistors are: electronic thermometers, room thermostats, duct sensors, electronic expansion valve sensors, and selected control circuits.

Vacuum measurement can also be obtained with thermistors. Two thermistors can be wired in a bridge circuit (*Figure 21*) so that any change in one thermistor will produce a reading on a current meter that is directly calibrated in microns. One thermistor is placed in the vacuum and the other thermistor is placed in the ambient air. With this configuration, the rate of heat loss can be directly converted into a vacuum reading.

Thermistors are used in temperature differential controls, such as the defrost control in some heat pumps. Two thermistors are used; one senses coil temperature and the other senses air temperature. They are wired in a circuit so that a 15°F to 25°F temperature differential will produce enough difference in resistance to turn a relay OFF or ON.

Another use of a thermistor is as a start-assist device for a compressor motor. A ceramic thermistor with a steep-slope positive temperature coefficient is wired in parallel with the run capacitor on a permanent split capacitor motor, increasing the starting torque by 200% to 300%.

Other uses of thermistors include sequence switching and current in-rush surge suppression.

### 4.4.1 Testing a Thermal-Electric Expansion Valve Sensor

When a thermistor is used as a sensing device for the operation of a thermal-electric (TE) expansion valve, both the expansion valve and the thermistor can be checked as follows. With the leads of the voltmeter placed across the terminals of the electric valve (*Figure 22*) and the system operating at or near peak load, the voltmeter reading should be within the range of 15V to 20V. With the system operating at lower loads, the reading should remain between 8V and 14V. The size of the expansion valve in relation to the capacity of the system will be directly related to the time that the valve registers the voltage limits. Therefore, the manufacturer's recommendations should be consulted. The readings should be observed for two or three minutes.

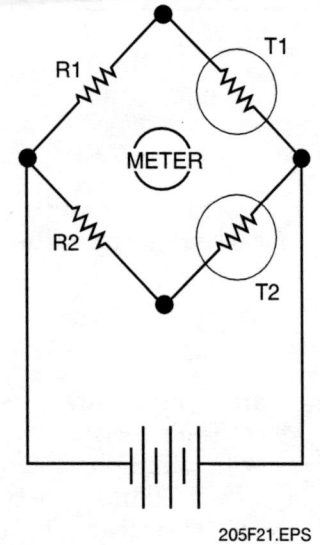

*Figure 21* ◆ Bridge circuit.

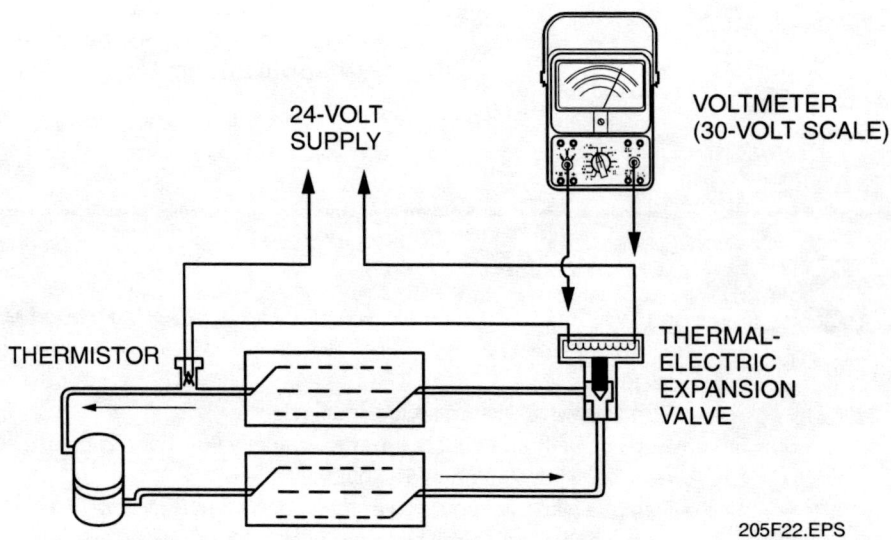

*Figure 22* ◆ TE expansion valve.

## 4.4.2 Testing Motor Protection Thermistors

Thermistors used on the internal windings for overload protection of three-phase motors can be checked by taking a resistance reading through them. An ohmmeter should be used and connected as illustrated in *Figure 23*.

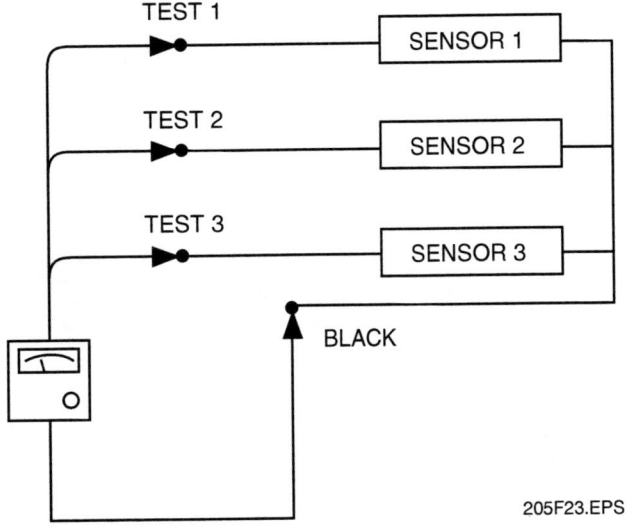

*Figure 23* ◆ Overload sensor test.

A short or ground in the thermistor will be indicated by a reading of zero or a value approaching zero. An open would be indicated by a reading of infinity. The procedure is as follows:

**Step 1** Shut off the disconnect switch and allow the equipment to cool.

**Step 2** Connect the ohmmeter leads as indicated in the illustration and record the resistance in ohms.

The values will change with a change in temperature, but at room temperature (75°F) the reading should be about 75Ω. Again, the resistance of the winding temperature sensors will vary by manufacturer and application, so unit specifications should be consulted before condemning or accepting the devices.

## 4.5.0 Cadmium Sulfide Detector

A cadmium sulfide flame detector (cad cell) is a device mounted in an oil burner that looks down the tube at the flame. Cadmium sulfide (*Figure 24*) is photoconductive; its electrical resistance is high in darkness but lower in the presence of visible light. The more intense the light, the lower the internal resistance. The size of the cadmium sulfide conductor determines the sensitivity of the cell to the light and the amount of current it can conduct. The resistance of the cell may exceed 100,000Ω in darkness but could drop to less than 1,500Ω in the presence of a flame. High internal resistance prevents current from flowing across the cell. This keeps the primary control from energizing, thus shutting off the flow of fuel.

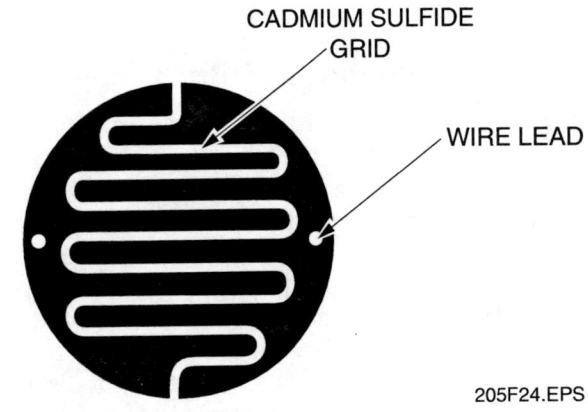

*Figure 24* ◆ Cadmium sulfide flame detector.

The cadmium sulfide flame detector can also be checked with an ohmmeter (*Figure 25*). Set the ohmmeter to the highest or intermediate scale. In darkness, the resistance of the cell should measure in excess of 100,000Ω. In the presence of an oil burner flame or other comparable light, the cell resistance should drop to less than 1,500Ω. Note that the resistance, in the presence of light or flame, will vary between makes and models. The manufacturer's instructions should specify the limits.

---

 **Cad Cell Maintenance**

A cad cell flame detector must detect light to prove that an oil burner is operating properly. If the face of the cad cell becomes coated with soot or a film of oily dirt, light can't be detected and the cad cell will shut the burner down. As part of any routine service of an oil burner, always remove and clean the face of the cad cell with a soft cloth to remove any soot or oil film.

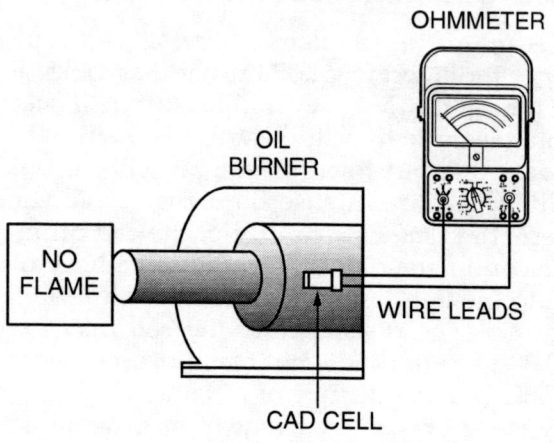
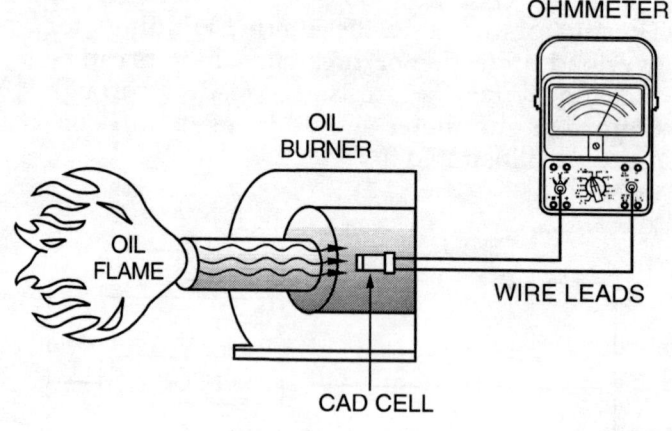

*Figure 25* ◆ Flame detector testing.

## 4.6.0 Electronically Commutated Motors

Electronically commutated motors (ECMs) are direct-current (DC) motors used in variable-speed applications. DC motors are considered far better for applications in which continuously variable control is needed. Like AC motors, DC motors rely on the interaction of the magnetic fields between two electromagnets, one fixed and the other rotating. Because DC voltage does not fluctuate, it is necessary to create a rotating magnetic field using a commutator, as shown in *Figure 26*. In many DC motors, the commutator is connected to the DC voltage source by brushes that remain in constant contact with the commutator and transfer power to it. Because brushes need periodic replacement, a brushless DC motor is used with ECMs, so that the motor can be used in non-reparable components, such as hermetic compressors.

The ECM receives its power from the system AC power source, and must convert that power to DC using a rectifier that is built into the ECM control. The electronic circuits in the ECM control convert the DC voltage to a voltage that is effectively three-phase AC; that is, three out-of-phase voltage waveforms. This voltage is applied to the three stator windings of the motor.

One application for the ECM is a variable-speed compressor motor. Rather than just cycling on and off in response to the thermostat, the ECM adjusts the speed of the compressor in proportion to the amount of cooling needed. This approach improves cooling efficiency and reduces energy costs.

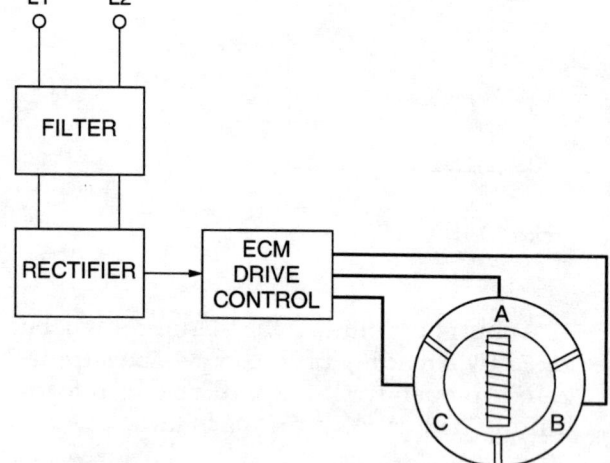

**SIMPLIFIED ECM CIRCUIT**

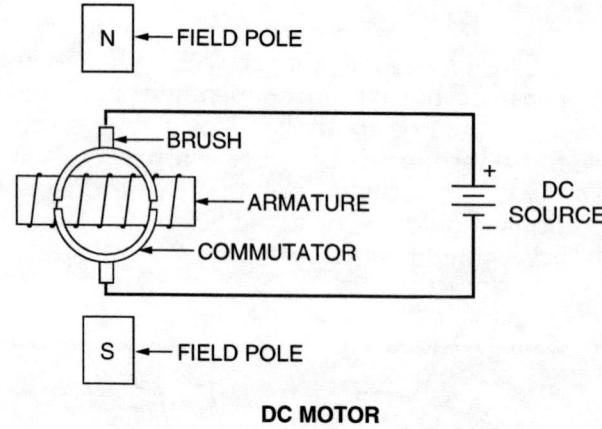

**DC MOTOR**

*Figure 26* ◆ Simplified ECM circuit.

## 4.7.0 Variable Frequency Drives

A variable frequency drive (VFD) is a motor control system that provides variable control of AC motors electronically. Most AC motors operate at a single speed. In the past, if a load required different speeds, it was common to mechanically adjust the load output using pulleys or to employ a method of throttling the load.

VFDs adjust the speed of the motor by adjusting the voltage and frequency of the electrical power supplied to the motor. The changes are made in response to system load requirements. The VFD is a typical closed-loop system. Sensors determine the load requirements and feed that information to the VFD control. A typical VFD contains a converter, an inverter, a control unit, and a sensor (*Figure 27*).

The sensor, which could be a temperature-sensing device, determines the instantaneous load requirement and feeds that information to the control unit. The rectifier converts the 60Hz system power signal into a DC voltage. The inverter converts the DC voltage into an adjustable-frequency, adjustable-voltage AC voltage. A technique known as *pulse width modulation (PWM)* is commonly used for the inversion process. PWM produces a current waveform that closely matches the power line waveform. This reduces the likelihood of motor overheating.

The control unit controls the amplitude and frequency of the voltage applied to the motor in response to the demand level indicated by the sensor.

## 5.0.0 ◆ PRINTED CIRCUIT BOARDS

Most of the electronic circuits you encounter will be mounted on printed circuit (PC) boards. In many HVAC systems, all the control circuits—relays, capacitors, diodes, etc.—are located on a single PC board (see *Figure 28*). The components are mounted on the top of the board; their electrical leads are inserted through holes in the board and soldered to terminal points on the bottom of the board. There is very little, if any, wiring on the PC board. Instead, a copper foil is bonded to the bottom of the board. The desired circuit is imprinted on the foil by a machine, and the copper is then chemically etched away from the unprinted areas. The printed copper acts as the conductor between the components on the circuit. Instead of a wiring harness and plug to connect the circuit to the outside, an edge connector is often built into the board. The edge connector then is plugged into a connector mounted on the hardware.

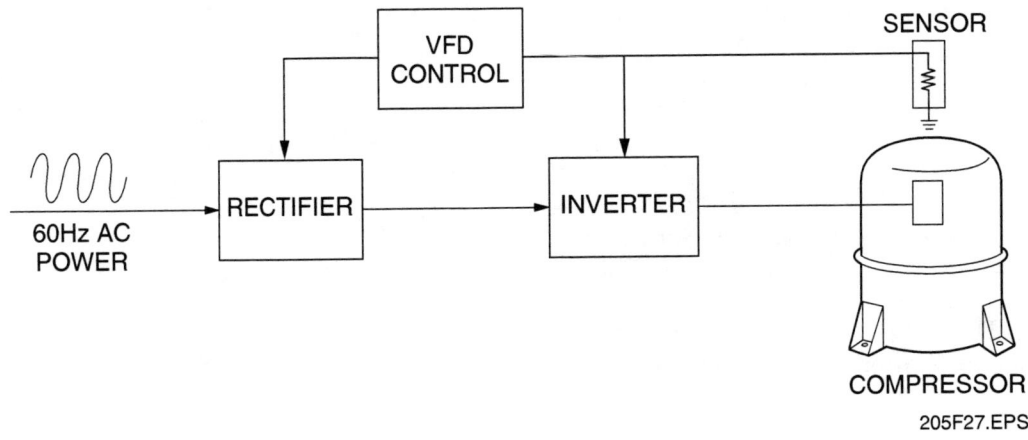

*Figure 27* ◆ Simplified VFD control circuit.

---

### Static Electricity

We've all experienced discharges of static electricity after walking across a carpeted floor and touching a light switch or doorknob. While seemingly harmless, these discharges can seriously damage or destroy delicate components on electronic devices. Before touching any electronic device, touch ground on the equipment chassis, a gas pipe, or a metal electrical conduit. This will ground any static charge in your body and prevent damage to the electronic device.

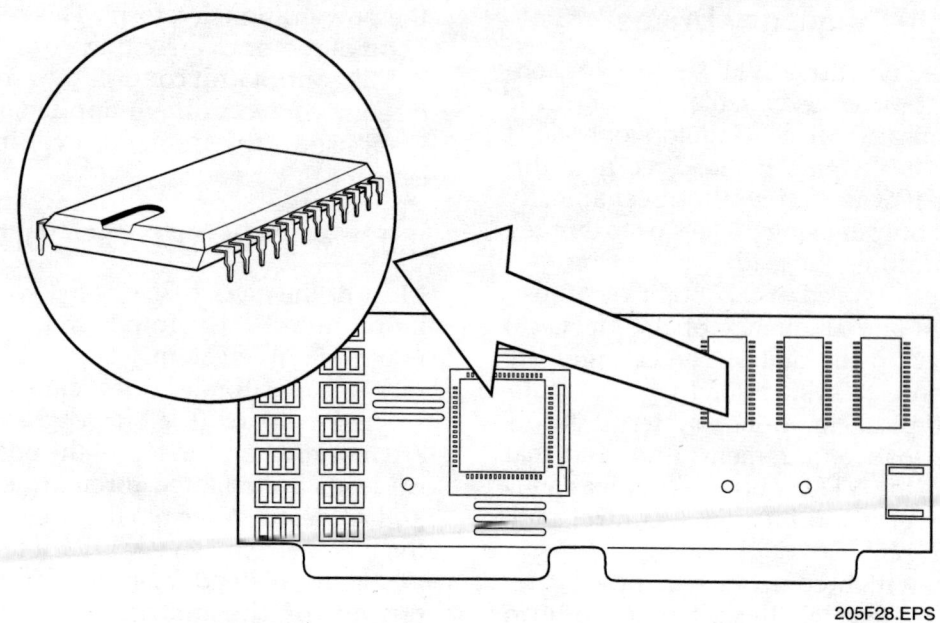

*Figure 28* ◆ Printed circuit board.

Some electronic circuits are packaged in sealed modules. This method is common with electronic control devices that perform a specific function and can be used in a number of different systems.

A very important feature that distinguishes packaged electronic controls from circuits built of discrete (separate) components is that the electronic circuit is treated as a black box; that is, if there is a control circuit failure, the entire board or module is replaced. In conventional circuits, on the other hand, you have to analyze the circuit and isolate the fault to the failed component; a bad relay, for example.

When an electronically controlled system is not working, it is tempting to just replace the control module or PC board without checking it. This will result in one of three outcomes, only one of which is desirable. The one good outcome is that it might fix the problem. A more likely outcome is that it won't. Electronic circuits are very reliable, and they have no moving parts, so it is not that common for an electronic control to fail. The worst possible outcome is that something external to the control caused it to fail, and will also cause its replacement to fail. It can be very embarrassing to explain to a customer why you charged them for a repair that didn't work.

Before replacing an electronic circuit, the troubleshooter must verify that the circuit has actually failed, and determine if an outside source caused the failure. This is done by first verifying that the printed circuit board or module is receiving the necessary supply voltages and control signals, and that they are at the proper levels. Once that is done, the outputs need to be verified. If the device is receiving the required inputs, but fails to produce the expected outputs, it can be assumed that the device has failed. Troubleshooting of electronic circuits is covered in detail in HVAC Level 3.

## 5.1.0 Integrated Circuit Chips

An integrated circuit chip (*Figure 29*) is a tiny wafer of semiconductor material containing microminiature electronic circuits designed to perform a specific function or functions. To get a perspective on what microminiature means, think about a multi-function digital wristwatch.

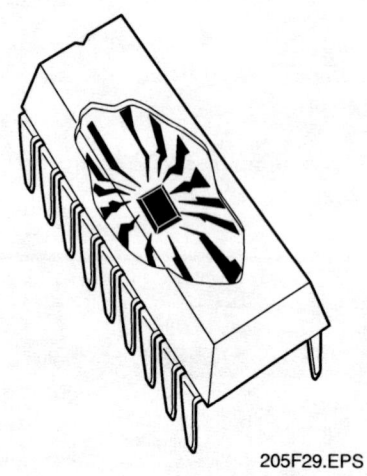

*Figure 29* ◆ An integrated circuit (IC) chip.

All the complex timekeeping, calendar, and display functions are contained on a single integrated circuit chip that you might have trouble finding if you looked inside the watch.

## 5.2.0 Microprocessors

Microminiaturization enables tiny devices smaller than the tip of your little finger to perform work that, in the early days of computers, used to take a roomful of electronic equipment to do. The semiconductor makes microminiaturization possible.

Semiconductors are materials in which the capacity to conduct electricity can be controlled by varying the voltage applied. In this case, we are talking about low-level DC voltages in the range of 5V to 15V. Heat, light, and pressure are also used to control current flow in semiconductors. Silicon and germanium are the two most widely used semiconductor materials.

Some integrated circuits can be programmed to perform complex tasks such as decision-making and mathematical calculations. These are known as *microprocessors* (*Figure 30*). They are the brains of the personal computer and are used to perform logical and analytical functions in many computerized HVAC systems. One of the most common applications is the programmable thermostat.

In microprocessor-controlled systems, integrated circuit chips and microprocessor chips are usually mounted with other components such as relays, resistors, and capacitors, on PC boards or encapsulated modules. The major advantage of a microprocessor-controlled system is its ability to provide very precise control. The microprocessor collects temperature, pressure, and humidity information from sensors located at strategic points in the system and the conditioned space. Information on the status of safety controls may also be supplied to the microprocessor. The microprocessor can evaluate the information and change system operation to meet changing conditions. Conventional controls are very limited in this sense; it would take many relays and hundreds of feet of wiring to accomplish even the most basic logic functions performed by a microprocessor.

## 5.3.0 Diagnostic Capability

Another important feature of microprocessor-controlled systems is their ability to recognize, isolate, and report faults. In small systems, the microprocessor receives sensor information such as temperatures and pressures and analyzes this information to locate a fault. The microprocessor is programmed to recognize patterns and relate those patterns to system components. Once the problem is isolated, the system will use a digital readout or a flashing light code to identify where the problem is located. Larger, more complex systems may have programmed tests that the technician can select to help isolate a malfunction.

## 5.4.0 Electrostatic Discharge Sensitivity

One important thing to remember about integrated circuits and microprocessors is that they can be damaged by static electricity. Be sure to ground yourself to something metal before handling them, and then don't touch the connector pins or wiring runs.

In later modules, we will discuss specific electronic control devices and circuits.

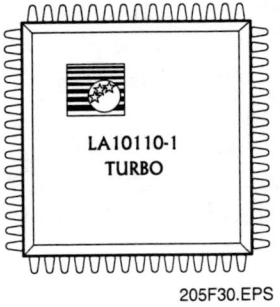

*Figure 30* ♦ A microprocessor chip.

### Troubleshooting an Electronic Furnace Control

A furnace contains an electronic control. Normal inputs to the control are 24 volts and Y, W, and G control signals from the room thermostat. Outputs from the control include 115 volts to operate the inducer motor and blower motors and 24 volts to operate the gas valve, blower relay, and cooling unit. On a no-heat service call, you find that 24 volts and all control signals from the room thermostat are present. A 24-volt output is available to the gas valve. The blower and inducer motor operate but there is no burner operation. Should you replace the electronic control in the furnace to correct the problem?

BASIC ELECTRONICS — TRAINEE MODULE 03205

## 6.0.0 ◆ INTRODUCTION TO COMPUTERS

Modern commercial buildings use computers to manage building systems such as HVAC, lighting, and security. In such buildings, the performance of these systems is monitored at a central computer workstation, which may be located in another building, and perhaps even another city. In addition to monitoring system functions, the computer is used as the control point for the building systems. Here's an example: If you needed to change the thermostat setpoint for all the offices in a 20-story building, you would no doubt prefer to do it once, rather than go from office to office changing all the thermostats. Computer-controlled building systems make this possible by electronically linking all the thermostats to a central control point.

Residential systems are now employing some of the same technology used in commercial systems. *Smart* homes, once a futuristic concept, are now a reality.

As an HVAC service technician, you can expect to work with computer-controlled systems. For that reason, it is essential that you become familiar with the terms and equipment associated with computers. This section reviews computer fundamentals for those who may have a limited knowledge of computers, and the special terminology associated with them.

### 6.1.0 Special Terms

Many special terms and abbreviations are used in the computer world. Here are some terms you may encounter in reading or talking about computers:

- *Bandwidth* – The speed at which data travels in transmission lines. Bandwidth is stated in cycles per second, which represents the number of bits of data per second. A typical telephone modem has a bandwidth of 56,000 bits per second (56Kbps), while a high-speed connection such as video cable has a bandwidth of 1.5 million bits per second. A file transferred on the 1.5Mbps line would move about 27 times faster than the same data transferred over a 56K line. A speed of 1.5Mbps is required to receive high-quality video.
- *Basic input/output system* – The **basic input/output system (BIOS)** is the first set of instructions to run when a computer is started (booted) up. The BIOS is stored in Read-Only Memory (ROM) located on the system board.
- *Binary digit (bit)* – The smallest unit of information in a computer system.
- *Byte* – One character, such as a number. It consists of 8 bits. A kilobyte (KB) is 1,024 bytes; a megabyte is approximately one million bytes; a gigabyte is approximately one billion bytes; and a terabyte is approximately one trillion bytes. These terms are used to define storage capacity.
- *Bus* – The wiring pathway between the internal elements of a computer. Buses are defined in terms of their width, which means how many bits of data they can carry, which in turn affects the processing speed of the computer. Two common busses are the ISA and PCI. The PCI bus is the faster of the two, and is able to handle 32-bit and 64-bit data. The ISA bus, in contrast, handles 16-bit data.
- *Cache* – Cache is a type of memory in which data is stored, or stock-piled ahead of time, so it is available for use when needed. Having key instructions and information readily available in cache allows the computer to work faster because it does not have to go the hard drive or other device to search for it.
- *DIP switch* – One of a set of tiny switches, located on a circuit board. DIP switches are used to configure the processor to perform certain functions. They are often used to select options.
- *Digital subscriber line (DSL)* – DSL is a method of providing high-speed communication over telephone lines. It is one of several such methods.
- *Handshake* – The process by which two computers initially establish communication. During the handshake, the computers determine if a connection is possible, then establish the best mode for the transmission.
- *Integrated drive electronics (IDE)* – IDE is a high-speed interface protocol associated with hard drives.
- *Integrated services digital network (ISDN)* – ISDN is a high-speed telecommunications connection. It has a bandwidth of 64Kbps, as compared with DSL which transfers at 150Kbps and up.
- *Network interface card (NIC)* – A NIC is a special printed circuit card or adapter that enables a workstation to connect to a computer network.
- *Parallel I/O* – I/O stands for input/output. A parallel I/O is one in which multiple data bits being transferred from one device to another are sent simultaneously on separate wires. Printers are typically connected to parallel I/O connections on computers.

- *Partition* – A section of a hard drive that is treated as a separate drive by the operating system. Partitions allow a single hard drive to have multiple formats. A computer user might put application software on one partition, and data files on another, or use one partition for one operating system, and a second one for a different operating system.
- *Plug and play* – A special process in which the BIOS recognizes peripheral devices such as printers, scanners, and drives, and automatically configures the system to interface with them. Before plug and play, the operator had to specifically configure the computer to handle each device as it was added.
- *Random Access Memory (RAM)* – RAM is the main temporary storage for information in a computer. Information stored in RAM can be accessed and changed very quickly by the computer. RAM is volatile storage. This means that the contents are erased if the power is shut off. RAM consists of memory chips located on a memory board connected to the main circuit board (motherboard). The memory modules are known as *SIMMs* (*single in-line memory modules*), which have memory chips on one side of the board, or *DIMMs* (*dual in-line memory modules*), which have memory chips on both sides of the board. The amount of RAM a computer has determines what applications it can use and how many applications it can have running at once. There are two types of RAM: DRAM (D is for dynamic) is the most common. It must be refreshed often by the computer to retain the information stored in it. SRAM (S is for static) retains information without being refreshed. It is more expensive, however, so it is only used where necessary, such as in video and cache applications. SDRAM and Rambus DRAM (RDRAM) are newer, faster versions of DRAM.
- *Read-only memory (ROM)* – ROM chips are pre-programmed with instructions or information for the computer in which they are used. One important allocation of ROM in a PC is the storage of the BIOS, which contains the boot-up instructions for the PC.
- *Small computer system interface (SCSI)* – Pronounced *scuzzy*. SCSI is an interface specification for connecting peripheral devices to a computer. It supports several high-speed devices through a single 50-pin or 68-pin cable.
- *Serial input/output (serial I/O)* – A method of transferring data between two devices one bit at a time. Modems and some printers are connected to a serial port.
- *T-1 line* – A high-speed communication line used for data transfer. It consists of 24 64Kbps channels, which can be used separately, combined into clusters, or combined into a single connection that will provide a 1.5Mbps data transfer rate.
- *T-3 line* – A very high-speed communication line consisting of 43 64Kbps channels, which can be combined into a single connection that will provide a 43Mbps transfer rate.
- *Virtual memory* – Hard disk space allocated to augment RAM. It is not as fast as RAM because of the disk access time, but there are some uses for which it is suitable. One of these is to serve as RAM when the PC is running multiple programs that require more RAM than the computer has available.

## 6.2.0 Mainframe Computers

Computers fall into two major classifications: mainframe computers and personal computers (PCs).

A mainframe computer (*Figure 31*) is intended for enterprise-wide applications where a large amount of processing is required. Large businesses and government entities use mainframe computers to handle their accounting and payrolls, keep track of inventory, and manage the flow of information and products. A single computer may be accessed by many people using personal computers or "dumb" terminals that consist of a monitor and keyboard. Clients of mainframe computers are linked to it in a network by cabling. External links are provided by telecommunications systems.

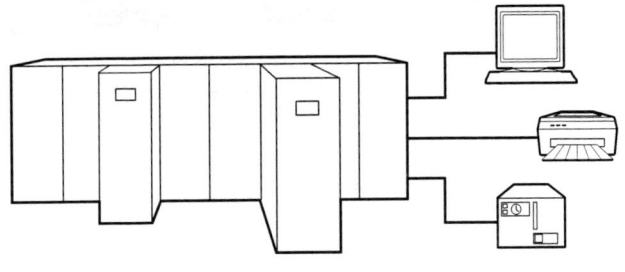

*Figure 31* ♦ Mainframe computer.

## 6.3.0 Personal Computers

Personal computers have become so commonplace that it is now unusual to meet someone who does not use one. People use them at home to play video games, obtain movies and music, correspond with friends and family, and do their shopping. At work, people use them to control their environments, manage their schedules, access information, and send information and correspondence around the world.

There are two basic types of PCs: The Apple Macintosh and the IBM PC Compatible. The latter, although originally developed by IBM, has been cloned by many companies and is the standard for business computing and most home computers. The Macintosh (or Mac) was the first PC to use a mouse, and has typically been favored by graphic designers and desktop publishers.

*Figure 32* shows a multimedia PC in the desktop configuration (that is, the central processing unit [CPU] is in a horizontal case with the monitor resting on it). Desktop PCs tend to be aimed at the low-cost home market, where expansion is less important.

If a computer capable of more expansion is needed, a tower configuration is chosen for the case, as shown in *Figure 33*. It has more internal expansion slots to accommodate additional special-purpose PC boards, more space for built-in storage drives, and more connectors (ports) to hook up peripherals such as scanners, printers, and game devices.

The components found in a typical system include:

- *Computer case* – This is the main box. It contains the processing circuits and other devices. It will be described further when we take a look inside.
- *Diskette (floppy) drive* – The floppy drive supports a 3½" magnetic diskette that will accommodate about 1.4 megabytes of data. It is therefore a convenient medium for storing and transferring text files outside the computer, but is not very useful for graphics files, which consume a lot more storage space than text. For many years, the floppy disk was a major means of transferring files from one computer to another, and a floppy disk drive was standard on every PC. With the advent of other read-write devices with much greater capacity (for example, ZIP™ disks and CD-R), the use of floppies has diminished.

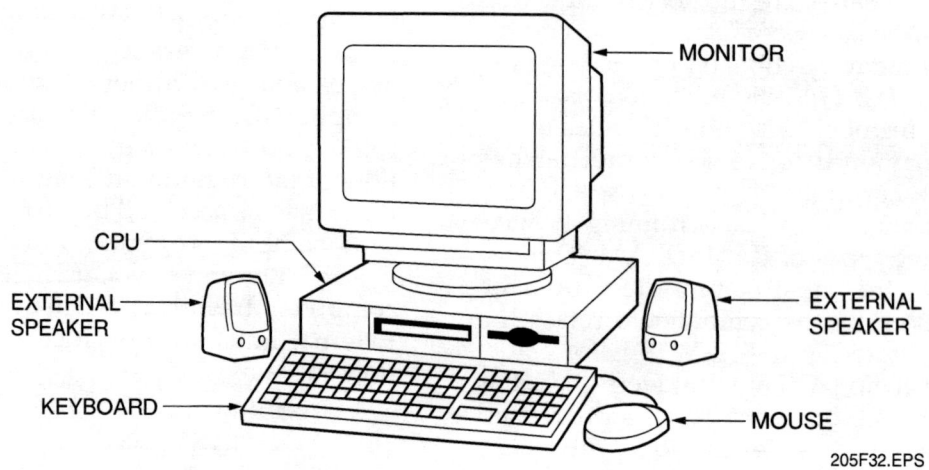

*Figure 32* ◆ Personal multimedia computer (desktop processor).

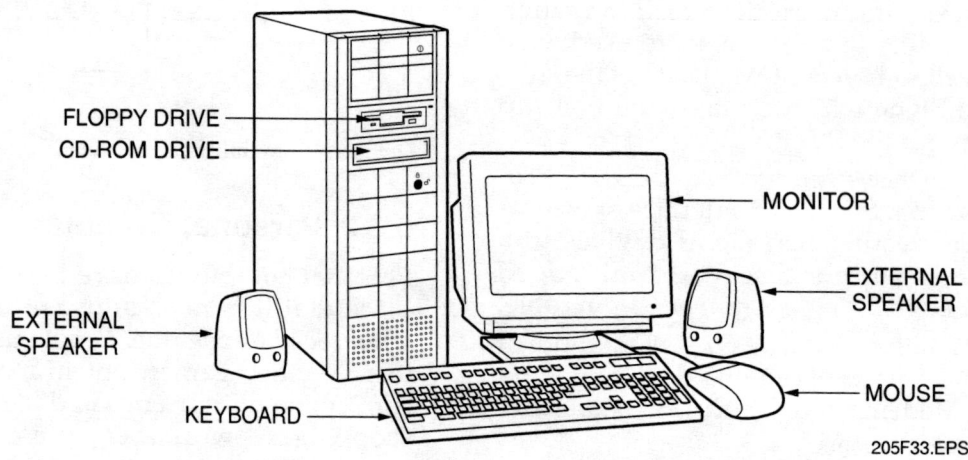

*Figure 33* ◆ Personal multimedia computer (tower configuration processor).

- *Compact Disc Read-Only Memory (CD-ROM)* – These disks are portable and can hold up to 650 megabytes of data. They look exactly like audio CDs. CD-ROMs are very inexpensive to reproduce, but need special equipment to manufacture. They are mainly used to mass-distribute software.
- *CD-Recordable (CD-R)* – The CD-R is a special blank CD which can be written to only once, using a special CD-R drive. Once it has been written to, it can then be read in most CD-ROM drives. The low cost and flexibility make this format ideal for backups and archiving. The availability of inexpensive CD-R drives has made this an extremely popular medium.
- *CD-Rewritable (CD-RW)* – The CD-R is another recordable CD format. These disks can be erased and written to repeatedly, but are less popular because they are more expensive and can only be read with special drives.
- *Digital versatile disk (DVD) drive* – The DVD is newer than the CD-ROM and has a much greater capacity to store information. It has four different storage modes, the lowest of which can store 4.7 gigabytes of data, which is about seven times the capacity of a CD-ROM. In its highest capacity mode (both sides, two layers per side), it can store more than 17 gigabytes.
- *Monitor* – The monitor is the display device. It receives information from the video card inside the CPU case.
- *Keyboard* – The keyboard allows the operator to enter alphabet characters and numbers. It also contains function and control keys that are used by computer programs to perform special functions. The use of function keys is less common in a mouse-driven system where interaction with the computer is done by clicking on graphic objects rather than entering keystrokes. However, the keys are still functional and can be used in place of the mouse for many tasks.
- *Speakers* – With the advent of multimedia, audio was introduced to the PC. Now, it is unusual to find a PC without a set of speakers. Speakers with a built-in amplifier provide the best quality sound. Powered speakers have their own power source; the sound level can be adjusted using a volume control on the main speaker, rather than doing it in the operating system. Speakers require an audio card or special audio circuits on the motherboard.
- *Mouse* – The mouse is so-called because of its shape and its long tail of cord. Trackballs and joysticks are other common input devices (*Figure 34*). The latter is used extensively for games.

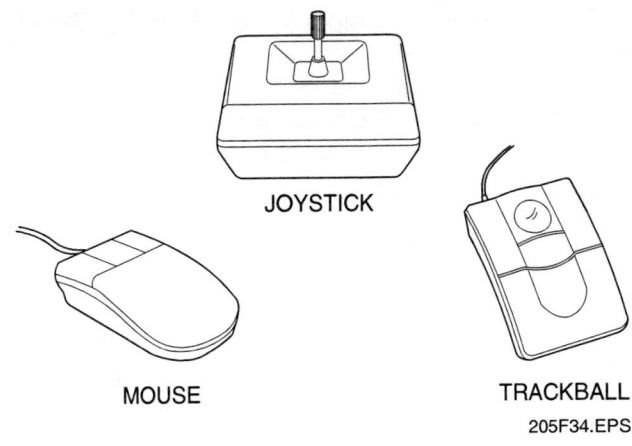

*Figure 34* ◆ Computer input devices.

### 6.3.1 Monitors

The monitor that appeared in the first PCs was a **monochrome** device capable of displaying only text and line drawings (simple plots, etc). Over the years since then, the monitor has evolved through several generations of devices capable of displaying color graphics, starting with the color graphic adapter (CGA) standard. CGA was followed by the enhanced graphic adapter (EGA) and video graphics array (VGA), which emerged in 1987. These were followed with super VGA (SVGA), XVGA, and Video Electronic Standards Association (VESA)-compliant monitors, which brought the screen image to new quality levels.

The quality of a monitor is measured in terms of its resolution and the number of colors it is capable of presenting. The screen image is made of up **pixels** (short for picture element), which are small dots that appear on the screen. Resolution is measured in terms of the number of pixels that appear, both horizontally and vertically. The more pixels, the higher the quality of the image.

Your computer must be set up to view the resolution of the subject image. This is done in Windows by going to the Display function on the Control Panel and resizing the window for a different resolution. Otherwise, the image on the screen will be undersized or oversized.

For example, if your monitor is set for 1280 × 1024 pixel resolution, and you are viewing a 800 × 600 pixel image, the image will be small in relation to the available viewing area. Changing the screen resolution will bring the image back to its normal size. Changing the resolution can be done from the display icon at the bottom of the screen on some Windows computers.

The number of colors (color depth) is another important measure of image quality. Early CGA monitors were capable of displaying 16 colors (4-bit color) with a resolution of 320 × 200 pixels. VGA brought that capability up to 256 colors (8-bit color) and 640 × 480 resolution, and VESA-compliant designs yielded 1280 × 1024 resolution with 16.8 million colors (24-bit color).

Monitors range in size from 14" to 21", with 15" to 17" being commonplace in homes and businesses. The size refers to the distance from one corner of the screen to the other (diagonally); the actual display area will be smaller. Although monitors typically look like the ones depicted earlier, flat screen displays are available and show signs of coming into general use.

### 6.3.2 Connections

On most PCs, the connection devices are located at the rear of the unit (*Figure 35*). In the computer world, connectors are referred to as *ports*.

- *Parallel ports* – A PC will have one or two parallel ports, which are designed to mate with 25-pin male connectors. They are designed to interface with peripheral devices such as printers, scanners, and tape drives. The computer designates parallel ports as LPT (line print terminal). If the PC has two parallel ports, they will be designated LPT 1 and LPT 2.

- *Serial ports* – A serial port is usually a 9-pin connector, but some are 25-pin connectors. They can be distinguished from other ports because they are male connectors. The serial port is used to connect such components as a mouse or external modem to the PC. Serial ports are designated COM ports.

- *Monitor port* – The monitor port is a 15-pin connector. The cable that carries video from the computer to the monitor is plugged in here.

- *Keyboard port* – The keyboard is connected to the computer with a round 9-pin connector. The connector is keyed to prevent the cable from being connected incorrectly.

- *Universal serial bus (USB) ports* – A USB port allows the user to connect a wide variety of peripheral devices to the PC. One of the problems with earlier PCs is that they did not have enough ports to connect all of the external peripherals a person might want to use. If you wanted to use a document scanner, for example, you might have to disconnect the printer to obtain a connection. On many PCs, the USB ports are located on the front of the computer to allow easy swapping of peripheral devices.

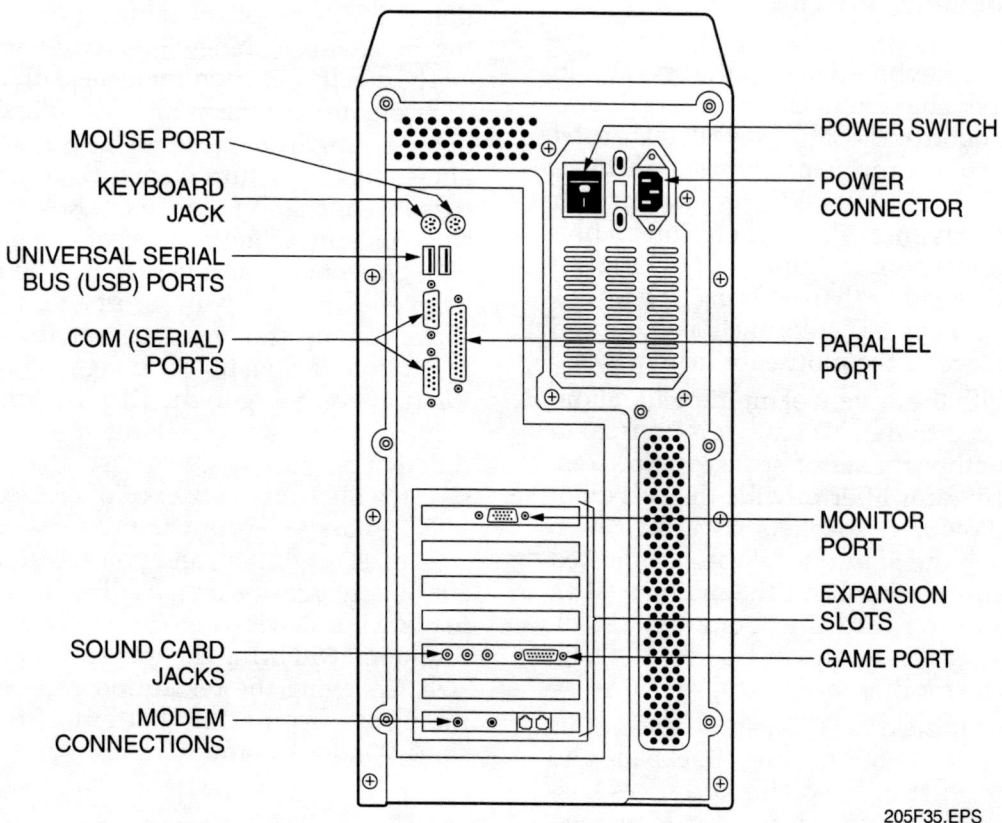

*Figure 35* ◆ Rear view of a tower configuration PC.

- *Game port* – The game port is a 15-pin connector used to connect a joystick or similar device used for games and simulations.
- *Phone jack* – This jack is used to connect the modem to the phone system. It is the same type of jack used in standard telephone circuits.
- *Mouse port* – Some versions of the mouse are connected with a 6-pin connector. Other versions are connected to a serial port.

### 6.3.3 Inside the PC

*Figure 36* shows the major components located inside the PC.

*System board* – The system board (commonly known as the *motherboard*) is the heart of the PC. It contains the CPU, or microprocessor, which performs calculations and manages the flow of information through the system. The speed at which a CPU processes information is stated in megahertz (MHz). Early Pentium PCs (around 1995) had a speed of 60MHz to 75MHz. By 2000, 750MHz was common. The added speed not only made data and graphic processing much faster, it opened the way for PCs to become true multimedia devices, delivering high quality video, audio, and animations.

One of the functions of the motherboard is to act as a base for other boards, known as *expansion boards*. Among these expansion boards are the memory module, which contains the RAM; the video board, which drives the monitor; and the modem. Inside the case, the motherboard is positioned so that its expansion slots are near the rear of the computer. The expansion boards contain connectors that protrude from openings in the rear of the case. These connectors are the ports we discussed earlier.

*Sound (audio) card* – The sound card is an expansion board that drives the speakers. The sound card contains connectors that are accessible at the back of the computer. In addition to the speaker jack, the sound card may also have jacks for other purposes, including: an audio input jack that allows a tape or CD player to play sound through the computer; an audio output jack that allows connection to an audio amplifier; and a microphone jack that allows users to record audio.

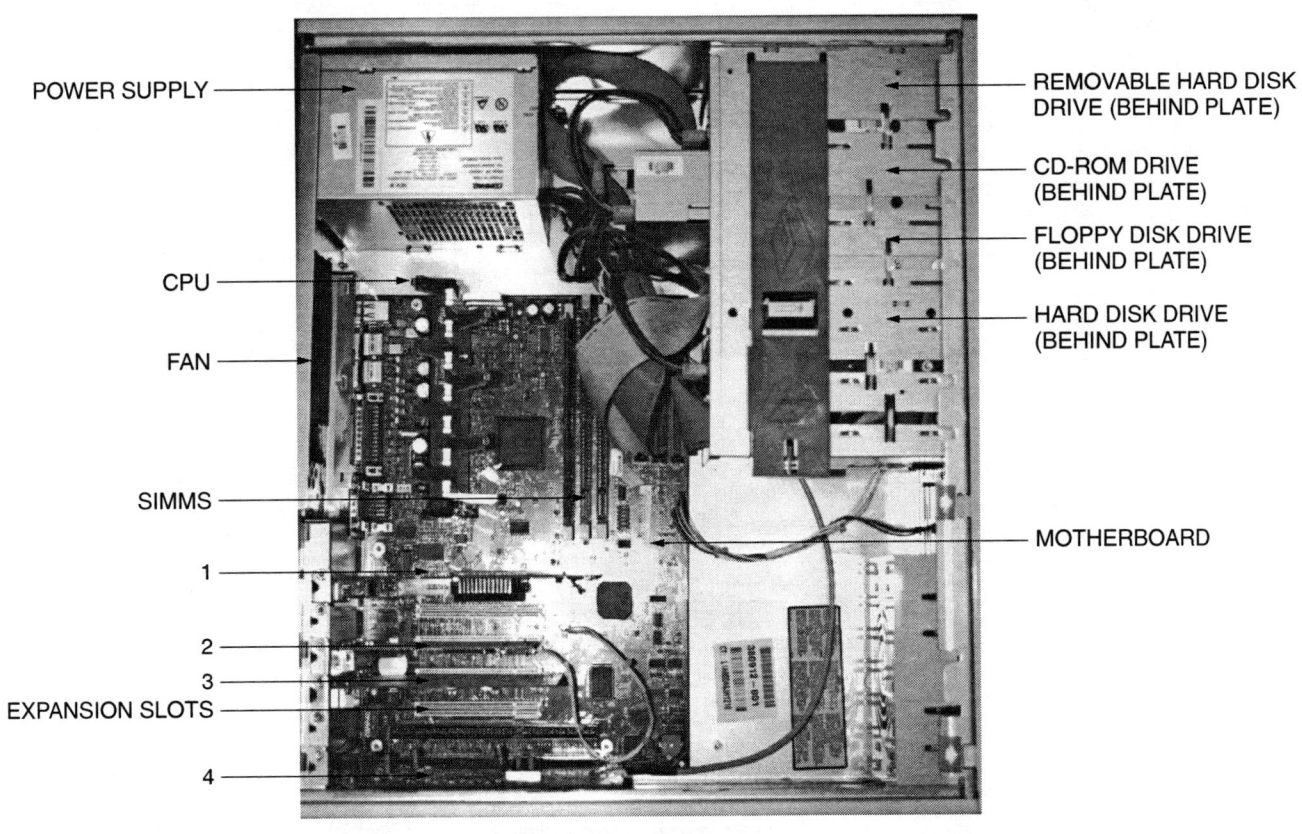

EXPANSION CARDS
1 VIDEO GRAPHICS CARD
2 NETWORK INTERFACE CARD
3 SPECIAL PURPOSE VIDEO CARD
4 AUDIO CARD

*Figure 36* ◆ Interior of a tower configuration PC.

To obtain high-quality sound, a card with at least a 16-bit sampling size and a 44.1kHz sampling rate is required.

*Video card* – A video card contains memory chips (modules) that store information before sending it to the monitor and microprocessors to interpret data from the CPU into displayed images. A PC needs at least 2 megabytes of memory in order to effectively display graphics.

A large, high-resolution graphic image can contain many megabytes of data. If your computer is not set up to handle such graphics, it will really slow your computer down. In order to eliminate this problem, many computers use AGP (advanced graphics port) video cards and special video RAM to display graphics. The AGP bus is designed to communicate directly with the computer's main memory in order to display complex graphics rapidly.

*Modem* – The term *modem* is an abbreviation for modulator-demodulator. It is the device that allows computers to communicate with each other over telephone lines. Its purpose is to translate incoming data from the analog form used on phone lines to the digital form required for the computer. It does the opposite for outgoing data. The modem must be plugged into a phone jack in order to work.

Modems communicate with other modems. The speed at which the data is transferred is determined by the slower of the two modems (e.g., if a 56Kbps modem is communicating with a 33.6Kbps modem, the data will transfer at the 33.6Kbps rate).

When you log on to the Internet, your modem contacts a modem at the site of your ISP (Internet service provider). The ISP then reaches out to other computers on the Internet to access the web site you select. The ISP uses high-speed transmission lines and special modems to make these connections rapidly. Businesses that transfer large amounts of information from one site to another also use high-speed lines to move the data faster.

*Hard drive* – The hard drive is the main storage device for the computer. It stores the software programs as well as the files created using those programs.

*Power supply* – The power supply converts the 120VAC source to the DC power that is needed to operate the solid-state electronics. The power supply will include a ventilating fan to dissipate the heat generated by the power supply and other devices. A PC will have one or more ventilating fans, depending on the amount and types of devices it contains.

### 6.4.0 Computer Storage Media

Since computer memory is erased when you shut the power off, computers need a way to keep information without power. This is called non-volatile storage. The computer can retrieve information from a non-volatile storage device and load it into RAM (reading) to work with it, then record information back to the device (writing). *Figure 37* shows some common storage media.

The most important form of non-volatile storage is the hard drive. Most computers contain

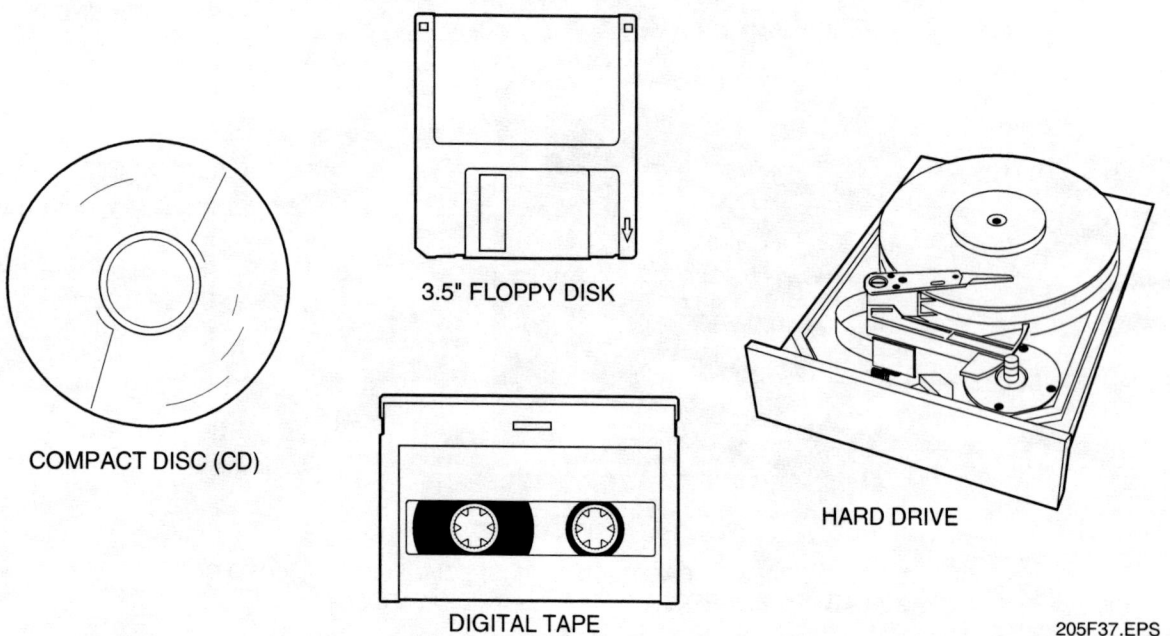

*Figure 37* ♦ Data storage devices.

some form of hard drive. This is a high-capacity, high-speed magnetic disk, mounted permanently inside the computer. The computer's operating system, applications, and other data are stored on the hard drive. The capacity of modern hard drives is measured in gigabytes, and is increasing rapidly. A computer can have more than one hard disk drive. Some computers do not contain internal hard drives, but store everything remotely on a network storage device.

The efficiency of a hard drive is determined by two critical factors:

- The rotational speed of the disk, measured in revolutions per minute (RPM). The faster it rotates, the faster it can find and retrieve information.
- Access (seek) time, measured in milliseconds (ms). Hard drives typically have access times ranging from 8 to 15 milliseconds. The shorter the time, the faster (and more expensive) the drive.

Transferring information to other computers, or making backup copies of important data from the hard disk is usually done with removable media. Removable media are storage devices that can be removed and replaced while the computer is running. There are several kinds of removable media available.

The CD-ROM is popular for distributing software to many computers, because it holds a lot of data and is very inexpensive to mass-produce. The disadvantage of CD-ROM is that it is read-only. Information on a CD-ROM cannot be changed, and the equipment for producing CD-ROMs is very expensive. To get around this limitation, CD-R technology was developed. CD-R (Compact Disc-Recordable) disks have a special coating that can be changed by exposure to a laser beam. These disks can be written to by special CD-R drives. Once they have been written to, these disks will function like normal CD-ROMs in most drives. The process of writing information to a CD-R is called *burning*. The availability of inexpensive CD-R drives and media has made this an extremely popular format for archiving and storing large files. Some of these drives also incorporate CD-RW technology, which makes it possible to erase and rewrite information on the same disk.

Because they are more expensive and cannot be read by normal CD-ROM drives, these disks are not as popular as CD-R disks.

DVD-ROM is similar to CD-ROM, but is a newer technology with a much larger storage capacity. Recordable and rewritable DVDs are also available.

The diskette, or floppy disk, is widely used, though it is becoming less popular as the need to store large files increases. A standard 3.5" floppy disk has a capacity of about 1.4 megabytes. This is enough to hold a few documents or small programs, but not enough for multimedia use. It would take hundreds of floppy disks to hold as much information as a single CD-ROM. Floppy disks are slow, and not reliable for long-term storage. Their biggest advantage is that they are the cheapest storage medium that can be easily written to, as well as read.

Digital magnetic tape drives can store large amounts of information reliably. There are many different kinds of digital tape. Currently, a large-capacity tape can hold hundreds of gigabytes, and as technology improves, this capacity is constantly growing. Digital tape is typically used to store a large amount of data at once. The most important use of tapes is making backups of the information on a hard disk in case the hard disk fails. Compared to other storage media, tapes are slow.

There are many other types of removable storage media, including ZIP™ disks, optical-magnetic devices, memory cards, and removable hard drives. Each type has advantages and disadvantages, and it is not unusual to find more than one kind of storage device within the same system.

## Summary

The science of electronics has made it possible to replace mechanical and electromechanical devices with electronic devices that take up less space, work much faster, and last longer. Electronic devices and circuits provide greater precision as well as a greater range of control. These capabilities add up to improved comfort control, greater operating efficiency, and easier servicing. Modular packaging of electronic controls, along with built-in diagnostic and testing capabilities, make it easier for service technicians to troubleshoot and repair HVAC systems.

**Review Questions**

1. An electrical charge occurs when an atom gains or loses _____.
   a. weight
   b. protons
   c. electrons
   d. protons

2. Which of the following is true of semiconductors?
   a. They are better conductors than insulators.
   b. They are better insulators than conductors.
   c. When mixed with impurities they become insulators.
   d. They are neither good insulators nor good conductors.

3. All semiconductor devices are _____.
   a. made from a combination of P-type and N-type material
   b. diodes
   c. forward-biased
   d. reverse-biased

4. For a diode to conduct current _____.
   a. the anode must be negative with respect to the cathode
   b. the anode must be positive with respect to the cathode
   c. the diode must be reverse-biased
   d. the diode must be used in a half-wave rectifier

5. A bridge rectifier might be used in place of a full-wave rectifier because _____.
   a. it is more effective
   b. it produces AC while the full-wave rectifier produces DC
   c. it does not require a center-tapped transformer
   d. it uses fewer diodes

6. Color-coding is normally used on _____ resistors.
   a. wire-wound
   b. carbon composition
   c. wire-wound and carbon composition
   d. variable

7. If a thermistor has a positive coefficient of resistance, its resistance will _____ as temperature increases.
   a. increase
   b. decrease
   c. remain the same
   d. decrease then return to its original value

8. A cadmium sulfide flame detector is sensitive to _____.
   a. light
   b. heat
   c. smoke
   d. vibration

9. A computer's BIOS is normally located _____.
   a. on the hard drive
   b. in RAM
   c. on the back of the unit
   d. in ROM

10. If it is necessary to open the computer to check internal connections, you should wear _____.
    a. a dust mask
    b. a hard hat
    c. a grounding wrist strap
    d. safety glasses

# GLOSSARY

## Trade Terms Introduced in This Module

*Anode:* The positive terminal of a diode.

*Basic input/output system (BIOS):* The basic method by which a computer exchanges information.

*Bridge rectifier:* A rectifier circuit that uses four diodes, two of which conduct current on each half-cycle. Has the advantage of not needing a center-tapped transformer.

*Cathode:* The negative terminal of a diode.

**Centrifugal force:** The force that makes rotating objects tend to move away from the center of rotation.

*Chip:* A common term used to describe an integrated circuit.

*Electromechanical components:* Electrical devices that contain moving parts.

*Electronics:* The science that deals with the behavior and effects of electron movement in conductors, insulators, and semiconductors.

*Free electrons:* Valence electrons that can easily be knocked out of orbit.

*Full-wave rectifier:* A rectifier circuit that uses two diodes.

*Half-wave rectifier:* A rectifier circuit that uses a single diode.

*Integrated circuit:* A plug-in circuit containing microminiature electronic circuits. Sometimes called a *chip*.

*Light-emitting diode (LED):* A diode that gives off light when current flows through it.

*Microminiaturization:* The technology that allows the manufacture of microscopic electronic circuits.

*Microprocessor:* An integrated circuit chip designed to perform computing functions. The microprocessor is the heart of a personal computer.

*Monochrome:* Able to display a single color.

*Photo diode:* A diode that conducts current when exposed to light.

*Pixel:* An abbreviation for picture element. A pixel is a single dot in a graphic image on a computer screen.

*Rectification:* The conversion of AC into DC using diodes.

*Semiconductor:* A material that contains four valence electrons and is used in the manufacture of integrated circuits.

*Valence electrons:* Electrons located in the outer orbit of an atom.

# ANSWER KEY

## Answers to Review Questions

| Answer | Section |
|---|---|
| 1. c | 2.0.0 |
| 2. d | 3.3.0 |
| 3. a | 4.0.0 |
| 4. b | 4.1.0 |
| 5. c | 4.1.1 |
| 6. b | 4.3.1 |
| 7. a | 4.4.0 |
| 8. a | 4.5.0 |
| 9. b | 6.1.0 |
| 10. c | 5.4.0 |

# REFERENCES

## Additional Resources

This module is intended to present thorough resources for task training. The following reference works are suggested for further study. These are optional materials for continued education rather than for task training.

*American Electrician's Handbook*, 1996. Terrell Croft and Wilfred I. Summers. New York, NY: McGraw-Hill.

*Solid State Fundamentals for Electricians*, 1993. Gary Rockis. Alsip, IL: ATP.

# NCCER CRAFT TRAINING USER UPDATES

The NCCER makes every effort to keep these textbooks up-to-date and free of technical errors. We appreciate your help in this process. If you have an idea for improving this textbook, or if you find an error, a typographical mistake, or an inaccuracy in the NCCER's Craft Training textbooks, please write us, using this form or a photocopy. Be sure to include the exact module number, page number, a detailed description, and the correction, if applicable. Your input will be brought to the attention of the Technical Review Committee. Thank you for your assistance.

*Instructors* – If you found that additional materials were necessary in order to teach this module effectively, please let us know so that we may include them in the Equipment and Materials list in the Instructor's Guide.

**Write:** Curriculum Revision and Development Department
National Center for Construction Education and Research
P.O. Box 141104, Gainesville, FL 32614-1104

**Fax:** 352-334-0932

**E-mail:** curriculum@nccer.org

Craft

Module Name

Copyright Date

Module Number

Page Number(s)

Description

(Optional) Correction

(Optional) Your Name and Address

# Module 03206-01

# *Electric Heating*

## COURSE MAP

This course map shows all of the modules in the second level of the HVAC curriculum. The suggested training order begins at the bottom and proceeds up. Skill levels increase as you advance on the course map. The local Training Program Sponsor may adjust the training order.

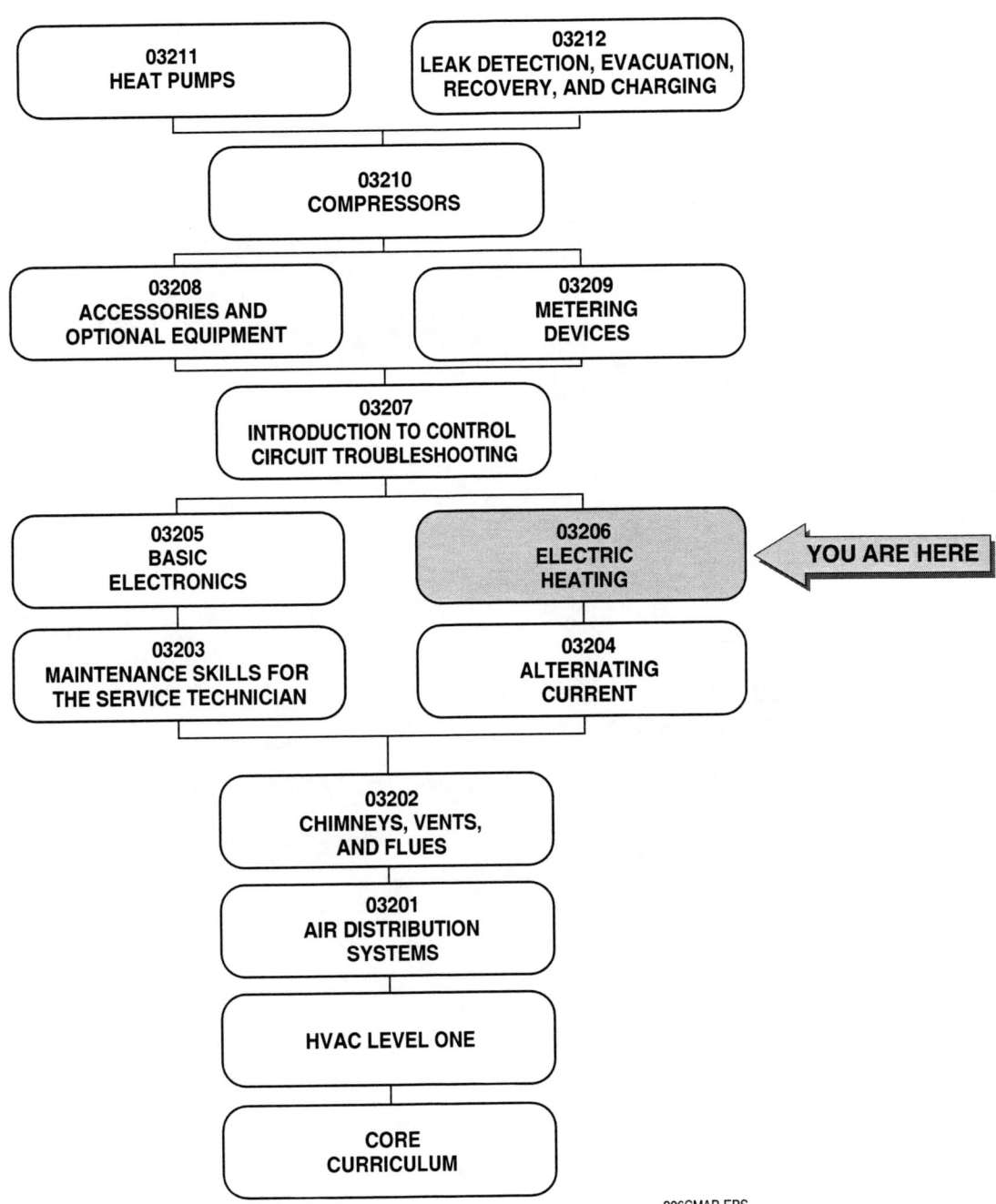

ELECTRIC HEATING — TRAINEE MODULE 03206     6.iii

## MODULE 03206 CONTENTS

- 1.0.0 INTRODUCTION .................................................. 6.1
- 2.0.0 GENERAL DESCRIPTION ...................................... 6.1
  - 2.1.0 Major Components ......................................... 6.3
  - *2.1.1 Heating Element* ......................................... 6.3
  - *2.1.2 Blower and Motor Assembly* ........................ 6.4
  - *2.1.3 Enclosure* .................................................. 6.4
  - *2.1.4 Accessories* ............................................... 6.4
  - *2.1.5 Power Supply* ............................................ 6.4
- 3.0.0 HEATING ELEMENTS ........................................... 6.4
- 4.0.0 FORCED-AIR ELECTRIC HEAT CONTROLS ............. 6.5
  - 4.1.0 Thermostat .................................................. 6.5
  - 4.2.0 Transformer ................................................ 6.6
  - 4.3.0 Heater Control ............................................. 6.6
  - 4.4.0 Blower Control ............................................. 6.6
  - *4.4.1 Alternate Blower Control Methods* ............... 6.6
  - 4.5.0 Safety Controls ............................................ 6.6
  - *4.5.1 Limit Switch* ............................................... 6.9
  - *4.5.2 Thermal Fuse* ............................................. 6.9
  - 4.6.0 Two-Stage Thermostat .................................. 6.9
- 5.0.0 TROUBLESHOOTING ........................................... 6.11
  - 5.1.0 Power Requirements .................................... 6.11
  - 5.2.0 Voltage Variations ....................................... 6.11
  - 5.3.0 Checking Loads ........................................... 6.11
  - *5.3.1 Voltage Check* ........................................... 6.12
  - *5.3.2 Current Check* ........................................... 6.12
  - *5.3.3 Resistance Check* ...................................... 6.12
  - 5.4.0 Replacing Resistance Wires .......................... 6.13
  - 5.5.0 Determining Btuh Output .............................. 6.13
  - 5.6.0 Safety With Electrical Circuits ....................... 6.13
  - *5.6.1 Safety Practices* ........................................ 6.15
- 6.0.0 AIRFLOW ........................................................... 6.15
  - 6.1.0 Calculating Airflow Volume ........................... 6.16
  - *6.1.1 Adjusting Blower Speed* ............................. 6.17
- 7.0.0 DUCT HEATERS ................................................. 6.17
- 8.0.0 OTHER ELECTRIC HEATING SYSTEMS .................. 6.18
  - 8.1.0 Baseboard Heaters ...................................... 6.18
  - 8.2.0 Space Heaters ............................................ 6.19
  - 8.3.0 Radiant Heating Panels ................................ 6.19
  - 8.4.0 Electric Boilers ........................................... 6.19

# MODULE 03206 CONTENTS (Continued)

**SUMMARY** .................................................. 6.20
**REVIEW QUESTIONS** ....................................... 6.21
**GLOSSARY** ................................................. 6.23
**ANSWERS TO REVIEW QUESTIONS** .......................... 6.24
**REFERENCES & ACKNOWLEDGMENTS** ......................... 6.25

## Figures

Figure 1   Fan coil with electric heat package ...................... 6.2
Figure 2   Fan coil components ................................... 6.3
Figure 3   Electric heating element .............................. 6.5
Figure 4   Room thermostat wiring diagram ....................... 6.5
Figure 5   Alternate blower control methods ...................... 6.7
Figure 6   Furnace control circuit with current sensing loop ...... 6.8
Figure 7   Limit switch wiring diagram ........................... 6.10
Figure 8   Voltage check ....................................... 6.12
Figure 9   Amperage check ..................................... 6.12
Figure 10  Resistance check ................................... 6.12
Figure 11  Ohm's law and derivative equations .................. 6.13
Figure 12  Air distribution ..................................... 6.16
Figure 13  Temperature check .................................. 6.16
Figure 14  Slip-in duct heater .................................. 6.18
Figure 15  Electric baseboard heater ........................... 6.18
Figure 16  Line voltage thermostat ............................. 6.18
Figure 17  Space heater ....................................... 6.19
Figure 18  Radiant heating applications ........................ 6.19
Figure 19  Wall-mounted electric boiler ......................... 6.20

## Tables

Table 1   Voltage Multipliers ................................... 6.13

# MODULE 03206

# Electric Heating

## Objectives

When you have completed this module, you will be able to do the following:

1. Describe and explain the basic operation of a fan coil equipped with electric heating elements.
2. Identify and describe the functions of major components of a fan coil equipped with electric heating elements.
3. Identify and describe the functions of electric heating controls.
4. Measure resistances and check components and controls for operation and safety.
5. Determine the cubic feet per minute (cfm) using the temperature rise method.
6. Describe and explain the basic operation of other electric heating systems.

## Prerequisites

Before you begin this module, it is recommended that you successfully complete the following modules: Core Curriculum; HVAC Level One; HVAC Level Two, Modules 03201 through 03204.

## Required Trainee Materials

1. Pencil and Paper
2. Appropriate Personal Protective Equipment

## 1.0.0 ♦ INTRODUCTION

Electric heating systems use electricity rather than combustible fuels, as an energy source. In all electric heat systems electrical current flows through resistive heating elements, causing them to generate heat. The resistive heating elements can be positioned in the airstream of a fan coil or packaged HVAC unit where heat is transferred into the supply air. The resistive elements can be part of a baseboard radiator which heats by convection and radiation or the elements can be embedded in a floor or ceiling to provide radiant heat. Resistive heating elements can also be immersed in water and used as the heat source for a boiler in a hydronic heating system.

## 2.0.0 ♦ GENERAL DESCRIPTION

The fan coil with electric heating elements (*Figure 1*) converts electrical energy to heat energy. This conversion takes place in resistance heaters that convert electrical energy into heat energy.

### Air Handlers

Furnaces which are entirely electric are relatively rare. Instead, manufacturers supply electric heat packages that are added to existing air handlers equipped with a heat exchanger (heat pump or cooling coil). Building codes in many areas prohibit the use of electric heat as a primary heat source. Heat pumps typically use electric resistance heaters to supplement compression (heat pump) heat since the heat output of the system drops as the outdoor temperature drops. The electric elements also add warmth to the supply air during heat pump defrost cycles.

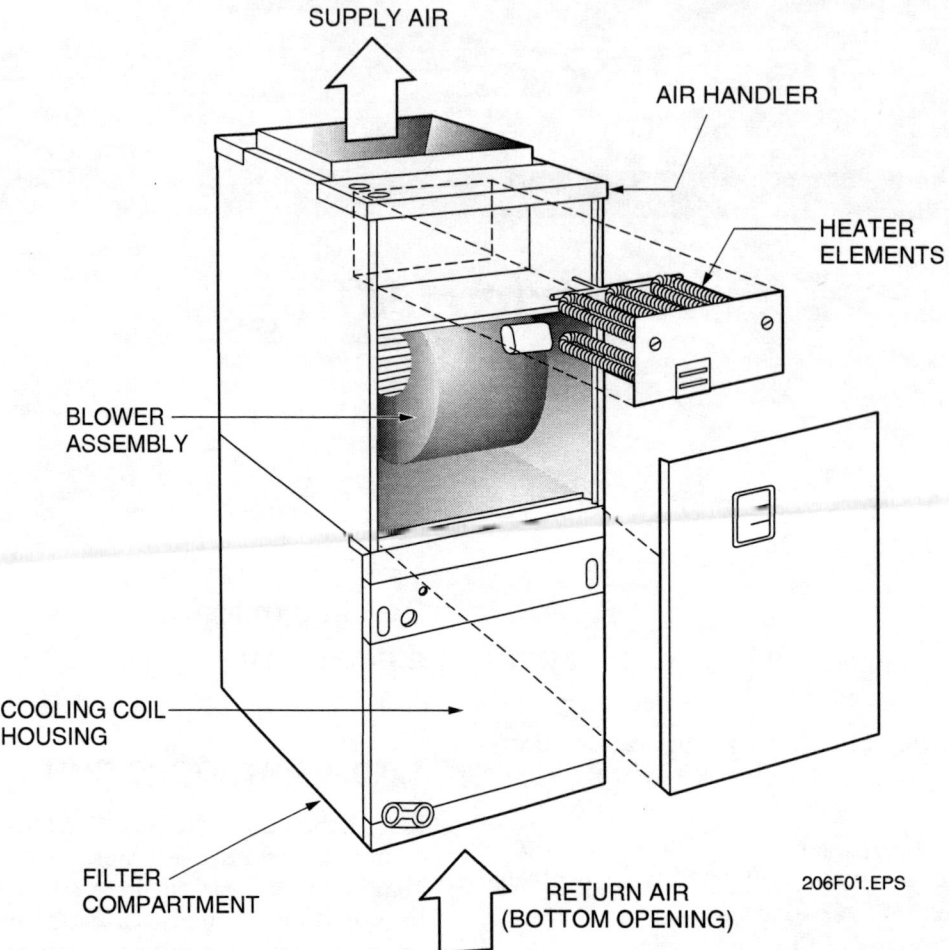

*Figure 1* ◆ Fan coil with electric heat package.

## The Rise and Fall of Electric Heating

In the 1960s, electricity was plentiful and inexpensive. The all-electric home was the ideal. Electric baseboard heaters were relatively inexpensive to install in comparison to forced-air systems, which required the installation of ductwork. Electric baseboard heat gave the added advantages of zoned heating and clean operation. In addition, utility companies were offering incentives to people who built homes with electric heat. In some parts of the country, buildings were heated with electric furnaces, which are forced-air systems using banks of heating coils instead of a fuel-burning heat source.

Over the next decade, the cost of electricity gradually rose, while the cost of fossil fuels such as natural gas and heating oil remained steady or declined. Heat pumps came into common use in many parts of the country. Eventually, the cost of heating homes with electricity far outweighed the cost of other methods, and the owners of all-electric homes began having trouble selling their homes. Conversion to forced-air heat was prohibitively expensive because of the difficulty in installing ductwork in a finished home.

Today, electric heat serves as the benchmark for measuring the efficiency of all other heating methods. A heat pump, for example, is 1.5 to 3 times more efficient than electric heat.

Electric heat is still in use because it is ideal for certain applications. One very common use is placing electric heaters under windows in commercial buildings that use forced-air heat. The electric heaters deal with the puddle of cold air that forms under a window in cold climates. These heaters are usually controlled by outdoor thermostats, so they are only used when the outdoor temperature is low.

Electric forced-air heating systems differ from fuel furnaces in that no special heat exchanger is required. Because no fuel is burned, neither a chimney nor a vent is needed to carry the products of combustion outdoors. This feature allows for greater safety and more installation flexibility. The return air from the conditioned space passes directly over the resistance heaters and into the supply air plenum. The amount of heat supplied by an electric forced-air heating system depends upon the number and size of the resistance heaters used.

## 2.1.0 Major Components

The major components of a fan coil with electric heat (see *Figure 2*), excluding the controls, are the heating elements; the blower and motor assembly; the heat pump or cooling coil; the furnace enclosure or cabinet; and the filter.

### 2.1.1 Heating Element

The function of the heating element is to provide the heat required for the conditioned space. Its wires are made of nickel and chromium (Nichrome™). The heating element wire is spiraled and threaded through a metal holding rack, which has ceramic insulators that prevent the resistance wires from shorting out on the frame of the rack. Calrod-type heating elements are sometimes used. These are similar to the heating elements used in electric ranges, burners, and oven elements.

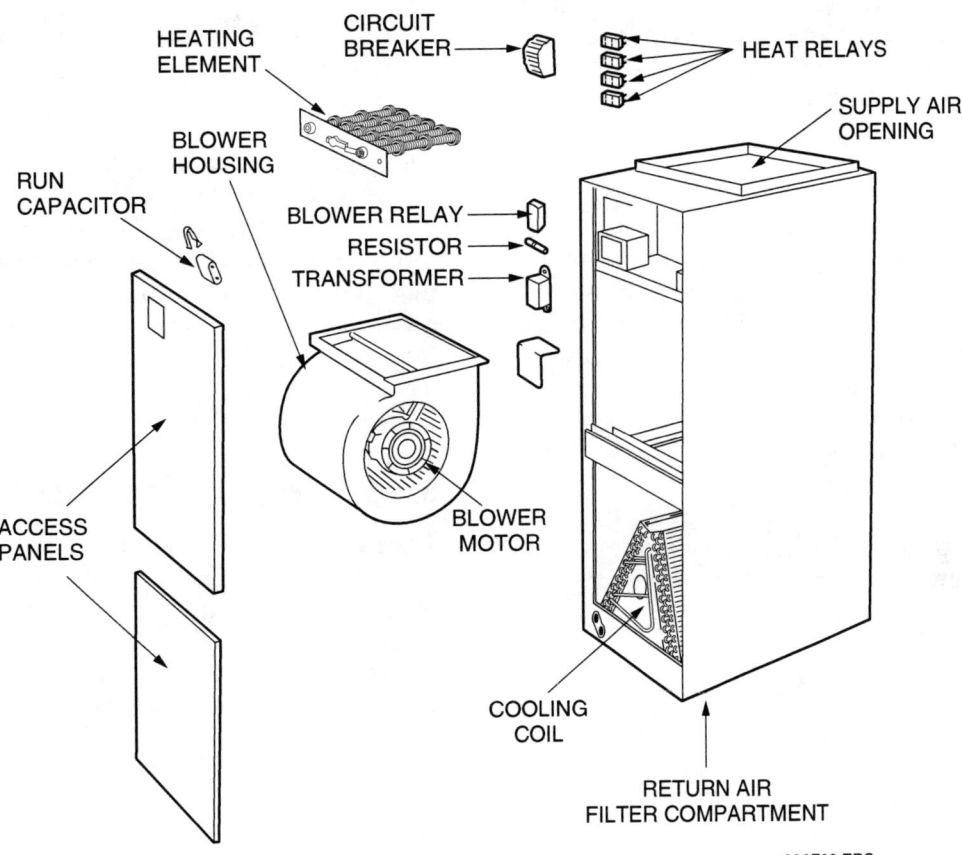

*Figure 2* ◆ Fan coil components.

### Limit Switches

In a correctly installed, properly maintained, normally operating system, a **limit switch** may never open during the life of the system. If you encounter a forced-air electric heating system that is cycling on the limit switch, there is probably something wrong. Restricted airflow, dirty filters, and an inadequate duct system are some problems that could cause a limit switch to open. Each of these conditions results in increased temperature due to reduced airflow.

Each element assembly contains a **thermal fuse** and a safety limit switch. The thermal fuse is a backup for the limit switch. It is set to open at a temperature slightly higher than that of the switch. A limit switch may be set to open at 160°F and close when the temperature drops to 125°F. The limit switch resets automatically. Once the thermal fuse opens, it must be replaced. A manual reset limit switch is sometimes used in place of a thermal fuse. It provides the same protection.

A forced-air electric heating system may have one or more banks of heating elements, depending on the required capacity. For example, we will assume there are four banks of heating elements. If each provides 5kW, the furnace capacity would be 20kW. This converts to 68,260 Btuh when multiplied by 3.413, which is the number of Btus per kilowatt. At 240V, each 5kW circuit draws about 21A (I = P/E). Because of the high current drawn by the resistive heating elements, the banks of elements are turned on in stages using sequencing relays with different time delays. This avoids large power surges that could blow fuses or affect other electrical appliances.

Electric heater packages typically range in capacity from about 17,000 Btuh (5kW) to 120,000 Btuh (35kW). At 240V, the latter will draw about 145A. The capacity of the required package is determined by calculating a heat loss in the structure. The equipment manufacturer will supply all components necessary to install the pre-wired heater packages. In most fan coils or packaged units, an access panel is removed to insert the actual heater. The sheet metal in the air handler cabinet is pre-drilled to accept the other components, such as the 24V transformer, relays, and terminal strip. New wiring diagrams and complete instructions are provided.

### 2.1.2 Blower and Motor Assembly

Like those used in fuel furnaces, the fans used in fan coils and packaged units equipped with electric heaters are usually direct-driven. Most are equipped with multi-speed blower motors and are sized to handle the airflow requirements of cooling, which tend to be greater than those of heating. Field adjustment of multi-speed blowers may be required.

### 2.1.3 Enclosure

The casing of an air handler is similar to that of a gas or oil furnace, but without the vent pipe connection. The interior of the cabinet is designed to permit the air to flow first over the cooling coil and then over the heating elements, which are usually insulated from the exterior casing by an air space.

### 2.1.4 Accessories

Filters and humidifiers are added to a fan coil in much the same manner as they are in gas and oil furnaces. Therefore, a fan coil with electric heaters can provide all of the climate-control features found in fuel-fired furnaces.

### 2.1.5 Power Supply

Fan coils with electric heaters usually require 208/240V single-phase, 60Hz AC. This type of heating unit is supplied by three wires: two hot and one grounded. The hot lines leading to the furnace contain fused disconnects. All wiring should be enclosed in conduit with the proper connectors, as specified by the National Electrical Code (NEC). Because these furnaces use 240V, every possible protection must be provided. The NEC also requires that the fan coil be grounded. A supply ground is provided for that purpose. Fuses or circuit breakers may be used at the cabinet terminal block. Check local codes for maximum temperature and clearance from combustible building components.

## 3.0.0 ♦ HEATING ELEMENTS

The function of the heating element (*Figure 3*) is to generate enough heat to satisfy the requirements of the conditioned space. Elements are manufactured in various Btuh or heating capacity sizes. One typical size is 5kW. It consumes 5,000W of power, or 50 times as much as a 100W light bulb. It delivers about 17,065 Btuh (5,000 × 3.413).

A typical heating element consists of a metal frame or holding rack, ceramic insulators, and resistance wire. The wire is spiraled and threaded through supports in the metal frame, which has ceramic insulators on the support framework. The insulators prevent electrical shorts between the wire and the framework.

---

### Converting Between Watts and Btuh
A heating element rated in watts can be converted to Btuh by multiplying the watts by 3.413. If given a rating in Btuh, how would you convert the Btuh to watts?

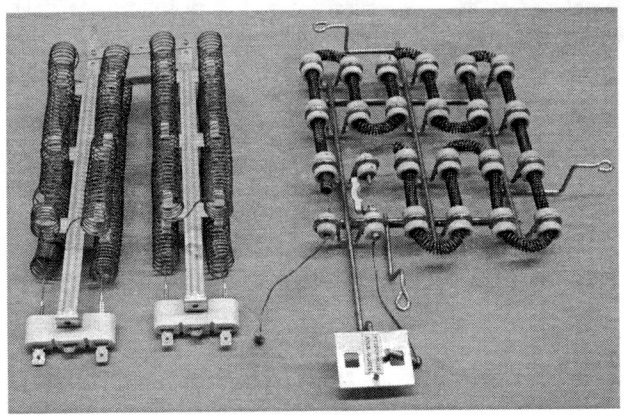

*Figure 3* ♦ Electric heating elements.

When line voltage is applied to the resistance wire, current will flow and the element will heat up because the wire spiral resists the flow of current. The blower in the fan coil or packaged unit forces return air over the heated element and the ductwork distributes the air to the space requiring heat.

Each leg of the heating element in the furnace requires a line fuse (or circuit breaker). The size of the fuse depends on the size of the element. The function of the fuse is to protect the circuit wiring.

Each heating element has a specific rating, such as 5kW at 220V. The heater relay sequences the heating elements in increments. Each heating element is protected by both a thermal fuse and a limit control. The thermal fuse has a cutoff temperature of a specific rating in degrees Fahrenheit, such as 333°. The thermal fuse also has a resistive interrupt current of a specified rating, such as 40A. The heating element limit control de-energizes the element in the event of excessive temperatures. The cut-out (opening) temperature for a given limit control may be 130°F, while the cut-in (closing) temperature of the same control may be 95°F.

## 4.0.0 ♦ FORCED-AIR ELECTRIC HEAT CONTROLS

This section covers various forced-air electric heat controls. Several components are involved, including the thermostat, transformer, heater control, blower control, safety controls, and two-stage thermostat.

### 4.1.0 Thermostat

To regulate the amount of heat produced in a forced-air electric heating system, a low-voltage (24V) thermostat must be used. Mounted in the conditioned space, the room thermostat senses the temperature and cycles the heating unit on and off to maintain the desired temperature. While a single-stage heat thermostat could be used with a single small (5kW) heater, most forced-air electric heating systems use a multi-stage room thermostat.

The room thermostat represented in *Figure 4* can control multi-stage compressor operation as well as multiple stages of heat. It can also function as a heat pump thermostat. Multi-stage room thermostats are used in heat pump applications.

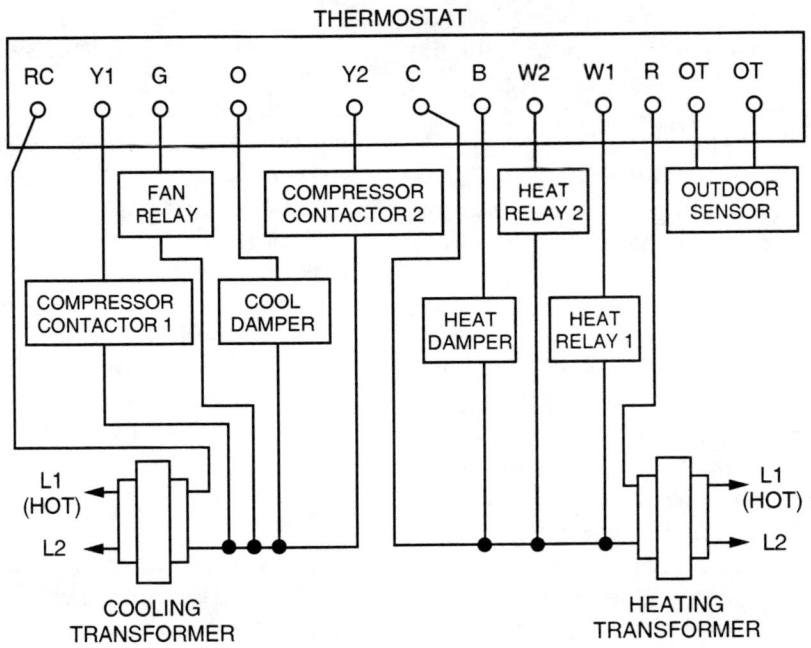

*Figure 4* ♦ Room thermostat wiring diagram.

Typically, the first stage of the room thermostat controls compression (heat pump) heat and the second and subsequent stages of the room thermostat control electric resistance heaters.

Since electric resistance heat is expensive, the use of multi-stage room thermostats can result in an energy savings because the heaters will be activated only when needed. In some areas, electric resistance heaters used with heat pumps must be controlled by an outdoor thermostat to conserve energy. This further ensures that the electric heaters will only be energized when the outdoor temperature is below a certain level and electric heat is really needed to heat the structure.

## 4.2.0 Transformer

Transformers are rated in volt/amperes (VA). The VA rating is the amount of electrical power (volts times amperes) that it can supply. The transformer must be of adequate size to handle the amperage required to operate the low-voltage loads connected to the secondary (low-voltage) side of the transformer. The transformer may be larger than the load requirement, but it can never be smaller. Short-circuit protection on the secondary side of the transformer is provided by a replaceable fuse, fusible link, or manually resettable external circuit breaker. If the fuse or fusible link is built into the transformer, the transformer requires replacement if the fuse or link burns out.

## 4.3.0 Heater Control

Some form of heater control, usually a **sequencer**, is used to bring on the various stages of electric heat. In *Figure 5*, the 1S sequencer controls the first stage of heating. A 24V first-stage heat signal from the room thermostat energizes the 1S sequencer. After 20 seconds the M1 – M2 contacts close, providing power to the 1HTR heater. Most sequencers contain a heating element that warms a bimetal strip used as a control. As the bimetal heats and warps, it will make contact, completing a current path. A sequencer timing graph is also shown in *Figure 5*.

If the room thermostat is not satisfied, another set of thermostat contacts will send a 24V signal to the 2S sequencer, bringing on the second stage 2HTR heater through the M1 – M2 contacts of 2S. Note that the A jumper wire must be removed for two-stage operation. If the jumper is left in place, both the 1S and 2S sequencers would energize on the call for first stage heat. As the room thermostat stages are satisfied, the 1S and 2S sequencers delay the shutdown of the heating elements to provide a gradual reduction.

## 4.4.0 Blower Control

The blower circuit must be designed to simultaneously energize the blower when the heating elements energize. This must be done to prevent overheating of the heating elements. If no airflow is present, the elements will quickly glow red hot and open a limit switch or overheat and burn out. *Figure 6* shows that the blower motor is energized through the normally closed 2R-2 blower relay contacts. In a normal call for heat, there is no 24V signal at G to energize the blower relay. As soon as the 1S, M1 – M2 sequencer contacts close, power is applied to the blower motor.

An exception to this sequence would be if the room thermostat fan switch was placed in the continuous position or the unit was operating in the heat pump compression heat mode. In both of these cases, the blower relay would be energized and the blower motor would already be operating when the heaters were energized.

### 4.4.1 Alternate Blower Control Methods

The sequence just described is commonly used in many air handlers equipped with electric heaters. That particular configuration does not always provide the best indoor comfort. Newer control schemes utilize electronic controls and variable or multi-speed blowers to provide better indoor comfort. For example, with a fixed blower speed (usually high speed) the supply air temperature will vary depending on the number of electric heaters energized. With a small amount (low wattage) of electric heat energized, the air coming from supply registers could feel cool with a high volume of air moving through the system. By matching the fan speed to the heat capacity, the supply air temperature would not experience wide (and possibly uncomfortable) temperature swings. Variable-speed motors have the capability to slowly come up to speed when the heaters are first energized and to slowly reduce speed as the heaters are de-energized, resulting in enhanced comfort.

## 4.5.0 Safety Controls

Safety controls must be included in any forced-air electric heating system. These include limit switches and thermal fuses.

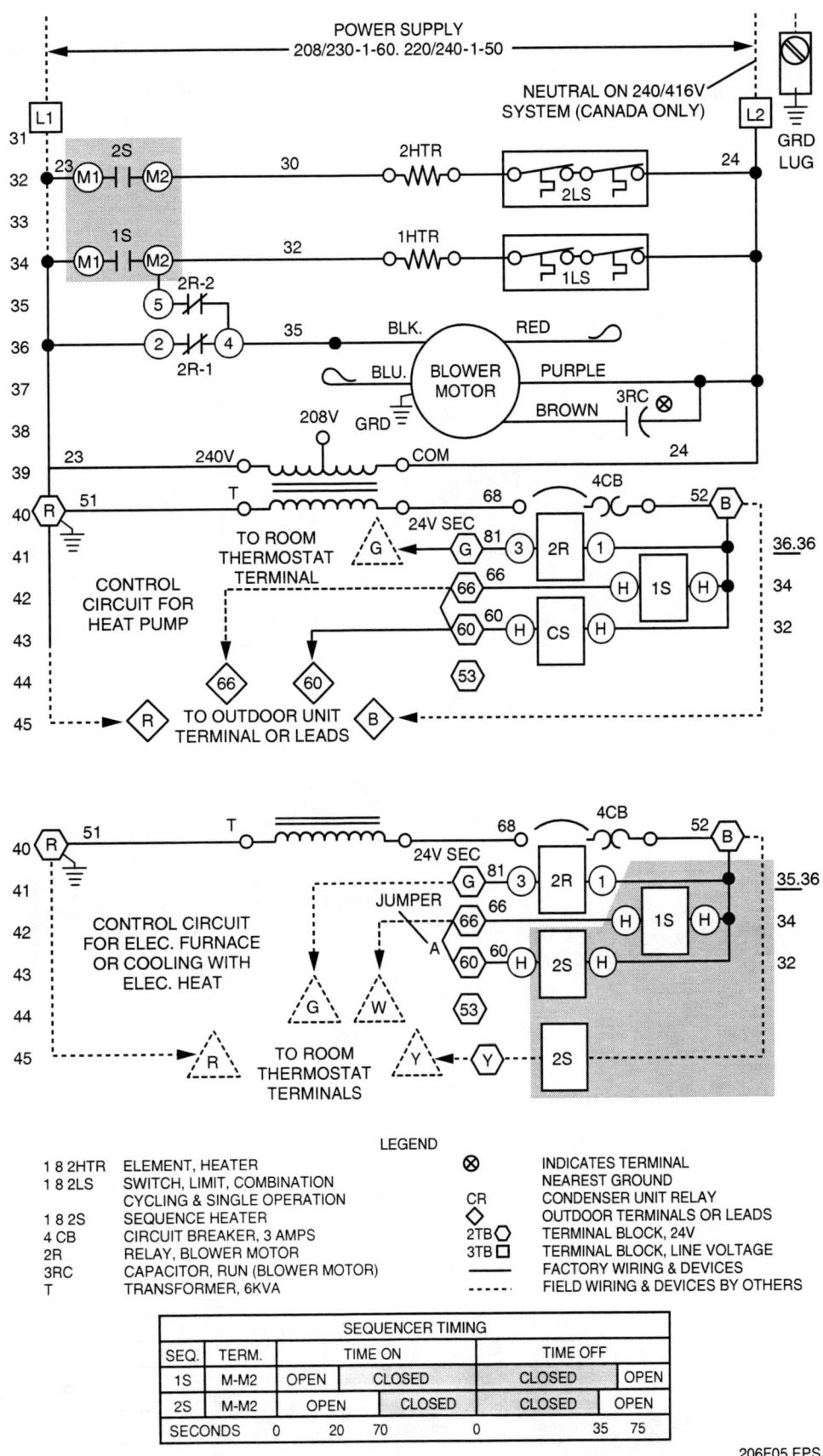

*Figure 5* ♦ Alternate blower control methods.

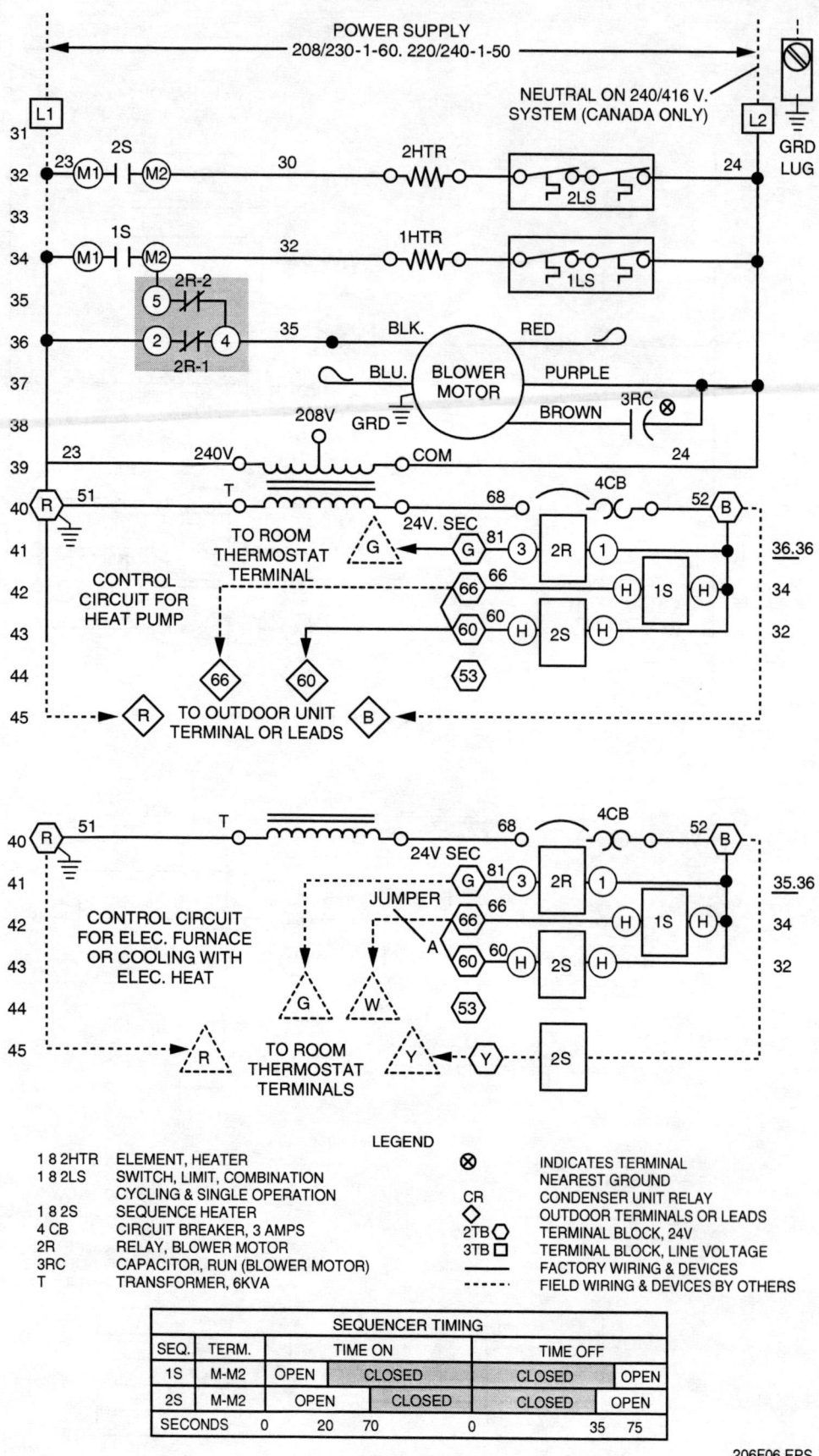

*Figure 6* ◆ Furnace control circuit with current sensing loop.

### 4.5.1 Limit Switch

The function of the limit switch is to de-energize the heating element branch circuit when the ambient temperature surrounding the element exceeds the limit switch setting. The limit switch protects the heating element and surrounding materials.

The limit switches (*Figure 7*) are wired in series with the heating elements 1HTR and 2HTR, and have a disc-type, **bimetal sensor** placed in a small metal housing. The position of the switch is such that the metal shell absorbs radiant heat from the element. As the blower moves air across the metal shell of the limit switch, heat is dissipated from the switch and it remains closed. If the air cannot be moved for some reason, the heat is absorbed by the shell. The bimetal sensor reacts to the excessive heat and will snap the contacts open, taking the element off-line. When the element cools down, the limit switch automatically closes and the heating element will begin to function again. Note that the limit switches 1LS and 2LS are combination devices. One part acts like a conventional cycling limit switch, while the other part is called *single-operation*. This means that the single-operation device may require a manual reset, or it may require replacement of the entire combination control if the single-operation portion opens.

As previously mentioned, there is a differential between the temperature at which the contacts open and the one at which they close. The close setpoint is usually 35°F cooler than the open setpoint. Limit switches have different open and close setpoints; therefore, only a limit switch with the same rating as the original may be used as a replacement for a defective switch.

### 4.5.2 Thermal Fuse

The thermal fuse is designed to open the line voltage circuit to the heating element if an abnormally high temperature exists around the element. It acts as a backup safety switch if the limit switch fails to operate. It has a higher temperature opening setting time than the limit switch. The thermal fuse must be replaced with one of the same rating. If both the limit switch and thermal fuse open, the cause should be determined and repaired before the unit is put back into service.

**NOTE**

A manual reset limit switch may be used in place of a thermal fuse. Since it must be manually reset, the manual reset device provides the same type of protection as the thermal fuse.

### 4.6.0 Two-Stage Thermostat

A two-stage thermostat provides improved efficiency by cycling the heating elements on in stages as the demand increases.

---

### Multi-Stage Heat Control

A heat pump with two stages of electric heat is really a three-stage heating system. The first stage is reversed cycle (compression) heat. When there are two stages of electric heat, the second stage is often controlled by an outdoor thermostat (ODT). This stage does not kick in until the outdoor temperature falls below a preset level (55°F is a common setting).

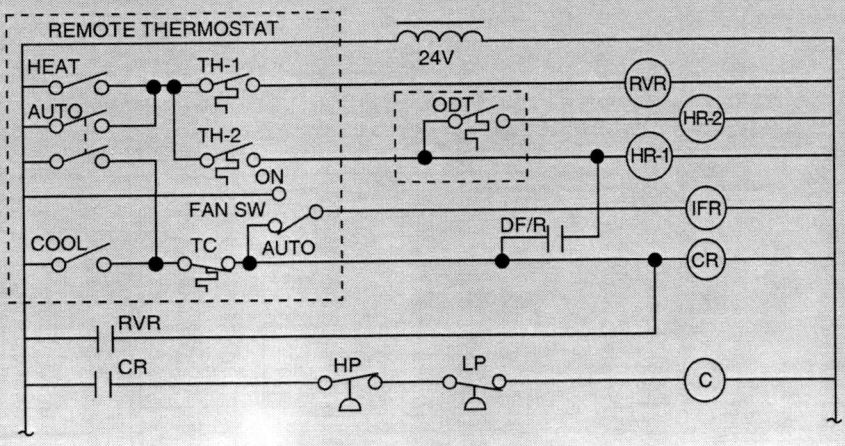

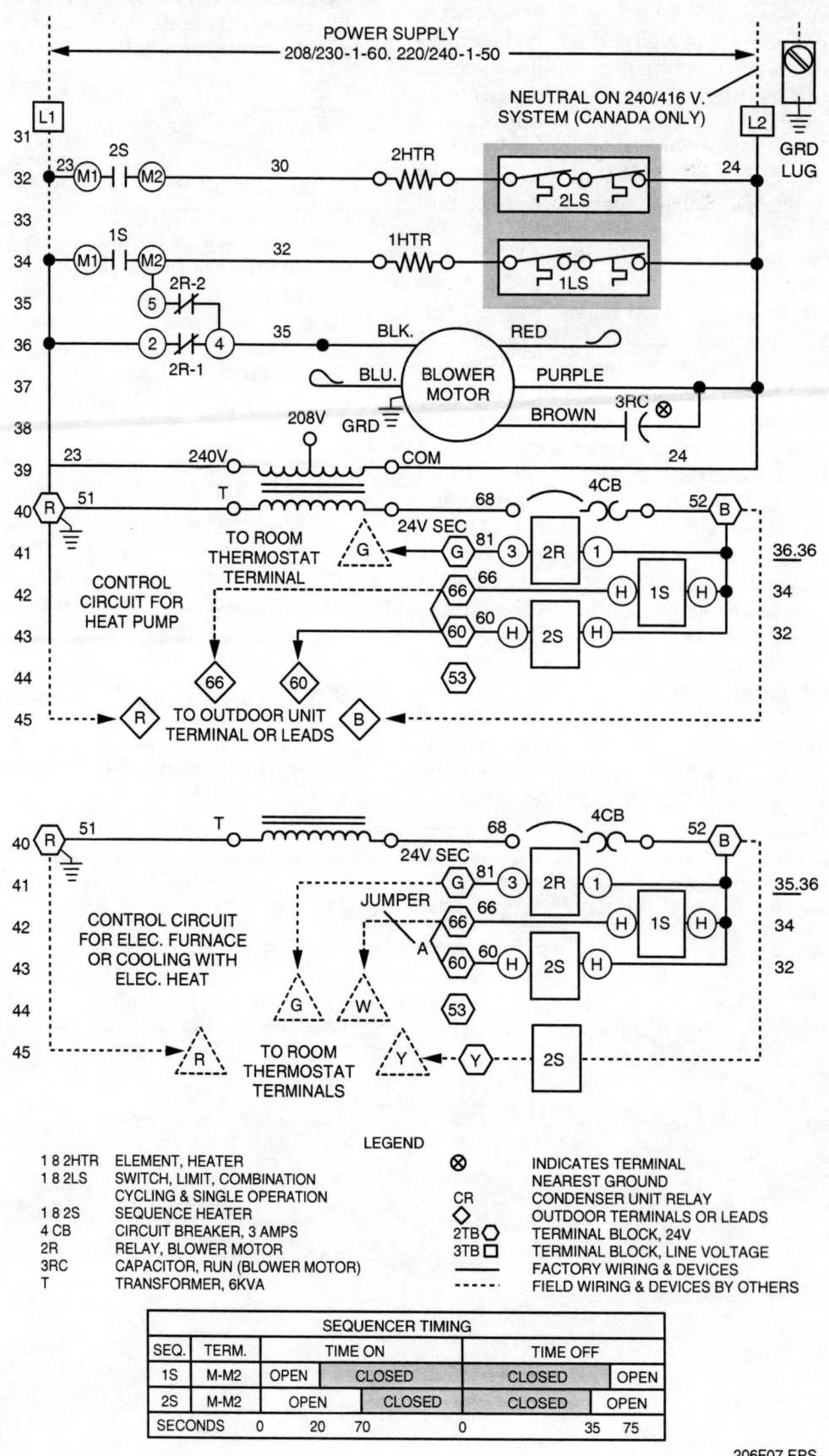

*Figure 7* ♦ Limit switch wiring diagram.

The first stage of the thermostat is set to close at a temperature a few degrees higher than the second stage. The first stage controls the fan and one or more of the heating elements. The second stage controls the rest of the heating elements. The elements come on in order. When the call for heat is met, the elements go off in the same order in which they came on. Consequently, it is very important that the heater relays match the control voltage.

An outdoor thermostat is sometimes used to control some or all of the heat stages. (Some utilities require that all electric heaters used with heat pumps be activated through an outdoor thermostat for energy conservation.) The outdoor thermostat monitors outdoor air temperature. The contacts open on a rise in temperature and close on a fall in temperature. The setpoint is manually adjusted to a temperature where the heat loss in the structure cannot be met by the first stage of heating.

The first bank of heating elements is normally cycled on and off by the indoor heating thermostat. As long as the outdoor temperature is above the setpoint, the first stage will handle the heating load.

Sequencing is similar to that of the two-stage thermostat. The circuit is designed so that unless the indoor thermostat is calling for heat, the outdoor thermostat cannot energize the heating elements.

## 5.0.0 ♦ TROUBLESHOOTING

This section covers troubleshooting. Troubleshooting is covered in detail in the HVAC Level Three Module, *Troubleshooting Electric Heating*.

### 5.1.0 Power Requirements

Electrical wires are run from the 240V branch circuit for the equipment to a fused manual disconnect switch located in the vicinity of the unit. The ground wire is grounded to the box, and the two hot wires are connected to the line side of the disconnect switch.

The disconnect switch provides a means of disconnecting the equipment for service. It is also a safety device in case of equipment malfunction.

The fuses inside the disconnect box must be sized in accordance with the maximum amperage stated on the Underwriters' Laboratories (UL) plate attached to the equipment.

Two hot wires are run from the load side of the disconnect switch to a set of terminals on the fuse block inside the cabinet. This supplies 240V, single-phase electric power to the furnace. When the service entrance, branch lines, disconnect switch, and furnace fuse block electrical connections have been made, the field wiring is complete. The remainder of the 240V circuits within the equipment are usually factory installed.

For a review of the generation and distribution of 240VAC power, refer to the HVAC Level Two Module, *Alternating Current*.

### 5.2.0 Voltage Variations

All load devices (resistance elements, motors, etc.) are designed to operate at the voltage indicated on the equipment nameplate. Most electrical equipment will tolerate variations from 10% above to 10% below the rated or specified voltage. Consequently, a motor rated at 120V will operate with voltages between 108V and 132V. Under-voltage or over-voltage is considered to be any voltage that is not within ±10%.

Under-voltage is a more common problem than over-voltage. Low voltage can cause motor failure and may cause resistance heating elements to become ineffective. It may also cause a drop in transformer voltage, which could result in the failure of some controls to operate.

All power supplies should be checked during peak load conditions to determine if proper voltage is being delivered to the service entrance. The peak load voltage check should be done at the service entrance, at the equipment disconnect, and at load devices. If there is a problem of overloaded circuits, the owner should be notified and an electrician's services obtained. If there is a problem in the furnace wiring, the service technician should look for defective wiring, improperly sized wiring, and improper transformer size and/or application.

---

### *Electrical Disconnects*

Packaged air conditioners and heat pumps equipped with electric heaters may require more than one electrical disconnect. For example, one disconnect would supply power to compressor and fan motor circuits while another disconnect would power the electric heaters. This is commonly seen in large-capacity equipment installed in commercial settings. On some commercial units, higher voltages and three-phase power circuits may be encountered. Exercise appropriate caution when working around energized electrical circuits.

## 5.3.0 Checking Loads

This section covers checking for proper voltage, current, and resistance loads. It is important to note all safety precautions here, as some testing is done with the system power on.

### 5.3.1 Voltage Check

To determine if proper voltage is available at the load, the voltmeter function of a multimeter (VOM/DMM) is used. A reading of 240V should be measured between the two hot legs (*Figure 8*), and 120V from either leg to ground. With proper voltage, the load should be operational.

> **CAUTION**
> The voltage test is made with the power on, so be sure to follow applicable safety precautions.

checking the current through the sequencer to be certain that the elements are being staged on at proper intervals. To conduct a current check, the jaws of the clamp-on ammeter are placed around one of the main supply lines (*Figure 9*). If there is no current reading, check the fuses. If the fuses are good, disconnect the power and check for continuity using an ohmmeter.

> **CAUTION**
> The current check is made with the power on, so be sure to follow applicable safety precautions.

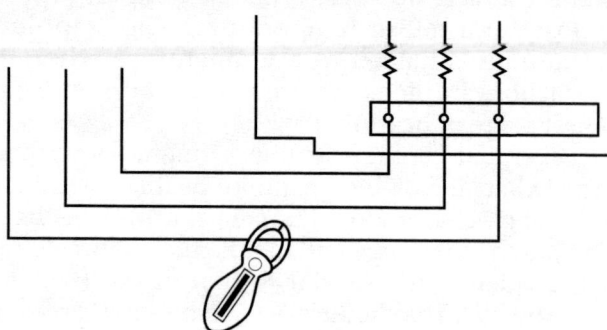

WITH THERMOSTAT CALLING FOR HEAT, CHECK THE CURRENT DRAW OF EACH HEATING ELEMENT. IF NO READING, CHECK FUSES WITH AN OHMMETER. IF FUSES ARE GOOD, MAKE CONTINUITY CHECK ON THE HEATER ELEMENTS WITH OHMMETER, **AFTER THE POWER IS DISCONNECTED.**

*Figure 9* ◆ Current check.

### 5.3.3 Resistance Check

The ohmmeter function of a VOM/DMM can be used to check heating elements for opens or shorts (*Figure 10*). The power must be turned off and the ohmmeter leads placed on the two leads of the heating element. At least one lead of the heating element must be disconnected from the rest of the circuit. If the ohmmeter reads infinity (∞), the element is open and the element or winding must be replaced. If the reading is within the values specified for the element, look elsewhere for the cause of the problem.

## 5.4.0 Replacing Resistance Wires

Some electric furnace manufacturers require replacement of the entire heating element if it is found to have an open resistance wire coil. Others furnish element replacement coil kits that can be installed in the old frame. When replacing an element or installing a coil kit, follow the instructions furnished with the kit or element.

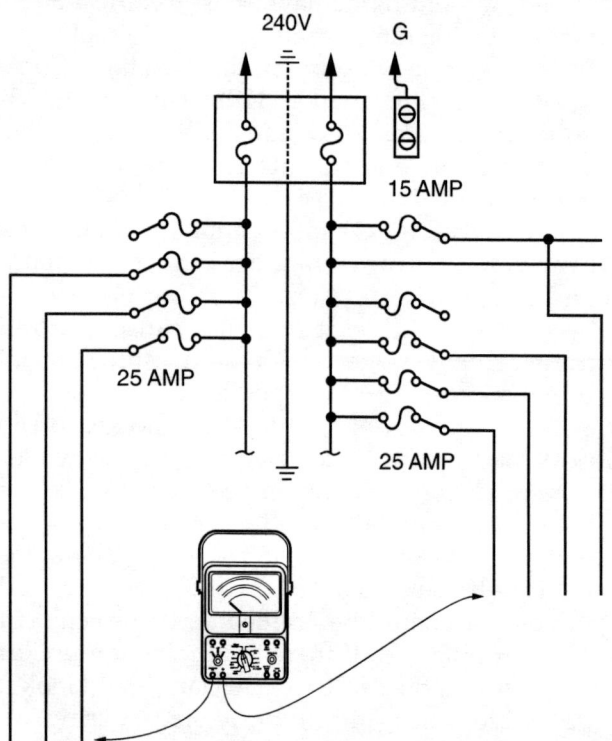

CHECK VOLTAGE TO EACH HEATING ELEMENT. IT SHOULD READ 240V. IF NO VOLTAGE, CHECK FUSES.

*Figure 8* ◆ Voltage check.

### 5.3.2 Current Check

A clamp-on ammeter is used to check the current draw of the equipment or any one of its loads. The ammeter test is performed with the power on. The meter readings are compared with the data on the unit nameplate. This test provides a means of

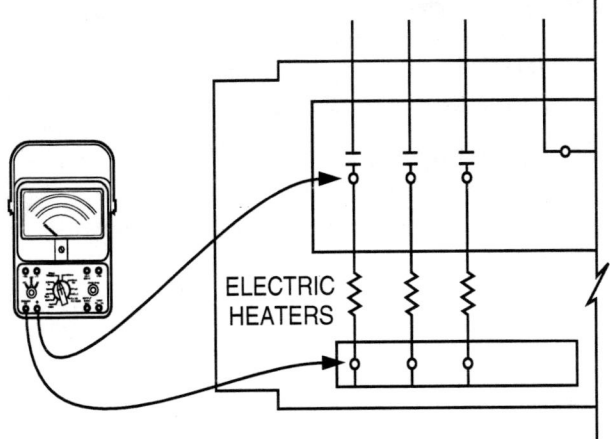

WITH THE POWER OFF, MAKE A RESISTANCE CHECK ON EACH HEATING ELEMENT WITH OHMMETER. IF INFINITY IS READ, REPLACE THE ELEMENT.

*Figure 10* ◆ Resistance check.

## 5.5.0 Determining Btuh Output

The heating capacity of electric heating elements is rated in thousands of watts or kilowatts (kW). As previously discussed, kilowatts can be translated into Btuh output. Recall that 1 watt equals 3.413 Btuh heat output. Thus, 1kW = 3,413 Btuh.

Most resistance elements are rated for full power at 240V input. This data is stamped directly on the frame of the heating element. As the voltage decreases, the kilowatt capacity also decreases. To determine the actual kilowatts at the unit for various voltages, use *Table 1* as a guide.

**Table 1** Voltage Multipliers

| Volts | Multiplier |
|---|---|
| 208 | Multiply output at 240 volts by 0.715 |
| 220 | Multiply output at 240 volts by 0.839 |
| 230 | Multiply output at 240 volts by 0.917 |

After using a VOM/DMM to find the resistance of an element in ohms, Ohm's law can be used to determine the actual kilowatt output of the element. *Figure 11* shows the various relationships between voltage, power, current, and resistance.

The following formula can be used to calculate the output of an element with a known resistance:

$$P = \frac{E^2}{R}$$

> **NOTE**
> The resistance changes as the coil heats up. Use a correction factor of 1.065 when making calculations on a cold coil.

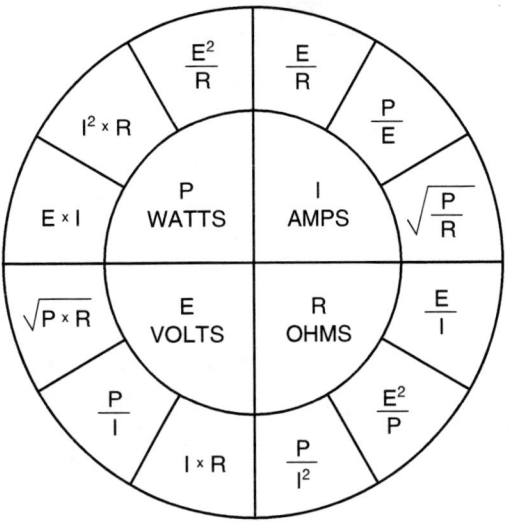

E = VOLTAGE
I = CURRENT
R = RESISTANCE
P = POWER

*Figure 11* ◆ Ohm's law and derivative equations.

### Study Example

Find the kW value of a heating element with a cold resistance of 12.26 ohms (Ω).

Using the applicable power formula, find P (in watts):

$$P = \frac{E^2}{R}$$

$$P = \frac{240^2}{12.26} \times 1.065$$

P = 5,004W or about 5kW

## 5.6.0 Safety With Electrical Circuits

Electricity is a powerful force. When under control, it can safely perform an endless variety of tasks. If it is not controlled, it can be a very destructive force. Control of electricity begins with the right kinds of wiring and the correct installation of equipment.

Most electrical equipment is designed and built for specific types of service. It will operate safely only when used for the purposes and under the conditions for which it is designed. When selecting equipment, always follow the recommendations of applicable codes and standards. Transformers, switches, motor starters, and other electrical equipment and controls should be installed in a way that eliminates or minimizes the possibility of accidental contact with energized conductors.

The severity of electrical shock is determined by the amount of current flow through the victim. An alternating current of 100 milliamperes at the frequency of 60Hertz (Hz) may be fatal if it passes through vital organs. It is estimated that a current of 16 milliamps is the average current at which a person can still release an object held with the hand. This current flow may be readily obtained upon contact with ordinary lighting or power circuits.

High voltage, 60Hz alternating current (600V or higher, according to the NEC) causes violent muscular contraction so severe that, on occasion, the victim is thrown clear of the circuit. Although low voltage (less than 600VAC) can also result in muscular contraction, the effect is not so violent.

Death or injury by electric shock may result from the following effects of current on the body:

- Contraction of the chest muscles, which may interfere with breathing
- Temporary paralysis of the nerve center
- Interference with normal heart rhythm
- Suspension of heart action due to muscular contraction
- Destruction of tissues, nerves, and muscles due to the heat generated by heavy current

Other types of injuries associated with electrical equipment include falls due to loss of muscular control, and mechanical injuries from the accidental starting of motors, blowers, or fans.

The safe current-carrying capacity of conductors and equipment is determined by their size, construction material, and insulation. If circuits overheat from carrying more than the maximum

### Using Power Tools

To minimize your risk of shock when using power tools, use an extension cord with a built-in ground-fault circuit interrupter (GFCI). If there is any current leakage to ground, the GFCI will trip long before a conventional circuit breaker trips, and before any potentially harmful levels of current are reached. A GFCI-equipped extension cord like the one shown here can save your life. If the power tool is equipped with a ground, it must be used. Never remove the ground plug.

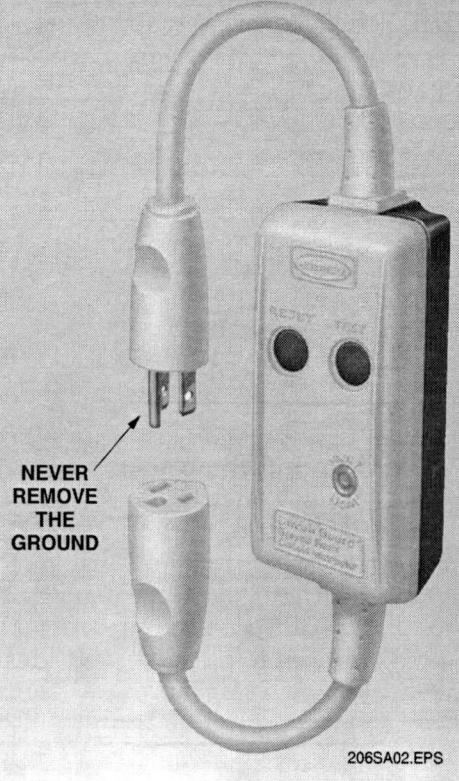

**NEVER REMOVE THE GROUND**

safe load, a fire hazard is created. Overcurrent devices, such as fuses and circuit breakers, provide circuit protection by opening the circuit automatically in the event of excessive current flow from accidental ground, short circuit, or overload.

Before fuses are replaced, the circuit should be opened or locked out, and the cause of the short circuit or overload should be determined. Blown fuses should be replaced by others of the same type and size. Fuses should never be installed in a live circuit. Circuit breakers should be selected for specific installation as determined by engineers or designers, and should be kept in good operating condition at all times.

The importance of using the proper tools and electrical testing equipment cannot be overemphasized. Terminal connections should be completed only with UL-approved fittings and connectors. Hand tools used for the installation of electrical equipment and controls should have insulated (dielectric) handles. Pay particular attention to applying the proper tool for each task.

### 5.6.1 Safety Practices

Electricians and technicians in many different trades work with potentially deadly levels of electricity every day. They can do so because good safety practices have become second nature to them. Here are some general safety practices to follow whenever you are working with electricity:

- Always wear appropriate personal protective equipment, such as rubber gloves.
- Use insulated tools.
- Use a voltmeter to verify that the power to the unit is actually off. Remember that even though the power may be switched off, there is still potential at the input side of the shutoff switch.
- Remove metal jewelry such as rings and watches.
- Keep one hand outside the unit whenever possible.

The *National Electrical Code*, when used together with the electrical code for your local area, provides the minimum requirements for the installation of electrical systems. Always use the latest edition of the Code as your on-the-job reference. It specifies the minimum provisions necessary for protecting people and property from electrical hazards.

In some areas, different editions of the Code may be in use, so be sure to use the edition specified by your employer. When in doubt, follow the most stringent code requirement.

## 6.0.0 ♦ AIRFLOW

Air handlers with electric heaters differ from gas or oil furnaces in that no heat exchanger is needed. Return air from the conditioned space passes directly over the resistance heaters and into the supply air plenum.

The air distribution system of a fan coil (*Figure 12*) is similar to, and serves the same function as, air distribution systems for gas and oil furnaces. Refer to the HVAC Level Two Module, *Air Properties and Distribution*, for information pertaining to air distribution.

### Safety
Unless it is essential to work with the power on, always shut off electricity at the source. Lock and tag the power switch in accordance with company or site procedures and good safety practices.

206SA03.EPS

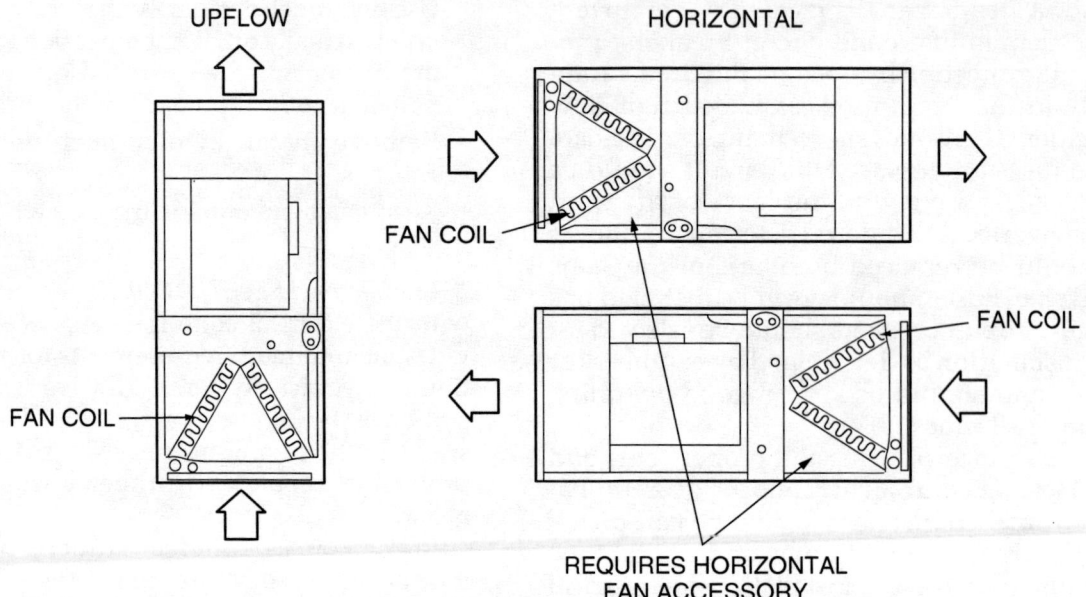

*Figure 12* ◆ Air distribution.

## 6.1.0 Calculating Airflow Volume

The airflow volume (in cfm) of the blower in a fan coil or packaged unit with electric heaters can be calculated using the **temperature rise**, which is the difference in temperature between the supply air and return air. In an electric furnace, the temperature rise is typically 45°F to 75°F. For example, the supply air might be 120°F and the return air 70°F. The correct range is stated on the furnace nameplate. When measuring temperature rise, all diffusers and registers should be in the normal operating position. The return air temperature should be obtained as near to the blower as possible. The supply air temperature can be taken in the supply air plenum. An average of several readings in each location should be taken for best results. The following procedure is recommended:

*Step 1* Drill access holes in the supply and return ducts (*Figure 13*). Insert a digital thermometer probe in each opening.

*Step 2* Turn up the thermostat to energize the heaters and operate the blower. Make sure to set the thermostat high enough to cause the unit to run continuously during the following measurements.

*Step 3* Allow the equipment to run for about 10 minutes in order for it to stabilize.

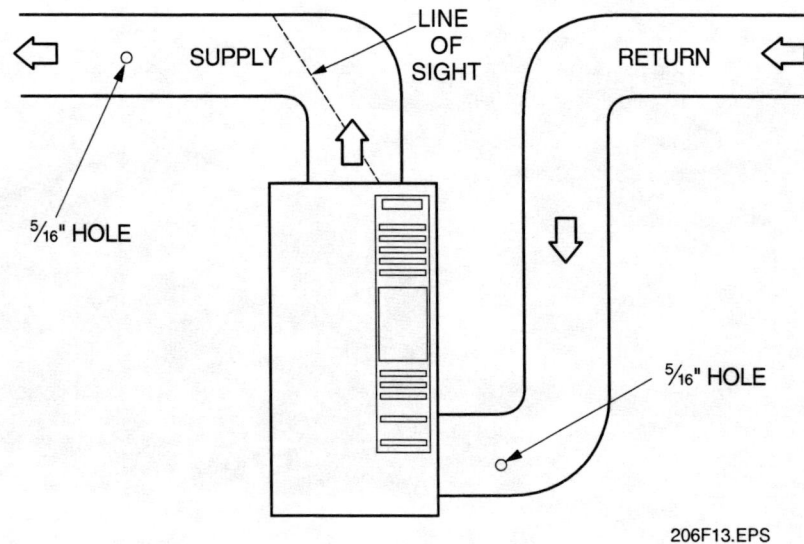

*Figure 13* ◆ Temperature check.

### Temperature Rise Measurement

When drilling a hole in the supply duct to measure the supply air temperature, drill the hole out of the line of sight of the heating elements to prevent radiant heat from affecting the readings.

*Step 4* Measure and record the return and supply air temperatures with a digital thermometer. When done measuring the temperatures, make sure to seal the holes in the supply and return duct.

*Step 5* Calculate the temperature difference (TD) across the furnace as follows:

TD = supply temp. − return temp.

Assuming the measured supply air temperature was 120°F and the return air temperature was 60°F, the furnace temperature rise equals 60°F (120°F − 60°F = 60°F).

*Step 6* Set a VOM/DMM to measure the equipment input AC voltage. At the unit disconnect box or terminal board, measure and record the input voltage. For example, assume the measured input voltage is 230V.

*Step 7* Set a clamp-on ammeter to measure AC current on the highest scale. At the disconnect box or terminal board, measure and record the total current draw. For example, assume the measured current is 90A.

*Step 8* Using the measured values of volts, amperes, and temperature rise, calculate the airflow in cfm using the following formula:

$$\text{Airflow in cfm} = \frac{\text{sensible heat}}{1.08 \times TD}$$

$$= \frac{\text{volts} \times \text{amps} \times 3.413}{1.08 \times TD}$$

Where:
 1.08 = constant
 TD = temperature difference (rise) across furnace
 Sensible heat in Btuh = watts × 3.413
 (watts = voltage × current)

For the example:

$$\text{Airflow cfm} = \frac{230 \times 90 \times 3.413}{1.08 \times 60}$$

= Approx. 1,090 cfm

*Step 9* Compare the calculated airflow to the manufacturer's recommended airflow for the equipment. Refer to the equipment service instructions for recommended values of airflow. If the airflow is within the recommended range, no adjustment of airflow is required. If the airflow is too low or too high, increase or decrease the blower speed, respectively, to change the airflow.

#### 6.1.1 Adjusting Blower Speed

The speed of belt-drive blowers can be changed by adjusting the belt-drive sheave. If a multi-speed direct-drive motor is used, a different speed can be selected to adjust the cfm. If a single-speed motor is used and the cfm needs to be changed, the motor (and possibly the blower) must be replaced. Discharge dampers may be adjusted to achieve the correct airflow.

### 7.0.0 ◆ DUCT HEATERS

Duct heaters are electric resistance heating packages installed directly into the ductwork of an air distribution system as a means of providing heat. Duct heaters and their control circuits are similar to the heating elements and controls of a heater package for a fan coil or packaged unit.

Duct heaters are inserted into the ductwork as a package (*Figure 14*). The heating elements are in the path of airflow and the control boxes are in a closed unit external to the duct. The air flowing over these heaters should be relatively clean to avoid the odor of burning dust particles.

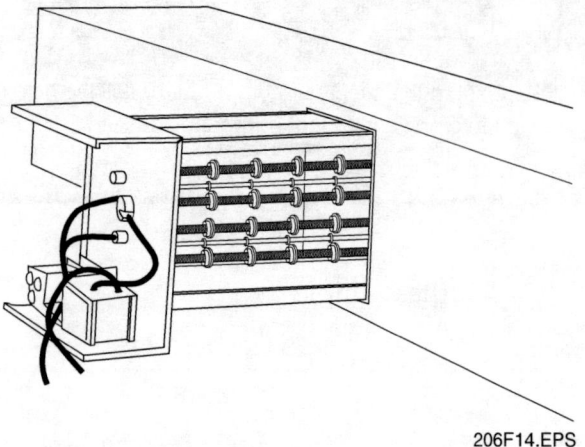

*Figure 15* ◆ Electric baseboard heater.

*Figure 14* ◆ Slip-in duct heater.

## 8.0.0 ◆ OTHER ELECTRIC HEATING SYSTEMS

There are several other types of heating systems that use electricity as the energy source. They include:

- Baseboard electric heaters
- Radiant heating panels
- Electric hot water boilers

### 8.1.0 Baseboard Heaters

Baseboard electric heaters consist of a low-density (250 watts per foot) heating element enclosed in a sheet-metal housing (*Figure 15*). The heaters are mounted a few inches above the floor. The individual units are sized to match the heat loss in individual rooms and come in a variety of lengths. The longer the unit is, the higher the heat output.

The heat elements are bonded to fins which help dissipate radiant heat. The sheet-metal housing allows air to flow over the finned elements by convection. The low heat density of the elements prevents them from getting very hot. This is necessary since baseboard units are installed near carpets, furniture, and drapes which are usually made of combustible materials.

Most baseboard electric heaters are powered by 240VAC and are controlled by a remotely mounted line voltage room thermostat (*Figure 16*) or an adjustable thermostat built into the baseboard unit.

A variation of the baseboard heater contains a fluid-filled reservoir in which the fluid is heated by an electric heating element that is immersed in the fluid. The heated fluid then circulates through a finned radiator within the baseboard unit. It provides heat to the structure in the same way as a conventional baseboard radiator in a hydronic system.

*Figure 16* ◆ Line voltage thermostat.

Advantages of baseboard heaters include low installed cost, zoned operation, and almost trouble-free operation. The main disadvantage is higher operating costs as compared to non-electric heating systems. Another disadvantage is that conventional line voltage room thermostats, which use a bimetal sensing element, do not provide precise temperature control. This is because the bimetal elements and switch contacts must be heavier to carry the larger current. Because they are heavier, they do not respond to temperature changes as quickly or easily as the smaller and lighter contacts in a low-voltage (24V) room thermostat.

Digital line voltage room thermostats that provide much more precise temperature control are now available.

## 8.2.0 Space Heaters

The most basic example of a space heater is the portable type readily available at hardware or discount stores. There are also space heaters designed for permanent installation to provide comfort in areas that are hard to heat (such as bathrooms and underneath a kitchen sink) or in an area where it might be impractical to run ductwork. Typically, these units are self-contained, forced-air heaters with a built-in thermostat and limit switch (*Figure 17*). Because of their small size and low heating capacity, space heaters are often powered by 115V circuits, but can be powered by higher voltages. The advantages of this type of system include low installed cost and simplicity of installation. The main disadvantage is operating cost.

## 8.3.0 Radiant Heating Panels

Radiant heating panels provide quiet, comfortable heat. A grid of low-density heating elements or cable is imbedded in the ceiling or under a floor (*Figure 18*). When energized by individual line voltage room thermostats, the radiant heat provides unsurpassed comfort. Since the grid of heating elements must be installed above the finished ceiling or below the finished floor surface, care must be taken when applying the finished floor or ceiling to prevent damage to the heating elements. Since the heating elements are concealed, repairs to such a system can be time-consuming and expensive.

Comfort is the major advantage of this system; installed cost and potential costs of repair are the major disadvantages. Like all electrically powered heating systems, radiant heating panels can be costly to operate.

## 8.4.0 Electric Boilers

Boilers used in hydronic heating systems are usually thought of as being fired by gas or oil, but electricity can also be used to heat water in a boiler. The elements are immersed in the water and provide heat when energized. All other controls on the boiler and hydronic system are similar to those found on gas- or oil-fired hydronic heating systems. Due to the compact nature of an electric boiler, it is often mounted on a wall (*Figure 19*).

A major advantage of this system is that it does not need a vent for disposal of combustion products. Compact size and relatively trouble-free operation in comparison to a gas- or oil-fired system are additional advantages. The main disadvantage is operating cost.

*Figure 17* ♦ Space heater.

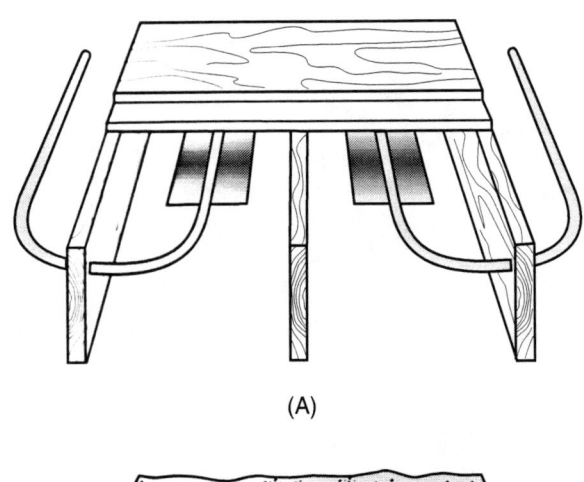

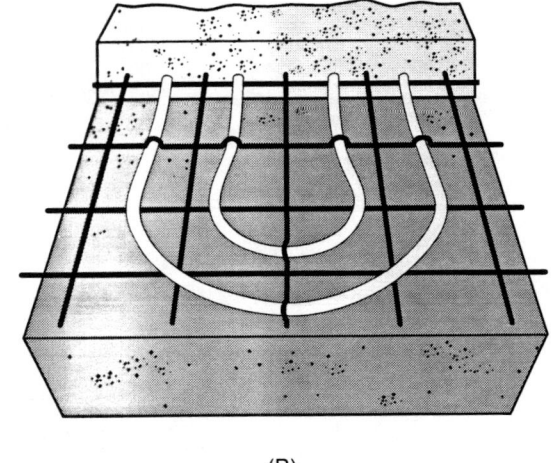

*Figure 18* ♦ Radiant heating applications.
(A) Wood floor. (B) Concrete floor.

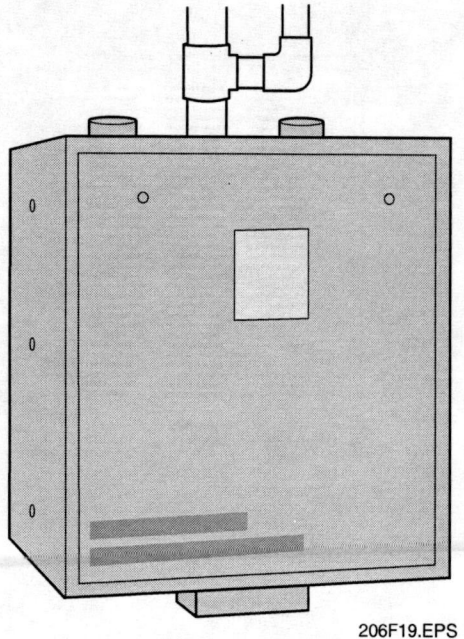

*Figure 19* ◆ Wall-mounted electric boiler.

## Summary

In some locations, especially where electrical power is inexpensive, electric heat is a good alternative to fossil-fuel heat. It is cleaner and offers certain safety advantages.

The heating elements in a forced-air heating system are usually made of resistive wire that gives off heat when current flows through it. A blower circulates building air over the heating elements in the same way that air is circulated over the heat exchangers of a forced-air gas or oil furnace.

Electrical controls and safety devices are different from those used on other types of forced-air heating systems. One big difference is that units with multiple heating elements are staged on by sequencing relays to avoid the power drain that could occur if all the elements came on at once.

Baseboard heaters, radiant elements imbedded in the floor or ceiling, and electrically powered hydronic boilers are other means through which electricity can provide comfort heating.

## Review Questions

1. Which of the following is *not* a component of a forced-air electric heating system?
   a. Heating element
   b. Burner manifold assembly
   c. Blower motor
   d. Heater relay

2. How much current will be drawn by a 240V furnace with three 5kW heating elements?
   a. 63A
   b. 120A
   c. 21A
   d. 51A

3. Heating elements are sequenced on over a period of several seconds in order to _____.
   a. avoid overheating the conditioned space
   b. prevent cold blow
   c. prevent the furnace from overheating
   d. avoid power surges

4. Which of the following safety controls is *not* likely to be found in a fan coil equipped with electric heaters?
   a. Safety limit switch
   b. Flame rollout switch
   c. Thermal fuse
   d. Circuit breaker

5. What type of blower fan is used on a fan coil equipped with electric heaters?
   a. Propeller fan and belt-drive
   b. Belt-drive only
   c. Direct-drive
   d. Propeller fan only

6. The purpose of energizing the blower at the same time the heaters are energized is to _____.
   a. prevent cold blow
   b. take advantage of residual heat
   c. prevent the heating elements from overheating
   d. save wear and tear on the blower

7. The purpose of the limit switch located on each heating element is to _____.
   a. keep the furnace from exceeding its capacity
   b. turn on the heat when the conditioned space gets too cold
   c. turn off the heat when the conditioned space gets too warm
   d. prevent the heat inside the fan coil from exceeding a safe level

8. What is the heating capacity of a furnace with four 5kW heating elements at 230V?
   a. 68,280 Btuh
   b. 20,000 Btuh
   c. 15,000kW
   d. 62,613 Btuh

9. How many watts of energy will a heating element with a cold resistance of 15Ω produce at 240V?
   a. 16
   b. 3,600
   c. 4,090
   d. 10,000

10. A fan coil with electric heat produces a supply air temperature of 130°F and has a return air temperature of 65°F. If the supply voltage is 208V and the fan coil current draw is 60A, the volume of airflow in cfm is _____.
    a. 607
    b. 641
    c. 700
    d. 822

11. When calculating temperature rise, the supply diffusers in the conditioned space should be _____.
    a. closed
    b. in their normal operating positions
    c. fully open
    d. open halfway

12. Baseboard electric heaters are usually controlled by a _____.
    a. built-in 24V thermostat
    b. line voltage thermostat
    c. 24V room thermostat
    d. cycling limit switch

13. Baseboard electric heaters operate at low temperatures in order to _____.
    a. prevent furniture and drapes from igniting
    b. conserve energy
    c. provide better temperature control
    d. reduce stress on the room thermostat

14. A unique feature of electric space heaters is that they _____.
    a. can be equipped with a cooling coil
    b. can provide central heating
    c. can operate on 12VDC
    d. have a built-in thermostat

15. A major disadvantage of floor- or ceiling-installed radiant electric heating systems is that _____.
    a. they provide poor comfort
    b. they require a special electronic room thermostat
    c. they can be expensive to repair
    d. most building codes prohibit their installation

# GLOSSARY

# Trade Terms Introduced in This Module

*Bimetal sensor:* A sensing element made of two dissimilar metals. It warps when exposed to heat. A bimetal sensor is used in control circuits to activate and deactivate a thermostat.

*Limit switch:* A heat-sensing safety switch that cuts off power if the temperature near the heating elements exceeds a set limit.

*Sequencer:* A relay with a built-in time delay of several seconds that allows the electric heating elements to be gradually staged on.

*Temperature rise:* The difference in temperature between the supply air and return air, as measured at the supply and return plenums.

*Thermal fuse:* A fuse that opens if the temperature around it gets too high (as opposed to a standard fuse, which opens if the current in the circuit gets too high).

# ANSWER KEY

## Answers to Review Questions

| Answer | Section |
|---|---|
| 1. b | 2.1.0 |
| 2. a | 2.1.1 |
| 3. d | 2.1.1 |
| 4. b | 2.1.1; 2.1.5 |
| 5. c | 2.1.2 |
| 6. c | 4.4.0 |
| 7. d | 4.5.1 |
| 8. d | 5.5.0 |
| 9. c | 5.5.0 |
| 10. a | 6.1.0 |
| 11. b | 6.1.0 |
| 12. b | 8.1.0 |
| 13. a | 8.1.0 |
| 14. d | 8.2.0 |
| 15. c | 8.3.0 |

# REFERENCES & ACKNOWLEDGMENTS

## *Additional Resources*

This module is intended to present thorough resources for task training. The following reference works are suggested for further study. These are optional materials for continued education rather than for task training.

*Modern Refrigeration and Air Conditioning*, 2000. A.D. Althouse, C.H. Turnquist, A.F. Bracciano. Tinley Park, IL: The Goodheart-Willcox Company, Inc.

*Refrigeration & Air Conditioning Technology*, 2000. William C. Whitman, William M. Johnson, John A. Tomczyk. Albany, NY: Delmar Publishers, Inc.

*Pocket Guide to Electrical Installations Under NEC 2002, Volumes I and II*, 2001. Quincy, MA: National Fire Protection Association.

## *Figure Credits*

| | |
|---|---|
| **Thomas P. Burke** | 206F03, 206F15, 206F16 |
| **Hubbell** | 206SA02 |
| **Carrier Corporation** | 206SA03 |
| **Radiante Co.** | 206F18 |

# NCCER CRAFT TRAINING USER UPDATES

The NCCER makes every effort to keep these textbooks up-to-date and free of technical errors. We appreciate your help in this process. If you have an idea for improving this textbook, or if you find an error, a typographical mistake, or an inaccuracy in the NCCER's Craft Training textbooks, please write us, using this form or a photocopy. Be sure to include the exact module number, page number, a detailed description, and the correction, if applicable. Your input will be brought to the attention of the Technical Review Committee. Thank you for your assistance.

*Instructors* – If you found that additional materials were necessary in order to teach this module effectively, please let us know so that we may include them in the Equipment and Materials list in the Instructor's Guide.

**Write:** Curriculum Revision and Development Department
National Center for Construction Education and Research
P.O. Box 141104, Gainesville, FL 32614-1104

**Fax:** 352-334-0932

**E-mail:** curriculum@nccer.org

Craft _____ Module Name _____

Copyright Date _____ Module Number _____ Page Number(s) _____

Description

_____
_____
_____
_____

(Optional) Correction

_____
_____
_____

(Optional) Your Name and Address

_____
_____
_____

# Module 03207-01

# *Introduction to Control Circuit Troubleshooting*

## COURSE MAP

This course map shows all of the modules in the second level of the HVAC curriculum. The suggested training order begins at the bottom and proceeds up. Skill levels increase as you advance on the course map. The local Training Program Sponsor may adjust the training order.

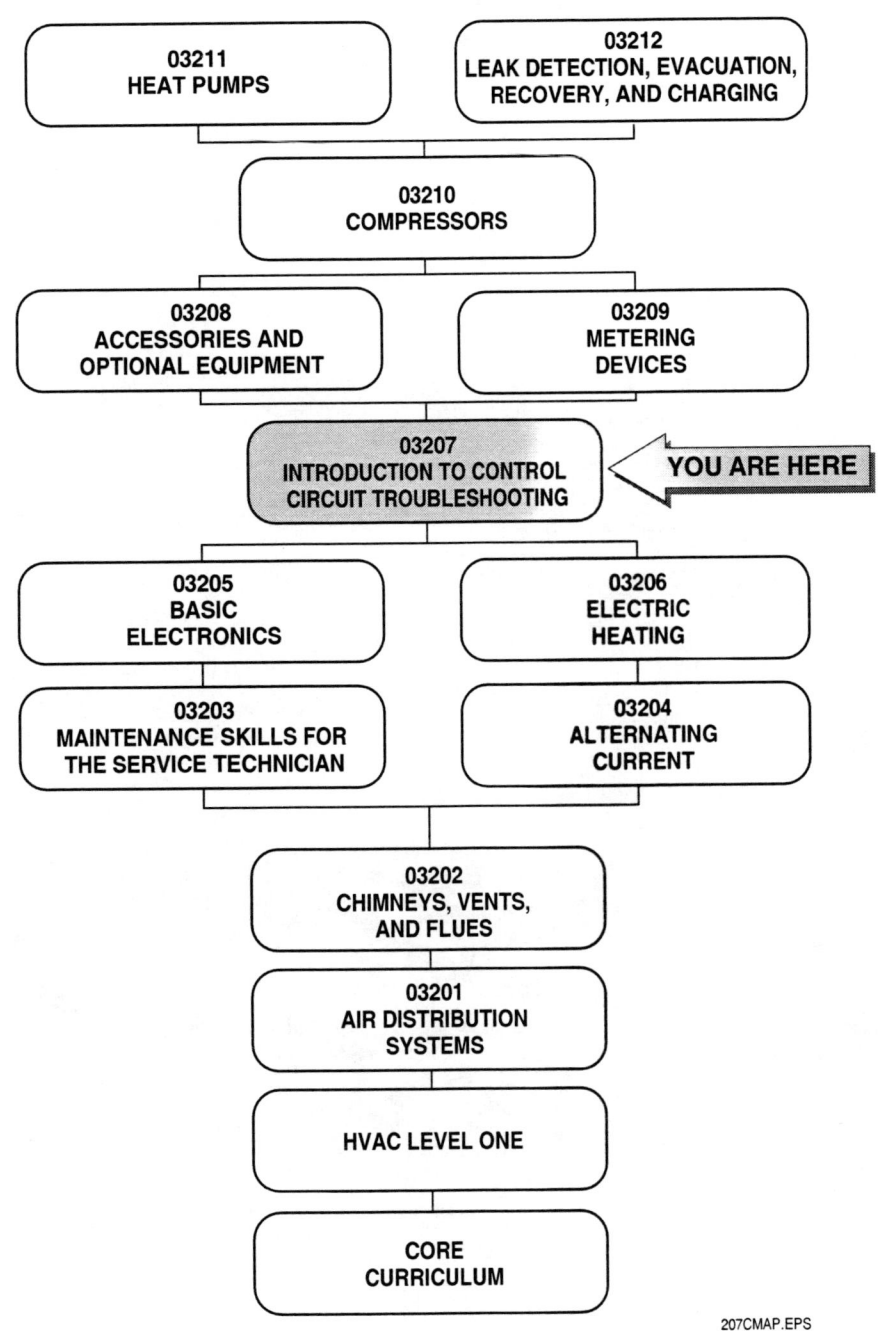

INTRODUCTION TO CONTROL CIRCUIT TROUBLESHOOTING — TRAINEE MODULE 03207

# MODULE 03207 CONTENTS

- **1.0.0 INTRODUCTION** .................................................. 7.1
- **2.0.0 THERMOSTATS** .................................................... 7.1
  - 2.1.0 Principles of Operation ........................................ 7.2
  - 2.2.0 Heating-Only Thermostats ....................................... 7.2
  - 2.3.0 Cooling-Only Thermostats ....................................... 7.3
  - 2.4.0 Heating-Cooling Thermostats .................................... 7.4
  - 2.5.0 Heating-Cooling Automatic Changeover Thermostats ............... 7.4
  - 2.6.0 Two-Stage Thermostats .......................................... 7.5
  - 2.7.0 Programmable Thermostats ....................................... 7.5
  - 2.8.0 Line Voltage Thermostats ....................................... 7.6
  - 2.9.0 Thermostat Installation ........................................ 7.6
    - *2.9.1 Installation Guidelines* .................................... 7.7
    - *2.9.2 Thermostat Wiring* .......................................... 7.8
    - *2.9.3 Checking Current Draw* ...................................... 7.8
    - *2.9.4 Adjusting Heat Anticipators* ................................ 7.10
    - *2.9.5 Cycle Rate* ................................................. 7.10
    - *2.9.6 Final Check* ................................................ 7.10
    - *2.9.7 Adjusting the Thermostat* ................................... 7.11
- **3.0.0 HVAC CONTROL SYSTEMS** ......................................... 7.11
  - 3.1.0 Relays, Contactors, and Starters ............................... 7.12
    - *3.1.1 Relays* ..................................................... 7.12
    - *3.1.2 Electronic Solid-State Relays* .............................. 7.15
    - *3.1.3 Contactors and Starters* .................................... 7.15
  - 3.2.0 Motor Speed Controls ........................................... 7.16
  - 3.3.0 Lockout Control Circuit ........................................ 7.17
  - 3.4.0 Time Delay Relay ............................................... 7.18
  - 3.5.0 Compressor Short-Cycle Timer ................................... 7.19
  - 3.6.0 Control Circuit Safety Switches ................................ 7.19
    - *3.6.1 Pressure Switches* .......................................... 7.19
    - *3.6.2 Freezestat* ................................................. 7.19
    - *3.6.3 Outdoor Thermostats* ........................................ 7.20
  - 3.7.0 Furnace Controls ............................................... 7.20
    - *3.7.1 Fan Control* ................................................ 7.20
    - *3.7.2 Limit Control* .............................................. 7.20
    - *3.7.3 Thermocouple* ............................................... 7.21
    - *3.7.4 Inducer Proving Switch* ..................................... 7.21
- **4.0.0 CONTROL CIRCUIT SEQUENCE OF OPERATION** ........................ 7.21

## MODULE 03207 CONTENTS (Continued)

**5.0.0 USING AN ORGANIZED APPROACH TO ELECTRICAL
TROUBLESHOOTING** .................................................7.26
    5.1.0   Customer Interviews .........................................7.26
    5.2.0   Physical Examination of the System ......................7.26
    5.3.0   Basic System Analysis ......................................7.27
    5.4.0   Use of Manufacturer's Troubleshooting Aids ...............7.27
        *5.4.1   Label Diagrams* .......................................*7.27*
        *5.4.2   Troubleshooting Tables and Fault Isolation Diagrams* .......*7.28*
        *5.4.3   Diagnostic Equipment and Tests* ...........................*7.28*
    5.5.0   Fault Isolation in the Equipment Problem Area ............7.28

**6.0.0 SAFETY** ................................................................7.30
    6.1.0   Safety Practices ............................................7.30
    6.2.0   OSHA Lockout/Tagout Rule .................................7.30
    6.3.0   Lockout/Tagout Procedure ..................................7.30
    6.4.0   Restoring Machines or Equipment .........................7.31

**7.0.0 HVAC SYSTEM TROUBLESHOOTING** .........................7.31

**8.0.0 HVAC EQUIPMENT INPUT POWER, LOAD,
AND CONTROL CIRCUITS** ....................................7.32
    8.1.0   Isolating to a Faulty Circuit via the Process of Elimination ....7.33
    8.2.0   Isolating to a Faulty Circuit Component ....................7.34

**9.0.0 ELECTRICAL TROUBLESHOOTING COMMON
TO ALL HVAC EQUIPMENT** ...................................7.35
    9.1.0   Input Voltage Measurements ...............................7.35
        *9.1.1   Effects of High and Low Voltage* .........................*7.35*
        *9.1.2   Voltage Phase Imbalance* ................................*7.36*
    9.2.0   Fuse/Circuit Breaker Checks ...............................7.36
        *9.2.1   Fuse Checks* ..........................................*7.37*
        *9.2.2   Circuit Breaker Checks* ..................................*7.38*
    9.3.0   Resistive and Inductive Load Checks ......................7.39
    9.4.0   Switch and Contactor/Relay Contact Checks ..............7.40
    9.5.0   Control Transformer Checks ...............................7.41
    9.6.0   Thermostat Checks ........................................7.42
        *9.6.1   Fan Switch Operation Checks* ............................*7.42*
        *9.6.2   Cooling Operation Checks* ...............................*7.42*
        *9.6.3   Heating Operation Checks* ...............................*7.42*

## MODULE 03207 CONTENTS (Continued)

- **10.0.0 MOTORS AND MOTOR CIRCUIT TROUBLESHOOTING** .......... 7.42
  - 10.1.0 Precautions for Motor Testing .......................... 7.44
  - 10.2.0 Start and Run Capacitor Checks ....................... 7.44
  - 10.3.0 Start Relay Checks ...................................... 7.45
  - 10.4.0 Start Thermistor Checks ................................ 7.46
  - 10.5.0 Identifying Unmarked Terminals of a PSC/CSR Motor ...... 7.47
  - 10.6.0 Open, Shorted, or Grounded Winding Checks ............ 7.47
    - *10.6.1 Open or Shorted Winding Checks* ................... 7.47
    - *10.6.2 Grounded Winding Check* ........................... 7.48
- **11.0.0 HYDRONIC CONTROLS** ...................................... 7.49
  - 11.1.0 Aquastat ................................................ 7.49
  - 11.2.0 Reset Controller ........................................ 7.50
  - 11.3.0 Low Water Cutoff ....................................... 7.50
  - 11.4.0 Circulator Pump ........................................ 7.50
  - 11.5.0 Zone Valves ............................................ 7.51
- **12.0.0 PNEUMATIC CONTROLS** .................................... 7.51
  - 12.1.0 Basic Components ...................................... 7.54
  - 12.2.0 Pneumatic Control System .............................. 7.56
  - 12.3.0 Airflow Control ......................................... 7.57
- **13.0.0 HVAC DIGITAL CONTROL SYSTEMS** ........................ 7.57
  - 13.1.0 Direct Digital Control .................................. 7.57
  - 13.2.0 Controlling Devices .................................... 7.58
  - 13.3.0 Example of a Digital Control System ................... 7.59

**SUMMARY** ........................................................... 7.60

**REVIEW QUESTIONS** ................................................. 7.61

**GLOSSARY** .......................................................... 7.64

**APPENDIX** .......................................................... 7.65

**ANSWERS TO REVIEW QUESTIONS** .................................. 7.67

**REFERENCES & ACKNOWLEDGMENTS** ................................ 7.68

## Figures

| | | |
|---|---|---|
| Figure 1 | Bimetal sensing elements | 7.2 |
| Figure 2 | Heating-only thermostat | 7.3 |
| Figure 3 | Cooling-only thermostat | 7.3 |
| Figure 4 | Cooling compensator | 7.3 |
| Figure 5 | Heating-cooling thermostat | 7.4 |
| Figure 6 | Heating-cooling contacts | 7.4 |
| Figure 7 | Automatic changeover thermostat | 7.5 |
| Figure 8 | Heat pump thermostat | 7.5 |
| Figure 9 | Programmable thermostat | 7.6 |
| Figure 10 | Electronic programmable thermostat | 7.6 |
| Figure 11 | Remote bulb thermostat | 7.6 |
| Figure 12 | Thermostat mounting | 7.7 |
| Figure 13 | Thermostat leveling | 7.8 |
| Figure 14 | Thermostat wiring | 7.8 |
| Figure 15 | Thermostat heat anticipator | 7.9 |
| Figure 16 | Gas valve electric ratings | 7.9 |
| Figure 17 | Amperage check | 7.9 |
| Figure 18 | Connection points | 7.10 |
| Figure 19 | Examples of plug-in relays | 7.12 |
| Figure 20 | Normally-open and normally-closed relay contacts | 7.13 |
| Figure 21 | Single-pole, single-throw (SPST) relay | 7.13 |
| Figure 22 | Single-pole, double-throw relay | 7.14 |
| Figure 23 | Double-pole double-throw (DPDT) relay | 7.15 |
| Figure 24 | Solid-state relay | 7.15 |
| Figure 25 | Typical contactor | 7.16 |
| Figure 26 | Autotransformer motor speed control | 7.17 |
| Figure 27 | Motor speed control using TRIACs | 7.17 |
| Figure 28 | Potentiometer damper control | 7.17 |
| Figure 29 | Electronic variable-speed furnace control | 7.18 |
| Figure 30 | Lockout relay used in an HVAC control circuit | 7.18 |
| Figure 31 | Four compressor control circuits with time delay relays | 7.19 |
| Figure 32 | Complete 24V furnace control circuit | 7.21 |
| Figure 33 | Basic cooling system control circuit | 7.22 |
| Figure 34 | Typical cooling system control circuit | 7.23 |
| Figure 35 | Circuit diagram of a cooling/gas heating system | 7.25 |
| Figure 36 | Typical label diagram | 7.27 |
| Figure 37 | Ladder diagram | 7.28 |

## Figures (Continued)

| | | |
|---|---|---|
| Figure 38 | Typical troubleshooting diagram | 7.29 |
| Figure 39 | Typical fault isolation diagram | 7.29 |
| Figure 40 | Lock out and tag HVAC equipment | 7.31 |
| Figure 41 | HVAC equipment functional circuit areas | 7.33 |
| Figure 42 | High-voltage and low-voltage circuits | 7.34 |
| Figure 43 | Isolating to a faulty circuit component | 7.35 |
| Figure 44 | Single-phase input voltage checks | 7.36 |
| Figure 45 | Three-phase voltage and current checks | 7.37 |
| Figure 46 | Fuse checks | 7.38 |
| Figure 47 | Circuit breaker checks | 7.38 |
| Figure 48 | Resistive and inductive load resistance checks | 7.41 |
| Figure 49 | Relay contact checks | 7.41 |
| Figure 50 | Control transformer checks | 7.41 |
| Figure 51 | Troubleshooting fan switch function of a heating/cooling thermostat | 7.42 |
| Figure 52 | PSC and CSR motors | 7.43 |
| Figure 53 | PSC multi-speed motor | 7.43 |
| Figure 54 | Three-lead, wye-connected, single-voltage, three-phase motor | 7.44 |
| Figure 55 | Capacitor checks | 7.45 |
| Figure 56 | Start relay check | 7.46 |
| Figure 57 | Start thermistor check | 7.46 |
| Figure 58 | Identifying unmarked terminals of a PSC/CSR motor | 7.47 |
| Figure 59 | Motor open or shorted winding check | 7.48 |
| Figure 60 | Grounded winding check | 7.48 |
| Figure 61 | Simple hydronic system | 7.49 |
| Figure 62 | Aquastat controls | 7.49 |
| Figure 63 | Aquastat used for low limit and circulator control in an oil-fired hydronic system | 7.50 |
| Figure 64 | Typical reset controller outside temperature sensor mounting | 7.50 |
| Figure 65 | Low water cutoff control | 7.50 |
| Figure 66 | Circulator pump | 7.51 |
| Figure 67 | Typical zone valve | 7.51 |
| Figure 68 | Pneumatic system | 7.54 |
| Figure 69 | Bleed-type thermostat | 7.55 |
| Figure 70 | Non-bleed thermostat | 7.55 |
| Figure 71 | Pneumatic actuator | 7.55 |

**Figures (Continued)**

Figure 72  Normally open damper ........................ 7.55
Figure 73  Normally closed damper ...................... 7.55
Figure 74  Modulating system with E-P relay ............. 7.56
Figure 75  Sail switch .................................. 7.57
Figure 76  Centralized building management system ....... 7.58
Figure 77  System controller module ..................... 7.58
Figure 78  Digital vs. analog signal .................... 7.59
Figure 79  Changes are small and occur gradually ........ 7.59
Figure 80  Typical constant volume HVAC system .......... 7.60

# MODULE 03207

# Introduction to Control Circuit Troubleshooting

## Objectives

When you have completed this module, you will be able to do the following:

1. Explain the function of a thermostat in an HVAC system.
2. Describe different types of thermostats and explain how they are used.
3. Demonstrate the correct installation and adjustment of a thermostat using proper siting and wiring techniques.
4. Explain the basic principles applicable to all control systems.
5. Identify the various types of electromechanical, electronic, and pneumatic HVAC controls, and explain their function and operation.
6. Describe a systematic approach for electrical troubleshooting of HVAC equipment and components.
7. Recognize and use equipment manufacturer's troubleshooting aids to troubleshoot HVAC equipment.
8. Exhibit competence in isolating electrical problems to faulty power distribution, load, or control circuits.
9. Identify the service instruments needed to troubleshoot HVAC electrical equipment.
10. Make electrical troubleshooting checks and measurements on circuits and components common to all HVAC equipment.

## Prerequisites

Before you begin this module, it is recommended that you successfully complete the following modules: Core Curriculum; HVAC Level One; HVAC Level Two, Modules 03201 through 03206.

## Required Trainee Materials

1. Pencil and Paper
2. Appropriate Personal Protective Equipment

### 1.0.0 ◆ INTRODUCTION

The first half of this module describes the operation of several common HVAC electrical control devices. You have been introduced to some of these devices in your HVAC Level One training. This module expands on this information and also covers several additional devices.

In an HVAC system, control devices such as relays, contactors, switches, and thermostats interact to control every aspect of system operation. Control devices are typically used in circuits that operate to stop and start HVAC system load devices such as motors, compressors, and heaters.

The second half of this module introduces the task of electrical **troubleshooting** with a focus placed on troubleshooting methods that are common to most types of HVAC equipment. Troubleshooting methods that are unique to the specific types of HVAC equipment, such as gas heating, cooling, and heat pumps are covered in individual modules in your HVAC Level Three training. Control devices used in hydronic and pneumatic control systems and digital control systems are introduced in this module.

### 2.0.0 ◆ THERMOSTATS

The room thermostat is the primary control in an HVAC system. It can be as simple as the single temperature-sensitive switch you learned about in HVAC Level One. It can also be a complex collection of sensing elements and switching devices that provide many levels of control. When we talk about thermostats in this module, we will generally be referring to the control devices that are mounted on a wall in the conditioned space.

Most residences, and many small commercial businesses such as retail stores and shops, have a single thermostat. Large office buildings, shopping malls, and factories will be divided into cooling and heating zones, and will have several thermostats—one for each zone. An important fact to remember about zoned heating and cooling is that each zone is independent of the others. Therefore, each thermostat must control either a separate system or the airflow from a common system.

## 2.1.0 Principles of Operation

Programmable electronic thermostats using electronic sensing elements are becoming increasingly popular; however, thermostats with **bimetal** sensing elements are still being manufactured and are still widely used. Although not quite as precise as the electronic thermostat, bimetal devices are considerably less expensive and have proven to be effective and reliable. Most bimetal thermostats will maintain the space temperature within ±2° of the setpoint. A good electronic thermostat will maintain the temperature within ±1°.

A bimetal element (*Figure 1*) is composed of two different metals bonded together; one is usually copper or brass. The other, a special metal called **Invar**®, contains 36% nickel. When heated, the copper or brass has a more rapid expansion rate than the Invar®, and changes the shape of the element. The movement that occurs when the bimetal changes shape is used to open or close switch contacts in the thermostat.

While bimetal elements are constructed in various shapes, the spiral-wound element is the most compact in construction and the most widely used. In *Figure 1*, for example, a glass bulb containing mercury (a conductor) is attached to a coiled bimetal strip. When the bulb is tipped in one direction, the mercury makes an electrical connection between the contacts and the switch is closed. When the bulb is tipped in the opposite direction, the mercury moves to the other end of the bulb and the switch is opened. Thermostat switching action should take place rapidly to prevent arcing, which could cause damage to the switch contacts. A magnet is used to provide rapid action to help eliminate the arcing potential in some bimetal thermostats.

Most residential and small commercial thermostats are of the low-voltage (24V) type.

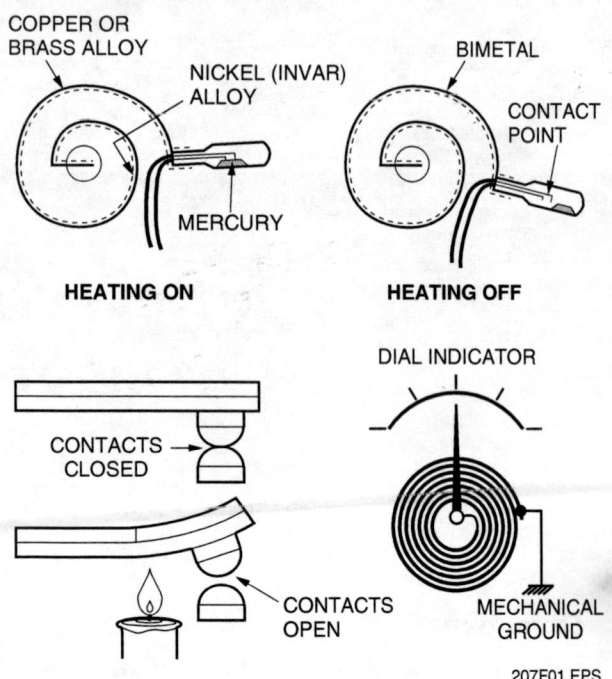

*Figure 1* ◆ Bimetal sensing elements.

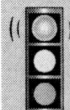 **WARNING!**
Mercury is toxic. Even short-term exposure may result in damage to the lungs and central nervous system. Mercury is also an environmental hazard. Do not dispose of thermostats containing mercury bulbs in regular trash. Contact your local waste management or environmental authority for disposal/recycling instructions.

With low-voltage control circuits, there is less risk of electrical shock and less chance of fire from short circuits. Low-voltage components are also less expensive and less likely to produce arcing, coil burnout, and contact failure. Some self-generating systems use a millivolt power supply that generates about 750mV to operate the thermostat circuit. These thermostats are very similar in construction and design to low-voltage thermostats; however, they are not interchangeable with 24V thermostats.

## 2.2.0 Heating-Only Thermostats

A wall-mounted heating-only thermostat typically contains a temperature-sensitive switch as shown in *Figure 2*. In this arrangement, the heating device (e.g., a furnace) will not come on unless the thermostatic switch is calling for heat.

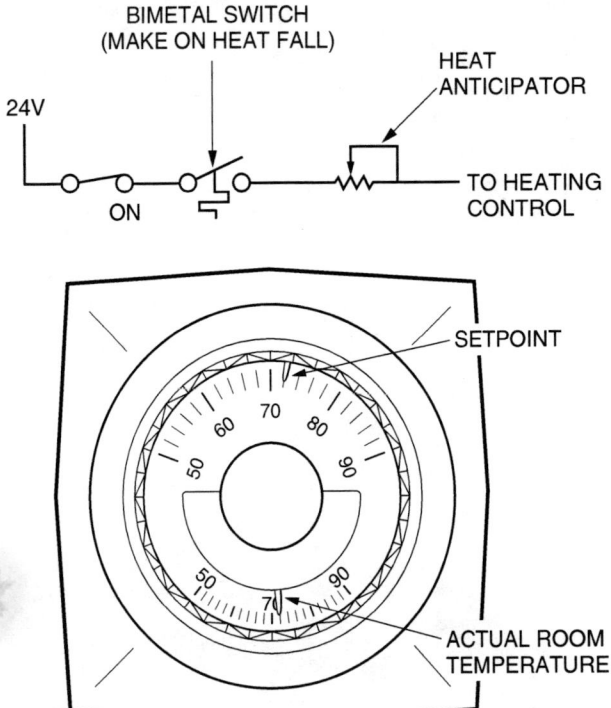

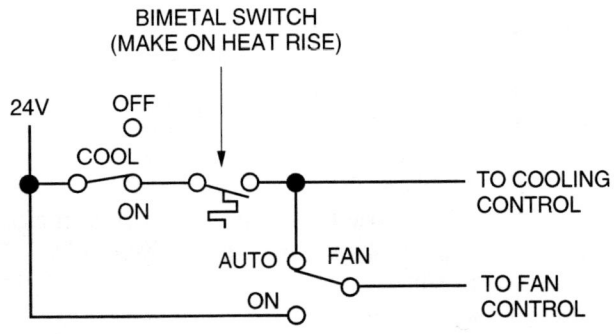

comfort. The cooling compensator (*Figure 4*) is a fixed resistance in parallel with the thermostatic switch. (Heating anticipators are adjustable.) No current flows through the compensator when cooling is on because it has a much higher resistance than the switch contacts. In this case, the contacts are essentially a short circuit. When the thermostat is open, however, a small current can flow through the compensator and the contactor coil.

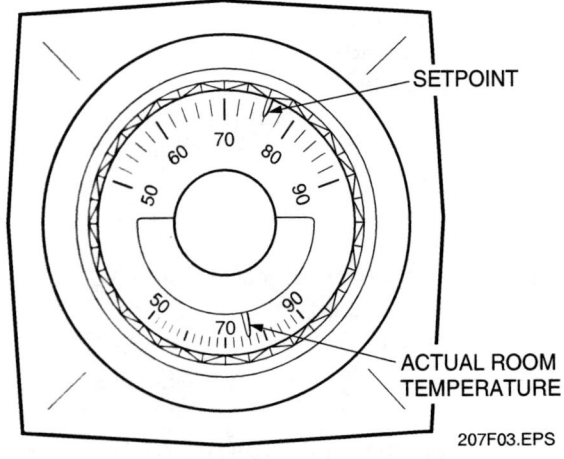

*Figure 2* ◆ Heating-only thermostat.

When the temperature in the conditioned space reaches the thermostat setpoint, the thermostatic switch will open. Because of the residual heat in the heat exchangers and the continued rotation of the fan as it slowly comes to a stop, the temperature will overshoot the setpoint. To avoid the discomfort that this condition might cause, the thermostat contains an adjustable heat anticipator that causes the thermostat to open before the temperature in the space reaches the setpoint.

As discussed in an earlier module, the heat anticipator is a small resistance heater in series with the switch contacts. The anticipator heats the bimetal strip, causing the contacts to open early.

## 2.3.0 Cooling-Only Thermostats

The cooling thermostat (*Figure 3*) is the opposite of a heating thermostat. When the bimetal coil heats up and unwinds, the mercury switch closes its contacts and starts the compressor. When the cooling thermostat is turned down to make the conditioned space cooler, it tips the mercury bulb so that the coil must cool more and wind tighter to turn the cooling system off.

Cooling thermostats contain a device called a **cooling compensator** to help improve indoor

*Figure 3* ◆ Cooling-only thermostat.

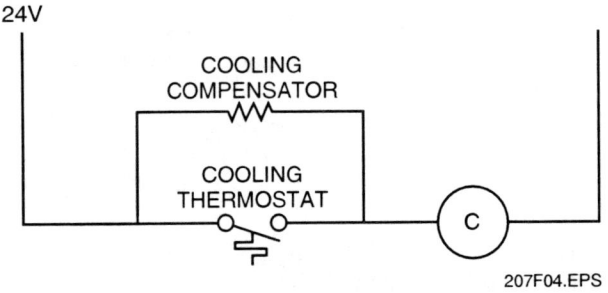

*Figure 4* ◆ Cooling compensator.

INTRODUCTION TO CONTROL CIRCUIT TROUBLESHOOTING — TRAINEE MODULE 03207          7.3

Because of the size of the compensator, the current is not enough to energize the contactor. The heat created by the current flowing through the compensator makes the thermostatic switch contacts close sooner than they would without the compensator. In this way, the cooling compensator accounts for the lag between the call for cooling and the time when the system actually begins to cool the space.

## 2.4.0 Heating-Cooling Thermostats

When heating and cooling are combined for year-round comfort, it is impractical to use a separate thermostat for each mode. Therefore, the two are combined into one heating-cooling thermostat (*Figure 5*). When a mercury bulb design is used, a set of contacts is located at one end of the bulb for heating and the other end of the bulb for cooling (*Figure 6*). The cooling contacts close the control circuit on a rise in temperature and open the circuit on a drop in temperature. The bulb with both sets of contacts is attached to a single bimetal element.

Unless a switch is provided in the heating-cooling thermostat, the thermostat will continuously switch back and forth from heating to cooling. In effect, heating and cooling will combat each other for control. A switch provides a means to direct the control to cooling, while disconnecting the heating control circuit. Likewise, when the switch is moved to heating, the switch connects the heating components, while electrically isolating the cooling circuit. When the switch is in the center or OFF position, neither the heating nor cooling control circuits are energized.

## 2.5.0 Heating-Cooling Automatic Changeover Thermostats

The disadvantage of a heating-cooling thermostat is that the building occupant must determine whether heating or cooling is needed at a particular time and set the thermostat switch accordingly. In some climates, that is very impractical; the need could change several times a day.

The **automatic changeover thermostat** automatically selects the mode, depending on the heating and cooling setpoints. The thermostat shown in *Figure 7* is essentially the same as that shown in *Figure 5*, with the exception that in *Figure 7* there is an AUTO position on the main control switch. The occupant can still select either heating or cooling. When the switch is in the AUTO position, however, the thermostat makes the selection. All that is necessary is for one of the thermostatic switches to close, indicating that the conditioned space is too warm or too cold.

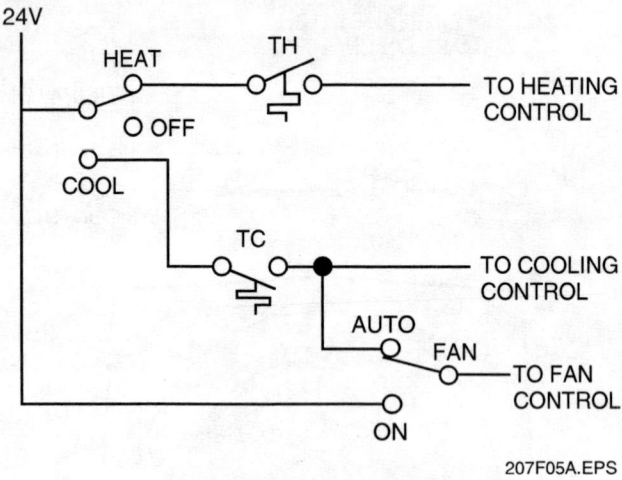

*Figure 5* ◆ Heating-cooling thermostat.

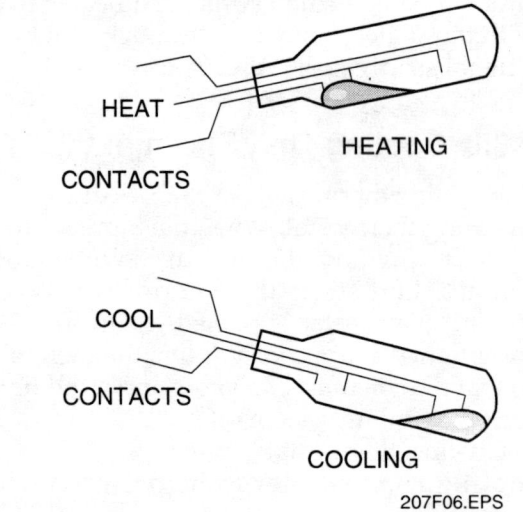

*Figure 6* ◆ Heating-cooling contacts.

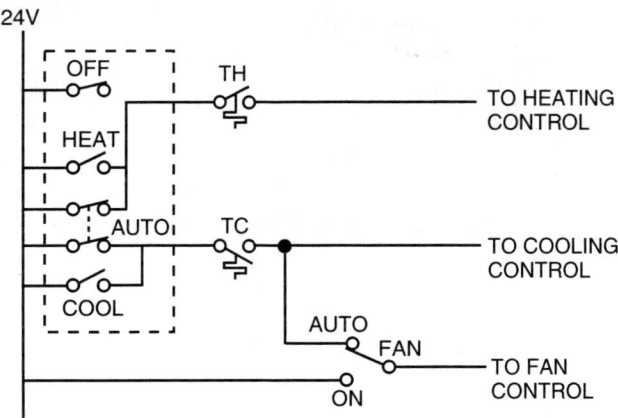

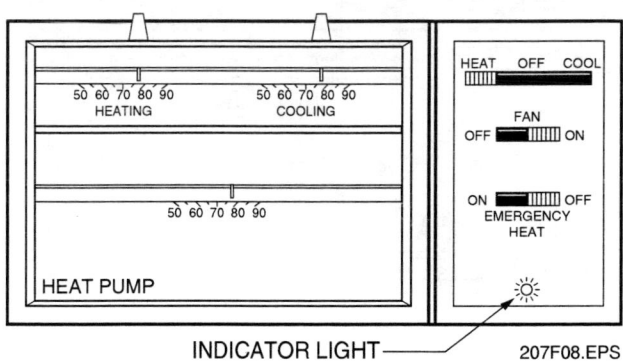

*Figure 8* ◆ Heat pump thermostat.

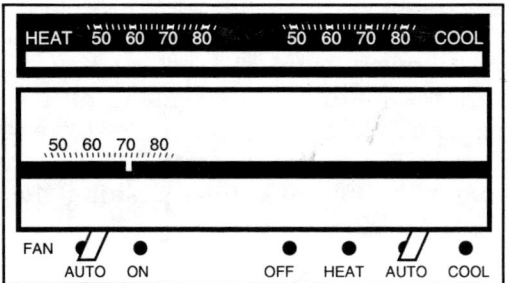

*Figure 7* ◆ Automatic changeover thermostat.

A thermostat contains a built-in mechanical **differential**, which is the difference between the cut-in and cut-out points of a thermostat. The differential is normally 2°F. For example, if the heating setpoint is 70°F, the furnace will turn on at 70°F and run until the temperature is 72°F.

Automatic changeover thermostats also have a minimum interlock setting, commonly known as the **deadband**. The deadband is a built-in feature that prevents the heating and cooling setpoints from being any closer together than 3°F.

## 2.6.0 Two-Stage Thermostats

A two-stage indoor thermostat is normally used to control heat pumps during the heating season. During the cooling season, it functions as a conventional air conditioning thermostat. During the heating season, the first stage controls the compressor of the heat pump. The second stage, usually preset 1°F to 2°F below the first stage, allows the supplementary electric heat to be energized if the indoor temperature continues to drop while the heat pump is operating. A two-stage thermostat is also used to control both cooling and heating on systems using two-speed compressors.

Heat pump thermostats usually have an emergency or auxiliary heat switch (*Figure 8*). If the heat pump becomes inoperative, this switch locks out the normal heat pump operation and heats the area with supplementary electric heat until the problem can be corrected. An indicator light, usually red, is mounted on the thermostat. It will come on when the selector switch is in the emergency heat position. As soon as the unit has been repaired, return the switch to the normal operating position. The operation of a two-stage heat pump thermostat will be discussed in more detail in the HVAC Level Two Module, *Heat Pumps*.

Because of the wide range of heating and cooling applications, multi-stage heating-cooling thermostats usually come in two pieces. The **sub-base** contains the wiring terminals and control switches. The **thermostat base** is selected for the specific heating and cooling needs of the application. The product lines offered by thermostat manufacturers have a variety of sub-bases and bases that can be combined to meet numerous applications.

## 2.7.0 Programmable Thermostats

Programmable thermostats are self-contained controls with the timer, temperature sensor, and switching devices all located in the unit mounted on the wall. Early programmables looked very much like conventional thermostats (*Figure 9*). This thermostat contains a motor-driven time clock driving a wheel containing cams that can be set to raise and lower the temperature settings at desired intervals. It is a self-contained version of the arrangement described previously, in which two thermostats and a timer were used.

Modern electronic programmable thermostats (*Figure 10*) use microprocessors and integrated circuits to provide a wide variety of control and energy-saving features. Their control panels use touch-screen technology and their indicators are digital readouts rather than analog needle positioning. Different thermostats offer different features; the more sophisticated (and expensive) the thermostat, the more features it offers.

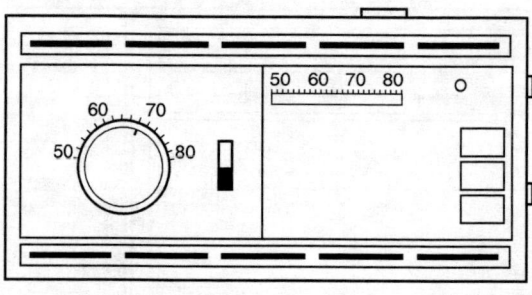

*Figure 9* ♦ Programmable thermostat.

*Figure 10* ♦ Electronic programmable thermostat.

Some of the features available on electronic thermostats are:

- *Override control* – This feature allows the occupant to override the program when desired. For example, you might override the night setback on Monday night so the thermostat isn't lowered before the football game is over.
- *Multiple programs* – This feature allows the occupant to design and select different schedules for different conditions. For example, you could program a special schedule for a vacation away from home.
- *Battery backup* – This feature prevents program loss in the event of a power failure.
- *Staggered start-up for multi-unit systems* – This feature avoids excessive current drain. This is an important feature in office buildings, shopping malls, and hotels.
- *Maintenance tracking* – This feature indicates when maintenance is to be performed (for example, when to replace filters).

The savings available from programmable thermostats are significant. For example, it is estimated that a setback of 10°F for both daytime and nighttime can result in a 20% energy savings. A 5°F setback will yield a 10% energy savings.

## 2.8.0 Line Voltage Thermostats

Just about all the thermostats you encounter will operate at low voltages (for example, 24V). There are also thermostats that operate at line voltages (for example, 240V). They are commonly used in controlling electric baseboard heat. Line voltage thermostats may be controlled by a bimetal sensing element or a hydraulic sensing bulb. The latter controls the thermostat by means of pressure. *Figure 11* shows a thermostat that is actuated by pressure on a bellows. The sensing bulb contains refrigerant which increases in pressure as the temperature increases. The increasing pressure acts to expand the bellows, thus pushing the switch toward the closed position.

Line voltage thermostats often suffer from a mechanical condition known as *thermal offset*, commonly called **droop**. Droop causes the thermostat control point to drift away from the selected setpoint. Because it is caused by heat, droop most often occurs in line voltage thermostats, which are subject to heat from high current. It may also occur in 24V thermostats that use anticipators.

## 2.9.0 Thermostat Installation

Even the best, most sensitive thermostat cannot perform correctly if it is poorly installed. Selecting the proper location for the thermostat is the first step in any installation procedure.

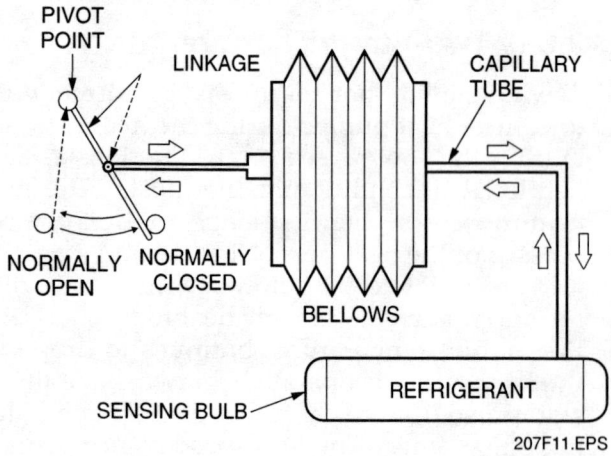

*Figure 11* ♦ Remote bulb thermostat.

## 2.9.1 Installation Guidelines

The thermostat should be installed in the space in which it will be called upon to control the temperature and other conditioning factors. The thermostat should be installed on a solid inside wall that is free from vibration that could affect operation by making the thermostat contacts chatter. For the same reason, it should not be located on a wall near slamming doors or near stairways. The following practices should be observed when installing a thermostat:

- The installer must be a trained, experienced technician.
- The manufacturer's instructions must be carefully read prior to installing. Failure to follow them could lead to a product damage or a hazardous condition.
- The rating should be checked in the instructions and on the unit to make sure the thermostat is suitable for the particular application.
- When the installation is complete, the thermostat must be operationally checked and adjusted as indicated in the installation instructions.

To replace an existing heating thermostat, first turn furnace power off. This de-energizes the 24V transformer in the furnace. Loosen the screws on the existing thermostat base and lift the thermostat away from the wall and wallplate or sub-base. Where applicable, remove the existing wallplate or sub-base from the wall. Disconnect and label each wire with the letter or number on the wiring terminal as each wire is removed, in order to avoid miswiring upon installation of the new thermostat.

Install the new wallplate or sub-base (*Figure 12*). Mount the wallplate or sub-base directly onto the wall with the screws enclosed in the package. If the wallplate or sub-base must be mounted on a vertical outlet box, use the proper adapter ring or cover plate. Mercury bulb thermostats must be carefully leveled (*Figure 13*). If they are not, the thermostat will not operate properly. The wallplate or sub-base of an electronic thermostat usually does not require leveling except for appearance.

> **CAUTION**
>
> Thermostats containing solid-state devices are sensitive to static electricity when not mounted on the wall; therefore, you must discharge body static electricity before handling the instrument. This can be accomplished by touching a metal doorknob or similar hardware. Touch only the front cover when holding the device.

When unpacking the new thermostat and wallplate or sub-base, handle with care. Rough handling can damage the thermostat. Save all instructional information, screws, and literature for later reference and use.

Locate the thermostat and wallplate or sub-base about five feet above the floor in an area with good air circulation at room temperature. Avoid locations that create the following conditions:

- Drafts or dead air spots
- Hot or cold air from ducts or diffusers
- Radiant heat from direct sunlight or hidden heat from appliances
- Concealed supply ducts and chimneys
- Unheated areas behind the thermostat, such as an outside wall or garage

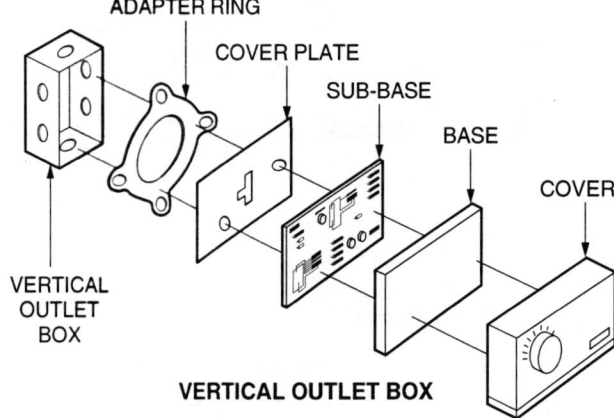
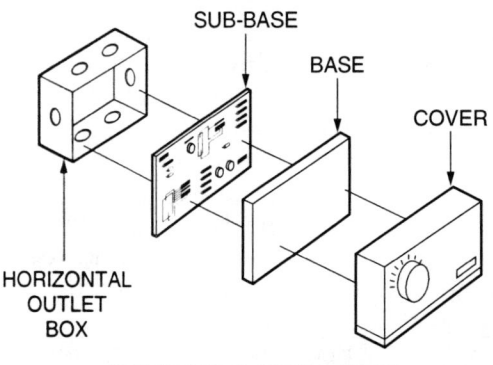

*Figure 12* ♦ Thermostat mounting.

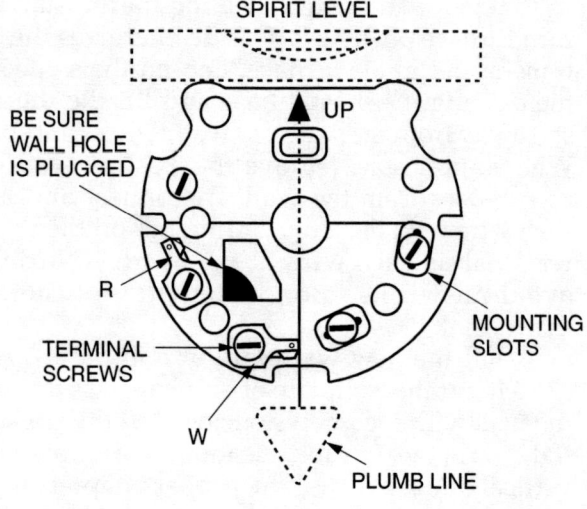

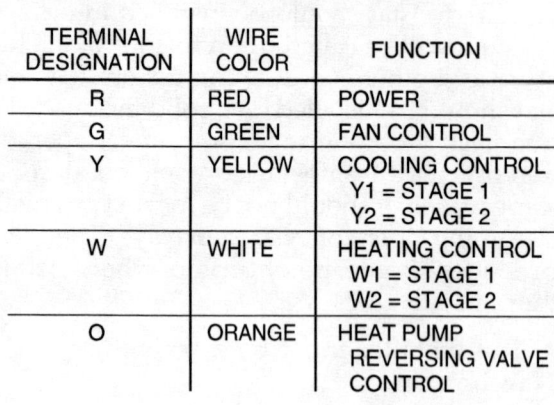

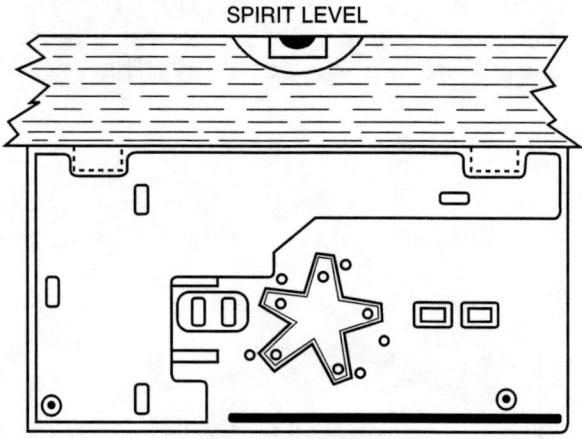

*Figure 13* ◆ Thermostat leveling.

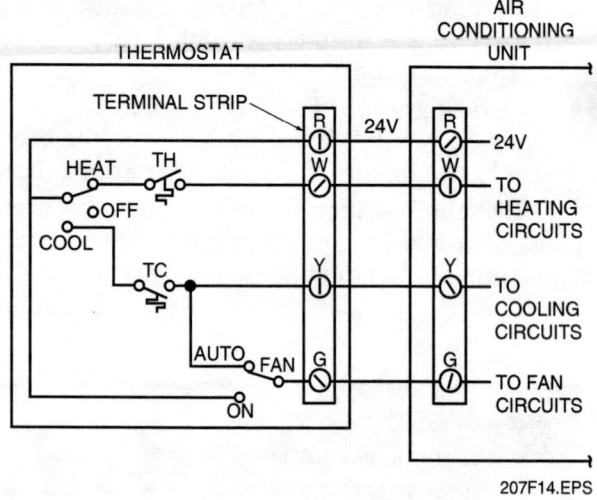

*Figure 14* ◆ Thermostat wiring.

## 2.9.2 Thermostat Wiring

The thermostat is connected through multi-conductor thermostat wire to a terminal strip or junction box in the air conditioning unit. A standard coding method is used in the HVAC industry to designate wiring terminals and wire colors. *Figure 14* shows the coding method and illustrates how the terminals of the heating-cooling thermostat we saw earlier would be designated. The terminal designation arrangement is fairly standard among thermostat and equipment manufacturers. It is not safe to assume, however, that the person who installed an existing system followed the color scheme in wiring the control circuits. There are additional codes for more complex thermostats; for example, the letter O designates orange and is connected to the reversing valve control in a heat pump. When there are multiple stages of heating or cooling, those terminals are designated with the appropriate letter plus a number.

Thermostat wiring should be done in accordance with national and local electrical codes. All wiring connections should be tight. To avoid damaging the control wire conductor, always use a stripping tool designed to strip small-gauge wire. Color-coded wiring should be used where possible for easy reference should system troubleshooting be required at a later date.

Wires should not be spliced, but if splicing is absolutely necessary, soldered splices are recommended. If wires are stapled to prevent movement, care must be taken to be sure that the staple does not go through the wire insulation. Seal the wire opening in the wall space behind the thermostat so it is not affected by drafts within the wall stud space.

## 2.9.3 Checking Current Draw

The next step is to determine the current draw. This may be done by locating the current draw of the primary control in the heating unit, or the heat

### Thermostat Wiring

Thermostat wires come in a variety of conductor configurations. Simple two-conductor wire can be used in a heating-only installation. Thermostat wires with three, four, five, and six conductors are readily available from any HVAC parts distributor for use in more complex installations. Normally, 18-gauge thermostat wire is adequate for most installations. However, long runs of thermostat wire can produce a voltage drop that might prevent equipment operation. If you have a run of thermostat wire that seems excessively long, use 16-gauge or heavier wire to reduce voltage drop.

anticipator setting on the existing thermostat (*Figure 15*). The current draw is usually printed on the furnace nameplate and/or a primary control such as the gas valve (*Figure 16*), the relay, or the oil burner control. It may also be found in the manufacturer's installation and service literature.

 **WARNING!**
To prevent electrical shock or equipment damage, make sure the power is off before connecting the wiring for the current draw adjustment.

If the current draw or heat anticipator setting cannot be found, shut off the power and connect a clamp-on ammeter of the appropriate range (0 to 2 amperes) around a wire connected between the R and W terminals of the existing wallplate or subbase (*Figure 17*). Since the current is so low, you will have to wrap 11 passes of wire (10 turns) around the jaws of the clamp-on ammeter to get a reading. Divide this reading by 10 to get the actual current draw of the circuit. Then operate the system for one minute, take the reading, and shut off the power.

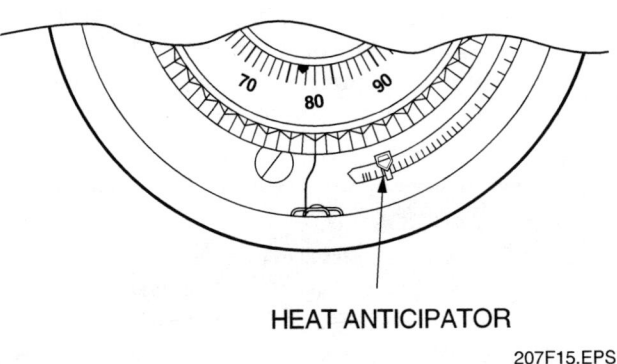

*Figure 15* ♦ Thermostat heat anticipator.

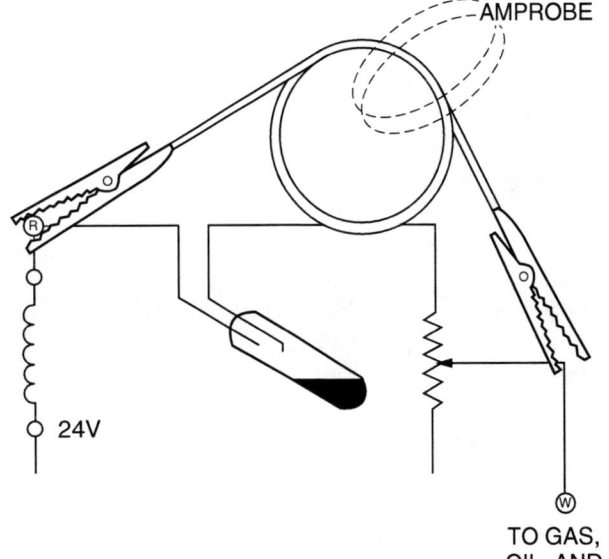

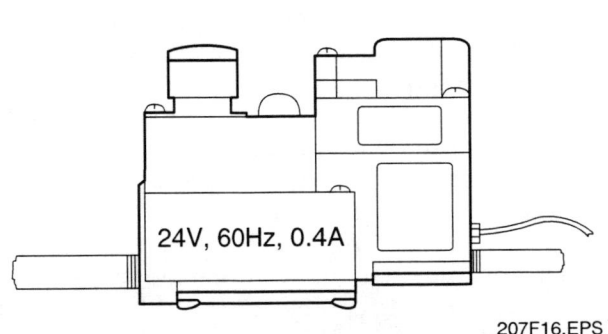

*Figure 16* ♦ Gas valve electric ratings.

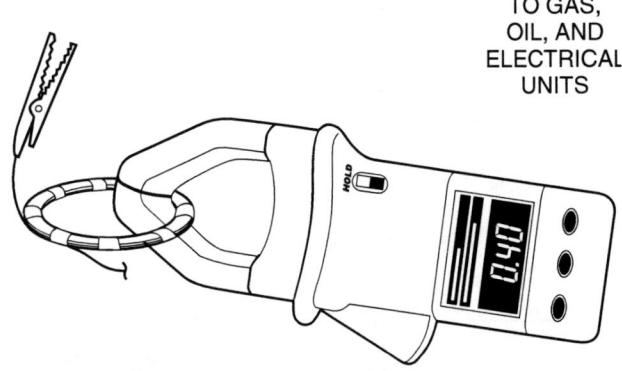

*Figure 17* ♦ Amperage check.

Some thermostats have an adjustable current draw feature to allow proper operation regardless of system current draw. An example is shown in *Figure 18*. To use this feature, connect the white wire to the W terminal. After the current has been determined, wire the red wire to R-Lo if the current draw is greater than 0.15A and less than 0.60A. If the current draw is greater than 0.60A and less than 1.2A, connect the red wire to the R-Hi terminal. This should help guarantee accurate temperature control.

### 2.9.4 Adjusting Heat Anticipators

The heat anticipator is adjustable and should be set at the amperage indicated on the primary control. Small variations from the required setting can be made by the service technician to improve performance on individual jobs.

Changing the setting of the anticipator changes the resistance of the wire resistor. This shortens or lengthens the heating cycle. Some heat anticipators have arrows to indicate the heating cycle adjustment.

Some heat anticipators have a fixed resistance, while others are equipped with an adjustable dial to change the resistance. The adjustable heat anticipator has a slide wire adjustment with the pointer scale marked in tenths of an ampere. This is used to set the anticipator to agree with the control amperage draw of the particular furnace. Furnaces are provided with an information sticker near the burner which states the amperage drawn by the control circuit of that particular furnace. This is the amperage at which the thermostat heat anticipator should be set under ideal conditions.

For example, if the amperage draw of a control circuit is shown as 0.45A, the installer should adjust the anticipator setting to 0.45A on the scale. The heat anticipator adjustment determines the length of the thermostat call for heat cycle by artificially heating the bimetal coil. As more heat is directed at the bimetal coil, a shorter heating cycle will occur. Conversely, as less heat is directed at the bimetal coil, a longer heating cycle will occur until the thermostat satisfies the call for heat.

When the control circuit amperage is high, less of the heater wire is needed; when the control amperage is low, more of the heater wire is needed. The control circuit amperage draw should be measured for each heating system as previously described.

### 2.9.5 Cycle Rate

Electronic programmable thermostats may require a change in cycle rate for correct equipment operation. The heating cycle rate can be adjusted by following the procedure as outlined in the thermostat manufacturer's instructions.

### 2.9.6 Final Check

The final step is to check the heating and/or cooling system when the thermostat is installed. With the thermostat in the heating mode, turn the power on, place the system switch at HEAT, and leave the fan switch in the AUTO position. Turn the setpoint dial to at least 5°F above the room temperature. The burner should come on within 15 seconds. The fan will start after a short delay. Then turn the setpoint dial to 5°F below the room temperature. The main burner should shut off within 15 seconds, but the blower may continue to run for one to two minutes. Next, set the thermostat to the cooling mode.

> **CAUTION**
> To avoid compressor damage, do not check cooling operation unless the outside temperature is at least 50°F and the crankcase heater (if so equipped) has been on for at least 24 hours.

If the outside temperature is at least 50°F, return power to the unit, set the thermostat to the COOL position, and the fan switch to AUTO. Leave the setpoint of the thermostat at 5°F below the room temperature. The cooling system should come on either immediately, or after any start delay, if the unit is so equipped. The indoor fan should come on immediately.

After the cooling system has come on, set the thermostat to at least 5°F above the room temperature. The cooling equipment should turn off within 15 seconds. Place the system switch to OFF. Move the setpoint dial to various positions. The system should not respond for heating or cooling.

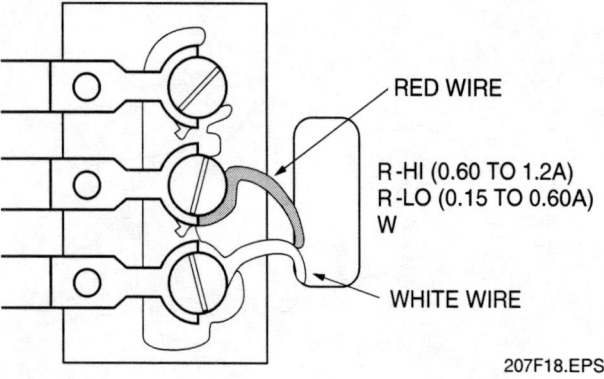

*Figure 18* ◆ Connection points.

### Compressor Short-Cycle Time Delay

Some systems are equipped with a compressor short-cycle time delay circuit to protect the compressor. This circuit, which is discussed later in this module, prevents the compressor from cycling back on for five minutes after it has turned off.

### 2.9.7 Adjusting the Thermostat

Thermostats are calibrated or preset at the factory for accurate temperature response and will not normally need recalibration. If a thermostat seems out of adjustment, the first thing to do is to check for accurate leveling. If that doesn't solve the problem, check and/or adjust the calibration as follows:

*Step 1* Move the temperature setting lever to the lowest setting. Set the system switch to HEAT and wait about ten minutes.

*Step 2* Remove the thermostat cover and move the thermostat temperature selector lever toward a higher temperature setting until the mercury switch just makes contact. This can be done by observing the mercury droplet in the bulb.

*Step 3* If the thermostat pointer and the setting lever read about the same at the instant the switch makes contact, no recalibration is needed. If recalibration is necessary, follow the manufacturer's instructions. A typical recalibration procedure is as follows:

a. With the system switch on HEAT, move the setting lever several degrees above room temperature.

b. Insert the end of a hex (Allen) wrench in the socket at the top center of the main bimetal coil. Use an open-end wrench to hold the hex nut under the coil.

c. Hold the setting lever so it will not move, and turn the Allen wrench clockwise until the mercury switch breaks contact. Remove the wrench.

d. Move the setting lever to a low setting. Wait at least five minutes until the thermostat loses any heat gained from your hands and its own operation.

e. Slowly move the setting lever up the scale until it reads the same as the thermometer.

f. Reinsert the Allen wrench. Holding the setting lever so it won't move, carefully turn the Allen wrench counterclockwise until the mercury just makes contact (turn it no further than necessary).

g. Recheck the calibration. Note that the calibration process may have to be repeated if the calibration is still off.

h. When you are satisfied with the calibration, replace the cover and set the thermostat lever switches for the desired operation.

### 3.0.0 ◆ HVAC CONTROL SYSTEMS

Most HVAC control systems are designed to automatically maintain the desired heating, cooling, and ventilation conditions set into the system. The controls for a small system such as a window air conditioner are very simple—a couple of control switches, a thermostat, and a couple of relays. As the system gets larger and provides more features, the controls become more complicated. For example, add gas heating to a packaged cooling unit, and the size and complexity of the control circuits will more than double. Make it a heat pump instead, and the control complexity will triple.

Large commercial systems may use pneumatic and electronic controls in conjunction with conventional electrical controls. These systems may have thirty or forty control devices, whereas a window air conditioner has just a handful.

The good news is that there are only a few different kinds of control devices. Once you learn to recognize them and understand the role each plays, it won't matter how many are used to control a particular system.

### Thermostat Calibration

When calibrating a room thermostat, you may have to repeat the steps several times because the heat from your hands and your breath can affect the thermostat operation. Try to avoid breathing on the thermostat and keep handling to a minimum to achieve a more accurate calibration.

All automatic control systems have the following basic characteristics in common:

- The sensing element (thermistor, thermostat, pressurestat, or humidistat) measures changes in temperature, pressure, and humidity.
- The control mechanism translates the changes into energy that can be used by devices such as motors and valves.
- The connecting wiring, pneumatic piping, and mechanical linkages transmit the energy to the motor, valve, or other devices that act at the point where the change is needed.
- The device then uses the energy to achieve some change. For example, motors operate compressors, fans, or dampers. Valves control the flow of gas to burners or cooling coils and permit the flow of air in pneumatic systems. Valves also control the flow of liquids in chilled-water systems.
- The sensing elements in the control detect the change in conditions and signal the control mechanism.
- The control stops the motor, closes the valves, or terminates the action of the component being used. As a result, the call for change is ended.

### 3.1.0 Relays, Contactors, and Starters

Some of the most common devices in HVAC control system circuits are relays, contactors, and motor starters. As you will see in the following sections, these devices are physically different in size and configuration, but their principles of operation are the same. Relays, contactors, and motor starters were covered briefly in the *Basic Electricity* module in HVAC Level One. Because of their extensive use in control and power distribution circuits, it is extremely important that you understand how these devices operate. Without this understanding, you will have difficulty in reading schematics and troubleshooting circuits. Refer to the appendix at the back of this module for common schematic symbols.

### 3.1.1 Relays

A relay operates to stop or permit the flow of electricity. Sometimes, a relay is used to reroute the flow of electricity in a different direction. Relays can be hard-wired into a circuit, or plug-in relays (*Figure 19*) can be used. Plug-in relays make troubleshooting and replacement much easier. Instead of having to connect and disconnect wiring, a new relay can simply be snapped into place.

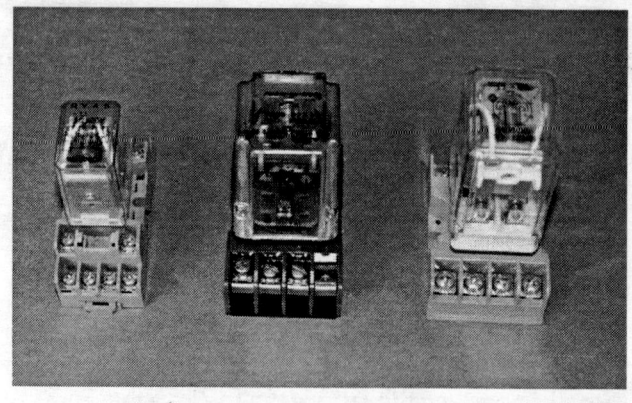

*Figure 19* ♦ Examples of plug-in relays.

The operation of a relay is sometimes difficult to grasp, because there seems to be a lot going on inside the relay's sealed enclosure that can't be seen. In its basic form, the relay consists of two parts: an electromagnetic coil and a set of contacts. When the coil is energized by the application of the proper voltage, it causes the position of the relay's contacts to change. Contacts are identified as being either *normally closed* or *normally open*. This refers to their position when no power is applied to the relay coil. With no power to the relay coil, the relay is referred to as *de-energized*. When power is applied to the coil, it is referred to as *energized*. All relay contacts shown on schematic diagrams are shown with the relay in the de-energized position.

*Figure 20* shows the open and closed contacts of a typical relay. It also shows the schematic symbol for normally closed and normally open relay contacts. Remember this is the position of the contacts when the relay coil is de-energized. As you can see, a normally closed set of contacts will allow electric current to flow through the contacts, while a normally open set of contacts does not allow the flow of electric current.

*Figure 21(A)* shows the schematic symbol for a simple relay consisting of the relay coil and a set of contacts. As shown, this relay is a normally closed relay. It is also classified as a *single-pole, single-throw (SPST) relay* because it only has one set of contacts and one current path. When the relay coil is de-energized as shown, current present at terminal 3 can travel through the closed contacts to terminal 4 for subsequent application to the remainder of control circuit. When voltage is applied across the coil of the relay via terminals 1 and 2, the coil energizes and the relay contacts open, preventing any current applied at terminal 3 from flowing through the contacts to terminal 4.

### Relays

The invention of the relay is credited to Samuel Morse, one of the inventors of the telegraph. Morse developed the relay to boost signal strength. An incoming signal activated an electromagnet, which closed a battery circuit, thereby transmitting the signal to the next relay.

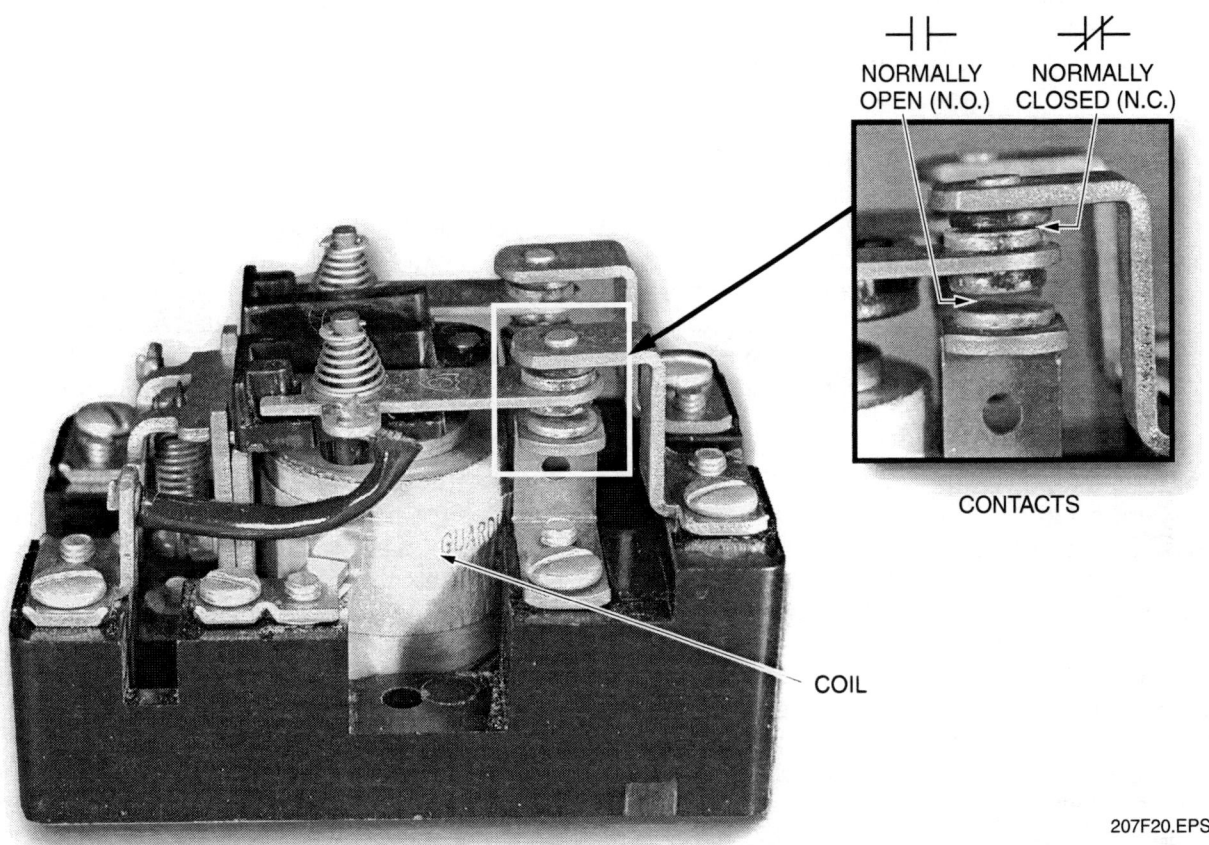

*Figure 20* ◆ Normally open and normally closed relay contacts.

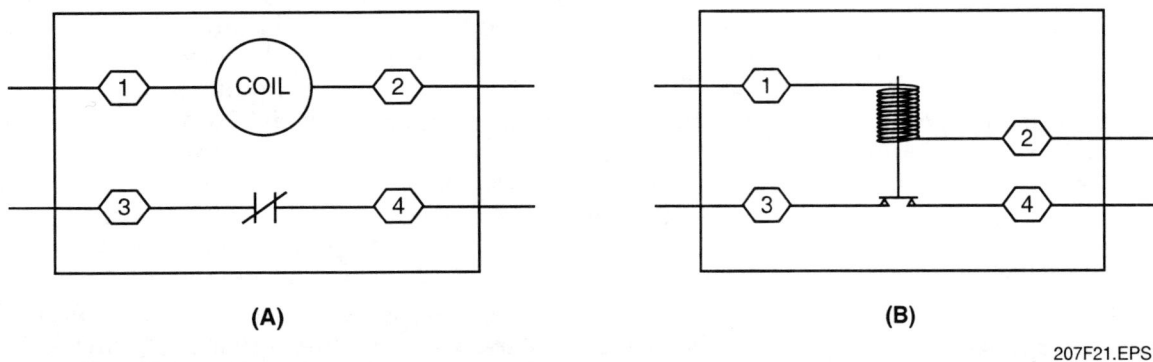

*Figure 21* ◆ Single-pole, single-throw (SPST) relay. (A) Schematic symbol. (B) Electromechanical presentation.

Remember, when the coil is energized, the normally closed contacts open. The last thing that needs to be determined about a relay is its coil voltage. Most relay coils used in HVAC control circuits operate on 24V. If such a relay needs to be replaced, use an SPST normally closed relay with a 24V coil.

It is important that you understand that no electrical connection exists between the coil of a relay and its contacts within the relay housing. Opening and closing of the relay contacts happens because of electromechanical linkage between the coil and contacts. *Figure 21(B)* shows a mechanical representation of the same relay we've been describing up to this point. As shown, terminals 1 and 2 are still the coil connections, and terminals 3 and 4 are still the contact connections. When power is applied to the relay coil, an electromagnetic field is created by the coil. This electromagnetic field pulls the plunger up, causing the normally closed contacts to open. There are other ways to accomplish this in relays, but this diagram should help you understand what's going on inside the relay mechanically.

Typically, the power applied to the coil might be 24V, while 120V may be applied through the contacts. This is one common situation but there are a number of other coil/contact voltage combinations. There are some applications in which the same voltage is applied to both the coil and the contacts of a relay. In this case, the connection between the two is always made using terminals outside of the relay. Again, it is important to remember that there is no electrical connection between the coil and contacts of a relay within the housing of the relay.

Now that you understand how a basic SPST normally closed relay works, let's move on to other types of relays. The counterpart of our first relay is the SPST normally open relay. This relay is similar to the SPST normally closed relay, except that when the coil is de-energized, the contacts are normally open. When the relay coil is energized, the contacts close.

The next relay is the single-pole, double-throw (SPDT) relay (*Figure 22*). In the SPST relay described earlier, the power applied at terminal 3 was applied to terminal 4, or not applied to terminal 4, depending on whether the contacts where closed or open, respectively. With a double-throw relay, we can allow power to be directed between two different terminals. As shown, with the coil de-energized, power coming in on the common terminal 3 is applied through the normally closed set of contacts to terminal 5. If the coil is energized, the positions of the relay contact sets change, causing the normally open contacts to close and the normally closed contacts to open. This causes the power applied from terminal 3 to be redirected from terminal 5 to terminal 4 for subsequent application to a different branch of the unit's control circuits.

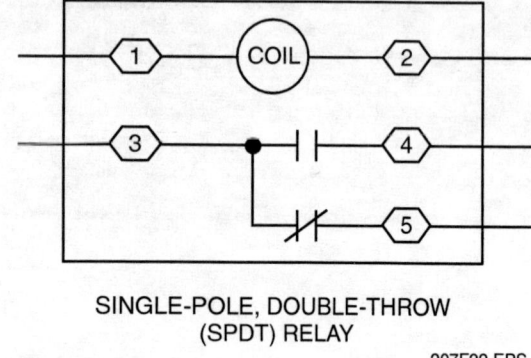

*Figure 22* ♦ Single-pole, double-throw relay.

The next relay we'll cover is the double-pole, double-throw (DPDT) relay. As shown in *Figure 23(A)*, another set of double-throw contacts is added to a basic SPDT relay to form this type of relay. When the coil is energized, it causes both sets of contacts to change position simultaneously. It is important to remember that there is no electrical connection between the first set of contacts and the second set of contacts, just as there is no connection between the coil circuitry and the contact circuitry. Again, if one or more relay terminals must be connected together, the connections are made at terminals external to the relay housing.

Now that you understand the concept of adding contact sets, we can add poles to our relay at will. You might find applications that require the use of a 3PDT (three-pole, double-throw) relay or even a 4PDT (four-pole, double-throw) relay. Regardless of the relay coil/contact configuration, each pole is electrically isolated from the coil and from all other poles. When the coil is energized, the position of all the poles change simultaneously.

Since each pole in a relay is electrically isolated from each other and from the coil, the poles from the same relay can be wired into different branches of the equipment's control circuits. Even though the coil and poles are physically housed in the same relay assembly, they are commonly shown in different areas of schematic and **ladder diagrams** used for troubleshooting. Schematic and ladder diagrams are covered later in this module.

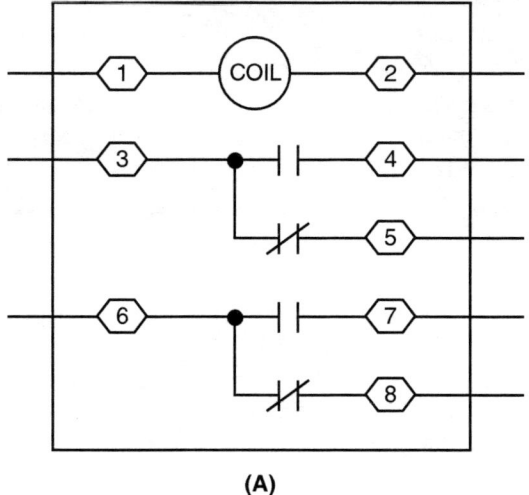
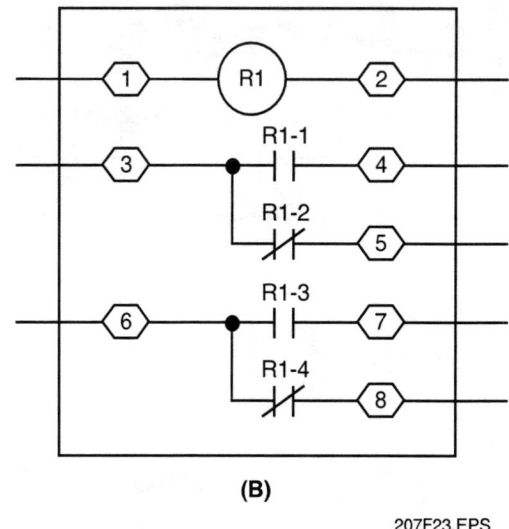

*Figure 23* ♦ Double-pole double-throw (DPDT) relay.

Because the relay coil and contacts are often shown in different areas of a schematic or ladder diagram, we need a method of designating which relay is associated with which contact set. An example is shown in *Figure 23(B)*, where the coil is designated R1 (relay 1) and all related contact sets are also identified with numbers that begin with R1. As shown, the first pole has contacts R1-1 and R1-2 (relay 1/contact set 1, and relay 1/contact set 2). The second pole has contacts R1-3 and R1-4. This method of relay designation is also used for three- and four-pole relays. For example, if you look at a schematic or ladder diagram and you see a normally open contact set designated R1-3, you know that the coil of relay R1 controls the contacts. If you energize the R1 coil, the R1-3 contacts will close. Likewise, all the other sets of R1 contacts are also going to change position.

**NOTE**
You will see many different designations for coils. For example, you may find CR (control relay), CC (compressor contactor), or FR (fan relay). There are dozens of different designations and they vary widely from manufacturer to manufacturer. Always consult the legend provided with the unit diagram.

### 3.1.2 Electronic Solid-State Relays

Unlike the electromechanical relays just discussed, solid-state relays have no moving parts. They use electronic devices such as TRIACs (a special type of current gateway) and light-emitting diodes (LEDs) to control switching action. In the example shown

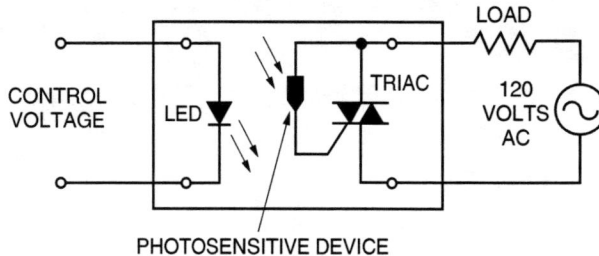

*Figure 24* ♦ Solid-state relay.

in *Figure 24*, the light created by current flow through the LED will activate a photosensitive device that triggers the TRIAC. Solid-state relays are used in general control applications, as well as in motor starting circuits.

### 3.1.3 Contactors and Starters

Contactors (*Figure 25*) are a type of heavy-duty relay. They are used to start and stop high-current, nonmotor loads or in motor circuits where overload protection is provided separately. A contactor does not provide overload protection.

INTRODUCTION TO CONTROL CIRCUIT TROUBLESHOOTING — TRAINEE MODULE 03207

### Electronic Solid-State Relays

Some advantages of solid-state relays are longer life, higher reliability, high-speed switching, and high resistance to shock and vibration. The absence of mechanical contacts eliminates contact bounce, arcing when contacts open, and hazards from explosives and flammable gases.

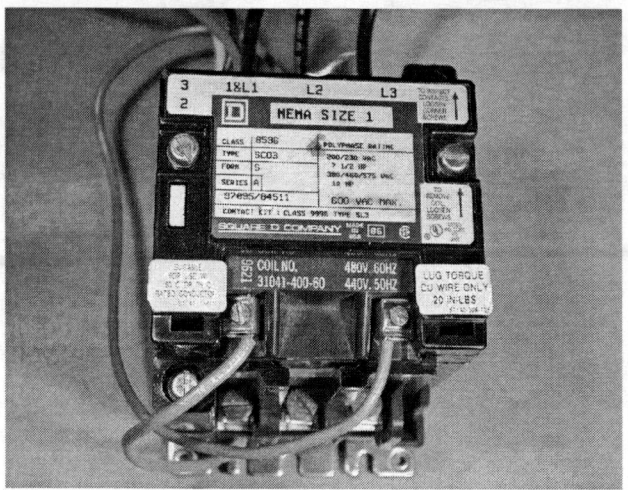

*Figure 25* ♦ Typical contactor.

A contactor is a normally open, single-throw device with one or more poles. It operates in the same way as a relay. When the coil is energized, the movable contacts are closed against the stationary contacts, thus completing the circuit. When power is removed from the coil, the contacts open, stopping the flow of electric current to the related load.

A motor starter is used to stop and start motors and provide overload protection. Standard starters include overload protection but they do not include a disconnection means or short circuit protection. Thermal overloads in the starter sense excessive current flowing to a motor and protect the motor from overload. If more current is flowing than the motor is designed to handle, the overload causes the motor to shut down. Starters usually have auxiliary contacts (additional poles). Combination starters are also available. They include a standard starter and a fused or nonfused switch or circuit breaker in the same enclosure.

### 3.2.0 Motor Speed Controls

Greater efficiency can be achieved by varying the speed of compressors and fan motors to adapt to changing heating and cooling loads. In this way, the equipment consumes only as much power as is needed to meet the demand. For example, HVAC equipment commonly uses variable-speed blowers. In addition, two-speed compressors are used in small systems, while larger systems use unloaders or multiple compressors to adapt to changing loads.

Some variable controls are connected or adjusted manually to vary the speed. For example, a furnace blower motor can be adjusted during installation to run at a speed that is optimum for the particular application. *Figure 26* shows an autotransformer motor speed control. The autotransformer acts as an impedance in series with the transformer stator winding. The greater the impedance, the slower the motor will run. While autotransformers once provided a common means of motor speed control, they are no longer widely used.

A potentiometer can also be used to adjust motor speed. A potentiometer is a resistor with a wiper arm. The position of the wiper arm determines how much resistance is offered by the potentiometer, and therefore determines how much current flows in the circuit. *Figure 27* shows a potentiometer with two TRIACs.

### Relays, Contactors, and Motor Starters

Like any switching device, relays, contactors, and motor starters have a limited life. Normal failure modes include contact sticking and improper operation. Be careful when selecting replacement contactors because they are rated for different uses. Some are rated for inductive loads, others are rated for resistive loads, and still others are rated for both types of loads.

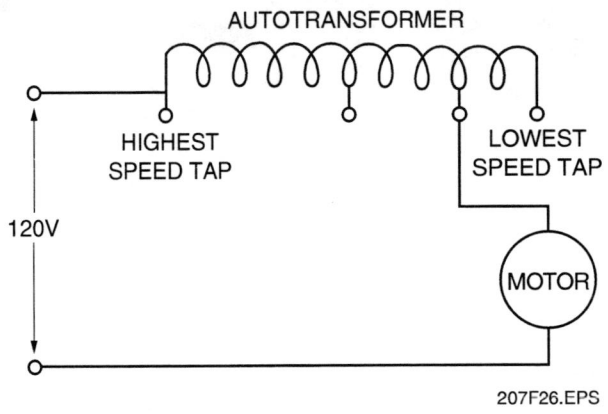

*Figure 26* ♦ Autotransformer motor speed control.

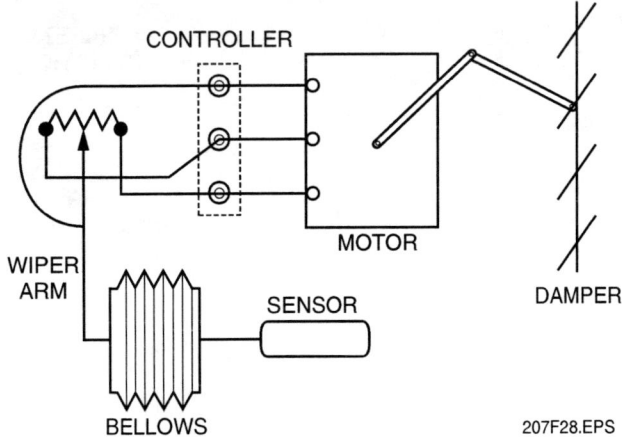

*Figure 28* ♦ Potentiometer damper control.

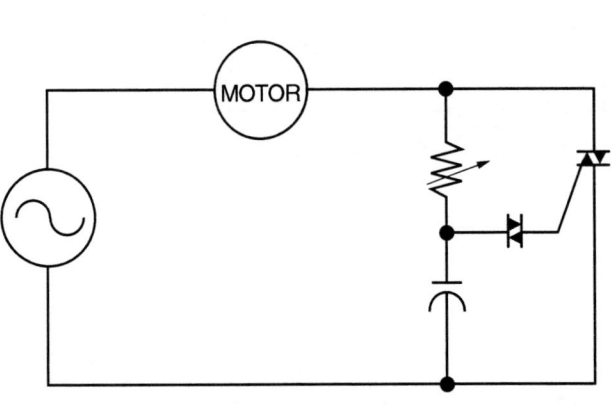

*Figure 27* ♦ Motor speed control using TRIACs.

The TRIAC is the most common motor control device. A knob on the control device is used to adjust a potentiometer; the setting of the potentiometer determines the motor speed. This circuit is similar to that of a light dimmer.

In *Figure 28*, the potentiometer is controlled by a bellows that responds to temperature changes detected by the sensor. As the bellows moves, the wiper arm of the potentiometer also moves. This changes the amount of current flowing to the damper motor, and thus changes the position of the damper to compensate for the temperature change. This is an example of a continuously variable motor control.

Modern electronic motor controls use microprocessor chips to achieve continuous control. The furnace control system shown in the simplified diagram in *Figure 29* controls the combustion system as well as the blower speed. The microprocessor monitors information such as the length of the last heating cycle and how often the furnace is cycling on and off. It then optimizes the heating cycle by selecting the appropriate fan speed and adjusting the length of the low-fire and high-fire cycles to match the conditions in the space. It is also able to sense any changes within the duct system such as a dirty filter or a closed zone valve, and vary the blower speed to maintain the correct CFM in the system.

### 3.3.0 Lockout Control Circuit

The purpose of the lockout relay in a control circuit is to prevent the automatic restart of the HVAC equipment. If the lockout relay has been activated, the system may be reset only by interrupting the power supply to the control circuit; for example, resetting the thermostat (in the case of an HVAC system), or turning the main power switch off and then on again.

In the circuit in *Figure 30*, the lockout relay coil, due to its high resistance, is not energized during normal operation. However, when any one of the safety controls opens the circuit to the compressor contactor coil, current flows through the lockout relay coil, causing it to become energized and to open its contacts. These contacts remain open, keeping the compressor contactor circuit open until the power is interrupted after the safety control has reset. Performance depends on the resistance of the lockout relay coil being much greater than the resistance of the compressor contactor coil.

If the lockout relay becomes defective, it should be replaced with an exact duplicate to maintain the proper resistance balance. This type of relay is sometimes called an *impedance relay*.

It is permissible to add a control relay coil in parallel with the contactor coil when a system

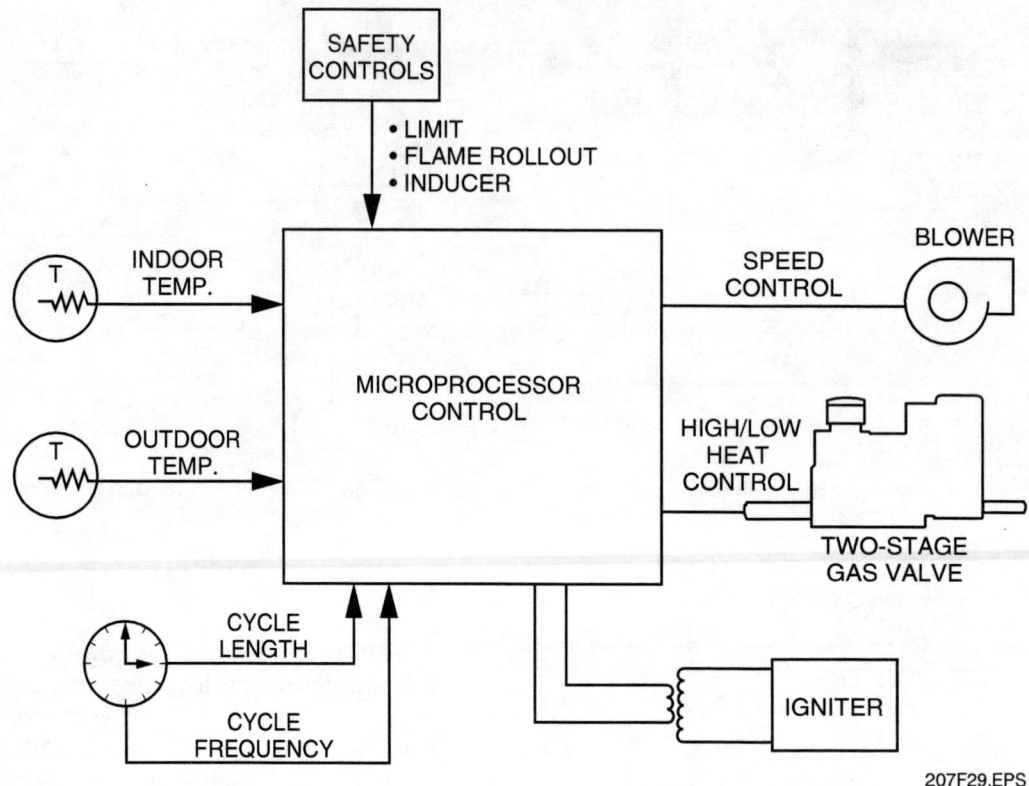

*Figure 29* ◆ Electronic variable-speed furnace control.

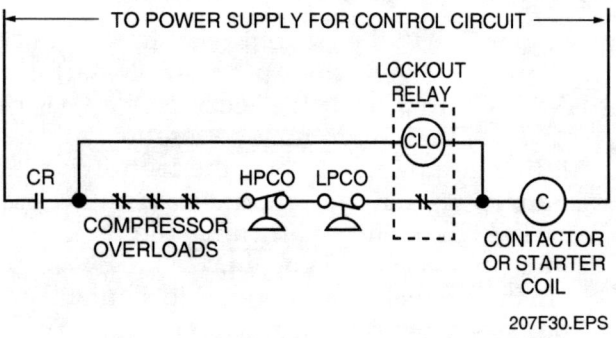

*Figure 30* ◆ Lockout relay used in an HVAC control circuit.

demands another control. The resistance of the contactor coil and the relay coil in parallel decreases the total resistance and does not affect the operation of the lockout relay.

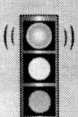 **WARNING!**
Never put additional lockout relays, lights, or other load devices in parallel with the lockout relay coil. Doing so might defeat the lockout and create a very hazardous situation.

### 3.4.0 Time Delay Relay

The purpose of a time delay relay is to delay the normal operation of a compressor or motor for a predetermined length of time after the control system has been energized. The length of the delay depends on the time built into the relay coil, and may vary from a fraction of a second to several minutes. A common use for a time delay relay is to delay the start-up or shutdown of a furnace blower to improve heating efficiency.

Electrical systems containing several motors may also use time delays to start the motors one at a time in order to limit the inrush current. For example, the schematic drawing in *Figure 31* shows four compressor control circuits with time delay relays for sequencing the starting of each motor. In this type of motor design, if all three stages of the thermostat are closed and electrical power is supplied to the units, compressor contactor coil 1 and time delay relay 1 become energized to start compressor 1. After the specified time delay, the contacts of time delay relay 1 close to energize compressor contactor coil 2, which starts compressor 2, and time delay relay 2, which starts compressor 3, and so on, until all four compressors are in operation.

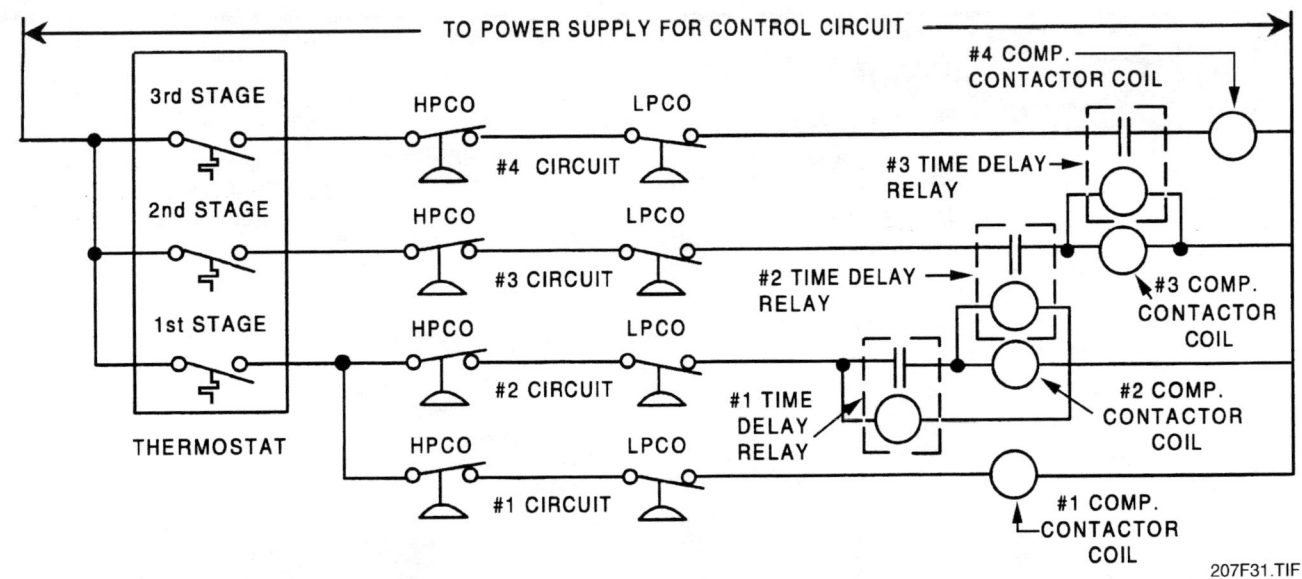

*Figure 31* ◆ Four compressor control circuits with time delay relays.

## 3.5.0 Compressor Short-Cycle Timer

Attempting to start a compressor against high head pressure can damage the compressor. When a refrigerant system shuts down, it should not be restarted until the pressures in the system have had time to equalize. Short-cycling can be caused by a momentary power interruption or by an occupant changing the thermostat setting.

A compressor short-cycle protection circuit contains a timing function that prevents the compressor contactor from re-energizing for a specified period of time after the compressor shuts off. Lockout periods typically range from 30 seconds to several minutes.

Short-cycle timers are available as self-contained modules that can be direct-wired into a unit. They are often sold as optional accessories.

Electromechanical versions use a motor-driven cam to run a timing mechanism, while electronic versions use an electronic timer.

## 3.6.0 Control Circuit Safety Switches

Refrigerant system compressor control circuits normally include several different types of safety switches. These include pressure switches, freezestats, and outdoor thermostats.

### 3.6.1 Pressure Switches

Many systems use one or more pressure switches in the compressor control circuit. These are safety devices designed to protect the compressor. A pressure switch is normally closed and wired in series with the compressor contactor control circuit. *Figures 30* and *31* show examples of low-pressure cutout (LPCO) and high-pressure cutout (HPCO) switches used in such circuits.

High-pressure switches are designed to open if the compressor head pressure is too high. Low-pressure switches are designed to open if the suction pressure is too low. Pressure switches use a bellows mechanism that presses against switch contacts that have preset open and close settings. When the system pressure begins to rise above or drop below the normal operating pressure, the related high-pressure or low-pressure switch will open, causing the compressor contactor to de-energize.

Some manufacturers use a type of low-pressure switch called a *loss of charge* switch that removes power to the equipment if the refrigerant charge is low.

### 3.6.2 Freezestat

Another type of safety switch is the freeze-up protection thermostat commonly called a *freezestat*. Its purpose is to prevent evaporator coil freeze-up. The freezestat switch is a normally closed bimetal switch that is usually attached to one of the endbells in the evaporator coil. It will open if the refrigerant temperature drops below a predetermined setpoint.

## Pressure Problems

High-pressure and low-pressure switches prevent system operation if the system pressure exceeds a preset range. Common causes of high head pressure are dirty condenser coils or a failed condenser fan motor. Common causes of low suction pressure include a loss of charge and low evaporator airflow. A dual-stage pressure switch capable of monitoring both high and low pressures is shown here.

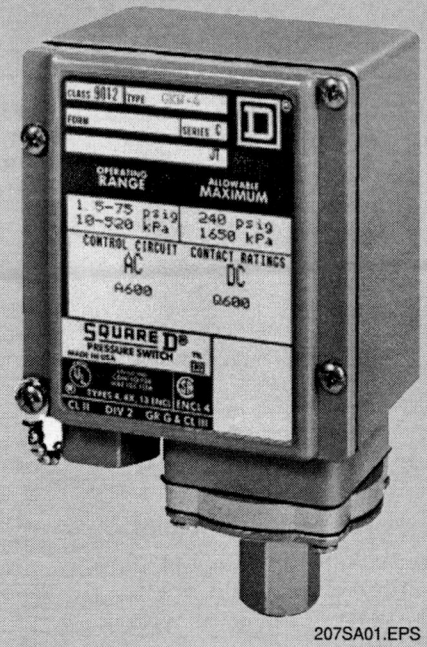

### 3.6.3 Outdoor Thermostats

Some equipment uses an outdoor thermostat to shut off the equipment when the ambient temperature drops below a predetermined setpoint (typically between 55°F and 65°F). This prevents equipment damage.

## 3.7.0 Furnace Controls

Common furnace controls include the fan control, limit control, thermocouple, and inducer proving switch. These controls are described in this section. Note that controls used with boilers are covered later in the hydronic controls section of this module.

### 3.7.1 Fan Control

A fan control (*Figure 32*) is a temperature-actuated control that, when heated, will close a set of contacts to start the indoor fan motor. The sensing element of the fan control is positioned inside one of the heat exchangers where the temperature is the highest.

The fan control is normally actuated by a bimetal element that opens or closes the contacts in response to a temperature change. The fan control may be set to bring the fan on at about 150°F and to stop the fan at about 100°F.

In operation, the burner or heating element provides heat to the heat exchangers for a few seconds to warm the heating chamber before the fan is started. This prevents blowing cold air into the conditioned space. When the thermostat is satisfied, the main burner or heating element stops providing heat, but the fan continues to operate until the temperature in the heating chamber has been reduced. This removes excess heat from the heat exchangers and improves efficiency by using residual heat.

### 3.7.2 Limit Control

A limit control is also a heat-actuated switch with a bimetal sensing element positioned near the heat exchangers. This is a safety control that is wired into the primary side of the transformer. If the temperature inside the heating chamber or

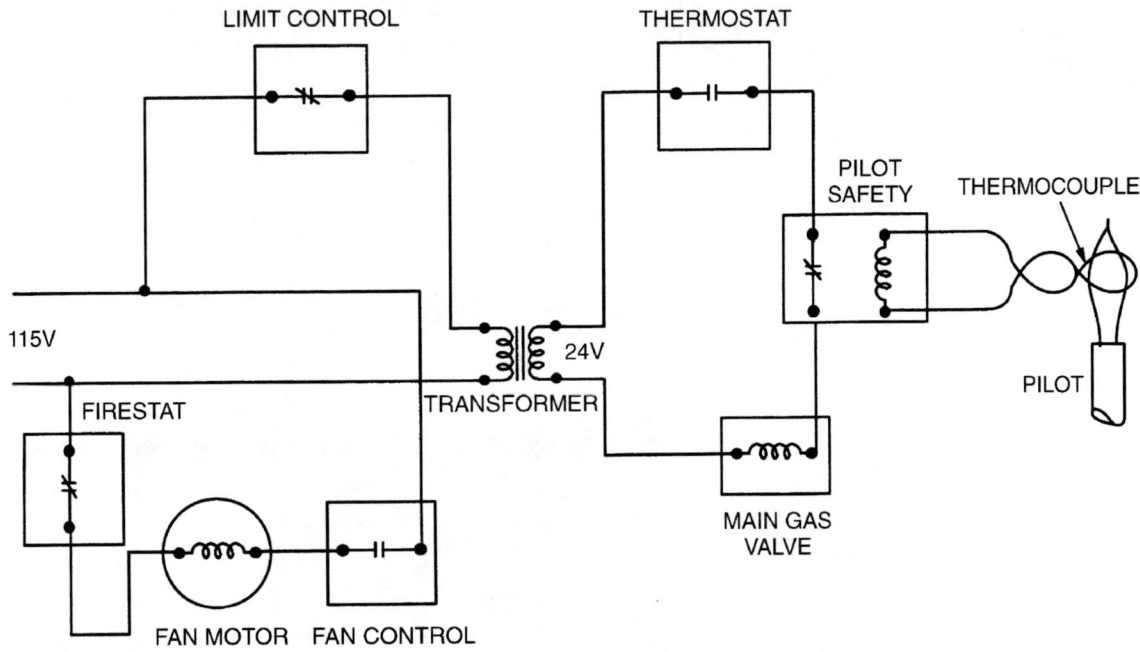

*Figure 32* ◆ Complete 24V furnace control circuit.

plenum reaches approximately 200°F, the power will be shut off to the transformer, which also stops all power to the temperature control circuit.

### 3.7.3 Thermocouple

A thermocouple is a device that uses dissimilar metals to control electron flow. The hot junction of the thermocouple is located in the pilot flame of a gas-fired furnace. When heat is applied to the welded junction, a small voltage is produced. This small voltage is measured in millivolts (mV) and is the power used to operate the pilot safety control. The output of a thermocouple is about 30mV. This simple device can cause many problems if the connections are not kept clean and tight or the pilot flame is inadequate to generate the correct voltage.

### 3.7.4 Inducer Switch

Newer furnaces that use an inducer motor have either a centrifugal switch or a pressure-operated switch in the heating control circuit to monitor the operation of the inducer motor. If the inducer motor fails to operate, the switch opens and prevents control voltage from being applied to the furnace gas valve.

The centrifugal switch has a set of contacts located near the inducer motor shaft. The centrifugal force created by the spinning shaft throws the contacts outward, closing a switch that allows control power to be applied to the furnace gas valve.

The disadvantage of a centrifugal switch is that it can stick in either the closed or open position. This can keep the unit from firing up, or in some cases, allow the unit to continue firing without a demand from the thermostat.

The pressure switch is the more common type of inducer switch. It has tubing connected to the housing, and in some cases, to the burner enclosure. When the inducer motor is operating, a negative pressure is created that causes the pressure switch to close, allowing control power to be applied to the furnace gas valve.

## 4.0.0 ◆ CONTROL CIRCUIT SEQUENCE OF OPERATION

Control circuits can be relatively simple, yet they are sometimes difficult for the new technician to understand. The function of most control circuits is simply to control the start and stop of motors. To begin to understand a basic control circuit, you need to know when the motor should start and when it should stop. This is called the *sequence of operation*.

The sequence of operation determines the types of devices used in the control circuits of each HVAC system. When sitting down to design a control circuit, the first thing an engineer must know is the required sequence of operation—that is, what motor is supposed to be running and when. Then, a control circuit sequence can be designed.

## Testing the Operation of an Inducer Motor Pressure Switch

To check the operation of an inducer pressure switch, disconnect the pressure tube from the switch and set the furnace thermostat to call for heat. The furnace burner should not ignite.

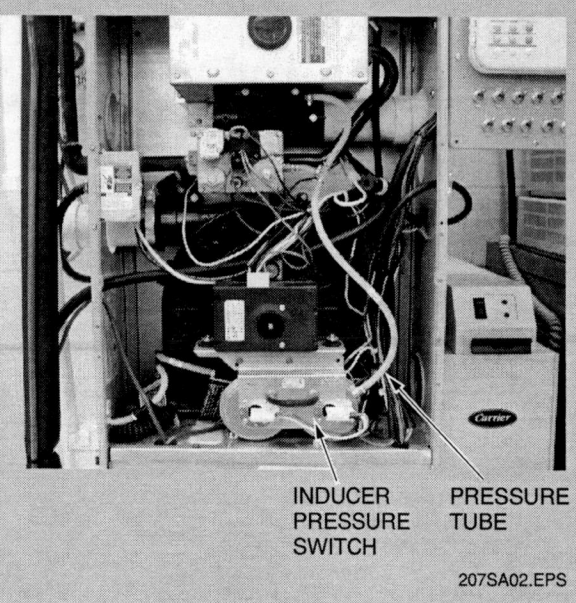

An automatic air conditioning circuit is simple in nature, and is a good place for you to begin your analysis of control circuits. *Figure 33* shows a basic cooling system control circuit.

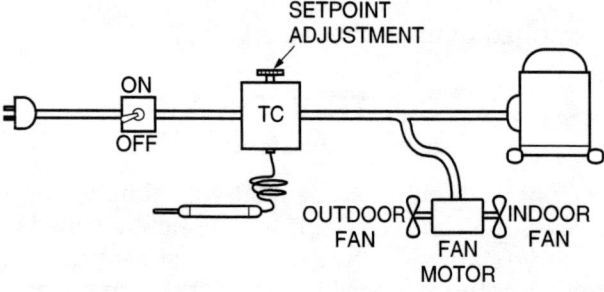

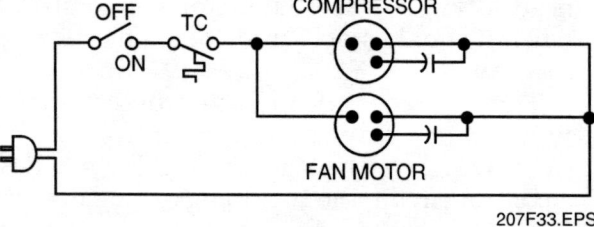

*Figure 33* ◆ Basic cooling system control circuit.

The outdoor unit of a split-system air conditioning system, called a *condensing unit*, has only two motors. These are the compressor and the condenser (outdoor) fan motor. The control sequence is simple because during cooling operation, both motors must be turned on at the same time. If the two motors were connected directly to the power supply, they would run continuously until someone turned the power off. Obviously, this is not a practical or economical method of operation. By adding a thermostat to the control circuit, the unit can come on when the building needs cooling, and shut off when it doesn't. When the thermostat is calling for cooling, both motors must be turned on. When the thermostat is satisfied, both must be turned off.

As shown in *Figure 33*, power from the power plug is applied to the ON-OFF switch. When the switch is in the ON position, power is passed to the cooling thermostat (TC). When the thermostat calls for cooling and closes, power is passed simultaneously to both the compressor and fan motor, causing both motors to turn on and run. When the thermostat is satisfied and opens, power is no longer applied to the motors, causing both motors to turn off. This describes the sequence of operation for this circuit. As you can see, it is very simple, but you must know what is supposed to be happening at any given time.

## Inside Track

### Split-System Condenser Units

This is an example of a typical outdoor (condensing) unit used in a split-system residential air conditioning system.

*Figure 34* shows a control circuit that is more typical of a basic cooling-only system. You can see that some additional features have been added to this circuit. A transformer has been added because we want to use 24V for the control circuit. This makes the control circuit components and wiring less expensive and easier to install. An indoor fan motor (IFM) has also been added to the circuit. Because the control circuit is now operating at 24V, the control relays (C and IFR) are added to provide control for the motors.

Let's trace this circuit to determine its control sequence. As shown, line voltage is applied at terminals L1 and L2, making it present at contactor (C), the indoor fan relay (IFR), and the primary of the system transformer at all times. With the transformer energized, the 24V output from its secondary is applied through one control circuit path to the VENT terminal of the FAN switch and by a second control circuit path to the ON-OFF switch. Remember that the power is applied to both control circuit paths at the same time. We now have two control circuit paths to analyze. Let's take them one at a time, starting with the simplest circuit, which is the one that controls the fan.

When the FAN switch is in the VENT position as shown, power is applied through the first control circuit path to the coil of the indoor fan relay (IFR), energizing the relay. This causes the related IFR contacts to close, applying power to the indoor fan motor (IFM), which energizes the fan motor. This mode of operation allows the occupant to use the

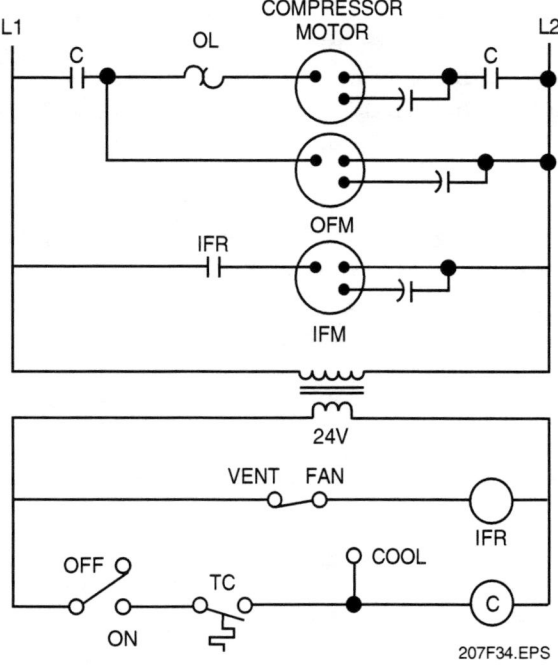

*Figure 34* ◆ Typical cooling system control circuit.

INTRODUCTION TO CONTROL CIRCUIT TROUBLESHOOTING — TRAINEE MODULE 03207

fan for ventilation without operating the compressor for cooling. If the FAN switch is placed in the COOL position, the control circuit path to the indoor fan relay remains open, causing the indoor fan relay to be de-energized.

Now look at the second control circuit path. With the ON-OFF switch closed, power is applied to the cooling thermostat (TC). When the thermostat calls for cooling, the thermostat contacts close and apply power simultaneously to the COOL terminal of the FAN switch and to the coil of contactor C. With power applied to the contactor coil, it energizes and closes its contacts. This allows power to be applied through C to the compressor motor and the outdoor fan motor (OFM), turning them both on. Assuming the FAN switch is still in the VENT position as shown, no power is applied through the FAN switch COOL terminal to the indoor fan relay (IFR) through this path. However, the IFR coil is already energized through the VENT terminal of the FAN switch, allowing the indoor fan motor (IFM) to operate as needed to circulate the cooled air.

With the ON-OFF switch closed, the thermostat calling for cooling, and the FAN switch placed in the COOL position, both the contactor coil (C) and the indoor fan motor coil (IFR) are energized through the operation of the second control circuit only. As before, this causes the compressor, outdoor fan, and indoor fan motors to be turned on simultaneously as required for the cooling mode of operation.

No matter how complex the control circuit appears to be, you will find the basic control arrangement shown in *Figure 34* (or something very much like it) at the heart of the circuit. Any additional circuits will represent special features used to improve equipment safety or operating efficiency. *Figure 35* illustrates the point. This diagram is for a combined cooling and gas heating unit. The circuit looks different and there are several more components, but if you trace out the cooling control function, you will see that it is essentially the same as the simple circuit discussed earlier. The additional features are not that complicated. First of all, the heating controls have been added near the bottom of the diagram, along with a heating-cooling thermostat. The cooling control has more extras, such as a compressor short-cycle protection circuit, high-pressure and low-pressure safety switches, a crankcase heater, and a current-sensitive overload device. Take away these components and you have a circuit that's identical to the one in *Figure 34*.

The diagram appears different because instead of drawing L1 and L2 and the two sides of the transformer secondary as the verticals on a ladder, they are shown emanating from common terminals. This reflects the way the circuit is actually wired, and is a common method used by manufacturers to draw control circuits. Ladder diagrams are helpful, but they are not supplied by all manufacturers.

Some of the features of the circuit shown in *Figure 35* are as follows:

- The unit has a two-speed indoor fan that runs on high speed for cooling and low speed for heating. In heating, the indoor fan is controlled by the time delay relay.

- Operation of the inducer fan must be proven before the gas valve is turned on. The inducer pressure switch (PS), located in the draft hood, will close when the induced-draft motor is running at the required speed. If the induced-draft fan stops, the stack pressure will drop and the switch will open, disabling the gas supply. The induced-draft motor (IDM) is energized by the induced-draft relay (IDR) as soon as the thermostat closes.

- The heating section has two additional safety devices in series with the gas valve. A flame rollout switch (RS) will open if the burner flame escapes from the burner box. This usually indicates insufficient air for combustion, or a leak in the burner box. The limit switch (LS) is a heat-sensitive switch that extends into the heat exchanger. If the heat is excessive, it will disable the gas valve. One cause of excessive heat is insufficient air flowing over the heat exchangers. This could be caused by a blower failure or a dirty filter.

### Analyzing a Circuit Diagram

Use the diagram in *Figure 35* to determine the answers to these questions.
1. Does the induced-draft motor turn off if the rollout switch opens?
2. If you connect a voltmeter from L2 to terminal 4 of IFR while the compressor contactor is de-energized, what voltage, if any, would the meter read?
3. Is the fan relay (IFR) energized or de-energized in the heating mode?

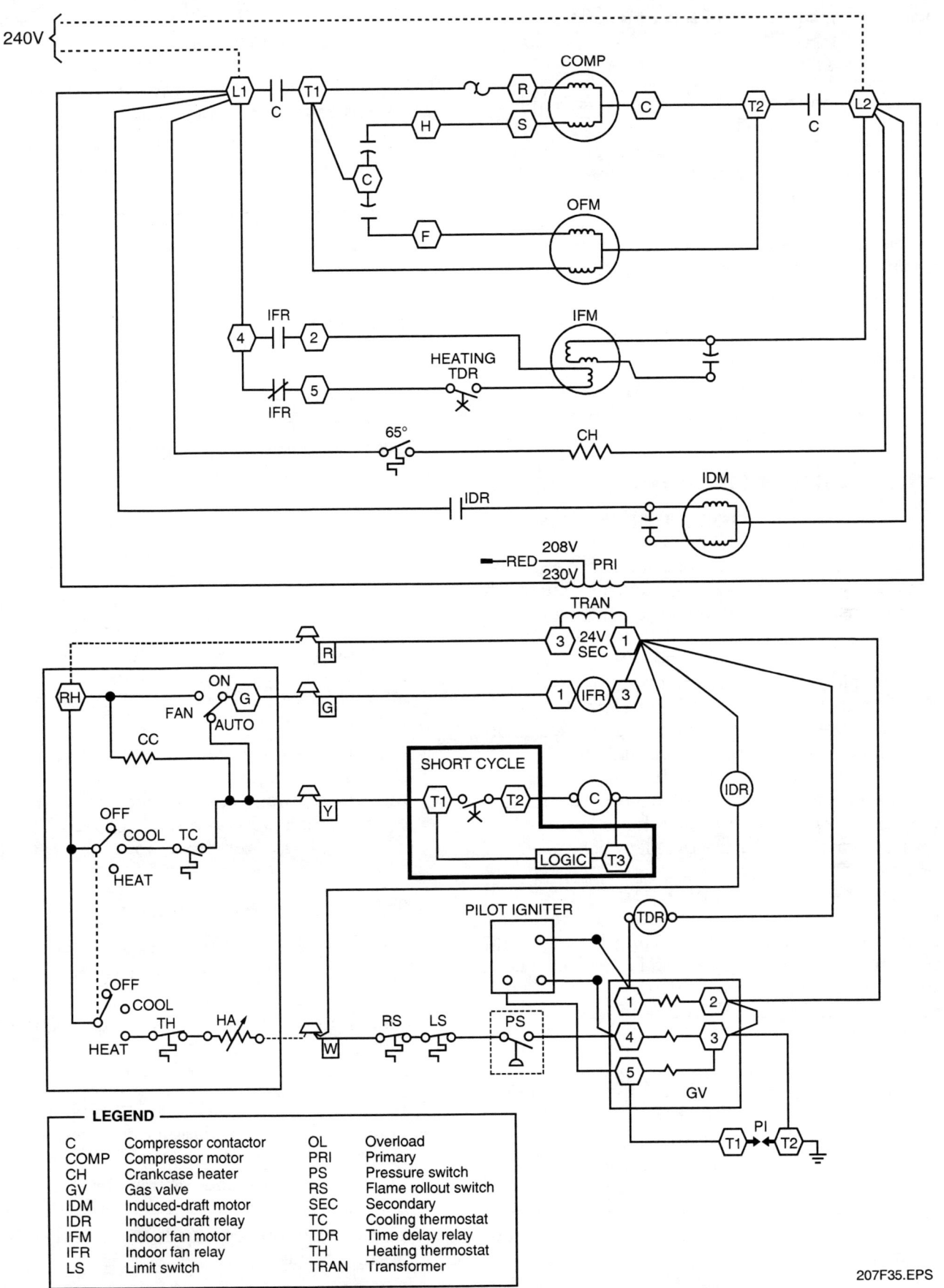

*Figure 35* ◆ Circuit diagram of a cooling/gas heating system.

INTRODUCTION TO CONTROL CIRCUIT TROUBLESHOOTING — TRAINEE MODULE 03207

**7.25**

## 5.0.0 ◆ USING AN ORGANIZED APPROACH TO ELECTRICAL TROUBLESHOOTING

Troubleshooting can be defined as a procedure by which the technician locates the source of a problem, then makes the repairs and/or adjustments to correct the cause of a problem so that it will not recur. Troubleshooting can be divided into the five basic elements listed here. The vast majority of problems can usually be found quickly if a systematic approach is used. This includes:

- Customer interviews
- Physical examination of the system
- Basic system analysis
- Use of manufacturer's troubleshooting aids
- Fault isolation in equipment problem area

### 5.1.0 Customer Interviews

The troubleshooting procedure should begin with the technician learning all that can be learned about the customer's complaint by talking to the customer and the service dispatcher. Talking with the customer prior to working on the equipment is always recommended because it can provide valuable information on equipment operation that can aid in the troubleshooting process. It can also identify the source of a problem that is not related to the HVAC equipment, thereby eliminating unnecessary equipment maintenance.

The first evidence of trouble with the HVAC system is often a complaint from an individual who is too cold, too hot, or who is bothered by drafts. In many cases, the problem behind such a complaint may not be an equipment or control system malfunction, but a personal comfort problem. If the system appears to be operating correctly but individuals are complaining about comfort problems, the technician should check for one or more of the following conditions before assuming the HVAC equipment is malfunctioning.

- *Air distribution and circulation problems* – Persons outside of the immediate area that is controlled by a thermostat may feel too hot or cold. The thermostat senses only the temperature at its particular location. Temperature levels in all the other areas controlled by the same thermostat are subject to variation that can be caused by poor air distribution and/or room air circulation.

- *False heat loads* – Direct sunlight on the thermostat or heat from lamps, appliances, and pipes can cause overcooling of a zone. Direct sunlight or artificial sources of radiant heat can also cause heating discomfort.

- *Covered grilles and diffusers* – Occupants frequently cover part or all of a discharge grille, causing improper heating or cooling.

- *Occupant locations* – Occupants located adjacent to outside walls or windows may be subject to air infiltration or radiant cooling or heating from a wall.

- *Insufficient conditioned air supply* – May be caused by poor air distribution design or fan speed, dirty filters, or lack of the proper amount or size of return air outlets.

- *Overcrowding* – Overheating will result if more people or mechanical equipment occupy a conditioned area than the space was designed to hold.

- *System size* – Extreme weather conditions may exceed the capacity of the heating or cooling equipment.

- *Drafts* – Forced-air systems rely on the movement of air to deliver the desired conditioning. To many people, even a slight air motion may be uncomfortable. This problem may be alleviated by moving the occupant's work station or redirecting the airflow.

- *Stale air* – A stuffy or smoky atmosphere usually results from an insufficient fresh air supply, air that is too humid, or air that has inadequate exhaust. Stale air can also be caused by overcrowding or low air circulation.

### 5.2.0 Physical Examination of the System

Many problems can be identified by simply using your senses to check the system. Conduct a preliminary, power-off visual system inspection followed, if possible, by a preliminary power-on system inspection using your senses.

- Look for evidence of leaks and physical damage.
- Look for dirt accumulation on filters and coils that can affect system operation.
- Listen for unusual sounds that could indicate a malfunction and possibly lead you to its source.
- Check for odors, especially the smell of overheating or gas.

## 5.3.0 Basic System Analysis

The proper diagnosis of a problem requires that you know what the unit should be doing when it is operating properly. If you are not familiar with how a particular unit should operate, you must first study the manufacturer's service literature to acquaint yourself with the equipment's modes and sequence of operation.

The second part of the diagnosis is to find out what the unit actually is doing, and what symptoms are exhibited by the improperly operating unit. Finding out what the unit is doing is accomplished both by carefully listening to the customer's complaints and by analyzing the operation of the unit yourself. As applicable, this means making electrical, temperature, pressure, and/or airflow measurements at key points in the system. The set of measured values can then be compared with a set of typical readings for a properly operating system as previously recorded on system operating logs or in the manufacturer's service literature. This process can often quickly pinpoint the system problem.

## 5.4.0 Use of Manufacturer's Troubleshooting Aids

To aid in the isolation of faults, many manufacturers provide troubleshooting information marked on the equipment or contained in the service instructions for a particular product. This information typically includes:

- **Label diagrams**
- **Troubleshooting tables**
- **Fault isolation diagrams**
- Diagnostic equipment and tests

### 5.4.1 Label Diagrams

Label diagrams (*Figure 36*) are usually placed in a convenient location inside the equipment, typically on the inside of a control circuit access panel. They normally show a component arrangement diagram, **wiring diagram**, legend, and notes pertaining to the equipment.

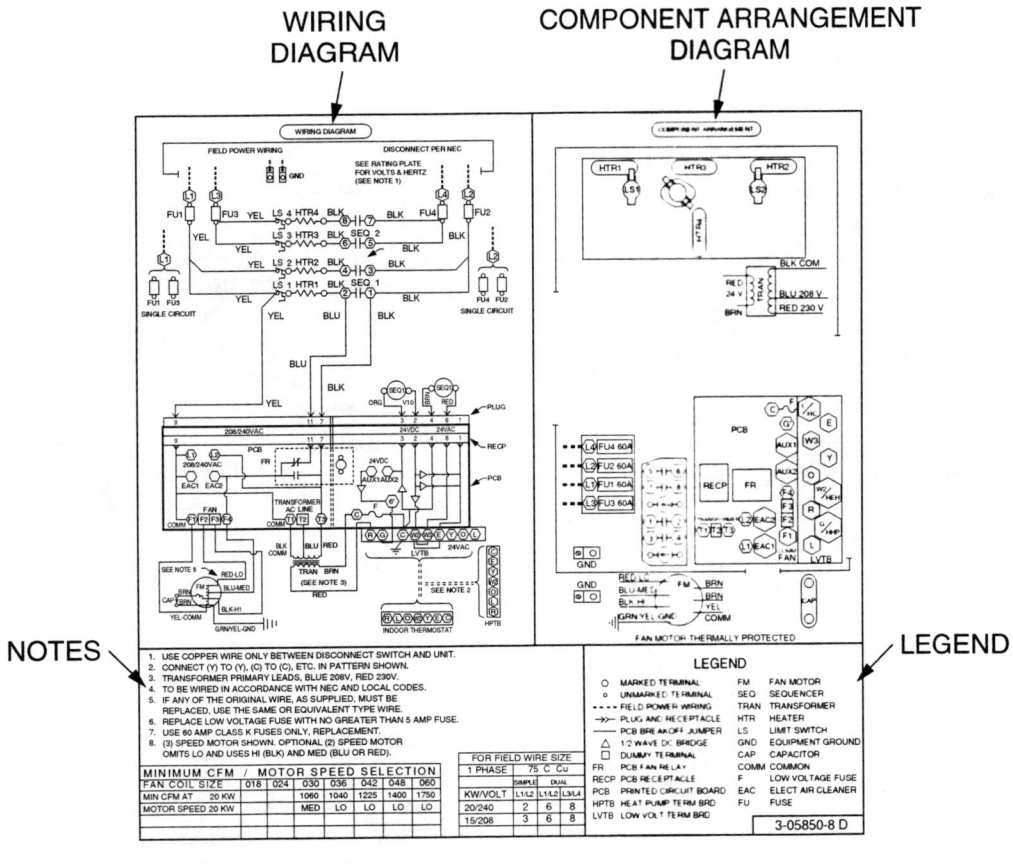

*Figure 36* ◆ Typical label diagram.

The component arrangement diagram shows where the components are physically located in the unit. It is useful because it helps you locate and identify the components shown on the wiring diagram.

The wiring diagram, sometimes called a *schematic*, provides a picture of what the unit does electrically and shows the actual external and internal wiring of the unit. Many label diagrams also contain a simplified schematic that is commonly called a *ladder diagram* (*Figure 37*). Ladder diagrams usually have the wire color and physical connection information eliminated from the diagram. This makes it more useful by focusing on the functional, not the physical, aspects of the equipment. Wiring and ladder diagrams are the primary troubleshooting aids used when isolating electrical problems.

The legend identifies the meanings of the abbreviations used on the label diagram.

### 5.4.2 Troubleshooting Tables and Fault Isolation Diagrams

Troubleshooting tables and/or fault isolation diagrams are usually contained in the manufacturer's Installation, Start-up, and Service Instructions for a particular product. As shown in *Figure 38*, troubleshooting tables are intended to guide you to a corrective action based on your observations of system operation. Fault isolation diagrams (*Figure 39*), also called *troubleshooting trees*, normally start with a failure symptom observation and take you through a logical decision-action process to isolate the failure.

### 5.4.3 Diagnostic Equipment and Tests

Many manufacturers incorporate electronically controlled or semi-automatic testing features in their equipment to help isolate malfunctions. Depending on the equipment, these built-in diagnostic devices can be simple or complex. Some units contain microprocessor controllers that can run a complete check of all system functions, then report back the results by means of a system numeric display or flashing LED display. Normally, these built-in test functions isolate a fault to a functional problem area. For example, if a test indicates a compressor failure, this means that the failure has been isolated to the compressor and its related control circuits and wiring. The technician must perform additional troubleshooting within the problem area to find out exactly where the fault is located.

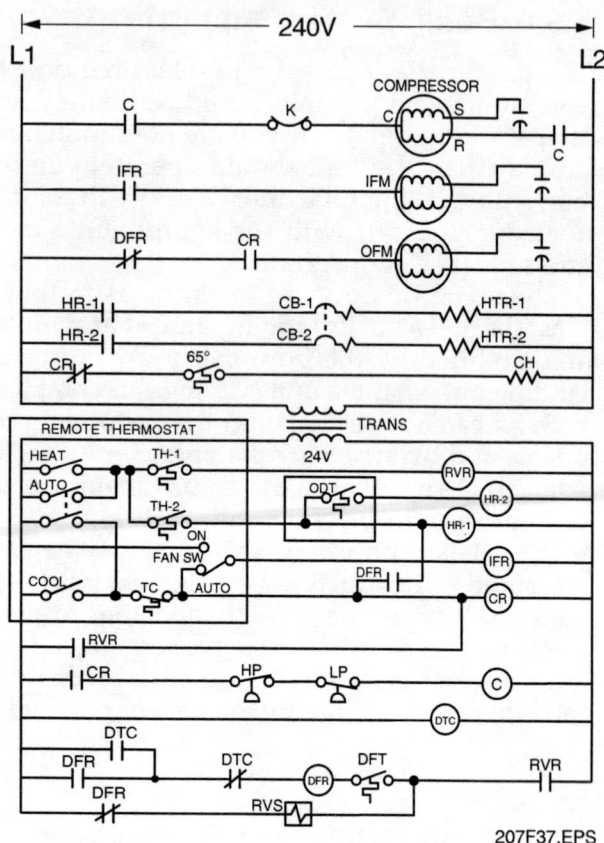

*Figure 37* ◆ Ladder diagram.

In addition to built-in diagnostic equipment, many manufacturers have developed a series of stand-alone electronic module testers that are available to troubleshoot the different electronic control modules commonly used in the manufacturer's product line. The module tester is usually plugged into the control module in the equipment under test.

Troubleshooting involves the testing of each module control circuit using a sequential troubleshooting process that is performed per the manufacturer's instructions provided with the module tester.

### 5.5.0 Fault Isolation in the Equipment Problem Area

Once troubleshooting aids have isolated a problem to a functional equipment area, it may be necessary to make additional measurements and use a step-by-step process of elimination to isolate the specific cause of the problem. This is usually the case for the more difficult problems you will encounter.

| Malfunction | Probable Cause | Corrective Action |
|---|---|---|
| Compressor motor and condenser motor will not start, but fan/coil unit (blower motor) operates normally | Check the thermostat system switch to ascertain that it is set to COOL. | Make necessary adjustments to settings. |
| | Check the thermostat to make sure that it is set below room temperature. | Make necessary adjustments. |
| | Check the thermostat to see if it is level. Most thermostats must be mounted level; any deviation will ruin their calibration. | Remove cover plate, place a spirit level on top of the thermostat base, loosen the mounting screws, and adjust the base until it is level; then tighten the mounting screws. |
| | Check all low-voltage connections for tightness. | Tighten. |
| | Make a low-voltage check with a voltmeter on the condensate float switch; the condensate may not be draining. | The float switch is normally found in the fan/coil unit. Repair or replace. |
| | Low air flow could be causing the trouble, so check the air filters. | Clean or replace. |
| | Make a low-voltage check of the antifrost control. | Replace if defective. |
| | Check all duct connections to the fan-coil unit. | Repair if necessary. |
| Compressor, condenser, and fan/coil unit motors will not start | Check the thermostat system switch setting to ascertain that COOL. | Adjust as necessary. |

*Figure 38* ◆ Typical troubleshooting diagram.

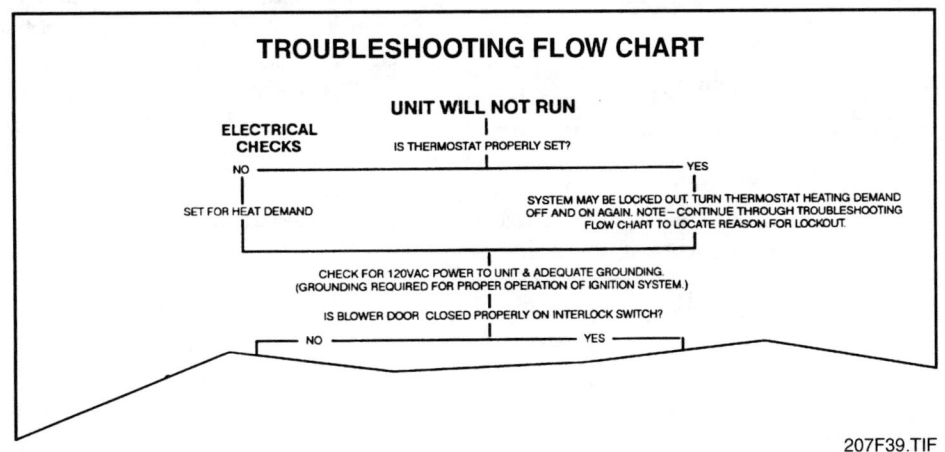

*Figure 39* ◆ Typical fault isolation diagram.

# 6.0.0 ◆ SAFETY

This section covers safety practices and procedures.

## 6.1.0 Safety Practices

Practically all electric shocks are due to human error, rather than equipment failure. Nearly all deaths due to electric shock are due to the worker's failure to observe safety precautions, failure to repair equipment for electrical defects, or failure to remedy all defects found by tests and inspections.

The following is a list of recommended safety practices:

- Always turn off power at the main disconnect before working on an electrical circuit or device. Lock and tag the power switch.

- Use a voltmeter to verify that the power to the unit is actually off. Remember that even though the power may be switched off, there is still potential at the input side of the shutoff switch.

- Never switch equipment on or off while standing in or touching a wet surface or area.

- Notify the power company or utility whenever a power line is touching the ground.

- If it becomes necessary to work on live electrical wiring, try to use one hand only. If you are shocked while using only one hand, current will probably flow through the hand and arm, then down the side and through the feet to the ground. If a shock is conducted through both hands and arms, the electrical path would be through the heart and lungs and would be more likely to be fatal.

- Use protective equipment such as rubber gloves and insulated boots.

- Use tools with dielectric insulation.

- Remove metal jewelry such as rings and watches.

The *National Electrical Code (NEC)*, when used together with the electrical code for your local area, provides the minimum requirements for the installation of electrical systems. Always use the latest edition of the code as your on-the-job reference. It specifies the minimum provisions necessary for protecting people and property from electrical hazards. In some areas, different editions of the code may be in use, so be sure to use the edition specified by your employer.

## 6.2.0 OSHA Lockout/Tagout Rule

In addition to these general electrical safety rules, OSHA released 29 CFR 1926 (Lockout/Tagout) in December 1991. This rule covers the specific procedure to be followed for the "servicing and maintenance of machines and equipment in which the unexpected energization or start-up of the machines or equipment, or releases of stored energy, could cause injury to employees. This standard establishes minimum performance requirements for the control of such hazardous energy."

The purpose of the OSHA procedure is to make sure that machinery is isolated from all potentially hazardous energy, and tagged and locked out before employees perform any servicing or maintenance activities where the unexpected energization, start-up, or release of stored energy could cause injury.

**WARNING!**
The OSHA procedure provides only the minimum requirements for the lockout/tagout procedure. Consult the lockout/tagout procedure for your company and the local area in which you are working. Remember, your life could depend on this procedure. It is critical that you use the correct procedure for your site.

## 6.3.0 Lockout/Tagout Procedure

To prepare for a lockout/tagout, make a survey to locate and identify all isolating devices (see *Figure 40*). You need to be certain which switch(es), valve(s), or other energy isolating devices apply to the equipment to be locked and tagged. More than one energy source (electrical, mechanical, or others) may be involved.

The following procedure outlines the general steps for lockout/tagout (note that this procedure is provided only as an example). Each employer will designate who is qualified to use the procedure.

*Step 1* Notify all affected employees of the lockout/tagout and why it is necessary. The authorized employee will know the type of energy that the machine or equipment uses and the associated hazards.

*Step 2* If the machine or equipment is operating, shut it down using the normal procedure. For example, press the stop button or open the toggle switch.

*Figure 40* ♦ Lock out and tag HVAC equipment.

*Step 3* Operate the switch, valve, or other energy isolating device(s) so that the equipment is isolated from its energy source(s). Stored energy must be dissipated or restrained by repositioning, blocking, or bleeding down. Examples of stored energy are springs, elevated machine members, rotating flywheels, hydraulic systems, and air, gas, steam, or water pressure.

*Step 4* Lock and tag the energy isolating devices with assigned individual lock(s) and tag(s).

*Step 5* After making sure that no personnel are exposed, operate the start button or other normal operating controls to make certain the equipment will not operate. This will confirm that the energy sources have been disconnected.

The equipment is now locked and tagged.

All equipment must be locked and tagged to protect against accidental operation when such operation could cause injury to personnel. Never try to operate any switch, valve, or energy isolating device when it is locked and tagged.

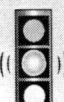 **CAUTION**
Return operating control(s) to their neutral or OFF position after the test.

## 6.4.0 Restoring Machines or Equipment

After service and/or maintenance is complete, the equipment is ready for normal operation. Use the following procedure to restore the machines or equipment to their normal operating condition.

*Step 1* Check the area around the machines or equipment to make sure that all personnel are at a safe distance.

*Step 2* Remove all tools from the machine or equipment and reinstall any guards.

*Step 3* Again making sure that all personnel are in the clear, remove all lockout/tagout devices.

*Step 4* Operate the energy isolating devices to restore energy to the machine or equipment.

## 7.0.0 ♦ HVAC SYSTEM TROUBLESHOOTING

Troubleshooting HVAC systems covers a wide range of electrical and mechanical problems. Obviously, not all problems fit easily into these two categories. For example, a loose or corroded wire may cause a compressor in a cooling system to cycle on and off intermittently. Although the problem is electrical, it looks like a mechanical refrigeration problem.

Another example is a compressor that fails from shorted windings because of the acids formed in a poorly evacuated system. This is an example of a mechanical refrigeration problem that looks like an electrical problem.

The methods used to troubleshoot mechanical problems in cooling, heating, and other HVAC equipment tend to be unique to the type of equipment being serviced. For this reason, no further information about troubleshooting mechanical problems is given in this module. Information about troubleshooting HVAC mechanical problems is described in detail in HVAC Level Three.

The methods used to troubleshoot electrical problems and components are very similar regardless of the type of HVAC equipment being serviced. The remainder of this module describes those procedures that are common to troubleshooting electrical problems in most types of HVAC equipment.

## Electrical Troubleshooting

Most electrical troubleshooting in HVAC equipment can be done using a multimeter (VOM/DMM) and an AC clamp-on ammeter. Some types of clamp-on instruments like the one shown here incorporate the functions of a clamp-on ammeter and multimeter into one instrument.

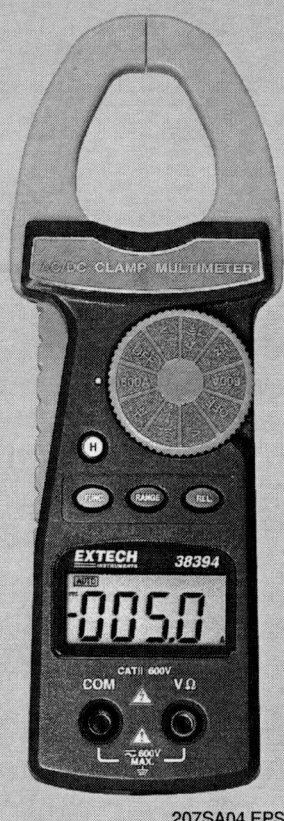

### 8.0.0 ◆ HVAC EQUIPMENT INPUT POWER, LOAD, AND CONTROL CIRCUITS

Troubleshooting electrical problems in HVAC equipment may appear complex. However, the process can be simplified if the unit's electrical components are divided into smaller functional circuit areas based on the operation they perform in the equipment.

Most HVAC equipment can be divided into three functional circuit areas (*Figure 41*):

- Input power distribution circuits
- Load circuits
- Control circuits

Input power distribution circuits serve as the power source for the entire unit. They operate at either single-phase or three-phase line voltages, and act to distribute the input power to the various loads in the unit. Power circuits usually consist of the field-installed power wiring from the main electrical service to a disconnect switch located near the unit, and from the disconnect switch to the unit. The input power and distribution circuits include protective devices such as fuses and/or circuit breakers.

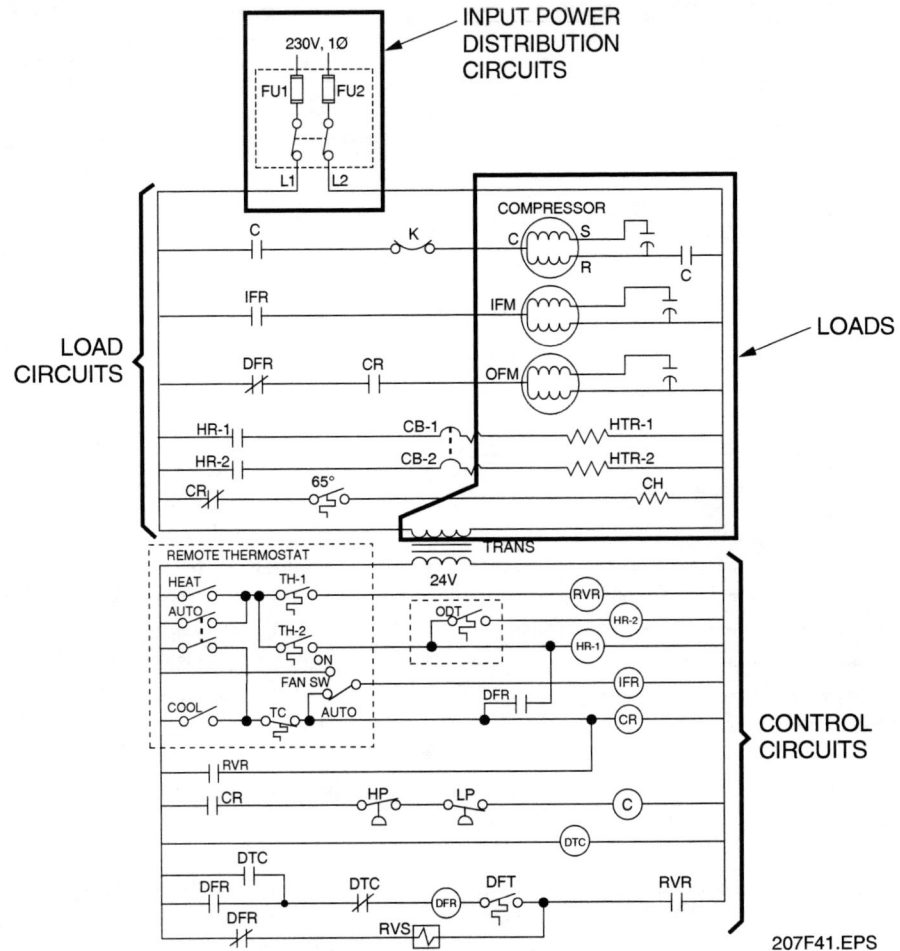

*Figure 41* ♦ HVAC equipment functional circuit areas.

Loads are devices that consume power to do work. Compressor motors, fan motors, heater elements, and the primary winding of transformers are all loads normally found in cooling and heating units. Because the input power distribution circuits and the load circuits are both energized and operate at the input voltage level, they are often called the *high-voltage circuits* (*Figure 42*).

Control circuits provide a link between loads and the input power. Control circuits start, stop, or otherwise control the operation of a load. They usually contain one or more control devices such as relays, switches, and thermostats that work to apply or remove power from the loads. The more complex the system, the more control devices it will have. When a load such as a compressor motor is not working, you have to determine whether the problem is in the load itself, or in the circuits controlling the load.

As discussed earlier, control circuits generally operate at 24V. This low voltage is obtained by using a control transformer to step down the line voltage. Because most control circuits operate at 24V, they are often called the *low-voltage circuits*. In some larger systems using three-phase line voltages of 240V and higher, a control voltage of 120V or higher may be used for some of the control devices. For this reason, you must always measure the control circuit voltage. Never assume the control circuit voltage is a low voltage.

## 8.1.0 Isolating to a Faulty Circuit via the Process of Elimination

Isolation to the faulty functional circuit (input power distribution circuit, load circuit, or control circuit) can be relatively easy based on an analysis of the equipment operation and a process of elimination. Also, talking to the customer prior to working on the equipment is always recommended because it can provide valuable information that can aid in the troubleshooting process.

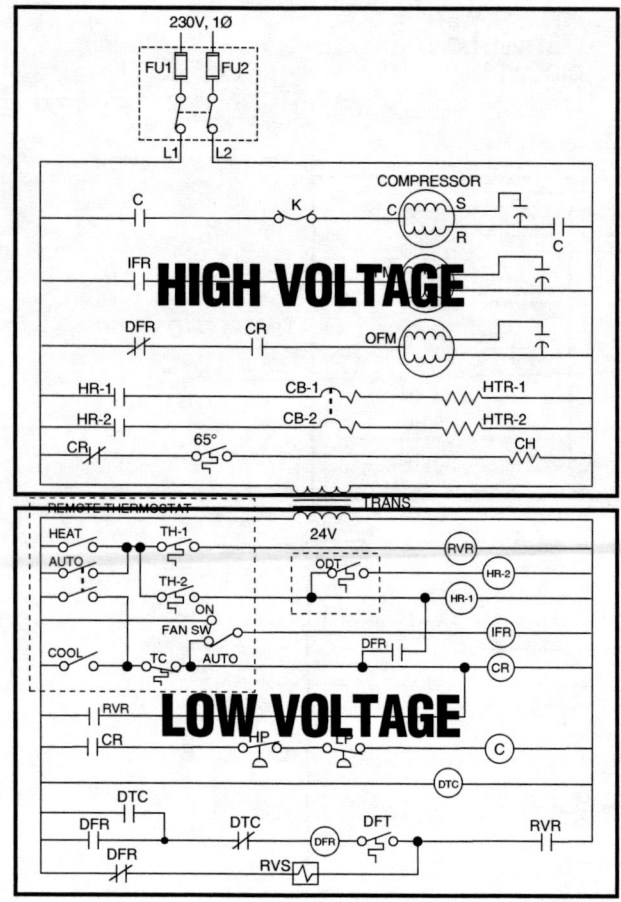

*Figure 42* ◆ High-voltage and low-voltage circuits.

For example, suppose you answer a service call on the heat pump system shown in *Figure 41*. The customer complains that the unit is running but is blowing warm air instead of cool air. Your preliminary check of system operation reveals that the compressor is not running when the thermostat is calling for cooling.

You begin fault isolation through the process of elimination. Because the indoor fan motor runs, you can immediately eliminate the input power distribution circuits as the source of the problem. Next, find out if the compressor will run in the reverse cycle heating mode by setting the thermostat to call for heating. (If it is too hot, thermostat TH-1 can be jumpered to simulate a call for heating.) For the purpose of this example, assume the compressor runs. Now you can eliminate the compressor load circuits and everything in the compressor contactor energizing path, including the control relay (CR).

This isolates the problem to the only devices left in the control circuit that are unique to the cooling mode. By studying the schematic (*Figure 41*), you identify these components to be the cooling thermostat (TC) and the related COOL control switch. The thermostat can be eliminated as the cause of the problem if the unit works when the control switch is set to the AUTO mode.

## 8.2.0 Isolating to a Faulty Circuit Component

Once the source of an electrical problem has been isolated to the malfunctioning load or control circuit, the next step is to make a series of voltage measurements across the components in the malfunctioning circuit to find the faulty component. As shown in *Figure 43*, the measurements can start from the line or control voltage side of the circuit and move toward the load device, such as a motor or a relay coil. Measurements are made until either no voltage is observed, or until the voltage has been measured across all the components in the circuit. Note that when there are many devices in the circuit under test, the measurements can be made by starting at the midpoint in the circuit (divide by two), then working towards either the source of line or control voltage or the load device, depending on whether voltage was or was not measured at the mid-point. As a result of taking these voltage measurements, one of the following situations should exist.

At some point within the circuit under test, no voltage will be indicated on the meter. This pinpoints an open component or set of switch contacts between the last measurement point and the previous measurement point. *Figure 43(A)* shows an example of this situation. In this case, the contacts of the low-pressure switch are open, preventing the compressor contactor coil (C) from energizing.

If the open is caused by a set of contactor or relay contacts, you must find out if the related contactor or relay coil is not being energized or is bad. *Figure 43(B)* shows an example of this situation. In this case, the contacts (CR) in the control circuit are open, preventing the compressor contactor (C) from energizing. These contacts close when the control relay coil (CR) is energized. Before assuming that the problem is caused by the open contacts (CR), you must troubleshoot the control circuit containing the related relay coil (CR) to find out if the coil is energized or de-energized.

If the coil is de-energized, you must further troubleshoot its control circuit to find out why. For example, if the thermostat cooling switch contacts are open, the relay coil (CR) will not be energized.

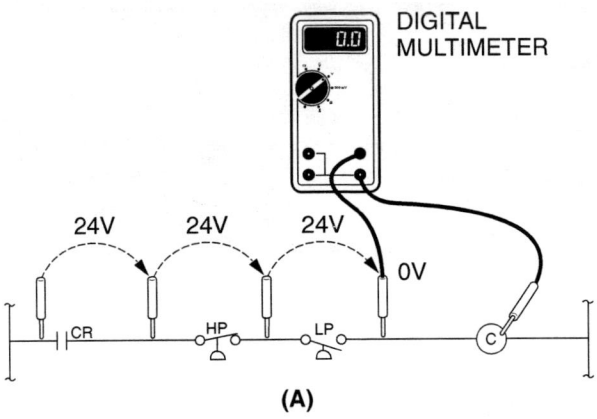

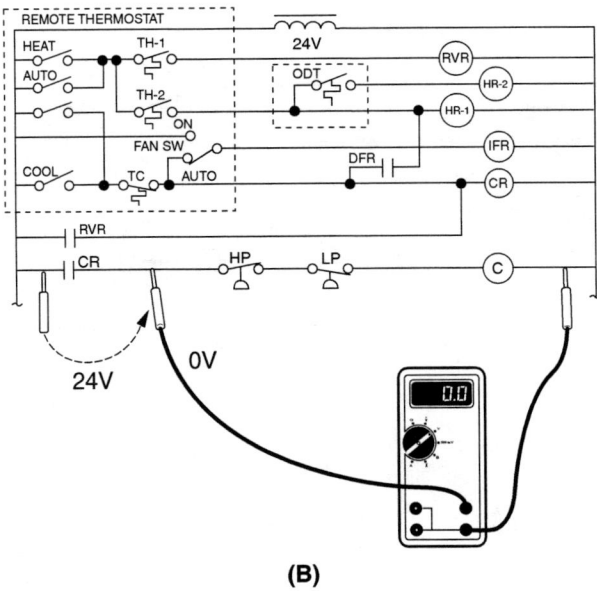

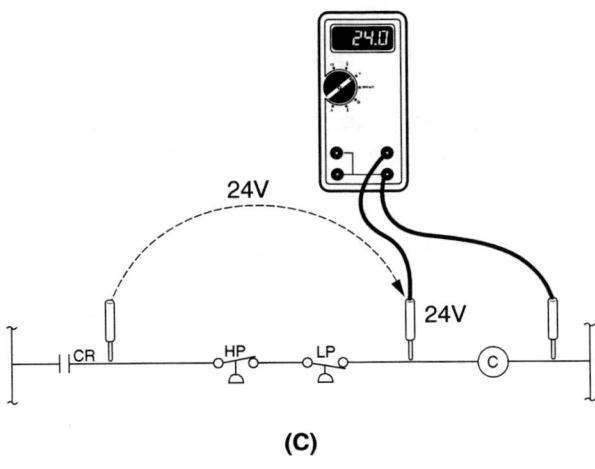

*Figure 43* ♦ Isolating to a faulty circuit component.
(A) Open low-pressure switch contacts.
(B) Open relay contacts – check related coil control circuit. (C) Load device (contactor coil) is bad.

If voltage is measured at the contactor coil, motor, or other load device, and the device is not working, the load device is most likely at fault. You should turn off the power to the unit, then disconnect the device from the circuit and test it to confirm that it is defective. *Figure 43(C)* shows an example of this situation in which 24V power is applied to the compressor contactor (C), but it is not energized. In this case, the contactor is probably bad.

## 9.0.0 ♦ ELECTRICAL TROUBLESHOOTING COMMON TO ALL HVAC EQUIPMENT

This section covers the electrical troubleshooting procedures common to all HVAC equipment. Specific procedures for motors, hydronic controls, and pneumatic controls are covered later in this module.

### 9.1.0 Input Voltage Measurements

All HVAC equipment is designed to operate within a specific range of system voltages including a safety factor, typically ±10%. This safety factor is added to compensate for temporary supply voltage fluctuations that might occur. Continuous operation of HVAC equipment outside the intended range of voltages can damage the equipment.

#### 9.1.1 Effects of High and Low Voltage

Too high or too low an operating voltage can cause overheating and possible failure of motors and other devices. The power supply voltage applied to the equipment should be measured and checked against the supply voltage indicated on the unit nameplate. See *Figure 44*. If low voltage exists at the equipment, the voltage should be measured at the electrical service entrance to make sure that a voltage drop does not exist in the branch circuit or feeder that supplies the HVAC equipment. If a voltage drop exists, it may be necessary to install larger wires between the service and the equipment.

Operating voltages applied to motors in the equipment must be maintained within limits from the voltage value given on the motor nameplate. The voltage tolerances used for most HVAC motors are as follows:

- *Single-voltage rated motors* – The input supply voltage should be within ±10% of the nameplate voltage. For example, a motor with a nameplate single voltage rating of 230V should have an input voltage that ranges between 207V (–10% of 230V) and 253V (+10% of 230V).

### Measuring Input Voltage

When the contactor closes in a cooling unit and applies power to loads such as the compressor and outdoor fan motor, the voltage level may drop about 3% from the measured open circuit voltage (contactor open). This is acceptable as long as the voltage does not drop below the manufacturer's stated minimum voltage.

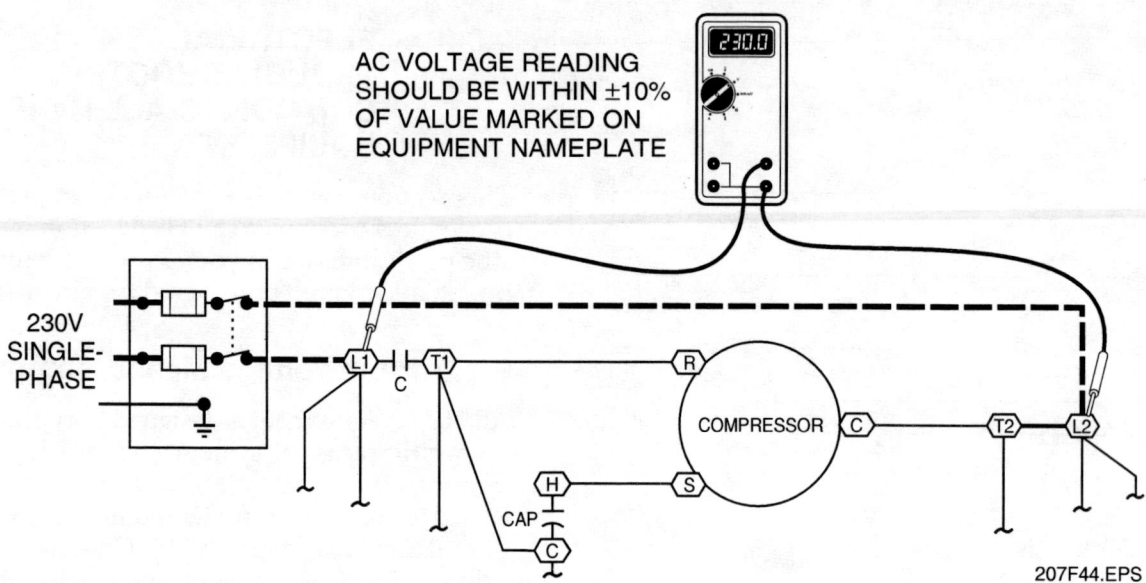

*Figure 44* ◆ Single-phase input voltage checks.

- *Dual-voltage rated motors* – The input supply voltage should be within ±10% of the nameplate voltage. For example, a motor with a nameplate dual voltage rating of 208V/230V should have an input voltage that ranges between 187V (−10% of 208V) and 253V (+10% of 230V).

#### 9.1.2 Voltage Phase Imbalance

Voltage imbalance becomes very important when working with three-phase equipment. A small imbalance in phase-to-phase voltage can result in a much greater current imbalance. This current imbalance increases the heat generated in the motor windings. Both current and heat can cause nuisance overload trips and may cause motor failure. For this reason, the voltage imbalance between any two legs of the supply voltage applied to a three-phase system or motor should not exceed 2%. If a voltage imbalance of more than 2% exists at the input to the HVAC equipment, correct the problem in the building or utility power distribution system before operating the equipment. *Figure 45* shows how the amount of voltage imbalance is determined in a three-phase system using the formula:

$$\% \text{ imbalance} = \frac{\text{maximum deviation from avg.}}{\text{average voltage}} \times 100$$

The current imbalance in any one leg of a three-phase system should not exceed 10%. A current imbalance may occur without a voltage imbalance. This can occur when an electrical terminal, contact, etc. becomes loose or corroded, causing a high resistance in the leg. Since current follows the path of least resistance, the current in the other two legs will increase, causing more heat to be generated in the devices supplied by those legs. The current imbalance in a three-phase system is determined in the same way as voltage imbalance, but average current is substituted for average voltage.

### 9.2.0 Fuse/Circuit Breaker Checks

Fuses and/or circuit breakers are normally the first components checked when a unit is totally inoperative.

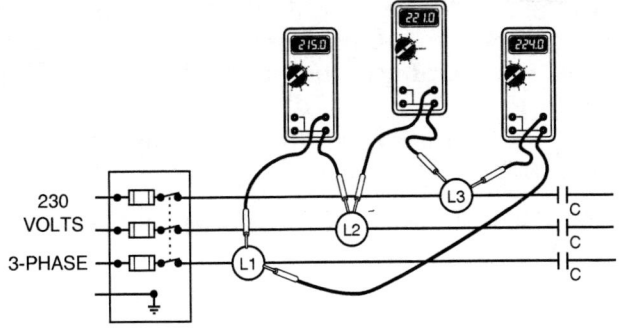

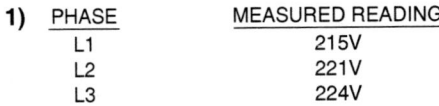

*Figure 45* ♦ Three-phase voltage and current checks. (A) Calculating voltage imbalance. (B) Calculating current imbalance.

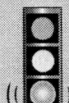

**NOTE**

The system must be operating when voltage and current imbalance tests are performed.

### 9.2.1 Fuse Checks

The best way to test a fuse is by measuring continuity (*Figure 46*). To check fuses, always open the unit disconnect switch, then remove the fuses using an insulated fuse puller. Test the fuses for continuity using an analog or digital multimeter (VOM/DMM). If a short (zero ohms) exists across the fuse, it is usually good. If an open (infinite resistance) exists across the fuse, it is bad. A blown fuse is usually caused by some abnormal overload condition, such as a short circuit within the equipment or an overloaded motor. Replacing a blown fuse without locating and correcting the cause can result in damage to the equipment.

Fuses can also be tested with the circuit energized. This is accomplished by shutting off power to the unit, then disconnecting the wires from the load side of each fuse. This eliminates the possibility of current being fed back to the meter through a short circuit within the unit. Set the multimeter to measure AC voltage on a range that is higher than the highest voltage expected. Turn on the power and place one of the multimeter test leads on the line side of a fuse. Touch the other test lead to the load side of another fuse. If voltage is measured on the load side of a fuse, the fuse is good; if not, the fuse is bad. Repeat this procedure so that all fuses are measured with one test lead on the load side and the other test lead on the input side of a different fuse. This method tests one fuse at a time. If the measurement was performed with both test leads on the load side of the fuses, and the multimeter showed no reading, you would know that a fuse was blown, but not which one.

### 9.2.2 Circuit Breaker Checks

At the power distribution panel, set the circuit breaker to OFF. If required, remove the panel that covers the circuit breaker to expose the body of the breaker and the wires connected to its terminals. Set the multimeter to measure AC voltage on a range that is higher than the highest voltage expected.

See *Figure 47*. Measure the voltage applied to the circuit breaker input terminals:

- A to neutral or ground (single-pole breaker)
- A to B (two-pole breaker)
- A to B, B to C, and C to A (three-pole breaker)

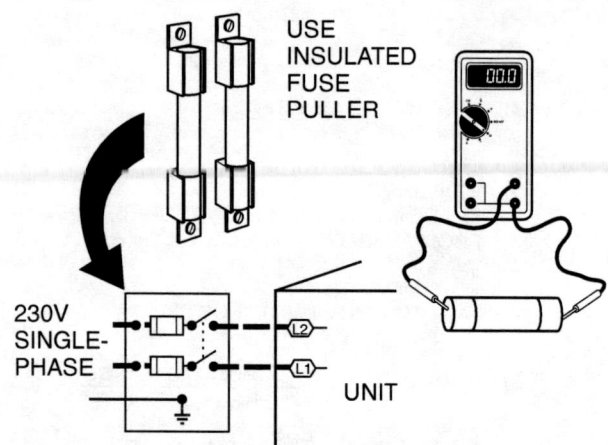

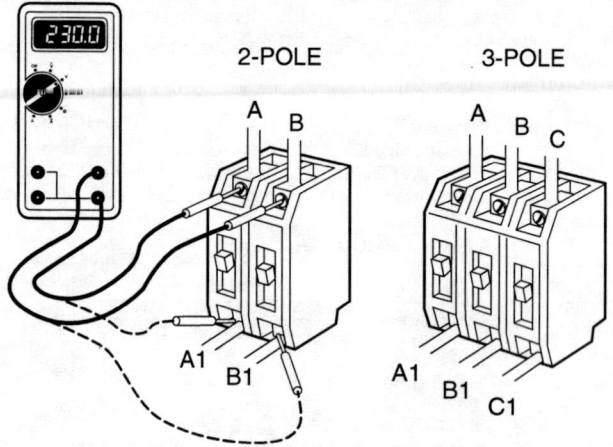

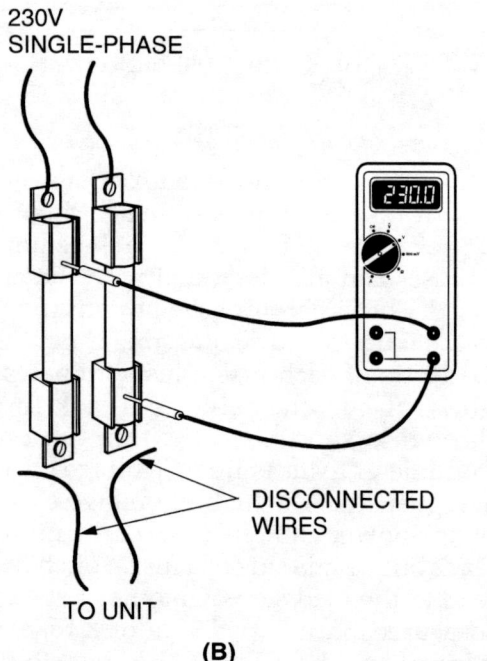

*Figure 46* ◆ Fuse checks. (A) Continuity check. (B) Voltage check.

*Figure 47* ◆ Circuit breaker checks. (A) Voltage check. (B) Current check.

Make sure that the breaker is closed by first setting it to the OFF position, then setting it to the ON position. Measure the voltage at the circuit breaker output terminals:

- A1 to neutral or ground (single-pole breaker)
- A1 to B1 (two-pole breaker)
- A1 to B1, B1 to C1, and C1 to A1 (three-pole breaker)

The measured input and output voltages should be the same. If the voltage is significantly lower than that measured at the input to the circuit breaker, visually inspect the circuit breaker for loose wires and terminals or signs of overheating. If none are found, the circuit breaker should be replaced.

If the circuit breaker shows signs of overheating, or trips when voltage is applied to the equipment, reset it, then check the current flow through the breaker using an AC clamp-on ammeter. Set up the AC clamp-on ammeter to measure AC current on a range that is higher than the highest current expected. Check the ampere rating marked on the breaker. It is usually stamped on the breaker lever or body. One wire at a time, measure the current flow in the wires connected to the circuit breaker output terminals:

- A1 (single-pole breaker)
- A1, B1 (two-pole breaker)
- A1, B1, and C1 (three-pole breaker)

If the circuit breaker trips at a current below its rating or is not tripping at a higher current, the circuit breaker should be replaced. Be sure that the breaker is not being tripped because of high ambient temperature.

## 9.3.0 Resistive and Inductive Load Checks

Electric crankcase heaters and electric heater elements are examples of resistive loads found in HVAC equipment. Inductive loads include contactor, relay, and motor starter coils. They also include control transformers, solenoid valves, and some gas valves. Compressor motors, fan motors, and other motors are also inductive loads. Because motors require special troubleshooting methods, they are covered as a separate category later on in this module.

---

### Circuit Breakers

Circuit breakers are available in a wide variety of sizes and types to suit various applications. Some types of circuit breakers are listed as heating, air conditioning, and refrigeration (HACR) circuit breakers. HACR breakers have a built-in time delay that allows a higher-than-rated current to momentarily flow in the circuit. This compensates for the large starting current drawn by such loads. The equipment being protected must also be marked by the manufacturer as suitable for protection by this type of breaker.

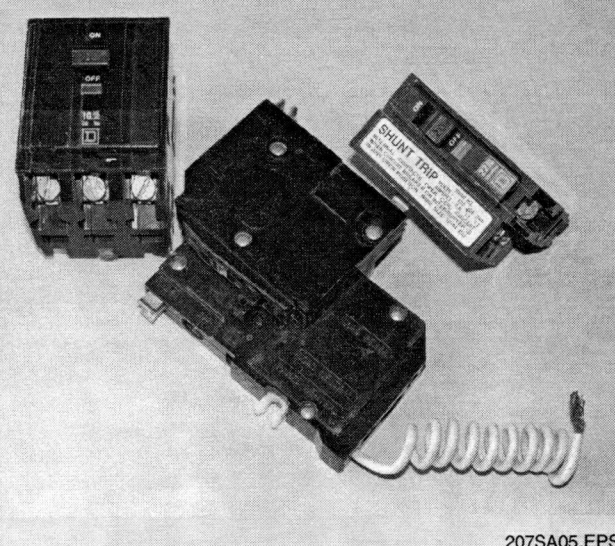

INTRODUCTION TO CONTROL CIRCUIT TROUBLESHOOTING — TRAINEE MODULE 03207

## *Inductive Load*

This is a solenoid valve used to control the flow of water to a humidifier. The solenoid coil is an inductive load.

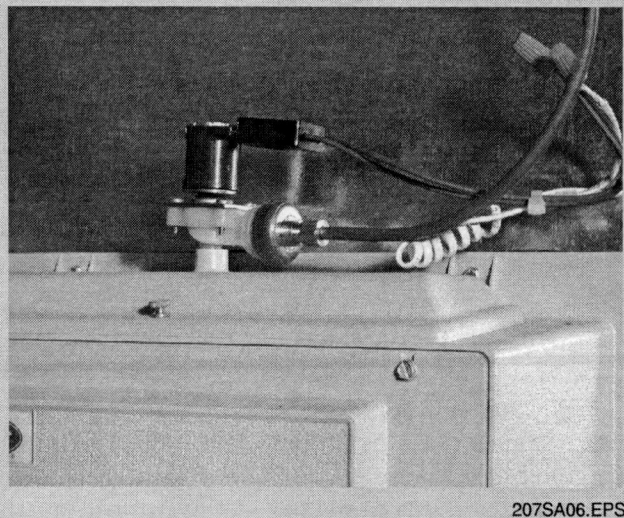

207SA06.EPS

Once a resistive or inductive load has been identified as the probable cause of an electrical problem, it should be tested to confirm that it is good or bad. The best way to test a resistive or inductive load is by measuring the resistance across the terminals of the device. Before measuring resistance, make sure to electrically isolate the component being measured by disconnecting at least one lead of the component from the circuit. This is important in order to achieve an accurate resistance reading. Otherwise, the meter will read the resistance of other components that are connected in parallel with the component to be measured. As shown in *Figure 48*, a reading of zero ohms indicates a shorted load, while a reading of infinite resistance indicates an open load. In either case, the device should be replaced. When the multimeter indicates a measurable resistance, it usually indicates that the device is good. If a low resistance is measured on a contactor, relay, or starter coil, place one meter probe to ground or to the unit frame. Touch the other probe to each coil terminal. If a resistance is measured from either terminal to ground, replace the device.

### 9.4.0 Switch and Contactor/Relay Contact Checks

Once a switch or contactor/relay contact has been identified as the probable cause of an electrical problem, the contacts should be tested to confirm their position. Switches and contactor/relay contacts can be tested by making a continuity measurement to determine whether the contacts are open or closed (*Figure 49*). If the switch contacts are open, the multimeter indicates an infinite resistance reading. If the switch contacts are closed, the multimeter indicates a short (zero ohms).

## *Measuring Resistance*

The actual resistance value measured for resistive and inductive loads can vary widely depending on the type of device. Ideally, the exact resistance value for the device can be found in the manufacturer's service literature. Another way to judge the resistance reading is by comparing the resistance of the device being tested with that of a similar device that is known to be good.

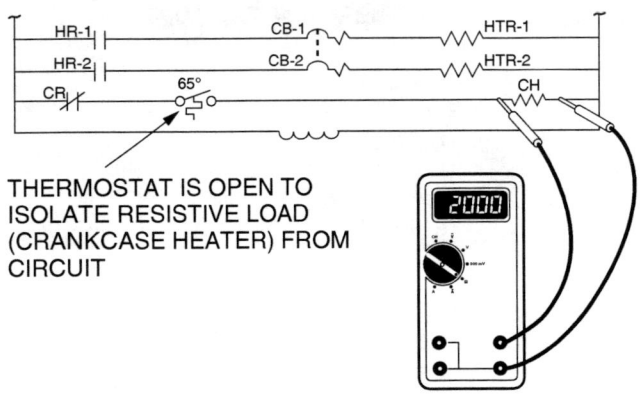

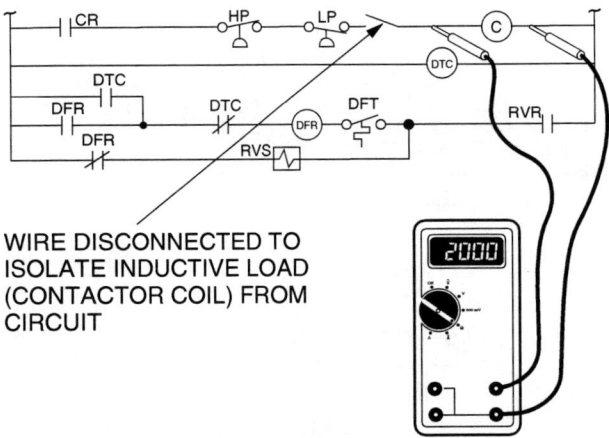

METER READS
MEASURABLE RESISTANCE = GOOD LOAD
ZERO RESISTANCE = SHORTED LOAD
INFINITE RESISTANCE = OPEN LOAD

*Figure 48* ◆ Resistive and inductive load resistance checks.

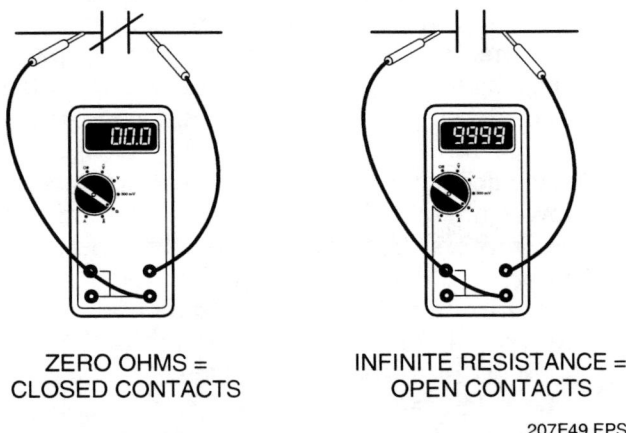

ZERO OHMS = CLOSED CONTACTS
INFINITE RESISTANCE = OPEN CONTACTS

*Figure 49* ◆ Relay contact checks.

> **NOTE**
> The power must be turned off to make the continuity check. As such, the check can only confirm the status of normally open and normally closed switch contacts. It does not reflect the status of the contacts when the system is powered up.

### 9.5.0 Control Transformer Checks

Control transformers are usually checked by measuring the voltages across the secondary and primary windings (*Figure 50*). Typically, the secondary winding is measured first. The multimeter should be set to measure AC voltage on a range that is higher than the control voltage expected. If the voltage measured across the secondary winding is within ±10% of the required voltage, the transformer is good. If no voltage is measured at the secondary winding, the voltage across the primary winding must be measured. If so equipped, also check the secondary fuse to see if it is blown.

If the voltage measured at the primary winding is within ±10% of the required voltage, the transformer most likely is bad. This can be confirmed by performing a continuity check of the transformer primary and secondary windings.

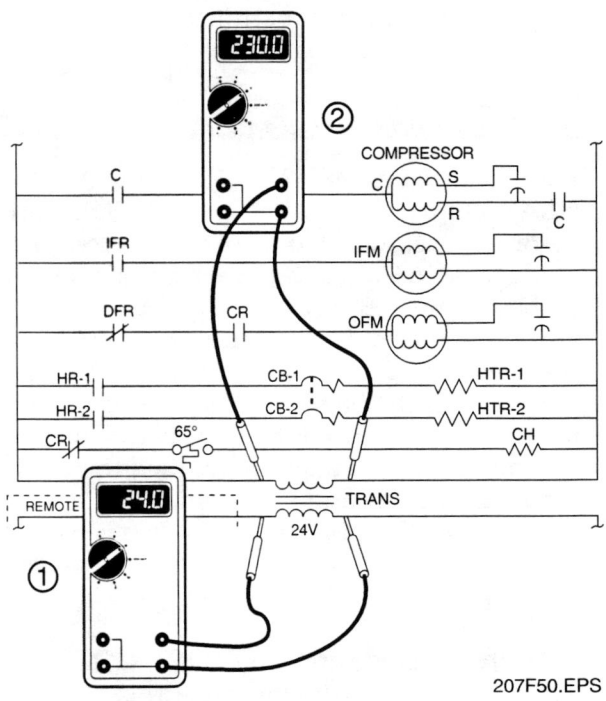

*Figure 50* ◆ Control transformer checks.

If no voltage or low voltage is measured across the primary winding, the power supply voltage to the equipment should be checked. If the power supply voltage is OK, troubleshoot the circuit wiring between the power supply and control transformer primary winding.

## 9.6.0 Thermostat Checks

Troubleshooting procedures for a typical non-electronic heating/cooling thermostat are covered here. They can easily be checked by using an insulated jumper wire connected across the thermostat's R (24V) terminal and fan (G), heat (W), or cool (Y) terminal, as applicable.

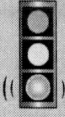

**NOTE**

Because of the diversity and complexity of electronic thermostat design, manufacturers normally provide troubleshooting aids in their service literature. Always follow the manufacturer's troubleshooting instructions.

### 9.6.1 Fan Switch Operation Checks

With the thermostat FAN switch set to the ON position, the indoor fan motor should be running. If not, connect the jumper wire across the thermostat's R and G terminals (*Figure 51*). If the indoor fan runs with the jumper in place, this proves that 24V control voltage is being applied to the thermostat; therefore, the thermostat FAN switch or related fan wiring is bad.

### 9.6.2 Cooling Operation Checks

With the thermostat FAN switch set to AUTO, the HEAT/OFF/COOL switch set to COOL, and the thermostat set to call for cooling, the compressor, outdoor fan motor, and indoor fan motor should all be running. If not, connect the jumper wire across the thermostat's R and Y terminals. If the compressor and outdoor fan motors run, the thermostat's HEAT/OFF/COOL switch, cooling thermostat (TC), or related wiring is bad. If the indoor fan motor fails to run, connect the jumper across the thermostat R and G terminals. If the indoor fan runs, the thermostat FAN switch is bad.

### 9.6.3 Heating Operation Checks

With the thermostat HEAT/OFF/COOL switch set to HEAT, and the thermostat set to call for heat, the furnace pilot should be ignited. If not, connect the jumper wire across the thermostat's R

| WIRE COLOR | | FUNCTION |
|---|---|---|
| R | RED | POWER |
| G | GREEN | FAN CONTROL |
| Y | YELLOW | COOLING CONTROL |
| W | WHITE | HEATING CONTROL |

THERMOSTAT WIRING CODES

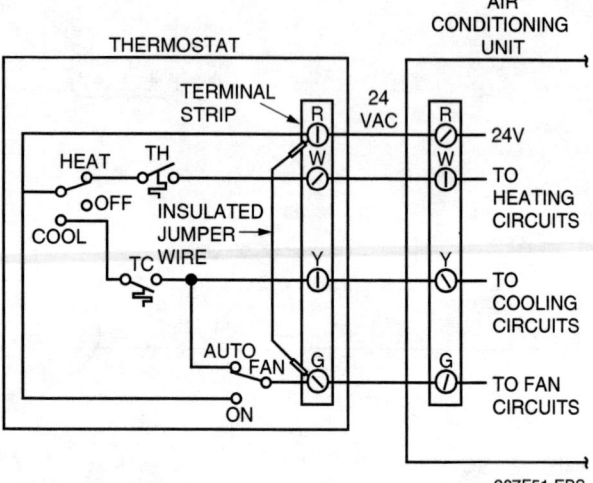

*Figure 51* ◆ Troubleshooting fan switch function of a heating/cooling thermostat.

and W terminals. If the furnace pilot ignites, the thermostat HEAT/OFF/COOL switch, heating thermostat (TH), or related wiring is at fault.

## 10.0.0 ◆ MOTORS AND MOTOR CIRCUIT TROUBLESHOOTING

The operation and uses of motors in HVAC equipment has been studied previously, but is reviewed briefly here. Five types of single-phase motors are commonly used in HVAC equipment: shaded-pole; split-phase; permanent split capacitor; capacitor start; and capacitor start, capacitor run. Permanent split capacitor (PSC) and capacitor start, capacitor run (CSR) motors are most often used in single-phase hermetic compressors because of their good running characteristics and high efficiency. Indoor and outdoor fan motors and blower motors are usually single- or multi-speed PSC motors. Use of electronically commutated motors (ECMs) is also on the increase in fan and blower motor applications. Shaded-pole motors are typically used in low-torque applications such as small direct-drive fan and blower motors.

Both PSC and CSR motors (*Figure 52*) have at least three external terminals leading to two internal windings. The main or run winding (R) contains relatively few turns of heavy wire. The start winding (S) contains a greater number of turns of lighter wire. The point where the two windings meet internally is called the *common (C)*. The arrangement of the motor windings used in both the PSC and CSR motors are the same. The configuration of the motor as a PSC or CSR motor is determined by the run and/or start circuit components used with the motor. The PSC motor has a run capacitor permanently connected across the run and start windings. The CSR uses an extra capacitor called a *start capacitor* to aid in starting. As shown, the start capacitor and the contacts of a start relay (SR) are connected in parallel with the run capacitor. When the motor is turned on and reaches about 75% of full speed, these contacts open and remove the start capacitor from the circuit. The start relay method of removing the start capacitor from the circuit is commonly used with hermetic and semi-hermetic compressor motors. A start relay can be used with all CSR motors; however, in non-compressor applications, a centrifugal switch is frequently used to disconnect the start capacitor from the circuit when the motor comes up to speed. Start capacitor failures are often the cause of compressor and other motor problems.

Multi-speed PSC motors (*Figure 53*) used to drive fans and/or blowers in HVAC equipment are capable of operating at two or more speeds. The motor's speed can be changed by switching the motor leads, terminal taps, or by the use of speed control switches or relays. In many heating/cooling units, the motor speed is selected automatically by the control circuits, as determined by the mode of operation. Normally, slower fan speeds are used with heating modes of operation, and higher speeds are used with cooling modes of operation. There are many types of multi-speed motors. As shown, the speed is changed by connecting the line voltage either to the low speed tap (LO), medium speed tap (MED), or high speed tap (HI) of the motor. The specific taps used are selected when installing the unit. A control relay with contacts located in the motor load circuit is normally used to prevent more than one motor winding speed tap from being energized at the same time, a condition that would destroy the motor.

Because multi-speed motors use tapped windings, series-connected winding sections, and/or other wiring configurations that enable operation at different speeds, they may fail in such a way that the motor will not run at one or more speed(s), but runs at other speed(s). When troubleshooting multi-speed motors, it is important to eliminate the speed selection circuits external to the motor as the cause of the problem before condemning the motor itself.

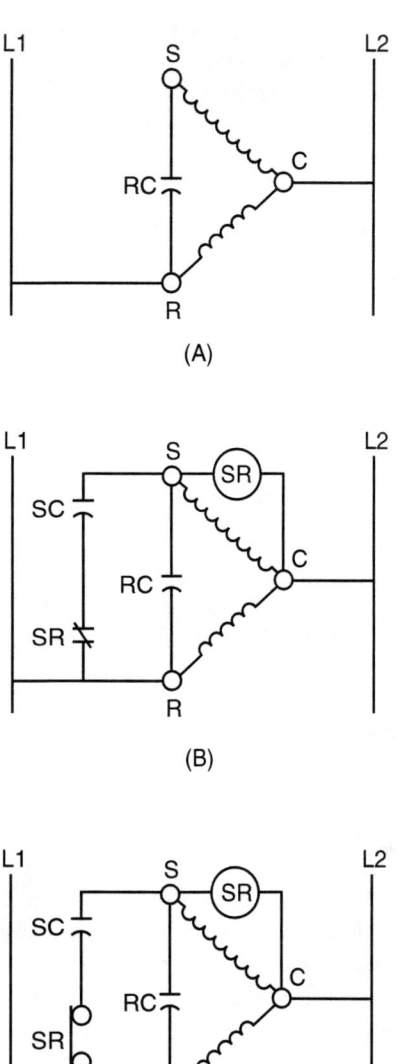

*Figure 52* ◆ PSC and CSR motors. (A) PSC. (B) CSR with start relay. (C) CSR with centrifugal switch.

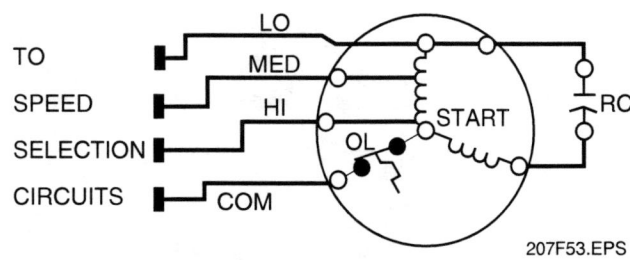

*Figure 53* ◆ PSC multi-speed motor.

Three-phase motors (*Figure 54*) are generally used when high starting torque is needed or when the motor requirements are greater than 7hp. All have at least three internal windings, with each winding having an equal resistance and the same number of wire turns. Six- and nine-lead, three-phase motors are also found in large applications where part winding start is necessary to reduce the initial inrush current at motor start-up. Three-phase motors have good starting and running characteristics and high efficiency. Three-phase motors require no external starting relays or capacitors.

### 10.1.0 Precautions for Motor Testing

In addition to the standard safety precautions that must be taken when working on electrical equipment, the following precautions must be adhered to when troubleshooting compressors and other motors.

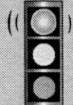

> **WARNING!**
>
> If damaged, the terminals of hermetic and semi-hermetic compressor motors have been known to blow out when disturbed in a pressurized system. To avoid injury, do not disconnect or connect wiring at the compressor terminals. When testing compressors, do not place test probes on the compressor terminals. Instead, use test points downstream from the compressor. To be safe, measurements and connecting/disconnecting wiring should only be done at the compressor terminals when the system pressure is at 0 psig.
>
> The capacitors used in motor circuits can hold a high-voltage charge after the system power is turned off. Always discharge capacitors before touching them.

### 10.2.0 Start and Run Capacitor Checks

The start and/or run circuits on single-phase motors use capacitors. Capacitors affect the wattage, amperage draw, torque, speed, and efficiency of a motor. Run capacitors are connected in the motor circuit at all times; therefore, they are referred to as *continuous-duty capacitors*. They are usually larger in physical size, but have lower capacitance ratings than start capacitors. Because run capacitors are in the circuit at all times, they are typically filled with a dielectric fluid that acts to dissipate heat. A shorted capacitor may provide a visual indication of its failure. The pop-out hole at the top of a start capacitor may appear bulged or blown. A run capacitor may be bulged and/or leaking. If a capacitor is found to be defective, always replace it with one specified by the manufacturer.

Testing capacitors is commonly done by making resistance checks using an analog multimeter (*Figure 55*). A capacitor analyzer can also be used, especially when it is necessary to measure the actual capacitance (MFD value) of a capacitor. A digital multimeter can be used to check for open and shorted capacitors and, if it has the capability, to measure the MFD value.

Capacitors are tested with the equipment power off, the capacitors discharged, and the capacitor(s) under test isolated from the remainder of the circuit. If testing a start capacitor, it is also recommended that one end of the bleeder resistor be disconnected. The multimeter is set to measure resistance on the $R \times 100\Omega$ or $R \times 1,000\Omega$ scale, then is connected across the capacitor terminals. If a capacitor is good, the multimeter will make a rapid swing toward zero and slowly return to infinity or some high value of measurable resistance. If the reading goes to zero or a low resistance and stays there, the capacitor is shorted. A reading of infinity indicates that the capacitor is open. In both cases, the capacitor must be replaced.

If testing a capacitor with a metal case, also check to see if the capacitor is grounded. Set up the multimeter to measure resistance on the $R \times 1,000\Omega$ or $R \times 10,000\Omega$ scale. Connect the multimeter between each one of the capacitor terminals and the metal case and measure the resistance. If the multimeter reads a measurable resistance, there is leakage to ground and the capacitor must be replaced.

If you must know the exact capacitance value of a capacitor, use a capacitor tester. Follow the tester manufacturer's instructions to perform the test. Typically, the measured MFD value for a start capacitor should be ±20% of the value shown on

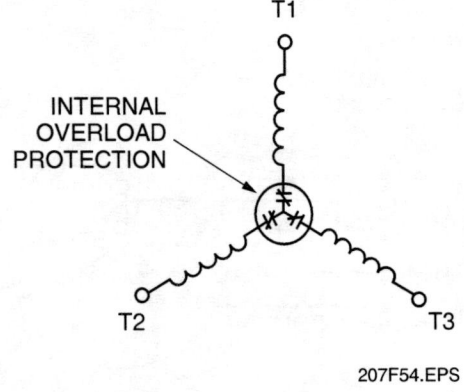

*Figure 54* ◆ Three-lead, wye-connected, single-voltage, three-phase motor.

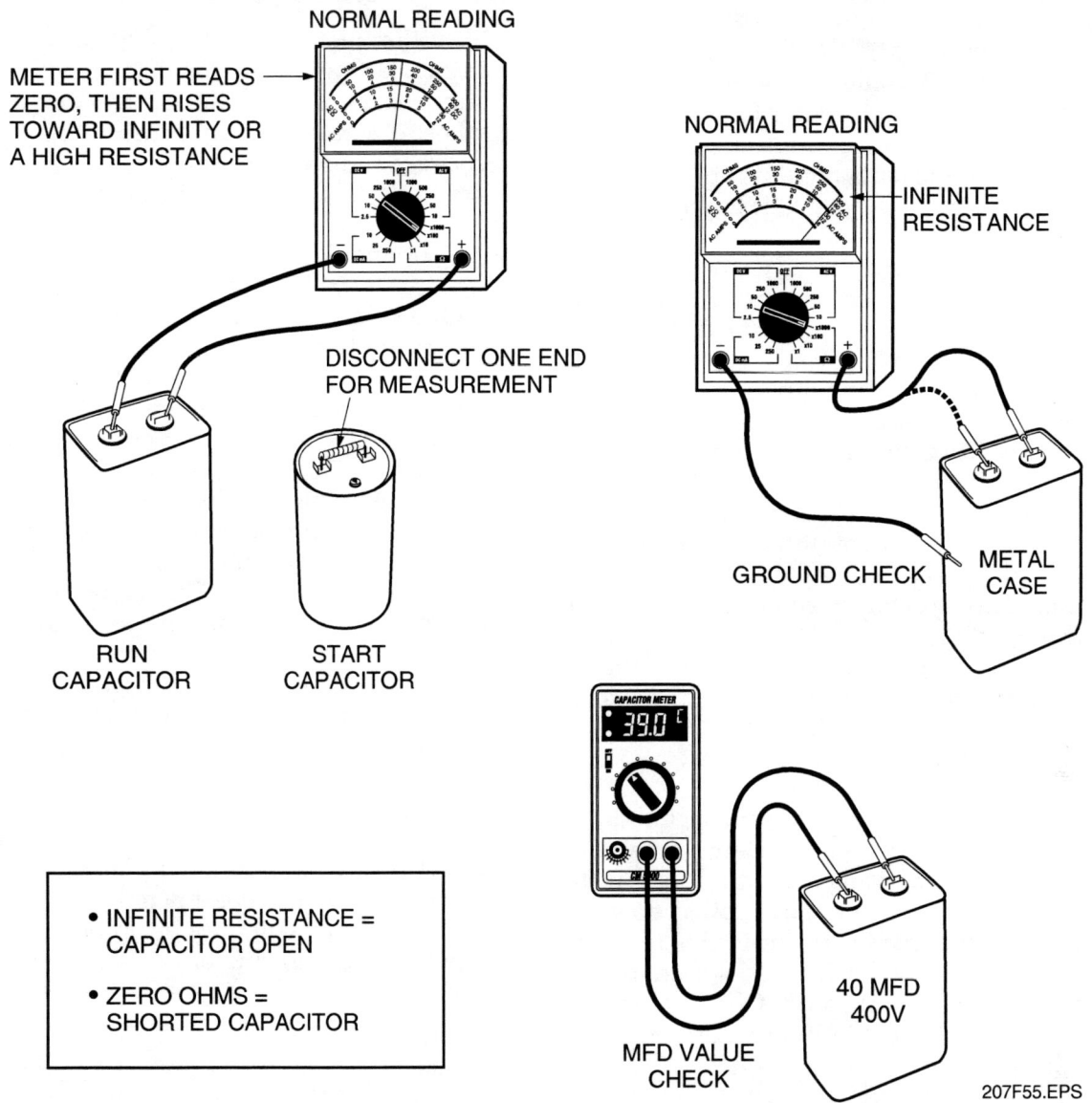

*Figure 55* ◆ Capacitor checks.

the capacitor label. If the measured value is outside the range of ±20%, replace the capacitor. For a run capacitor, the measured value should be ±10% of the value shown on the capacitor. If the measured value is outside the range of ±10%, replace the capacitor.

## 10.3.0 Start Relay Checks

The start relay is used to remove the start capacitor from the motor starting circuit when the motor reaches about 75% to 80% of its operating speed. Start relays are made that can be actuated by either current or voltage. The start relays used with HVAC equipment motors are normally voltage-actuated relays (potential relays); therefore, the remainder of this discussion will cover the testing of a voltage-actuated start relay.

Start relays tend to fail with their contacts closed. This results in the start capacitor remaining in the start circuit, causing the motor's start winding to overheat and fail. It may also result in failure of the start capacitor. When it is necessary to replace a start relay, an identical replacement must be used. Substitution of a relay with a different pickup voltage can cause damage to the start capacitor or motor start winding. Also, the replacement relay must be positioned and wired exactly as the original.

Start relays can be tested by measuring the motor start winding current with a clamp-on ammeter. The use of an analog clamp-on ammeter is recommended because it is easier to observe the current reading. If using a digital clamp-on ammeter, one with a MIN/MAX current capability must be used. Testing begins by first finding the full load amps (FLA) rating for the motor as marked on the compressor or motor nameplate. The clamp-on ammeter is then set up to measure AC current on a range scale that is higher than the motor FLA.

With the power turned off, the clamp-on ammeter jaws are placed around the wire that connects the motor start capacitor to the start relay contacts, as shown in *Figure 56*. Power is turned on to the unit while watching the clamp-on ammeter indication to observe the current flow in the start capacitor circuit as the motor starts.

When the start relay is operating properly, the clamp-on ammeter current indication should momentarily indicate current flow, then fall back to zero as the motor comes up to speed. This shows that the start relay is good because its contacts have opened.

If current continues to be read on the clamp-on ammeter after the motor is up to speed, the relay contacts are stuck closed. This means the relay is bad and should be replaced. The start capacitor and/or motor start winding can be damaged when the start relay contacts are stuck closed.

If no current is shown on the clamp-on ammeter, the relay contacts may be stuck open, the related start capacitor may have failed open, or the related wiring may be open. In this instance, the relay contacts can be checked for continuity with the unit power turned off. Contacts that are stuck open will measure infinite resistance.

## 10.4.0 Start Thermistor Checks

Start thermistors can be used to provide additional starting torque for PSC compressors. The start thermistor is a temperature-sensitive device whose electrical resistance changes as a result of a change in temperature. Positive temperature coefficient (PTC) thermistors increase their resistance with an increase in temperature. PTC thermistors are commonly used in the start circuits of PSC motors. *Figure 57* shows a PSC compressor motor with a PTC start thermistor. As shown, the PTC thermistor is placed across the run capacitor.

At room temperature, the PTC thermistor resistance is very low, about 25Ω to 50Ω. When the compressor is turned on, the application of voltage provides an initial surge of high current through the start winding, because the low resistance of the PTC thermistor is effectively bypassing (shorting) the run capacitor. This surge results in increased motor starting torque. The temperature increase created by the high current causes the PTC thermistor resistance to increase very rapidly to several thousand ohms, blocking current flow and effectively removing the thermistor from across the run capacitor. The motor then runs as a normal PSC motor. A small leakage current through the thermistor keeps the thermistor heated and its resistance high while the motor is running. Circuit operation remains this way until the motor is turned off. After a cool-down period, the thermistor's resistance will once again be the low value needed to start the motor.

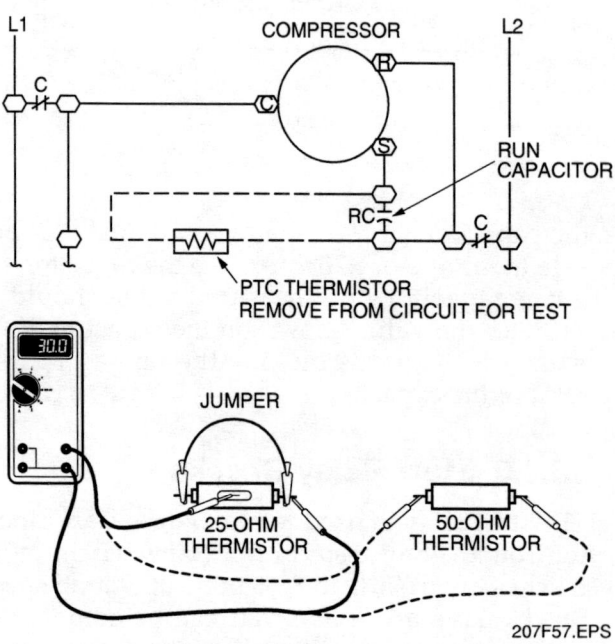

*Figure 57* ◆ Start thermistor check.

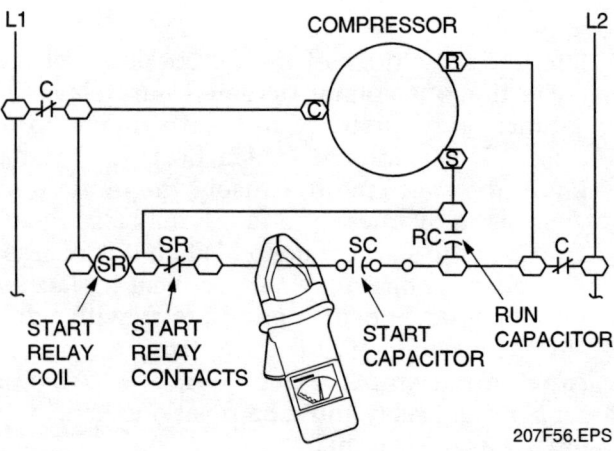

*Figure 56* ◆ Start relay check.

The thermistor is tested with the equipment power off, the capacitors discharged, and the thermistor under test isolated from the remainder of the circuit. Testing the thermistor is done by making a resistance measurement of the thermistor. Before attempting to measure the resistance, you should wait at least 10 minutes to allow the thermistor to cool to ambient temperature. The cold resistance of any PTC thermistor should be about 100% to 180% of the thermistor ohm rating. For example, a thermistor rated at 25Ω should have a cold resistance of 25Ω to 45Ω. If the PTC thermistor resistance is much lower or more than 200% higher than its rating, the thermistor should be replaced.

## 10.5.0 Identifying Unmarked Terminals of a PSC/CSR Motor

Sometimes the terminals on a single-phase motor are not marked, or are hard to identify. The terminals can be identified by using a multimeter to measure the resistance of the motor windings. First, the multimeter is used to find the two terminals across which the highest resistance is measured (*Figure 58*). These are the run (R) and start (S) terminals. The remaining terminal is the common (C) terminal. Next, put one lead of the multimeter on the common (C) terminal, and find which of the remaining terminals gives the highest resistance reading. This is the start winding (S) terminal. The remaining terminal is the run winding (R) terminal.

## 10.6.0 Open, Shorted, or Grounded Winding Checks

Motors are tested for open, shorted, and/or grounded windings with the equipment power off, the capacitors discharged, and the motor leads disconnected from all the related components, including the run and start capacitors and the start relay.

### 10.6.1 Open or Shorted Winding Checks

Testing for shorted or open windings is done by measuring the resistance of the windings with a multimeter. Be sure to use a multimeter that has an accurate low range (R × 1Ω) scale because the resistance of some undamaged windings can be as low as ½Ω. Perform the test with the multimeter set to measure resistance on the R × 1Ω scale. Make sure that the meter is zeroed. One lead of the multimeter is connected to one of the motor leads,

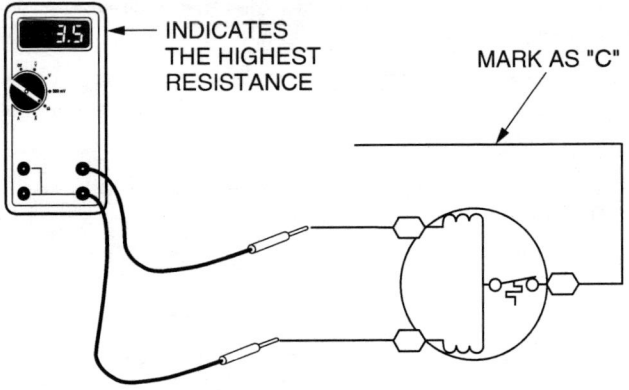

IDENTIFY THE COMMON (C) TERMINAL

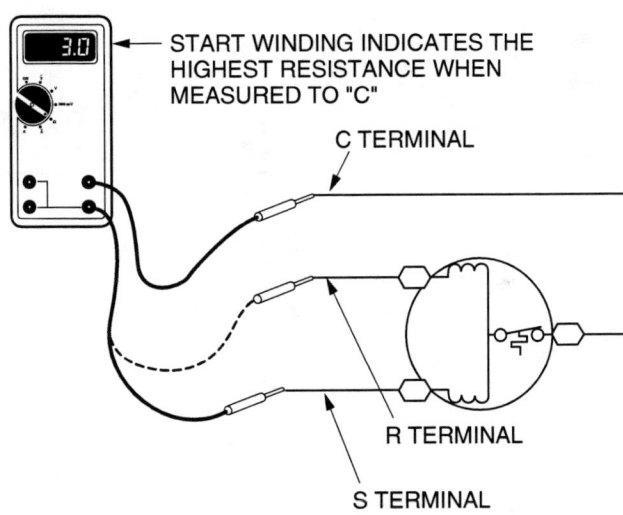

IDENTIFY THE START (S) AND RUN (R) TERMINALS

*Figure 58* ♦ Identifying unmarked terminals of a PSC/CSR motor.

as shown in *Figure 59*. Touch the other lead to the remaining motor leads, one lead at a time, and observe the meter indication. If the multimeter reads a measurable resistance, the windings are probably good.

If the multimeter reads zero resistance at one or more leads, the motor has a shorted winding; if it reads an infinite resistance, it has an open winding.

When checking a motor with an internal motor protection device, always make sure that the motor has had adequate time to cool off so that the protective device has time to reset. (This may take an hour or more.)

## Checking for Open or Shorted Motor Windings

Some rules of thumb commonly used to judge the condition of windings are as follows:

- If testing a single-phase PSC/CSR compressor motor, the resistance of the start winding is typically three to five times that of the run winding. For non-compressor motors, the resistances can vary widely depending on the design of the motor.
- When testing a single-phase multi-speed motor run winding, the highest resistance is normally measured between the common lead and the low (LO) speed lead, and the lowest resistance between the common lead and the high (HI) speed lead. The resistance measured between the common lead and the medium speed lead (MED) should be somewhere between that measured for the LO and HI leads.
- If testing a three-phase motor, the motor windings are usually judged to be good if the resistance measured across each winding is nearly identical to the other two windings.

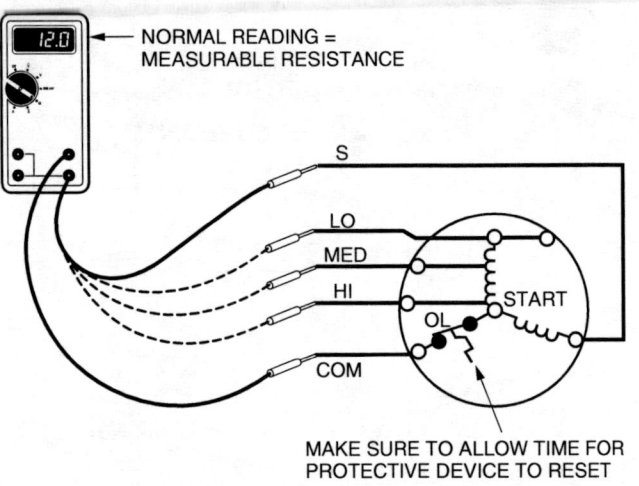

*Figure 59* ◆ Motor open or shorted winding check.

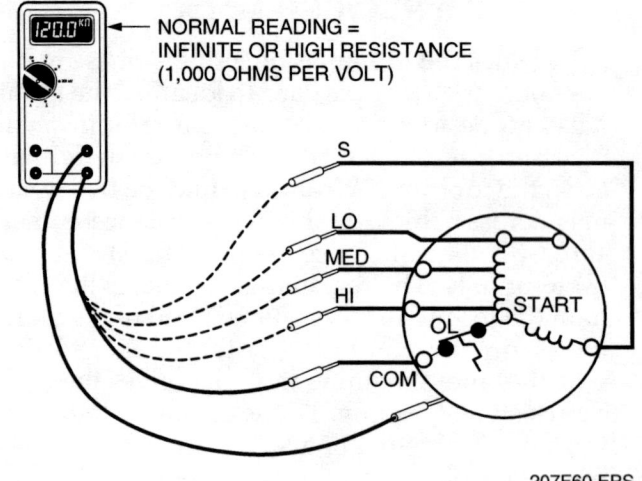

*Figure 60* ◆ Grounded winding check.

### 10.6.2 Grounded Winding Check

Testing a motor for grounded windings is done by measuring the resistance of the windings with a multimeter. Perform the test with the multimeter set up to measure resistance on the R × 10,000Ω scale. Connect one lead of the multimeter to a good ground connection, such as the motor or compressor frame or compressor discharge/suction line. Poor electrical contact because of a coat of paint, layer of dirt, or corrosion can cause an inaccurate measurement and hide a grounded winding. The other meter lead is then touched to each of the motor leads, one lead at a time, while watching the meter indication (*Figure 60*). An infinite or high resistance should be measured from each lead to ground. If a high resistance reading is indicated, it should not be less than 1,000Ω per volt. For example, on a 230V motor, the resistance should not be less than 230,000Ωs (230V × 1,000Ω/V = 230,000Ω). This indicates that the motor winding is not grounded.

If you measure a low or zero resistance, or a measurable resistance that is less than 1,000Ω per volt, this usually indicates that the motor winding is grounded. However, if testing a hermetic or semi-hermetic compressor motor, note that erroneous resistance readings to ground can be measured if liquid refrigerant is present in the compressor shell. In this instance, it is recommended that the refrigerant be recovered from the system, then the compressor be retested before condemning it.

## 11.0.0 ♦ HYDRONIC CONTROLS

In many parts of the United States, especially New England, the Middle Atlantic States, and the Upper Midwest, hot water or hydronic heat is very popular. With this type of heat, hot water (or in some cases, steam) is circulated through pipes to radiators in different parts of the building. Boilers are pressurized vessels and have unique safety and installation requirements. The nature of hydronic heat dictates specialized components and controls not found in other areas of HVAC.

A simple hydronic system, such as the one shown in *Figure 61*, contains a hot water boiler where the water is heated, a circulating pump to move the heated water throughout the system, a tank to absorb water as it expands when heated, a device called an *aquastat* that controls the water temperature, and a relief valve to bleed excess pressure so that boiler pressures do not reach explosive levels. The water may be heated by an oil or gas burner. These burners and many of the controls associated with them are similar to the burner controls on gas or oil-fired warm air furnaces.

### 11.1.0 Aquastat

The aquastat performs several important functions. In its simplest form, it is nothing more than a limit switch set to prevent water from exceeding a certain temperature. An aquastat, such as the one shown in *Figure 62(A)*, is usually mounted on the boiler and contains a temperature-sensing element that is inserted in a well in the boiler jacket. With such an aquastat, the water in the boiler will stay cold until there is a call for heat. The burner will then ignite and start to warm the water. The burner will operate until the room thermostat is satisfied or the water in the boiler reaches the high limit temperature setting.

A more complex aquastat, such as the one shown in *Figure 62(B)*, provides additional functions. In addition to a high limit setting, it contains a low limit setting that prevents the water in the boiler from getting too cold. This feature is necessary if domestic hot water is provided by a heat exchanger (called a *tankless coil*) that is inserted in the boiler jacket. This type of aquastat also contains a circulator control that prevents circulator operation until the water is warm enough to provide heat from the radiators.

Other specialized aquastats, especially those used on oil-fired boilers, have the oil burner primary control built into the aquastat. *Figure 63* shows how an aquastat with circulator and high and low limits would operate on an oil-fired boiler. If the water temperature is too low, contacts R-B remain closed, allowing the water to heat up but with no circulation. As the water heats up, R-W closes and R-B opens. If there is a call for heat, power from the switching relay feeds through the closed high limit switch to keep the burner running and through R-W to power the circulator pump.

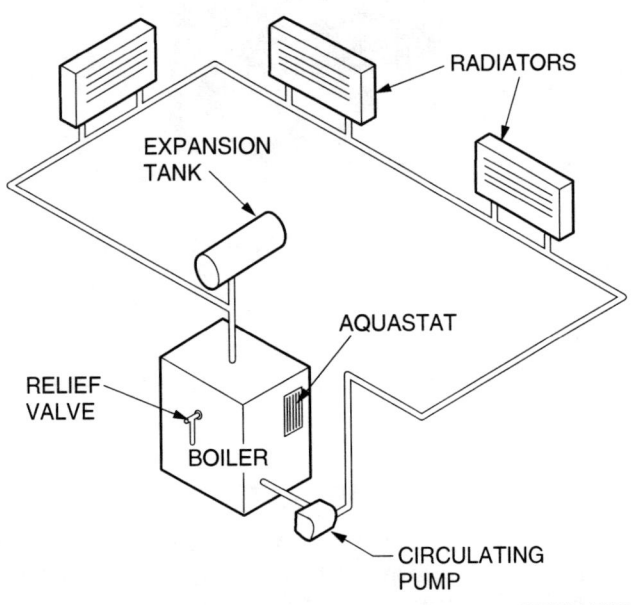

*Figure 61* ♦ Simple hydronic system.

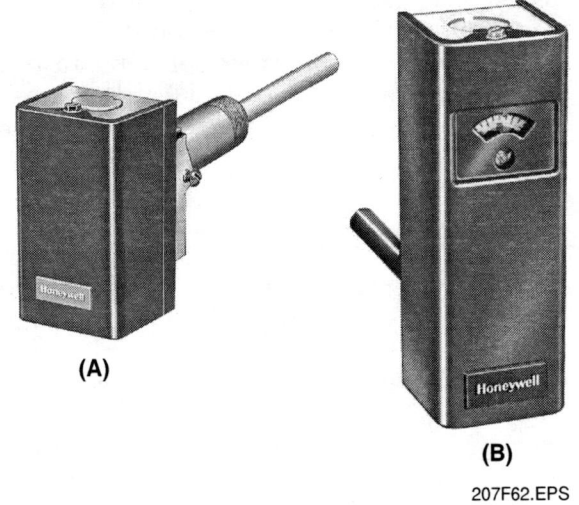

*Figure 62* ♦ Aquastat controls.

INTRODUCTION TO CONTROL CIRCUIT TROUBLESHOOTING — TRAINEE MODULE 03207

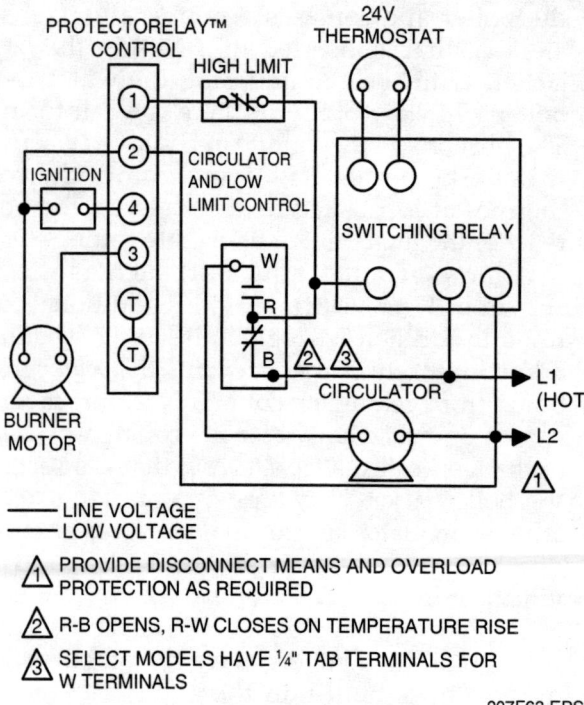

*Figure 63* ◆ Aquastat used for low limit and circulator control in an oil-fired hydronic system.

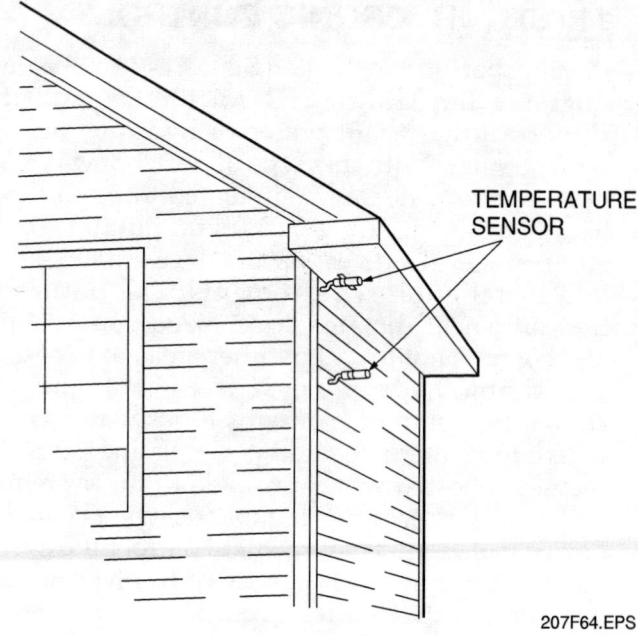

*Figure 64* ◆ Typical reset controller outside temperature sensor mounting.

## 11.2.0 Reset Controller

To achieve maximum comfort from a hydronic heating system, some systems incorporate an aquastat device called a *reset controller*. This device monitors the outdoor temperature and adjusts the boiler water temperature for maximum comfort and energy savings. It allows the boiler water temperature to rise on cold days to supply more heat but limits the boiler water temperature to lower levels on milder days when less heat is required. In addition to a temperature sensor in the boiler jacket, the reset controller contains an outside temperature sensor (*Figure 64*) that is typically mounted on the north side of the structure away from direct sunlight.

## 11.3.0 Low Water Cutoff

Loss of water in a boiler can have catastrophic results. The boiler may produce large amounts of steam, overheat, and build up explosive pressures. To prevent this, a low water cutoff device may be installed in the boiler. The simplest devices are nothing more than a float that activates a switch if the water level drops. The switch shuts off burner operation to prevent overheating and/or activates an alarm or warning light. Electronic versions (*Figure 65*) have a probe that is inserted in the boiler to monitor the water level.

*Figure 65* ◆ Low water cutoff control.

These devices also shut off burner operation and/or sound an alarm when the water level drops too far.

## 11.4.0 Circulator Pump

The circulator pump in a hydronic system performs a job similar to the blower in a forced air furnace. The major difference is that the pump moves water instead of air. Circulating pumps can be mounted on the floor or directly in the return water line. *Figure 66* shows a typical circulator pump. The pump motor is direct coupled to the pump's centrifugal impeller through a shaft seal in the impeller housing. Circulator operation can be controlled by the aquastat or by the zone valves (if so equipped). Most residential hydronic systems have one circulator pump. In some zoned installations, each zone has its own pump. Some older circulator pumps require periodic lubrication of the motor and pump, while newer models are sealed units with little or no maintenance required.

## 11.5.0 Zone Valves

Zoned control is easily accomplished with hydronic heating systems due to the simplicity of the zone valve. This device is simply an electrically operated valve. It is placed in the supply line to the zone and opens when the thermostat in the zone calls for heat. *Figure 67* shows a typical zone valve.

In residential applications, zone valves are powered by 24VAC. In commercial applications, higher voltages may be used. Most valves are motorized so that on a call for heating from the zone, the valve opens. Some valves are equipped with a switch that closes as the valve opens. This switch closure starts the circulator pump so that water can flow to the zone that is calling for heat. There are a number of ways to control burner and circulator operation on a gas or oil-fired boiler. Much of it depends on the aquastat used but the installer does have some control over the sequence of operation.

## 12.0.0 ♦ PNEUMATIC CONTROLS

Pneumatic control systems use compressed air to supply energy for the operation of valves, motors, relays, and other pneumatic control components. They are used primarily in commercial air conditioning systems, although they may sometimes be used in residential systems. Pneumatic control circuits consist of air piping, valves, orifices, and similar mechanical devices.

Thermostats control an air line; the air in the line can, in turn, operate the pneumatic motors, which in turn operate dampers, valves, and switches.

The following advantages are available through the use of pneumatic control systems, especially in commercial and industrial applications:

- Pneumatic equipment is adaptable to modulating operation, but two-position or positive operation can also be provided.
- A great variety of control sequences and combinations are available while using relatively simple pieces of equipment.
- Pneumatic equipment is relatively free of operating difficulties.
- It is very suitable for controlling explosion hazards.
- Installation and maintenance costs may be less than for electrical controls, particularly if building codes require the use of electrical conduit.

*Figure 66* ♦ Circulator pump.

*Figure 67* ♦ Typical zone valve.

## Wiring Diagram Exercise

Using the schematic diagram of the packaged heat pump shown on the follwoing page, answer the following questions:

1. Where would your voltmeter leads be placed to check for the correct incoming voltage to the unit?

2. You place voltmeter leads across terminals R and X of the low-voltage terminal strip and read 24V. What does this tell you?

3. During defrost, the defrost relay (DR) is energized. During a normal defrost, you place voltmeter leads across terminal 2 of the defrost relay and terminal X on the low-voltage terminal strip. What should the voltmeter read?

4. Assume a room thermostat is connected to the low-voltage terminal strip. Across which terminals on the low-voltage terminal strip would you have to place voltmeter leads to confirm that a call for second-stage heat was coming from the room thermostat? What voltage would you expect to measure?

5. During defrost, the defrost relay (DR) is energized. Knowing this, what happens to the outdoor fan motor during defrost?

6. You wish to know if the demand defrost control is receiving the correct input power. Across which terminals of the demand defrost control would you check for power and what is the voltage you would expect to read?

7. The compressor in this unit occasionally won't start and trips the power supply circuit breaker. You wish to check the compressor's current draw during startup. What test instrument would you use to do this and where in the circuit would you attach or apply the instrument?

8. The blower control (BC) has three power leads and requires two inputs to operate properly. What are the two input signals and at what points are they applied to the control?

9. If the compressor fails in this heat pump, it is still possible to heat the home. How is this done? At what terminals on the low-voltage terminal strip would you expect to measure a signal that would make this happen?

10. Since this unit energizes the defrost relay during heat pump defrost operation, what does this fact tell you about the state of the reversing valve solenoid (RVS) during cooling operation? Is the RVS energized or de-energized during the cooling mode?

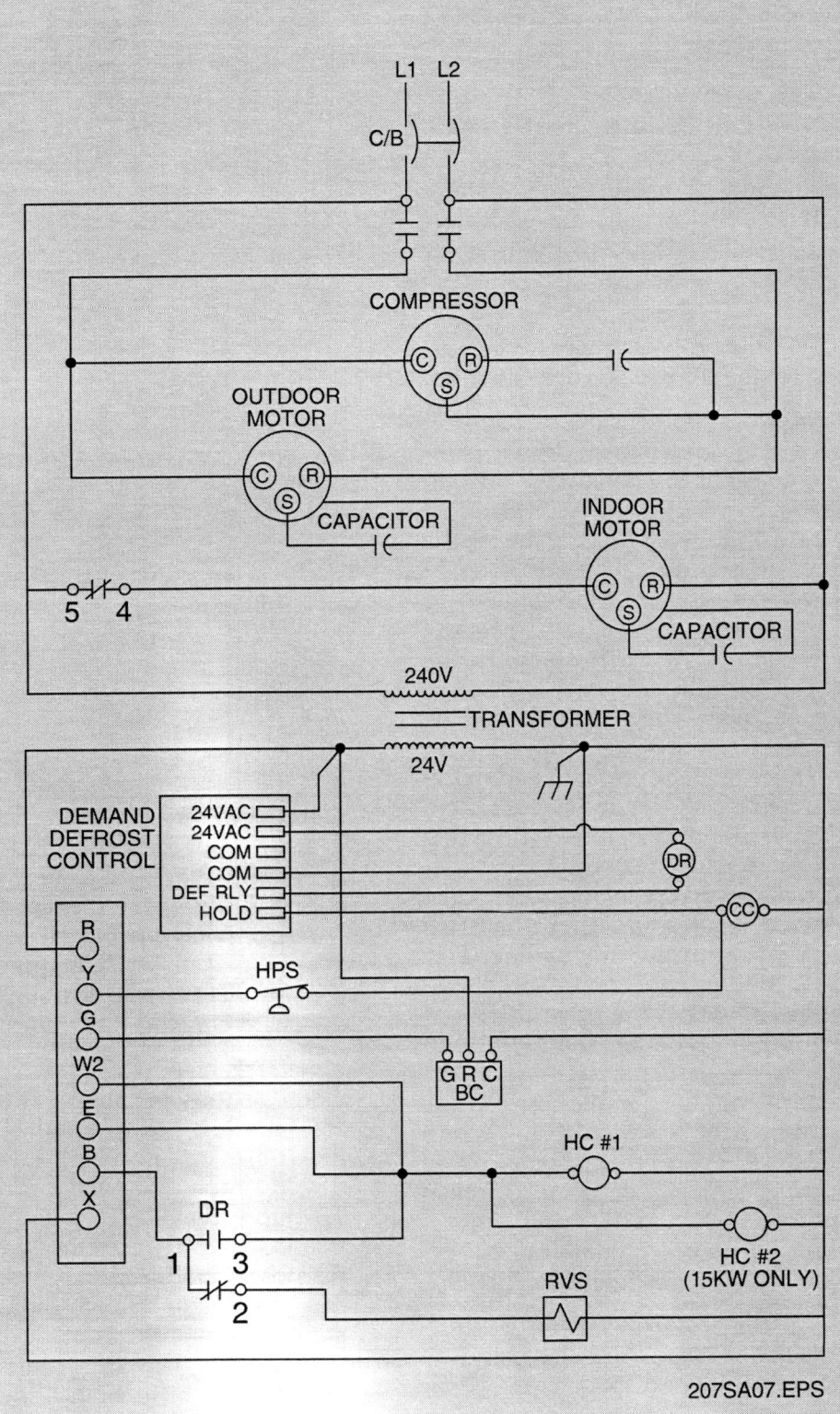

## 12.1.0 Basic Components

Pneumatic control systems (*Figure 68*) consist of the following components:

- A source of clean, dry, compressed air
- Air lines called *mains* deliver air from the supply to the controlling devices
- Controlling devices, such as thermostats, humidistats, relays, and switches or controllers
- Branch circuits or branch lines that deliver air from the controlling devices to the controlled devices
- Operators or **actuators**, such as valves and motors, which are the controlled components of the system

The air source is an electrically driven compressor that is connected to the storage tank in which the pressure is maintained between fixed limits (usually 20 to 30 psi). Air leaving the tank is filtered to remove oil and dust. In many installations, a small refrigeration unit is included to condense out any entrained moisture. Pressure-reducing valves control the air pressure to the main that feeds the controller (thermostat, etc.).

The controller regulates the positioning of the controlled device. It does this by taking air from the supply main at a constant pressure and delivering it through the branch line to the controlled component at a pressure that is varied according to the change in the measured condition.

For example, a change in the temperature of the conditioned space causes the thermostat to change the air pressure in the branch line. This change in pressure then causes the controlled component to move toward the open or closed position, depending on whether the room temperature has increased or decreased. When the valve of the controlled device moves toward the open position, more heat is added to the space; when the valve moves toward the closed position, less heat is added. A similar procedure applies to the cooling position.

There are four types of controllers or thermostats:

- A direct-acting thermostat increases the pressure of its branch line when the air temperature increases in the conditioned space.
- A reverse-acting thermostat decreases the pressure of its branch line when the air temperature increases in the conditioned space.
- A graduate thermometer gradually changes its branch line pressure when the air temperature changes in the space. This type of thermometer can maintain any pressure from 0 psi to 15 psi.

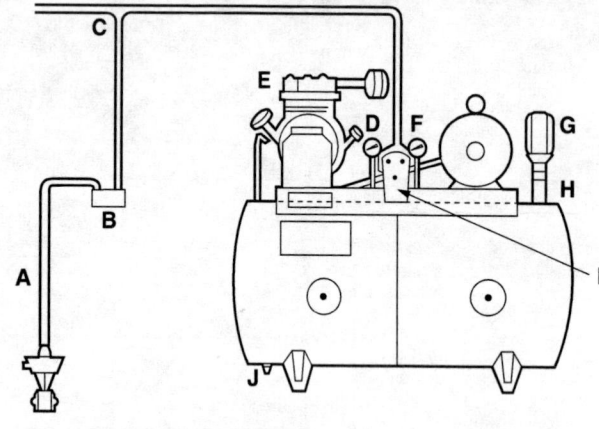

A. TO ACTUATOR
B. THERMOSTAT
C. TO OTHER PARTS OF SYSTEM
D. HIGH-PRESSURE GAUGE
E. COMPRESSOR UNIT
F. LOW-PRESSURE GAUGE
G. PRESSURE SWITCH
H. RELIEF VALVE
I. PRESSURE REGULATOR AND FILTER
J. DRAIN

*Figure 68* ♦ Pneumatic system.

- A positive-acting thermostat abruptly changes its branch line pressure when the room temperature changes. In this instance, the branch line pressure is either 0 psi or 15 psi. The thermostat does not maintain pressures between these values. It is only a two-position control, either fully open or fully closed. Humidity and pressure controllers operate in a similar manner.

A bleed thermostat control (*Figure 69*) has a bimetal element which reacts to temperature and positions the vane near or away from the air nozzle. Thus, the pressure in the branch is relative to how much air is bled off. **Bleed controls**, if used directly, do not provide a wide range of control. Therefore, they are frequently coupled to a relay that is separately furnished with air for activating purposes; the bleed thermostat simply controls the relay action. Bleed controls maintain a constant drain on the compressed air source.

Non-bleed thermostat controllers (*Figure 70*) use air only when the branch line pressure is being increased. The air pressure is regulated by system valves. The valves eliminate the constant bleeding of air that is present in the bleed thermostat. The exhaust and air main valves are controlled by the action of the bellows resulting from changes in space temperature. The exhaust or bleeding action is relatively small and occurs only on a pressure increase.

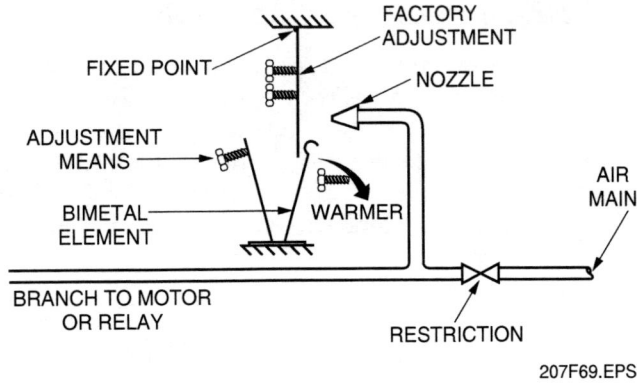

*Figure 69* ♦ Bleed-type thermostat.

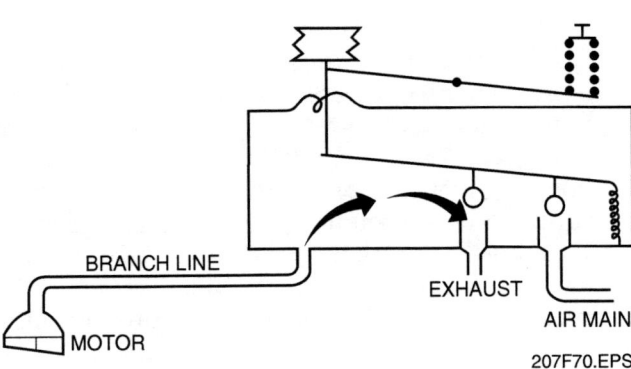

*Figure 70* ♦ Non-bleed thermostat.

Relays are installed between a controller and a controlled device. They are used to perform a function that cannot be accomplished by the controller. A diverting relay, for example, can supply branch air in one position and exhaust it in another position. It can also supply air to either of two branches without exhausting the other branch, or it can shut off the branch air without exhausting the branch.

The controlled devices (actuators) are mostly pneumatic damper motors and valves. The motor moves according to changes in the branch line pressure. A pneumatic motor contains one of three mechanisms: a bellows (*Figure 71*); a diaphragm; or a cylinder and piston. The damper linkage or valve stem is connected to the bellows and when the branch line pressure decreases, the bellows expand due to the internal spring pressure. As the branch line pressure increases, the bellows contract as a result of the increase in air pressure. The expansion and contraction of the bellows, piston, or diaphragm moves the linkage so that the damper opens and closes.

A normally open damper (*Figure 72*) is installed so that it moves toward the open position as the air pressure in the damper motor decreases. A normally closed damper (*Figure 73*) is a damper that is installed so that it moves toward the closed position when the air pressure in the damper motor decreases.

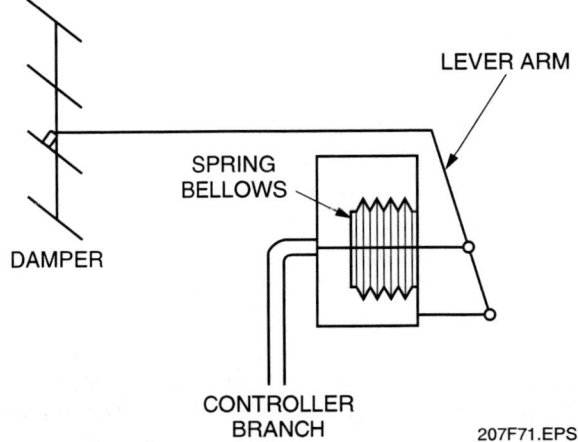

*Figure 71* ♦ Pneumatic actuator.

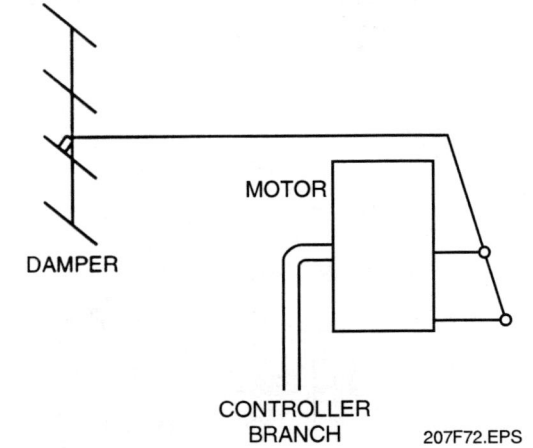

*Figure 72* ♦ Normally open damper.

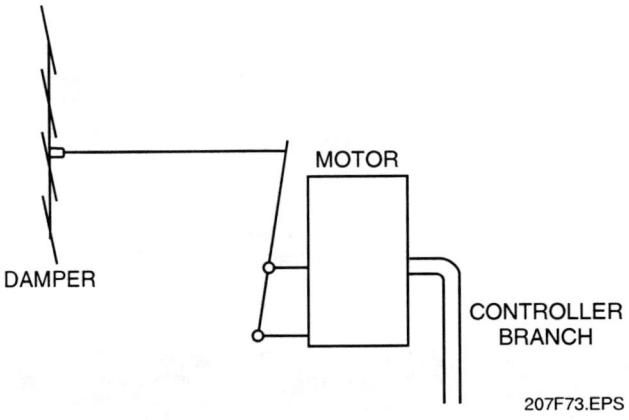

*Figure 73* ♦ Normally closed damper.

As the branch line pressure changes, the movement of the bellows actuates the lever arm or valve stem. The spring exerts an opposing force so that a balanced, controlled position can be stabilized. The motor arm can be linked to a variety of functions.

There will always be some crossover between the air devices and the electrical system. The devices most widely used for this process are **pneumatic-electric (P-E) relays** and **electric-pneumatic (E-P) relays**.

P-E relays are simply pressure switches in which a pneumatic signal causes an electrical change. P-Es are used to turn on electrical devices such as pumps and electric heaters. All P-Es have switch differentials (the pressure change that causes the switch to make and break).

E-P relays are three-way solenoid valves (see *Figure 74*). They are generally used as interlocking devices where a circuit is enabled under certain conditions which are signaled electrically.

When the E-P is de-energized, the common and normally open ports are connected, and the normally closed port is blocked. When the E-P is energized, the common port is connected with the normally closed port, and the normally open port is blocked.

## 12.2.0 Pneumatic Control System

The components of a typical modulating air conditioning system are shown in *Figure 74*. The system controls both the room temperature and the temperature of the air as it enters the room. It also provides adjustment of the quantity of outside air used for ventilation. This system may be used in any structure that has a demand for good air circulation to all areas.

Solid-state electronics have made possible the accurate modulation of temperature, humidity, and/or air volume. The modulator is usually controlled by a thermostat, a remote bulb and bellows or diaphragm, and/or pressurestats or humidistats. Most modulating systems have continuously running blowers.

These controls are usually not used for residential and small commercial applications. They are more often used for large commercial or industrial buildings. They are required for large structures having complex air handling needs. Control of the volume, temperature, and relative humidity of air supplied to the conditioned space of large buildings must be regulated to suit the space load and varying occupant requirements. The regulating system is generally composed of proportioning

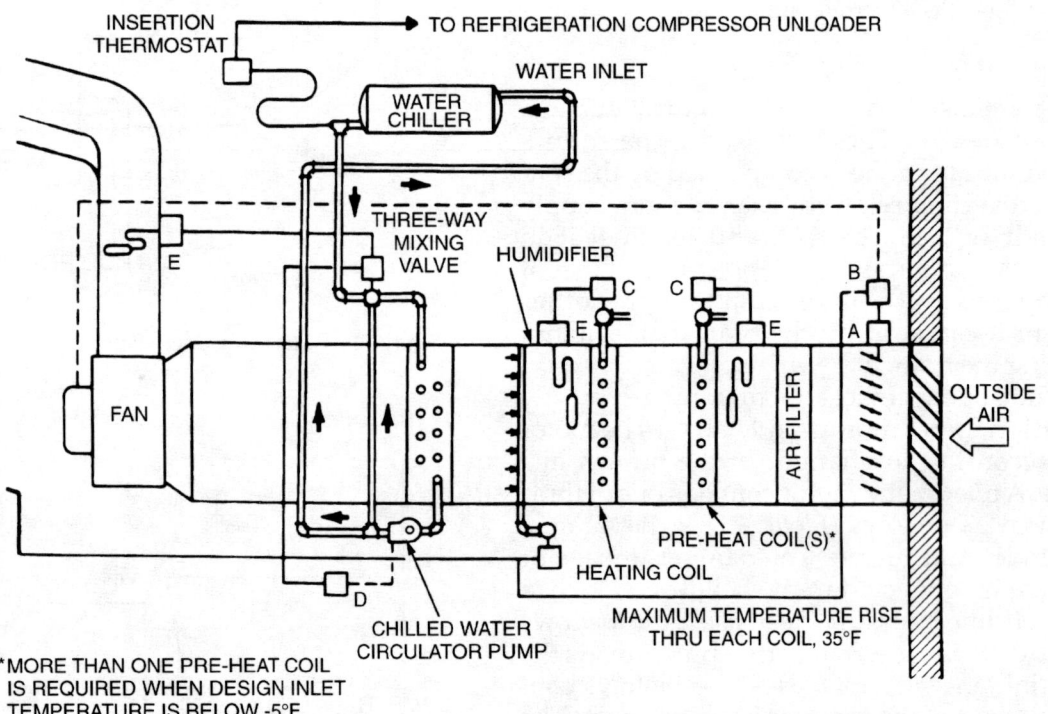

*Figure 74* ◆ Modulating system with E-P relay.

thermostats, supply and return air controllers, damper and valve motors, plus other controls to start and stop fans and to position air dampers.

Modulating controls contain potentiometers that provide electrical signals to actuators when the temperature, humidity, air volume, or air pressure being controlled deviates from the setpoint. The strength of the electrical signal is directly proportional to the amount of deviation. In order to make a correction, the controller signals the actuator to assume a new position to correct the deviation. Modulating controls operate on low voltages (usually 24V).

Modulating systems have been developed to fit the machine capacity very closely to the heating or cooling loads.

For example, in a cooling application, two or more compressors that are connected in parallel may be used. Each compressor is operated by a motor control. During operation, if the load increases and the temperature starts to rise, one compressor will begin to run. If the temperature keeps rising, the second compressor will start to operate. Additional compressors may cut in until enough capacity is obtained. The control contains a special switching device that rotates the service of the various compressors so that each compressor will be used about the same amount of time. Some modulating refrigeration systems may use a multiple cylinder compressor, with each cylinder equipped with an unloader device. Variable-speed motors are also used to provide a modulated cooling capacity.

## 12.3.0 Airflow Control

An airflow switch (*Figure 75*), also known as a *sail switch*, is often installed in ductwork as a safety device. It is used in electric heating systems to guarantee that air is circulating through the air distribution system before the heating elements are turned on. It is also used in cooling systems to verify that there is air flowing across the condenser and evaporator before the compressor is turned on and to activate an electronic air cleaner.

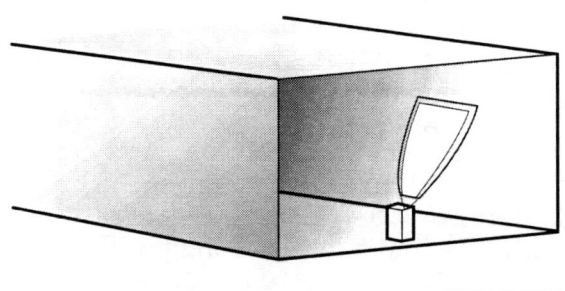

*Figure 75* ♦ Sail switch.

## 13.0.0 ♦ HVAC DIGITAL CONTROL SYSTEMS

The increasing size of modern buildings and building complexes, and the difficulty of monitoring and controlling the operation of up to several hundred elements of the heating, ventilating, and air conditioning systems has led to a revolution in the way manufacturers design HVAC controls. With this revolution in controls, occupants are enjoying a very high level of comfort and control. The microprocessor has led to this revolution in today's HVAC control systems.

The microprocessor is a high-speed instruction executer. You give it a list of things to do (instructions) and it simply executes or carries out those instructions over and over.

A goal of today's HVAC control designs is to build an integrated building system that is able to collect information and process that information to achieve maximum building comfort, energy usage, indoor air quality, and building safety and security.

As *Figure 76* illustrates, controlling the HVAC equipment is only one aspect of today's control systems, but it is an important one. When correctly integrated, it will prolong equipment life, increase customer satisfaction, aid in the repair of the equipment, and sometimes even communicate a problem before it can damage the equipment.

## 13.1.0 Direct Digital Control

HVAC, lighting, and other building systems controlled by traditional pneumatics, timers, switches, and thermostats may appear to function well. But slow response, calibration shifts, mechanical wear, and the inability to coordinate operation with other systems results in wasted energy and an inconsistent level of comfort.

Direct digital control (DDC), on the other hand, connected to a central computer, can integrate all building systems. It can perform complex control sequences automatically, improving the building environment and lowering costs at the same time. The greatest strength of DDC is its ability to communicate with the real world and to understand and sort the signals it receives. This communication helps it to reject obviously erroneous signals that older controllers were unable to detect.

Great care goes into the design of the controller input/output (I/O) function of the DDC system because it is like a nerve system to the controller (*Figure 77*). It receives all of the signals from the system and directs them to the microprocessor to be used based on the instructions stored within its memory.

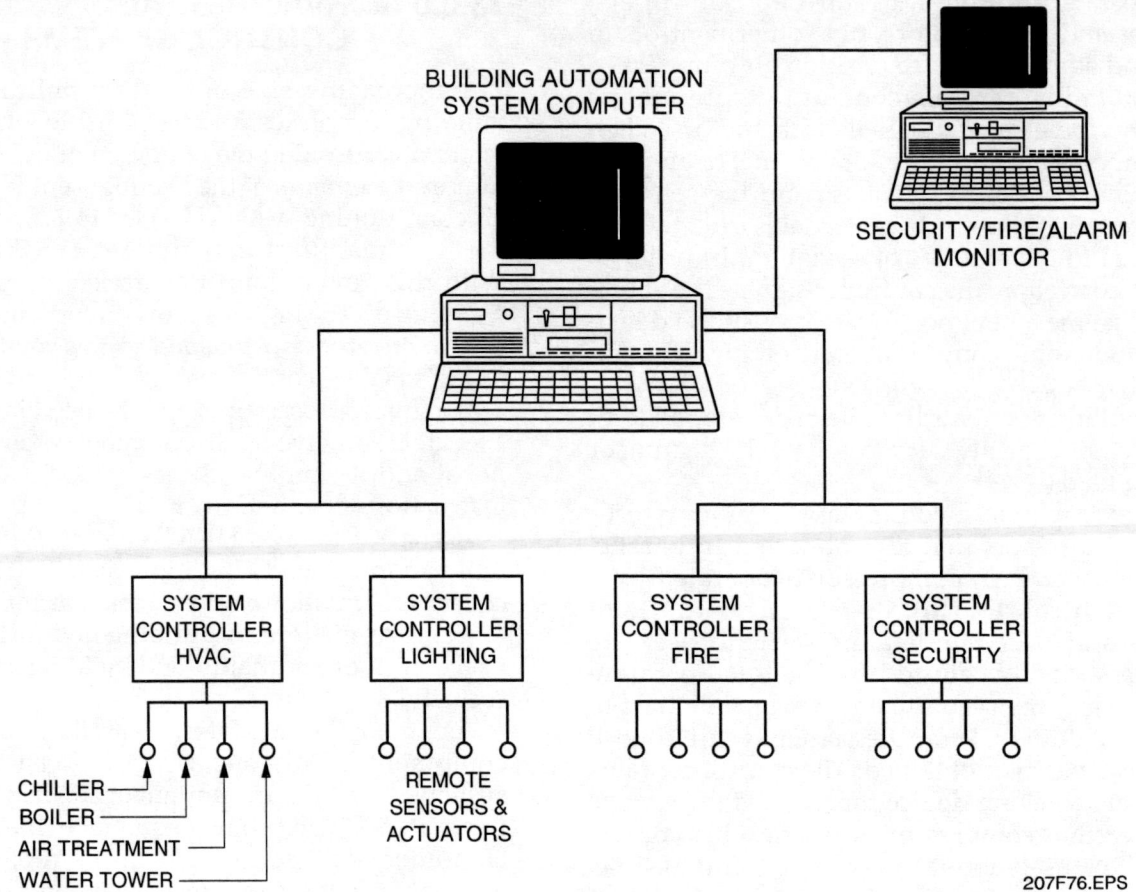

*Figure 76* ◆ Centralized building management system.

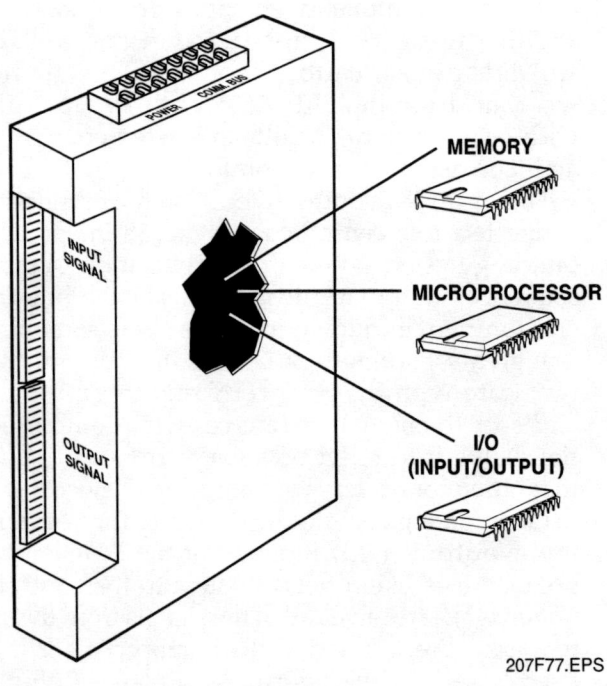

*Figure 77* ◆ System controller module.

## 13.2.0 Controlling Devices

All control systems require some means of communication between devices. This is accomplished by two basic methods: digital signals and analog signals (*Figure 78*). In the HVAC industry, these signals are called *points*.

The first category of control devices includes external digital devices. These are things such as relays, switches, lights, and other devices that can be operated in either a full on or full off (binary) condition.

The second category includes external analog devices. The types of devices in this group include thermistors, photocells, and DC motor controls. The problem of interfacing to an analog device is somewhat more complicated. In this case, we are attempting to take a signal with an infinite number of values (analog) and convert it to a form that can be represented and manipulated by two-state (binary) devices (digital). Most real-world processes are continuously changing (*Figure 79*).

Physical quantities such as pressure, temperature, liquid levels, and fluid flow tend to change

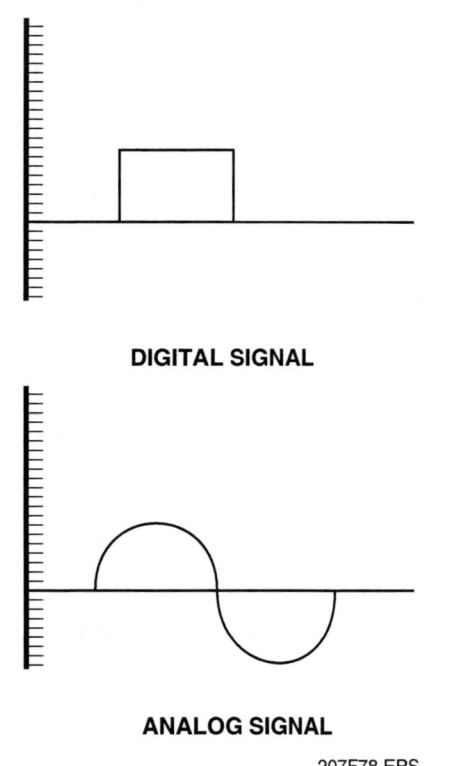

*Figure 78* ◆ Digital vs. analog signal.

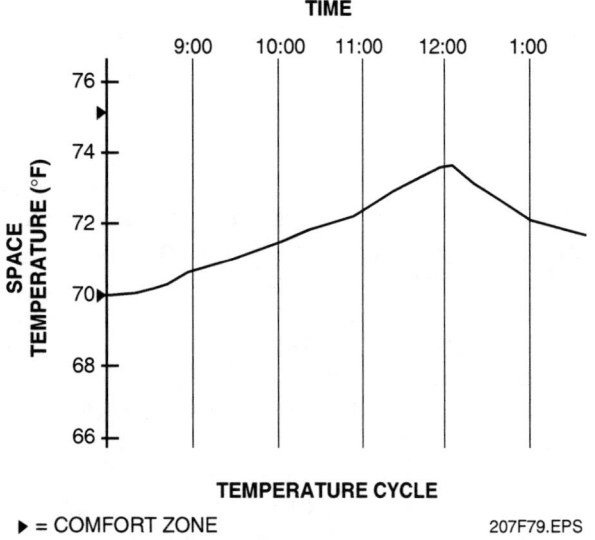

*Figure 79* ◆ Changes are small and occur gradually.

value rather gradually. Changes of this type produce a large number of discrete values before ever reaching a final state. Converting from analog form to digital form requires the use of a circuit called an **analog-to-digital converter**.

The analog and digital signals can be further divided into inputs and outputs. Analog In (AI) signals are from sensors such as temperature, pressure, and humidity. Analog Out (AO) signals are analog commands, such as reset of system setpoints. Digital In (DI) signals are contact closures or openings, showing status or alarm conditions in a two-position mode. Digital Out (DO) signals are two-position commands like start/stop or open/close states.

### 13.3.0 Example of a Digital Control System

The key to a successful control system is the integration of all the sensors and unit controllers. Let's look at the integration of a typical constant volume HVAC system. A constant air volume system is one in which the volume of supply air remains constant and the temperature of the air is varied to achieve the desired comfort conditions. It is a single-zone system. If there is more than one zone, each must have its own dedicated unit. The constant air volume system contrasts with a variable air volume system in which the supply air temperature is held constant and the volume of air changes to meet the changing demand. In a variable air volume system, a single unit can serve several zones.

A schematic of a constant volume system is shown in *Figure 80*. The control system is made up of several control loops: economizer control of mixed air; heating-cooling sequencing; and humidification-dehumidification sequencing.

When the unit fan is energized, as sensed by a static pressure sensor in the supply duct, the damper control system becomes activated. A mixed-air sensor maintains the mixed-air temperature by modulating the outdoor air, return air, and exhaust dampers. When the outdoor air temperature exceeds the setting of the outdoor air sensor, the outdoor and exhaust air dampers return to the minimum open position, as programmed, to provide ventilation. The return air damper takes the corresponding open position. In large buildings, it is often required that some outside ventilation air be provided at all times. Therefore, the damper has a minimum open position.

A space temperature sensor, also through the controller, maintains the space temperature by modulating the heating coil valve in sequence with the chilled water coil valve. A space humidity sensor, also through the controller, maintains space humidity. Upon a drop in space relative humidity, the humidifier steam valve modulates toward the open position, subject to a duct-mounted high-limit humidity sensor. With a rise in space relative humidity, the humidifier steam valve modulates to the closed position, followed

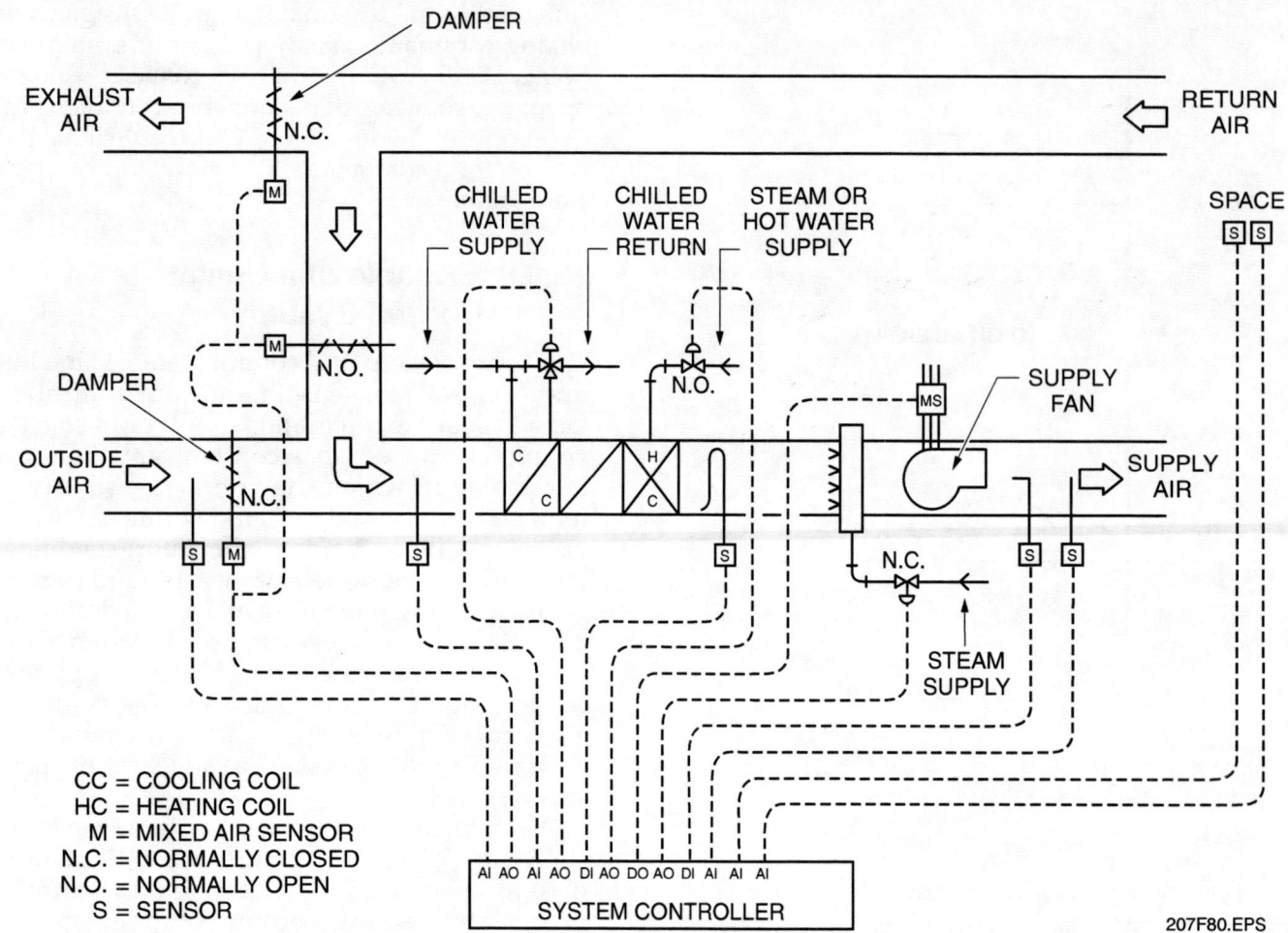

*Figure 80* ◆ Typical constant volume HVAC system.

by the opening of the chilled water coil valve to provide dehumidification. During the dehumidification cycle, the space temperature sensor modulates the heating coil valve to maintain space temperature conditions.

A low-temperature controller, with its capillary located on the discharge side of the heating coil, will de-energize the unit fan, close the outdoor and exhaust dampers, and open the return damper if the discharge air temperature drops below its setting. Whenever the unit fan is de-energized, as sensed by the supply duct static pressure sensor, the damper control system will be de-energized, closing the outdoor and exhaust dampers, along with the humidifier steam control valve.

## Summary

HVAC control systems are made up of a variety of electrical, electronic, and pneumatic controls. These controls perform two basic functions: they automatically turn functions on and off, and they protect the system from damage.

The ability to analyze HVAC control systems is a critical skill for the service technician because a large percentage of the problems that occur in HVAC systems are control circuit faults. No matter how complex the control system, it consists of individual components that are combined to form control functions such as heating, cooling, fan control, and defrost. If you know how the components work, you can figure out how they fit together and how the system functions as a whole. Once you have these skills, you can troubleshoot any system.

Effective troubleshooting is a process by which the HVAC technician listens to a customer's complaint, performs an independent analysis of a problem, and then initiates and performs a systematic step-by-step approach to troubleshooting that results in the correction of the problem. The HVAC technician must understand the purpose and principles of operation of each component in the equipment being serviced. You must be able to tell whether or not a given device is functioning properly and to recognize the symptoms arising from the improper operation of any part of the equipment.

## Review Questions

1. Low-voltage (24V) control circuits are used in HVAC systems because _____.
   a. step-down transformers produce only 24V output
   b. 24V power is safer and cheaper
   c. 24V power is readily available from the electric company
   d. it is the same voltage used to operate the compressor and fan motors

2. For what purpose is a heat anticipator used in heating thermostats?
   a. It turns on the heat just before the temperature reaches the thermostat setpoint.
   b. It keeps the system warm in cold weather.
   c. It turns on a light to let the occupants know when the furnace is about to come on.
   d. It opens the thermostat just before the heat in the conditioned space reaches the thermostat setpoint.

3. Which of the following is *not* true regarding a cooling compensator?
   a. It compensates for the lag between the thermostat call for cooling and the time when the system actually begins cooling the conditioned space.
   b. It is a variable resistor.
   c. It causes the cooling thermostat to close sooner than it otherwise would.
   d. It is usually placed in parallel with the cooling thermostat switch.

4. The differential in a thermostat is _____.
   a. the difference between the cut-in and cut-out points of the thermostat
   b. the difference between the cooling and heating setpoints
   c. normally at least 3°F
   d. the difference between the settings of the heat anticipator and the cooling compensator

5. Which of the following is *not* a common feature on a programmable electronic thermostat?
   a. Battery backup to prevent loss of programs
   b. Motor-driven heat anticipator
   c. Program override
   d. Digital readouts

6. A good location for a thermostat is _____.
   a. anywhere that it will receive direct sunlight
   b. about five feet above the floor on an outside wall
   c. about five feet above the floor on an inside wall
   d. in a corner where there is minimum air circulation

7. In a standard thermostat wiring scheme, the R terminal would be connected to _____.
   a. the fan control
   b. 24V power
   c. the reversing valve
   d. the cooling circuits

8. The correct current draw setting for a thermostat heat anticipator may be obtained from _____.
   a. the thermostat label
   b. the wallplate
   c. measurements taken across the thermostat R and G terminals
   d. the thermostat literature

9. The purpose of a motor starter is to _____.
   a. jump-start the compressor if the power fails
   b. vary the speed of a motor as the conditions change
   c. check the motor current to make sure the motor is running
   d. control power application to the motor and provide motor overload protection

10. The purpose of a lockout relay is to _____.
    a. make sure no one turns on the power when you want it off
    b. make sure the compressor doesn't turn on until the fans are running
    c. protect the compressor motor from start-up surges
    d. keep a motor from operating out of sequence when an automatic trip device resets itself

11. The purpose of a compressor short-cycle timer is to _____.
    a. prevent the compressor from restarting before system pressures can equalize
    b. make sure the compressor runs for at least five minutes once it cycles on
    c. keep track of how often the compressor turns on
    d. run the compressor every five minutes to make sure it doesn't freeze up

12. A limit control is provided in a furnace to _____.
    a. make sure the furnace puts out as much heat as possible
    b. control the amount of heat the furnace generates
    c. prevent the furnace from overheating
    d. sense the presence of carbon monoxide

13. In the electrical control circuit of a combined heating/cooling system, the outdoor fan motor _____.
    a. usually has a continuous-on mode for ventilation
    b. runs at low speed in the heating mode
    c. is on whenever the compressor is on
    d. is off whenever the compressor is off

14. Listening to the customer's complaint does *not* provide _____.
    a. valuable information about the nature of an equipment problem
    b. information on what the unit should do
    c. information that may eliminate the HVAC equipment as the cause of the problem
    d. all of the information needed to make an accurate problem diagnosis

15. A troubleshooting aid normally given on a label diagram is the _____.
    a. troubleshooting tree
    b. troubleshooting table
    c. wiring diagram
    d. fault isolation diagram

16. When troubleshooting an electrical problem by the process of elimination, which statement is *not* true?
    a. If the problem is only with heating in a combined heating/cooling unit, the cooling components can be eliminated.
    b. If a component operates properly, it is probably not the problem.
    c. Areas of the diagram that show optional equipment that is not installed can be eliminated.
    d. If an outdoor fan motor runs but a related compressor does not, the compressor control circuit can be eliminated.

17. When troubleshooting an energized compressor load circuit, the multimeter indicates voltage on both sides of a set of switch contacts. This indicates that the switch contacts are _____.
    a. open
    b. bad
    c. closed
    d. designated as normally closed (NC) contacts

18. When troubleshooting in an energized fan motor load circuit, the multimeter indicates voltage on one side of a set of relay contacts but not on the other. This indicates that the _____.
    a. relay contacts are shorted
    b. relay contacts are closed
    c. relay contacts are open
    d. related relay coil is energized

19. The power supply voltage measured across the output terminals of a two-pole circuit breaker is lower than was measured across the input terminals of the same breaker. The problem can be _____.
    a. loose wires and/or corroded circuit breaker terminals
    b. the breaker has been subjected to extremely cold temperatures
    c. there is no problem
    d. the breaker is set to OFF

20. When testing a relay coil, the multimeter indicates a measurable resistance. This means that the coil is _____.
    a. shorted
    b. probably good
    c. open
    d. partially shorted

21. When troubleshooting the operation of a compressor start relay circuit by measuring the start winding current draw with a clamp-on ammeter, no current is measured on the clamp-on ammeter when the compressor is started. This means that the _____.
    a. start relay contacts are stuck open
    b. start relay contacts are stuck closed
    c. start relay coil is open
    d. start capacitor is shorted

22. On a single-phase PSC compressor, the highest resistance should be measured between _____.
    a. the common and start terminals
    b. the run and start terminals
    c. the common and run terminals
    d. depends on the winding characteristics of the PSC compressor being measured

23. When checking a 440V motor for grounded windings, the minimum resistance that should be measured between any lead and ground is _____.
    a. 1,000Ω
    b. 230,000Ω
    c. 440,000Ω
    d. infinity

24. A pneumatic-electric relay is one that _____.
    a. uses air pressure to open and close electrical relay contacts
    b. uses a solenoid to close a pneumatic valve
    c. supplies power to the air compressor
    d. uses hydraulic fluid pressure to open and close relay contacts

25. The form in which temperature information is likely to be received by a computer-controlled system is _____.
    a. binary
    b. analog
    c. digital
    d. audio

# GLOSSARY

## Trade Terms Introduced in this Module

*Actuator:* The portion of a regulating valve that converts one type of energy, such as pneumatic pressure, into mechanical energy (for example, opening or closing a valve).

*Analog-to-digital converter:* A device designed to convert analog signals such as temperature and humidity to a digital form that can be processed by logic circuits.

*Automatic changeover thermostat:* A thermostat that automatically selects heating or cooling.

*Bleed control:* A valve with a tiny opening that permits a small amount of fluid to pass.

*Cooling compensator:* A fixed resistor installed in a thermostat to act as a cooling anticipator.

*Deadband:* A temperature band, usually 3°F, that separates heating and cooling in an automatic changeover thermostat.

*Differential:* The difference between the cut-in and cut-out points of a thermostat.

*Droop:* A mechanical condition caused by heat that affects the accuracy of a bimetal thermostat.

*Electric-pneumatic (E-P) relay:* A three-way pneumatic solenoid valve.

*Fault isolation diagram:* A troubleshooting aid usually contained in the manufacturer's Installation, Start-Up, and Service Instructions for a particular product. Fault isolation diagrams are also called *troubleshooting trees*. Normally, fault isolation diagrams begin with a failure symptom then guide the technician through a logical decision-action process to isolate the cause of the failure.

*Invar®:* An alloy of steel containing 36% nickel. It is one of the two metals in a bimetal device.

*Label diagram:* A troubleshooting aid usually placed in a convenient location inside the equipment. It normally depicts a wiring diagram, a component arrangement diagram, a legend, and notes pertaining to the equipment.

*Ladder diagram:* A troubleshooting aid that depicts a simplified schematic diagram of the equipment. The load lines are arranged like the rungs of a ladder between vertical lines representing the voltage source. Normally, all the wire color and physical connection information is eliminated from the diagram to make it easier to use by focusing on the functional, not the physical, aspects of the equipment.

*Pneumatic-electric (P-E) relay:* A pressure switch in which a pneumatic signal causes an electrical change.

*Sub-base:* The portion of a two-part thermostat that contains the wiring terminals and control switches.

*Thermostat base:* The portion of a two-part thermostat that contains the heating and cooling thermostats.

*Troubleshooting:* A procedure by which the technician locates the source of a problem, then makes the repairs and/or adjustments to correct the cause of a problem so that it will not recur.

*Troubleshooting table:* A troubleshooting aid usually contained in the manufacturer's Installation, Start-Up, and Service Instructions for a particular product. Troubleshooting tables are intended to guide the technician to a corrective action based on observations of system operation.

*Wiring diagram:* A troubleshooting aid, sometimes called a *schematic*, that provides a picture of what the unit does electrically and shows the actual external and internal wiring of the unit.

# APPENDIX

## Schematic Symbols

| SWITCHES | | | | | | | |
|---|---|---|---|---|---|---|---|
| DISCONNECT | MAGNETIC CIRCUIT BREAKER | THERMAL CIRCUIT BREAKER | LIMIT | | | | MAINTAINED POSITION |
| | | | SPRING RETURN | | | | |
| | | | NORMALLY OPEN | NORMALLY CLOSED | NEUTRAL | | |
| | | | HELD CLOSED | HELD OPEN | | | |

| Liquid Level | | Vacuum & Pressure | | Temp. Actuated | | Air or Water Flow | |
|---|---|---|---|---|---|---|---|
| NORMALLY OPEN (1) | NORMALLY CLOSED (2) | NORMALLY OPEN (1) | NORMALLY CLOSED (2) | NORMALLY OPEN (1) | NORMALLY CLOSED (2) | NORMALLY OPEN (1) | NORMALLY CLOSED (2) |

| CONDUCTORS | | FUSES | COILS | | | |
|---|---|---|---|---|---|---|
| NOT CONNECTED | CONNECTED | OR | RELAYS, TIMERS, ETC. | OVERLOAD THERMAL | SOLENOID | TRANSFORMER |

| PUSHBUTTONS | | | | |
|---|---|---|---|---|
| SINGLE CIRCUIT | | DOUBLE CIRCUIT | MUSHROOM CIRCUIT | MAINTAINED CONTACT |
| NORMALLY OPEN | NORMALLY CLOSED | | | |

| TIMER CONTACTS | | | | GENERAL CONTACTS | | |
|---|---|---|---|---|---|---|
| CONTACT ACTION IS RETARDED WHEN COIL IS: | | | | STARTERS, RELAYS, ETC. | | |
| ENERGIZED | | DE-ENERGIZED | | OVERLOAD THERMAL | NORMALLY OPEN | NORMALLY CLOSED |
| NORMALLY OPEN | NORMALLY CLOSED | NORMALLY OPEN | NORMALLY CLOSED | | | |

(1) Make on rise
(2) Make on fall

207A01.EPS

INTRODUCTION TO CONTROL CIRCUIT TROUBLESHOOTING — TRAINEE MODULE 03207

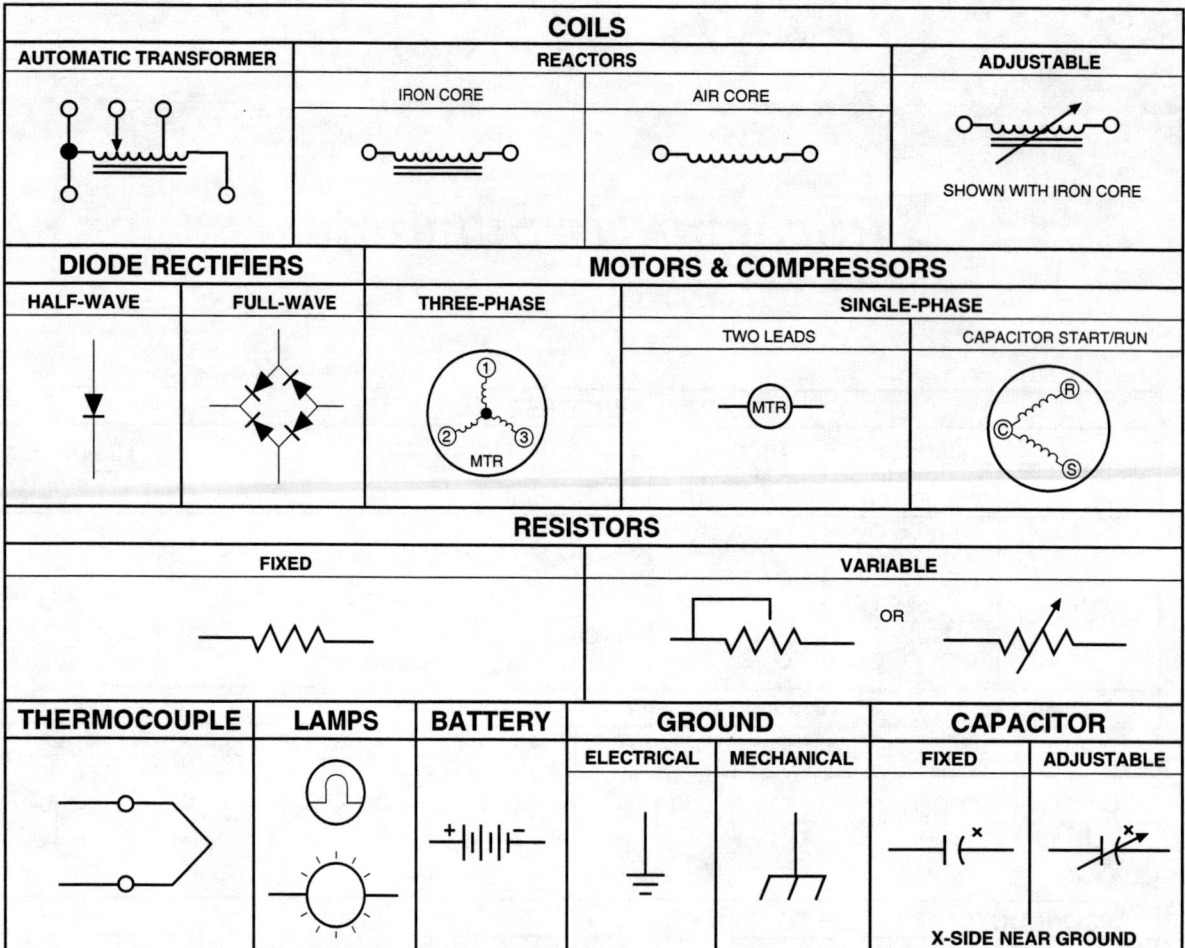

# ANSWER KEY

## Answers to Review Questions

| Answer | Section |
|---|---|
| 1. b | 2.1.0 |
| 2. d | 2.2.0 |
| 3. b | 2.3.0 |
| 4. a | 2.5.0 |
| 5. b | 2.7.0 |
| 6. c | 2.9.1 |
| 7. b | 2.9.2 |
| 8. d | 2.9.3 |
| 9. d | 3.1.3 |
| 10. d | 3.3.0 |
| 11. a | 3.5.0 |
| 12. c | 3.7.2 |
| 13. c | 4.0.0 |
| 14. d | 5.1.0 |
| 15. c | 5.4.1 |
| 16. d | 8.1.0 |
| 17. c | 8.2.0 |
| 18. c | 8.2.0 |
| 19. a | 9.2.2 |
| 20. b | 9.3.0 |
| 21. a | 10.3.0 |
| 22. b | 10.5.0 |
| 23. c | 10.6.2 |
| 24. a | 12.1.0 |
| 25. b | 13.2.0 |

# REFERENCES & ACKNOWLEDGMENTS

## Additional Resources

This module is intended to present thorough resources for task training. The following reference works are suggested for further study. These are optional materials for continued education rather than for task training.

*Electricity & Controls for Heating, Ventilating, & Air Conditioning*, 2001, Albany, NY: Delmar Publishers.

*HVAC Servicing Procedures*, 1995. Syracuse, NY: Carrier Corporation.

*Modern Refrigeration and Air Conditioning*, 2000. A.D. Althouse, C.H. Turnquist, A.F. Bracciano. Tinley Park, IL: The Goodheart-Willcox Company, Inc.

*Pocket Guide to Electrical Installations Under NEC 2002, Volumes I and II*, 2001. Quincy, MA: National Fire Protection Association.

*Refrigeration & Air Conditioning Technology*, 2000. William C. Whitman, William M. Johnson, John A. Tomczyk. Albany, NY: Delmar Publishers, Inc.

## Figure Credits

| | |
|---|---|
| **Thomas P. Burke** | 207F05B, 207SA03, 207SA06 |
| **Gerald Shannon** | 207F10, 207F20, 207F25, 207SA02 |
| **Veronica Westfall** | 207F19, 207SA05 |
| **Tyco Electronics (Potter & Brumfield)** | 207F24 |
| **Square D/Schneider Electric** | 207SA01 |
| **Carrier Corporation** | 207F35, 207F36, 207F37, 207F40, 207F41, 207F42 |
| **Extech** | 207SA04 |
| **Honeywell** | 207F62, 207F63 |
| **McDonnell & Miller** | 207F65 |
| **Taco, Inc.** | 207F66, 207F67 |

# NCCER CRAFT TRAINING USER UPDATES

The NCCER makes every effort to keep these textbooks up-to-date and free of technical errors. We appreciate your help in this process. If you have an idea for improving this textbook, or if you find an error, a typographical mistake, or an inaccuracy in the NCCER's Craft Training textbooks, please write us, using this form or a photocopy. Be sure to include the exact module number, page number, a detailed description, and the correction, if applicable. Your input will be brought to the attention of the Technical Review Committee. Thank you for your assistance.

*Instructors* – If you found that additional materials were necessary in order to teach this module effectively, please let us know so that we may include them in the Equipment and Materials list in the Instructor's Guide.

**Write:** Curriculum Revision and Development Department
National Center for Construction Education and Research
P.O. Box 141104, Gainesville, FL 32614-1104

**Fax:** 352-334-0932

**E-mail:** curriculum@nccer.org

---

Craft _____ Module Name _____

Copyright Date _____ Module Number _____ Page Number(s) _____

Description
_____
_____
_____
_____

(Optional) Correction
_____
_____
_____

(Optional) Your Name and Address
_____
_____
_____

Module 03208-01

# *Accessories and Optional Equipment*

## COURSE MAP

This course map shows all of the modules in the second level of the HVAC curriculum. The suggested training order begins at the bottom and proceeds up. Skill levels increase as you advance on the course map. The local Training Program Sponsor may adjust the training order.

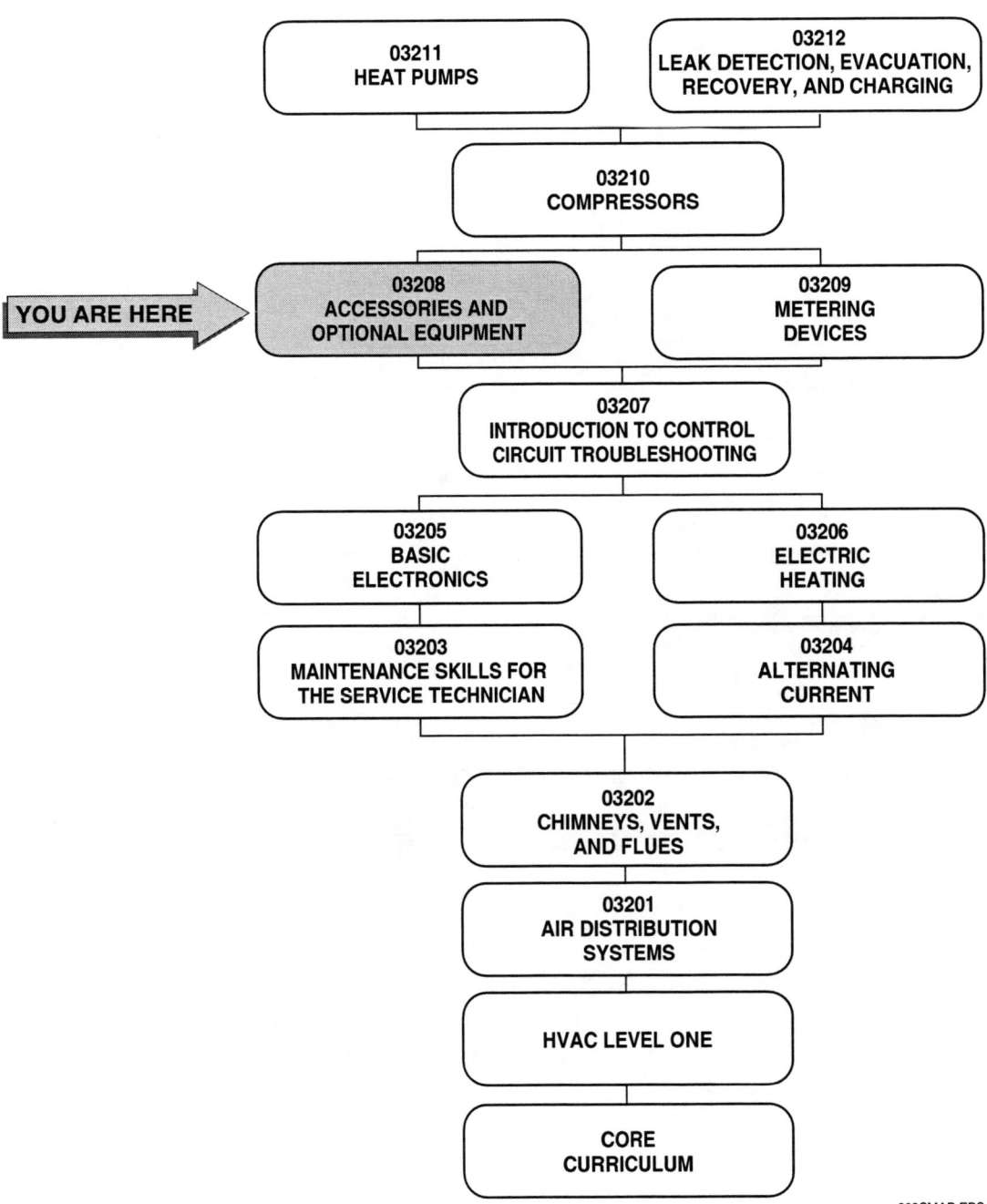

Copyright © 2001 National Center for Construction Education and Research, Gainesville, FL 32614-1104. All rights reserved. No part of this work may be reproduced in any form or by any means, including photocopying, without written permission of the publisher.

# MODULE 03208 CONTENTS

1.0.0 **INTRODUCTION** .................................................. 8.1
2.0.0 **PROCESS AND COMFORT AIR CONDITIONING** ............... 8.1
    2.1.0 Comfort Air Conditioning ................................. 8.3
    2.2.0 Maintaining Body Comfort ................................ 8.3
    *2.2.1 Conduction* ............................................... 8.3
    *2.2.2 Convection* ............................................... 8.3
    *2.2.3 Radiation* ................................................ 8.4
    *2.2.4 Evaporation* .............................................. 8.4
3.0.0 **HUMIDITY CONTROL** .......................................... 8.4
    3.1.0 Humidification ........................................... 8.5
    3.2.0 Dehumidification ......................................... 8.5
    3.3.0 Humidifiers .............................................. 8.6
    *3.3.1 Plate-Type Humidifiers* ................................... 8.6
    *3.3.2 Wetted-Element Humidifiers* .............................. 8.6
    *3.3.3 Atomizing Humidifiers* ................................... 8.8
    *3.3.4 Infrared Humidifiers* .................................... 8.9
    *3.3.5 Steam Humidifiers* ....................................... 8.9
    *3.3.6 Humidifier Capacity* ..................................... 8.9
4.0.0 **INTRODUCTION TO INDOOR AIR QUALITY** ................... 8.10
    4.1.0 Mechanical Air Filters .................................. 8.11
    *4.1.1 Conventional Disposable Filters* ........................ 8.11
    *4.1.2 Extended-Surface Disposable Filters* .................... 8.11
    *4.1.3 Bag-Type Disposable Filters* ............................ 8.12
    *4.1.4 Activated Carbon Disposable Filters* .................... 8.12
    *4.1.5 Electrostatic Permanent Filters* ........................ 8.12
    *4.1.6 Steel/Aluminum Mesh Filters* ............................ 8.12
    4.2.0 Electronic Air Cleaners ................................. 8.12
    *4.2.1 Gas-Phase Air Filtration* ............................... 8.13
    4.3.0 Filter and Electronic Air Cleaner Installation and Servicing ............................................ 8.13
5.0.0 **AIR CONDITIONING ENERGY CONSERVATION EQUIPMENT** ................................................... 8.15
    5.1.0 Energy and Heat Recovery Ventilators .................... 8.15
    5.2.0 Economizers ............................................. 8.15
    5.3.0 Evaporative Pre-Coolers ................................. 8.18
    5.4.0 Zoned Control ........................................... 8.18
6.0.0 **FIRE AND SMOKE DAMPERS** .................................. 8.19
7.0.0 **ULTRAVIOLET LIGHT AIR PURIFICATION SYSTEMS** ......... 8.20
8.0.0 **CARBON MONOXIDE AND CARBON DIOXIDE MONITORS** ..... 8.20

## MODULE 03028 CONTENTS (Continued)

**SUMMARY** .................................................. 8.21
**REVIEW QUESTIONS** ......................................... 8.22
**GLOSSARY** ................................................. 8.23
**ANSWERS TO REVIEW QUESTIONS** .............................. 8.24
**REFERENCES & ACKNOWLEDGMENTS** ............................. 8.25

### Figures

| | | |
|---|---|---|
| Figure 1 | Comfort zone | 8.2 |
| Figure 2 | Transfer of body heat and ways to control it | 8.3 |
| Figure 3 | Comparison of outside and inside air with the same specific humidity | 8.5 |
| Figure 4 | Recommended indoor relative humidity in winter | 8.5 |
| Figure 5 | Plate-type humidifier | 8.6 |
| Figure 6 | Wetted-element evaporative humidifiers | 8.7 |
| Figure 7 | Atomizing humidifiers | 8.9 |
| Figure 8 | Steam humidifiers | 8.9 |
| Figure 9 | Humidifier sizing chart | 8.10 |
| Figure 10 | Airborne particles | 8.11 |
| Figure 11 | Mechanical filters | 8.12 |
| Figure 12 | Electronic air cleaner | 8.13 |
| Figure 13 | Typical filter and electronic air cleaner placement | 8.14 |
| Figure 14 | Recovery ventilators | 8.16 |
| Figure 15 | Basic economizer unit | 8.17 |
| Figure 16 | Evaporative pre-cooler mounted on an air conditioning unit | 8.18 |
| Figure 17 | Typical forced-air zoned system | 8.19 |
| Figure 18 | Example of a UVC air purification unit | 8.20 |
| Figure 19 | Typical carbon monoxide (CO) monitor | 8.21 |
| Figure 20 | Typical carbon dioxide ($CO_2$) monitor | 8.21 |

# MODULE 03208

# Accessories and Optional Equipment

## Objectives

When you have completed this module, you will be able to do the following:

1. Explain how heat transfer by conduction, convection, radiation, and evaporation relates to human comfort.
2. Explain why it is important to control humidity in a building.
3. Recognize the various kinds of humidifiers used with HVAC systems and explain why each is used.
4. Demonstrate or describe how to install and service the humidifiers used in HVAC systems.
5. Recognize the kinds of air filters used with HVAC systems and explain why each is used.
6. Demonstrate or describe how to install and service the filters used in HVAC systems.
7. Use a manometer or differential pressure gauge to measure the friction loss of an air filter.
8. Identify accessories commonly used with air conditioning systems to improve indoor air quality and reduce energy cost, and explain the function of each.

## Prerequisites

Before you begin this module, it is recommended that you successfully complete the following modules: Core Curriculum; HVAC Level One; HVAC Level Two, Modules 03201 through 03207.

## Required Trainee Materials

1. Pencil and Paper
2. Appropriate Personal Protective Equipment

## 1.0.0 ♦ INTRODUCTION

Air conditioning is the process of treating indoor air to control its temperature, humidity, cleanliness, and distribution. It must do this in both the summer and in winter. Air conditioning is done to provide comfort for humans and in support of controlled manufacturing processes and materials.

Control of the temperature involves automatic control of the heating and cooling system(s) to maintain the best temperature range in any weather. Control of humidity usually involves adding moisture to the conditioned air in the winter, or dehumidifying it in the summer to remove moisture.

Methods used to control the cleanliness of air, such as with filtering devices, are usually the same for all seasons. The methods and devices used to control the temperature and air distribution systems of heating and cooling units are described in detail in several other modules.

This module deals with the HVAC equipment, called *accessories*, used to control the humidity, cleanliness, and overall indoor quality of air.

## 2.0.0 ♦ PROCESS AND COMFORT AIR CONDITIONING

Air conditioning in support of a manufacturing process or refrigeration of materials is based on the nature of the process or material, and will vary from system to system. On the other hand, air conditioning in support of human comfort is well known and basic requirements rarely vary from system to system. Control of the air for human comfort is sometimes called *comfort air conditioning*.

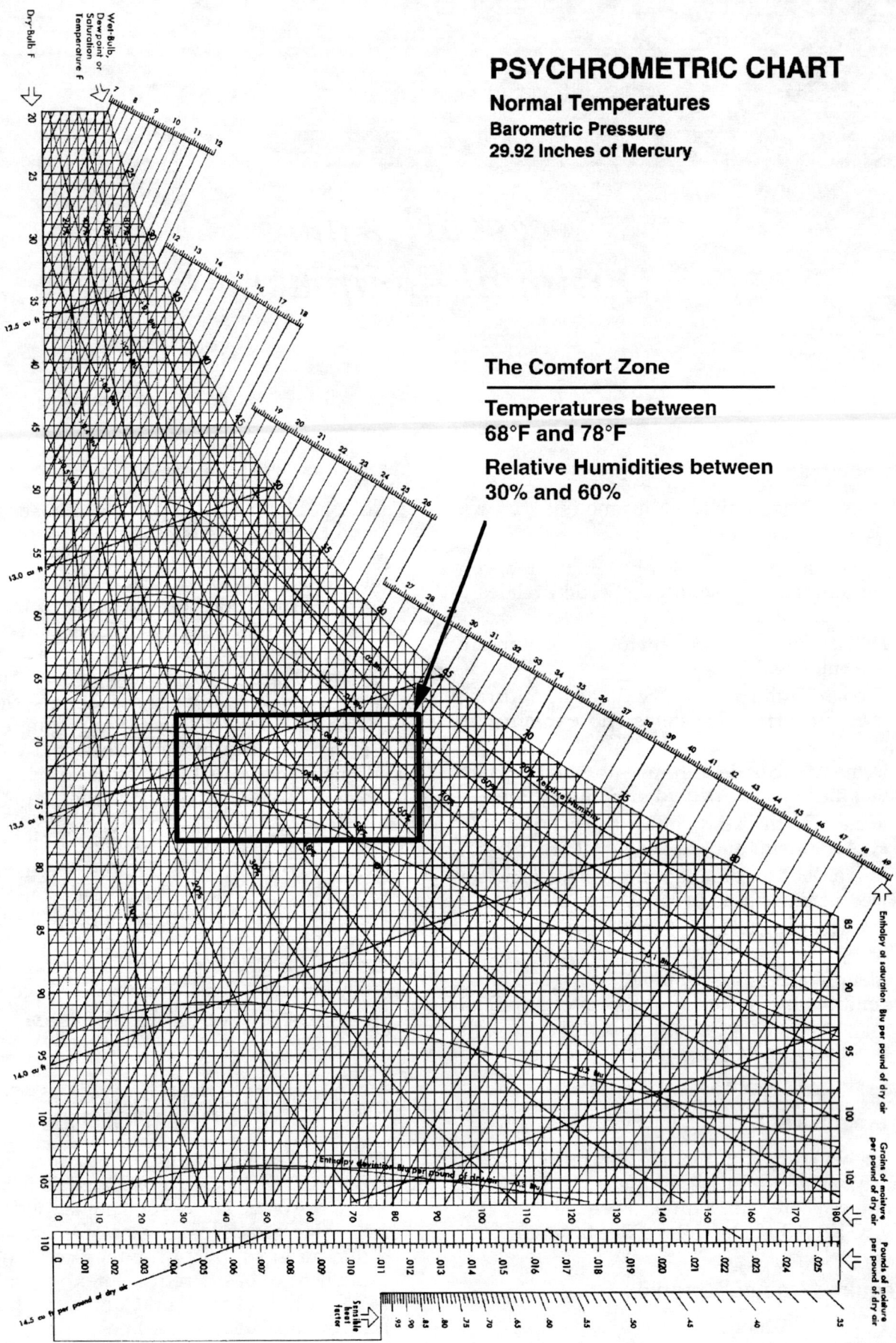

*Figure 1* ◆ Comfort zone.

>
> **DID YOU KNOW?**
> *Psychrometric Chart*
>
> In 1911, Dr. Willis Carrier presented his *Rational Psychrometric Formulas* to the American Society of Mechanical Engineers. This formula led to the development of the psychrometric chart. A psychrometric chart gives a graphical representation of the interrelationships that exist for all properties of air.

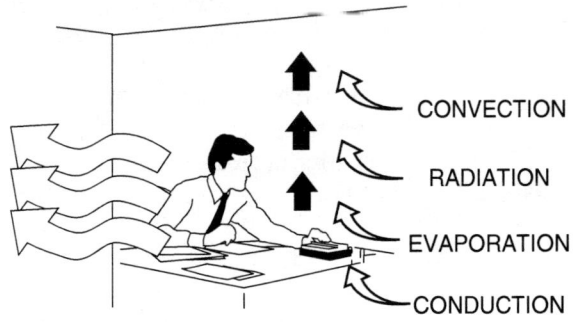

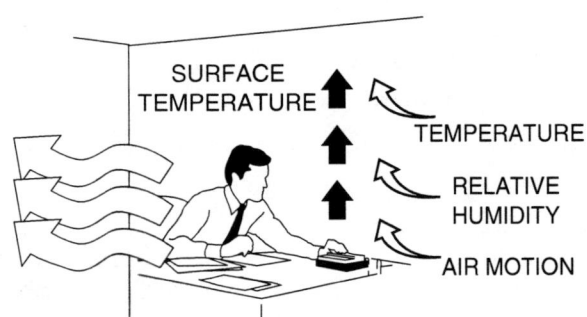

*Figure 2* ♦ Transfer of body heat and ways to control it.

Comfort air conditioning performs the following functions:

- Heating and cooling
- Humidification and dehumidification
- Ventilation
- Filtration
- Air circulation

## 2.1.0 Comfort Air Conditioning

Normally, a comfort air conditioning system must maintain the temperature and humidity of the conditioned rooms within an acceptable range called the *comfort zone*. *Figure 1* shows the comfort zone plotted on a psychrometric chart.

As shown in *Figure 1*, the comfortable dry-bulb temperatures range from about 68°F to 78°F. The comfortable relative humidity range is from about 30% to 60%. The lowest temperatures and humidities typically apply to winter, while the highest temperatures and humidities apply to summer.

## 2.2.0 Maintaining Body Comfort

The temperature of the inner human body is normally 98.6°F. In order to maintain this temperature, the body must reject any excess heat. In a conditioned room, the body always generates more heat than is needed to maintain its internal temperature. This excess heat is constantly being transferred to the room air. You will recall that heat is transferred from one place to another by three methods: conduction, convection, and radiation. Excess body heat is also transferred via these three methods. In addition, **evaporation** in the form of perspiration is another way that the body rejects heat. *Figure 2* shows the four ways that bodies transfer heat. It also shows the properties of air and conditions in a building that must be controlled in order to provide comfort. As shown, these include the room air and surface temperatures, air motion, and relative humidity.

### 2.2.1 Conduction

Conduction is the transfer of heat from a higher-temperature area to a lower-temperature area via direct contact. It can occur through a substance in contact with the higher- and lower-temperature materials, or directly from one substance to another. Heat generated by a person is conducted to objects the person touches and to the air surrounding them. Air temperature controls conduction; therefore, an acceptable temperature must be maintained in a conditioned space. The cooler the air surrounding a person, the faster heat leaves the body. When the room temperature is too low, occupants will complain about feeling cold. When the temperature is too warm, they will complain about feeling hot.

### 2.2.2 Convection

Convection is heat transfer by the movement of a fluid (gas or liquid) from one place to another. In a room, convection air currents carry heat to the body or away from it. As body heat is conducted into the air close to a person, the air next to the person becomes warmer than the air which is farther away. Since warm air is lighter, or less dense, than cool air, the warm air rises and is replaced by cooler air in a continuing process.

The velocity of airflow in a room is important to comfort. If the room air moves too slowly, the temperatures within the room become uneven and the occupants feel stuffy. If it moves too fast, the occupants may complain about draftiness and feeling cold. In a properly conditioned room, the air motion is typically between 15 and 45 feet per minute (fpm). The natural movement of convection air currents in a room is usually too slow to maintain comfort. This is one reason why forced-air systems are used. The blower in a forced-air system causes the air to move more rapidly than it would by natural means.

### 2.2.3 Radiation

Radiation is the movement of heat in the form of invisible rays or waves, similar to light. Radiant heat travels from a warmer object to a colder object without heating up the area in between. Radiant heat flows from a body to the cooler surfaces around it. If a surface is warmer than the body, radiant heat will flow from the surface to the body.

Even when the room air temperature is fine, the temperature of surrounding surfaces can cause discomfort because of radiant heat flow. The surface temperatures of ceilings, walls, floors, and windows is determined mainly by how well insulated they are. The closer these temperatures are to the room air temperature, the more comfortable the room will be. This reduces the discomfort caused by radiant heat loss from a body in the winter. It also reduces the discomfort caused by radiant heat gain in the summer. Typically, ideal room comfort exists when all room surfaces are between 70°F and 80°F. To eliminate the radiation effect, the surface temperature must be within 1° of the room air temperature.

### 2.2.4 Evaporation

Evaporation is the condition in which the heat absorbed by a liquid causes it to change into a vapor. Evaporation transfers heat from bodies to the surrounding air. Moisture, in the form of perspiration, is given off through the pores in the skin. As body heat causes this moisture to evaporate, it turns into water vapor. Evaporation from our bodies goes on constantly, whether perspiration is visible or not. Actually, when perspiration is visible on the skin, the body is producing more heat than it can reject at the normal rate. Evaporation is the most important factor for keeping the body cool. Two things affect evaporation: air motion and relative humidity.

If there were no air movement in a room, the layer of air close to the body would rapidly absorb all the water vapor it could hold, causing evaporation of moisture from the body to stop. Normally, convection currents constantly move the air that has absorbed evaporated moisture away from the body, replacing it with drier air. This drier air allows the evaporation process to continue. The faster the air movement in a room, the more rapid the evaporation of moisture from a body. This causes the body to feel a greater sensation of coolness.

The relative humidity (RH) of room air greatly affects evaporation of moisture from the body. Relative humidity is a measure of how much moisture is in the air, expressed as a percentage. The lower the percentage, the lower the moisture content in the air and the greater the air's ability to absorb more moisture. For each pound of water evaporated from the skin, about 1,000 Btus of heat are removed from the body. The relative humidity of the air around a body affects the rate of evaporation. The lower the relative humidity, the faster evaporation occurs. The higher the relative humidity, the slower evaporation occurs.

## 3.0.0 ◆ HUMIDITY CONTROL

Proper humidity control improves comfort and health conditions in all seasons. Also, it often reduces the cost of equipment operation. The recommended RH levels are about 30% in winter and 50% in summer. RH levels that are too low or too high can cause problems.

Low RH levels can cause:
- Dry, itchy skin
- Static electricity shocks
- Clothing static cling
- Sinus problems
- A chilly feeling
- Sickly pets and plants
- Loose furniture joints

High RH levels can cause:
- Condensation on windows and inside exterior walls in cold weather
- Moist environments in warm weather that can lead to:
  – Development of bacteria, viruses, fungi, and mite infestations
  – The warping of wood
  – Loss of personal comfort

### Think About It

*Humidity in a Home*

Compare the relative humidity level within the average home during the winter to the humidity level of a desert.

## 3.1.0 Humidification

In the fall or winter, heated buildings tend to be dry, making the occupants feel uncomfortably cool or cold. This is because the drier indoor air accelerates the body's process for rejecting excess heat by evaporation. Most of the moisture within a building comes in from the outside by infiltration or ventilation. Because of its cold temperature in the winter, outside air brought into a building contains less moisture. For example, if the outdoor air is 20°F and has 50% RH, its specific humidity is only about 7.5 grains/lb (*Figure 3*). The specific humidity is low because at 20°F, the air can hold very little water vapor before it becomes saturated. When this same air enters a building heated to 72°F, its moisture content is still only 7.5 grains/lb, but its RH drops to 11%. This makes the air much drier. *Figure 4* shows some examples of winter outdoor air temperatures and

| OUTDOOR TEMPERATURE (°F) | OUTDOOR RH (%) | INDOOR RH (%) | RECOMMENDED (SAFE) INDOOR RH (%) |
|---|---|---|---|
| -10 | 30 TO 70 | 2 | 20 |
| 0 | 30 TO 70 | 5 | 25 |
| 10 | 30 TO 70 | 7 | 30 |
| 20 | 30 TO 70 | 11 | 35 |
| 30 | 30 TO 70 | 15 | 40 |

BASED ON AN INDOOR TEMPERATURE OF 72°F

*Figure 4* ◆ Recommended indoor relative humidity in winter.

RH related to the corresponding indoor RH for the same air. It also shows the recommended indoor RH levels that should be maintained to keep the building occupants comfortable for the same conditions. To achieve the recommended RH levels in the winter, a **humidifier** is almost always installed in air conditioning systems used in cold climates. A humidifier is a device used to add and control humidity.

## 3.2.0 Dehumidification

In the summer when the RH is above 60%, people complain about feeling hot and sticky, regardless of the room temperature. This is because the moisture-laden room air slows down the body's process for rejecting excess heat by evaporation. To reduce the high humidity levels that occur in the summer, the conditioned air is normally dehumidified by the operation of the system

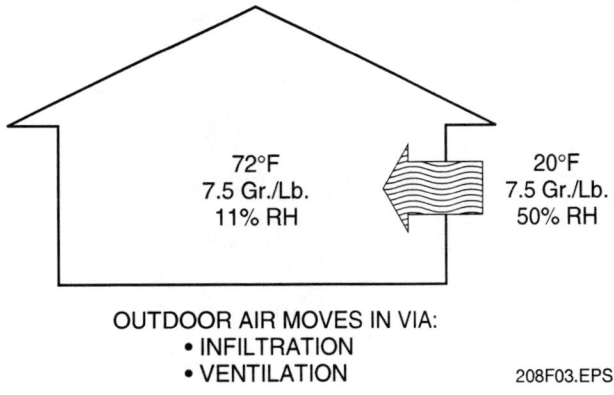

*Figure 3* ◆ Comparison of outside and inside air with the same specific humidity.

### Humidifier Settings

Although 40% to 60% relative humidity is considered the optimum comfort zone, humidifiers are designed to operate in the 35% relative humidity range. This is because most homes cannot tolerate wintertime humidity levels in the 40% to 60% range for any length of time without causing condensation water damage.

## Dehumidification

To increase dehumidification in the cooling mode, it is acceptable to lower the evaporator blower speed to produce a slightly colder coil. The colder coil and lower volume of air moving across it allow more moisture to be removed. Many manufacturers offer a kit that includes a humidistat and relay that automatically puts the cooling unit in dehumidification mode when the humidity is high.

cooling unit. For buildings that have only a heating system, the air can be dehumidified on a room-by-room basis using individual **dehumidifiers**. A dehumidifier is a device used to remove moisture from the air.

### 3.3.0 Humidifiers

All humidifiers introduce water vapor into a building's conditioned air at a predetermined rate. All must be connected to a water supply. Some humidifiers require a drain connection. Others are motor-driven and controlled by a humidistat. There are five basic types of humidifiers:

- Plate-type
- Wetted-element
- Atomizing
- Infrared
- Steam

#### 3.3.1 Plate-Type Humidifiers

The plate or pan-type humidifier (*Figure 5*) is an evaporative humidifier. It uses heat from the air to vaporize water, allowing the vapor particles to mix with the heated air. A plate-type humidifier consists of a pan and a series of porous, mineral wool plates mounted in a rack. The bottoms of the plates extend down into water that is contained in the pan. The plates absorb water from the pan by wick action until they are completely wet. The pan, with

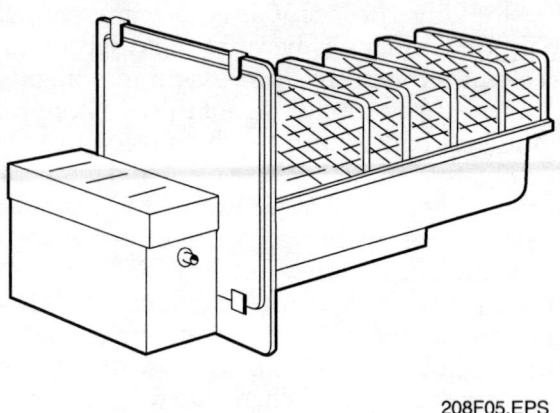

*Figure 5* ◆ Plate-type humidifier.

the plates, is installed into the furnace plenum. There, the warm air picks up moisture from the plates as it passes over them. A float device regulates the supply of water to maintain a constant level in the pan. Because a plate-type humidifier has a limited capacity, wetted-element humidifiers are normally used in new installations.

#### 3.3.2 Wetted-Element Humidifiers

Wetted-element humidifiers (*Figure 6*) are evaporative units commonly used in forced-air systems. They use a porous media that has been moistened with water. Air passes over the media by the action of either an air pressure differential or a

## Plate-Type Humidifiers

Plate-type humidifiers have been responsible for the failure of many plate-type furnace heat exchangers. As the water evaporates, scale and mineral deposits build up and can cause the float device that regulates water flow to the pan to stick open. This causes the water in the pan to overflow onto the heat exchanger located below the humidifier, resulting in rust and early heat exchanger failure. Plate-type humidifiers, like all humidifiers, must be cleaned on a regular basis to prevent this type of problem.

self-contained fan. The air current picks up moisture from the media and carries it into the duct system. Wetted-element humidifiers can be mounted in many ways, depending on the type and manufacturer. Always follow the manufacturer's instructions for the type of humidifier being installed.

There are three common types of wetted-element humidifiers: rotating drum, bypass, and fan-powered.

- *Rotating drum* – A rotating drum humidifier consists of a drum covered with a screen or sponge pad. The drum is motor-driven and rotates very slowly through a pan of water. The bottom part of the pad is immersed in the water and, as the pad rotates, the entire pad becomes wet. Air is drawn across the wet pad via a bypass duct because of the air pressure difference between the supply and return sides of the system. The water level in the pan is controlled by a float valve. A humidistat controls the operation of the drum. Typically, the humidistat is located near the thermostat or in the return air duct. If mounted in the return duct, it must be out of the path of any radiant heat. The humidifier can be wired so the drum operates only when the furnace is on and the humidistat senses a drop in humidity.

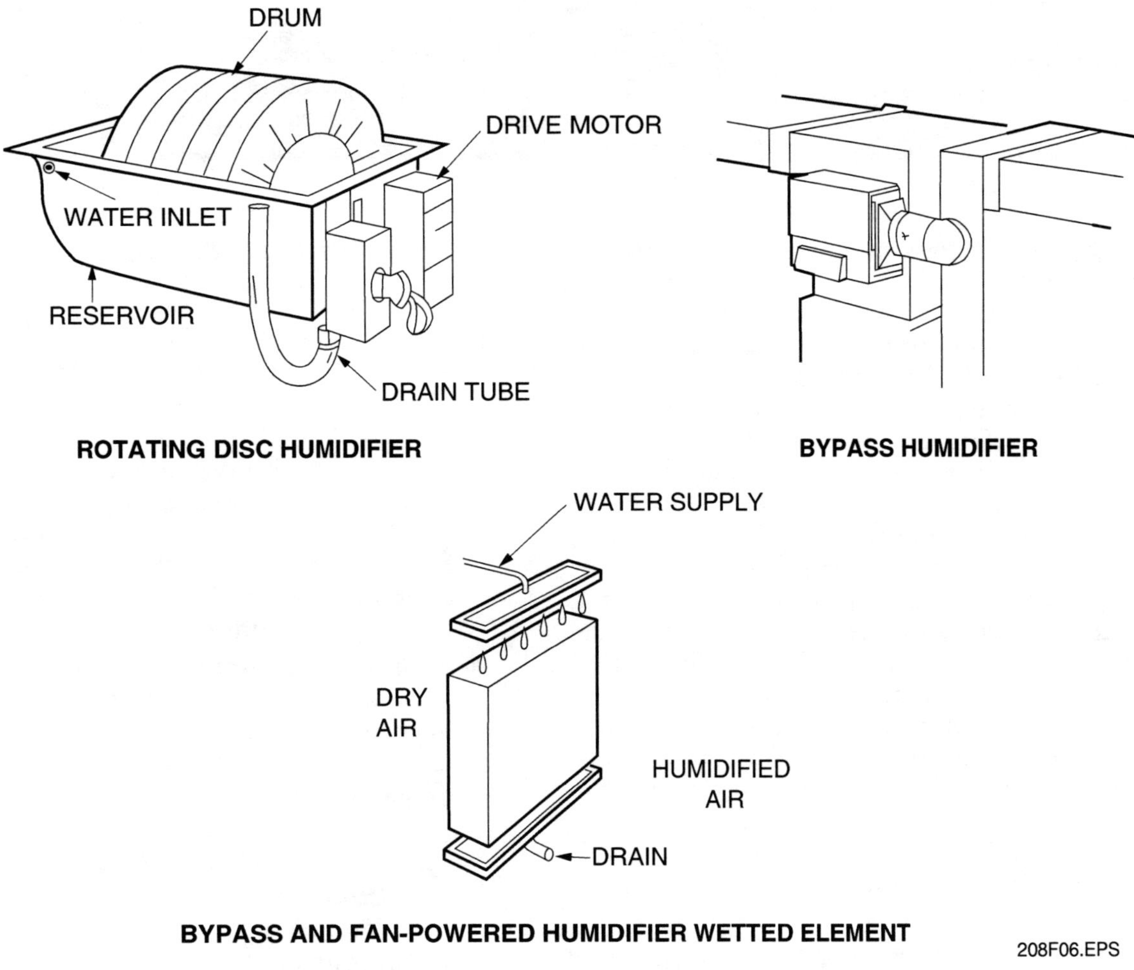

**ROTATING DISC HUMIDIFIER**

**BYPASS HUMIDIFIER**

**BYPASS AND FAN-POWERED HUMIDIFIER WETTED ELEMENT**

*Figure 6* ◆ Wetted-element evaporative humidifiers.

### Bypass Humidifier

A bypass humidifier (or any humidifier in which the excess water is drained out at the bottom) greatly reduces the amount of mineral deposits that build up over time as the result of water evaporating.

### Bypass Humidifier Installation

This is an example of a typical bypass humidifier installation.

- *Bypass* – The bypass humidifier uses a porous wetted element that is mounted vertically. Water applied to the top of the element moves evenly across the entire top surface. The porous material allows the water to migrate slowly from the top to the bottom to maintain a uniform wet surface. A pan with a drain is placed under the element to catch any water that does not evaporate. This humidifier is normally mounted on the supply air plenum or duct, with a bypass pipe connected to the return air plenum or duct. It may also be mounted on the return duct with a bypass duct connected to the supply. Warm air is drawn through the wetted element by the difference in air pressure between the supply and return sides of the system. The moving air picks up moisture by evaporation, then returns to the duct system via the bypass pipe. Water flow applied to the wet element is controlled by a humidistat, which allows water to flow only when it senses a drop in humidity.
- *Fan-powered* – The fan-powered humidifier mounts on the furnace plenum and operates basically the same as the bypass type. However, airflow through the wet element is caused by a separate fan built into the humidifier. Air bypass openings in the humidifier allow the fan to draw warm air directly from the plenum and then push it through the wet element and back into the plenum. This eliminates the need for a bypass pipe.

### 3.3.3 Atomizing Humidifiers

Atomizing humidifiers (*Figure 7*) convert water into small droplets for release into the air stream. They do this in one of three ways: using a spinning disc, a high-pressure spray nozzle, or ultrasonic frequencies.

- *Spinning disc* humidifiers use a circular wheel or cone that rotates at a fairly high speed. Water is fed into the spinning disc and centrifugal force converts it into small droplets. This type of atomizing humidifier is commonly used in self-contained units.
- *High-pressure spray nozzle* humidifiers either blow water through a metered orifice into the duct air stream, or spray water onto an evaporative pad where it is absorbed into the air stream as vapor. They can be mounted on the plenum, on the side of the duct, or under the duct.
- *Ultrasonic* humidifiers contain a crystal known as a *piezoelectric crystal*. This crystal vibrates at a high frequency (above 16,000 Hertz) when an electric current is applied to it. Water dripping onto the vibrating crystal is atomized and injected into the air stream.

Certain precautions must be taken when using atomizing humidifiers. They should not be used with hard water because it contains minerals such as lime and iron that leave the water vapor and settle throughout the building. Also, if all of the water spray does not vaporize, it can settle in the duct system and cause rust and corrosion.

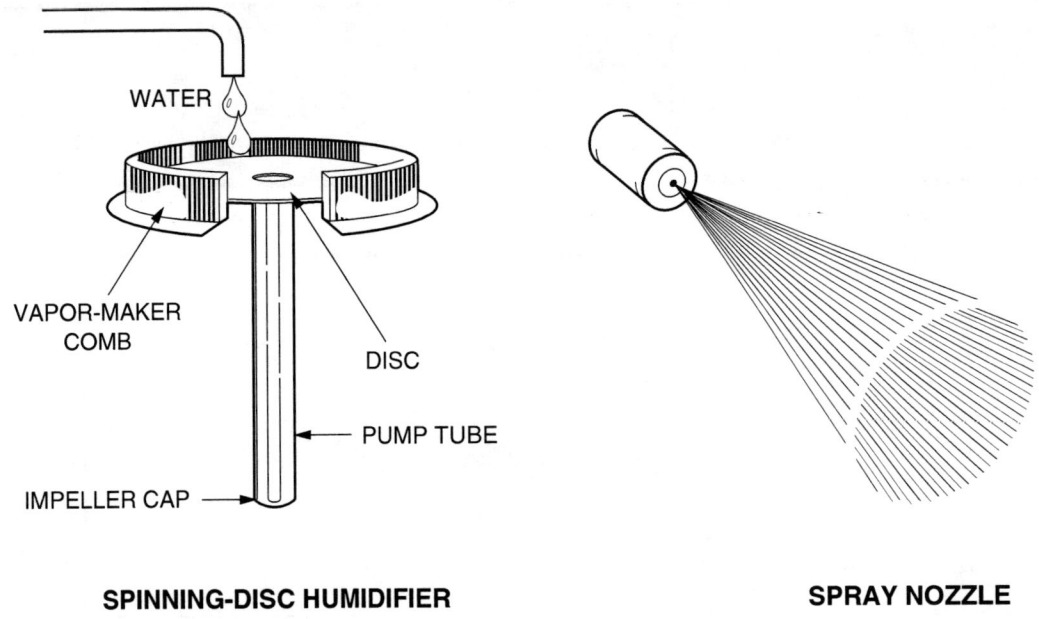

*Figure 7* ◆ Atomizing humidifiers.

### 3.3.4 Infrared Humidifiers

The infrared humidifier consists of a horizontal water reservoir and infrared lamps with reflectors. Infrared humidifiers do not use heat from the heating system to vaporize water. Instead, energy from the infrared lamps is reflected into the water, where the radiant heat evaporates the water rapidly into the air stream. Because infrared humidifiers do not use heat generated by the system, they can be installed almost anywhere in the ductwork.

### 3.3.5 Steam Humidifiers

In this type of humidifier, water is heated and converted into steam. Steam humidifiers (*Figure 8*) can be installed almost anywhere in the ductwork because they do not use heat generated by the system to vaporize the moisture.

### 3.3.6 Humidifier Capacity

Humidifier capacity is typically rated in gallons of water per day. The required capacity depends on the volume of the building or area in square feet (ft.$^2$). It also depends on the building's construction relative to its air tightness. *Figure 9* shows a typical graph used for the selection of residential humidifiers. Similar graphs and charts are available for commercial and industrial humidifiers.

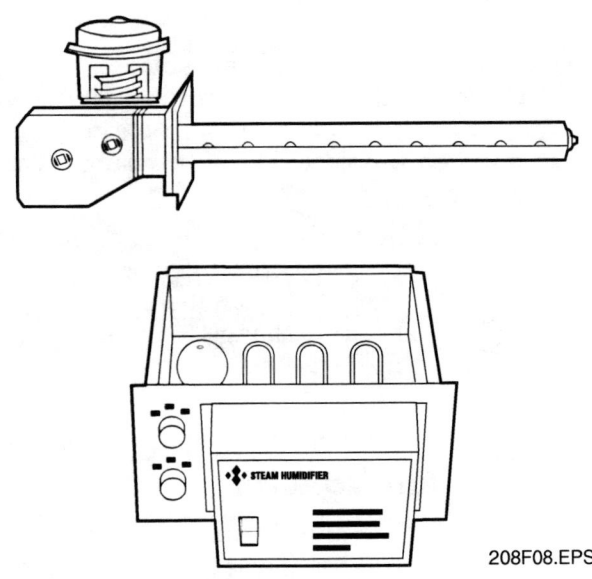

*Figure 8* ◆ Steam humidifiers.

Loose, average, and tight houses are defined as follows:

- *Loose* – The building has little insulation, no vapor barriers, and no storm doors or windows. In homes, this can also mean there is an undampered fireplace. The air exchange rate is about 1.5 changes per hour.

### Humidifier Capacity

A wetted-element humidifier normally is connected to a cold water source. An increase in capacity can be obtained for the same humidifier by connecting it to a hot water source.

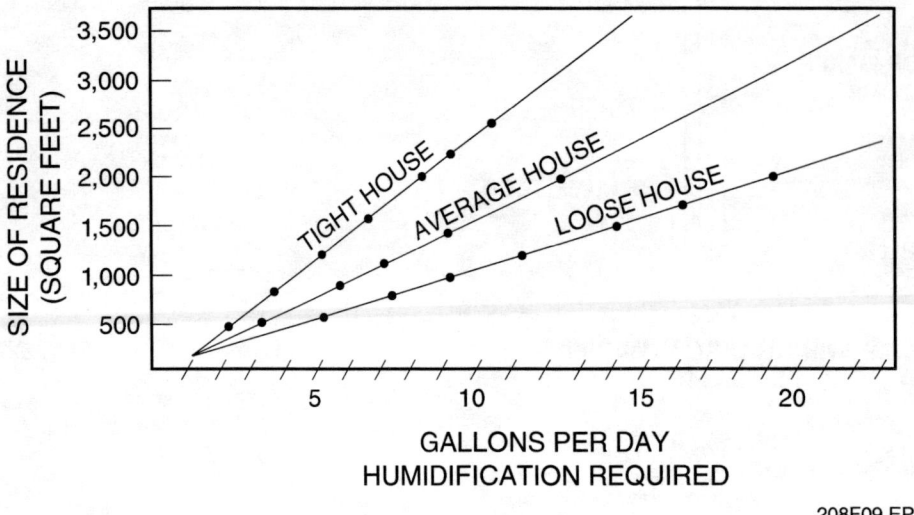

*Figure 9* ♦ Humidifier sizing chart.

- *Average* – The building is insulated, has vapor barriers, and has loose storm doors or windows. In homes, this can also mean there is a dampered fireplace. The air exchange is about 1.0 change per hour.
- *Tight* – The building is well insulated, has vapor barriers, and tight storm doors or windows. In homes, this can also mean there is a dampered fireplace. The air exchange is about 0.5 change per hour.

## 4.0.0 ♦ INTRODUCTION TO INDOOR AIR QUALITY

As energy conservation became more of a concern, the immediate response was to tighten up the construction of homes and commercial buildings to retain the heated or cooled environment inside. Energy was saved, but the tight construction trapped contaminants within the buildings, causing a variety of health problems. These included allergic reactions and respiratory problems.

Investigators found that the closed environment of some buildings and homes encouraged the growth of molds and spores. The stale air contributed to the spread of disease. Second-hand cigarette smoke became such an issue that smoking is now banned in most public buildings in the United States. Even some of the building materials, such as adhesives, carpets, and wall coverings, emitted substances that caused allergic reactions. Clearly, something had to be done to improve the quality of indoor air. The American Society of Heating, Refrigeration and Air Conditioning Engineers (ASHRAE) has established standards governing the design of HVAC systems to ensure that indoor air quality (IAQ) standards are met and maintained. The factors that cause poor IAQ and methods used to achieve good IAQ are covered in detail in the HVAC Level Four Module, *Indoor Air Quality*.

Properly designed and maintained HVAC systems play a key role in maintaining good IAQ. Humidifiers discussed earlier in this module work to maintain the correct moisture levels in the air. Air filters and cleaners remove unwanted particles from the air. **Energy recovery ventilators (ERVs)**, **heat recovery ventilators (HRVs)**, and **economizers** ensure that outdoor air is constantly being brought into the building as stale, contaminated air is being discharged from the building. These items of equipment, and others, are covered in the remainder of this module.

*Figure 10* shows some of the common particles that contaminate air. As shown, these particles have diameters that range in size from less than 0.01 **micron** to more than 100 microns. A micron is a unit of length that is one-millionth of a meter, or about $\frac{1}{25,400}$ of an inch. The size of about 99% of

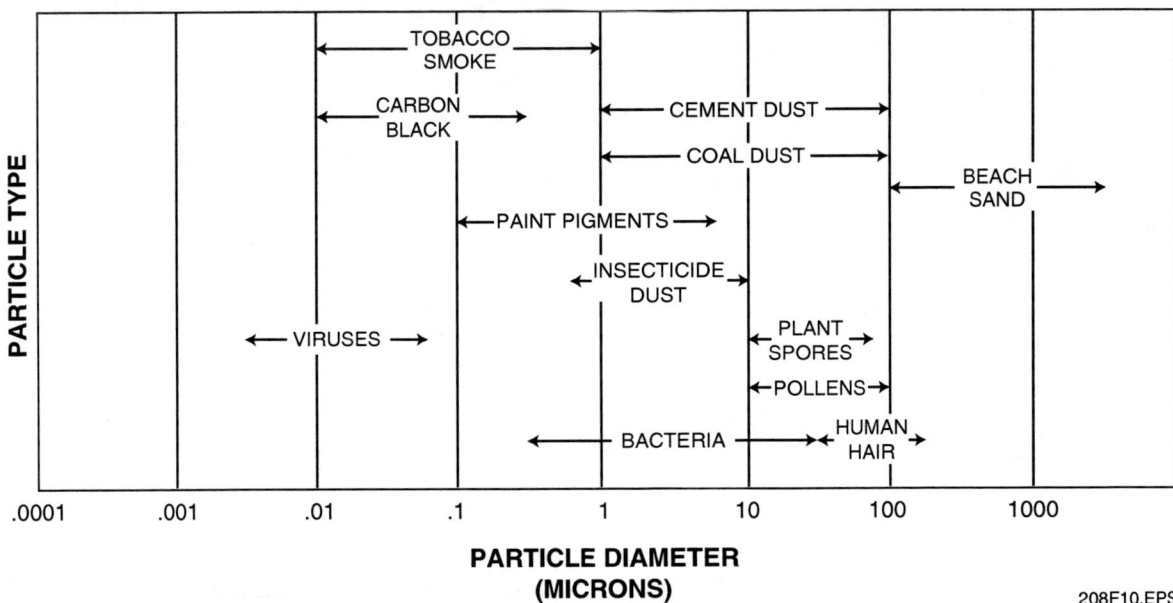

*Figure 10* ◆ Airborne particles.

the airborne particles is less than 1 micron. The remaining 1% consists of the larger, heavier particles such as dust, lint, and pollen. Several types of filtration devices are used to remove contaminants from the air, making it cleaner and healthier to breathe. Both mechanical filters and electronic air cleaners are in common use.

## 4.1.0 Mechanical Air Filters

There are many kinds of mechanical filters. Some of the more common ones include:

- Conventional disposable
- Extended-surface disposable
- Bag-type disposable
- Activated carbon disposable
- Electrostatic permanent
- Steel/aluminum mesh permanent

### 4.1.1 Conventional Disposable Filters

Conventional disposable filters (*Figure 11*) are available in bulk rolls of filtering material or in frames. They are typically made from fiberglass, hog-hair, or open-cell foam material. The filtering material is coated with a nondrying, nontoxic, adhesive coating that catches airborne particles. Even when the filter is coated with dust, this adhesive remains effective. The filter material gets progressively more dense as the air passes through it. For this reason, this type of filter must be placed in the air stream correctly. Normally, framed versions of these filters are marked with an arrow that points in the direction of airflow when installed in the duct system. Conventional filters do a good job of protecting the furnace and air distribution system from accumulating dust and dirt. However, they do not remove smaller particles, such as pollen, spores, and smoke.

### 4.1.2 Extended-Surface Disposable Filters

Extended-surface disposable filters are used when a higher level of cleanliness is required, such as in computer or electronic equipment rooms. They are also used when conditions prohibit the use of fiberglass or hog-hair filters. Extended-surface filters use a pleated, non-woven, cotton-synthetic material reinforced with a metal backing. The filter material is installed in a rigid frame. Extended-surface filters typically have a usable life four times that of conventional disposable filters.

---

### Filter Replacement

Always replace an air filter with a similar or identical one. Different filters have different resistances to system overflow. If a filter with a high resistance is used to replace a filter with a much lower resistance, system performance may suffer and equipment damage or failure could result.

an ultra-fine fiberglass material that is reinforced on the leaving-air side by nylon backing material within the bags.

### 4.1.4 Activated Carbon Disposable Filters

Activated carbon disposable filters provide the dual air cleaning functions of particle filtration and odor removal. The filter media and a honeycomb cell of odor-absorbing carbon are enclosed in a frame for installation in the system. Activated carbon acts like a porous sponge that absorbs odors.

### 4.1.5 Electrostatic Permanent Filters

Electrostatic filters clean the air using an electrostatic charge that is generated by the airflow as it passes over layered pairs of woven synthetic (plastic) filter material. Electrostatic action weakly attracts small, naturally charged particles, such as smoke particles, and causes them to collect and settle on the filter material. Electrostatic filters can be cleaned by washing and/or vacuuming.

### 4.1.6 Steel/Aluminum Mesh Filters

Steel/aluminum mesh filters are permanent filters used in commercial and institutional buildings such as restaurants, hotels, and schools. The steel/aluminum filter mesh can be cleaned by washing.

## 4.2.0 Electronic Air Cleaners

Electronic air cleaners normally outperform mechanical filters in trapping airborne particles and odors. They can be standalone units or may be mounted in the air conditioning system. Electronic filters have a high-voltage, solid-state power supply. The high voltage produced by the power supply, which can range from 6,000 to 10,000 volts DC, is used to electrically charge (ionize) all particles in the air that pass through the filter. *Figure 12* shows a typical electronic air cleaner.

As shown, the filter portion consists of a prefilter, ionizer section, collector section, and charcoal filter section. As the air is moved through the filter, larger particles are trapped by the prefilter section. Smaller particles pass through the prefilter to the ionizing section. This section consists of a fine, tungsten-wire grid connected to the high-voltage power supply. An ionizing field created by the high voltage on the wire grid charges the particles in the air stream as they pass through the grid. These charged particles are then drawn into the collector section.

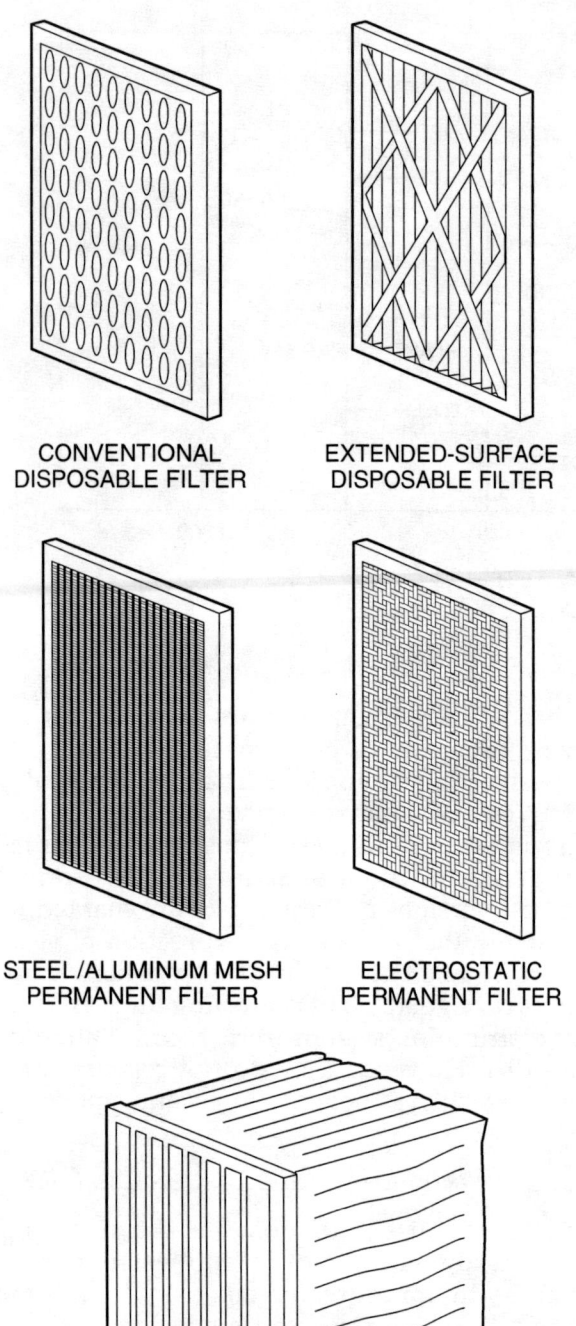

*Figure 11* ◆ Mechanical filters.

### 4.1.3 Bag-Type Disposable Filters

Bag-type disposable filters are typically used where tiny airborne particles cannot be tolerated, such as in hospital operating and recovery rooms, or pharmaceutical process rooms. These filters use

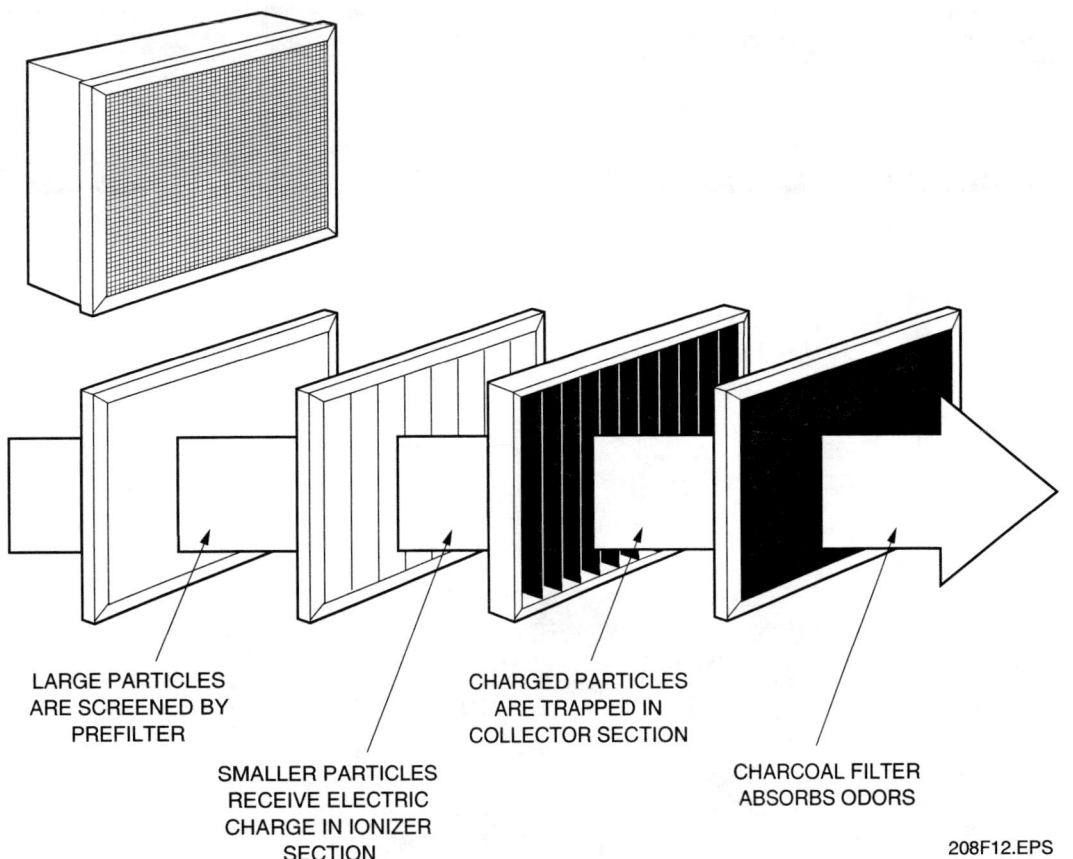

*Figure 12* ◆ Electronic air cleaner.

The collector section consists of a series of equally spaced, parallel collector plates. These plates are connected to the high-voltage power supply so that the even and the odd numbered plates are at a positive and negative DC voltage, respectively. As the ionized particles flow between the plates, they are attracted and held on oppositely charged collector plates. The air, cleaned of pollutants, then passes through the charcoal filter section where any odors are absorbed. From there it is passed on to the conditioned space. The pollutants remain held in the collector section until they are removed when the filter is cleaned. Electronic filter operation is sensitive to airflow. When operated at airflow rates above those recommended by the manufacturer, they can become quite inefficient. When the airflow is reduced below the manufacturer's recommended minimum, enough ozone can be generated to cause an annoying odor.

### 4.2.1 Gas-Phase Air Filtration

Gas-phase air filtration filters are made to filter the air for both particulate and odor contaminants. The system uses dry scrubbing, gas-phase filtration media. These types of filters are used mainly in commercial applications, such as office buildings, museums, airports, hospitals, and hotels. Gas-phase filtration systems are typically custom designed to meet the unique requirements for a specific application. This is because the control media used for a specific filter differs depending on the type of odor contaminant gas involved. The gas-phase media used removes the airborne gaseous contaminants from the air either by adsorption, where the gaseous molecules are captured and held to the media surface, or chemisorption, where the media reacts with gaseous molecules to change their chemical form to a nontoxic end product. Typically, the gas-phase filtration media is made in a dry, pellet bulk form used to fill filtration modules, or it is supplied as a coating on pleated fiber filters.

### 4.3.0 Filter and Electronic Air Cleaner Installation and Servicing

The location and installation of mechanical and electronic filtration devices should be as directed in the manufacturers' instructions. Typically, both types can be placed in similar locations within a duct system. *Figure 13* shows typical mounting locations.

### Air Filter Installation

Most air filters have an airflow directional arrow printed or stamped on their frames. Always install the filter so that the tip of the arrow points in the direction of the airflow.

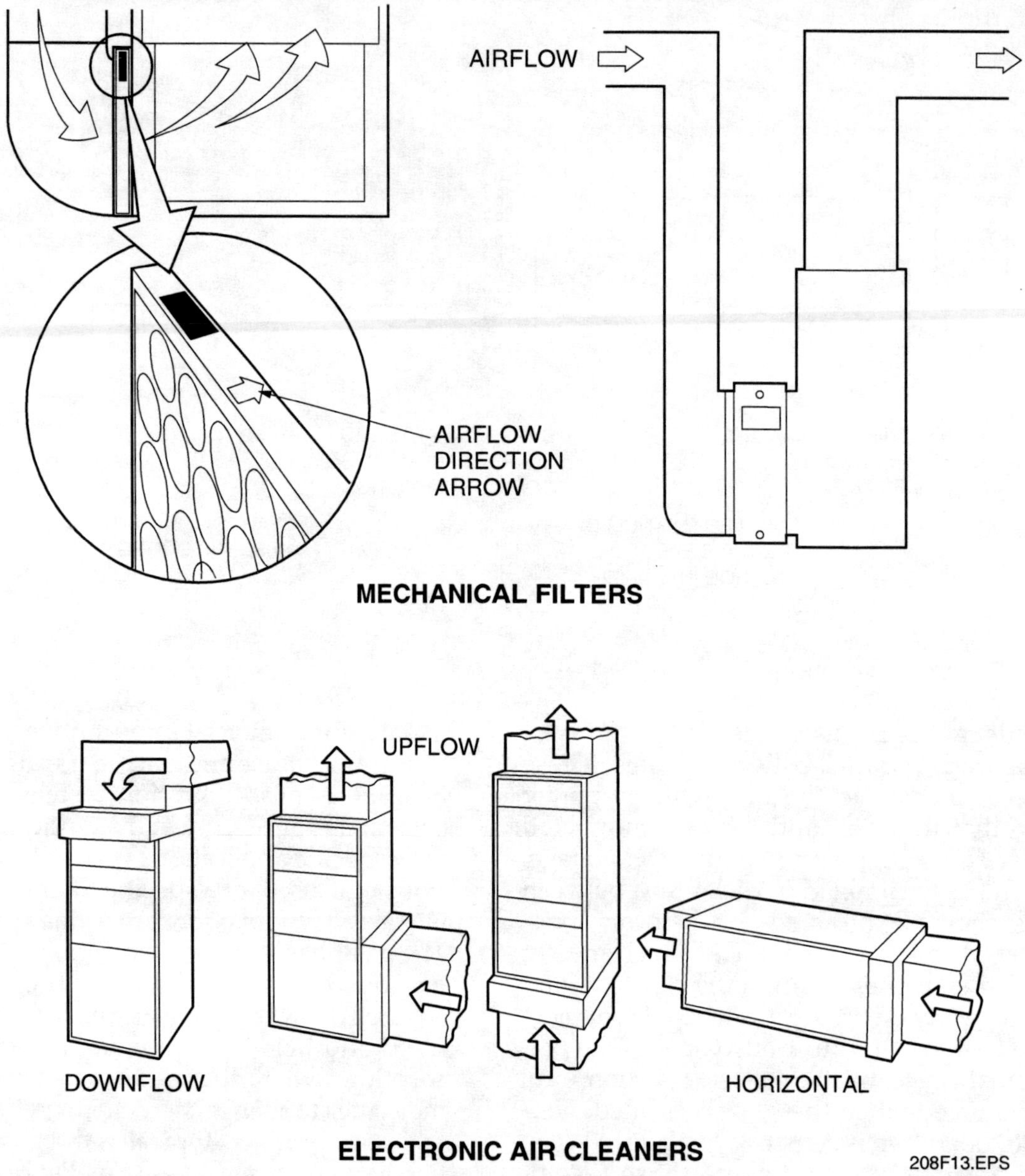

*Figure 13* ♦ Typical filter and electronic air cleaner placement.

Disposable filters should be replaced, and permanent ones cleaned, when they lose their efficiency or become so clogged that they produce a pressure drop that is too high. Visual inspection is one way to determine if a filter needs replacement. If it has turned black, the frame is bent or warped, or the filtering material is ripped or punctured, replace the filter. Filter cleanliness can be checked by placing a strong light on one side of the filter and looking through the filter from the other side to see how much light can be seen and how uniform the pattern is. Another method is to use a manometer or differential pressure gauge to determine if a filter has a pressure drop that is too high. The two ports on the manometer are connected to measure the airflow on the two opposite sides of the filter. Typically, if the pressure drop exceeds more than 25% of the pressure drop across the fan, it is too high. If possible, check the manufacturer's specifications to find out what the normal pressure drop of the filter should be.

Permanent mechanical filters are cleaned by vacuuming and/or washing with a mixture of mild detergent and warm water, followed by a rinse. Normally, the collector filter in electronic air cleaners can be cleaned in the same way. Be sure to disconnect the power before attempting any cleaning. Follow the manufacturer's instructions for cleaning the unit.

## 5.0.0 ◆ AIR CONDITIONING ENERGY CONSERVATION EQUIPMENT

Many kinds of energy-conservation equipment are being incorporated into air conditioning systems. Three popular conservation methods are:

- Energy and heat recovery ventilators
- Economizers
- Zoned control

### 5.1.0 Energy and Heat Recovery Ventilators

Energy-efficient homes and buildings do a good job of keeping conditioned heated or cooled air in, but they also seal in air that has been recirculated within the building many times. This causes the air to become stale and contaminated with airborne particles. Without proper ventilation, there is more of an opportunity for mold to grow, moisture can condense on windows, people may become sick more often, and a musty smell may be noticed. For this reason, more and more states now require mechanical ventilation in every new residential building. For the healthiest living environment, ASHRAE standards recommend that a building's indoor air be exchanged for fresh outdoor air at a rate of .35 air changes per hour. An alternate method recommended by ASHRAE calls for an exchange rate of 15 cfm per person, 20 cfm per bathroom, and 25 cfm per kitchen. Ventilators are one type of HVAC equipment that can be used to help solve poor indoor air quality problems within a building by bringing a controlled amount of outside air into the building. In addition to helping maintain good indoor air quality, ventilators also help to conserve energy.

There are two types of ventilators: energy recovery ventilators (ERVs) and heat recovery ventilators (HRVs). ERVs are used to supply fresh air and recover both heating and cooling energy year-round. ERVs are used in most localities in the United States. HRVs are used to supply fresh air and recover heat energy during the heating season. They typically are installed in homes in colder climates that have longer heating seasons, such as those in the northern part of the United States and Canada.

There are several manufacturers and designs for ERV and HRV equipment. The residential ERV and HRV units of one major manufacturer described here are similar both in construction and operation. Both have a heat exchanger central core and one or more blowers to push air through the unit. According to the U.S. Department of Energy, most models are capable of recovering about 60% to 80% of the energy from the exiting air and delivering the energy to the incoming fresh air. Typically, an ERV/HRV improves the indoor air by changing the air about every three hours.

Air from the living space is passed through the ERV or HRV and exhausted outside (*Figure 14*). At the same time, fresh air is brought in from the outside and sent through the unit. When the two air streams pass through the heat exchanger core, most of the heat or cooling from the outgoing indoor air is transferred to the incoming fresh outdoor air. The core design allows this transfer of heat and cooling between the entering and leaving air streams to occur without mixing the two air streams. The result is a constant stream of fresh air being delivered to the living space.

The main difference between an ERV and HRV is the way the heat exchanger core works. In the HRV, only sensible heat is transferred. That's why HRVs are used mainly in colder climates. In the ERV, the core has the capability of transferring both sensible and latent heat, allowing it to transfer heat in the winter and remove moisture from the air during the summer cooling season. This makes the use of ERVs popular in humid climates, such as in the southeast. Upon installation of an ERV or HRV, balancing of the air distribution system is critical to make sure that the amounts of incoming and outgoing air are equal.

### 5.2.0 Economizers

An economizer is an accessory often found on commercial rooftop packaged heating/cooling units. The economizer mixes outdoor air with conditioned air in a proportion that depends on the outdoor air temperature and humidity. The controlled use of outdoor air reduces the amount of conditioned air needed, and thus reduces the operating cost of the system. In addition to using outdoor air for cooling, the economizer also controls building ventilation by introducing a minimum amount of outdoor air into the indoor environment.

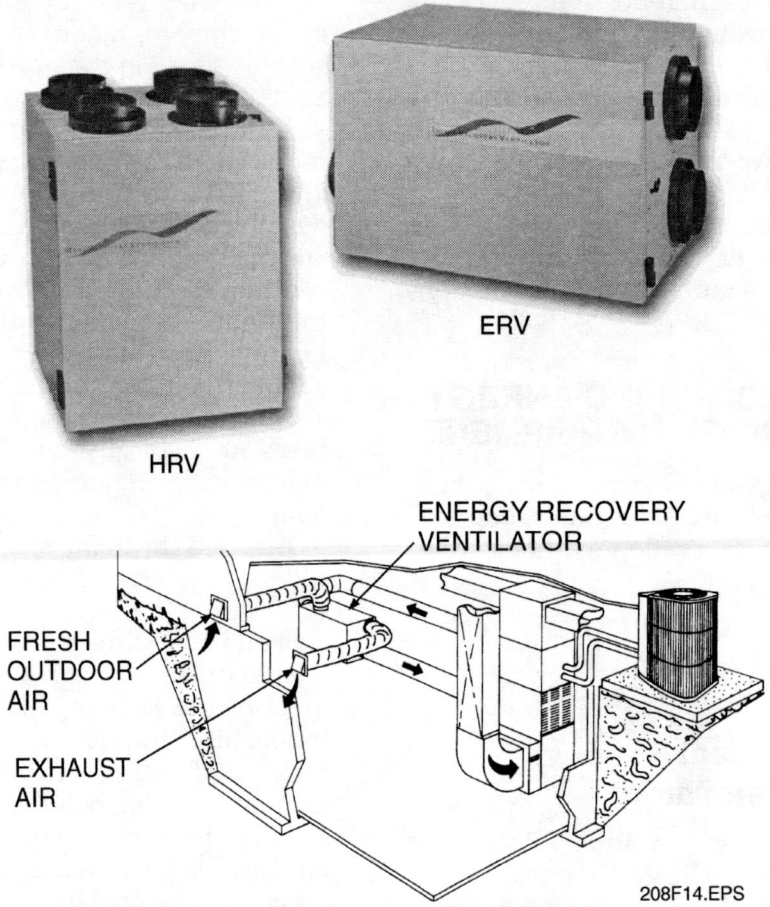

*Figure 14* ◆ Recovery ventilators.

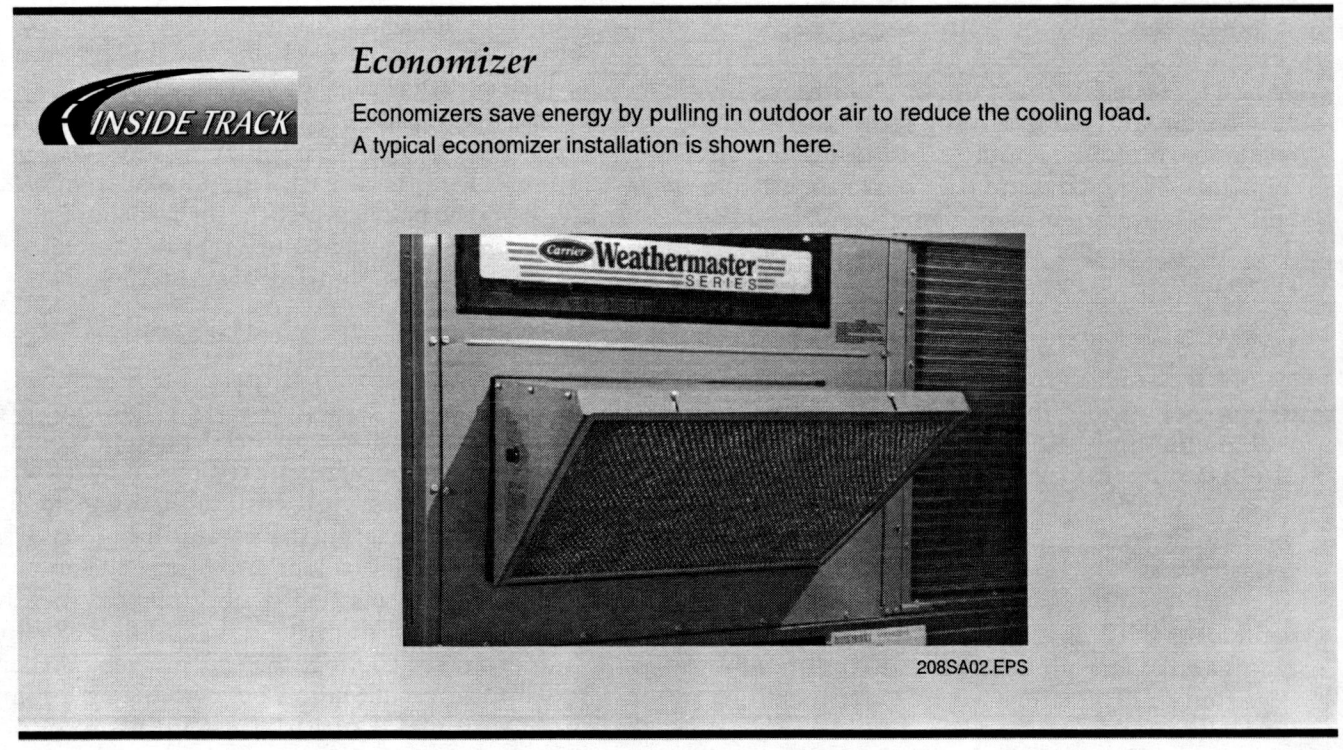

## Economizer

Economizers save energy by pulling in outdoor air to reduce the cooling load. A typical economizer installation is shown here.

The benefit of using an economizer is related mainly to the cooling mode of operation. Four conditions are used to control operation of an economizer: outside air, return air, mixed air temperature, and the required ventilation air.

Outdoor air conditions are sensed by either an outdoor air thermostat that senses the outdoor dry-bulb temperature or an **enthalpy** controller (sensor) that senses both the temperature and humidity. Enthalpy-based controls are typically used in high humidity areas. When the outdoor air sensor detects that the outdoor air is above its setpoint, cooling for the building is provided in the conventional way by the air conditioner compressor (**mechanical cooling**). When the outdoor air falls below the sensor setpoint, the economizer control system acts to cool the building by turning off the system compressor and using the indoor fan to bring outside air into the building through motor-actuated dampers. This mode of operation is called **free cooling**. The outdoor sensor setpoint is normally selected by the system designer and is based on both comfort and economy.

*Figure 15* shows a basic economizer. As shown, it consists of a mixed air thermostat (MAT) located in the conditioned space; damper motor and control circuit; and outdoor and return air dampers. During the free cooling mode, the MAT monitors the average air temperature on the face of the indoor coil and compares it to a predetermined setpoint. This thermostat is called a *mixed air* thermostat because it senses the temperature of the building air, which is a mixture of both outdoor and return air coming from the conditioned space. The MAT provides a voltage input to the damper motor and control unit.

The function of the damper motor and its control circuit is to control the position (open, closed, or somewhere in between) of the system outdoor air and return air dampers. It does this through linkage rods connected to the damper motor's shaft. The damper motor has two windings that determine the direction of shaft rotation. When energized, one of the windings will cause the shaft to turn clockwise and the other winding will cause it to turn counterclockwise. Only one winding can be

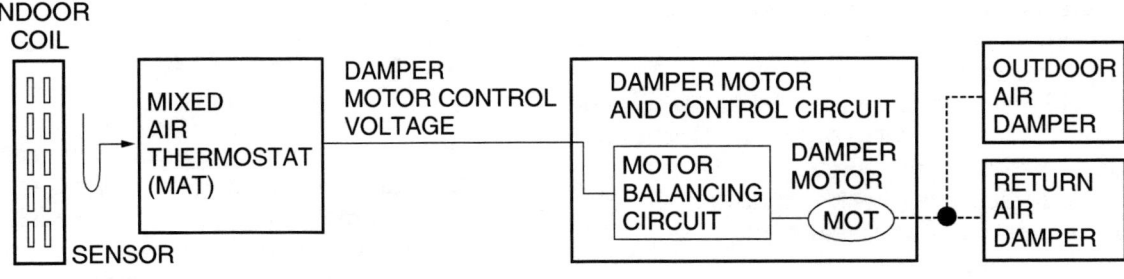

| MODE OF OPERATION | OUTDOOR AIR DAMPER | RETURN AIR DAMPER |
|---|---|---|
| OFF | CLOSED | WIDE OPEN |
| FAN ONLY AND MECHANICAL COOLING | OPENS TO MINIMUM POSITION FOR VENTILATION | MODULATES TO COMPLEMENT OUTDOOR AIR DAMPER |
| FREE COOLING | MODULATES TO PROVIDE THE PROPER MIXED AIR TEMPERATURE | MODULATES TO COMPLEMENT OUTDOOR AIR DAMPER |
| HEATING | OPENS TO MINIMUM POSITION FOR VENTILATION | MODULATES TO COMPLEMENT OUTDOOR AIR DAMPER |

*Figure 15* ◆ Basic economizer unit.

energized at a time. The direction of rotation is determined by the voltage level applied to the damper motor from the MAT. This voltage level changes in response to changes in the condition of the indoor air sensed by the thermostat. The correct damper motor winding is energized by the action of a motor balancing circuit, which is part of the motor damper and control unit. This causes the damper motor to rotate and reposition the vanes of the outdoor air and return air dampers as needed. The damper motor continues driving the dampers until turned off by the motor balancing circuit. In this way, the predetermined mixed air temperature level is maintained in the building.

The damper motor and control unit circuitry also controls the operation of the dampers to provide for ventilation during all operating modes of the air conditioning system. The damper positions for all modes of operation are summarized in *Figure 15*.

## 5.3.0 Evaporative Pre-Coolers

Evaporative pre-coolers are used to pre-cool air that is used to cool equipment such as heat exchangers. A typical example of how an evaporative pre-cooler can be used is with a conventional air conditioner condenser coil (*Figure 16*). The condenser coil is used to dissipate heat to the outside air during the cooling cycle of the air conditioner. Without pre-cooling, the condenser coil uses outside air to dissipate the heat. If that heat is 120°F, then the efficiency of the heat transfer is greatly reduced. With evaporative pre-cooling, the 120°F air temperature is reduced to about 90°F, which greatly improves the efficiency of the heat transfer of the condenser coil. At lower temperatures across the condenser coil, the air conditioner efficiency improves, the energy used is reduced, and the useful life of the compressor is increased. The air conditioner produces colder air at a lower energy cost.

As shown, the pre-cooler is placed over the air intake of the condenser unit. The pre-cooler unit has wet and dry sections and a pump and sump-type water distribution system with a metering valve and bleed-off line. Water is delivered to the evaporative cooler from its source by PVC Schedule 40 pipe. The wet section contains a rigid media over which the water is circulated. The dry section contains one or more blowers used to force air across the media pad in the wet section. Direct evaporative cooling results from warm air being blown over the water-soaked media. In the wet section, the water is discharged upwards against a water deflector which redirects it back evenly over the full width of the media. The water absorbs the heat from the air, bringing the dry-bulb tempera-

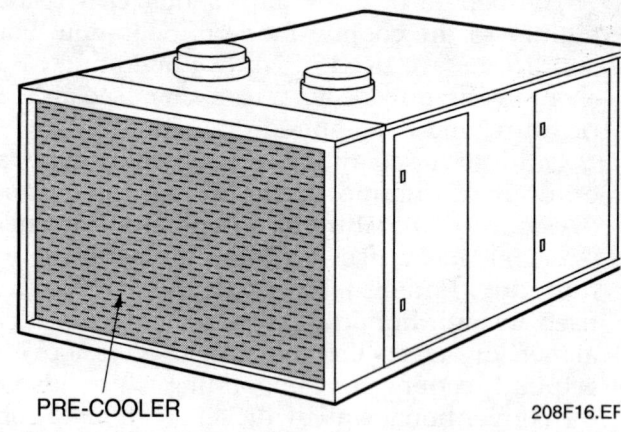

*Figure 16* ◆ Evaporative pre-cooler mounted on an air conditioning unit.

ture closer to the wet-bulb temperature. It is the difference between the wet- and dry-bulb temperatures that determines the efficiency of the evaporative cooling system. The greater the wet-bulb/dry-bulb temperature differential, the greater the transfer of heat from the air to the water.

## 5.4.0 Zoned Control

Zoned control is the division of a building into a number of separately controlled spaces, or zones, where different heating or cooling temperatures can be maintained at the same time. This provides a way to overcome variations in cooling and heating loads that occur in different areas of the building at different times. Zoned control allows the occupants to set conditions in each zone independently. Energy can be saved when a zone is unoccupied by lowering or raising the thermostat setpoint temperature to control cooling or heating, respectively. When climatic conditions allow, cooling or heating in any zone can be turned off while comfort conditions are maintained in the rest of the zones.

Zones are set up by grouping one or more rooms in a building where there is little variation in heating or cooling needs, allowing them to be controlled by one thermostat. The selection of zones is usually based on either usage or exposure. When based on usage, the factors include:

- The zone is occupied only during the day or night. For example, bedrooms are only occupied at night. When unoccupied during the day, the thermostat can be set to save energy. Similarly, business offices are typically occupied only during the day, allowing the thermostat to be set to save energy during the night.

- The zone is occupied on an irregular basis (for example, recreational areas, and meeting and conference room areas).
- The zones are occupied at specific times of the day by the same group of people (for example, classroom, shop, cafeteria, and laboratory areas in a school).

When setting up zones by exposure, the factors involved include:

- Portions of the building have large areas of glass, or poorly shaded areas, with cooling loads caused mainly by solar heat gain through the glass.
- The building is occupied by a small amount of people so the internal heat gain from people is small compared to the overall cooling requirements for the building.

There are several types of zoned systems in use. *Figure 17* shows a typical forced-air zoned system. As shown, it consists of four zones supplied from a central cooling/heating unit. Supply of the conditioned air to each zone is through a damper controlled by the zone thermostat.

Some forced-air zoned systems use separate heating and cooling systems for each zone. Other types of zoned systems include:

- Hydronic (hot water) systems that use separate hot water loops to each zone controlled by zone valves or separate circulator pumps.
- Electric heat perimeter systems in which each area or room is typically set up as its own zone.
- Ductless, split systems in which each zone uses a separate indoor fan coil with its own thermostat. A single outdoor condensing unit is connected to two or more indoor coils with parallel refrigerant circuits controlled by solenoid valves.

## 6.0.0 ◆ FIRE AND SMOKE DAMPERS

Openings for ducts in walls and floors with fire-resistance ratings must be protected by fire dampers as required by local codes. Air transfer openings should also be protected. Smoke dampers are used for smoke management (smoke containment) or for smoke control. In smoke containment, a smoke damper stops the passage of smoke under forces of buoyancy, stack effect, and wind. In a smoke control system, a smoke damper stops the flow of air that may or may not contain smoke. They may be used to shut off air to a smoke zone or shut off exhaust air from a non-smoke zone.

At locations requiring both fire and smoke dampers, combination dampers that meet the requirements can be used. Fire and smoke dampers are covered in detail in the *Air Distribution Systems* module studied earlier in this level.

Smoke detection systems monitor the smoke density that exists in a chimney or duct system. One method uses a photoelectric cell (photocell) and beam tube installed in the chimney or duct. A signal generated by the photocell indicates the amount of smoke. It is amplified for display on a local indicator or can be recorded using a recording device. If the amount of smoke detected exceeds a preset level, an alarm may be given. In addition to the alarm, the signal may also be applied to a control circuit where it initiates the shutdown of the system. Photocells and other electronic sensors are also used to control smoke dampers in duct runs. If the amount of smoke detected exceeds a preset level, a damper-holding device is tripped, causing the smoke damper to close.

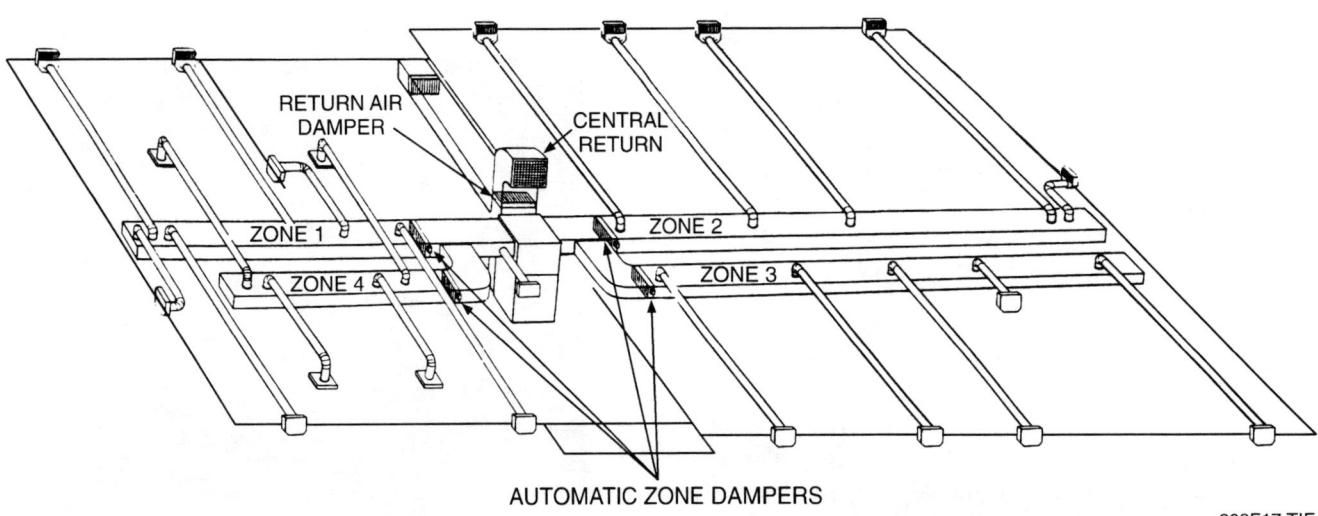

*Figure 17* ◆ Typical forced-air zoned system.

> **NOTE**
> 
> The location of fire and smoke dampers must be determined by a qualified architect or engineer. See NFPA-90A, *Installation of Air Conditioning and Ventilating Systems*.

Photocells and other types of sensing devices normally need periodic cleaning to remove soot buildup. They should be cleaned according to the sensor manufacturer's instructions.

## 7.0.0 ◆ ULTRAVIOLET LIGHT AIR PURIFICATION SYSTEMS

Ultraviolet (UV) light air purification equipment can be used in HVAC air distribution systems to help prevent the growth of bacteria and other microorganisms that are known to cause indoor air problems and musty, mold-related odors. There are many manufacturers and designs of UV air purification equipment, but they all share the same basic principles of operation. C-band UV light (UVC) energy in the 240- to 280-nanometer wavelength range destroys microorganisms by penetrating their cell walls and damaging the protein structure of the cells and/or chemically altering the cell's DNA. Once this occurs, the organisms die or cannot reproduce. Germicidal effectiveness (killing power) is directly related to the UV dose applied, which is a function of time and intensity.

HVAC system air purification by UVC light is done in one of two ways: purification of a fixed object, or purification of the moving air stream. In fixed-object purification, the HVAC supply side evaporator/indoor coil and drain pan is continuously irradiated with light rays generated by stationary quartz UVC lamps, probes, etc. (emitters). Over time, the UVC rays destroy bacteria and viruses that are present on the fixed object. The time required to destroy microorganisms on fixed objects depends on a number of factors, including the distance at which the UVC emitter is mounted from the fixed object, the size and intensity or killing power of the UVC emitter, and the temperature of the air and UVC emitter.

In UV purification of the moving air stream, the air in a duct system is irradiated as it moves past a stationary UVC emitter. Achieving air purification using this method is much more difficult because of the short time (dwell time) that the air moving past the emitter is irradiated. Typically, the air moves past the UVC emitter at a speed of about 600 fpm or faster, spending only about 20 milliseconds in front of the probe/emitter. The intensity or killing power of the UVC emitter, how fast the UV ray intensity decreases with distance as the air moves away from the emitter, and how far into the air stream the UV rays penetrate determine the UV purification efficiency on the moving air. Because of the short dwell time that a portion of the moving air is subjected to the UV light rays, purification of the air stream normally requires the use of multiple UV light sources and reflectors that are capable of producing much stronger UV light rays than needed for a fixed object. For this reason, fixed object air purification systems are more widely used.

*Figure 18* shows an example of a typical UVC air purification unit. It is designed to protect coils, drain pans, and humidifiers from mold and bacterial growth while killing some airborne microorganisms. It consists of a housing, power supply, and emitters. The components are incorporated into one assembly that is mounted outside the equipment at the cooling coil ductwork, with the emitters protruding into the center of the coil and air stream.

## 8.0.0 ◆ CARBON MONOXIDE AND CARBON DIOXIDE MONITORS

Carbon monoxide and carbon dioxide monitors are widely used HVAC accessories. The deadly effects of carbon monoxide (CO) are well known. Carbon monoxide gas results from incomplete combustion of carbon fuel. Carbon monoxide is colorless, odorless, and tasteless. A carbon monoxide (CO) detector is both a safety and IAQ monitor. Early detection of high CO is almost impossible without a CO detector. Stationary CO monitors (*Figure 19*) are made for use in automated systems. They are installed in strategic

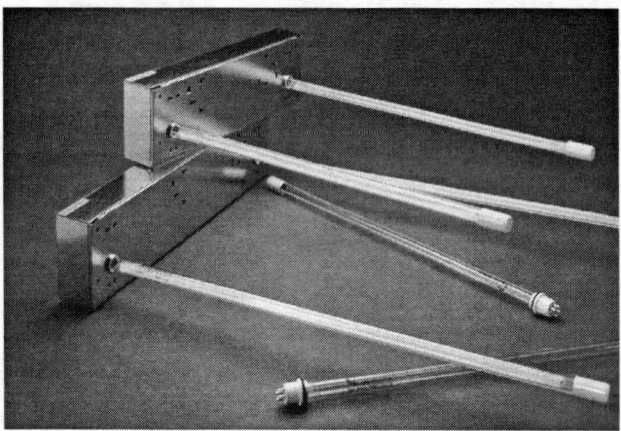

*Figure 18* ◆ Example of a UVC air purification unit.

locations throughout a building and normally activate a contact closure and sound an alarm when a high level of CO is detected. Portable CO detectors are also available for use mainly when testing the operation of HVAC combustion equipment.

Carbon dioxide ($CO_2$) is also a colorless, odorless, tasteless gas derived from combustion and metabolic processes such as human breathing. The concentration level of $CO_2$ in a building is commonly used as one indicator of IAQ. Low concentrations of $CO_2$ produced by people are always present in buildings. Concentrations below 1,000 parts per million (ppm) generally indicate that the building's ventilation is adequate to deal with the routine by-products of human occupancy. $CO_2$ levels above 1,000 ppm can indicate a ventilation problem. At higher building concentrations, some loss of the occupants' mental awareness is noted.

$CO_2$ sensors (*Figure 20*) are used to monitor $CO_2$ levels in a building. Some models can be used as a ventilation controller in demand-based ventilation control systems. When used as a ventilation controller, the $CO_2$ sensor determines the need for ventilation based on the $CO_2$ concentration and modulates the position of the building's dampers to maintain acceptable ventilation. If a space is unoccupied, the $CO_2$ controller will set the air intake volume at a minimum setting that allows established ventilation rates to be maintained while reducing over-ventilation.

In addition to CO and $CO_2$ monitors, there are many other types of gas monitors used to monitor the quality of air. These monitors are covered in detail in HVAC Level Four.

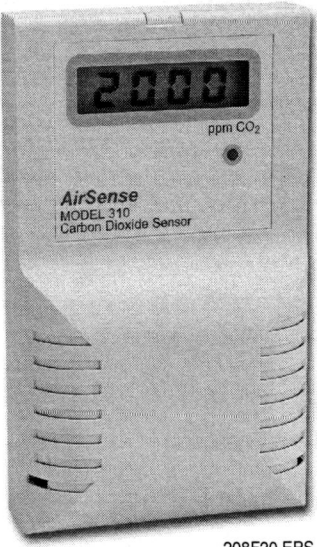

*Figure 20* ♦ Typical carbon dioxide ($CO_2$) monitor.

## Summary

Air conditioning is the process of treating indoor air to control indoor air quality, including its temperature, humidity, cleanliness, and distribution. Air conditioning is done to provide comfort for humans and to support controlled manufacturing processes and materials.

Air conditioning for supporting a manufacturing process or refrigerating materials is based on the nature of the process or material and will vary from system to system. Air conditioning in support of human comfort is well-known, and basic requirements rarely vary from system to system. Control of the air for human comfort is sometimes called *comfort air conditioning*. It provides:

- Heating and cooling
- Humidification and dehumidification
- Ventilation
- Filtration
- Air circulation

Several accessories are used with the basic air conditioning system to enhance human comfort and/or conserve energy in the conditioned space. Common accessories include:

- Humidifiers
- Mechanical filters and electronic air cleaners
- Energy and heat recovery ventilators
- Economizer equipment
- Evaporative pre-coolers
- Zoned air conditioning systems
- Fire and smoke dampers and smoke detection equipment

*Figure 19* ♦ Typical carbon monoxide (CO) monitor.

**Review Questions**

1. Human comfort can be maintained by controlling the room _____.
   a. air temperature and humidity
   b. convection
   c. radiation
   d. evaporation

2. Heat is transferred between humans and surrounding surfaces such as windows or walls by _____.
   a. conduction
   b. convection
   c. radiation
   d. evaporation

3. The relative humidity in a room affects the transfer of heat from a body mainly through _____.
   a. conduction
   b. evaporation
   c. radiation
   d. convection

4. Too much humidity can cause _____.
   a. dry, itchy skin
   b. static electricity shocks
   c. growth of bacteria
   d. sinus problems

5. Which humidifier uses heat from the system to vaporize the water?
   a. Ultrasonic
   b. Infrared
   c. Steam
   d. Fan-powered

6. Which humidifier is *not* an evaporative humidifier?
   a. Spinning disc
   b. Rotating disc
   c. Rotating drum
   d. Bypass

7. A 1,500-square foot tight house needs a humidifier that supplies about _____ gallons per day.
   a. 5
   b. 7
   c. 10
   d. 12

8. Which type of disposable filter would most likely be used in a hospital operating room?
   a. Extended surface
   b. Bag-type
   c. Conventional
   d. Electronic

9. A device that uses a heat exchanger to transfer heat from the furnace exhaust air to ventilation air entering a building is a(n) _____.
   a. economizer
   b. zone damper control
   c. heat recovery ventilator
   d. mixed air thermostat

10. When dividing a building into zones based on exposure, you must consider _____.
    a. areas in the building with greater amounts of glass
    b. if the zone is occupied at specific times of the day
    c. if the zone is occupied only during the day or night
    d. if the zone is occupied on an irregular basis

## GLOSSARY

# Trade Terms Introduced in this Module

*Dehumidifier:* A device used to remove moisture from the air.

*Economizer:* An HVAC device that substitutes outdoor air for the cooled air produced by the air conditioning system, when outdoor air conditions permit. It also controls the amount of outdoor air used to ventilate a building.

*Energy recovery ventilator (ERV):* HVAC equipment used to supply fresh air and recover both heating and cooling energy year-round.

*Enthalpy:* The total heat content of a substance. In HVAC, the total heat content of the air and water vapor mixture as measured from a predetermined base or point.

*Evaporation:* The condition in which the heat absorbed by a liquid causes it to change into a vapor.

*Free cooling:* A mode of economizer operation. It is the cooling provided by outside air rather than the compressor.

*Heat recovery ventilator (HRV):* HVAC equipment that saves energy by using a heat exchanger to transfer heat from the heating system exhaust air to the cold ventilation air that is entering the building.

*Humidifier:* A device used to control humidity.

*Mechanical cooling:* A mode of economizer operation. It is the cooling provided in the conventional manner by the compressor.

*Micron:* One-millionth of a meter (about $\frac{1}{25,400}$ of an inch). It is also a precise measurement of pressure used with electronic vacuum measuring instruments and vacuum pumps. One inch of mercury equals 25,400 microns.

# ANSWER KEY

## Answers to Review Questions

| Answer | Section |
|---|---|
| 1. a | 2.2.0 |
| 2. c | 2.2.3 |
| 3. b | 2.2.4 |
| 4. c | 3.0.0 |
| 5. d | 3.3.2 |
| 6. a | 3.3.3 |
| 7. b | 3.3.6; Figure 9 |
| 8. b | 4.1.3 |
| 9. c | 5.1.0 |
| 10. a | 5.4.0 |

# REFERENCES & ACKNOWLEDGMENTS

## *Additional Resources*

This module is intended to present thorough resources for task training. The following reference works are suggested for further study. These are optional materials for continued education rather than for task training.

*Modern Refrigeration and Air Conditioning*, 2000. A.D. Althouse, C.H. Turnquist, A.F. Bracciano. Tinley Park, IL: The Goodheart-Willcox Company, Inc.

*Refrigeration & Air Conditioning Technology,* 2000. William C. Whitman, William M. Johnson, John A. Tomczyk. Albany, NY: Delmar Publishers, Inc.

## *Figure Credits*

| | |
|---|---|
| **Thomas P. Burke** | 208SA01 |
| **Carrier Corporation** | 208F01, 208F14, 208SA02 |
| **Steril-Aire, Inc.** | 208F18 |
| **Brooks Equipment Company** | 208F19 |
| **Digital Control Systems** | 208F20 |

# NCCER CRAFT TRAINING USER UPDATES

The NCCER makes every effort to keep these textbooks up-to-date and free of technical errors. We appreciate your help in this process. If you have an idea for improving this textbook, or if you find an error, a typographical mistake, or an inaccuracy in the NCCER's Craft Training textbooks, please write us, using this form or a photocopy. Be sure to include the exact module number, page number, a detailed description, and the correction, if applicable. Your input will be brought to the attention of the Technical Review Committee. Thank you for your assistance.

*Instructors* – If you found that additional materials were necessary in order to teach this module effectively, please let us know so that we may include them in the Equipment and Materials list in the Instructor's Guide.

**Write:** Curriculum Revision and Development Department
National Center for Construction Education and Research
P.O. Box 141104, Gainesville, FL 32614-1104

**Fax:** 352-334-0932

**E-mail:** curriculum@nccer.org

Craft _____ Module Name _____

Copyright Date _____ Module Number _____ Page Number(s) _____

Description
_____
_____
_____
_____

(Optional) Correction
_____
_____
_____

(Optional) Your Name and Address
_____
_____
_____

Module 03209-01

*Metering Devices*

## COURSE MAP

This course map shows all of the modules in the second level of the HVAC curriculum. The suggested training order begins at the bottom and proceeds up. Skill levels increase as you advance on the course map. The local Training Program Sponsor may adjust the training order.

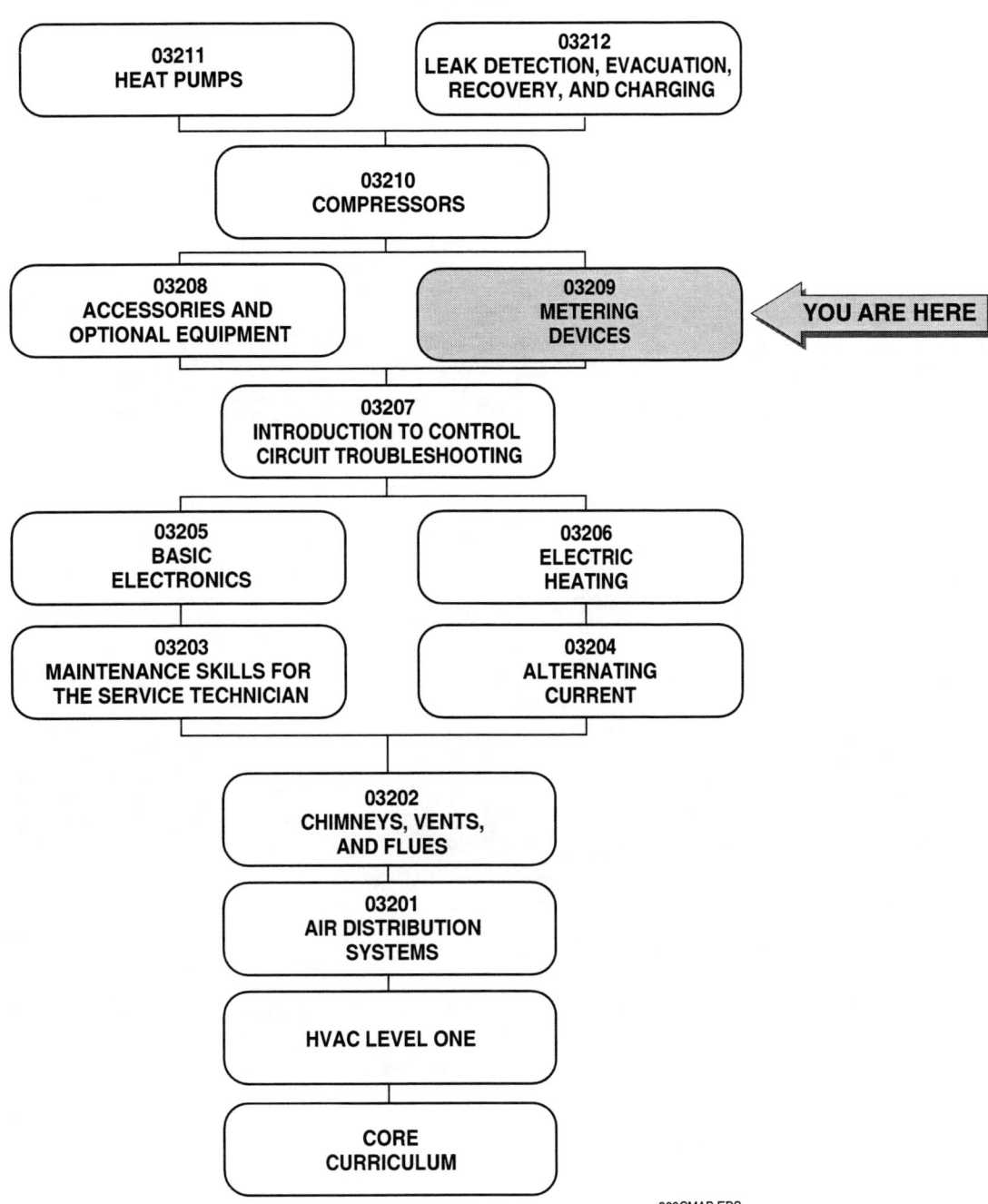

## MODULE 03209 CONTENTS

1.0.0 INTRODUCTION . . . . . . . . . . . . . . . . . . . . . . . . . . . . . . . . . . . . . . . . .9.1
2.0.0 BASIC OPERATION . . . . . . . . . . . . . . . . . . . . . . . . . . . . . . . . . . . . .9.1
    2.1.0 Functions . . . . . . . . . . . . . . . . . . . . . . . . . . . . . . . . . . . . . . . . .9.3
    2.2.0 Adapting to Load Changes . . . . . . . . . . . . . . . . . . . . . . . . . . . .9.3
3.0.0 FIXED METERING DEVICES . . . . . . . . . . . . . . . . . . . . . . . . . . . . . .9.4
    3.1.0 Capillary Tubes . . . . . . . . . . . . . . . . . . . . . . . . . . . . . . . . . . . . .9.4
    3.2.0 Fixed-Orifice Metering Devices . . . . . . . . . . . . . . . . . . . . . . . .9.6
4.0.0 EXPANSION VALVES . . . . . . . . . . . . . . . . . . . . . . . . . . . . . . . . . . . .9.7
    4.1.0 Manual Expansion Valves . . . . . . . . . . . . . . . . . . . . . . . . . . . .9.7
    4.2.0 High-Side Float Valves . . . . . . . . . . . . . . . . . . . . . . . . . . . . . .9.7
    4.3.0 Low-Side Float Valves . . . . . . . . . . . . . . . . . . . . . . . . . . . . . .9.8
    4.4.0 Automatic Expansion Valves . . . . . . . . . . . . . . . . . . . . . . . . . .9.8
    4.5.0 Thermal Expansion Valves . . . . . . . . . . . . . . . . . . . . . . . . . . .9.9
    *4.5.1 Operating Principles* . . . . . . . . . . . . . . . . . . . . . . . . . . . . . . .9.9
    *4.5.2 Equalizers* . . . . . . . . . . . . . . . . . . . . . . . . . . . . . . . . . . . . . .9.11
    4.6.0 Thermal-Electric Expansion Valves . . . . . . . . . . . . . . . . . . . .9.13
    4.7.0 Electronic Expansion Valves . . . . . . . . . . . . . . . . . . . . . . . .9.14
    4.8.0 Hunting . . . . . . . . . . . . . . . . . . . . . . . . . . . . . . . . . . . . . . . . .9.14
5.0.0 DISTRIBUTORS . . . . . . . . . . . . . . . . . . . . . . . . . . . . . . . . . . . . . . .9.15
6.0.0 TXV REPLACEMENT . . . . . . . . . . . . . . . . . . . . . . . . . . . . . . . . . .9.16
    6.1.0 Selection . . . . . . . . . . . . . . . . . . . . . . . . . . . . . . . . . . . . . . .9.16
    6.2.0 Placement . . . . . . . . . . . . . . . . . . . . . . . . . . . . . . . . . . . . . .9.16
    6.3.0 Sensing Bulb . . . . . . . . . . . . . . . . . . . . . . . . . . . . . . . . . . . .9.16
    6.4.0 TXV Adjustment . . . . . . . . . . . . . . . . . . . . . . . . . . . . . . . . . .9.18
7.0.0 METERING DEVICE PROBLEMS . . . . . . . . . . . . . . . . . . . . . . . . . .9.18
SUMMARY . . . . . . . . . . . . . . . . . . . . . . . . . . . . . . . . . . . . . . . . . . . . . . .9.19
REVIEW QUESTIONS . . . . . . . . . . . . . . . . . . . . . . . . . . . . . . . . . . . . . .9.20
GLOSSARY . . . . . . . . . . . . . . . . . . . . . . . . . . . . . . . . . . . . . . . . . . . . . .9.22
ANSWERS TO REVIEW QUESTIONS . . . . . . . . . . . . . . . . . . . . . . . . . .9.23
REFERENCES & ACKNOWLEDGMENTS . . . . . . . . . . . . . . . . . . . . . . .9.24

## Figures

Figure 1    Metering device location ....................... 9.2
Figure 2    Metering device operation ..................... 9.2
Figure 3    Capillary tube ................................. 9.4
Figure 4    Fixed-orifice device ........................... 9.6
Figure 5    Manual expansion valve ....................... 9.7
Figure 6    High-side float valve used with flooded evaporator ... 9.7
Figure 7    Low-side float valve ........................... 9.8
Figure 8    Automatic expansion valve ..................... 9.8
Figure 9    Refrigerant flow in the evaporator ............. 9.9
Figure 10   Thermal expansion valve ....................... 9.10
Figure 11   Expansion valve with internal equalizer ......... 9.11
Figure 12   Expansion valve with external equalizer ......... 9.11
Figure 13   Internally equalized TXV ....................... 9.12
Figure 14   Externally equalized TXV ....................... 9.13
Figure 15   Thermal-electric expansion valve ............... 9.14
Figure 16   Electronic expansion valve (EEV) ............... 9.14
Figure 17   Suction temperature variations indicate hunting ... 9.15
Figure 18   Multi-circuit evaporator fed by a distributor ....... 9.15
Figure 19   Distributor .................................... 9.16
Figure 20   Sensing bulb positioning ....................... 9.17
Figure 21   Trap in a suction line riser ..................... 9.18

# MODULE 03209

# Metering Devices

## Objectives

When you have completed this module, you will be able to do the following:

1. Explain the function of metering devices.
2. Describe the operation of selected metering devices and expansion valves.
3. Identify types of thermal expansion valves (TXVs).
4. Describe problems associated with replacement of TXVs.
5. Describe the procedure for installing and adjusting selected TXVs.

## Prerequisites

Before you begin this module, it is recommended that you successfully complete the following modules: Core Curriculum; HVAC Level One; HVAC Level Two, Modules 03201 through 03208.

## Required Trainee Materials

1. Pencil and Paper
2. Appropriate Personal Protective Equipment

## 1.0.0 ♦ INTRODUCTION

The metering device in a refrigeration system performs two important functions. First, it matches the rate of refrigerant flow entering the evaporator with the rate at which the evaporator will boil liquid refrigerant into vapor. Second, it provides a pressure drop that separates the high side of the system from the low side, where the actual cooling occurs. A variety of metering devices are used in modern systems to control the flow of refrigerant. These devices include **capillary tubes**, precision **orifices**, pressure-operated valves, **thermal expansion valves (TEVs or TXVs)**, float valves, **electronic expansion valves (EEVs)**, and **distributors**. The metering device is also called a *flow control device* or *refrigerant control*.

## 2.0.0 ♦ BASIC OPERATION

The metering device exists because of the need for a refrigeration system to operate at both high and low pressures. The refrigerant entering the condenser must be at a high pressure, because its temperature must be higher than that of the outdoor air in order for heat transfer to take place. The refrigerant entering the evaporator, on the other hand, must have a low pressure and temperature, because the temperature of the refrigerant entering the evaporator must be lower than that of the air entering the evaporator in order for heat transfer to take place. To accomplish this, a metering device is installed between the condenser and the evaporator, usually at the evaporator input (*Figure 1*). In order for a metering device to operate, there must be a minimum pressure drop across the orifice.

In a metering device, high-temperature, high-pressure, subcooled refrigerant from the condenser is forced through a tiny opening or orifice, which reduces the pressure at the output (see *Figure 2*). In addition, some of the refrigerant turns into vapor, known as **flash gas**, which helps to cool the remaining refrigerant. If the pressure of the refrigerant entering the metering device is 300 psig, the pressure at the evaporator entry might be 69 psig. If the temperature at the metering device input is 110°F, the leaving temperature might be about 40°F. This is, by the way, the temperature at which HCFC-22 boils at 69 psig.

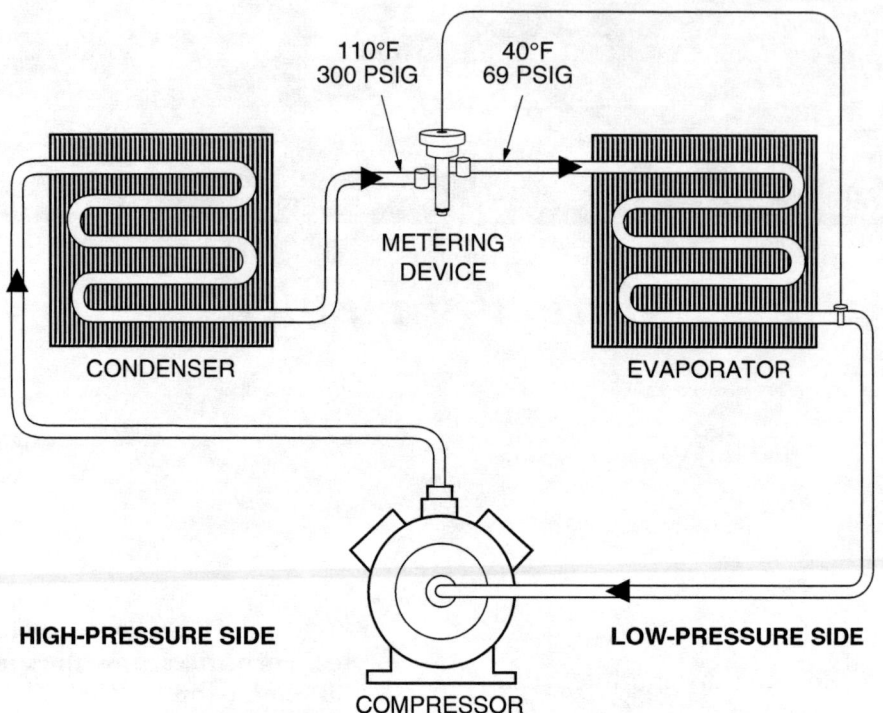

*Figure 1* ◆ Metering device location.

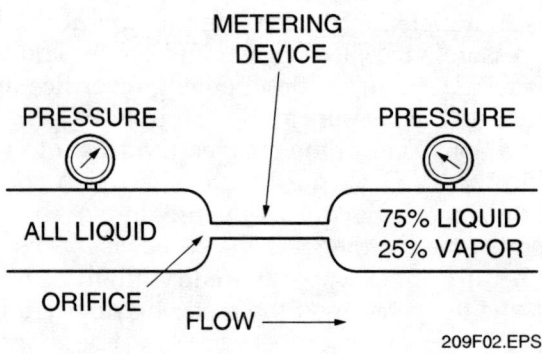

*Figure 2* ◆ Metering device operation.

In many systems, the flash gas enters the evaporator along with the liquid refrigerant. Thus, the refrigerant entering the evaporator is a mix of about 75% to 80% liquid and 20% to 25% vapor. This applies to systems of 100 tons or less that use **direct-expansion (DX) evaporators**. Large systems containing centrifugal or absorption chillers use **flooded evaporators**, in which only liquid refrigerant flows through the evaporator. In these systems, a **surge chamber** at the evaporator inlet is used to separate the liquid from the flash gas. The liquid then reenters the evaporator, while the vapor is bypassed directly to the compressor suction line.

This module focuses on DX systems; large commercial systems using flooded evaporators are covered later in your training.

## 2.1.0 Function

The metering device controls (meters) the amount of liquid refrigerant that enters the evaporator. Only as much liquid as is needed for system operation is allowed to pass through the metering device. It must pass enough liquid to provide the cooling requirements of the evaporator (the liquid needed for evaporation), while at the same time preventing a liquid surplus.

If not enough liquid enters the evaporator, it evaporates too quickly and much of the coil surface becomes ineffective. In some cases, the coil may freeze. Conversely, if too much liquid enters the evaporator, some of it will not boil into a gas and will pass in liquid form through the coil and into the suction line. If this liquid floods back through the suction line into the compressor, it will cause liquid **slugging** and possible damage. Even if the compressor is not slugged, the returning liquid will dilute the refrigerant and oil. This causes increased wear and shortens the life of the compressor. Thus, the metering device must:

- Restrict the refrigerant flow to maintain the pressure drop needed to produce the low-temperature refrigerant required for cooling.
- Regulate the quantity of refrigerant that can pass into the evaporator according to the cooling demand.

## 2.2.0 Adapting to Load Changes

The expansion valve metering device adapts to changing loads by increasing or decreasing the amount of refrigerant flowing to the evaporator. Fixed metering devices, on the other hand, are selected to meet the maximum cooling load or **design load**. This means that the flow capacity of the device is equal to the pumping capacity of the compressor.

Fixed metering devices are suitable for small systems that operate under a fairly constant load (such as refrigerators, freezers, and air conditioners of five tons or less). Such systems can, to some extent, adapt themselves to changing loads because system pressures vary in response to changes in the amount of heat absorbed by the refrigerant as it flows through the evaporator.

As the cooling load reduces, for example, the compressor pumps less refrigerant through the system. The refrigerant flow through a fixed metering device depends on the condenser pressure and the size of the metering device's orifice.

In the case of a capillary tube, the refrigerant flow also depends on the length of the tubing.

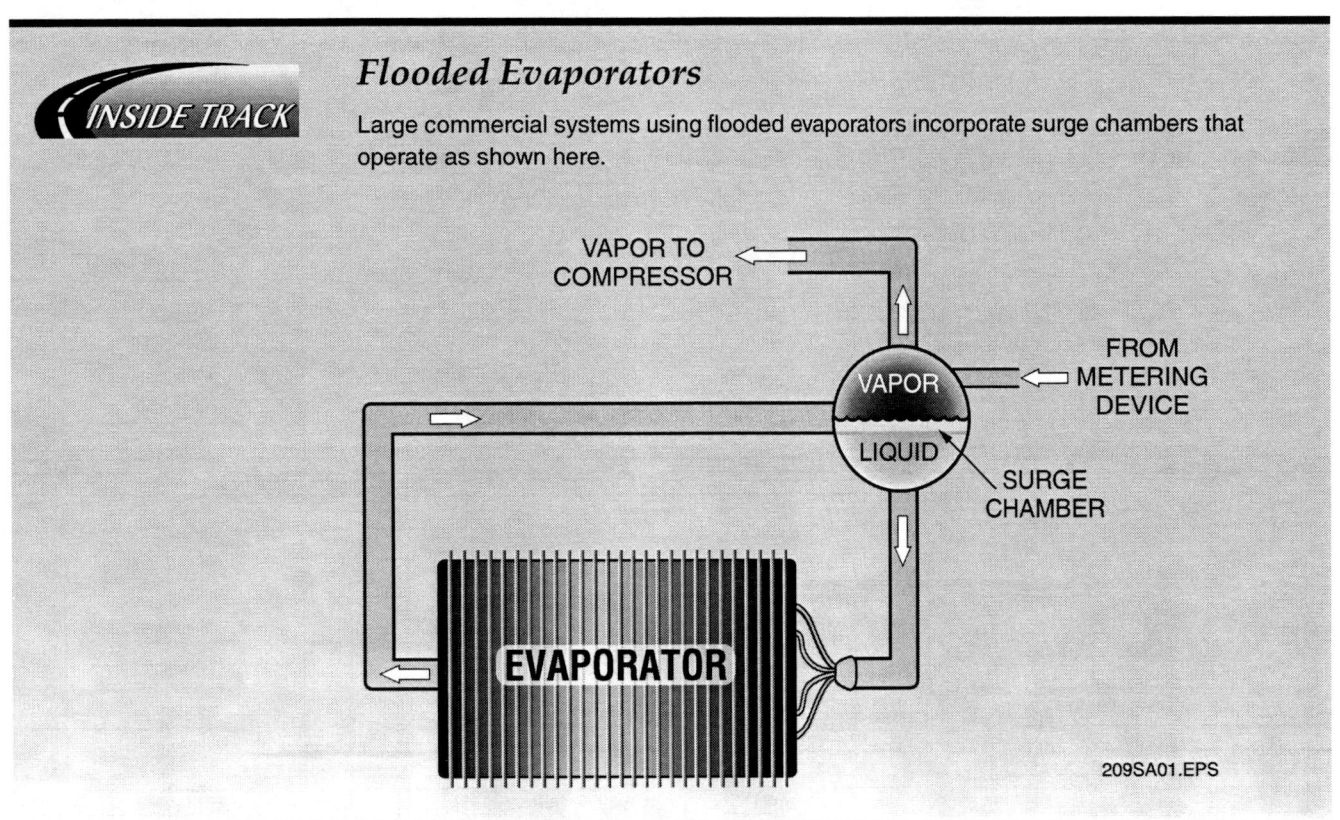

*Flooded Evaporators*

Large commercial systems using flooded evaporators incorporate surge chambers that operate as shown here.

## 3.0.0 ◆ FIXED METERING DEVICES

This category includes capillary tubes and **fixed-orifice metering devices**. These devices have a fixed opening through which the refrigerant from the condenser passes on its way to the evaporator. The outside temperature affects the condensing pressure achieved with fixed metering devices. On a hot day, the condensing pressure will be high, driving more refrigerant through the fixed metering device. On a mild day, the condenser pressure will be lower, reducing the amount of refrigerant flow. **Superheat** will be lower on hot days and higher on mild days.

### 3.1.0 Capillary Tubes

The capillary tube or *cap tube* (*Figure 3*) is a long copper tube with an inside diameter of 1/16" to 1/8". It may range in length from 20" to 140". It is usually coiled to save space and protect it from damage. The combination of diameter and length are selected to match the pressure drop with the pumping capacity of the compressor at design load.

The cap tube is usually selected and installed at the factory because its length and diameter are critical. If the tube is too long or the diameter too small, the evaporator will be starved for refrigerant, and excess liquid will build up in the condenser. The effect will be high head pressure, high superheat, high subcooling, and inadequate cooling. The same effect would be caused by a restricted tube. Therefore, a liquid strainer is sometimes provided at the cap tube input. A liquid line filter drier is always recommended to protect any metering device.

If the tube is too short or the diameter too great, excess liquid will be fed to the evaporator. This could cause liquid slugging at the compressor.

One of the drawbacks of the cap tube is that it doesn't stop refrigerant flow during the off-cycle. Thus, liquid refrigerant from the condenser can migrate to the evaporator. When the system cycles on, a slug of liquid will be drawn into the compressor. If the liquid migrates to the compressor crankcase during the off-cycle, it will mix with the compressor oil and affect lubrication. Over time, this could seriously damage the compressor. At minimum, system performance will be affected. To prevent liquid from condensing in the oil, a crankcase heater can be used to keep the compressor oil warm.

> **NOTE**
> Capillary tubes should be secured so they do not rub against each other or the case. Leaks can occur over time.

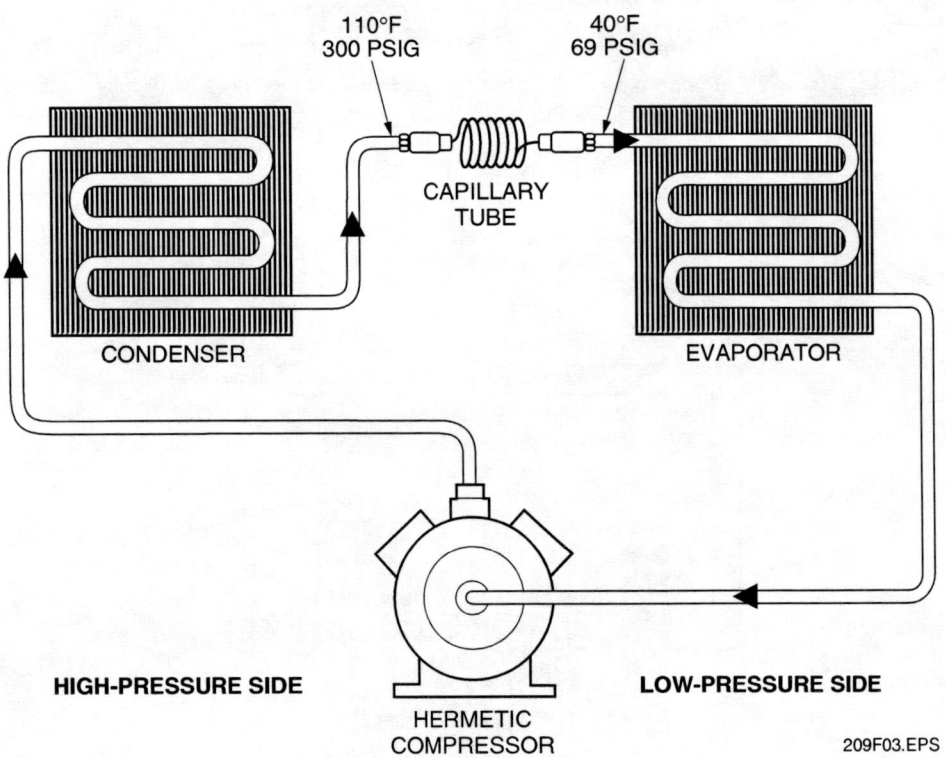

*Figure 3* ◆ Capillary tube.

## *Capillary Tubes*

In the past, capillary tubes were widely used in systems of up to five tons. With the introduction of fixed-orifice metering devices, cap tubes have fallen out of use and now are mainly found in room air conditioners and packaged terminal air conditioners (PTACs). Cap tubes are also commonly used in domestic refrigerators and freezers.

Because capillary tubes are small and relatively fragile, they must be handled with extreme care. Because of the risk of damage, cap tubes must not be cut with regular tubing cutters or hacksaws. They may, however, be cut with special cutters like those shown here.

Cap tubes may also be cut by using a file to score the tube, and then filing the tube until it easily breaks apart. Be careful not to pinch the cap tube, as this may change the inside diameter (I.D.) of the tube, which will affect system performance. Always remember to examine the tube ends for any burrs or debris before performing maintenance or repair.

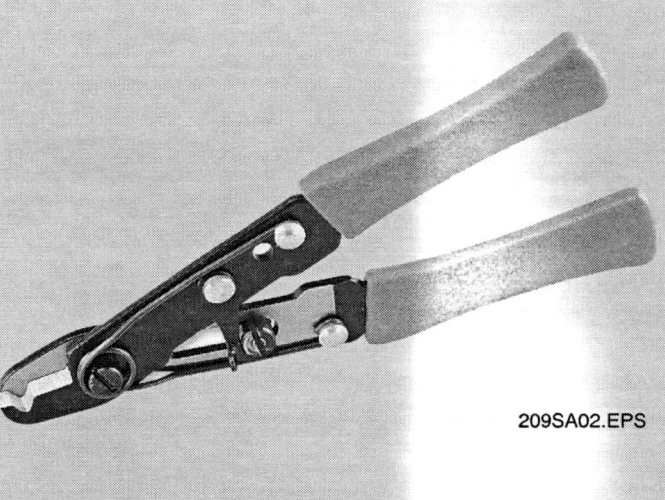

209SA02.EPS

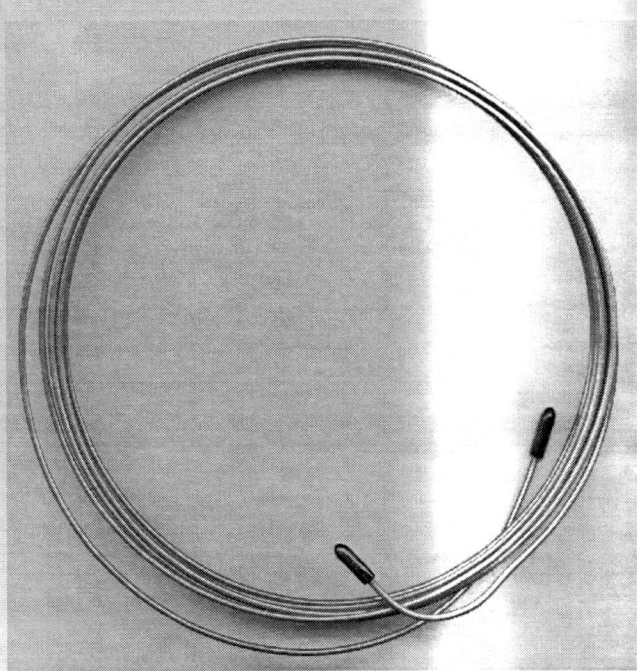

209SA03.EPS

## Servicing Capillary Tubes

When servicing a system equipped with a capillary tube, is it permissible to simply cut out a damaged or blocked section of the tube?

The advantage of the capillary tube is that it equalizes system pressure to reduce wear on the compressor. During the off-cycle, the high and low pressures equalize and allow the compressor to start more easily. This reduces the need for starting components such as start capacitors and potential relays.

### 3.2.0 Fixed-Orifice Metering Devices

One popular metering device uses a removable piston that contains a fixed orifice (*Figure 4*). It has the advantage of being much smaller and more rugged than the cap tube and is less prone to damage.

In a split system, the condenser capacity determines the orifice size. Therefore, it is common for the manufacturer to ship a metering piston with the condensing unit.

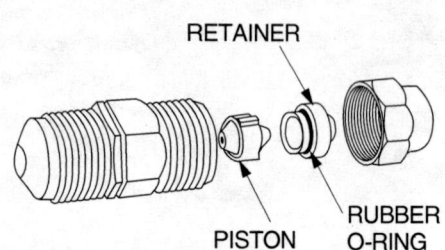

*Figure 4* ◆ Fixed-orifice device.

## Individual Feeder Tube Metering Devices

Some systems have a distributor that feeds refrigerant to the evaporator circuits. In these systems, it is possible to equip each evaporator feeder tube with a fixed-orifice metering device as shown here. Because there are no moving parts, refrigerant is distributed with greater consistency and reliability.

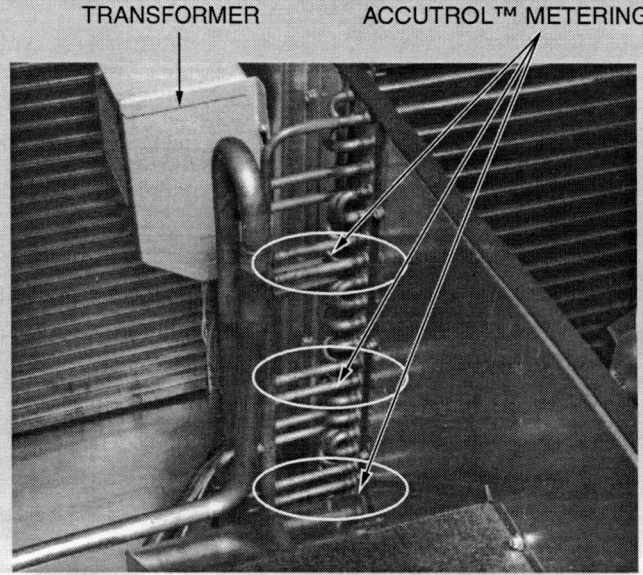

### Cleaning Fixed-Orifice Metering Devices

Once removed, fixed-orifice metering devices are easy to clean, but use caution when cleaning the device. If the orifice is enlarged or even scratched, system operation will be affected. After disassembling the device, use a small wire to pull the piston from the body of the device.

The evaporator is also shipped with a piston installed in a metering device. In most cases, the piston shipped with the evaporator should be replaced with the piston supplied with the condensing unit. For long line applications, refer to the installation instructions to determine the correct piston to use.

Specially designed fixed-orifice devices are popular in heat pumps because they can have a built-in check valve to control reverse refrigerant flow.

An advantage of the fixed-orifice metering device is that it can be removed easily from the system and replaced with a device of a different size to fine-tune system performance.

## 4.0.0 ◆ EXPANSION VALVES

Cooling demand varies according to changes in temperature and the quantity of indoor and/or outdoor air. When the air in the conditioned space becomes warmer, it increases in quantity and more heat is available to be absorbed. The ideal metering device reacts and allows more refrigerant to flow into the evaporator. When the conditioned air becomes cooler or decreases in quantity, less heat is available to be absorbed and the metering device allows less refrigerant flow.

There are several types of automatic expansion devices. There are also manual expansion valves, which are adjusted by the installer to match the pressure drop to the design load.

### 4.1.0 Manual Expansion Valves

A manual expansion valve (*Figure 5*) is a special hand valve with needle-pointed valve stems. The valve orifice and adjusting needle control refrigerant flow.

Manual expansion valves are adjusted by hand to match the design load, so they are only suitable in systems with a fairly constant load. If they are used in systems with variable loads, and not re-adjusted as the load changes, system performance will be affected (i.e., the evaporator will be either starved or flooded as the load increases or decreases).

### 4.2.0 High-Side Float Valves

The high-side float valve (*Figure 6*) reacts to the level of liquid from the condenser, opening and closing the orifice in response to changes in the liquid level. The amount of liquid leaving the condenser is driven by the amount of vapor produced by the evaporator. This, in turn, is a function of the cooling load on the evaporator.

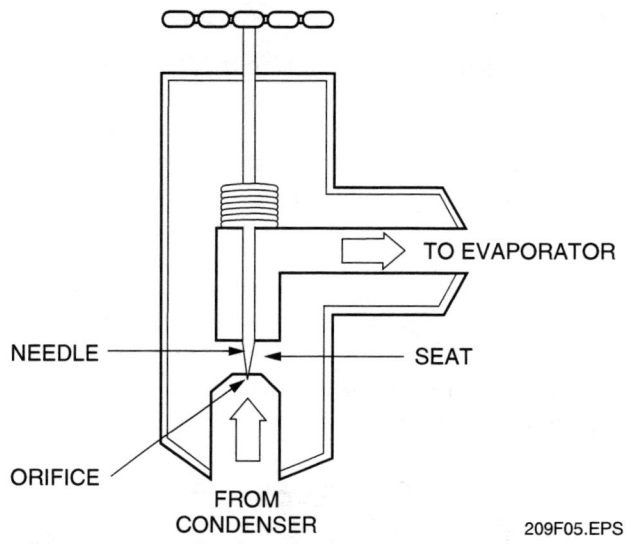

*Figure 5* ◆ Manual expansion valve.

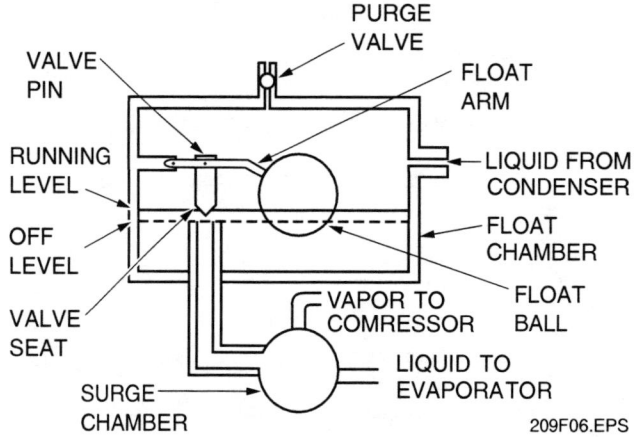

*Figure 6* ◆ High-side float valve used with flooded evaporator.

As the cooling load increases, more liquid leaves the condenser. This extra liquid raises the float, and in doing so, increases the opening in the orifice to send more refrigerant to the evaporator.

In a system with a high-side float valve, most of the refrigerant stays in the evaporator when the compressor is not running. System charging is therefore critical. If the system is overcharged, the excess charge in the evaporator will flood the compressor with liquid when the system cycles on. High-side float valves are commonly used in the flooded cooler of a centrifugal chiller; however, they are also used with DX evaporators. When used with a flooded evaporator, the metering device output is sent through a surge chamber. This bypasses flash gas to the compressor, leaving only liquid at the evaporator input. High-side float valves are used on systems with a constant load.

## 4.3.0 Low-Side Float Valves

The low-side float valve (*Figure 7*) is the best control device for use with flooded evaporators. It may be installed directly in the evaporator or in a separate float chamber outside the evaporator. Note that it is installed on the low side of the metering orifice, while the high-side float valve is installed on the high side. When the load on the evaporator increases, more liquid is boiled into vapor, reducing the amount of liquid in the float chamber. The lowered liquid level lowers the float, opening the metering orifice, and supplying more refrigerant to the evaporator.

## 4.4.0 Automatic Expansion Valves

The automatic expansion valve (*Figure 8*) maintains a constant evaporator pressure. The pressure maintains the temperature in the evaporator. It relies on a spring-loaded diaphragm and valve that react to pressure in the evaporator.

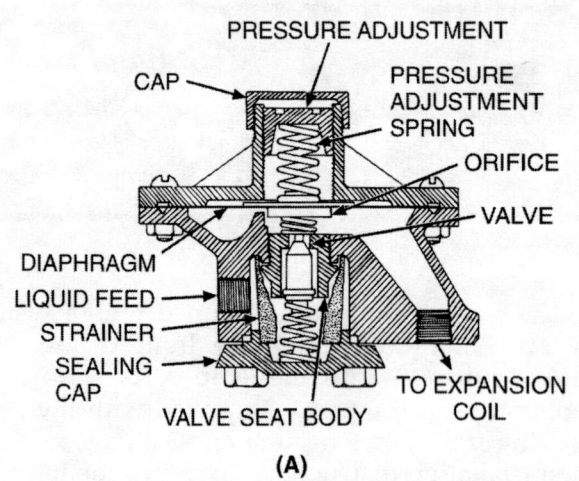

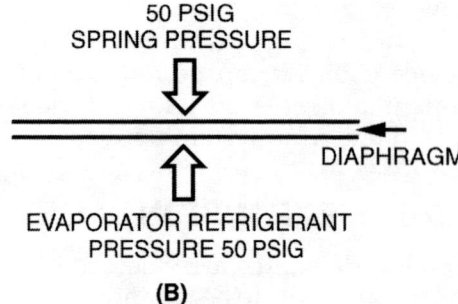

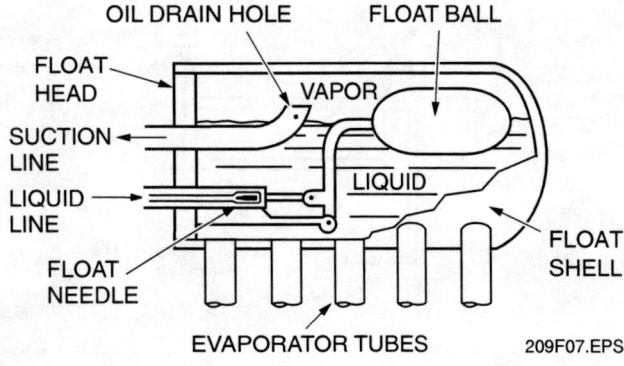

*Figure 7* ◆ Low-side float valve.

*Figure 8* ◆ Automatic expansion valve. (A) Cross section. (B) Valve in equilibrium. (C) Exterior.

At a preset pressure, the force applied by the spring and the force provided by the evaporator cancel each other out. If the evaporator pressure changes in response to a changing load, the diaphragm will raise or lower, depending on whether the pressure is increased or decreased. Movement of the diaphragm causes the valve to expand or reduce the orifice opening. Automatic expansion valves are best suited for small capacity equipment with a fairly constant load (e.g., refrigerators and freezers).

The automatic expansion valve responds to load changes in a manner opposite of that you might expect. When there is an increase in the heat load being handled by the related evaporator coil, the suction pressure starts to rise. This causes the automatic expansion valve to start to reduce, or *throttle*, the amount of refrigerant being fed to the evaporator by closing the valve enough to maintain the constant suction pressure setpoint. This has the effect of starving the coil slightly. A larger increase in the heat load will cause even more starving of the coil.

Similarly, when there is a decrease in the heat load being handled by the evaporator, the suction pressure goes down. This causes the automatic expansion valve to open and feed more refrigerant to the coil. If the load is reduced too much, it is possible for liquid refrigerant to leave the evaporator coil and flood into the suction line going to the compressor.

## 4.5.0 Thermal Expansion Valves

The thermal expansion valve is a metering device with an external sensing bulb that senses the refrigerant temperature at the evaporator outlet. The bulb sends this information, which represents the amount of superheat produced by the evaporator, back to the metering device, where it is used to adjust the flow of refrigerant to match the load. This type of device is alternately referred to as a *thermal expansion valve* or *thermostatic expansion valve*, and may be abbreviated *TEV* or *TXV*. The terms are interchangeable, as are the abbreviations. It tries to maintain a constant superheat at the evaporator outlet. This device should not be confused with the **thermal-electric expansion valve (TEEV/ THEV)** or the electronic expansion valve (EEV), both of which use a thermistor as a temperature sensor. We will cover the TXV first.

### 4.5.1 Operating Principles

The TXV is a device in which control is based chiefly on the amount of superheat in the refrigerant when it leaves the cooling coil and enters the suction line (see *Figure 9*). The DX evaporator coil in a system running at design conditions carries a mix of low-temperature liquid and vapor refrigerant. This refrigerant is boiling, and the air flowing through the evaporator is supplying the heat to change the refrigerant from a liquid to a vapor. When the refrigerant becomes a saturated vapor, its sensible heat begins to increase, creating superheat.

As the cooling load fluctuates, the point at which the liquid becomes a gas will move back and forth in the evaporator. The expansion valve must open and close automatically to adjust refrigerant flow to meet the load.

When the load is constant, the diaphragm is balanced by the sensing bulb pressure at the top and a combination of spring pressure and evaporator inlet pressure at the bottom. (See *Figure 10*.) When the cooling load changes, the pressure at the evaporator outlet will also change. The diaphragm will move up or down, opening or closing the orifice to change the amount of refrigerant applied to the evaporator.

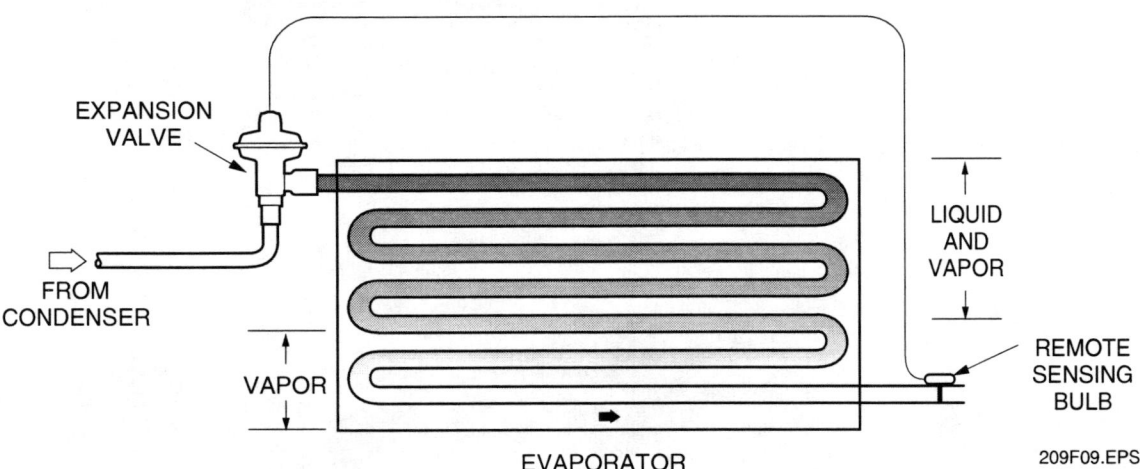

*Figure 9* ◆ Refrigerant flow in the evaporator.

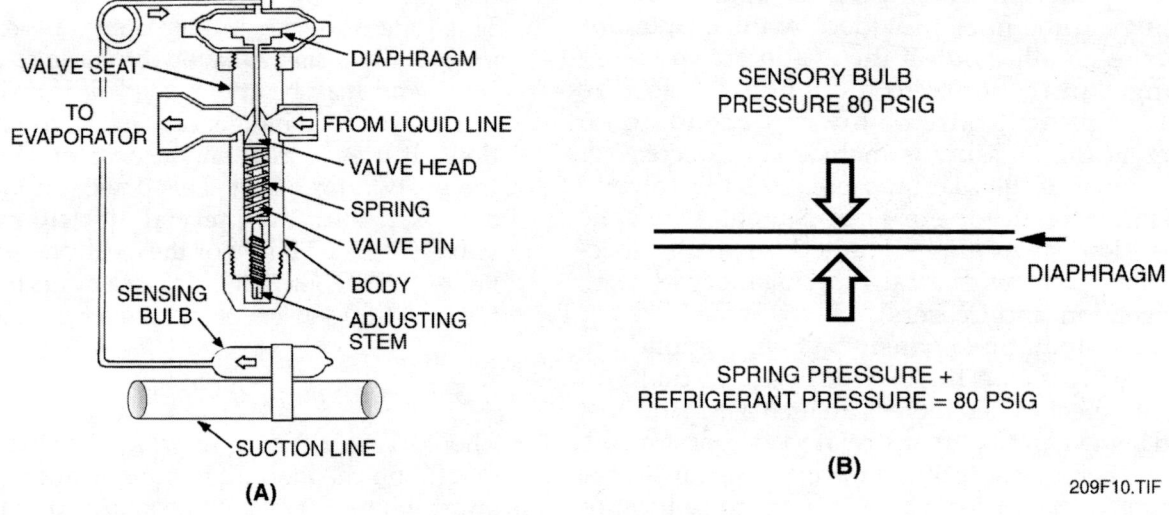

*Figure 10* ◆ Thermal expansion valve. (A) Thermostatic expansion valve component. (B) Valve in equilibrium.

If the cooling load increases, the evaporator will be starved for refrigerant and superheat will begin earlier. The suction line temperature will increase because of the increased evaporator superheat. The increased superheat causes an increase in the pressure of the refrigerant in the sensing bulb.

The pressure increase is transmitted back to a diaphragm (or bellows). This opens the refrigerant needle valve to a greater degree. The thermostatic fluid used in the sensing bulb is often the same refrigerant used in the system. This refrigerant does not mix with the system refrigerant. For air conditioning, a TXV is commonly selected for 15°F to 20°F superheat at the factory. In most cases, the TXV should not be adjusted.

Constant-pressure or automatic-diaphragm expansion valves have spring-loaded diaphragms activated by evaporator pressure. As the pressure lowers due to insufficient refrigerant, the pressure on the diaphragm allows the spring to open the

## Converting to Thermal Expansion Valves

Because thermal expansion valves offer superior control of refrigerant, they are often found on more expensive lines of residential air conditioners and heat pumps. Manufacturers of these products frequently offer thermal expansion valve kits for converting evaporator coils equipped with fixed-orifice metering devices.

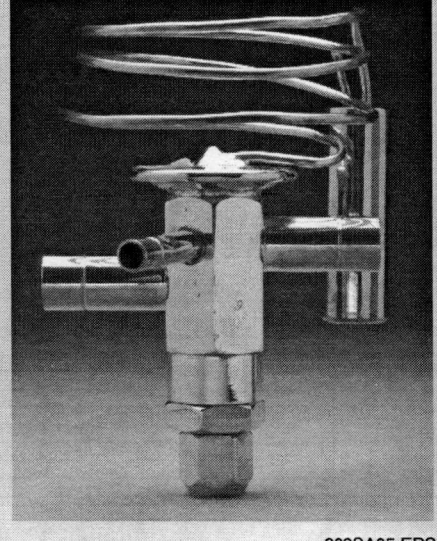

### Adjusting Thermal Expansion Valves

Most of the expansion valves used in residential and light commercial HVAC equipment are not adjustable. The manufacturer selects a thermal expansion valve with the correct superheat and installs it at the factory, or makes it available for field installation. If a thermal expansion valve is adjustable, do not adjust it unless absolutely necessary.

valve, increasing the refrigerant flow. This kind of valve operates to keep the suction pressure almost constant.

### 4.5.2 Equalizers

Along with spring pressure, the evaporator inlet pressure, which is also the pressure at the outlet of the TXV, acts on the bottom of the diaphragm. The sensing bulb pressure acts on the top of the diaphragm. If there is a significant pressure drop across the evaporator, the pressure created by the sensing bulb will be less than that sensed at the evaporator inlet. This unbalances the TXV, and too much superheat is produced.

In systems with little or no pressure drop across the evaporator, an internal equalizer is used. (See *Figure 11*.) The allowable pressure drop will vary with the refrigerant and the evaporating temperature, and may range from 0.75 psig to 3 psig. In an internally equalized TXV, evaporator inlet pressure is fed to the bottom of the diaphragm through an internal passage. Internal equalization is used only with evaporators having a very small pressure drop. These include refrigerators, food freezers, and small air conditioners.

If there is a larger pressure drop across the evaporator, an external equalizer is used. The use of a refrigerant **distributor line** or a long suction line are factors that can cause an excessive pressure drop. When a distributor is used, an externally equalized expansion valve will also be used.

The external equalizer (*Figure 12*) isolates the TXV from the evaporator inlet pressure. The evaporator outlet pressure is sensed at a location immediately after the TXV sensing bulb. It is then fed back through the equalizer fitting to the bottom of the diaphragm. Because the pressures at the top and bottom of the diaphragm are sensed at the same point, the TXV is equalized.

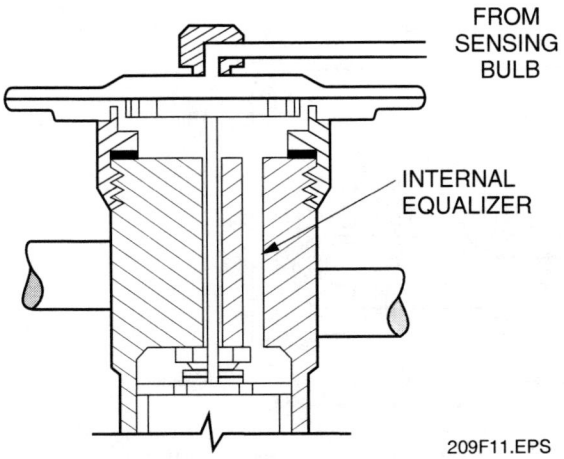

*Figure 11* ◆ Expansion valve with internal equalizer.

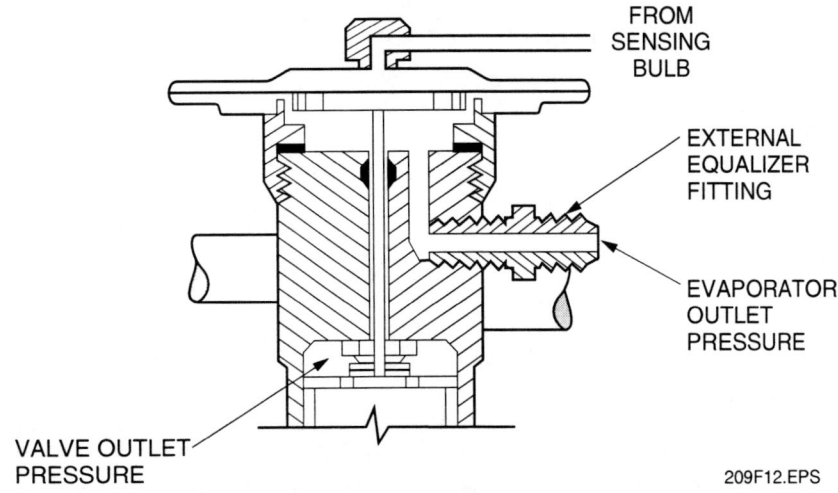

*Figure 12* ◆ Expansion valve with external equalizer.

METERING DEVICES — TRAINEE MODULE 03209

Let's examine a typical R-22 system (*Figure 13*). Assume that we want to maintain 10°F of superheat at the evaporator outlet to achieve optimum system performance. If the pressure at the TXV outlet is 76 psig and the spring pressure is equal to 18 psig, the combined pressure on the underside of the diaphragm is 94 psig. With zero pressure drop across the evaporator, the superheated vapor at the evaporator outlet will produce a suction line temperature of 55°F. In the TXV sensing bulb, the 55°F temperature corresponds to a pressure of 94 psig, which is fed back to the top of the diaphragm. Since the pressures on the top and bottom of the diaphragm are equal, the valve maintains a constant refrigerant flow. With a small pressure drop, the TXV may fluctuate for a while, but will eventually stabilize. In this situation, the internally equalized TXV is suitable for the conditions.

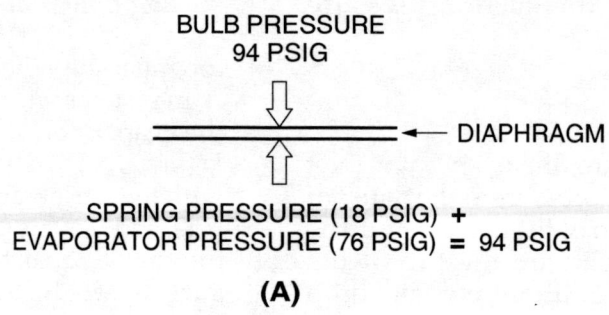

(A)

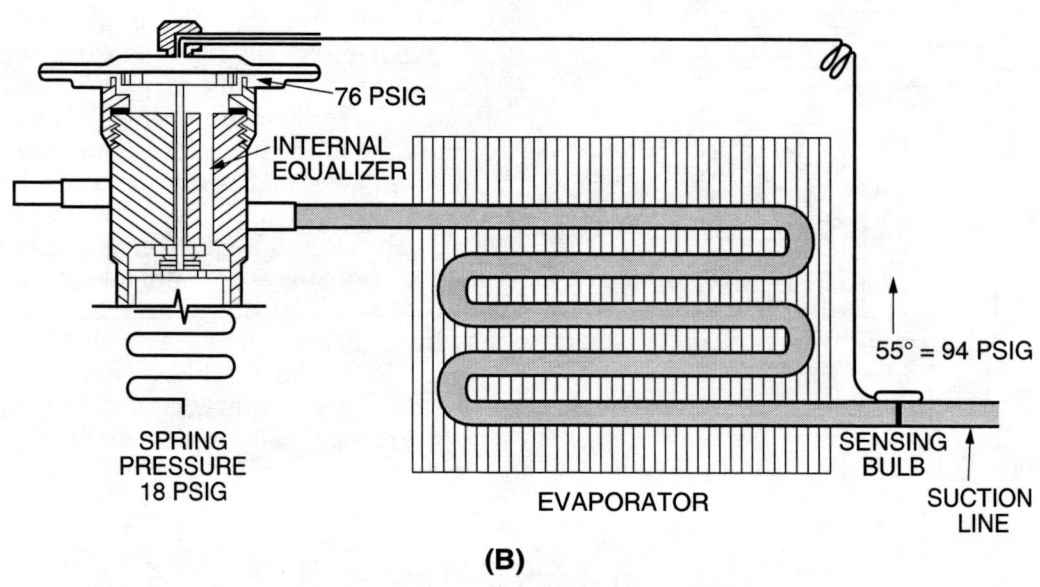

(B)

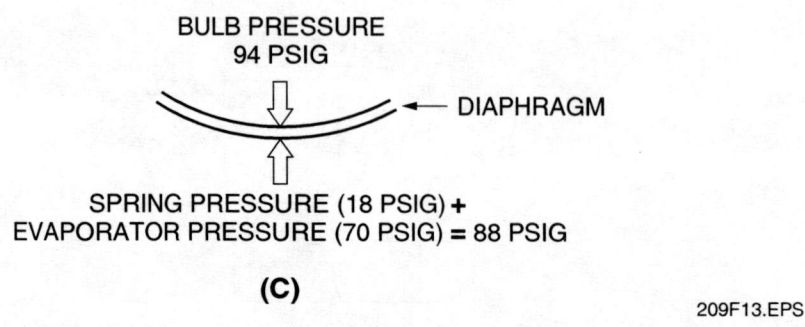

(C)

*Figure 13* ◆ Internally equalized TEV. (A) Valve in equilibrium. (B) Internally equalized TXV components. (C) Valve diaphragm deflected down with lower superheat.

If the evaporator had a 6 psig pressure drop, the effect with an internally equalized TXV would be very different. The evaporator outlet pressure would be 6 psig less than the inlet pressure. When the bulb sensed this pressure drop, the valve would modulate, reducing the amount of refrigerant to the evaporator. This would cause the evaporator to operate at about 15°F of superheat, which would adversely affect system performance.

With an externally equalized TXV, this problem is eliminated (*Figure 14*). The pressure in the suction line is tapped and fed back to the bottom of the diaphragm. Since the pressure at the top of the diaphragm is also a function of the suction line pressure, the forces on the two sides of the diaphragm remain equal as long as the load remains the same.

Many valve manufacturers recommend externally equalized valves for improved performance. When replacing a defective internally equalized valve, it may be necessary to replace the valve with an externally equalized valve. The equalizer line is usually ¼" soft copper tubing run from the valve to the suction line near the evaporator outlet. In these cases, an external equalizer line must be installed on the system. Make sure not to cap off the external equalizer fitting.

### 4.6.0 Thermal-Electric Expansion Valves

The thermal-electric expansion valve (it may be abreviated as either TEEV or THEV) senses the temperature of the refrigerant leaving the evaporator. (See *Figure 15*.) A thermistor placed in the suction line acts as the sensor. As the cooling load changes, the resulting changes in the refrigerant temperature will reduce or increase the resistance offered by the thermistor. Inside the valve, there is a bimetal needle that reacts to the change in current by modulating the valve to reduce or increase refrigerant flow. The valve will maintain a constant superheat in the system.

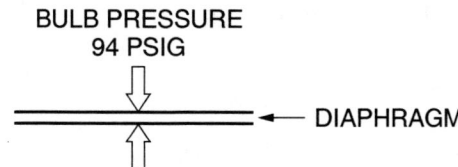

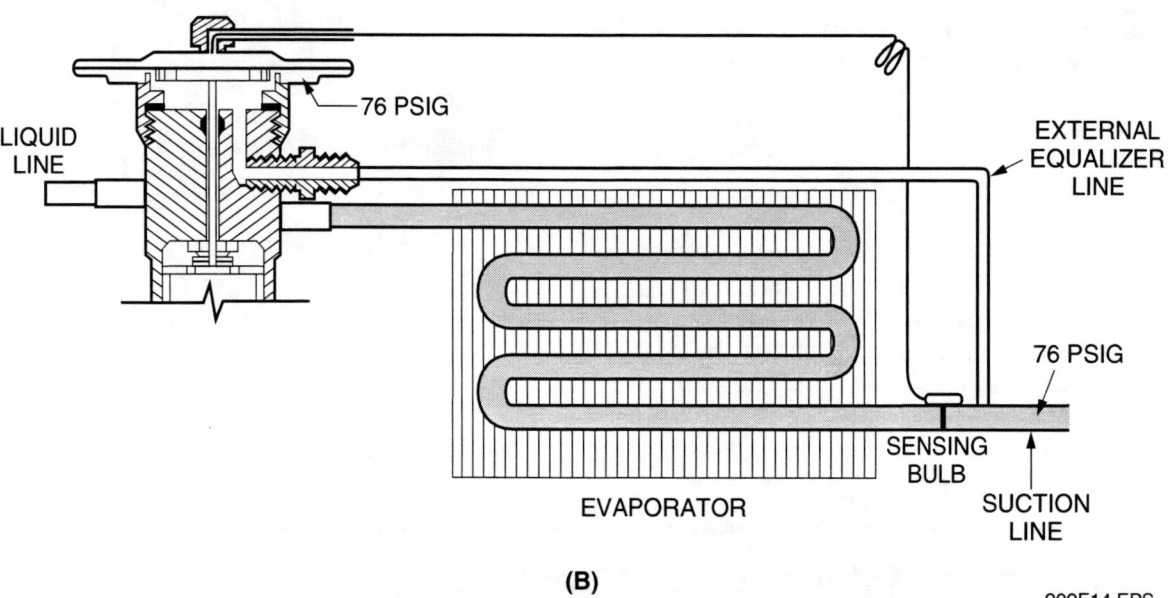

*Figure 14* ◆ Externally equalized TXV. (A) Valve in equilibrium. (B) Externally equalized TXV components.

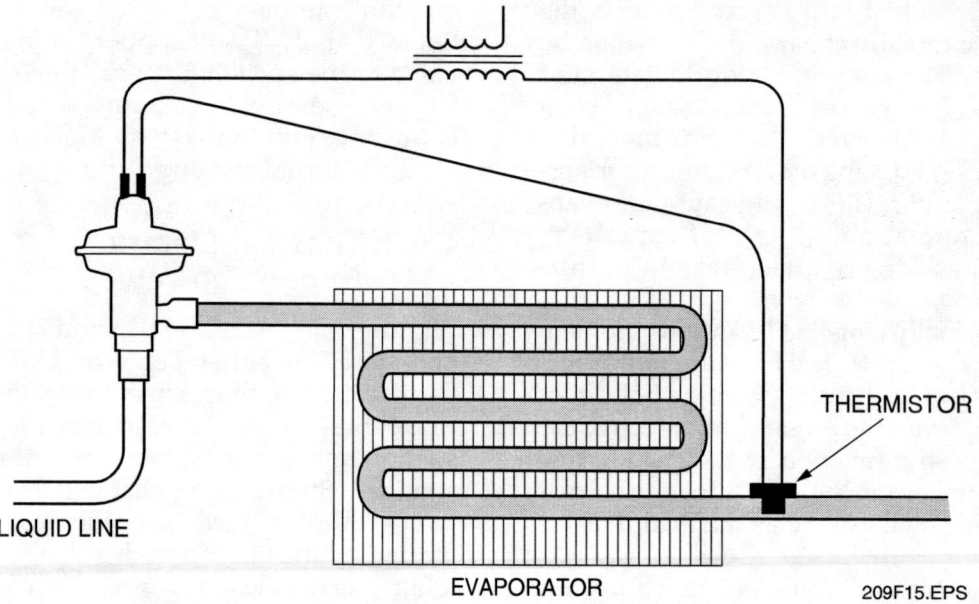

*Figure 15* ♦ Thermal-electric expansion valve location.

### 4.7.0 Electronic Expansion Valves

An electronic expansion valve (EEV) (*Figure 16*) is a microprocessor-controlled device that is driven by a precision DC motor known as a *stepper motor*. This motor can provide 760 discrete steps of orifice control in the metering device. Like the TEEV, the EEV responds to temperature changes sensed by a thermistor and reacts to these changes to maintain a constant superheat at the evaporator outlet. Some EEVs are driven by a pulse system instead of a stepper motor.

### 4.8.0 Hunting

**Hunting** is a term commonly used to describe the changes in refrigerant flow as the expansion device adapts to changing conditions. As the load increases, for example, the valve may open too wide, allowing too much refrigerant into the evaporator. When this happens, the system may overcompensate in the other direction. Hunting occurs because there is a time lag between when the expansion valve modulates and when the sensing bulb senses the change. Proper selection of the expansion valve and adjusting for the correct superheat can minimize hunting. The position of the sensing bulb and how firmly the bulb is clamped to the suction line can also affect hunting.

Some devices include anti-hunting features such as thermal ballast. The thermal ballast is included inside the sensing bulb. It delays the response of the sensing tube, and thus prevents the valve from overfeeding or underfeeding

*Figure 16* ♦ Electronic expansion valve (EEV).

---

### Non-Electric TXVs

Non-electric TXVs operate in a narrow range of 3% to 5% of the set superheat rating. This is not hunting.

before the system can balance. It also acts as a safety device by quickly closing the valve when it senses a liquid slug in the suction line.

Hunting can only be verified by making several suction line temperature or pressure measurements over a period of time. If there is a repetitive pattern such as that shown in *Figure 17*, hunting is occurring. It may be necessary to select a smaller valve to correct the situation.

## 5.0.0 ◆ DISTRIBUTORS

For best system performance, the mixture of liquid and gas coming from the TXV must be distributed evenly throughout the evaporator coil. It is critical that the pressure drop within the evaporator be held within reasonable limits. Additionally, refrigerant must be regulated to guarantee proper oil return to the compressor. These needs are usually met by adding multiple refrigerant circuits in the cooling coil.

For some applications, each circuit can be served by its own TXV. However, normal practice is to employ a **distributor** that evenly proportions refrigerant flow from the expansion valve to all circuits (*Figure 18*). Ideally, a distributor should provide even refrigerant distribution with a minimal pressure drop to ensure stable system control by the TXV.

*Figure 19* shows one type of distributor. The orifices in the distributor are precision machined to assure equal distribution of the liquid/vapor refrigerant mixture to the evaporator circuits. Some distributors are designed to drop the refrigerant pressure further and could be considered extensions of metering devices.

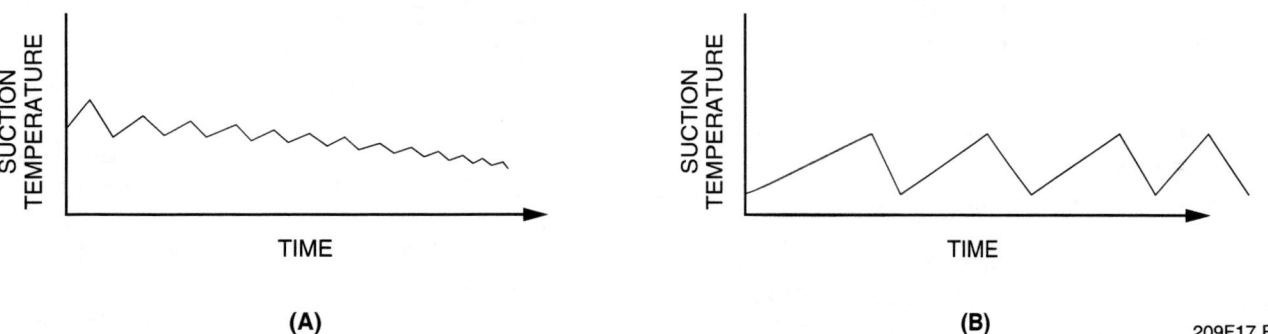

*Figure 17* ◆ Suction temperature variations indicate hunting. (A) Normal valve operation. (B) Hunting.

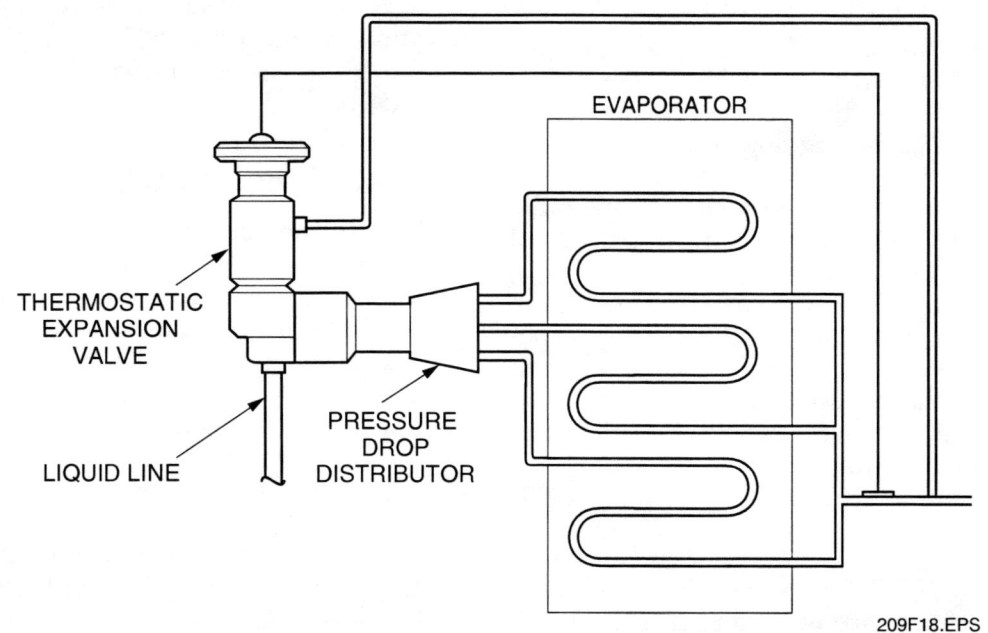

*Figure 18* ◆ Multi-circuit evaporator fed by a distributor.

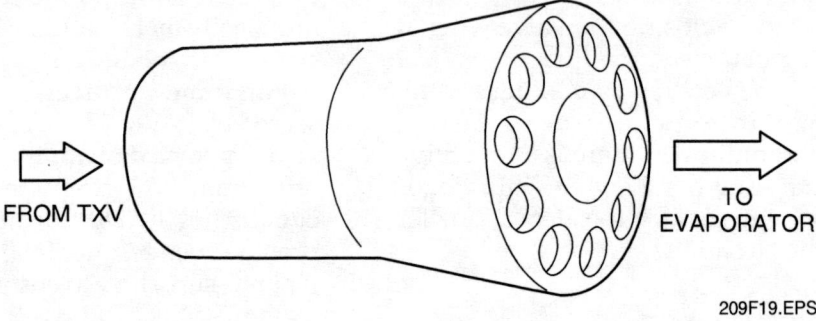

*Figure 19* ♦ Distributor.

## 6.0.0 ♦ TXV REPLACEMENT

This section covers TXV replacement.

### 6.1.0 Selection

It is important to select a TXV with the correct capacity for peak system performance. Most manufacturers print selection tables. Correct expansion valve selection is vitally important; therefore, it is advisable to review a selection procedure prior to obtaining a replacement valve.

Valve capacity must be compatible with system capacity. For example, if a load is calculated at 17.5 tons and the only valves available have capacities of 12 tons and 18 tons, the 18-ton valve should be selected. The smaller-capacity valve would starve the evaporator at full load, whereas the slightly oversized valve would be acceptable. TXV selection tables usually contain data in relation to system capacity in tons, condensing temperature, evaporator temperature, and pressure drop across the valve.

If an extremely oversized valve is selected, it will not be able to control closely at part-load conditions and may cause slugging of the compressor.

During selection, consider such factors as correct capacity, selective charge of the sensing bulb, and whether it should have an external or internal equalizer. According to some manufacturers, TXVs may be mounted in any position; however, it is advisable to place them as close to the evaporator inlet as possible and to follow each manufacturer's installation instructions. If a distributor is used, it should be mounted directly to the valve outlet. Common valve mountings are brazed, soldered, flared, or flanged.

### 6.2.0 Placement

The best performance is obtained if the valve feeds vertically up or down into the distributor. If a hand valve is located on the outlet side of the expansion valve, it should not have a restricted port, nor should any restrictions appear between the TXV and the evaporator (refrigerant distributor excepted). To minimize refrigerant migration on some types of valves, it is recommended that the valve diaphragm case be insulated so that it remains warmer than the bulb.

If TXVs are located in corrosive atmospheres, they must be protected with appropriate materials to prevent early failure.

It is usually not necessary to disassemble solder-type valves when soldering to the connecting lines. Any of the commonly used solders such as 95-5, silfos, easy-flow, or phos-copper may be used. However, exercise caution to keep the flame of the torch directed away from the valve body to avoid excessive heat on the valve diaphragm. A wet cloth wrapped around the valve during the soldering process is an extra precaution that is well worth the effort.

A nitrogen purge will prevent copper oxide contamination during the brazing process. Nitrogen will also help keep the valve cool. Replace the liquid line filter drier after installing the new valve.

### 6.3.0 Sensing Bulb

Locate the sensing bulb with care and follow accepted piping standards to allow for precise valve control. The sensing bulb should be attached to a horizontal suction line, but it may be attached to a vertical line if absolutely necessary. The bulb should not be mounted in a trap or pocket in the suction line because refrigerant boiling out of the trap will falsely influence the temperature of the bulb, resulting in improper valve control. The bulb should never be installed on unions or fittings because the thermal conductance of these materials will delay response to changes in suction line temperature.

On suction lines up to and including ⅝" outside diameter (O.D.), the bulb should be installed on top of the suction line (see *Figure 20*). On lines with a ⅞" O.D. or larger, the bulb should be clamped near the bottom of this line. Locating the bulb on the bottom of the line is not recommended because oil-refrigerant mix is usually present at this point.

For difficult installations, the proper bulb location may be determined by trial and evaluation. Good thermal contact between the bulb and the suction line is essential; therefore, the bulb should be fastened securely to a clean, straight section of the suction line with two bulb straps. The straps should be a copper alloy to prevent reaction with the copper. The sensing bulb should be insulated to improve performance. Good piping practices usually include attaching the TXV to a horizontal suction line leaving the evaporator. This line is pitched downward.

When a vertical riser follows the horizontal line, a short trap (*Figure 21*) is placed immediately ahead of the vertical line. This will collect any liquid refrigerant and/or oil passing through the suction line that might influence the temperature of the sensing bulb. On multiple evaporator installations, the piping and sensing bulb should be arranged so the flow from any one valve cannot influence the others.

On commercial and low-temperature applications, the bulb should be clamped on the suction line at a point where the bulb temperature will be the same as the evaporator temperature during the off-cycle. The insulation used to wrap the sensing bulb must not be water absorbent.

On brine tanks and water coolers, the bulb should be attached below the surface of the liquid. If the bulb is located in a brine tank, the bulb and cap tubing must be painted with corrosion-resistant paint or pitch. If the bulb has to be located where its temperature will be higher than the evaporator during the off-cycle, a solenoid valve must be used ahead (upstream) of the TXV.

When an externally equalized TXV is installed in the system, the equalizer connection should be made at a point immediately downstream from the sensing bulb. If evaporator pressure or temperature control valves are located in the suction line near the evaporator outlet, the equalizer line must be connected on the evaporator side of these valves.

Many valve failures are due to the presence of dirt, sludge, and moisture. Therefore, filter-driers should be installed in the system whenever the expansion valve is replaced. It may also be desirable to install a sight glass and moisture indicator at the same time. As an added precaution, most replacement TXVs are equipped with built-in inlet screens for removing scale or other particles that could obstruct the closure of the valve and seat.

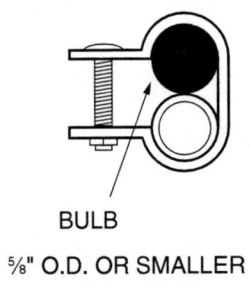

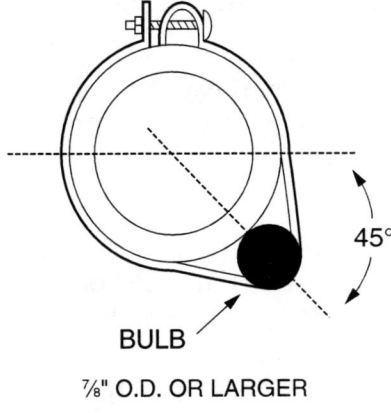

*Figure 20* ◆ Sensing bulb positioning.

---

### TXV Installation

Proper installation of the sensing, or feeler, bulb is critical. It must be securely attached to a clean section at the bottom of the suction line near the evaporator outlet. The bulb must be insulated to ensure that it only reacts to the temperature at the suction line. If the bulb is loose or not properly insulated, the system will not operate correctly. A loose or poorly insulated TXV bulb will result in low superheat, along with high discharge and suction pressures and low subcooling. The opposite conditions will exist if the TXV feeler bulb loses its charge.

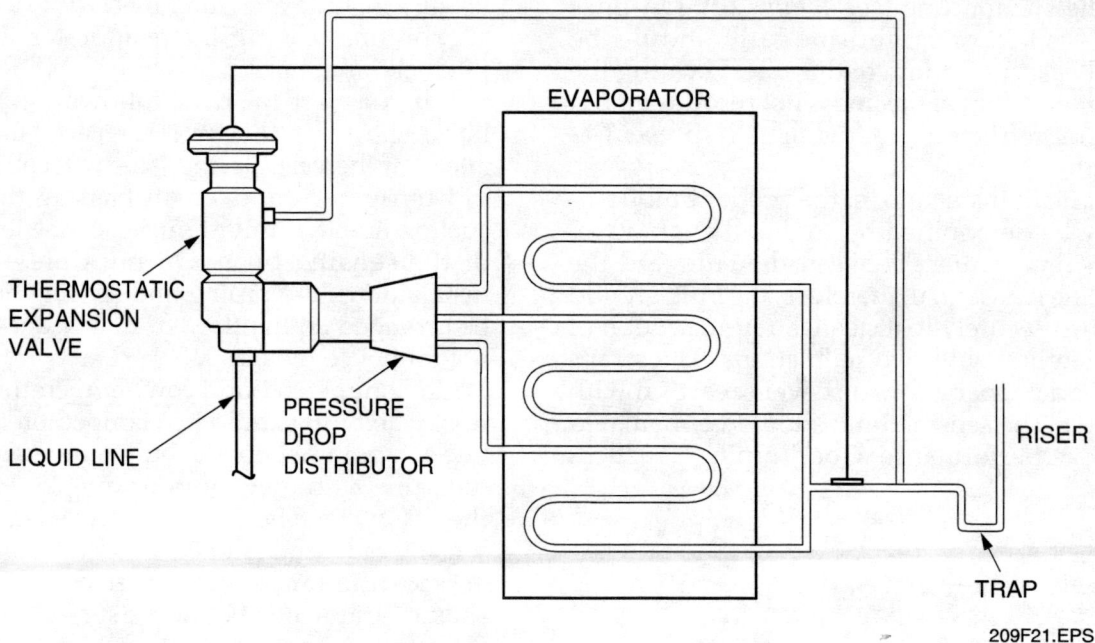

*Figure 21* ◆ Trap in a suction line riser.

## 6.4.0 TXV Adjustment

All TXVs are tested and set at the factory prior to shipment. The factory superheat setting will be correct in relation to the size of the valve; no further adjustment is required in most cases. If operating conditions require a different setting, consult the equipment manufacturer's literature and install the specified replacement valve.

The amount of superheat for a given evaporator may be obtained using the following procedure:

*Step 1* Record the temperature of the suction line at the position where the sensing bulb is attached.

*Step 2* Obtain the suction pressure in the suction line at the bulb location. Place a gauge in the external equalizer line (if the unit is so equipped) or record the gauge pressure at the suction valve of the compressor. If the pressure is obtained at the compressor, an estimated line pressure drop must be added to the gauge reading. The sum of the gauge reading and estimated pressure drop should be about equal to the suction line pressure at the sensing bulb.

*Step 3* Convert the recorded pressure to saturated refrigerant temperature using a pressure-temperature chart. Some compound gauges have the corresponding temperatures for this conversion.

*Step 4* Subtract the two temperatures (bulb location temperature and converted suction line pressure-temperature). The difference is the superheat.

To find the system superheat rather than the evaporator superheat, measure the suction line temperature near the compressor and follow Steps 2 through 4. Most manufacturers require the system superheat reading.

Acceptable superheat readings may range from as low as 10°F to as high as 20°F. The manufacturer's recommendations should be followed for the correct valve size on factory equipment.

## 7.0.0 ◆ METERING DEVICE PROBLEMS

Metering devices are often blamed for system problems that can be caused by any number of other faults. For example, if the temperature leaving the evaporator is too high, the metering device may be blamed for not feeding enough refrigerant. The symptoms are obvious (not cooling properly), but the cause may be more difficult to find.

**NOTE**
Before troubleshooting any system, make sure the system is properly charged and has the correct level of subcooling.

### Before Troubleshooting a TXV

Before troubleshooting a TXV, establish a baseline superheat reading. The operation of a TXV can be checked by removing the sensing bulb and holding it, or holding the bulb up to a hot light source. The system should respond to the increase in the bulb's temperature by lowering the superheat.

Next, place the sensing bulb in a glass of ice water. The system superheat should increase, because the drop in temperature of the sensing bulb will cause the TXV to close. If the TXV does not respond, it is defective and must be replaced.

---

Causes of poor system performance include:

- Moisture due to water and oil frozen in the valve port or working parts when operating below 32°F
- Dirt or other contaminants that may not be trapped by the strainer and obstruct the flow of refrigerant through the valve port
- Wax that precipitates at low temperature and builds up on the needle valve and seat
- Refrigerant shortage due to a low state of charge
- Gas in the liquid line due to long or undersized lines or improperly piped systems
- Misapplication of an internally equalized valve in place of an externally equalized valve, or improper location of the externally equalized valve line
- Restricted, plugged, or capped external equalizer tube
- Undersized valve
- High superheat adjustment
- High pressure drop through the evaporator

Other problems, such as refrigerant overfeeding, feeding too much on start-up, improper valve feeding, or hunting and cycling could be due to some of the same causes.

Keep in mind that many system malfunctions may be eliminated if the system is kept clean, dry, and in a proper state of charge. Check these conditions before changing the metering device.

Neither TXVs nor distributors function well at very low loads. Low-load conditions may be caused by oversizing. If a distributor load falls below 50%, the liquid-vapor mixture entering the evaporator may separate. The liquid will go to the bottom circuits and the vapor will go to the top circuits. Excessive superheat and liquid floodback to the compressor will result. An easy way to check for this problem is to look for high temperatures at the top of the evaporator and low temperatures at the bottom. If this problem occurs, consult the equipment manufacturer to determine the proper corrective action.

### Summary

The purpose of the TXV or any other refrigerant metering device is to maintain the pressure drop needed to produce low-temperature liquid refrigeration for cooling and to regulate the quantity of refrigerant that can pass into the evaporator in response to the cooling load. The importance of the TXV cannot be overemphasized. If it is to operate properly, it must be carefully selected and precisely installed in a clean, dry, properly charged system.

## Review Questions

1. A metering device is a refrigeration system control that _____.
   a. converts high-pressure, high-temperature liquid refrigerant to a low-temperature, low-pressure pure vapor refrigerant
   b. converts high-pressure, high-temperature liquid refrigerant to a low-temperature, low-pressure mix of liquid and vapor refrigerant
   c. is located at the evaporator outlet
   d. always contains a feature to automatically adjust the refrigerant flow

2. A significant difference between a DX evaporator and a flooded evaporator is that _____.
   a. the DX evaporator is used primarily on large commercial systems, while the flooded evaporator is used more on small systems
   b. in the DX evaporator, the refrigerant is converted to vapor in the evaporator, whereas it remains in liquid form in the flooded evaporator
   c. DX evaporators are always used with a surge chamber, while flooded evaporators seldom are
   d. there is more flash gas at the inlet of a flooded evaporator

3. Which of the following is true of a capillary tube?
   a. It is always the same length.
   b. It is made of PVC pipe.
   c. It has no sensing bulb.
   d. It is used only with flooded evaporators.

4. In which of these systems would a manual expansion valve be a good choice?
   a. A system with a fairly constant load.
   b. A system that experiences frequent load changes.
   c. A lobby in a busy hotel.
   d. A movie theater.

5. The main difference between high-side and low-side float valves is that the _____.
   a. high-side float valve is located at the compressor outlet, while the low-side float valve is installed at the evaporator inlet
   b. low-side float valve is always used with DX evaporators, while the high-side float valve is used with flooded evaporators
   c. low-side float valve is installed at the bottom of the evaporator, while the high-side float valve is installed at the top of the evaporator
   d. low-side float valve is installed on the low side of the metering orifice, while the high-side float valve is installed on the high side of the metering orifice

6. A distinguishing feature of the TXV is that it _____.
   a. has a sensing bulb filled with refrigerant or other thermostatic fluid
   b. needs to be manually adjusted as the cooling load changes
   c. uses a thermistor as a sensing device
   d. uses a float to maintain a constant liquid level in the evaporator

7. Which of the following is best suited for a system with a large pressure drop across the evaporator?
   a. An internally equalized TXV.
   b. An externally equalized TXV.
   c. A high-side float valve.
   d. A manual expansion valve.

8. The primary purpose of a distributor in an air conditioning system is to _____.
   a. provide high voltage to the spark plugs
   b. feed equal amounts of refrigerant to the evaporator circuits
   c. make sure the evaporator gets enough refrigerant
   d. distribute refrigerant from the condenser to the metering devices

9. Which of the following is true of TXV selection?
   a. Any TXV will do because it can be adjusted to fit the system.
   b. It may be undersized, but not oversized.
   c. If too large a TXV is selected, it may cause compressor slugging.
   d. If the TXV is too large, it will starve the evaporator.

10. The correct location for a TXV sensing bulb is _____.
    a. at the condenser outlet
    b. on the suction riser
    c. on the compressor side of the external equalizer tube
    d. at the evaporator outlet

## GLOSSARY

# Trade Terms Introduced in This Module

*Capillary tube:* A copper tube with a fixed length and fixed diameter, usually with an inside diameter of 1/16" to 1/8". Used as a metering device.

*Design load:* The maximum load at which a system is designed to operate.

*Direct-expansion (DX) evaporator:* An evaporator in which liquid is completely converted to vapor; also known as *dry-expansion*.

*Distributor:* Small tubes containing orifices that distribute liquid refrigerant evenly to the evaporator circuits.

*Distributor line:* Tubing between the distributor and evaporator or the line between the metering device and evaporator.

*Electronic expansion valve (EEV):* A microprocessor-controlled expansion valve in which the orifice is controlled by a precision DC motor. The EEV maintains a constant superheat.

*Fixed-orifice metering device:* A device in which the metering orifice is located in a replaceable piston. The term *fixed-orifice metering device* may also be used to refer to a capillary tube.

*Flash gas:* Vapor refrigerant that is formed as the liquid refrigerant is squeezed through the metering orifice. Flash gas is produced when some of the liquid refrigerant boils off to cool the remaining liquid as the liquid passes through the metering device.

*Flooded evaporator:* An evaporator in which the refrigerant remains in liquid form as it leaves the evaporator.

*Hunting:* A condition in which an expansion valve alternately underfeeds and overfeeds the evaporator.

*Orifice:* A tiny opening designed to pass liquid refrigerant.

*Slugging:* A condition in which a slug of liquid refrigerant and/or oil enters the compressor, causing a hammering sound. Slugging can result in compressor damage or even failure.

*Superheat:* Heat added to the refrigerant above the refrigerant's boiling point.

*Surge chamber:* A device that separates liquid and vapor refrigerant. Used with flooded evaporators. Liquid is recirculated back to the chiller or evaporator and vapor returns to the compressor.

*Thermal expansion valve (TEV or TXV):* A metering device with an external sensing bulb that senses the refrigerant temperature at the evaporator outlet. Also referred to as a *thermostatic expansion valve*. The valve maintains superheat around a setpoint. The terms *TEV* and *TXV* are used interchangeably.

*Thermal-electric expansion valve (TEEV or THEV):* An expansion valve in which a thermistor senses liquid line temperature and adjusts the orifice by changing the current flow through a bimetal needle. The terms *TEEV* and *THEV* are used interchangeably.

# ANSWER KEY

## Answers to Review Questions

| Answer | Section |
|---|---|
| 1. b | 2.0.0 |
| 2. b | 2.0.0 |
| 3. c | 3.1.0 |
| 4. a | 4.1.0 |
| 5. d | 4.2.0; 4.3.0 |
| 6. a | 4.5.1 |
| 7. b | 4.5.2 |
| 8. b | 5.0.0 |
| 9. c | 6.1.0 |
| 10. d | 6.3.0 |

# REFERENCES & ACKNOWLEDGMENTS

## Additional Resources

This module is intended to present thorough resources for task training. The following reference works are suggested for further study. These are optional materials for continued education rather than for task training.

*Air Conditioning Systems: Principles, Equipment, and Service*, 2001. Joseph Moravek. Upper Saddle River, NJ: Prentice Hall, Inc.

*Modern Refrigeration and Air Conditioning*, 2000. A.D. Althouse, C.H. Turnquist, A.F. Bracciano. Tinley Park, IL: The Goodheart-Willcox Company, Inc.

*Refrigeration & Air Conditioning Technology*, 2000. William C. Whitman, William M. Johnson, John A. Tomczyk. Albany, NY: Delmar Publishers, Inc.

## Figure Credits

| | |
|---|---|
| Thomas P. Burke | 209F01 |
| Ritchie Engineering Co., Inc. | 209SA02 |
| A-1 Components Corp. | 209SA03 |
| Gerald Shannon | 209F04, 209F16 |
| Carrier Corporation | 209SA04 |
| ALCO Controls | 209SA05, 209F08 |

# NCCER CRAFT TRAINING USER UPDATES

The NCCER makes every effort to keep these textbooks up-to-date and free of technical errors. We appreciate your help in this process. If you have an idea for improving this textbook, or if you find an error, a typographical mistake, or an inaccuracy in the NCCER's Craft Training textbooks, please write us, using this form or a photocopy. Be sure to include the exact module number, page number, a detailed description, and the correction, if applicable. Your input will be brought to the attention of the Technical Review Committee. Thank you for your assistance.

*Instructors* – If you found that additional materials were necessary in order to teach this module effectively, please let us know so that we may include them in the Equipment and Materials list in the Instructor's Guide.

**Write:** Curriculum Revision and Development Department
National Center for Construction Education and Research
P.O. Box 141104, Gainesville, FL 32614-1104

**Fax:** 352-334-0932

**E-mail:** curriculum@nccer.org

---

Craft _____  Module Name _____

Copyright Date _____  Module Number _____  Page Number(s) _____

Description
_____
_____
_____
_____

(Optional) Correction
_____
_____
_____

(Optional) Your Name and Address
_____
_____
_____

# Module 03210-01

# *Compressors*

## COURSE MAP

This course map shows all of the modules in the second level of the HVAC curriculum. The suggested training order begins at the bottom and proceeds up. Skill levels increase as you advance on the course map. The local Training Program Sponsor may adjust the training order.

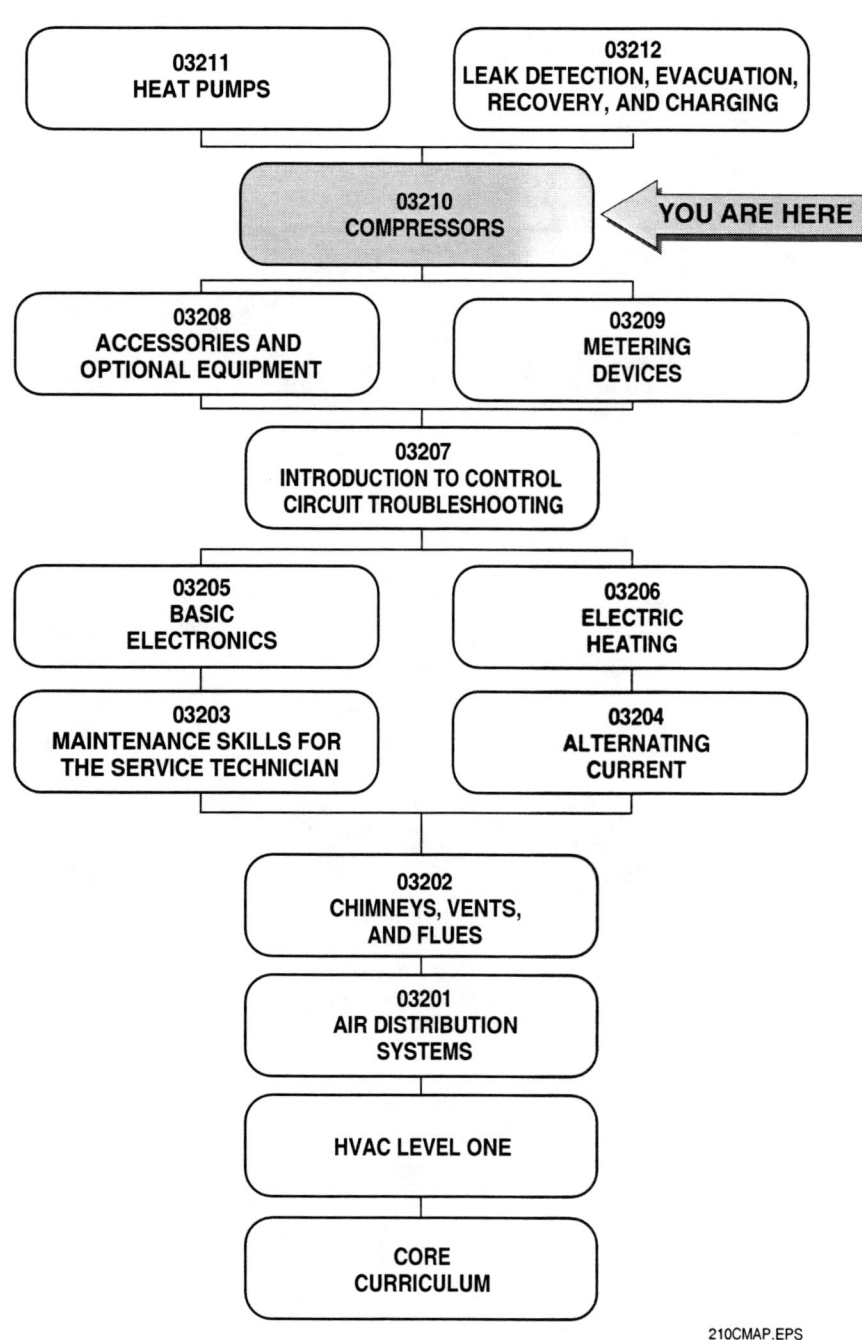

# MODULE 03210 CONTENTS

| | | |
|---|---|---|
| 1.0.0 | INTRODUCTION | 10.1 |
| 2.0.0 | THE ROLE OF THE COMPRESSOR | 10.1 |
| 3.0.0 | OPEN, HERMETIC, AND SEMI-HERMETIC COMPRESSORS | 10.3 |
| 4.0.0 | TYPES OF COMPRESSORS | 10.4 |
| | 4.1.0 Reciprocating Compressors | 10.4 |
| | *4.1.1 Piston and Piston Rings* | *10.5* |
| | *4.1.2 Compressor Cylinder Valves* | *10.6* |
| | *4.1.3 Open Compressor Crankshaft Seals* | *10.7* |
| | *4.1.4 Refrigerant Oils* | *10.7* |
| | *4.1.5 Lubrication Systems* | *10.8* |
| | 4.2.0 Rotary Compressors | 10.9 |
| | *4.2.1 Stationary Vane Compressors* | *10.9* |
| | *4.2.2 Rotating Vane Compressors* | *10.10* |
| | 4.3.0 Scroll Compressors | 10.10 |
| | 4.4.0 Screw Compressors | 10.11 |
| | 4.5.0 Centrifugal Compressors | 10.12 |
| 5.0.0 | CAPACITY CONTROL OF COMPRESSORS | 10.13 |
| | 5.1.0 Compressor Speed Control | 10.13 |
| | 5.2.0 On/Off Cycling | 10.14 |
| | 5.3.0 Cylinder Unloading | 10.14 |
| | 5.4.0 Hot Gas Bypass | 10.14 |
| | 5.5.0 Intake Slide Valve | 10.14 |
| | 5.6.0 Inlet Guide Vane | 10.14 |
| 6.0.0 | COMPRESSOR ELECTRIC DRIVE MOTORS | 10.15 |
| | 6.1.0 Compressor Motor Cooling | 10.15 |
| | 6.2.0 Compressor and Drive Motor Shaft Alignment | 10.16 |
| | 6.3.0 Input Power | 10.16 |
| | 6.4.0 Compressor Motor Overload Protection | 10.16 |
| | *6.4.1 External Line Break Overloads* | *10.17* |
| | *6.4.2 Internal Line Break Overloads* | *10.18* |
| | *6.4.3 Motor Thermostat Overloads* | *10.18* |
| | *6.4.4 Electronic Overloads* | *10.18* |
| | *6.4.5 Three-Phase Motor Overloads* | *10.20* |
| | *6.4.6 Current Monitoring Devices* | *10.20* |
| 7.0.0 | OTHER COMPRESSOR PROTECTION DEVICES | 10.21 |
| | 7.1.0 Pressure Protection | 10.21 |
| | 7.2.0 Evaporator and Condenser Airflow Protection | 10.21 |
| | 7.3.0 Lockout Protection | 10.22 |
| | 7.4.0 Short Cycling Protection | 10.22 |
| | 7.5.0 Electronic Head Pressure Control | 10.22 |

## MODULE 03210 CONTENTS (Continued)

**8.0.0 REDUCED-VOLTAGE MOTOR STARTING** .....................10.22
**9.0.0 CAUSES OF COMPRESSOR FAILURE** .......................10.24
    9.1.0    Slugging ..................................................10.25
    9.2.0    Flooding ..................................................10.25
    9.3.0    Flooded Starts ..........................................10.26
    9.4.0    Contamination ..........................................10.26
    *9.4.1    Air Contamination* .....................................10.26
    *9.4.2    Moisture Contamination* ..............................10.27
    *9.4.3    Acid Contamination* ...................................10.27
    *9.4.4    Dirt Contamination* ....................................10.27
    *9.4.5    Eliminating Contaminants* ...........................10.28
    9.5.0    Electrical .................................................10.28
    *9.5.1    Compressor Nameplate Information* .................10.28
    *9.5.2    Motor Operating Voltage Ranges* ...................10.28
    *9.5.3    Voltage Imbalance* ...................................10.29
    *9.5.4    Current Imbalance* ...................................10.29
    9.6.0    Compressor Heating ..................................10.30
**10.0.0 SYSTEM CHECKOUT FOLLOWING COMPRESSOR FAILURE** ...10.30
    10.1.0 Preliminary Inspection ................................10.30
    10.2.0 Analyzing System Operating Conditions .............10.31
    *10.2.1 System Operation Checks* ...........................10.31
    *10.2.2 Analyzing System Conditions* .......................10.32
    10.3.0 Final Compressor Checks ............................10.33
    *10.3.1 Electrical Reasons for Failure* ......................10.33
    *10.3.2 Mechanical Reasons for Failure* ....................10.33
**11.0.0 COMPRESSOR CHANGEOUT** ............................10.33
    11.1.0 Compressor Replacement Due to Mechanical Failure ......10.34
    11.2.0 Compressor Replacement Due to Electrical Failure .......10.35
    *11.2.1 Replacement Procedure After a Mild Burnout* ...........10.35
    *11.2.2 Replacement Procedure After a Severe Burnout* .........10.35
**12.0.0 HYDRONIC PUMPS** .....................................10.36
**SUMMARY** .....................................................10.37
**REVIEW QUESTIONS** ..........................................10.37
**GLOSSARY** ...................................................10.39
**ANSWERS TO REVIEW QUESTIONS** .............................10.40
**REFERENCES & ACKNOWLEDGMENTS** ...........................10.41

## Figures

| | | |
|---|---|---|
| Figure 1 | Typical refrigeration system | 10.2 |
| Figure 2 | Compressors | 10.3 |
| Figure 3 | Reciprocating compressor | 10.5 |
| Figure 4 | Compressor valves | 10.6 |
| Figure 5 | Bellows-type crankshaft seal | 10.7 |
| Figure 6 | Typical compressor lubrication systems | 10.9 |
| Figure 7 | Stationary vane rotary compressor | 10.10 |
| Figure 8 | Rotary vane compressor | 10.10 |
| Figure 9 | Scroll compressor | 10.11 |
| Figure 10 | Screw compressor | 10.12 |
| Figure 11 | Centrifugal compressor | 10.12 |
| Figure 12 | Suction bypass and suction cutoff unloaders | 10.15 |
| Figure 13 | Pilot and line duty devices | 10.17 |
| Figure 14 | Examples of external line break overloads | 10.17 |
| Figure 15 | Example of an internal line break overload | 10.18 |
| Figure 16 | Examples of motor thermostat overloads | 10.19 |
| Figure 17 | Sensor location in a three-phase motor | 10.19 |
| Figure 18 | Wiring connections for an overload module | 10.20 |
| Figure 19 | Control circuit with an overload module | 10.20 |
| Figure 20 | Line duty three-phase overloads | 10.21 |
| Figure 21 | Typical current-sensing transformer. | 10.21 |
| Figure 22 | Compressor pressure and airflow protection devices | 10.23 |
| Figure 23 | Compressor lockout relay protection | 10.23 |
| Figure 24 | Compressor anti-short cycle simplified circuit | 10.23 |
| Figure 25 | Typical electronic low ambient temperature head pressure control | 10.24 |
| Figure 26 | Typical compressor nameplate | 10.29 |
| Figure 27 | How to check voltage imbalance in a three-phase system | 10.29 |
| Figure 28 | Circulating pumps | 10.36 |

## Tables

| | | |
|---|---|---|
| Table 1 | Compressor Comparison Chart | 10.4 |
| Table 2 | Typical Operating Conditions for HCFC-22 System at 90°F Outdoors | 10.32 |

# MODULE 03210

# Compressors

## Objectives

When you have completed this module, you will be able to do the following:

1. Identify the different kinds of compressors.
2. Demonstrate or describe the mechanical operation for each type of compressor.
3. Demonstrate or explain compressor lubrication methods.
4. Demonstrate or explain methods used to control compressor capacity.
5. Demonstrate or describe how compressor protection devices operate.
6. Perform the common procedures used when field servicing open and semi-hermetic compressors.
   - Shaft seal removal and installation
   - Valve plate removal and installation
   - Unloader adjustment
7. Demonstrate the procedures used to identify system problems that cause compressor failures.
8. Demonstrate the system checkout procedure performed following a compressor failure.
9. Demonstrate or describe the procedures used to remove and install a compressor.
10. Demonstrate or describe the procedures used to clean up a system after a compressor burnout.

## Prerequisites

Before you begin this module, it is recommended that you successfully complete the following modules: Core Curriculum; HVAC Level One; HVAC Level Two, Modules 03201 through 03209.

## Required Trainee Materials

1. Pencil and Paper
2. Appropriate Personal Protective Equipment

## 1.0.0 ♦ INTRODUCTION

Compressors typically are the most expensive and, functionally, the most important part of a complete refrigeration system. Replacing a compressor is one of the most critical service procedures performed on refrigeration systems. Many compressors that are replaced are not defective. In many cases, this is because compressor operation and its relationship to the whole system are not fully understood by the HVAC technician. The result is needless service time and customer expense, as well as the failure to repair the equipment. Because understanding how a compressor works is so important, this module reviews some basic information about compressors previously covered in HVAC Level One. This is enhanced with new information that will help you accurately diagnose a compressor failure, understand why it failed, and correctly install a replacement compressor.

## 2.0.0 ♦ THE ROLE OF THE COMPRESSOR

The compressor is the heart of the refrigeration system. *Figure 1* shows the compressor in a typical system. Its role in the basic refrigeration cycle deserves review.

As shown in *Figure 1*, the refrigerant flows through the components of the system in the direction indicated by the arrows. Our review will begin with the evaporator.

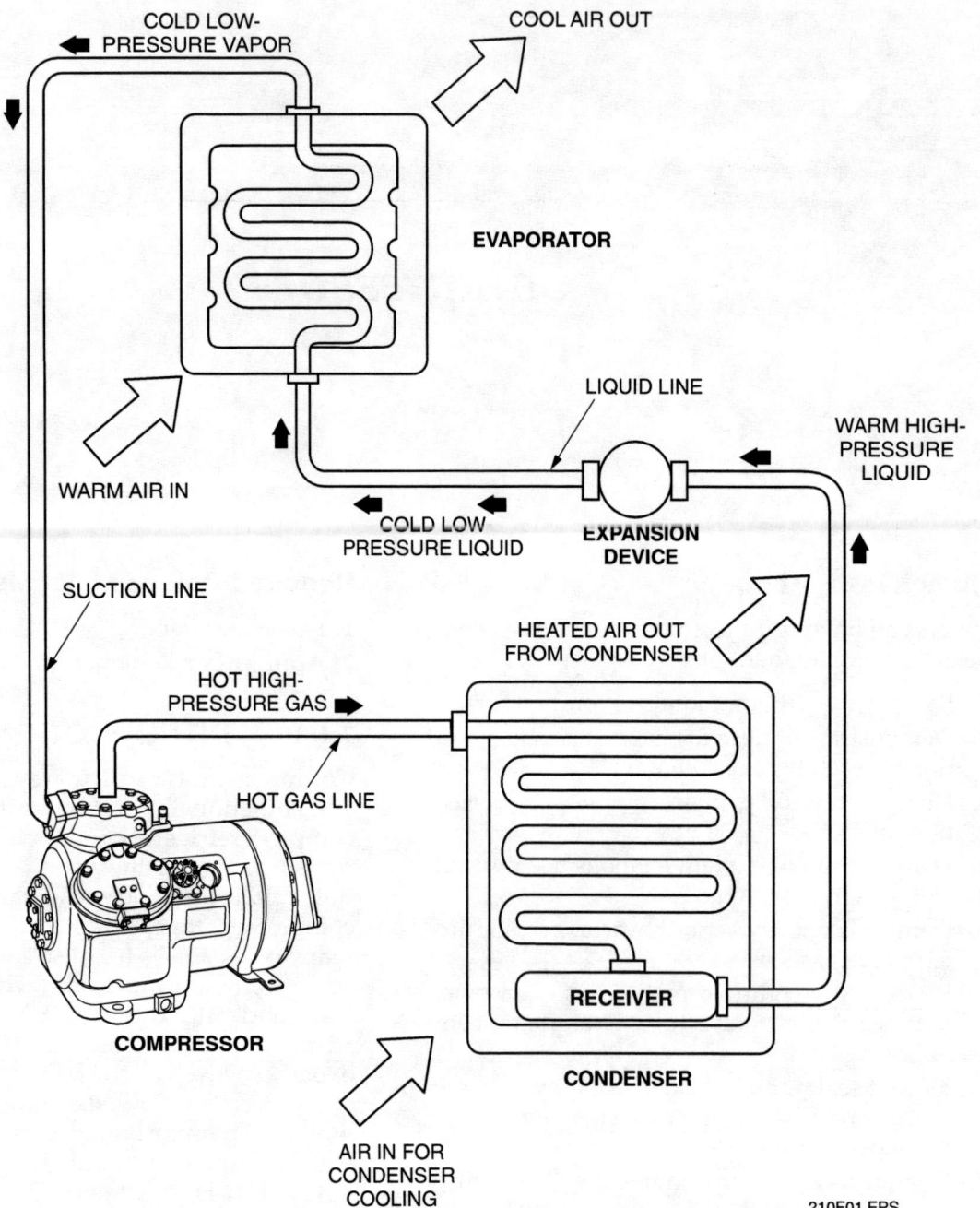

*Figure 1* ♦ Typical refrigeration system.

The evaporator receives low-temperature, low-pressure liquid refrigerant from the expansion device. The evaporator is mainly a series of tubing coils which expose the cooler liquid refrigerant to the warmer air passing over them. Heat from the warmer air is transferred through the evaporator tubing into the cooler refrigerant, causing it to boil or vaporize. Superheated, low-temperature, low-pressure refrigerant vapor at the output of the evaporator flows through the suction line to the suction input of the compressor. There, through the process of **compression**, the compressor converts the low-temperature, low-pressure vapor into a high-temperature, high-pressure vapor that flows to the condenser via the hot gas line.

Like the evaporator, the condenser is mainly a series of tubing coils through which the refrigerant flows. As cooler outside air moves across the condenser tubing, the hot refrigerant vapor gives up superheat and cools. As the refrigerant continues to give up heat to the outside air, it cools further, and the temperature drops to the saturated refrigerant vapor point, where the refrigerant begins to change from a vapor into a liquid. As more cooling takes place, called *subcooling*, the refrigerant is cooled below the saturation temperature. This medium-temperature, high-pressure liquid flows through the liquid line to the input of the expansion device.

The expansion or metering device regulates the flow of refrigerant to the evaporator. It also decreases the pressure, and therefore the temperature, of the refrigerant applied to the evaporator. Through the use of a built-in restriction, such as a small hole or orifice, it converts the medium-temperature, high-pressure refrigerant from the condenser into the low-temperature, low-pressure refrigerant needed to absorb heat in the evaporator.

The compressor performs two operations: it draws the refrigerant from the evaporator coil (suction cycle) and it forces the refrigerant into the condenser (discharge cycle). During this process, certain conditions are created:

- The pressure and temperature of the refrigerant in the evaporator are lowered, allowing the refrigerant to boil and absorb heat from its surroundings.
- The pressure and temperature of the refrigerant in the condenser are raised, allowing the refrigerant to give up heat at existing temperatures to whatever medium (air or water) is used to absorb the heat.
- A pressure difference is maintained between the high- and low-pressure sides of the system. It is this pressure difference that causes the refrigerant to flow through the system.

## 3.0.0 ◆ OPEN, HERMETIC, AND SEMI-HERMETIC COMPRESSORS

Compressors are usually driven by an electric motor. Very large compressors can be driven by internal combustion engines or steam turbines. Compressors are divided into three groups based on the way they are joined to their motors or engines (*Figure 2*).

- *Open-drive compressor* – This compressor is separate from its motor. One end of its horizontally mounted shaft extends outside the case. A mechanical seal on the rotating shaft prevents leakage of the refrigerant. The compressor motor drives the compressor by a belt (belt drive) or a flexible coupling (direct drive). Belt-driven arrangements allow the motor to run at one speed, while the compressor can run at another. The proper combination of pulleys, also called *drives,* produces the desired speed of the compressor. Most direct-drive systems use an electric motor to drive the compressor. This means that the compressor also runs at the speed of the drive motor.

- *Hermetic (welded hermetic) compressor* – In this unit, the compressor and motor have a common drive shaft. Hermetic reciprocating, rotary, and scroll compressors have the motor and drive shaft mounted vertically in the compressor casing. The reciprocating and rotary compressors usually have the motor mounted above the compressor. In the scroll compressor, the motor is usually mounted below the compressor. They are sealed in a welded steel enclosure or shell. Hermetic compressors (sometimes called *tin cans*) are more compact, less noisy, and require less maintenance than open-type compressors because they have no belts or couplings to break or wear out. Because they are sealed, the entire unit must be replaced when it fails.

- *Semi-hermetic (serviceable hermetic) compressor* – Similar to the hermetic compressor, the compressor and motor share the same housing and a common, horizontally mounted drive shaft. When they fail, access to the compressors or motors for limited repair is possible by removing the heads and/or bottom and end plates.

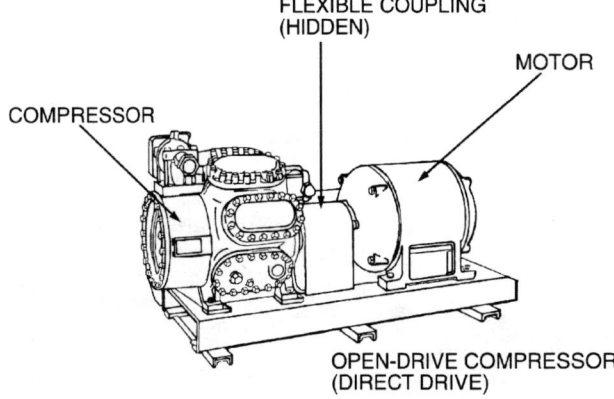

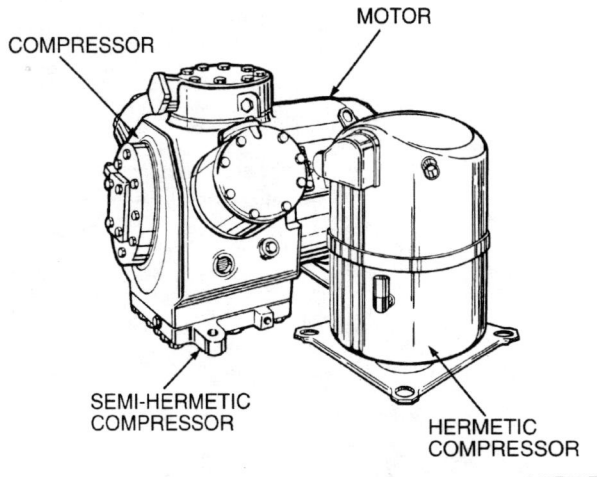

*Figure 2* ◆ Compressors.

### Open-Drive Compressors

Open-drive compressors are commonly used in many larger stationary air conditioning and refrigeration systems. Open-drive compressors of various sizes are also widely used in mobile applications such as in air conditioned vehicles and refrigerated trailers. In these applications, the belt-driven open-drive compressor is coupled to the vehicle's engine through a magnetic clutch mechanism located in the compressor drive pulley hub. The magnetic clutch, controlled by the vehicle's thermostat, engages or disengages the compressor belt-drive pulley and the compressor shaft. The clutch is operated by forcing a clutch disk, mounted to the compressor shaft, against the belt pulley. When the thermostat in the passenger compartment or refrigerated space of the vehicle calls for cooling, the clutch is engaged and causes the compressor to run. When the temperature reaches the desired cooling level, the thermostat signal releases the clutch. This disengages the compressor from the vehicle's motor and causes the compressor to stop running.

## 4.0.0 ◆ TYPES OF COMPRESSORS

The variety of refrigerants, and the size, location, and application of the systems are some of the factors that create the need for many types of compressors. The following compressors are commonly used in mechanical refrigeration systems:

- Reciprocating compressors
- Rotary compressors
- Scroll compressors
- Screw compressors
- Centrifugal compressors

*Table 1* summarizes the uses and characteristics of each type of compressor.

### 4.1.0 Reciprocating Compressors

Reciprocating compressors come in many types and can have from one to as many as ten pistons moving back and forth within a cylinder or cylinders. The main parts include a cylinder, piston, connecting rod, crankshaft, cylinder head, and suction and discharge valves enclosed in a crankcase. Piston movement is synchronized with the opening and closing of the suction and

**Table 1** Compressor Comparison Chart

|  | Reciprocating | Rotary | Scroll | Screw | Centrifugal |
| --- | --- | --- | --- | --- | --- |
| Use | Refrigeration and air conditioning, heat pumps, and transportation | Refrigerators, room air conditioners, and small central systems | Small central systems for refrigeration, air conditioning, and heat pumps | Refrigeration, air conditioning, and heat pumps | Refrigeration, air conditioning, and heat pumps |
| Size/Range | Fractional tonnage through 150 tons | Smallest: 5 tons or less | 1.5 to 70 tons | 50 to 750 tons | Largest: 100 to over 10,000 tons |
| Types | Open, serviceable hermetic, and hermetic | Sliding vane and rolling piston type, hermetic only | Compliant and noncompliant hermetic | Rolling rotor, open, and hermetic | Single and multi-stage open and hermetic |
| Displacement | Positive | Positive | Positive | Positive | Positive |
| Typical Capacity Control | Two-speed, on-off, and cylinder unloaders | On-off | On-off | Variable speed and intake slide valve | Variable speed and inlet guide vanes |
| Suction Valves | Yes | No | No | No | No |
| Discharge Valves | Yes | Yes | No | No | No |

discharge valves. These valves control the intake and discharge of the refrigerant into and from the cylinder. The crankcase contains the crankshaft and stores oil that is used to lubricate the moving compressor components.

*Figure 3* shows the intake (suction) and compression (discharge) strokes for one cylinder of a reciprocating compressor. The following events take place inside a reciprocating compressor cylinder during the intake stroke of operation:

1. At the start of the intake stroke, both the discharge and suction valves are closed. As the piston starts down, a low pressure is formed under the suction valve. When this low pressure becomes less than the suction line pressure, the suction valve opens and the cylinder begins to fill with gas supplied from the suction line.
2. As the piston continues to the bottom of the intake stroke, the cylinder becomes nearly full. At the bottom of the stroke there is a very slight time lag as the crankshaft rotates the connecting rod to complete the stroke. During this time lag, the cylinder continues to fill with suction gas.
3. When the piston reaches its lowest point of travel, called *bottom dead-center*, a spring on the suction valve closes the valve.

The following events take place inside a reciprocating compressor cylinder during the compression stroke of operation (*Figure 3*):

1. At the start of the compression stroke, both the suction and discharge valves are closed. The piston begins its upward stroke, compressing the gas in the cylinder. This increases the pressure and saturation temperature of the gas. The superheat of the gas is also increased. The suction valve remains held shut by the pressure in the cylinder while the discharge valve remains shut by the pressure in the discharge line.
2. As the piston continues its upward travel, the discharge valve remains held shut by the pressure in the discharge line until the piston gets close to the top of the cylinder. At this point in travel, the pressure in the cylinder becomes greater than the pressure in the discharge line, and the discharge valve opens. This allows the high-temperature, high-pressure gas to pass into the discharge line on its way to the condenser.
3. At the top of its travel, called *top dead-center*, the piston stops moving and leaves gas in a clearance area at the top of the cylinder. This area is called the **clearance volume**. The gas, which is at the discharge pressure, will re-expand when the piston starts back down at the start of the next intake stroke.

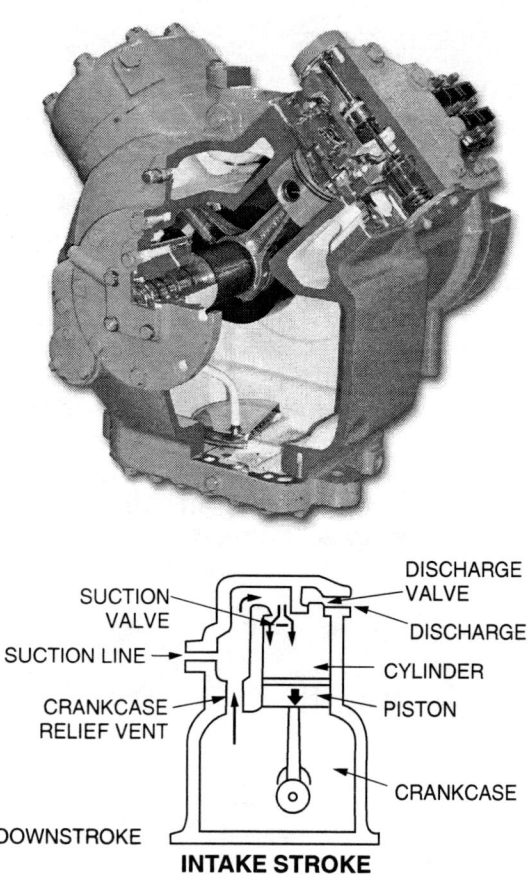

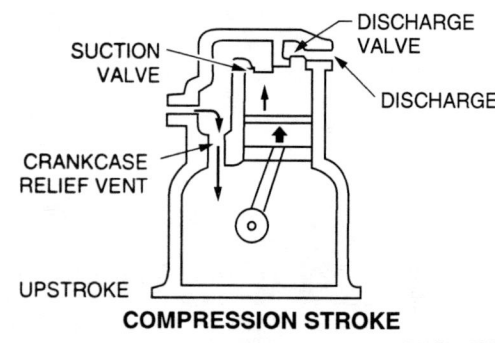

*Figure 3* ◆ Reciprocating compressor.

### 4.1.1 Piston and Piston Rings

The piston is exposed to the high-pressure gas during the compression stroke. During the upstroke, pistons have high-pressure gas on top and suction or low-pressure gas on the bottom. The piston must slide up and down in the cylinder in order to pump the gas. Two types of piston rings are mounted on the piston. The upper rings are called *compression rings*. They prevent the high-pressure discharge gas from leaking past the piston into the crankcase. Rings mounted on the piston below the compression rings are called *oil rings*. They function to control the oil flow past the piston. Some small hermetic compressors have no rings. In these compressors, the refrigerant oil acts as the seal.

## 4.1.2 Compressor Cylinder Valves

Two types of compressor valves are in common use. They are the reed (flapper) valve and the ring valve (*Figure 4*). Both types can be used either as a suction or discharge port valve. As shown in *Figure 4*, the reed valve is a thin, flexible piece of spring steel mounted on a valve plate. The valve plate mounts between the head of the compressor and the top of the cylinder wall. Some portion of the valve covers the suction or discharge port. The natural spring tension of the reed valve material works to keep the valve closed. It is forced open by pressures that exceed the valve spring tension. Reed valves have many different shapes, with each manufacturer having an individual version.

Ring valves are also mounted on valve plates. The assembly consists of valve seats, circular ring valves, and valve guides. The valve plate is mounted at the top of the cylinder, with the suction valve below the plate and the discharge valve above the plate. The ring valves are held in the closed position by a number of small springs installed in equally spaced holes around the guide. When the pressure exerted under either the discharge or suction valve is greater than the tension of the valve springs, the valve lifts from its seat, allowing gas to pass.

The operation, seating, and tightness of the valves is important. Broken valves, bent valves, or leaking valve plate gaskets affect the seal between the high-pressure and low-pressure sides of the compressor. High-pressure, high-temperature gas will leak into the suction side of the compressor. If this happens, the suction pressure will rise and the discharge pressure will be reduced. This can result in the compressor overheating. If the leak is small, such as might occur with a slightly bent or cracked valve, the pressure differences from normal operation will also be small. Continuous running of the compressor, low capacity, or upper cylinder head overheating may indicate that a compressor has worn or broken valves, or blown gaskets.

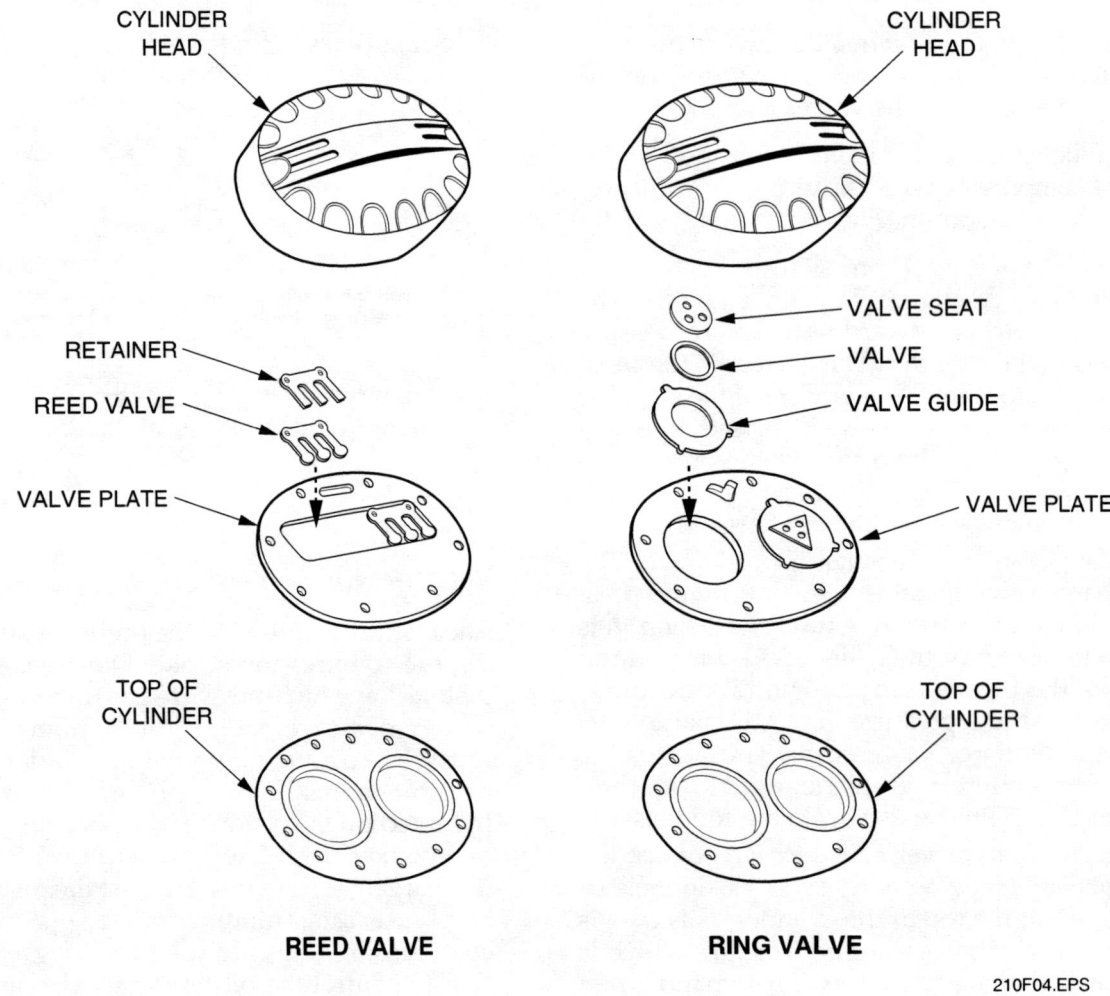

*Figure 4* ◆ Compressor valves.

> ### Compressor Valve Damage
>
> Under normal operation, all the refrigerant applied to a compressor is vaporized. However, if a problem occurs, such as can result from a faulty metering device, a slug of liquid refrigerant may be drawn into the compressor. This can have the same effect in a compressor as when birds are sucked into a jet engine; it is a disaster. Liquid refrigerant does not compress. Trying to compress a slug of refrigerant in a compressor generates pressures of over 1,000 psi. When applied to the compressor's valves or connecting rods, this pressure can damage (and even break) the valves or rods.
>
> Overheating of the compressor resulting from high suction superheat, high compression ratios, or air trapped in the refrigeration system can also damage the compressor by carbonizing the discharge valves and/or valve guides.

### 4.1.3 Open Compressor Crankshaft Seals

Open compressors must use a leakproof seal where the crankshaft exits the compressor crankcase. This seal prevents the refrigerant from leaking out of the compressor, regardless of whether or not the shaft is rotating. In certain low-temperature systems such as those using CFC-12 (R-12) refrigerant, the compressor crankcase pressure can be in a vacuum (below 0 psig). With this condition, a leaking seal can allow air and water vapor to enter the refrigerant system. A leaking seal usually causes the loss of the system refrigerant, resulting in constant running of the compressor and poor cooling.

All seals use two rubbing surfaces. One surface turns with the crankshaft and is sealed to the shaft with an O-ring or synthetic material. The other surface is stationary and is mounted on the housing with leak-proof gaskets. Most compressors use a bellows seal or a variation of the bellows seal. The bellows seal (*Figure 5*) consists of a bellows assembly, carbon ring, and cover plate installed on the compressor crankshaft in the same order as listed. The sealing surface is between the nose of the bellows and a shoulder on the crankshaft. Other seals in use are the packing gland, diaphragm, and rotary.

If installed correctly, crankcase seals can last for years without wearing out. For long seal life, it is important that the compressor shaft be properly aligned with the mating drive motor (or engine) shaft. In a direct-drive unit, the shafts should be aligned according to the manufacturer's instructions. In a belt-drive unit, the belts must be adjusted for the proper tension.

### 4.1.4 Refrigerant Oils

Refrigerant compressors use special oil. This oil must have excellent lubrication qualities and chemical stability. These characteristics are needed to prevent breakdown from long periods of use or from contact with contaminants outside or inside the system. If a substitute is used, system problems and eventual compressor failure may result. Oil in a refrigeration system performs the following functions:

- Minimizes friction and prevents wear of moving parts
- Provides a fluid seal which helps to separate the high side from the low side of the system
- Acts as a coolant
- Dampens mechanical noise made by the moving parts

Systems using CFC and HCFC refrigerants have used and continue to use mineral oils for lubrication. However, the new refrigerants, which do not contain chlorine, require lubricating oils with special characteristics. For example, R-410A is the refrigerant that appears likely to replace R-22 in most residential and light commercial air conditioners. Since R-22 is so widely used in HVAC, and R-410A is its likely replacement, let's look at the lubricant needed with this new refrigerant.

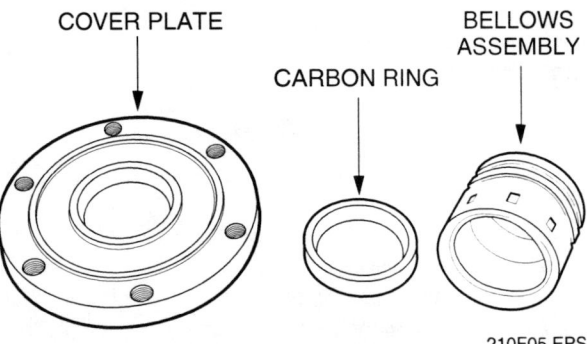

*Figure 5* ◆ Bellows-type crankshaft seal.

This new HFC refrigerant requires a synthetic polyolester (POE) oil that has the following unique characteristics:

- The oil is not compatibile with mineral oils used in R-22 systems.
- The oil is hygroscopic (it readily attracts water).
- The oil irritates the skin.

Since the oil does not mix with mineral oil, most of the remaining mineral oil must be removed from a system before R-410A refrigerant can be installed. This is important if you replace an R-22 split system condenser with an R-410A condenser. Tests have shown that the maximum amount of mineral oil that can remain in the system without causing problems is 5%. Since this low level may be difficult to achieve in a field service environment, it is probably easier to replace the entire system, including the evaporator coil and interconnecting tubing. In fact, some manufacturers recommend doing just that.

POE oil is very hygroscopic. It readily attracts water, and neither water nor moisture belong in an air conditioning system. So strong is the attraction that the oil must be kept in a sealed metal container. It will even absorb moisture through the walls of a plastic container. When adding POE oil to a system, use a pump; do not pour it. Use oil from a fresh, sealed, metal container and discard any oil that may be left over in the container. Do not open a system containing POE oil on a foggy or rainy day. Split systems containing this new oil and refrigerant must have a special liquid line filter-drier installed. If moisture does get into the POE oil, it can only be removed by the same type of special filter-drier.

There are many new refrigerants available today to replace CFC and HCFC refrigerants. Some will stand the test of time, while others will fall out of favor. Many of these new refrigerants have special oils that must be used with them. We have shown you an example of one such refrigerant and the special requirements of the oil it uses. When dealing with any of the new refrigerants, be sure to follow the equipment manufacturer's installation and service instructions.

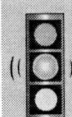

**CAUTION**

POE oil can cause skin irritation. If any gets on your skin, wash it off with warm, soapy water. This same oil can damage certain rubberized roof coatings. If a roof spill occurs, the oil will still damage the roof, even if the spill is cleaned up. To prevent roof damage, lay down protective plastic sheets to prevent any oil from coming into contact with the roof.

### 4.1.5 Lubrication Systems

Two lubrication methods are used with reciprocating compressors to lubricate the moving parts. Some compressors use a **splash lubrication system**. In the splash lubrication system (*Figure 6*), crankcase oil is splashed onto the cylinder walls and bearing surfaces of the compressor during each revolution of the crankshaft while the compressor is running. Some compressor connecting rods have little dips or scoops attached to the lower end that scoop up the oil and splash it around to the other parts.

The second and most popular method of lubrication is the **pressure (force-feed) lubrication system**. This method is used in all sizes of compressors. It uses an internal oil pump mounted on the end of the crankshaft. This pump forces oil through the crankshaft to the bearing surfaces. In some cases the connecting rods are drilled so the oil under pressure is also supplied to the piston pins.

---

### POE Compatibility With Older Lubricants

HFC refrigerants will not mix with mineral oil or alkylbenzene lubricants used with older CFC and HCFC refrigerants. For this reason, when retrofitting an older system to use a HFC refrigerant, it is necessary to flush the old oil out of the system and replace it with POE oil. If the residual level of non-POE oil remaining in the system after flushing exceeds 5%, pockets of oil may drop out of the refrigerant and coat the tubing, take up space, or clog openings inside the system.

There are concerns over the chemical reactivity of POEs in the presence of residual oils and contaminants left in the system. POEs are better solvents than older lubricants. Systems with residue on the inside of the piping or components may be cleaned by the POE, which can cause the released contaminants to circulate to the system valves and/or compressor where they can cause damage.

## Inside Track

### System Loading and Compressor Lubrication

A certain amount of compressor lubricating oil is normally entrained in the refrigerant being circulated in a system. The uniform movement of this oil through the system piping and back to the compressor is dependent on a reasonably high refrigerant velocity. However, at minimum load, the movement of refrigerant within the system can be greatly reduced, resulting in the trapping of oil within the system evaporator and discharge and suction piping. Excessive trapping of oil can reduce the oil level in the compressor, eventually resulting in bearing failure.

An oil screen may be located in the bottom of the crankcase to filter the oil supplied from the crankcase to the oil pump. An oil pressure regulator in some systems prevents the buildup of excessive oil pressure, which could result in high power consumption, loss of oil, and damage to the compressor. *Figure 6* shows typical pressure (force-feed) lubrication systems. The arrows show the direction of oil flow through the compressor.

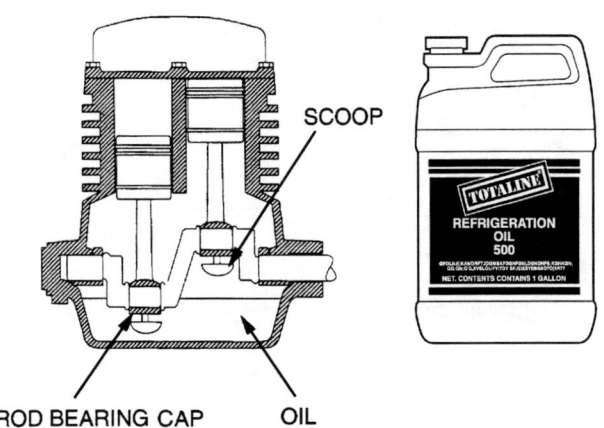

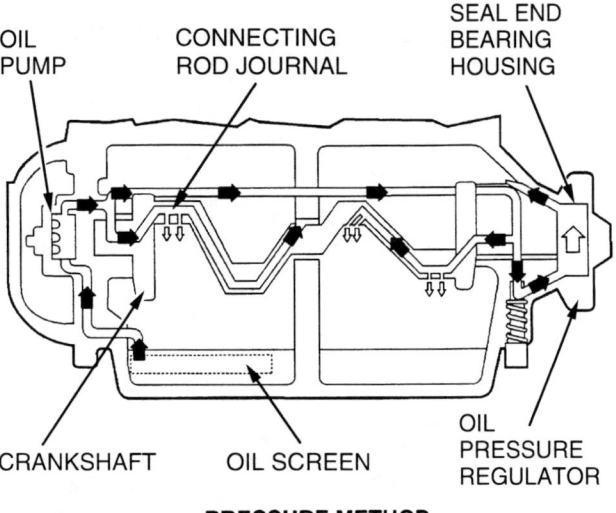

*Figure 6* ◆ Typical compressor lubrication systems.

### 4.2.0 Rotary Compressors

Rotary compressors are usually welded hermetic compressors. The motor and drive shaft are mounted vertically in the compressor housing with the motor above the compressor. Rotary compressors are usually of two types: stationary vane and rotary vane. In both types, the vanes, under spring tension, slide in and out of their retaining slots to provide a continuous seal for the refrigerant vapor within the cylinder. Rotary compressors do not have suction valves, but use a discharge valve to prevent backflow of refrigerant into the compressor when it is turned off. Normally, they have a check valve installed in the intake passage to prevent the compressor oil from being forced back into the suction line and into the evaporator. Lubrication in the rotary compressor is provided either by a splash system or force-feed system. Regardless of the method, proper operation of the rotary compressor depends on maintaining a continuous film of oil on the cylinder, roller, and vane surfaces.

#### 4.2.1 Stationary Vane Compressors

In the stationary vane compressor (*Figure 7*), a shaft with an attached off-center (eccentric) rotor rotates or rolls around the cylinder. As it rotates, one point on its circumference is always in contact with the cylinder wall. A stationary vane that is under spring tension is mounted in the compressor housing. It slides in and out and follows the out-of-round motion of the rotor as the rotor moves within the cylinder. This vane also isolates the suction and discharge sides of the cylinder. As the shaft turns, the rotor rolls around the cylinder, drawing suction gas in the intake opening while at the same time compressing the gas against the cylinder wall on the discharge or compression side. This process continues as long as the compressor is running. An exhaust valve mounted on the discharge port keeps the compressed gas from leaking back into the cylinder and into the suction side during the off cycle.

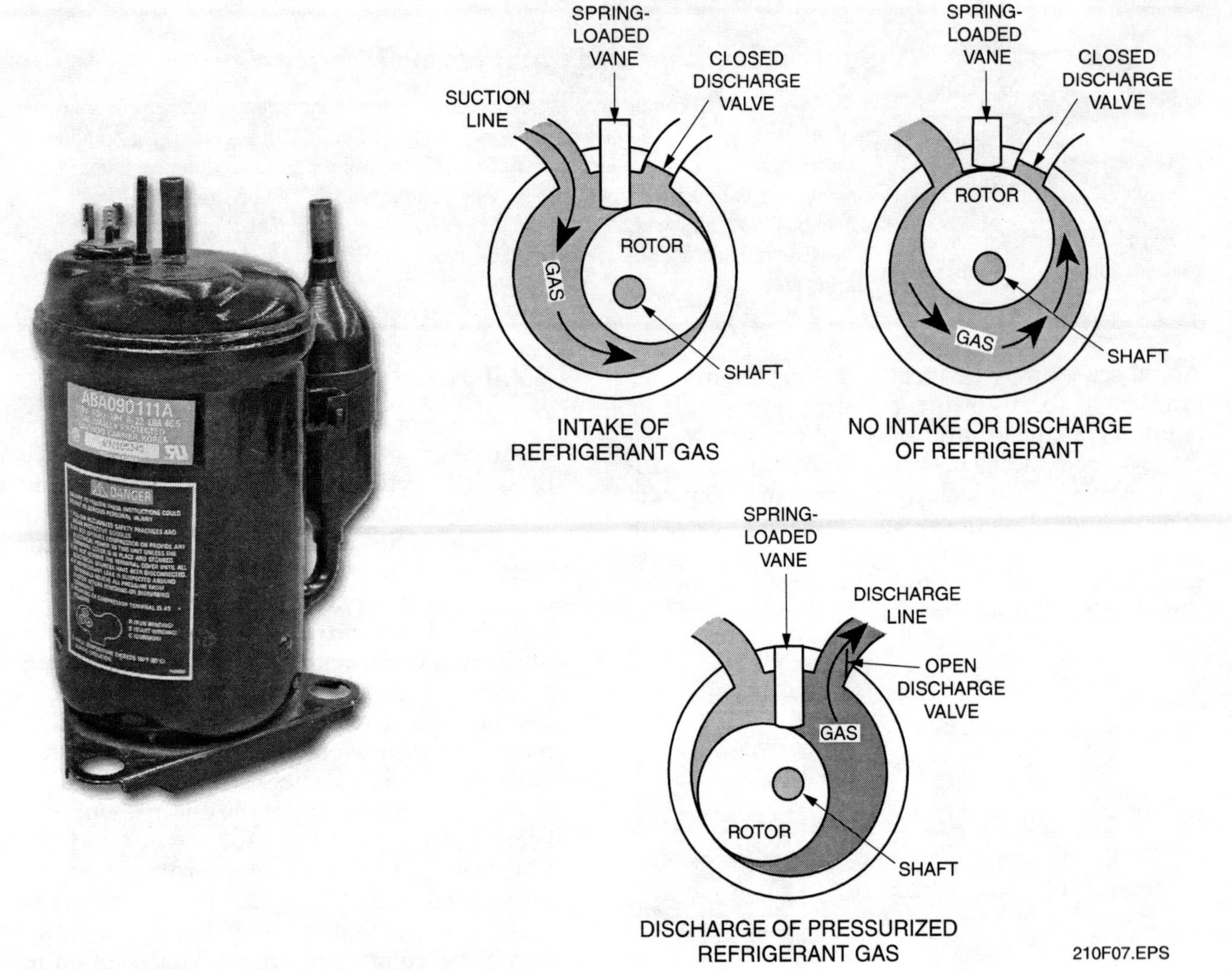

*Figure 7* ♦ Stationary vane rotary compressor.

### 4.2.2 Rotary Vane Compressors

Rotary vane compressors (*Figure 8*) have a rotor centered on the drive shaft. However, the drive shaft is positioned off-center in the cylinder. Mounted on the rotor are two or more vanes that slide in and out to follow the shape of the cylinder. As the rotor turns, low-pressure suction gas from the suction line is drawn into the cylinder behind the vanes. The trapped vapor ahead of the vanes is compressed against the cylinder wall until it is forced out of the discharge opening. The vanes also keep the compressed gas from mixing with the incoming low-pressure gas.

### 4.3.0 Scroll Compressors

Scroll compressors are usually welded hermetic compressors. The motor and drive shaft are mounted vertically in the compressor housing with the motor below the compressor. This compressor has neither suction nor discharge valves because its construction resists refrigerant backflow.

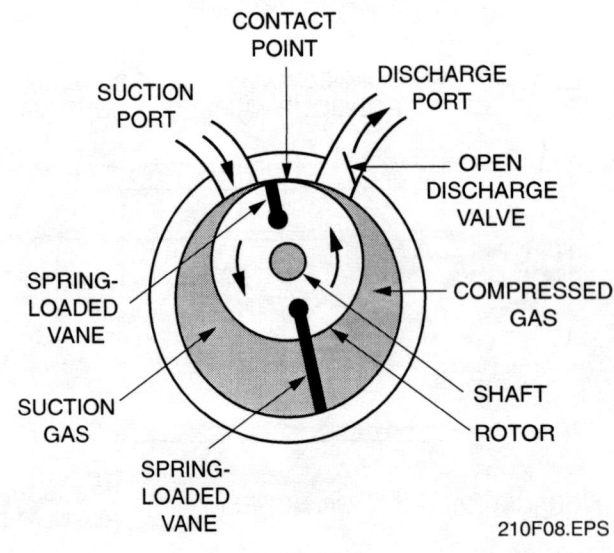

*Figure 8* ♦ Rotary vane compressor.

### Rotary Vane Compressors

Rotary vane compressors must be installed in clean systems. If small particles of debris are present, they can jam the rotor and stop the compressor. For this reason, rotary compressors are mainly used in packaged room air conditioners and packaged terminal air conditioners (PTACs) where the cleanliness of the sealed system can be controlled at the factory. In the past, rotary compressors were used in some split air conditioner systems. However, dirt and contamination introduced during installation caused a high rate of compressor failure, prompting manufacturers to discontinue their use in those systems.

Of all the compressor types, the scroll compressor has the fewest working parts. It operates efficiently even in applications that have large changes in refrigerant pressures, such as with commercial refrigeration and heat pumps.

The scroll compressor (*Figure 9*) achieves compression through the use of two spiral-shaped parts, called *scrolls*. The upper scroll is fixed; the lower scroll is driven by the motor and moves with an orbiting action inside the fixed scroll. There is contact between the two. Refrigerant gas enters the suction port at the outer edge of the scroll and, after compression, is squeezed out a separate discharge port at the center of the stationary scroll. The orbiting action draws gas into pockets between the two spirals. As this action continues, the gas opening is sealed off and the gas is compressed and forced into smaller pockets as it progresses toward the center.

A version of the scroll compressor, called a **compliant scroll compressor**, allows the orbiting scroll to temporarily shift from its normal operating position if liquid refrigerant enters the compressor.

### 4.4.0 Screw Compressors

Screw compressors are used in large commercial and industrial applications requiring capacities from 50 to 750 tons. They are made in both open and hermetic styles.

Screw compressors use a matched set of screw-shaped rotors (*Figure 10*), one male and one female, enclosed within a cylinder. The male rotor is driven by the compressor motor. In turn, it drives the female rotor. Normally, the driven male rotor turns faster than the female rotor because it has fewer lobes than the female rotor. Typically, the male has four lobes, and the female has six. As these rotors turn, they mesh with each other and compress the gas between them. Oil keeps them from actually touching. The screw threads form the boundaries separating several compression chambers, which move down the compressor at the same time. In this way, the gas entering the compressor is moved through a series of progressively smaller compression stages until the gas exits at the compressor discharge in its fully compressed state.

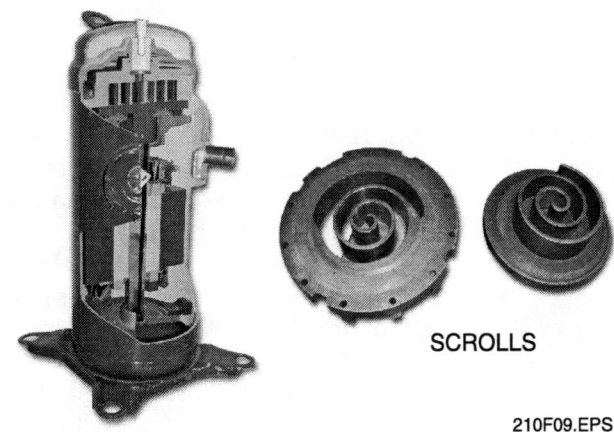

*Figure 9* ◆ Scroll compressor.

### Scroll Compressors

Scroll compressors make unique sounds when starting up, running, and shutting down. If you are not aware that these sounds are normal, you may think there is something wrong with the compressor. Various compressor manufacturers and equipment manufacturers have training programs and materials available that can help you recognize the normal and abnormal sounds made by scroll compressors.

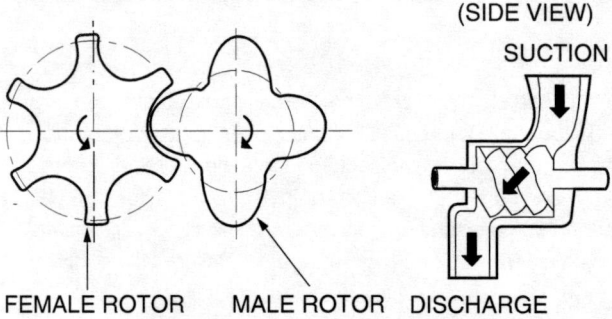

SCREW COMPRESSOR ROTORS

210F10.EPS

*Figure 10* ◆ Screw compressor.

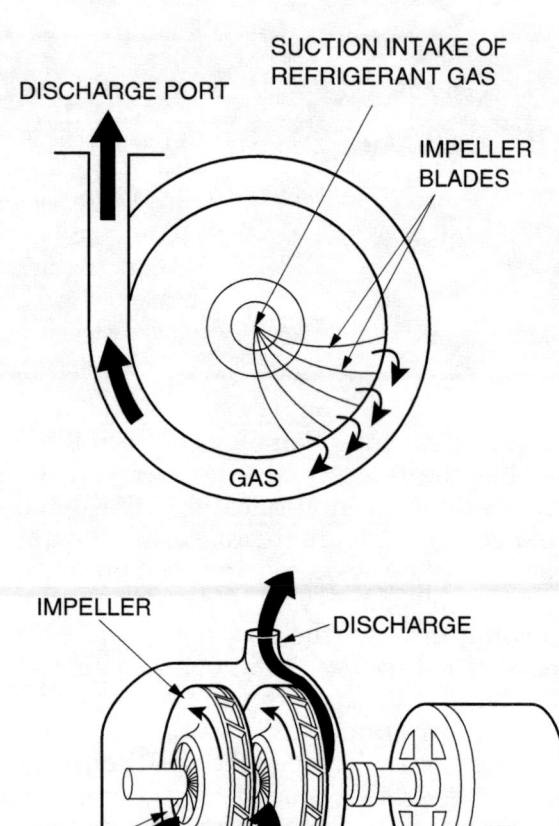

TWO-STAGE COMPRESSOR

## 4.5.0 Centrifugal Compressors

Centrifugal compressors are made in open and hermetic designs. They are typically used on commercial and industrial refrigeration and air conditioning systems using compressors larger than 100 tons. Standard models range up to 10,000 tons of capacity; custom-built models exceed 20,000 tons.

Centrifugal compressors use a high-speed impeller with many blades that rotate in a spiral-shaped housing (*Figure 11*). The impeller is driven at high speeds (typically 10,000 rpm) inside the compressor housing. Refrigerant vapor is fed into the housing at the center of the impeller. The impeller throws this incoming vapor in a circular path outward from between the blades and into the compressor housing. This action, called *centrifugal force*, creates pressure on the high-velocity gas and forces it out the discharge port. Often, several impellers are put in series to create a greater pressure difference and to pump a sufficient volume of vapor. A compressor that uses one impeller is a single-stage machine; one which uses two impellers is a two-stage machine, and so on. When more than one stage is used, the discharge from the first stage is fed into the inlet of the next stage.

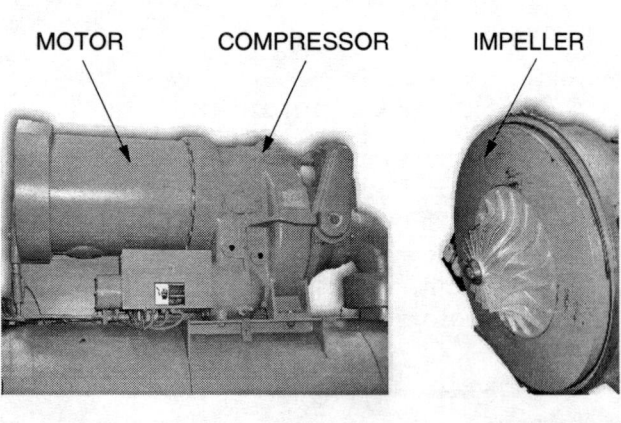

210F11.EPS

*Figure 11* ◆ Centrifugal compressor.

## 5.0.0 ◆ CAPACITY CONTROL OF COMPRESSORS

When a cooling system is operating properly, it undergoes a reduction in load whenever there is less heat for the system to dissipate. Loads handled by commercial cooling systems change rapidly and vary more than those usually encountered by residential systems. Commercial systems typically operate at 50% of the design (peak) cooling load. If the system compressor operated at its full capacity under these conditions, it would lower the evaporator pressure and the dew point of the evaporator coil, causing the relative humidity and temperature within the building to drop below a comfortable level. If the dew point dropped low enough, frost would build up on the evaporator coil, restricting airflow and resulting in compressor **short cycling** and erratic system operation. If the condition continued, it might even cause a compressor motor failure.

Common methods of controlling compressor capacity include:

- Multiple-speed motors
- On/off cycling (multiple compressors)
- Cylinder unloading
- Hot gas bypass
- Intake slide valves
- Inlet guide vanes

**Capacity control** of compressors is one of several methods used to change the pumping capacity of a compressor in order to match changes in the system load.

## 5.1.0 Compressor Speed Control

All compressors, except the centrifugal compressor, are **positive-displacement compressors**. Positive-displacement compressors are those in which the pumping action is created by moving pistons or moving chambers. When they are operated at a constant speed, they pump a constant volume of gas. The higher the compressor displacement, the higher its capacity. For a reciprocating compressor, the displacement is calculated as follows:

$$\text{Compressor displacement} = \text{piston displacement} \times \text{rpm}$$

$$\text{Compressor displacement (cubic inches)} = (pD^2Ln \div 4) \times \text{rpm}$$

*Where:*

p = 3.1416
D = cylinder bore (in inches)
L = length of stroke (in inches)
n = number of cylinders
rpm = revolutions per minute

The displacement, and therefore the capacity, of a compressor is proportional to the speed of its drive motor. For this purpose, multiple-speed motors are often used to drive compressors, especially reciprocating, scroll, and rotary compressors. Electronic controls that automatically select the proper motor speed to match load conditions make their use practical.

---

### Factors That Determine the Capacity of a Compressor

When a system is designed, a compressor is selected that has the capacity to handle the peak system load. The capacity of an off-the-shelf reciprocating compressor is based on several factors. Some of these factors are inherent to the compressor's design (mechanical design factors), while others are external to the compressor (application factors). Note that the application factors are determined by the conditions under which a compressor is operated and are changeable within limits.

Mechanical design factors that determine a compressor's capacity include:

- Piston displacement – A function of bore, stroke, and the number of cylinders
- Piston clearance – The space between the piston head and the end of the cylinder when the piston is at the top of its stroke
- The size of the suction and discharge valves

Application factors that affect the capacity of a compressor include:

- Speed in revolutions per minute (rpm)
- Suction pressure
- Discharge pressure
- Type of refrigerant

Variable-speed motors have been used to drive compressors. However, the complexity and expense of variable-speed motors have limited their use.

## 5.2.0 On/Off Cycling

On/off cycling is a method of capacity control used in multi-compressor systems. The capacity of the system is controlled by cycling (turning) on and off individual compressors within the system.

## 5.3.0 Cylinder Unloading

Cylinder unloading is a popular method of capacity control used with open and semi-hermetic reciprocating compressors, usually above 20 tons of capacity. A control valve internal to the compressor blocks the flow of refrigerant gas to some of the cylinders of a multi-cylinder compressor. The cylinders continue to move up and down but do not pump refrigerant into the system. Three types of cylinder unloading methods are in common use:

- Suction bypass unloading
- Suction cutoff unloading
- Hydraulic unloading

The suction bypass unloading method uses a special cylinder head and a pressure actuated or electrically actuated unloader control valve. When the system cooling load drops, the pressure and temperature at the compressor suction input also drops. The unloader control device senses this decrease in pressure (pressure actuated valve) or temperature (electrically actuated valve) and operates to actuate the unloader. A piston in the unloader retracts at a preset point and unloads the cylinder. This causes the suction refrigerant gas to be recirculated between the discharge and suction sides of the unloaded cylinder. *Figure 12* shows a pressure actuated bypass unloader operating in the unloaded position. The suction cutoff unloading method is actuated in a similar manner. With a suction cutoff unloader, the suction gas is prevented from entering the unloaded cylinder, rather than being recirculated through it.

Normally, both the suction bypass and suction cutoff types of unloader valves have adjustments to set the operating point at which the control valve will load and unload the cylinders. One adjustment, called the *control setpoint*, adjusts the point where the cylinder loads. Another adjustment is made to provide a pressure differential between the cylinder load and unload point.

The hydraulic unloader achieves unloading in compressors by holding open the suction valve, thus preventing compression of the gas. The unloader valve lifting mechanism is powered by the compressor lubricating system. When capacity reduction is needed, oil pressure is relieved from a piston located in the capacity reduction unit. This piston operates a mechanical linkage to lift the suction valve from its seat. The compressor piston is then no longer able to compress refrigerant within the cylinder and just travels up and down within the cylinder.

Like the other unloaders, the hydraulic unloader has adjustments to set the operating point at which the cylinders are loaded and unloaded.

## 5.4.0 Hot Gas Bypass

The hot gas bypass method reduces compressor capacity by routing some of the compressor hot gas discharge through a bypass line back into the suction line. The flow of hot gas through the bypass line is controlled by a solenoid stop valve in the line.

During full capacity operation, the solenoid valve is closed and blocks the line. When the related pressure or temperature sensor calls for a reduction in capacity, the solenoid valve is actuated and the bypass line is opened. This allows some of the hot gas discharged from the compressor to be returned to the suction line. In this method, the amount of capacity reduction is determined by the amount of gas bypassed.

## 5.5.0 Intake Slide Valve

The intake slide valve method of capacity control is used in screw compressors. It uses a sliding seal mechanism that varies the compressor capacity by changing the point where the suction entraps the vapor and starts to compress it as it moves through the rotors.

## 5.6.0 Inlet Guide Vane

The inlet guide vane method of capacity control is used in centrifugal compressors. It uses inlet guide vanes mounted in front of the impeller inlet. These vanes open and close while directing the refrigerant vapor into the impeller, changing compressor performance. The more the vanes are closed, the lower the capacity of the compressor.

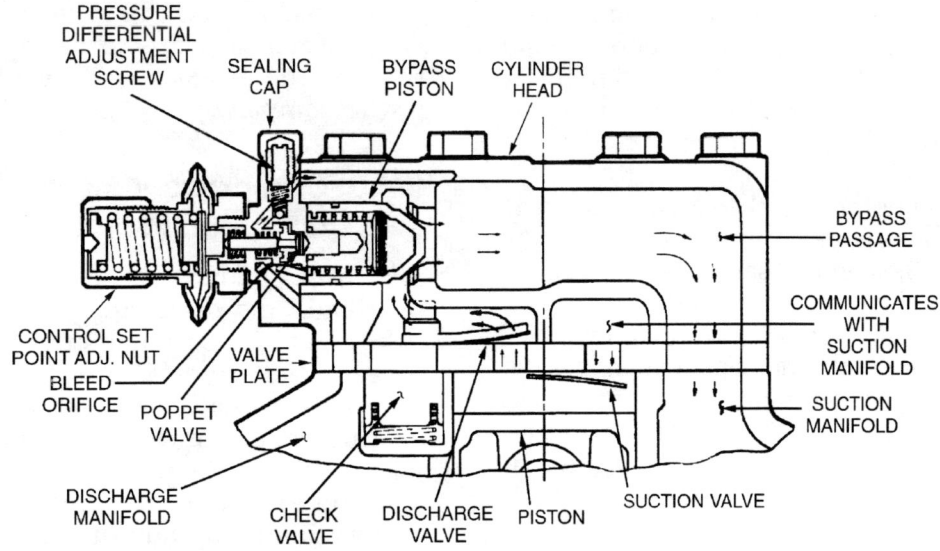

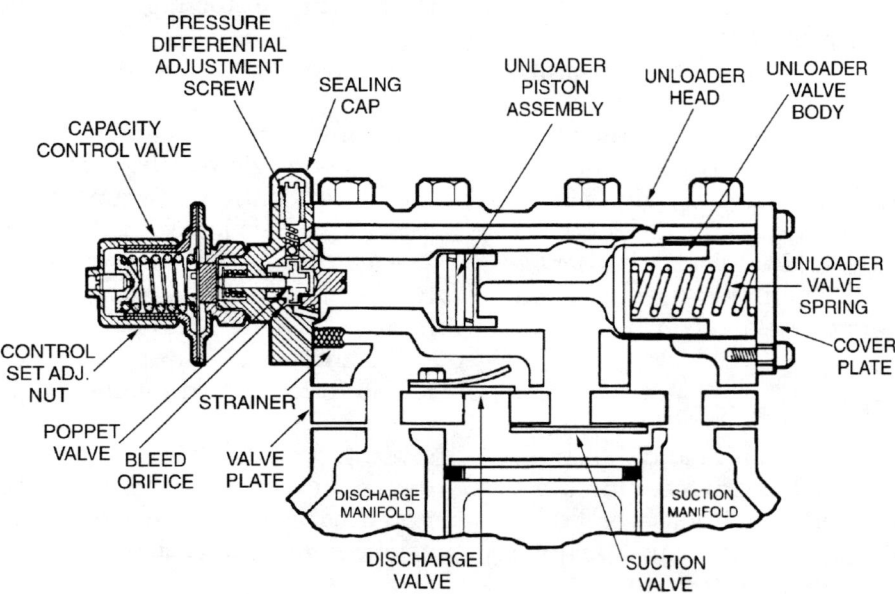

*Figure 12* ♦ Suction bypass and suction cutoff unloaders.

## 6.0.0 ♦ COMPRESSOR ELECTRIC DRIVE MOTORS

The electric motors used to drive compressors operate in the same way as other motors used in refrigeration systems. You have studied the principles of operation for electric motors previously in the HVAC Level Two Module, *Alternating Current*. This section describes cooling, mechanical, and electrical considerations specific to compressor motors. Also described are methods commonly used to protect compressors and compressor motors.

### 6.1.0 Compressor Motor Cooling

Unless specially designed, open compressors and their drive motors must be located in a well-ventilated area to aid in motor cooling. Electric motors used to drive open compressors are cooled by the surrounding air. Some have ventilation openings on their end bells or sides to pass external cooling air over and around the windings. This type of motor must be used in relatively clean locations to prevent the entrance of dust and other foreign material into the motor housing. Other motors used to drive open compressors are totally enclosed to prevent free air exchange between the inside and outside. They depend on radiation for cooling the windings. Often, the enclosed motor has external fins and/or a fan connected to its shaft to aid in cooling.

In hermetic compressors, the electric motor is enclosed in the same housing as the compressor. As the motor turns the compressor, electric energy is changed into mechanical energy and heat energy. This heat must be removed or it will build up, causing compressor and motor damage. In serviceable-hermetic and welded-hermetic compressors, cool refrigerant from the evaporator passes over and cools the motor windings before the refrigerant enters the suction side of the compressor.

## 6.2.0 Compressor and Drive Motor Shaft Alignment

Electric motors used to drive open compressors are coupled to the compressor crankshaft through pulleys and belts or a flexible coupling. Proper bearing wear depends on achieving the best possible alignment of the compressor and the drive motor pulleys or flexible couplings. The compressor and motor mountings should be firm and rigid to eliminate vibration and bearing wear. Refer to the HVAC Level Two Module, *Maintenance Skills for the Service Technician* for the methods used to align compressor and motor pulleys and couplings.

## 6.3.0 Input Power

Depending on the size of the motor and the application, electric motors used to drive compressors usually operate on either single-phase or three-phase AC power. Wiring to a motor must be done in accordance with the latest edition of the *National Electrical Code (NEC)* and local code requirements. Wire size should be based on the motor nameplate full-load amperage (FLA) or rated-load amperage (RLA), any listed NEC derating factor, and any increase in wire size needed to prevent voltage drop on long runs.

To prevent damage to motor windings, make sure the operating voltage, frequency, and phase stamped on the motor nameplate are compatible with the input electric power source. In three-phase systems, also make sure that the input voltage phase imbalance is no greater than 2%. Should the imbalance exceed 2%, contact your local power company.

## 6.4.0 Compressor Motor Overload Protection

The most common causes of motor failure are overloads and overheating. An overload condition is produced when current to the motor exceeds the motor's normal operating current flow. An overload condition may result in melted conductors or burned insulation on the motor's wiring. Considerable damage may result if the compressor motor overheats. Overheating can occur without the current draw becoming excessive. Heat can be caused by a defective start relay, excessive load, or loss of refrigerant gas cooling in a hermetic compressor. Another cause is operation at too high or too low a voltage. In three-phase motors, a leg-to-leg voltage imbalance exceeding 2% will shorten the life of the motor.

Another problem common only to three-phase motors is **single phasing**. Single phasing is when the motor continues to run after one of the three input phases is lost while the motor is operating. The resulting imbalance increases the temperature in the remaining two windings, causing the motor to overheat.

It is important to protect a motor from both current overloads and overheating. To accomplish this protection, current-sensing and/or heat-sensing devices are used to open the circuit before damage is caused to the motor. Electrical protection devices used with compressors are classified as either **pilot duty devices** or **line duty devices** (*Figure 13*).

- Pilot duty devices sense current overload or temperature within the motor and open the motor contactor control circuit to remove power from the motor.
- Line duty devices sense current flow and temperature in the motor winding and open the winding circuit to remove the line voltage when an overload occurs.

### *Pilot and Line Duty Devices*

Pilot duty devices are rarely used with welded hermetic compressors. Instead, this type of compressor uses a line duty device for motor protection. Pilot duty devices are still used on semi-hermetic compressors.

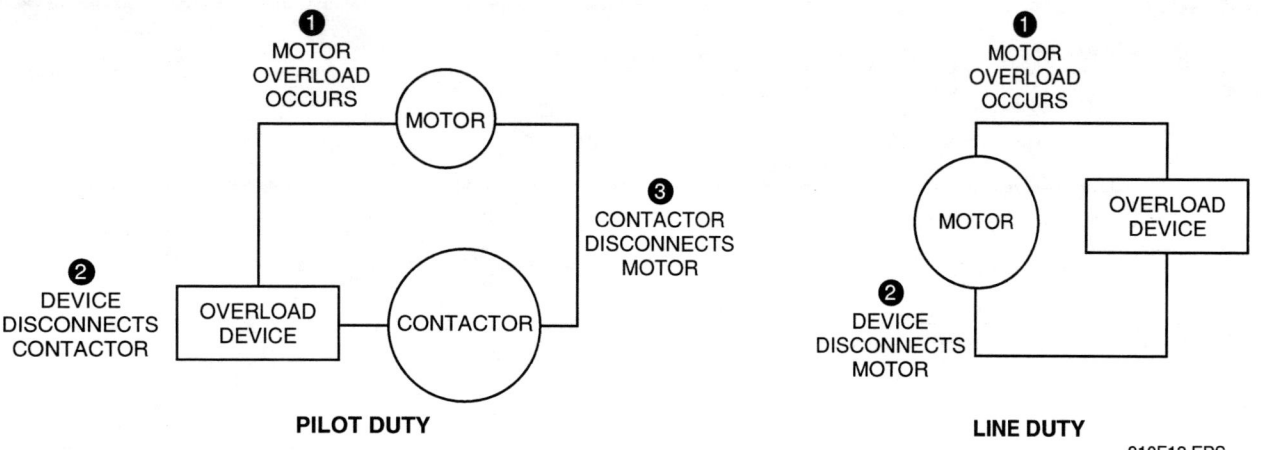

*Figure 13* ◆ Pilot and line duty devices.

Once an overload device has opened or tripped, it must be reset manually or automatically before the motor can operate again. A manual-reset overload device must be physically reset once it has tripped. An automatic-reset overload device will automatically reconnect the power to the motor after the overcurrent or over-temperature condition has passed. Reset time of automatic devices is based on recovery from the out-of-tolerance condition, and can vary from seconds to hours. Many types of motor protection devices are used, including:

- External line break overloads
- Internal line break overloads
- Motor thermostat overloads
- Electronic overloads
- Three-phase overloads
- Current monitoring devices

The operation of the devices described in this section are typical of those you will encounter in the field.

### 6.4.1 External Line Break Overloads

The external line break overload, often called a *klixon*, is a common single-phase motor overload protector. Some types provide current protection, while others are sensitive to both temperature and current. These devices use a bimetal warped-disc switch. Generally, the external line break is located in a motor terminal box and connected for line duty in series with the motor's common terminal. Some versions are used in pilot duty applications. In this case, the contacts are connected in series with the contactor circuit. Both devices reset automatically after recovery from the overload condition. *Figure 14* shows two examples of an external line break overload.

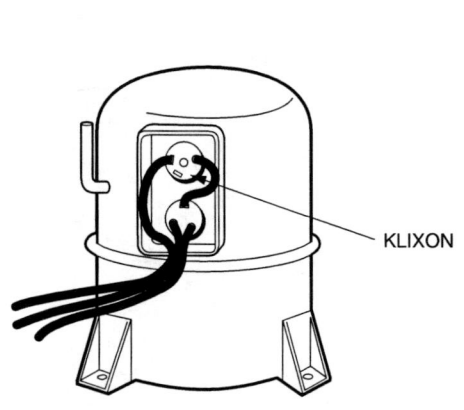
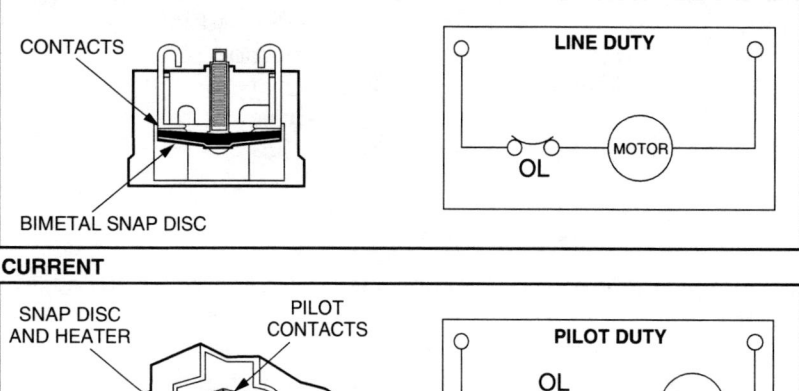

*Figure 14* ◆ Examples of external line break overloads.

COMPRESSORS — TRAINEE MODULE 03210   10.17

> ### Inside Track
> **Line Duty External Overloads**
> Line duty external overloads are commonly used on the compressors used in room air conditioners.

### 6.4.2 Internal Line Break Overloads

The internal line break overload is used often in single-phase hermetic compressors. This protector is a line duty device located inside the motor (*Figure 15*). It is placed in series with the terminal motor common and trips on either winding current or temperature, or a combination of both. It resets automatically after recovery of the overload condition. Because this type of protector is located inside the compressor motor, it is impossible to remove and inspect. The complete motor must be tested in order to isolate a faulty protector.

### 6.4.3 Motor Thermostat Overloads

Motor thermostats protect motors from overheating. They are either externally mounted or internally mounted directly on the motor windings. Three thermostat overloads are shown in *Figure 16*. The external shell device is mounted on the motor shell. When the shell overheats, the disc warps upward to disconnect power. The internal device is wound into the motor windings and will open the winding circuit when an over-temperature condition occurs. The hermetic motor thermostat shown on the bottom in *Figure 16* has rate-of-rise compensation. The case and internal strip expand at different rates. It trips early if the temperature rises rapidly, such as in a locked rotor condition. If the temperature rise is gradual, it will trip at its normal setting. This method prevents a rapidly rising temperature from overshooting the trip-out point and damaging the motor. The reset of all types is automatic.

### 6.4.4 Electronic Overloads

A negative temperature coefficient (NTC) thermistor is an electronic form of motor protection used to control the operation of an overload relay. The thermistor is encased in a capsule within the motor.

As the motor temperature increases, the resistance of the NTC thermistor decreases. When the motor temperature reaches an unsafe level, the increased current flow through the thermistor causes the overload relay coil to energize and the contacts to open the motor circuit. After the motor and thermistor cool back to a safe level, the overload relay will de-energize and its contacts close the motor circuit to start the motor. Some electronic overloads will automatically take a three-phase compressor off-line in the event of a phase loss. Thermistors imbedded in the motor winding will change resistance as the motor temperature changes. These resistance changes can be monitored by an electronic control, which will open the control circuit to the compressor contactor if motor temperature becomes excessive. Heat is the main enemy of a hermetic compressor motor.

Some of the input to an electric motor is converted to heat as well as to useful work, but excessive heat can damage the motor. Therefore, it is essential that the hermetic motor be effectively protected. One type of overload protection uses a single-module, three-sensor system. The sensors are inserted in the hermetic motor winding (*Figure 17*). The resistance of the sensors increases as the temperature rises. This action causes the module to react and break a set of contacts to open the control circuit of the compressor when the temperature reaches a preset level.

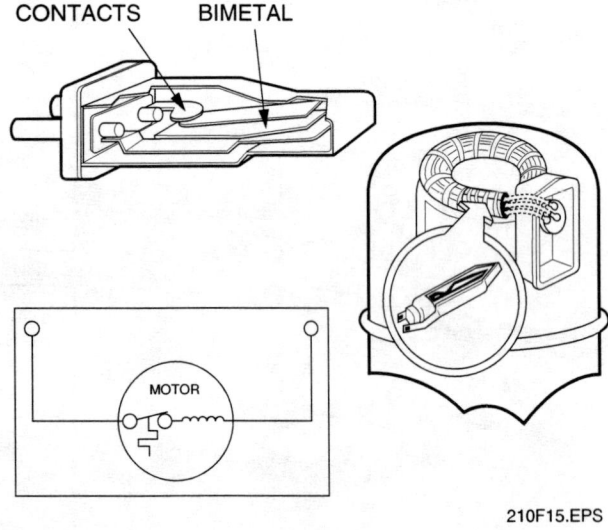

*Figure 15* ◆ Example of an internal line break overload.

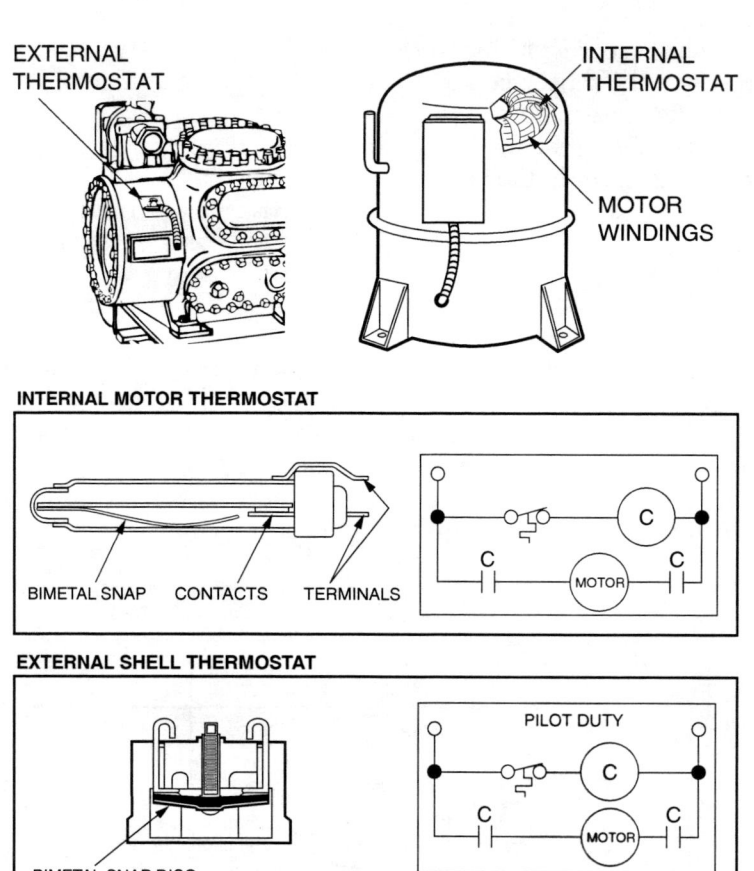

*Figure 16* ◆ Examples of motor thermostat overloads.

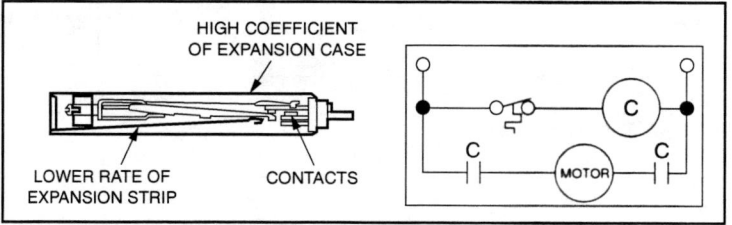

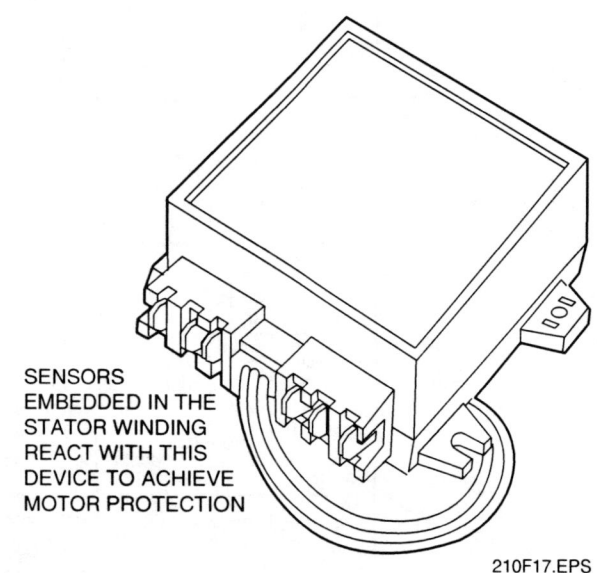

*Figure 17* ◆ Sensor location in a three-phase motor.

COMPRESSORS — TRAINEE MODULE 03210

An overload module is shown in *Figure 18*. The module requires a power supply, sensor connections, control circuit contacts, and manual reset connections. *Figure 19* shows the connections that are required between the module and other components in the system.

### 6.4.5 Three-Phase Motor Overloads

Three-phase connected motors in the smaller horsepower range can also be protected by line duty current-temperature devices like the ones shown in *Figure 20*. They can be internal or external to the motor when used with open compressors, but are always internal when used with hermetic compressors. When an overload occurs, the bimetal disc warps, breaking the electrical circuit. Reset is automatic when the disc cools.

### 6.4.6 Current Monitoring Devices

Electronic controls have been developed that monitor compressor current to ensure that preset levels are not exceeded. If current levels are exceeded, the control shuts the compressor off. These devices use either current-sensing loops (transformers) on the compressor power leads (*Figure 21*), or are wired in series with the compressor leads. Controls for single-phase compressors monitor current only in the common lead while three-phase controls monitor current in all three legs.

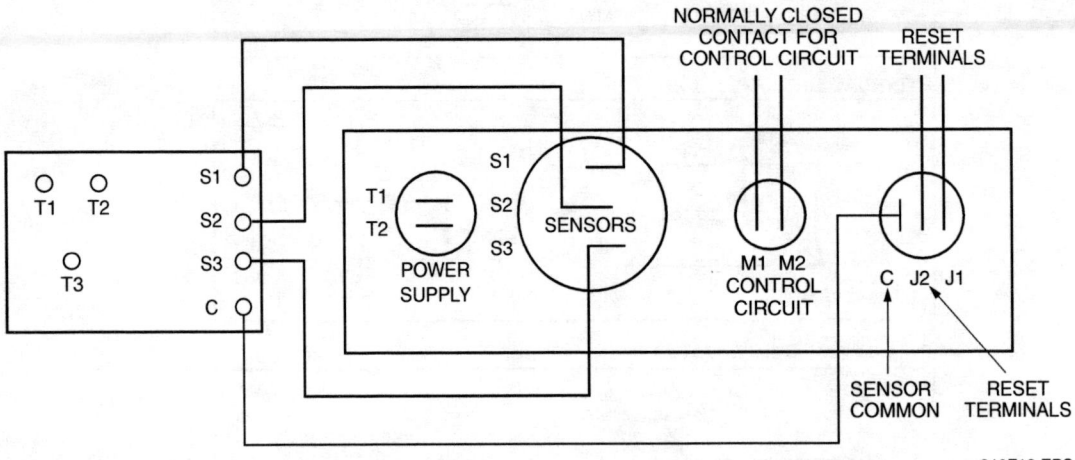

*Figure 18* ◆ Wiring connections for an overload module.

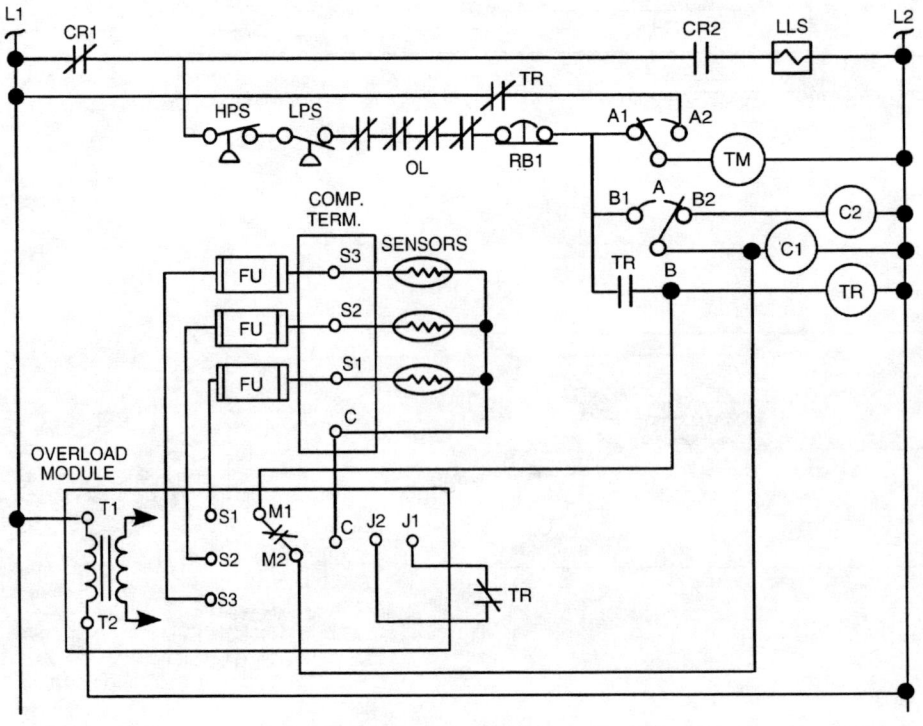

*Figure 19* ◆ Control circuit with an overload module.

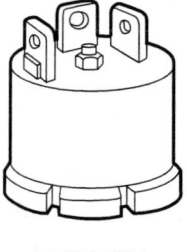
EXTERNAL CURRENT AND TEMPERATURE
LINE DUTY

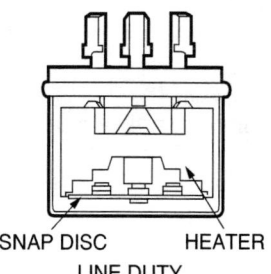
INTERNAL CURRENT AND TEMPERATURE
SNAP DISC   HEATER
LINE DUTY

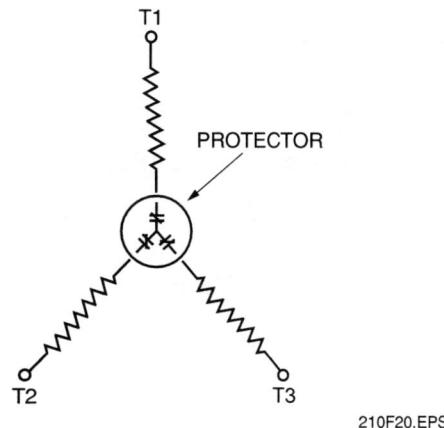

*Figure 20* ◆ Line duty three-phase overloads.

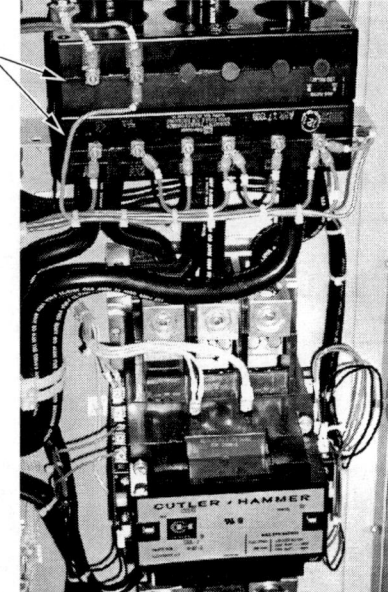

*Figure 21* ◆ Typical current-sensing transformer.

## 7.0.0 ◆ OTHER COMPRESSOR PROTECTION DEVICES

In addition to its electric drive motor overload devices, a compressor is often protected from damage by other devices. Typically, these protection devices include:

- High-pressure and low-pressure protection
- Loss of evaporator or condenser airflow protection
- Operational sequence protection
- Short cycling protection
- Head pressure control

### 7.1.0 Pressure Protection

The use of high-pressure and low-pressure safety cutout switches is the most common method of protecting the compressor when a pressure problem exists. High discharge or condensing pressure is one of the most harmful conditions affecting the compressor. High condensing temperature raises the temperature of the refrigerant vapor and oil moving across the compressor discharge valves. This could result in oil and refrigerant breakdown. In hermetic compressors, motor cooling depends on the amount and temperature of the returning refrigerant in the suction line. If suction pressure is too low, the motor may overheat and cause compressor failure.

As shown in *Figure 22*, the high-pressure and low-pressure switches are usually wired in series with the compressor contactor coil and any other protective devices. A sensor tube from the high-pressure switch is attached to the discharge line. The pressure in the discharge line acts against a diaphragm in the switch. The contacts in the switch are kept closed by a calibrated spring. If the pressure in the system exceeds the switch cutout setting, as determined by the spring tension, the contacts open and stop the compressor. Operation of the low-pressure switch is similar to that of the high-pressure switch, except the sensor tube is connected to the suction line. If the pressure in the suction line falls below the suction cutout level, the spring opens the contacts and stops the compressor. The cutout points for the switches are normally preset at the factory.

### 7.2.0 Evaporator and Condenser Airflow Protection

Another form of high-pressure and low-pressure protection can be provided by airflow switches mounted in the evaporator and condenser fan air stream. Restricted or no airflow at the evaporator can cause low suction pressures, while restricted or no airflow at the condenser can cause high compressor discharge pressures. Like the pressure switches, the contacts of the airflow switches are usually wired in series with the compressor contactor coil and any other protective devices (*Figure 22*). The switches are operated by the force of air created by the condenser fan or the evaporator fan. With the fans operating, and with normal airflow, the switch contacts are closed, allowing the compressor to operate.

### 7.3.0 Lockout Protection

Compressor lockout protection is used to prevent the compressor from operating out of sequence due to the opening and closing of the control circuit by any of the automatic reset safety controls. *Figure 23* shows a lockout relay (sometimes called an *impedance relay*) used for this purpose. The relay coil, due to its high resistance, is not energized during normal operation. However, when any one of the safety controls opens the circuit to the compressor contactor coil, current flows through the lockout relay coil, causing it to energize and open its contacts. The relay remains energized, keeping the compressor contactor circuit open until the power is interrupted from the control circuit (either by the thermostat or the main power switch) after the safety control has been reset.

Proper operation of this circuit depends on the resistance of the lockout relay coil being much greater than the resistance of the compressor contactor coil. If the lockout relay becomes faulty, it should be replaced with an exact replacement to maintain the proper circuit balance.

Another type of compressor lockout device uses a current-sensing loop to monitor current in the common leg of a single-phase compressor. This device must detect a compressor current within a few seconds of the compressor contactor energizing or the lockout will open the control circuit and lock it out until manually reset.

### 7.4.0 Short Cycling Protection

Before restarting a compressor after it has been turned off, enough time should be allowed to pass so that the pressures in the system can equalize. Short cycling is a condition in which the compressor is restarted immediately after it has been turned off. This causes the compressor to restart against a high discharge (head) pressure, which can cause damage to the compressor or motor windings. It can also cause the circuit breaker or overload to open. Short cycling can result when there is a momentary interruption of power to the compressor, such as might occur during severe thunder storms, or when the thermostat setting is manually changed in a manner which first opens then closes its contacts. Short cycling can also be caused by erratic operation of a marginal system. Short cycling protection can be provided by placing a motor-driven, cam-operated timer, or solid-state electronic timer, in the compressor contactor control circuit (*Figure 24*). When power is applied, the anti-short cycling device must first time out before power can be applied to the compressor contactor coil. Timers typically have delays ranging from 30 seconds to 5 minutes. The specific time delay used depends on the application.

### 7.5.0 Electronic Head Pressure Control

Commercial air conditioners are required to operate over a much wider outdoor temperature range than residential equipment. For example, a cooling unit may need to run in a crowded restaurant even if the outdoor temperature is below freezing. Operating the unit at this low ambient temperature will result in low head pressure and possible liquid floodback to the compressor.

To maintain a workable head pressure, electronic low ambient temperature head pressure controls (*Figure 25*) are available which control condenser fan motor speed. These controls use thermistors attached to pre-determined points on the condenser coil to monitor the coil temperature. If the thermistors detect the coil is getting too cold, that input is read by the control that then signals the condenser fan to slow down. By slowing down the fan motor, less heat is rejected by the coil and the pressures and temperatures in the coil rise as a result. A motor equipped with special ball bearings must be used with this control to maintain motor bearing lubrication at low motor speeds.

### 8.0.0 ◆ REDUCED-VOLTAGE MOTOR STARTING

When starting larger compressor motors, the starting current can sometimes be almost six times the rated load current. The motors are built to handle this current, but there may be a large voltage drop in the power system. In these cases, reduced-voltage starting methods are used to control the starting current and limit the voltage drop to a tolerable value. All reduced starting methods use specially designed motor starters and are controlled by adjustable timers between the start and run functions.

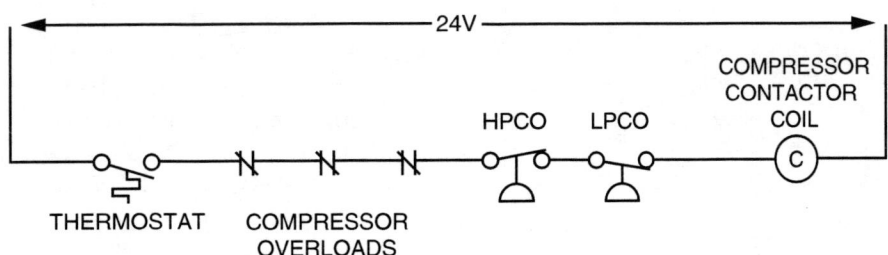

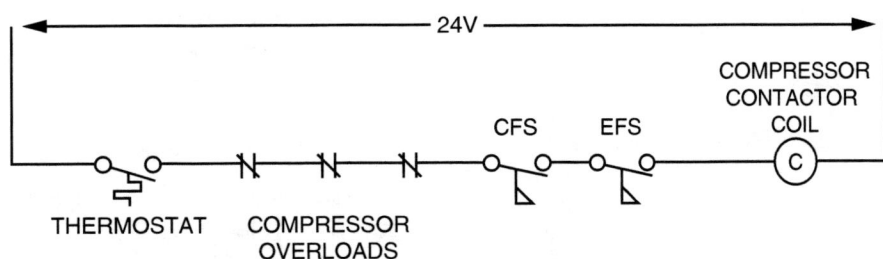

*Figure 22* ◆ Compressor pressure and airflow protection devices.

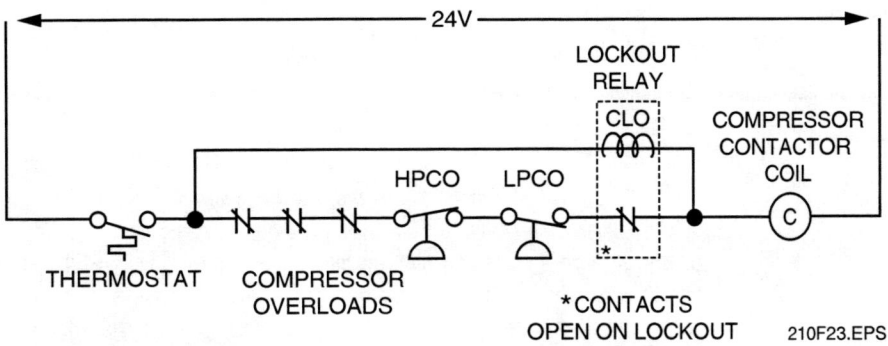

*Figure 23* ◆ Compressor lockout relay protection.

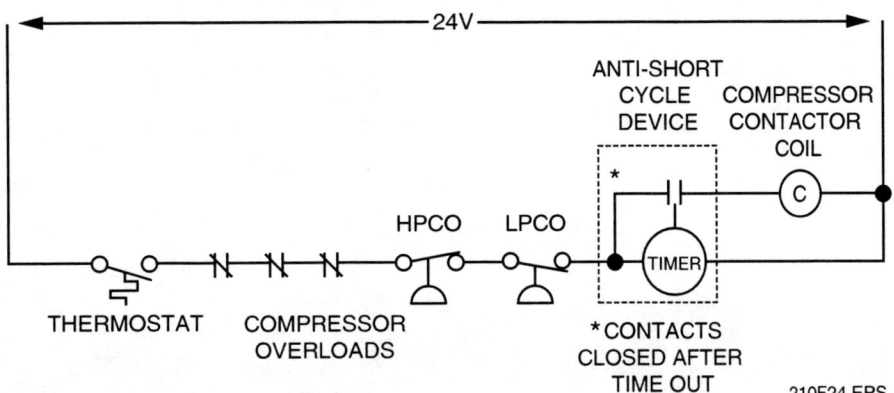

*Figure 24* ◆ Compressor anti-short cycle simplified circuit.

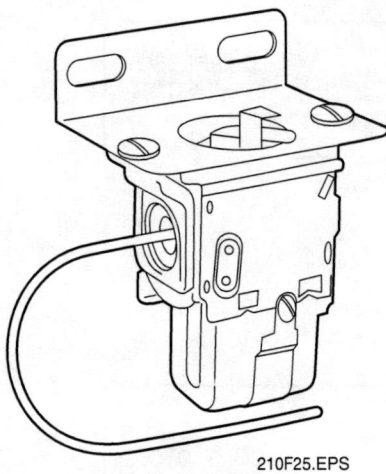

*Figure 25* ◆ Typical electronic low ambient temperature head pressure control.

Common methods of reduced-voltage starting include:

- *Primary resistor or reactor* – This method uses a series resistance or reactance to reduce the current on the first step. After a preset interval, the motor is connected directly across the line. This method can be used with any standard motor.
- *Autotransformer* – An autotransformer is used to directly reduce voltage and current on the first step. After a preset time interval, the motor is connected directly across the line. This method can be used with any standard motor.
- *Wye-delta* – This method induces the voltage across the wye-connection to reduce voltage and current on the first step. After a preset time interval, the motor is connected in delta, allowing full current. This method requires a motor capable of wye-delta connection.
- *Part-winding* – This method uses a motor with two separate winding circuits. When starting, only one winding circuit is energized, allowing the current to be reduced. After a preset time interval, both winding circuits of the motor are connected directly across the line. To avoid overheating and possible damage to the windings, the time between the connection of the first and second windings must be limited. Typically, this time is about four seconds.

There are also electronic soft-start kits that are used with large motors.

## 9.0.0 ◆ CAUSES OF COMPRESSOR FAILURE

If not corrected, most mechanical refrigeration system problems will result in compressor failure. If you assume the cause of the compressor failure lies within the compressor, it is likely you will make the first of many unnecessary compressor replacements. If the compressor fails and the cause has not been found and eliminated, the replacement compressor will also fail.

### Solid-State Reduced-Voltage Starters

Today, solid-state reduced-voltage starters are widely used with large compressor motors and other types of motors in new industrial and commercial HVAC equipment installations. They are also being used to replace older electromechanical-type reduced-voltage starters in retrofit work. Solid-state reduced-voltage starters provide for smooth start and acceleration. For this reason, they are referred to as *soft-start controllers*. This photo shows a family of solid-state reduced-voltage starters produced by one manufacturer for use with various motors.

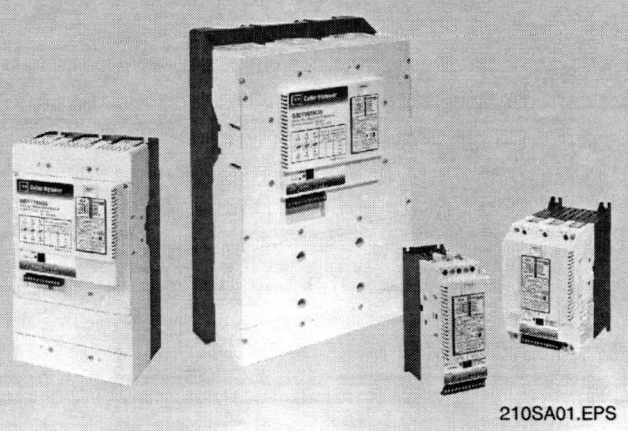

Problems that can cause compressor failure include:

- Slugging of liquid refrigerant and/or oil in the compression area of the compressor
- **Flooding** of liquid refrigerant into the crankcase of the compressor
- **Flooded starts** when the oil in the crankcase is mixed with a quantity of liquid refrigerant when it migrates back to the compressor at shutdown
- Loss of lubrication by loss of oil or by refrigerant diluting the oil
- Contamination of the refrigeration system with air, moisture, and dirt
- Overheating of the compressor cylinder components and/or the hermetic motor windings
- Electrical irregularities in voltage and current
- Incorrect installation of any system components, piping, or accessories

## 9.1.0 Slugging

Slugging occurs when a compressor tries to compress liquid refrigerant, oil, or both instead of superheated gas. If slugging occurs, it will occur at start-up or during rapid changes in system operating conditions. It can sometimes be detected by a periodic knocking noise at the compressor. When a compressor tries to compress a liquid refrigerant or oil, extremely high pressures that may exceed 1,000 psi can be reached in the cylinder. These pressures can result in blown cylinder heads and/or valve plate gaskets and damage to the compressor pistons and discharge valves. Slugging may result when:

- There is an overcharge of refrigerant.
- The crankcase heater has failed.
- There is an oversized or damaged thermostatic expansion valve or loose sensing bulb.
- Condensed refrigerant in any cold part of the system, such as the evaporator, during the off cycle. Buried refrigerant lines or lines passing through cold spots can allow the refrigerant to condense back into a liquid at shutdown.

- Slugs of oil are trapped in the suction line because the suction gas does not have enough velocity to return the oil to the compressor. Normally, the oil and refrigerant mix. The oil is circulated through the system in very small drops as it is being swept along by the velocity of the refrigerant vapor. If it gets trapped in the system piping and returns all at once, it can cause slugging. This condition tends to be found in the suction line of built-up systems. It also occurs in systems that use compressors with unloaders, especially when the compressor runs unloaded for long periods of time.
- The system has an overcharge of oil.

## 9.2.0 Flooding

Flooding is the continuous return of liquid refrigerant in the suction vapor to the compressor during operation. Flooding usually dilutes the oil, resulting in crankcase foaming and overheating of bearing surfaces. If severe enough, it can result in damage to the pistons, rings, and valves because the refrigerant washes the oil off the bearing surfaces. Flooding may happen if:

- The thermostatic expansion valve is oversized.
- The thermostatic expansion valve sensing element is broken, mislocated, in poor contact, or improperly insulated.
- The superheat setting is too low.
- There is a low load on the evaporator caused by low airflow. Reduced airflow often causes frosting of the coil, which adds to the problem. Possible causes of restricted airflow are dirty filters, air restriction, and dirty fan wheels.
- There is an overcharge of refrigerant in systems that use fixed-orifice metering devices. Since fixed metering devices do not react to load change, an overcharge of refrigerant can raise the head pressure, which can increase the flow rate to a point where there is more flow than available heat transfer.

---

### Refrigerant Piping and Slugging

Refrigerant piping installations in which the evaporator is located lower than the condensing unit require a properly sized and routed oil trap at the evaporator coil connection. Piping installations in which the evaporator is located higher than the condensing unit should include a reverse trap at the evaporator coil.

## 9.3.0 Flooded Starts

Flooded starts are a result of the oil in the compressor crankcase absorbing refrigerant. This condition usually occurs during shutdown. Oil will absorb refrigerant under most conditions. The amount absorbed depends on the temperature of the oil and the pressure in the crankcase. The lower the temperature and the higher the pressure, the more refrigerant is absorbed.

On start-up, this refrigerant-rich oil mixture is pumped through the oil pump of the compressor, resulting in marginal or inadequate lubrication of the bearings. This results in the bearing surfaces being overheated and scored because of the inadequate lubrication. As the crankcase pressure drops after start-up, the refrigerant will flash from a liquid to a gas, causing foaming. This foaming can restrict the oil passages and cause pressure to build. It can also cause a hydraulic slug of the oil and liquid mixture to enter the cylinder, resulting in compressor damage.

The problem of flooded starts can be minimized by following the manufacturer's specifications. It is important to maintain the correct refrigerant-to-oil ratio by making sure the system has a proper refrigerant charge and the correct amount of oil in the crankcase. Crankcase heaters are normally used to raise the temperature of the oil during shutdown and to prevent refrigerant from migrating to the compressor crankcase during the off cycle.

## 9.4.0 Contamination

Refrigeration systems are intended to contain only refrigerant and oil. Anything else in the closed refrigerant system is considered a contaminant. Contaminants in a system, such as air, moisture, acid, and dirt are major causes of compressor failure.

- Air, moisture, and dirt usually enter the system accidentally during installation or servicing.
- Acid, soot, varnish, hard carbon, and copper plating are created in the system either by chemical reactions with contaminants or as a result of compressor motor **burnout**.

### 9.4.1 Air Contamination

Air not only contains moisture but is noncondensable. Since air is noncondensable, it can accumulate in the condenser, taking up space needed for condensing the refrigerant. This

### Crankcase Heaters

Compressor crankcase heaters are used to prevent liquid refrigerant from migrating into the compressor and mixing with the oil when the compressor is off. All heaters evaporate refrigerant from the oil. Heaters used with semi-hermetic compressors are typically fastened to the bottom of the crankcase. Another type, called an *immersion heater*, is inserted directly into the compressor crankcase. Wrap-around (bellyband) crankcase heaters like the one shown here are widely used to encircle the outside shell of welded hermetic compressors.

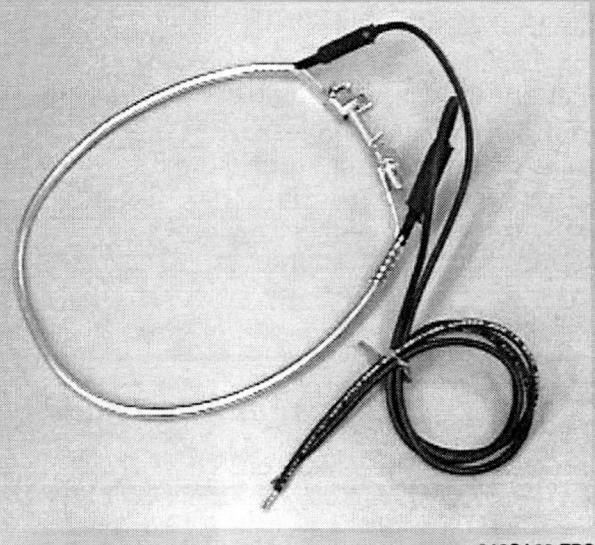

210SA02.EPS

### Installing Split Systems

Split systems are at risk for dirt contamination because the two sections of the system have to be connected with field-installed refrigerant tubing. Here is how to minimize the chances of system contamination when installing a split system:

- Use only ACR or refrigeration grade tubing, or use the refrigerant line sets specified by the equipment manufacturer.
- Keep the ends of the tubing plugged until just before making the connections to the equipment. Install a filter-drier in the liquid line.
- Evacuate the interconnecting tubing between the units to 500 microns before opening the service valves.

---

increases the condensing temperature and makes the system work harder. It also promotes the creation of acids by chemical reaction with the oil and refrigerant mixture. These acids erode machined surfaces, which can create copper plating, causing copper from the system to be deposited on the heated bearing surfaces in the compressor.

#### 9.4.2 Moisture Contamination

Moisture is a common contaminant. Under the heat of compression, it will react with the refrigerant to form hydrochloric and hydrofluoric acid. These acids cause corrosion of metals and breakdown of the insulation on the motor windings and other motor wiring. Moisture in the refrigerant can also cause oil sludge, which reduces the lubricating properties of the oil and plugs oil passages and screens in the compressor. Moisture can also freeze at the expansion device, especially in heat pumps where they operate at lower temperatures than those found in cooling-only systems. The presence of moisture in a system can be determined by testing the refrigerant with an acid/moisture test kit.

#### 9.4.3 Acid Contamination

Acid is not introduced into a system. It is formed inside an improperly operating system by the reaction of air or moisture with the refrigerant. Acid is produced in great quantities in a system that has experienced a severe motor burnout. Acid creates sludge and varnish, which plug oil passages and restrict the strainers in the lubrication system. The presence of acid in a system can be determined by testing the refrigerant with an acid/moisture test kit or the oil with an oil acid test kit.

#### 9.4.4 Dirt Contamination

Foreign material such as filings, chips, or lint can get into the system during installation or servicing. These particles can easily clog small oiling passages in the compressor, preventing its normal lubrication and forming restrictions in the system that cause pressures to increase or decrease.

Copper oxide is a black, flaky, solid contaminant that is formed around brazed joints. It forms when the copper tubing is heated to over 1,000°F

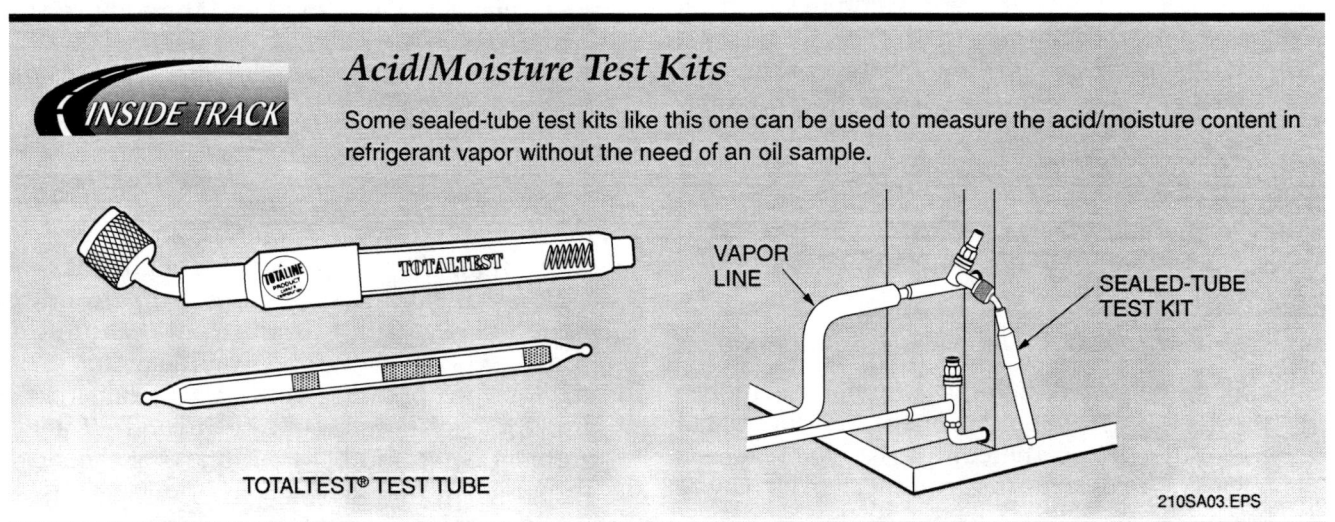

### Acid/Moisture Test Kits

Some sealed-tube test kits like this one can be used to measure the acid/moisture content in refrigerant vapor without the need of an oil sample.

in the presence of oxygen. It is easily washed from the tubing surfaces when the system is operated, allowing it to circulate with the refrigerant and oil. These solid particles of copper oxide circulating through the system can cause many problems that eventually lead to compressor failure. It is critical to prevent the formation of copper oxides within the tubing while brazing. You must practice proper brazing techniques when installing or servicing a system. This includes purging the refrigerant lines with nitrogen gas in order to displace the oxygen in the system when brazing the joints. Review the method for nitrogen purging previously described in HVAC Level One.

### 9.4.5 Eliminating Contaminants

Contaminants must be eliminated when installing new systems or when servicing existing equipment. Contaminants can be eliminated as follows:

- *Air* – Evacuate the system.
- *Moisture* – Evacuate and dehydrate the system.
- *Acid* – Recover and replace or recycle the system refrigerant, replace system filter-driers, and replace the compressor oil.
- *Chips and dirt* – Work carefully and install strainers and filters in the system.

You will study the methods used to recover and recycle refrigerant, and to evacuate and dehydrate a system in the HVAC Level Two Module, *Leak Detection, Evacuation, Recovery, and Charging*.

## 9.5.0 Electrical

Compressor motor electric failures typically occur as a result of other compressor problems such as contamination, overheating, and the loss of lubrication. Some electrical problems in compressors are caused by a failed component in the electrical start or run circuit, such as the capacitors, potential relay, or start switch. Review the material on electric motors, their components, and the procedures used to test them previously studied in the Level Two module, *Alternating Current*. The remainder of this section describes electrical problems external to the compressor motor that will result in compressor failure if not corrected.

### 9.5.1 Compressor Nameplate Information

Valuable information about a hermetic compressor's operating frequency, voltage, and current can be obtained from the compressor information plate (*Figure 26*). This stamped metal plate may be welded or riveted to the compressor. In addition to specific electrical information, it also gives information such as refrigerant type and model number, which may be needed if the compressor must be replaced. Unfortunately, many compressor nameplates are hard to see because the compressor is mounted deep within the unit. Fortunately, the important compressor electrical information is normally also included in a visible location on the unit nameplate.

An important piece of information given on the nameplate is the RLA or FLA value. RLA stands for rated load amps and FLA for full load amps. Nameplates for hermetic compressors built after 1972 are marked with the RLA. The RLA for a hermetic compressor is determined by the manufacturer by placing the compressor under actual operation at rated refrigerant pressure and temperature, voltage, and frequency. To determine a compressor's RLA, the manufacturer first determines the compressor's maximum continuous current (MCC). This is the maximum current value that the compressor's motor protector will carry without opening, thus stopping the compressor. Any additional current draw above the MCC will cause the protector to open. Sometimes the MCC value is also marked on the compressor's nameplate.

Once the MCC is determined, the RLA is then calculated. According to the Underwriters' Laboratory (UL), this is done by dividing the MCC by 1.56 to establish the minimum RLA; however, some compressor manufacturers may use a different divisor than 1.56 (such as 1.4). The RLA value is important because it can be used to size wire and contactors used with the compressor. In refrigeration low-temperature compressor applications, compressors are typically equipped with a crankcase pressure regulator (CPR) to prevent the pressure in the compressor crankcase from rising to a level that will overload the compressor. In this case, the RLA value is also used when adjusting the CPR so that the compressor amp draw does not exceed its RLA value.

### 9.5.2 Motor Operating Voltage Ranges

Too high or too low an operating voltage can cause overheating of a compressor motor and may cause compressor failure. Operating voltages must be maintained within minimum-maximum limits from the voltage value given on the motor's nameplate. If the operating voltage falls outside these limits, the system should be turned off and the problem corrected before restarting the system. The problem may be with the building distribu-

```
SERIAL 9370052                         MODEL PAC036010
                                       OUTDOOR FAN MOTOR
        FACTORY CHARGED R-22           VOLTS AC  208/230
LBS    7.0            Kg 3.2           HP 1/4          FLA 1.4
                                       PH 1            HZ 60
POWER SUPPLY     208/230     VOLTS        INDOOR FAN MOTOR
PH  1                HZ 60             VOLTS AC 208/230
        PERMISSIBLE VOLTAGE AT UNIT    HP 1/3          FLA 2.8
MAX    253           MIN 187           PH 1            HZ 60
        SUITABLE FOR OUTDOOR USE         DESIGN/TEST PRESSURE GAUGE
              COMPRESSOR               HI    PSI 300    kPA 2068
VOLTS   AC        208/230              LO    PSI 250    kPA 1034
PH 1              HZ 60                MINIMUM CIRCUIT AMPS 26.7
RLA 18.0          LRA 96               MAX FUSE40    40MAX CKT-BKR (*)
                                             HEATER PACKAGE
                                       VOLTS   AC   209/230
                                       HEATER AMPS      26.5
```

*Figure 26* ◆ Typical compressor nameplate.

tion system or the electric utility supply to the building. The voltage tolerances used for motors are as follows:

- *Single-voltage rated motors* – The input supply voltage should be within ±10% of the nameplate voltage. For example, a motor with a nameplate single voltage rating of 230V should have an input voltage that ranges between 207V and 253V (±10% of 230V).

- *Dual-voltage rated motors* – The input supply voltage should be within ±10% of the nameplate voltage. For example, a motor with a nameplate dual voltage rating of 208/230V should have an input voltage that ranges between 187V (–10% of 208V) and 253V (+10% of 230V).

### 9.5.3 Voltage Imbalance

The voltage imbalance between any two legs of the supply voltage applied to a three-phase motor should not exceed 2%. A small imbalance in the input voltage results in a considerable amount of heat being generated in the motor windings. For example, if the voltage imbalance were to exceed 2%, the temperature rise generated in the motor windings would increase to 8% over the safe level. With only a 5% imbalance, the winding temperature can increase to 50% over the safe level. *Figure 27* shows an example of how to calculate the voltage imbalance in a three-phase system.

The percent of voltage imbalance is calculated using the following formula:

$$\% \text{ voltage imbalance} = \frac{\text{maximum voltage imbalance}}{\text{average voltage}} \times 100$$

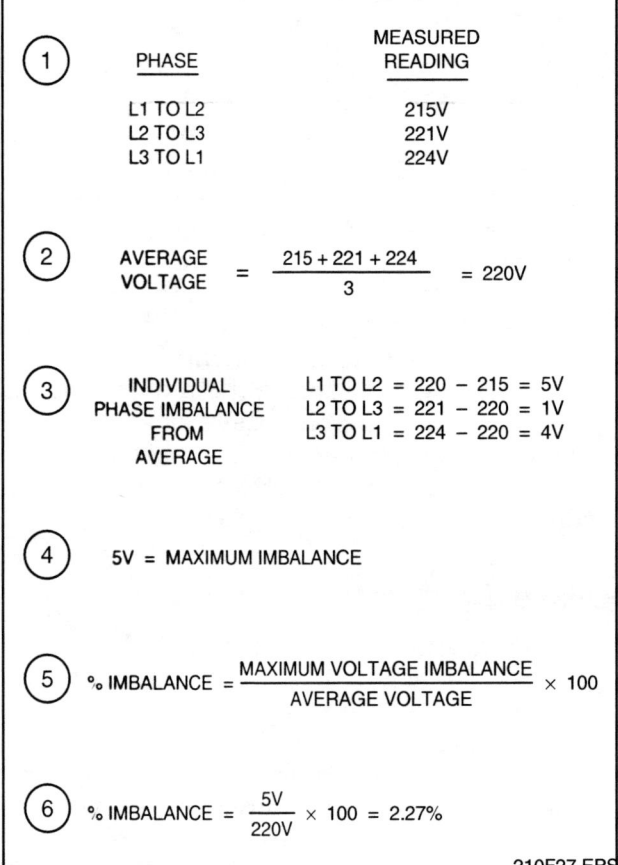

*Figure 27* ◆ How to check voltage imbalance in a three-phase system.

### 9.5.4 Current Imbalance

Current imbalance between any two legs of a three-phase motor should not exceed 10%. Voltage imbalance will always produce current imbalance, but a current imbalance may occur without

### Voltage Imbalance

The voltages measured between the terminals of a three-phase motor are L1 to L2 = 218V, L2 to L3 = 219V, and L3 to L1 = 221V. What is the percent of voltage imbalance, and is it within tolerance?

---

a voltage imbalance. This can happen when an electrical terminal or contact becomes loose or corroded, causing a high resistance in the leg. Since current follows the path of least resistance, the current in the other two legs will increase, causing more heat to be generated in the windings. Current imbalance is calculated using the following formula:

$$\% \text{ current imbalance} = \frac{\text{maximum current imbalance}}{\text{average voltage}} \times 100$$

Single-phasing of a three-phase motor is an example of a severe case of current imbalance. If the compressor is operating and one phase opens, the motor can continue to run. The other two phases will attempt to carry the load, resulting in a current increase of about 1½ times the normal running current. If the compressor is loaded, this will probably cause an overload device to trip. If unloaded, the overload might not trip, allowing the motor to continue running and causing the motor windings to overheat. Normally, once the motor is stopped, it cannot be restarted again because it will continuously trip the overload protectors. This can eventually cause a compressor failure.

### 9.6.0 Compressor Heating

Compressors normally generate heat and are designed to handle it. One indication of possible compressor overheating is a high compression ratio. When the compressor is overheated, the cause must be determined. As discussed previously, high superheat, lubrication problems, condenser problems, or electrical problems are all possible causes. High temperatures between 275°F and 300°F in the hot gas discharge line cause oil and refrigerant to break down with the potential for compressor failure. Laboratory tests have shown that for each 18°F rise above normal in the discharge temperature, the chemical reaction between the refrigerant-oil mixture and moisture and acid in the system doubles.

### 10.0.0 ◆ SYSTEM CHECKOUT FOLLOWING COMPRESSOR FAILURE

Before condemning the compressor, system checkout procedures should be performed to make sure the compressor is bad. Even if the compressor proves to be bad, these checks must be made to find out why it failed. Remember that if a compressor fails and the cause has not been determined and corrected, the replacement compressor will also fail. The system checkout procedure consists of three phases:

- Preliminary inspection
- Analyzing system operating conditions
- Final compressor tests

### 10.1.0 Preliminary Inspection

Preliminary inspection of the system is performed using your senses of sight, sound, touch, and smell to identify problems.

**WARNING!**
Be sure all electrical power to the equipment is turned off. Open, lock, and tag disconnects. Watch out for pressurized or hot components. Follow all safety instructions labeled on the equipment and given in the manufacturer's service manual for the equipment.

---

### Current Imbalance

The current measurements at the terminals (T1, T2, and T3) of a three-phase motor are T1 = 20 amps, T2 = 25 amps, and T3 = 30 amps. What is the percent of current imbalance, and is it within tolerance?

The preliminary inspection is performed as follows:

*Step 1* Turn off the equipment. Lockout and tag equipment so that no one can start it.

*Step 2* Look for an evaporator or condenser mounted above the compressor which might dump liquid refrigerant into the compressor.

*Step 3* Look for the following piping problems:
- Long or uninsulated suction line, which might develop excessive superheat
- Liquid line running through an unconditioned space (hot or cold), which might affect subcooling
- Buried lines, which might cause refrigerant to condense
- Extremely long liquid line, which might hold an excessive amount of refrigerant

*Step 4* At the evaporator and condenser, check for the following:
- Fin collars corroded
- Fins or coils dirty or damaged
- Supply plenum dirty
- Filters dirty or missing
- Fan belts at improper tension
- Blowers and fans dirty
- Evaporator shows signs of freezing up

*Step 5* As applicable, at the compressor:
- Check that the service valves are fully open.
- Check that the hold-down bolts are loosened or unloosened per the manufacturer's instructions.
- Check that the crankcase heater is working.
- Inspect the cylinder heads to see if they are scorched or blistered from excessive heat.
- Inspect for rust streaks, indicating condensation from cold return gas.
- Check that the oil level is at the proper height in the sight glass.

## 10.2.0 Analyzing System Operating Conditions

In order to tell if the system or compressor has a problem, the actual conditions that exist in the system must be known.

### 10.2.1 System Operation Checks

If the compressor is operable, system operation should be monitored and the critical parameters measured. This is necessary so the actual conditions can be compared against a set of previously recorded normal system operating parameters. Check the system operating conditions and make sure to record the values for system parameters as described in this section. It cannot be stressed enough how important it is to properly record the operating conditions and measured values for a system. The record may serve as a historical document that can be used for comparison purposes by a technician when troubleshooting in the future. The recorded data can also serve as a system commissioning record in order to verify compliance with contractual system specifications.

> **WARNING!**
> Watch out for rotating, pressurized, or hot components. Follow all safety instructions labeled on the equipment and given in the manufacturer's service manual for the equipment. Failure to do so could result in personal injury.

*Step 1* Install a gauge manifold set on the system gauge ports. Install thermometers on the suction line at the compressor input, on the hot gas discharge line, and on the liquid line at the input to the expansion device.

*Step 2* Turn on the system and adjust the thermostat to call for cooling.

*Step 3* Start the system and monitor its operating characteristics.
- Listen for excessive vibration of the compressor, piping, motors, and fans.
- Listen for compressor knocks or rattles, which may indicate liquid refrigerant is being drawn into the cylinders. If this condition continues for more than a few seconds, shut down the system and look for the cause of excess liquid return.

*Step 4* Check the compressor oil sight glass (if applicable). Heavy foaming at the sight glass should clear 5 to 10 minutes after startup. If not, there may be excessive refrigerant in the oil.

*Step 5* Measure and record the input voltage and current at the compressor contactor as follows. Current in the compressor motor leads can be measured with a clamp-on ammeter.
- The measured voltage should be within ±10% of the motor nameplate value.
- In a three-phase motor, the voltage imbalance between phases should not exceed 2%. Any current imbalance between legs should not exceed 10%.

>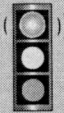
> **WARNING!**
> Danger exists if the compressor terminals are damaged and the system is pressurized. Disturbing the terminals to make measurements could cause them to blow out, causing injury. When making voltage, current, or continuity checks on a hermetic or semi-hermetic compressor in a pressurized system, always make measurements at terminal boards and points of test away from the compressor. Once the refrigerant has been recovered and the system is no longer under pressure, measurements can be made at the compressor terminals.

*Step 6* Check that all fans, motors, and pumps are operational and moving the proper amounts of air or water.

*Step 7* As applicable, measure or calculate and record the following operating parameters:

- Suction pressure
- Saturated suction temperature
- Suction line temperature
- Superheat
- Discharge pressure
- Saturated discharge temperature
- Discharge line temperature
- Subcooling
- Liquid line temperature entering the expansion device
- Oil pressure
- Hot gas discharge line temperature differential above the saturated condensing temperature (head temperature)
- Compressor temperature at the bottom of the cylinder heads
- Compressor temperature at top and bottom of the motor barrel
- Crankcase temperature
- Compression ratio
- Evaporator capacity
- Condenser capacity

### 10.2.2 Analyzing System Conditions

Once the actual conditions that exist in a system are known, they can be compared against a set of normal system operating parameters in order to determine if there is a problem and where it is. Remember, readings vary because of equipment application, operating conditions, and type of refrigerant used. The data in *Table 2* is typical of an HCFC-22 air conditioning system at 90°F outdoor temperature. Always refer to the manufacturer's service manual or system maintenance log to find typical readings for the specific equipment you are servicing.

Based on the analysis of system operation, repair the system, repair or replace the compressor, or both. Some common problems and their causes are listed below.

- Failure to start:
  - Thermostat setting too high
  - Power circuit voltage inadequate or missing
  - Control circuit voltage inadequate or missing
  - Safety switches open or defective

- Overheated compressor:
  - High superheat caused by low charge or defective metering device
  - Electrical problems such as low voltage or voltage imbalance
  - Lubrication problems such as low oil or poor oil return
  - Condenser problems such as dirty or plugged coil, or failed condenser fan motor

- Reduced system capacity:
  - Check load
  - Check evaporator flow
  - Check refrigerant flow
  - Check refrigerant charge

**Table 2** Typical Operating Conditions for HCFC-22 System at 90°F Outdoors

| System Operation | Parameter |
|---|---|
| Suction pressure | 68 psig |
| Saturated suction temperature | 40°F |
| Suction line temperature | 50°F |
| Superheat | 12°F |
| Discharge pressure | 160 psig |
| Saturated discharge temperature | 120°F |
| Subcooling | 10°F |
| Liquid line temperature entering the expansion device | 110°F |
| Hot gas discharge line temperature differential above the saturated condensing temperature (head temperature) | 70°F |
| Oil pressure | 40 psig |
| Compressor temperature at the bottom of the cylinder heads | 80°F to 120°F (max) |
| Compressor temperature at the top and bottom of cylinder heads | 90°F |
| Crankcase temperature | 100°F |
| Compression ratio | 3.32 to 1 |

## 10.3.0 Final Compressor Checks

Before replacing a compressor, some final checks should be made on the compressor to be sure it is defective. Compressor failures result from either an electrical failure or a mechanical failure.

### 10.3.1 Electrical Reasons for Failure

The methods used to troubleshoot motors were studied in the HVAC Level Two Module, *Alternating Current*. Review the material about electric motors, their components, and the procedures used to test them. The most likely causes of electrical failure include:

- A grounded, open, or shorted motor
- Open internal overload (stuck open)
- Defective (open) start relay
- Open or shorted start or run capacitor
- Electrical connections
- Relays
- Improper input voltage
- Malfunctioning contactor

### 10.3.2 Mechanical Reasons for Failure

Mechanical reasons for compressor failure include:

- Loss of charge or overcharge
- Physical damage
- Broken valves or rings
- Loss of lubrication

Leaks that exist at the weld or terminals of welded hermetic compressors cannot be repaired and the compressor must be replaced. If a leak exists at the stubs, recover all refrigerant and repair the leak with silver brazing alloy using the necessary precautions. If the leak cannot be repaired, the compressor must be replaced. System leaks can also contribute to compressor failure and should be repaired.

If a base unit or compressor is mishandled, internal damage can occur. A broken spring or bent line can cause excessive running noise. Compressors with these problems should be replaced.

Compressor valves and rings provide a seal between the high-pressure and low-pressure sides. If damaged, the compressor must be replaced or repaired (open or serviceable hermetic). Suction and discharge pressure checks can be used to test for this condition. Bad valves or rings may exist if the suction will not pull down or discharge builds up with the system properly charged.

Another check can be made by measuring the running current with an ammeter. If, under loaded conditions, the running current is considerably lower than normal, faulty valves or rings should be suspected. If the compressor has unloaders, make sure they are not activated. On heat pumps, make sure the reversing valve and check valves are not stuck open.

Many compressors that are replaced are incorrectly judged to be seized. Seized or tight motors usually hum but will not run. Also, when the current is measured, the motor is drawing locked rotor current. Before replacing a compressor with these symptoms, make sure the following conditions do not exist:

- Unequal system pressures (common with PSC motors)
- Low supply voltage
- Contactor not making good contact on all poles
- Defective (open) start relay
- Start or run capacitor open or shorted

## 11.0.0 ◆ COMPRESSOR CHANGEOUT

The procedures used to change out a compressor differ depending on whether the cause of the compressor failure is electrical or mechanical. There are many types and variations of compressors. The procedures in this section describe the methods used to replace a welded hermetic-type compressor. Regardless of the type of compressor being replaced, the guidelines given in the procedures normally apply. You must always consult the system and replacement compressor manufacturer's service literature for the specific system being serviced and the compressor being used.

### Internal Line Break Overloads

Internal line break overloads can take an hour or more to reset under some conditions. If you are on a service call, you don't want to waste your valuable time and the customer's money waiting for the overload to reset. Here's how you can speed up the reset time:

- Shut off all power to the unit.
- Protect any exposed electrical components and motors from water entry.
- Run water from a garden hose over the compressor to help cool it down.

## 11.1.0 Compressor Replacement Due to Mechanical Failure

When replacing a compressor because of a mechanical problem, use the following guidelines.

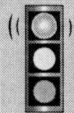

**WARNING!**
Be sure all electrical power to the equipment is turned off. Open, lock, and tag disconnects. Watch out for pressurized or hot components. Follow all safety instructions labeled on the equipment and given in the manufacturer's service manual. Failure to do so may result in personal injury.

*Step 1* Turn off the equipment. Lockout and tag equipment so that no one can start it.

*Step 2* Recover the refrigerant, regardless of its condition. Do not vent the refrigerant to the atmosphere. The methods used to recover refrigerant are described later in the HVAC Level Two Module, *Leak Detection, Evacuation, Recovery, and Charging*.

*Step 3* Disconnect and tag all wiring from the compressor.

**WARNING!**
Danger exists if the compressor terminals are damaged and the system is pressurized. Disturbing the terminals to make measurements could cause them to blow out, causing injury. When making voltage, current, or continuity checks on a hermetic or semi-hermetic compressor in a pressurized system, always make measurements at terminal boards and points of test away from the compressor. Once refrigerant has been recovered and the system is no longer under pressure, measurements can be made at the compressor terminals.

*Step 4* Use a tubing cutter to cut the system refrigerant tubing connected to the compressor. Do not use a hacksaw; it can introduce harmful chips into the system. Tape the open lines in the equipment to prevent dirt or moisture from entering.

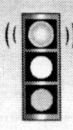

**WARNING!**
Never use a torch to cut the compressor lines because oil vapor in the lines can flare up and cause severe burns.

*Step 5* Remove the holddown bolts, then remove the old compressor. Be sure to get help to avoid injury caused by heavy lifting.

*Step 6* Unpack the new compressor and read any manufacturer's literature that accompanies it. Replacement compressors often have new features or devices that should be used. Compare nameplates between the old and new compressors to be sure the new compressor is the correct type.

*Step 7* Test the new compressor motor for open windings, grounds, or shorts.

*Step 8* On the old compressor, scratch matching marks on the stubs for alignment purposes with the new compressor. Then remove the stubs with a torch. Be sure to have a wet quenching cloth and fire extinguisher nearby in case the oil vapor flares up.

*Step 9* Scratch alignment marks on the new compressor corresponding to those marked on the old one. Align the old stubs on the new compressor, then braze on the stubs.

*Step 10* Remove the mounting grommets from the old compressor and install them on the new one.

*Step 11* Mount the new compressor in the equipment and bolt it down. Be sure to get adequate help to avoid injury caused by heavy lifting.

*Step 12* Braze the system refrigerant lines to the compressor using sweat couplings. Clean any flux from the joints and paint them for protection.

*Step 13* Remove the existing liquid line filter-drier and replace it with one that is one size larger. If the system is not equipped with a filter-drier, install one.

*Step 14* Leak test the system, then deep evacuate to a level of 500 microns. The methods used to leak test and evacuate a system are described in the HVAC Level Two Module, *Leak Detection, Evacuation, Recovery, and Charging*.

*Step 15* While evacuating the system, consult the equipment wiring diagram and connect all wiring to the compressor.

*Step 16* Recharge the system with the correct type and weight of refrigerant. The methods used to charge a system are described later in *Leak Detection, Evacuation, Recovery, and Charging*.

*Step 17*  Start the system and allow it to run in order to stabilize the system pressures. Refer to the equipment manufacturer's service instructions and follow them to make any adjustments in the refrigerant charge as deemed necessary.

## 11.2.0 Compressor Replacement Due to Electrical Failure

Electrical failures involving motor windings are often called *burnouts*. A burnout is the breakdown of the motor winding insulation, which causes the motor to short out or ground electrically. Burnouts are classified either as mild or severe.

- A mild burnout usually occurs suddenly, causing the motor to stop before the contaminants created by the burnout leave the compressor. Few contaminants, if any, are produced and little or no chemical reaction of the refrigerant and oil has occurred.
- Severe burnouts (cookouts) usually occur over a long period of time. Considerable contaminants are produced and may be pumped through the system while the compressor is still running.

**WARNING!**
When working on a system suspected of having a compressor burnout, wear appropriate personal protective equipment, including rubber gloves and eye protection. Contaminated refrigerant oil may contain heavy concentrations of acid. Do not allow contact with the skin or eyes as severe burns may result.

Before replacing a compressor that has failed because of a burnout, you must determine if the burnout was mild or severe. To do this, you must check the system for acid or moisture, using one of several acid/moisture test kits or oil test kits that are readily available. Acid/moisture test kits typically can be connected to a system service port to obtain a sample of the refrigerant. Oil test kits require that a sample of the system oil be taken. For systems with hermetic compressors, an oil sample may be obtained by inserting an oil trap in the suction line. In either case, follow the test kit manufacturer's instructions for using the kit and for determining the amount of contamination in the system.

Another way to check out the type of burnout in a hermetic compressor is to cut the suction and discharge lines and check for carbon by running a clean, lint-free swab into the lines. In a semi-her-

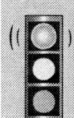

**WARNING!**
Do not cut into or attempt to unbraze a line with any system pressure in it. The system pressure must be 0 psig before repairs are attempted.

metic compressor, the easiest way to tell the type of burnout is to remove the cylinder head. If you find carbon using either method, it is a good indication that a severe burnout has occurred. With a severe burnout, the refrigerant will have a very strong rotten-egg smell.

### 11.2.1 Replacement Procedure After a Mild Burnout

Since most contaminants produced by a mild burnout are contained within the compressor, the procedure for changing the compressor is the same as that used with a mechanical failure, including the installation of an oversized filter-drier. One exception is that after the compressor installation is completed, the system should be triple evacuated before charging it with refrigerant. The triple evacuation method is described later in *Leak Detection, Evacuation, Recovery, and Charging*.

### 11.2.2 Replacement Procedure After a Severe Burnout

Servicing a system after a motor burnout requires not only compressor replacement, but also a thorough cleanup of the entire system to remove all harmful contaminants left by the burnout. Successive burnouts on the same system can generally be traced to improper cleanup. Unless the cleanup is performed correctly, a repeat burnout usually occurs in a short period of time. The following is an overview of the procedure for replacing a compressor after a severe burnout:

*Step 1*  Recover the system refrigerant and remove the compressor.

*Step 2*  Remove the liquid line filter-drier.

*Step 3*  Purge the system piping with dry nitrogen in the direction opposite to normal refrigerant flow.

*Step 4*  Remove, clean, and/or replace metering devices, accumulators, reversing valves, and related components, if contaminated.

*Step 5*  Install the new compressor.

*Step 6*  Add or replace the liquid line filter-drier. Use the next larger size.

*Step 7* Add a suction line filter-drier with input and output pressure taps.

*Step 8* Triple evacuate the system.

*Step 9* Recharge the system.

*Step 10* Run the system for one hour. Stop the system and change the suction line filter-drier any time the pressure drop across it is just below or at 3 psig.

*Step 11* After one hour has elapsed, stop the system and change the liquid and suction line filter-driers. If using a semi-hermetic compressor, replace the oil.

*Step 12* Run the system for two more hours.

*Step 13* After two hours have elapsed, stop the system and test for acid or moisture contamination to make sure the system is clean. If clean, change the liquid line filter-drier. Remove the suction line filter-drier from the system.

*Step 14* If acid or moisture is still present in the system at Step 13, change the oil (semi-hermetic) and repeat Steps 12 and 13 as necessary to achieve a clean system.

## 12.0.0 ◆ HYDRONIC PUMPS

Up to this point, this module has focused on refrigeration system compressors. This section briefly introduces circulating pumps used in water (hydronic) systems. More detailed information about centrifugal pumps is studied later in your HVAC Level Three training when studying hydronic systems.

A circulating pump is typically used in hot water heating systems to force or circulate the hot water from the boiler through the system piping and heat radiation terminals back to the boiler. Most circulating pumps are centrifugal-type pumps. A centrifugal pump works somewhat like a centrifugal compressor. In the centrifugal pump, the rotating action of an impeller in a scroll housing generates pressure that forces the water through the piping system. The pressure and volume developed depends on the pump size and rotational speed.

In small systems, in-line pumps are commonly used, while base-mounted pumps are used in large systems (*Figure 28*). Shut-off valves are normally installed on the inlet and outlet piping of the pump so that the pump can be isolated for repair or replacement.

A circulating pump causes circulation in a piping system by creating a pressure differential between its suction and discharge openings. Water flows in the system in an attempt to equalize the difference. As long as the pump runs, the pressure difference remains and the water keeps flowing. A circulating pump is sized to overcome the system pressure drop and to supply each terminal device with the necessary amount of water (in gallons per minute) at the proper temperature. The term *head pressure*, usually expressed in feet of water, is used to give the capacity of a circulating pump. It is just another way of expressing pressure drop. The maximum head of a pump is actually the maximum pressure drop against which the pump can cause water to flow.

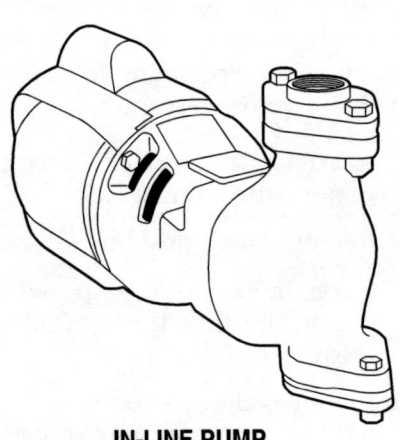
**IN-LINE PUMP**

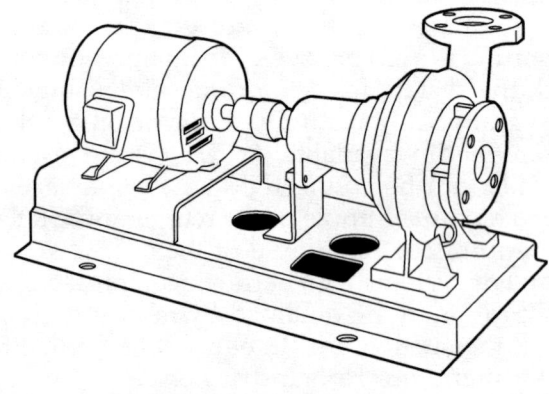
**BASE-MOUNTED PUMP**

*Figure 28* ◆ Circulating pumps.

## Summary

Compressors used in refrigeration systems fall into three groups: open compressors, welded hermetic compressors, and semi-hermetic compressors. Within each group, five different types of compressors can be found depending on the application. These are: reciprocating, rotary, scroll, screw, and centrifugal. Welded hermetic compressors are used in the majority of smaller commercial and residential systems, while open and semi-hermetic compressors are typically used for large commercial or industrial applications.

Because of the expense and time involved, compressor changeouts are considered to be one of the critical service procedures performed on cooling systems. The best way to prevent problems is to do a clean, professional installation. Ideally, the installation should be followed by periodic routine maintenance and service checks that enable early detection and correction of potential compressor problems. If not corrected, most mechanical refrigeration system problems will result in compressor failure. If you assume the cause of the compressor failure lies within the compressor, it is likely you will make the first of many unnecessary compressor replacements. If a compressor fails and the cause has not been found and corrected, the replacement compressor will also fail.

When a compressor fails, a systematic approach must be followed to correctly diagnose the cause of the failure. This includes examining the rest of the system for abnormal system conditions which may have caused the compressor to fail. These conditions include:

- Incorrect installation
- Refrigerant flooding
- Loss of lubrication
- System contamination
- Overheating
- Electrical problems
- Compressor incorrectly sized for the load

---

## Review Questions

1. A compressor that has a piston that travels back and forth in a cylinder is _____.
   a. reciprocating
   b. rotary
   c. centrifugal
   d. screw

2. In a scroll compressor _____.
   a. both scrolls move
   b. only the upper scroll moves
   c. only the lower scroll moves
   d. neither scroll moves

3. The capacity of a compressor is *not* related to the _____.
   a. horsepower of the driving motor
   b. compressor displacement
   c. number of cylinders
   d. length of the stroke in inches

4. The on/off method of capacity control is normally used with _____.
   a. reciprocating, rotary, and scroll multi-compressor systems
   b. rotary compressors
   c. scroll compressors
   d. reciprocating compressors

5. The cylinder unloader method of capacity control is used mainly with _____.
   a. centrifugal compressors
   b. screw compressors
   c. rotary compressors
   d. open and semi-hermetic reciprocating compressors

6. A compressor electrical motor pilot duty protection device _____.
   a. resets automatically
   b. senses current flow and/or temperature in the motor winding and opens the winding circuit to remove line voltage from the motor
   c. senses a current overload and/or temperature within the motor and opens the motor contactor control circuit to remove power from the motor
   d. must be reset manually

7. A compressor protection device that prevents the compressor from operating out of sequence is a _____.
   a. high-pressure switch
   b. short-cycle timer
   c. lockout relay
   d. airflow switch

8. Reduced-voltage starting of a compressor is used to _____.
   a. boost the motor starting current
   b. limit the supply voltage drop when the motor starts
   c. control the capacity of the compressor
   d. decrease the compressor motor rpm

9. Compressing a slug of liquid refrigerant can result in pressures applied to the compressor valves and pistons that exceed _____ psi.
   a. 40
   b. 100
   c. 1,000
   d. 2,000

10. Excessive foaming in the oil sight glass shows the oil is mixed with _____.
    a. refrigerant
    b. air
    c. acid
    d. moisture

11. A clean refrigeration system may contain _____.
    a. moisture
    b. air
    c. acid
    d. oil

12. The nameplate on a three-phase compressor motor is marked 460 volts. For proper motor operation, the input supply voltage must range between _____.
    a. 414V and 506V
    b. 430V and 590V
    c. 437V and 483V
    d. 451V and 469V

13. Voltage imbalance for a three-phase electrical supply must not exceed _____.
    a. 1%
    b. 2%
    c. 5%
    d. 10%

14. When determining the type of burnout in a system with a hermetic compressor, the first thing you would do is _____.
    a. turn off the power and remove the wiring from the compressor
    b. recover the refrigerant charge
    c. check the moisture and acid content of the system
    d. evacuate the system

15. A metering device should be cleaned or replaced after a _____.
    a. mild compressor burnout
    b. capacitor failure
    c. severe compressor burnout
    d. compressor mechanical failure

# GLOSSARY

# Trade Terms Introduced in This Module

*Burnout:* A condition in which the breakdown of the motor winding insulation causes the motor to short out or ground electrically.

*Capacity control:* Methods used in cooling systems to adjust system operation to match changes in the system cooling load.

*Clearance volume:* The amount of clearance between a piston at the top dead-center position of travel and the cylinder head.

*Compliant scroll compressor:* A version of the scroll compressor that allows the orbiting scroll to temporarily shift from its normal operating position if liquid refrigerant enters the compressor.

*Compression:* The reduction in volume of a vapor or gas by mechanical means.

*Flooded starts:* A condition in which slugging, foaming, and inadequate lubrication occur at compressor start-up as a result of the oil in the compressor crankcase having absorbed refrigerant during shutdown.

*Flooding:* A condition in which there is a continuous return of liquid refrigerant in the suction vapor being returned to the compressor during operation.

*Line duty device:* A motor protection device that senses current flow and temperature in the motor winding. If an overload occurs, it opens the motor winding circuit to remove the line voltage.

*Pilot duty device:* A motor protection device that senses current overload or temperature within the motor. If an overload occurs, it opens the motor contactor control circuit to remove power from the motor.

*Positive-displacement compressor:* Any compressor where the pumping action is created by pistons or moving chambers.

*Pressure (force-feed) lubrication system:* A method of compressor lubrication that uses an oil pump mounted on the end of the crankshaft to force oil to the compressor main bearings, lower connecting rod bearings, and piston pins.

*Short cycling:* A condition in which the compressor is restarted immediately after it has been turned off.

*Single phasing:* A condition in which a three-phase motor continues to run after losing one of the three input phases while operating.

*Splash lubrication system:* Method of compressor lubrication in which the crankcase oil is splashed onto the cylinder walls and bearing surfaces during each revolution of the crankshaft while the compressor is running.

# ANSWER KEY

## Answers to Review Questions

| Answer | Section |
|---|---|
| 1. a | 4.1.0 |
| 2. c | 4.3.0 |
| 3. a | 5.1.0 |
| 4. a | 5.2.0 |
| 5. d | 5.3.0 |
| 6. c | 6.4.0 |
| 7. c | 7.3.0 |
| 8. b | 8.0.0 |
| 9. c | 9.1.0 |
| 10. a | 9.2.0 |
| 11. d | 9.4.0 |
| 12. a | 9.5.2 |
| 13. b | 9.5.3 |
| 14. c | 11.2.0 |
| 15. c | 11.2.2 |

# REFERENCES & ACKNOWLEDGMENTS

## Additional Resources

This module is intended to present thorough resources for task training. The following reference works are suggested for further study. These are optional materials for continued education rather than for task training.

*Capacity Control–General Training Compressors*, 1990. Syracuse, NY: Carrier Corporation.

*Clean-Up After Burnout–General Training Compressors*, 1985. Syracuse, NY: Carrier Corporation.

*Compressors–General Training Air Conditioning*, 1991. Syracuse, NY: Carrier Corporation.

*Modern Refrigeration and Air Conditioning*, 2000. A.D. Althouse, C.H. Turnquist, A.F. Bracciano. Tinley Park, IL: The Goodheart-Willcox Company, Inc.

*Refrigeration & Air Conditioning Technology*, 2000. William C. Whitman, William M. Johnson, John A. Tomczyk. Albany, NY: Delmar Publishers, Inc.

## Figure Credits

| | |
|---|---|
| **Gerald Shannon** | 210F03, 210F07, 210F11, 210F21 |
| **Carrier Corporation** | 210F06, 210F17, 210F25, 210SA03 |
| **Thomas P. Burke** | 210F09, 210F10 |
| **Cutler-Hammer** | 210SA01 |
| **Springfield Wire** | 210SA02 |

# NCCER CRAFT TRAINING USER UPDATES

The NCCER makes every effort to keep these textbooks up-to-date and free of technical errors. We appreciate your help in this process. If you have an idea for improving this textbook, or if you find an error, a typographical mistake, or an inaccuracy in the NCCER's Craft Training textbooks, please write us, using this form or a photocopy. Be sure to include the exact module number, page number, a detailed description, and the correction, if applicable. Your input will be brought to the attention of the Technical Review Committee. Thank you for your assistance.

*Instructors* – If you found that additional materials were necessary in order to teach this module effectively, please let us know so that we may include them in the Equipment and Materials list in the Instructor's Guide.

**Write:** Curriculum Revision and Development Department
National Center for Construction Education and Research
P.O. Box 141104, Gainesville, FL 32614-1104

**Fax:** 352-334-0932

**E-mail:** curriculum@nccer.org

---

Craft _____ Module Name _____

Copyright Date _____ Module Number _____ Page Number(s) _____

Description
_____
_____
_____
_____

(Optional) Correction
_____
_____
_____

(Optional) Your Name and Address
_____
_____
_____

Module 03211-01

# Heat Pumps

## COURSE MAP

This course map shows all of the modules in the second level of the HVAC curriculum. The suggested training order begins at the bottom and proceeds up. Skill levels increase as you advance on the course map. The local Training Program Sponsor may adjust the training order.

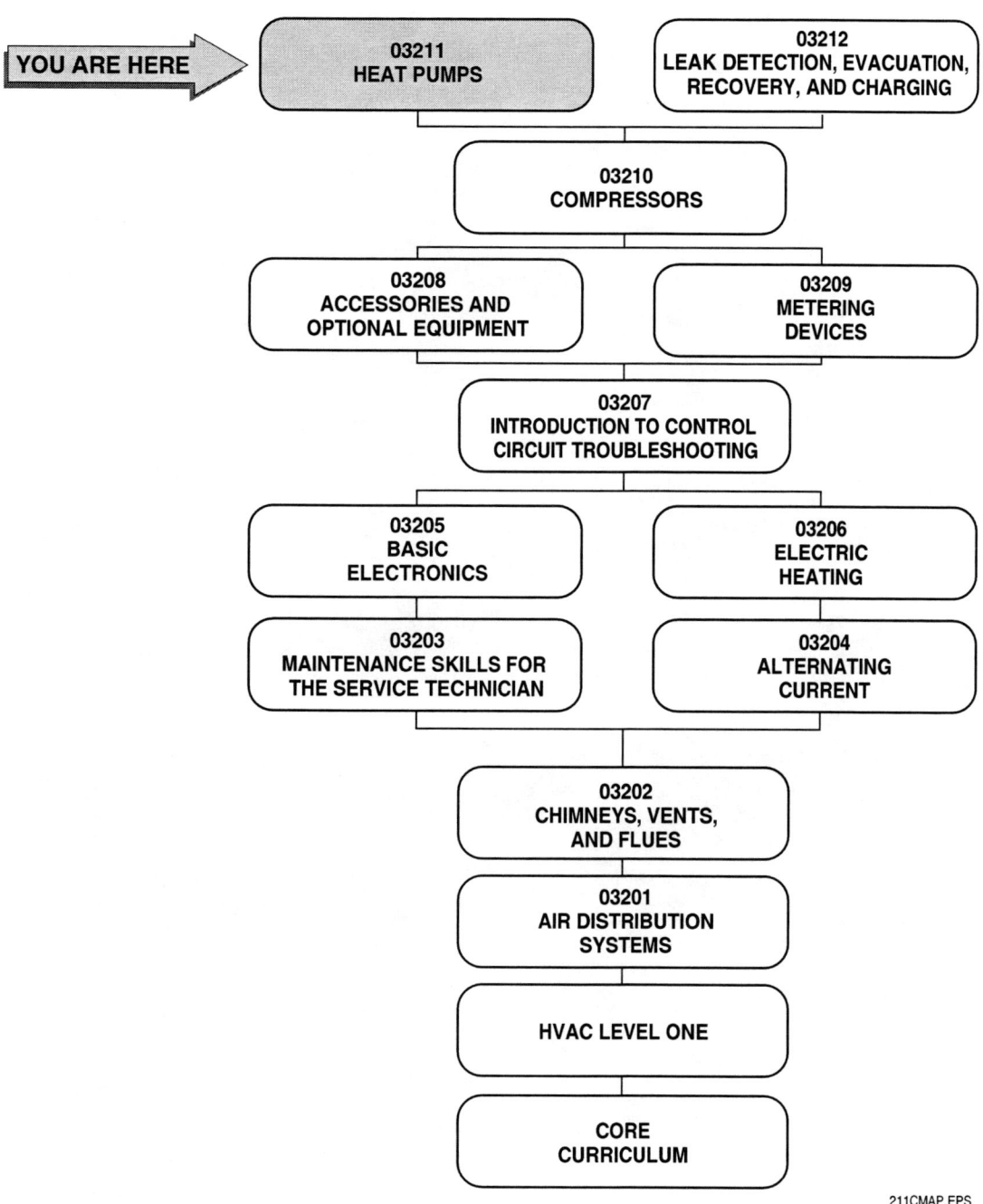

## MODULE 03211 CONTENTS

- **1.0.0 INTRODUCTION** .................................................. 11.1
- **2.0.0 HEAT PUMP CLASSIFICATIONS** ................................. 11.1
  - 2.1.0 Air-to-Air Heat Pumps ......................................... 11.2
  - 2.2.0 Water-to-Air Heat Pumps ...................................... 11.2
  - 2.3.0 Water-to-Water Heat Pumps ................................... 11.4
  - 2.4.0 Air-to-Water Heat Pumps ...................................... 11.5
  - 2.5.0 Ground-Source Heat Pumps ................................... 11.5
  - 2.6.0 Special Heat Sources .......................................... 11.5
- **3.0.0 HEAT PUMP OPERATION** ....................................... 11.6
  - 3.1.0 Cooling Cycle .................................................. 11.6
  - 3.2.0 Heating Cycle .................................................. 11.6
  - 3.3.0 Defrost Cycle .................................................. 11.6
    - *3.3.1 Electromechanical Defrost Controls* ....................... 11.8
    - *3.3.2 Electronic Defrost Controls* ............................... 11.9
- **4.0.0 HEAT PUMP COMPONENTS** ..................................... 11.10
  - 4.1.0 Reversing Valves .............................................. 11.10
  - 4.2.0 Metering Devices .............................................. 11.12
  - 4.3.0 Liquid Accumulators ........................................... 11.12
  - 4.4.0 Thermostats ................................................... 11.13
  - 4.5.0 Pressure Controls ............................................. 11.13
  - 4.6.0 Crankcase Heaters ............................................ 11.15
- **5.0.0 HEAT PUMP PERFORMANCE** .................................... 11.15
  - 5.1.0 Coefficient of Performance .................................... 11.15
  - 5.2.0 Performance Factor ........................................... 11.16
  - 5.3.0 Seasonal Energy Efficiency Ratio ............................. 11.16
  - 5.4.0 Heat Pump Efficiency Recommendations ..................... 11.17
- **6.0.0 BALANCE POINT AND SUPPLEMENTARY HEAT** ................. 11.17
- **7.0.0 INSTALLATION** ................................................. 11.18
  - 7.1.0 Split Systems .................................................. 11.18
    - *7.1.1 Location* .................................................... 11.19
    - *7.1.2 Mounting* .................................................... 11.20
    - *7.1.3 Indoor Section* .............................................. 11.20
  - 7.2.0 Packaged Units ................................................ 11.20
  - 7.3.0 Add-On Systems ............................................... 11.21
  - 7.4.0 Installation Checklist .......................................... 11.22
- **8.0.0 SERVICE** ....................................................... 11.23
- **9.0.0 HEAT PUMP CONTROLS** ........................................ 11.23

## MODULE 03211 CONTENTS (Continued)

**SUMMARY** .................................................... 11.25
**REVIEW QUESTIONS** ........................................... 11.25
**GLOSSARY** ................................................... 11.27
**ANSWERS TO REVIEW QUESTIONS** ................................ 11.28
**REFERENCES & ACKNOWLEDGMENTS** ............................... 11.29

### Figure

| | | |
|---|---|---|
| Figure 1 | Schematic of an air-to-air heat pump (shown in cooling mode) | 11.3 |
| Figure 2 | Water-to-air heat pump | 11.3 |
| Figure 3 | Water-to-air closed loop system | 11.4 |
| Figure 4 | Ground-source heat pump | 11.5 |
| Figure 5 | Refrigerant flow in an air-to-air heat pump | 11.7 |
| Figure 6 | Electromechanical defrost control | 11.8 |
| Figure 7 | Electronic defrost control simplified schematic | 11.9 |
| Figure 8 | Reversing (four-way) valve | 11.11 |
| Figure 9 | Refrigerant flow in a reversing valve | 11.11 |
| Figure 10 | Reversing valve position – cooling cycle | 11.11 |
| Figure 11 | Refrigerant flow – cooling cycle | 11.12 |
| Figure 12 | Reversing valve position – heating cycle | 11.12 |
| Figure 13 | Refrigerant glow – heating cycle | 11.12 |
| Figure 14 | Use of thermostatic expansion valves in a heat pump | 11.13 |
| Figure 15 | Heat pump thermostat | 11.14 |
| Figure 16 | Pressure switches used to protect the compressor | 11.14 |
| Figure 17 | Crankcase heater control | 11.15 |
| Figure 18 | Heat pump balance point | 11.18 |
| Figure 19 | Split system installation | 11.19 |
| Figure 20 | Ductless split system | 11.19 |
| Figure 21 | Packaged unit installation | 11.21 |
| Figure 22 | Packaged unit locations | 11.21 |
| Figure 23 | Add-on heat pump installation | 11.22 |
| Figure 24 | Heat pump ladder diagram | 11.24 |

# MODULE 03211

# Heat Pumps

## Objectives

When you have completed this module, you will be able to do the following:

1. Describe the principles of reverse-cycle heating.
2. Identify heat pumps by type and general classification.
3. List the components of heat pump systems.
4. Demonstrate heat pump installation and service procedures.
5. Identify and install refrigerant circuit accessories commonly associated with heat pumps.
6. Analyze a heat pump control circuit.

## Prerequisites

Before you begin this module, it is recommended that you successfully complete the following modules: Core Curriculum; HVAC Level One; HVAC Level Two, Modules 03201 through 03210.

## Required Trainee Materials

1. Pencil and Paper
2. Appropriate Personal Protective Equipment

## 1.0.0 ♦ INTRODUCTION

A heat pump is a combination heating and cooling unit. It produces cooling in the same manner as a conventional cooling unit, then reverses the cycle to produce heat. Both systems have the same components:

- Compressor
- Condenser fan
- Condenser and evaporator coils
- Pressure and temperature controls
- Refrigerant and refrigerant lines
- Blower fan
- Service valves

In addition to the typical air conditioning components listed, the heat pump requires a **reversing valve** and additional metering and control devices.

Since heat pumps perform both heating and cooling functions, the coils are identified by location: the evaporating coil and condensing coil on an air conditioner are identified as the **indoor coil** and **outdoor coil** on a heat pump in the cooling mode.

Why are heat pumps used? Heat pumps operate for about half the cost of electric resistance heat. Electric utilities offer customers, contractors, and developers rebates to install heat pumps. This is because electricity demand is low in the winter months, and heat pumps encourage customers to switch from oil or gas to electric utilities. Air Conditioning and Refrigeration Institute (ARI) statistics indicate that over one million heat pumps are sold each year.

Besides being efficient to operate in the heating mode, a heat pump provides cooling. For heating purposes, a heat pump may or may not be less costly to operate than gas or oil heat, depending on the climate and fuel costs. However, it is likely to be more cost effective for combined heating and cooling than other methods.

## 2.0.0 ♦ HEAT PUMP CLASSIFICATIONS

Heat pumps are classified according to their heat source and the medium to which the heat is transferred (**heat sink**). A water-to-air heat pump, for example, picks up heat from water flowing through a coil and transfers it to the air flowing over another coil.

The more common types of heat pumps are: air-to-air, water-to-water, air-to-water, and water-to-air. In the air-to-air system, the external heat source is air, and the medium (heat sink) that

### Indoor and Outdoor Coils

When a heat pump switches from the cooling mode to the heating mode, the indoor and outdoor coils switch functions. Like cooling equipment, a heat pump can consist of a single packaged unit, or it can be a matched split system.

comes in contact with the indoor refrigerant coil is also air. Some larger heat pump systems may have more than one heat source and may also use both water and air as heat sinks.

An air-to-water system also uses air as the heat source, but the heating and cooling are provided by a water distribution system within the structure. The air-to-water system uses convection devices to transfer the heat between the water and the air.

Ground water, from the standpoint of temperature, is an excellent heat source, but it is not available in all areas and in some areas is in limited supply during certain periods. Therefore, air is the predominant heat source in residential and small commercial installations. On occasion, heat pump systems may use a coil buried in the earth as a heat source.

Other heat sources besides air and water are sometimes used. Some of these are waste heat from selected industrial processes, exhaust air from ventilation, solar energy, and heat extracted from refrigerated spaces. These sources are commonly used in addition to, rather than as replacements for, the basic heat source.

### 2.1.0 Air-to-Air Heat Pumps

The most common heat source is air. The refrigeration circuit schematic in *Figure 1* represents an air-to-air heat pump in the cooling cycle. Outdoor air is used as the heat source in air-to-air units. The heat is then delivered to, or removed from, the indoor air through the use of refrigerant flowing through a coil.

Air-to-air heat pumps are generally smaller-tonnage units. Packaged heat pumps are available for residential and commercial applications up to 30 tons. However, built-up systems of any size may also be installed. Current air-to-air models are either self-contained in one unit (**packaged unit**) or **split systems**, which are divided into two sections. Depending on the type of application, both kinds offer definite advantages.

When air is used as the heat source, efficiency improves in mild climates. A heat pump is generally selected to match the cooling load.

If the heating load exceeds the heating capacity of the selected heat pump unit, electric resistance heaters may be required to supply supplementary heat during periods of higher heat load.

Auxiliary heat is also required for heating during the **defrost** cycle. Auxiliary heat should be sized for 100% of the building's heat load. This ensures that there will be adequate heat in the event of a **compression heat** failure.

Air-to-air heat pumps offer several advantages:

- They do not use water, so the problems related to piping, corrosion, and condenser water disposal are nonexistent.
- They operate as in-space units supplying filtered, conditioned air at a nominal cost.
- They are available as packaged (integral) or remote (split) systems.
- They operate at half the operating cost of electric resistance heat.

### 2.2.0 Water-to-Air Heat Pumps

Water-to-air heat pumps use well water, lake water, or another available water supply as a heat source and transfer the heat indoors through the use of a heat exchanger. These units are similar in size to the air-to-air type and are marketed in the range from 2 to 50 tons.

### Use of Heat Pumps

Recent changes in ventilation standards require that high volumes of outside air be used in high-occupancy applications such as restaurants, churches, and auditoriums. Heat pumps are not well suited for these types of applications because of the low air temperatures entering the indoor coil.

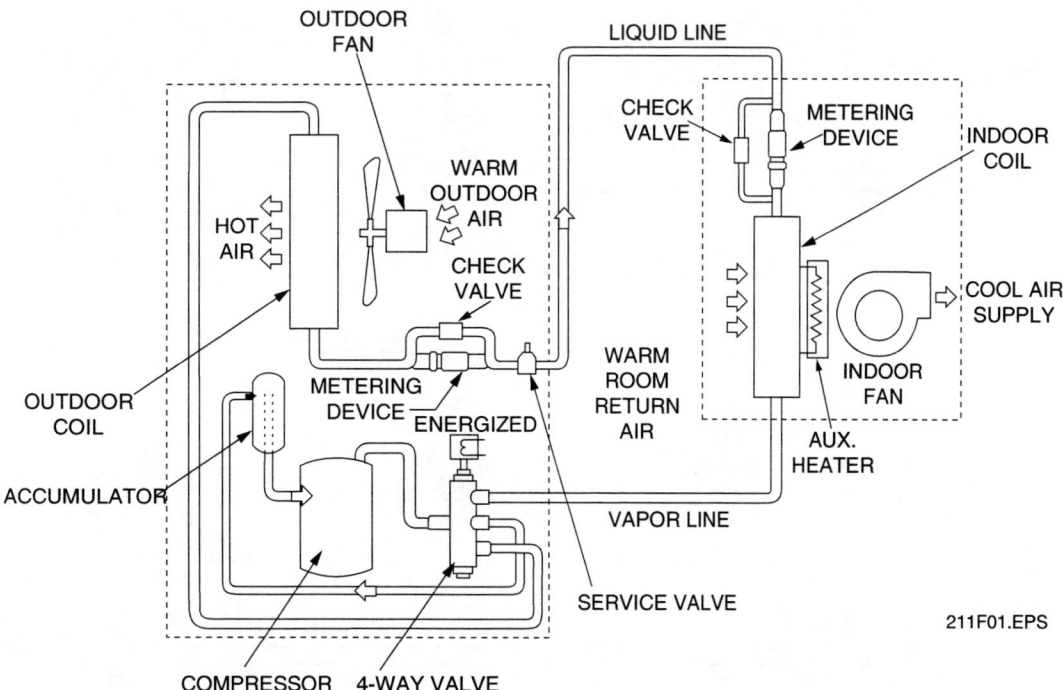

*Figure 1* ◆ Schematic of an air-to-air heat pump (shown in cooling mode).

Water-to-air heat pumps (*Figures* 2 and 3) have some of the same basic problems as water-to-water types, such as water availability and disposal. However, water quality is not as much of a problem because only the outdoor coil, which is immersed in water, is exposed. This coil can be made of a special copper-nickel alloy that is corrosion-resistant.

When surface water is not available, the contractor must drill for water. The drilling cost can significantly increase the cost of the installation.

Some of the advantages of water-to-air heat pumps include:

- They generally operate more efficiently than air-source units.
- Self-contained or integral water-to-air units have greater application flexibility than integral air-to-air units because they are not restricted to the use of an outdoor air supply.
- They may be installed as free-standing, in-space conditioning units.
- They do not require defrost.
- They generally do not require supplemental electric heat.
- They operate at half the operating cost of electric resistance heat.

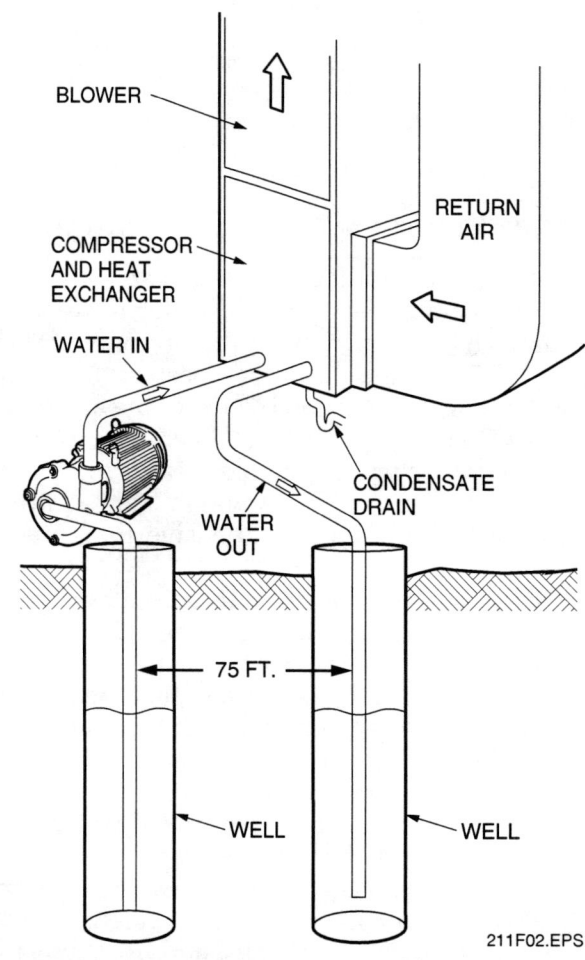

*Figure 2* ◆ Water-to-air heat pump.

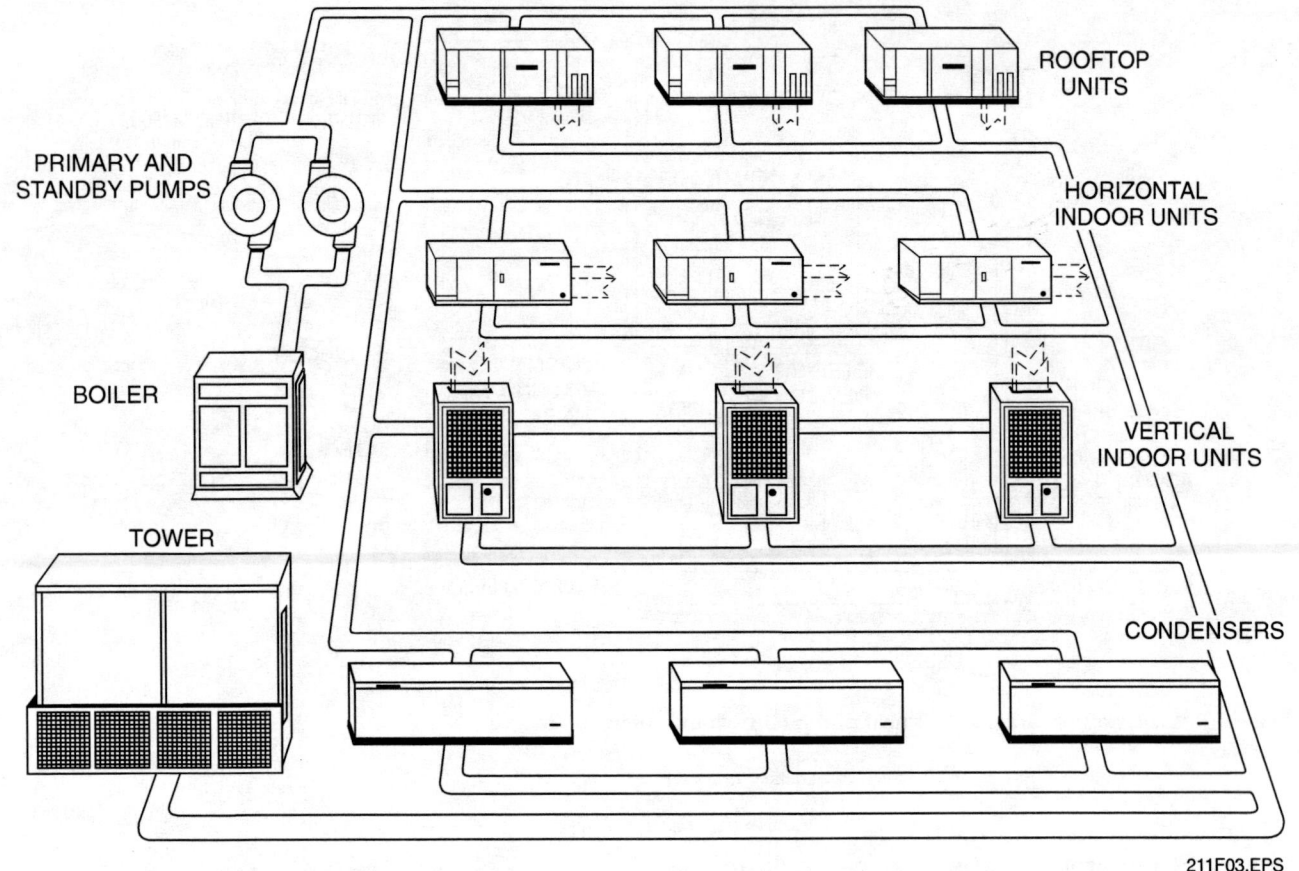

*Figure 3* ♦ Water-to-air closed loop system.

## 2.3.0 Water-to-Water Heat Pumps

Water-to-water heat pumps use water from wells, lakes, or other heat sources. Water-to-water heat pumps have been in use in the United States since the 1930s, but the growth has been relatively slow and has been surpassed by that of the air-to-water type. This slow growth has been due to the fact that there are few populated areas with generous supplies of cheap, clean water. The equipment itself is generally more expensive than air-to-air heat pumps.

Well or lake water has an advantage over air when used as a heat source because it is at a more constant temperature and transfers heat more efficiently. This is a benefit if the peak heating load occurs when the temperature of the water is considerably higher than the temperature of outdoor air. This temperature difference also makes the water-source heat pump more efficient than a comparable air-source heat pump.

The disadvantage of this type of heat pump is that water contains soluble minerals such as calcium salts, magnesium salts, and iron, which will form deposits on the surface of the heat exchanger and slow heat transfer. Water without mineral content (soft water) can also be corrosive. A neutral water supply is seldom found in large quantities.

Because of the large amounts of water involved, treatment of the supply water from the heat source is impractical. Thus, use of unsuitable water will result in excessive maintenance of heat exchanger surfaces and parts. In some areas, the sand content in well water can greatly shorten the lives of pumps, heat exchangers, and valves, thereby increasing maintenance costs.

The water-to-water heat pump has three major cost advantages:

- When compared to fossil fuel units, the water-to-water heat pump has substantially lower yearly electrical operating costs.
- It has a more efficient refrigeration cycle than that of the air-source heat pump. As a result, less electricity is usually required.
- The initial cost of the dual system will generally be lower than that of a conventional heating system combined with a second system to provide summer cooling.

Water-to-water heat pumps are sometimes used in conjunction with closed circuit, flat-plate solar collectors. The solar collectors are used as a heat source, and a water storage tank is used as a heat sink in these types of applications.

## 2.4.0 Air-to-Water Heat Pumps

An air-to-water heat pump is a system that uses the outdoor air as a heat source. It delivers the heat to the conditioned space through the use of a secondary medium which, in this case, is water. These units are often used to heat domestic water and swimming pools. Air-to-water installations can be designed to compete with conventional heating and cooling systems of all sizes. Heat is generally supplemented by electric resistance heaters if the load exceeds the heating capacity of the chosen heat pump. Air-to-water systems provide the following benefits:

- They can eliminate the need for large supply and return duct systems because warm and cold water are piped through the building.
- Rooms and areas may be thermostatically controlled more easily in large, central installations than with large supply and return air systems.
- They can be used as hot water heaters.

## 2.5.0 Ground-Source Heat Pumps

A **ground-source (geothermal) heat pump** (*Figure 4*) is a heat pump with a coil buried in the ground. The availability of heat in the earth for a heat pump system on peak-demand days is constant. A non-toxic antifreeze solution is pumped through the ground loop where heat is absorbed. A heat exchanger transfers this heat through the refrigeration process to the indoor coil, where the heat is distributed throughout the structure. This type of heat pump has gained in popularity in recent years.

## 2.6.0 Special Heat Sources

Economical, special heat sources for heat pump systems may be available from operations or processes that exist in nearby areas.

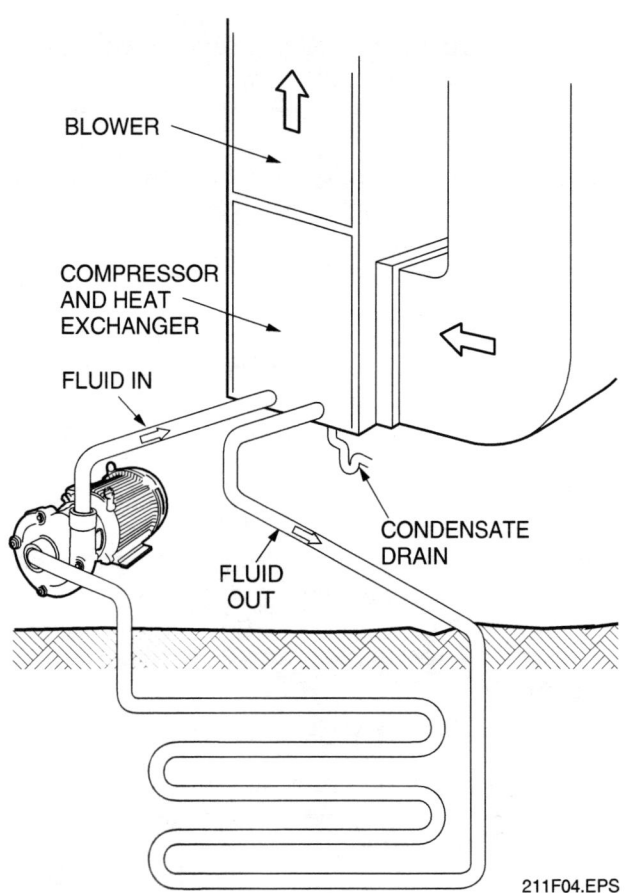

*Figure 4* ◆ Ground-source heat pump.

On large commercial and industrial buildings, one potential source of extracted heat is the central-point exhaust unit. Direct-expansion coils located in the exhaust air stream may be used to remove this heat and return it to other portions of the building.

Another method of extracting heat is the coil energy recovery loop (runaround loop) system. The coil energy recovery loop system incorporates coils filled with an antifreeze solution located in the ventilation supply and exhaust air ducts. Heat removed from the exhaust air is absorbed by the incoming ventilation air, thereby reducing the heating requirements of the mechanical equipment.

### Heat Pumps and Ground Water

Environmental laws are becoming very strict with regard to thermal pollution of any type. For this reason, it is important to always check the local codes concerning the use of ground water for heat pump applications.

Coil energy recovery loop systems are covered in more detail in the HVAC Level Four Module *Energy Conservation Equipment*.

A third method of extracting heat and reducing compressor motor electrical consumption is to place a refrigerant-liquid subcooling coil in the ventilation air plenum. The subcooling coil provides preheat to the incoming fresh air, while subcooling the liquid refrigerant as it flows to the evaporator.

The use of a subcooling coil is limited to buildings where a large portion of the ventilation supply air is introduced to the building at a central point close to the compressor room. It is impractical to run long refrigerant lines to widely dispersed coils.

Other sources from which heat may be recovered include computer rooms, industrial plants where heat is removed from process work, and refrigeration condensing units or packaged air coolers in supermarkets and warehouses.

## 3.0.0 ◆ HEAT PUMP OPERATION

To understand how the heat pump works, it is necessary to understand that, regardless of how cold it is outdoors, there is still some heat in the air. This is true as long as the temperature is above **absolute zero** (–460°F), which is 0°K on the **Kelvin scale**. At warmer temperatures, there is more heat available. As temperatures drop, there is less heat available for the heat pump to extract. Most heat pumps begin losing efficiency rapidly when the outdoor temperature falls below 35°F. When the temperature falls below 20°F, they can be quite ineffective. For that reason, heat pumps used in cold climates are usually equipped with auxiliary electric heaters that cycle on automatically when the heat pump alone cannot meet the heating demand.

In order to provide heating, the refrigerant must flow in the opposite direction from that of the cooling cycle. The outdoor coil becomes the evaporator and the indoor coil becomes the condenser. Heat from the outdoor air flows over the outdoor coil and is transferred to the low-temperature refrigerant flowing through that coil. After being compressed, the heated refrigerant flows through the indoor coil, and its heat is transferred to the relatively cooler indoor air.

### 3.1.0 Cooling Cycle

*Figure 5* shows a schematic diagram of the refrigerant cycle of a typical air-to-air heat pump. The reversing valve, also known as a **four-way valve**, is the key to the heat pump's dual nature.

In most heat pumps, the reversing valve is energized when the thermostat calls for cooling and de-energized when it calls for heating. This guarantees that heat is available even if there is a failure in the reversing valve control circuits.

When the unit is in the cooling mode (*Figure 5, View A*), the refrigerant flow (see arrows) is the same as that of any cooling unit. Notice the path through the reversing valve. The cold, low-pressure refrigerant flowing through the indoor coil (evaporator) absorbs heat from the conditioned space and is boiled into a superheated vapor. Hot, high-pressure refrigerant gas leaving the compressor is pumped through the outdoor coil (condenser), where the heat is rejected. The refrigerant pressures and temperatures shown are typical of those we have seen in the discussion of cooling systems.

### 3.2.0 Heating Cycle

In the heating mode (*Figure 5, View B*), the reversing valve changes position. The refrigerant leaving the compressor is routed in the opposite direction from that of the cooling cycle. Instead of flowing through to the outdoor coil, the hot, high-pressure refrigerant vapor leaving the compressor flows to the indoor coil, which is now acting as a condenser. Heat is extracted from the refrigerant at that point because the air in the conditioned space is cool in relation to the refrigerant in the coil (heat transfer from warm to cool). The condensed refrigerant then flows to the outdoor coil where it encounters cold outdoor air. This causes the refrigerant to boil. Heat provided in this way is known as **reverse cycle heat** or *compression heat*. Note the significant differences in the pressures and temperatures at key points in the system as compared with the cooling mode.

### 3.3.0 Defrost Cycle

When heat is transferred from the cold outdoor air to the refrigerant in the outdoor coil, moisture condenses on the coil. Because of the temperature relationships, this moisture will freeze on the coil, even at outdoor temperatures above 40°F. Over time, the frost will build up on the coil, blocking airflow and preventing effective heat transfer. This is more likely to be a problem at outdoor temperatures between 28°F and 40°F than it is at lower temperatures, because the colder the air, the less moisture it contains.

To prevent ice buildup on the outdoor coil, most heat pumps have a defrost function. In air-to-air heat pumps, the heating cycle is reversed, placing the unit in the cooling mode (*Figure 5, View A*).

**VIEW A. COOLING MODE**

**VIEW B. HEATING MODE**

211F05.EPS

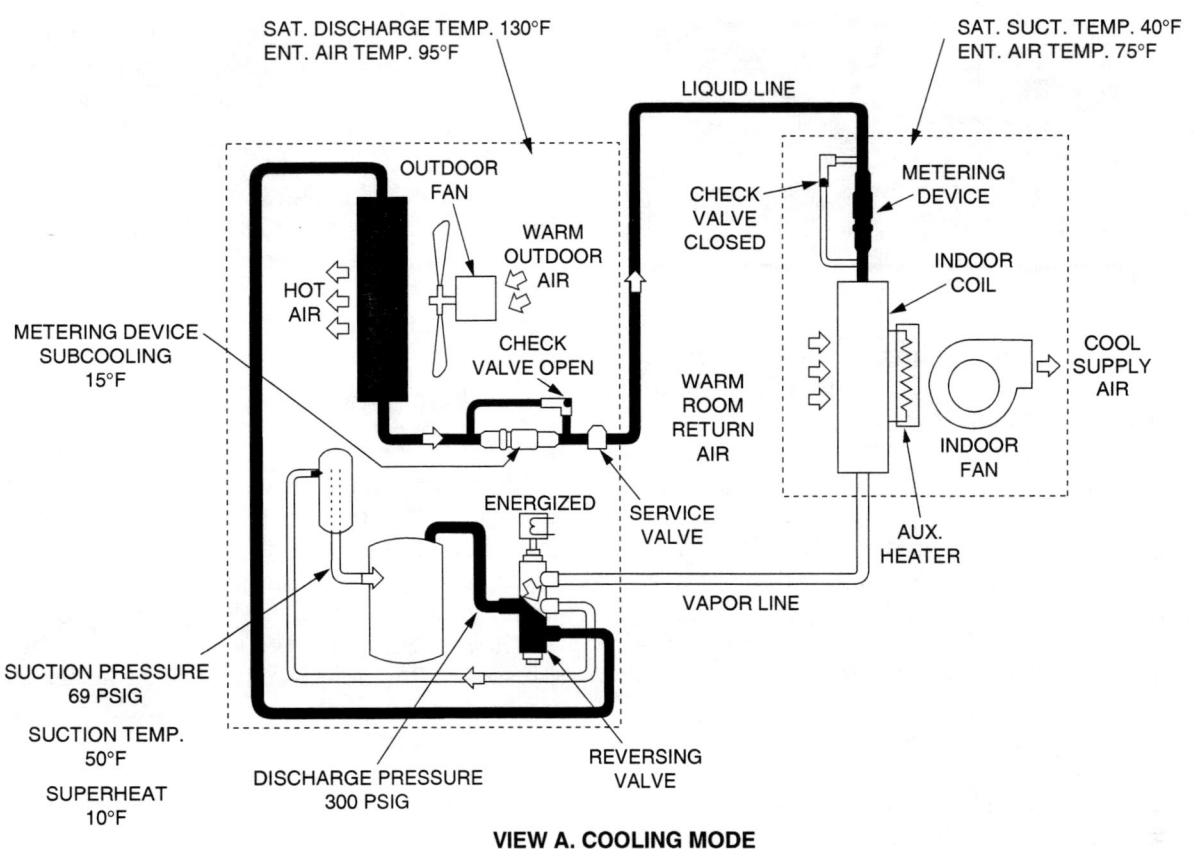

*Figure 5* ◆ Refrigerant flow in an air-to-air heat pump.

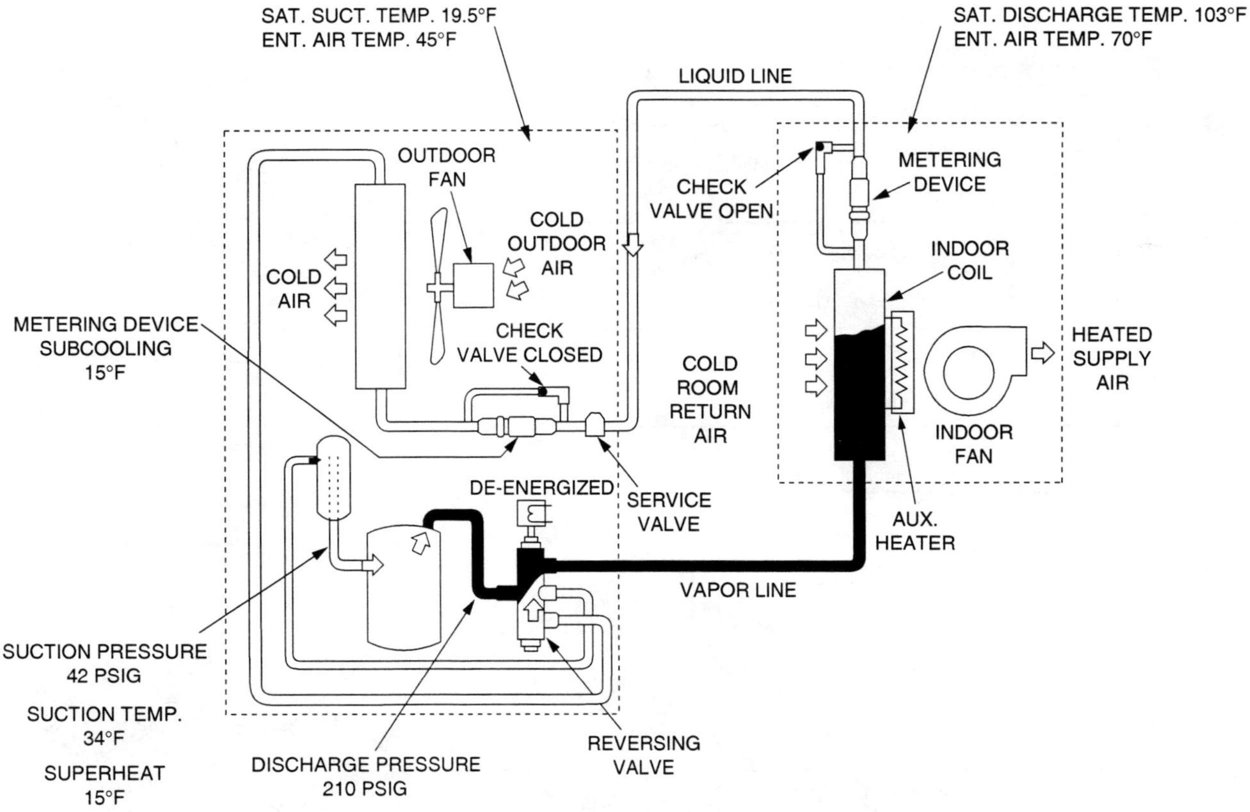

> **NOTE**
>
> For efficient transfer to occur, the outdoor coil must be 10°F to 20°F cooler than the outdoor air. Even at 60°F ambient temperature, the coil will be below freezing and moisture will freeze on the coil.

Hot refrigerant then flows through the outdoor coil to melt the frost buildup. This usually lasts about 10 minutes. During that time, the heat pump is not providing any heat. In fact, cool air is blowing off the evaporator into the conditioned space, just like it would on a hot day. To prevent discomfort to building occupants during the defrost cycle, an electric heater located between the indoor blower and the indoor coil is automatically switched on.

There are many different ways to control the defrost function. The most popular method uses a combination of time and temperature. The logic behind this method is that defrost is needed only if the unit has been operating long enough for frost to form on the coil (for example, 90 minutes), and the temperature is low enough for moisture to freeze on the coil. If both conditions don't exist, defrost is unnecessary. Most timers can be set for 45, 60, or 90 minutes of operation before defrost will occur.

This method requires a timer to keep track of the system's operating time and a means of sensing outdoor temperature. The latter is easy; a small thermostat or thermistor on the outdoor coil will do the job. Keeping time is easily accomplished with an electronic control.

### 3.3.1 Electromechanical Defrost Control

In some heat pumps, especially older models, the timing function is partly mechanical. One method uses a motor-driven timing device, which is switched on whenever the compressor is running. *Figure 6* shows the electrical schematic of such an arrangement. Whenever the compressor contactor is energized, power is supplied to the compressor motor and the defrost timer. As the defrost timer motor runs, it turns a cam that closes a set of contacts for 10 seconds every 90 minutes. Because the unit does not always run for 90 minutes at a time, it may take several on-cycles to accumulate 90 minutes.

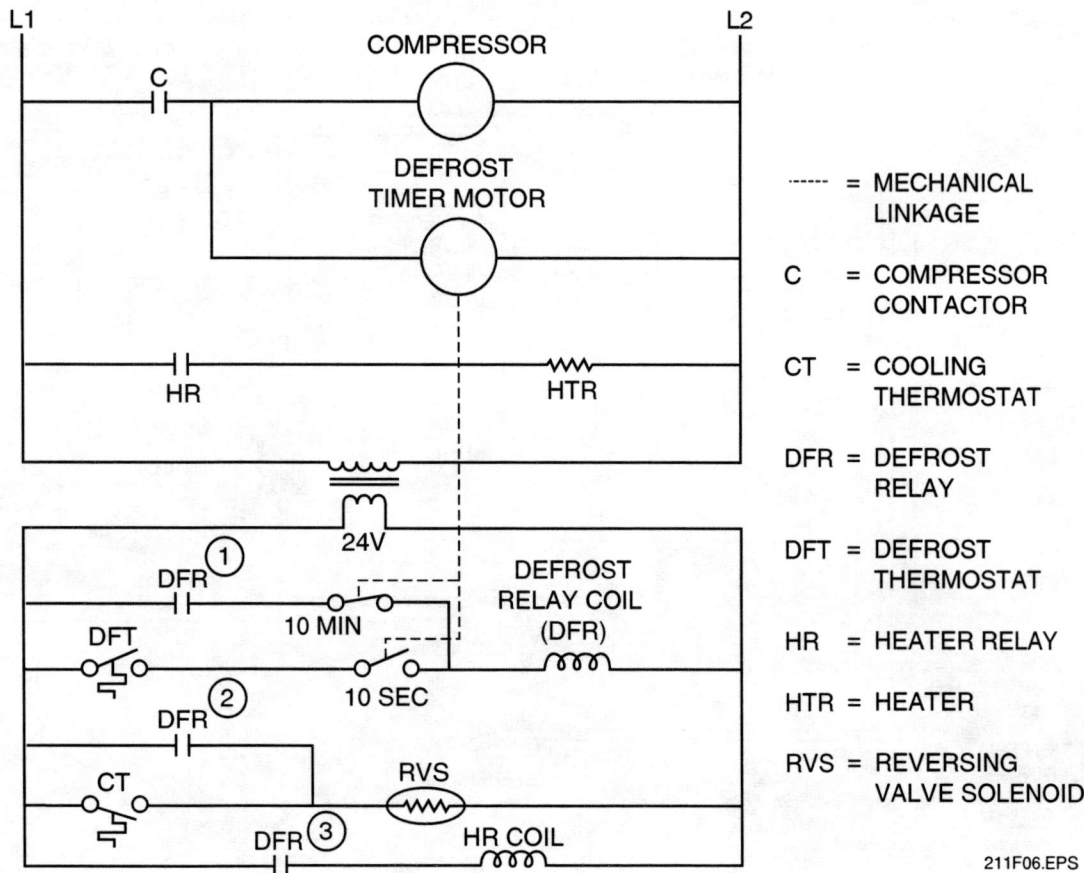

*Figure 6* ◆ Electromechanical defrost control.

11.8     HVAC LEVEL TWO — TRAINEE MODULE 03211

The 10-second closure of the mechanical contacts is not enough to start defrost. The defrost thermostat (DFT) must also be closed, indicating that the outdoor temperature is below 45°F (or some other selected setting). When both are closed, the defrost relay (DFR) will energize. One set of normally open DFR contacts (1) closes to act as a holding circuit for the DFR coil. This is needed because the mechanical contacts will open after 10 seconds.

Another set of normally open DFR contacts (2) closes, causing the reversing valve to energize and placing the system in the cooling mode. After 10 minutes (or some other selected period) of defrost operation, the other contacts that are mechanically linked to the timer motor will open momentarily, stopping the defrost process and restarting the 90-minute timing cycle.

If the heating thermostat opens during the defrost cycle, defrost will be suspended. When the thermostat cycles the unit on again, it will pick up where it left off, as long as the defrost thermostat has remained closed.

Notice that a third set of normally open DFR contacts (3) controls the heater relay (HR). When defrost cycles on, HR energizes, and its contacts in the heater circuit close, bringing on the electric heater to prevent cold air from blowing into the conditioned space.

In addition to mechanical timers, pressure switches that sense airflow through the outdoor coil and thermal controls may also be used to start and stop defrost. A pressure switch used to terminate defrost would be located in the liquid line, and would open the defrost circuit when the pressure at the outdoor coil reached 275 psi. This equates to a coil temperature of about 124°F, which is high enough to indicate that the coil is free of ice.

### 3.3.2 Electronic Defrost Controls

Most modern heat pumps use electronic defrost controls in which all the defrost circuitry, including the relays, is mounted on a printed circuit board. The timer is electronic, and therefore less likely to fail (see *Figure 7*). A temperature sensor, a defrost thermostat, or a thermistor sends temperature information from the outdoor coil to the board. Instead of keeping the operating time mechanically, an electronic clock in the timing logic receives a signal as long as the heating thermostat is closed. If defrost is interrupted by the opening of the heating thermostat, the timing logic will remember where it left off. If the coil temperature still indicates the presence of frost when the unit cycles on again, the unit will automatically go back into the defrost mode.

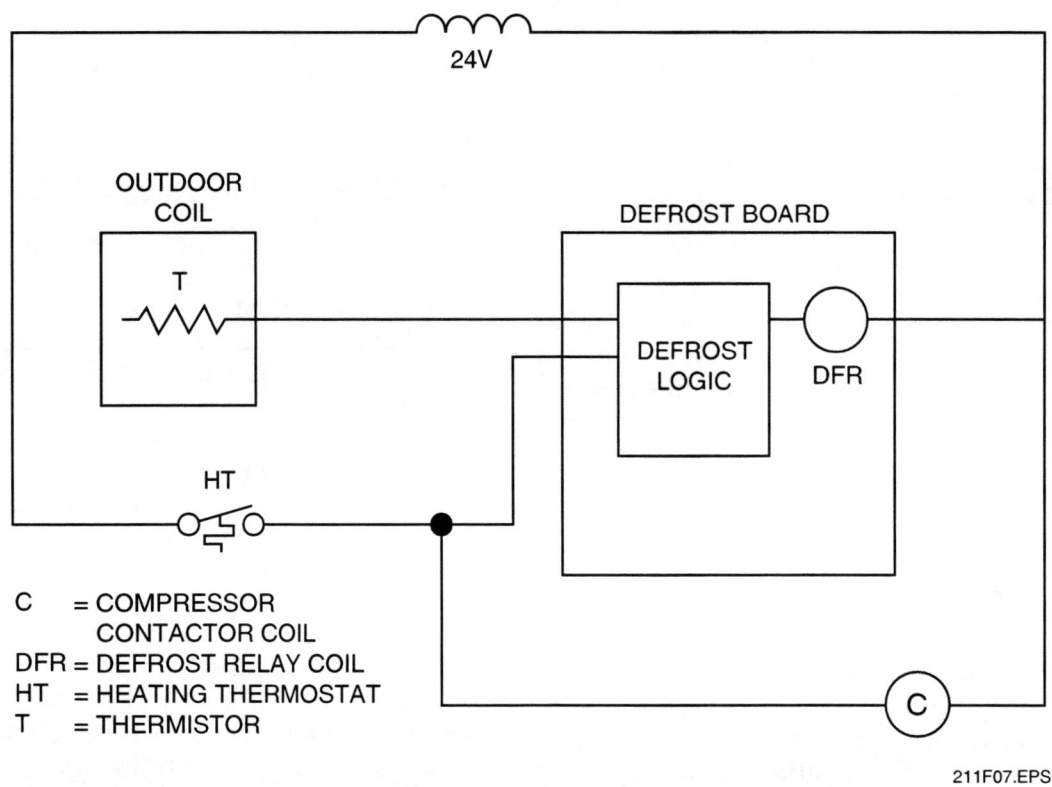

*Figure 7* ◆ Electronic defrost control simplified schematic.

>
>
> ### Electronic Defrost Controls
>
> An electronic defrost control is a solid-state control. Solid-state controls are generally more reliable than controls in which the timing function is partially or completely mechanical.
>
>
>
> DEFROST CONTROL BOARD

A jumper or other selection device on the board allows the installer to select the defrost frequency (30, 50, or 90 minutes, for example) based on local conditions.

An important feature of many of these boards is a built-in defrost test cycle. On one such control the technician can, by jumpering across two terminals on the board, artificially force a defrost cycle to determine whether the system is working properly in the defrost mode. The 90-minute defrost cycle is over-ridden and the usual 10-minute running cycle can be reduced to just a few seconds if desired. A major advantage of the electronic defrost control is that it is completely self-contained. If the technician finds that there is a defrost control problem, it is not necessary to isolate the failed component. Instead, the entire board is replaced.

Electronic defrost controls may be equipped with **demand defrost**, which combines time, temperature, and pressure information to determine if there is enough frost buildup to require defrost. It also terminates defrost when it senses that the coil is frost-free, regardless of whether the timed cycle is complete. Some electronic defrost controls contain a short-cycle timer.

## 4.0.0 ◆ HEAT PUMP COMPONENTS

Although a heat pump and a cooling-only unit are schematically similar, they are physically quite different. The heat pump not only has more components, but the components it has in common with the cooling unit are different because they perform a dual role. In a cooling-only unit, the evaporator can be smaller than the condenser because it handles less heat. In a heat pump the evaporator must be capable of rejecting the heat generated in the heat mode. The heat pump requires a heavy-duty compressor because it runs year-round. The heat pump compressor must be able to withstand compression ratios of 8:1, where the cooling-only compressor might reach only 3:1. Piping is also different. In a cooling-only unit, the vapor line carries only low-pressure suction gas from the evaporator. In a heat pump, this same line must also be able to carry hot, high-pressure gas from the compressor discharge.

### 4.1.0 Reversing Valves

There are several types of reversing valves, but they all operate in basically the same way. One particular type of valve has a main valve body containing a slide and a piston, plus a three-way pilot valve that is actuated by an electrically energized solenoid.

The reversing valve has four piping connections (*Figure 8*). Flow of refrigerant through two of these connections never changes. These two connections are the hot gas discharge from the compressor and the suction line back to the compressor. The remaining two ports connect to the indoor and outdoor coils.

The refrigerant gas is directed to either coil depending on the position of the piston in the valve. See *Figure 9*. When the piston shifts, the path of the discharge flow is changed from one coil to the other. When this occurs, the suction gas is also reversed in the coils.

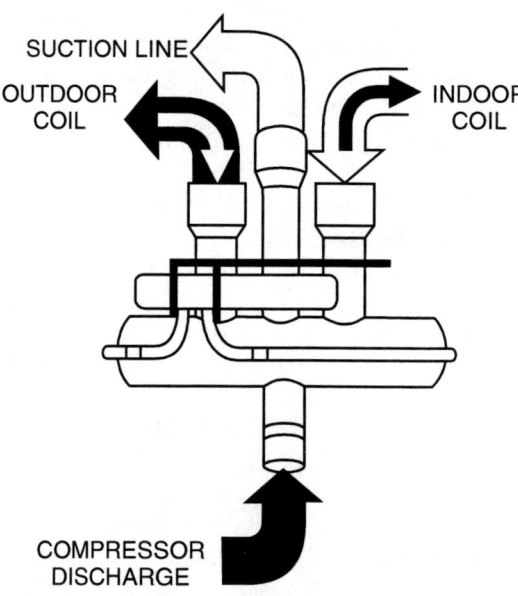

(A) TYPICAL REVERSING VALVE

(B) REVERSING VALVE IN ACTION

*Figure 8* ♦ Reversing (four-way) valve.

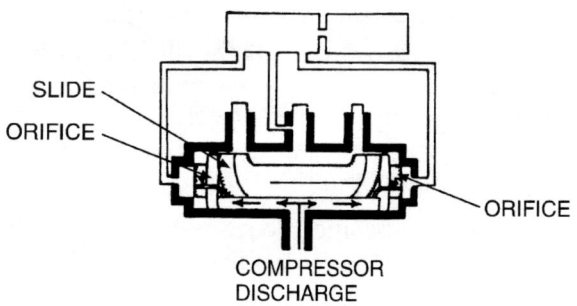

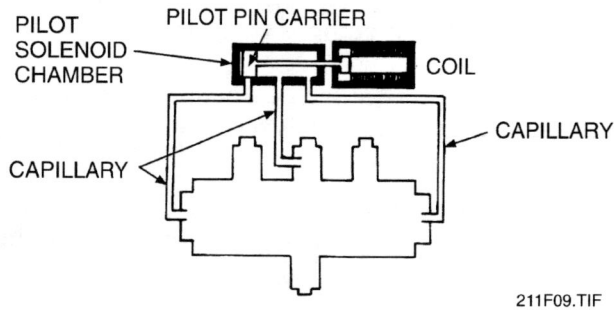

*Figure 9* ♦ Refrigerant flow in a reversing valve.

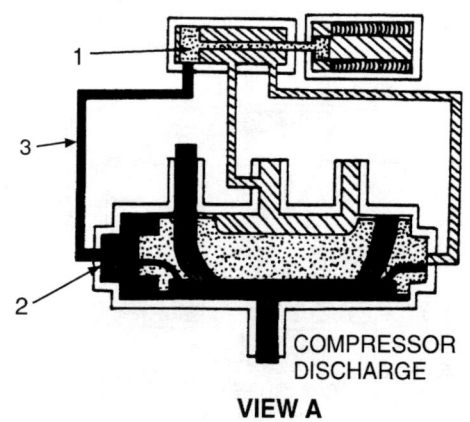

VIEW A

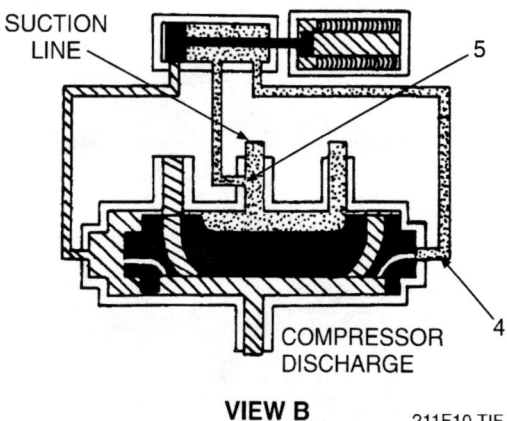

VIEW B

*Figure 10* ♦ Reversing valve position – cooling cycle.

The valve shifts due to the pressure difference within the valve body. The pressure difference is controlled by a 24V solenoid-actuated pilot valve. The pilot valve voltage may also be 120V or 240V. During the cooling cycle, the valve operates as shown in *Figure 10, View A*:

- The solenoid is de-energized and the pilot solenoid pin carrier is at the far left of the pilot solenoid chamber (point 1).
- The compressor discharge gas bleeds behind the left side of the slide (point 2).
- At point 3, the discharge gas is trapped in the capillary tube. The pressure builds until it reaches the discharge pressure.

- At point 4 (*Figure 10, View B*), the pressure behind the right side of the slide equals the compressor suction pressure, which then passes through the pilot solenoid chamber, down through the center capillary to the suction line port at point 5.

The pressure behind the right side of the piston is less than the pressure behind the left side of the piston; therefore, the reversing valve piston is moved to the right. With the piston in this position, the flow of refrigerant is directed to the outdoor coil and the heat pump is operating in the cooling cycle (*Figure 11*).

In order to place the heat pump in the heating mode, the solenoid valve is energized and the action is as follows:

- Upon energizing the solenoid valve, the pilot pin carrier moves to the right to point 1 (*Figure 12*).
- Compressor suction pressure is now able to pass through the center capillary and the pilot chamber down to the left side of the piston at point 2.
- At point 3, the compressor discharge pressure bleeds through the orifice behind the right side of the piston, where it is trapped.

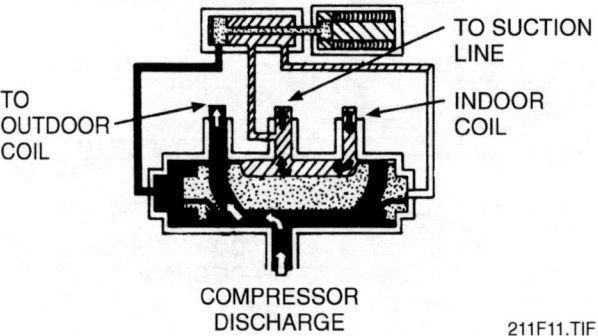

*Figure 11* ◆ Refrigerant flow – cooling cycle.

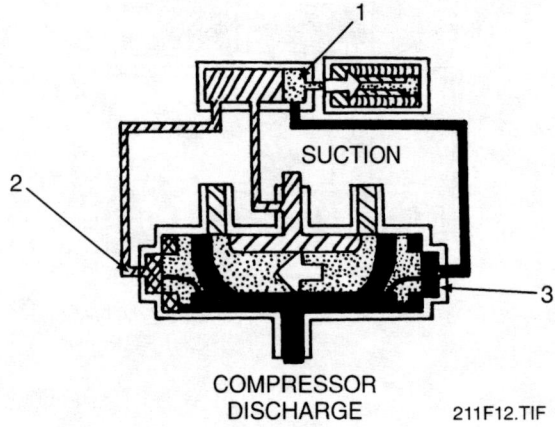

*Figure 12* ◆ Reversing valve position – heating cycle.

- In this position, the pressure behind the left side of the piston is less than the pressure behind the right side and the piston will move to the left.
- With the piston in this position (*Figure 13*), the hot gas from the compressor is now directed to the indoor coil and the unit is operating in the heating cycle.

## 4.2.0 Metering Devices

There are a number of methods of metering refrigerant in a heat pump. Most of them involve the use of two metering devices.

In one arrangement, a fixed-orifice metering device is installed at each coil. A check valve in parallel with each device either blocks or allows refrigerant flow, depending on the flow direction. The metering device piston and the check valve oppose each other. In the cooling mode, for example, the metering device at the output of the indoor coil passes the refrigerant flowing to the indoor coil. The check valve is positioned to prevent flow in this direction. The piston in the other metering device faces in the opposite direction, and is therefore forced into the orifice to block flow. The check valve in this location is positioned to pass refrigerant flowing toward the indoor coil, so the refrigerant flows through the check valve and is not metered. In the heating mode, the refrigerant flows in the other direction and the opposite occurs; that is, refrigerant is metered by the metering device at the input to the outdoor coil.

A similar arrangement can be made using thermostatic expansion valves (abbreviated as *TEVs* and *TXVs*) and check valves, as shown in *Figure 14*. TXVs with built-in check valves are also available.

## 4.3.0 Liquid Accumulators

A liquid accumulator is a large cylinder or container in series in the refrigerant suction line and is often necessary on many heat pumps. It is

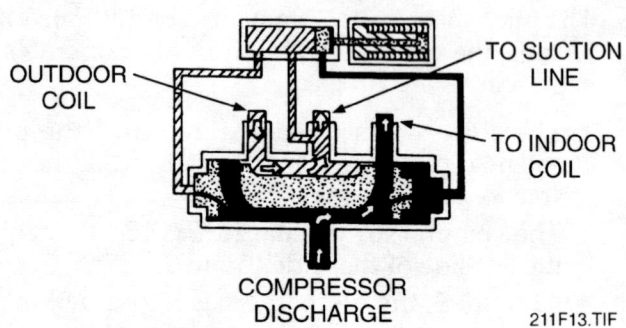

*Figure 13* ◆ Refrigerant flow – heating cycle.

### Check Valves and Fixed-Orifice Metering Devices

Today, many fixed-orifice metering devices combine the check valve and expansion device valve functions in the same assembly. Special heat pump TXVs are also available with a built-in check valve and bypass circuit.

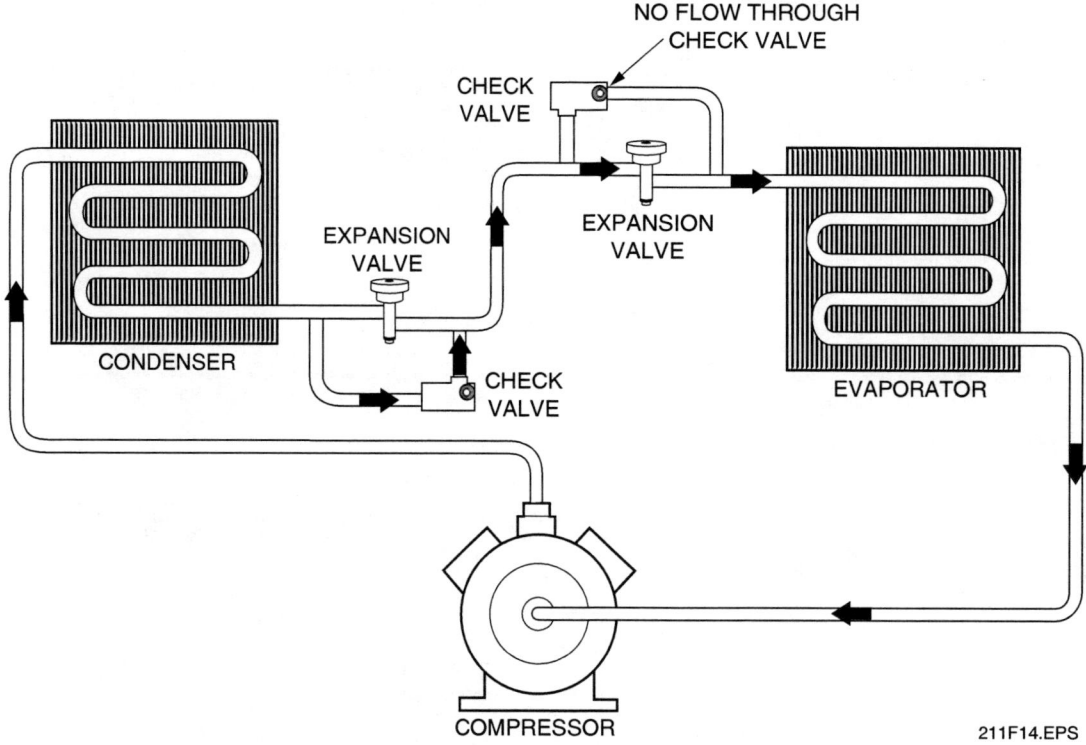

*Figure 14* ◆ Use of thermostatic expansion valves in a heat pump.

usually placed close to the compressor to help eliminate compressor damage due to liquid slugging. It collects liquid refrigerant and prevents it from entering the compressor.

Another function of the accumulator is to adjust the charge between the two cycles. During the heating mode, more refrigerant is available than is required for operation, so it has to be stored. Without the accumulator, this liquid could spill into the compressor crankcase and cause mechanical failure. A metering device built into the accumulator feeds the refrigerant oil to the compressor at a controlled rate.

### 4.4.0 Thermostats

A two-stage heat, one-stage cool, manual changeover thermostat (*Figure 15*) is commonly used to control residential heat pumps equipped with supplemental heat. These thermostats are identified by indicator lights that notify the owner of the operation of the auxiliary and emergency heat.

### 4.5.0 Pressure Controls

Both high-pressure and low-pressure safety controls are frequently installed on heat pumps. These two pressure controls are in addition to the defrost termination pressure switch. Physically, the high-pressure control is located in the compressor discharge line between the compressor and the reversing valve.

The high-pressure control protects the system against an excessive pressure buildup in both the heating and cooling cycle. Electrically, this control is connected in series with the coil of the compressor contactor (*Figure 16*).

The low-pressure control is located in the liquid line and has the primary function of protecting the system against a loss of charge. Therefore, it is often referred to as the *loss-of-charge switch*. The low-pressure control is wired in series with the compressor contactor coil and will open on a drop in pressure. Typically, the control contacts are adjusted to open at 10 psig and close at 30 psig.

### Liquid Accumulators

The liquid accumulator converts liquid refrigerant into a vapor and returns it to the compressor.

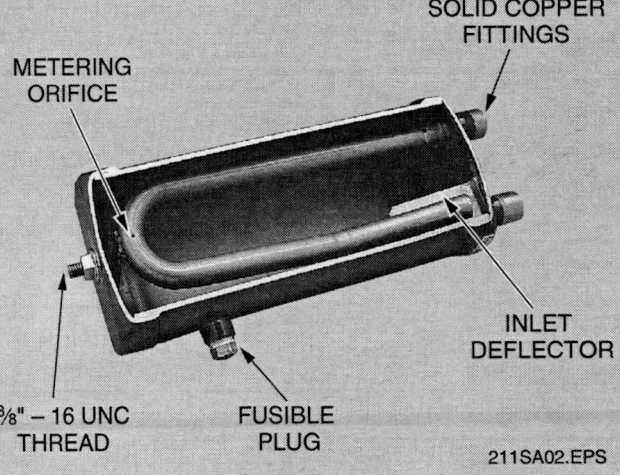

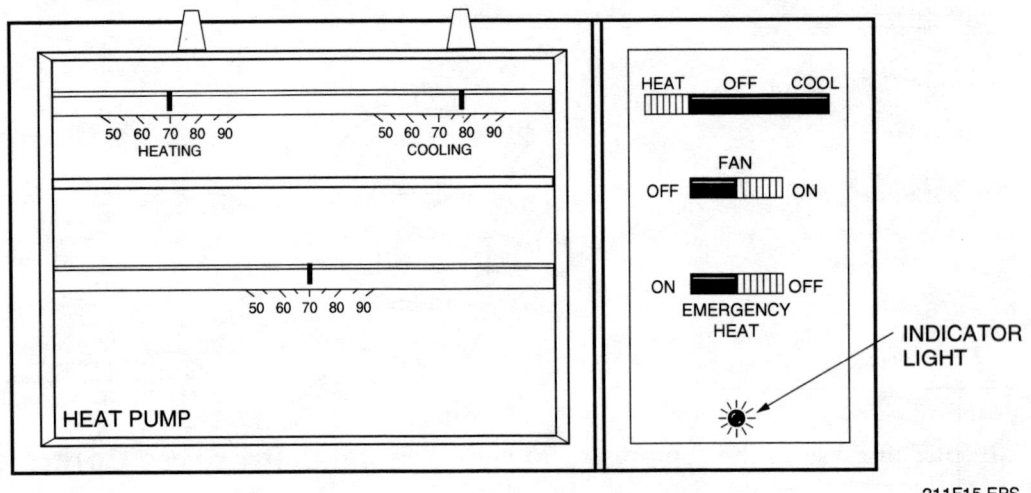

*Figure 15* ◆ Heat pump thermostat.

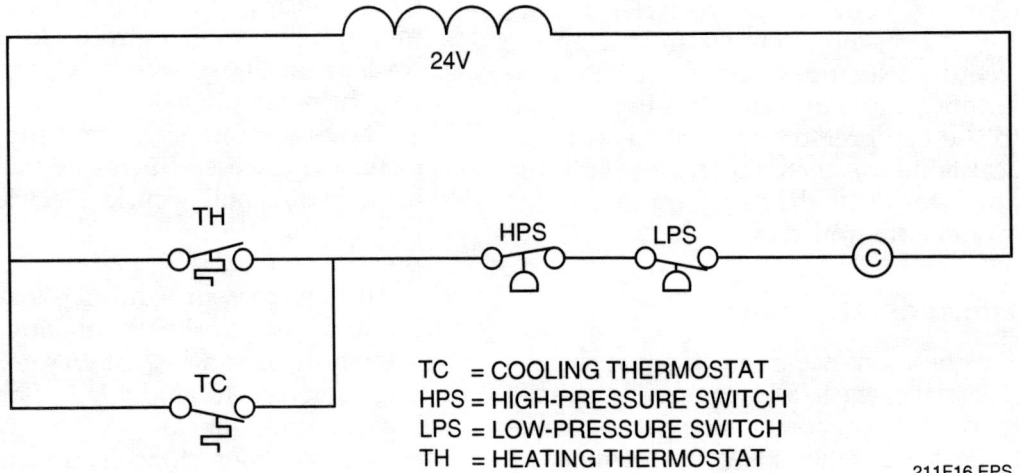

TC = COOLING THERMOSTAT
HPS = HIGH-PRESSURE SWITCH
LPS = LOW-PRESSURE SWITCH
TH = HEATING THERMOSTAT

*Figure 16* ◆ Pressure switches used to protect the compressor.

## 4.6.0 Crankcase Heaters

Heat pump compressors are often equipped with crankcase heaters. The crankcase heater prevents compressor damage as a result of liquid slugging that can occur on cold start-up. It is usually wired into the circuit through a crankcase heater relay so it is on whenever the compressor is off and the main disconnect is on. A thermostat may also be used with the crankcase heater relay (*Figure 17*) to turn on the heater when the outdoor temperature falls below a preset level. As shown, energizing voltage is available to the heater whenever the compressor contactor is de-energized.

If the main switch is disconnected for a long period of time, no attempt should be made to start the unit for 24 hours after the switch is reconnected. This allows all of the liquid refrigerant to be driven out of the compressor by the crankcase heater. A warning label pertaining to supplying power to the crankcase heater should be provided with the installation instructions for the outdoor unit. Attach this label to the disconnect switch where the owner or occupant can see it.

## 5.0.0 ◆ HEAT PUMP PERFORMANCE

The electric heat pump, when operating as a cooling unit, removes more heat from the conditioned space than the electrical energy required to run it. Likewise, when operating as a heating unit, it provides more heat energy to the conditioned space than the electrical energy it consumes. Heat pump economy can be measured using one of several rating methods.

## 5.1.0 Coefficient of Performance

The ratio of useful heat delivered to the equivalent heat consumed in operating the entire system is known as the **Coefficient of Performance (COP)**. This ratio is measured in Btus per hour (Btuh) or kilowatts and is determined under specific operating conditions. Occasionally, the COP may be specified with respect to only the compressor or to any given component or portion of the system.

Heat pump efficiency for heating is defined as follows:

$$COP = \frac{\text{Btuh output}}{\text{kW input} \times 3{,}413 \text{ Btu/kW}}$$

or

$$COP = \frac{\text{Btuh output}}{\text{watts} \times 3.413}$$

When electric heat is used, it generates 3,413 Btus of heat for each kilowatt of power consumed. Therefore, the COP of straight electric resistance heat is 1 and can never be any higher.

To calculate the COP of a three-ton heat pump delivering 36,000 Btuh, and with a total input including fan motor wattage of 4.8kW, proceed as follows:

$$\text{Cooling COP} = \frac{\text{Btuh output}}{\text{kW input} \times 3{,}413}$$

$$\text{Cooling COP} = \frac{36{,}000}{4.8 \times 3.413} = 2.197$$

Cooling COP = 2.2 (rounded off)

In order to calculate the COP of the heat pump in the heating mode for the same size unit, assume that the heating capacity of the same unit is 33,480 Btuh with a total input, including fan motor wattage, of 3.768kW.

$$\text{Heating COP} = \frac{\text{Btuh output}}{\text{kW input} \times 3{,}413}$$

$$\text{Heating COP} = \frac{33{,}489}{3.768 \times 3.413}$$

Heating COP = 2.6

**Study Examples**

1. Calculate the heating COP of a heat pump with a heating capacity of 55,800 Btuh, and with a total input of 6.28kW.

$$\text{Heating COP} = \frac{\text{Btuh output}}{\text{kW input} \times 3{,}413}$$

$$\text{Heating COP} = \frac{55{,}800}{6.28 \times 3.413}$$

Heating COP = 2.6

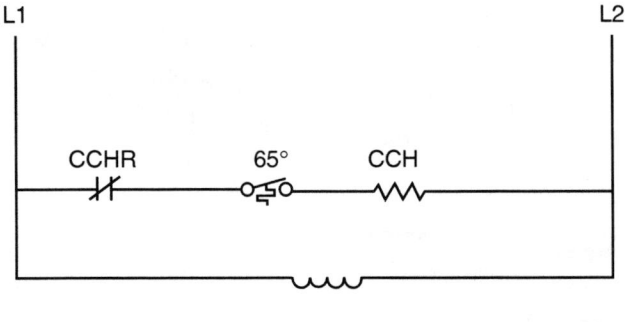

*Figure 17* ◆ Crankcase heater control.

### Computing COP
What is the heating COP of a heat pump with a heating capacity of 100,000 Btuh and a total input of 8kW?

---

2. Calculate the cooling COP of a heat pump with a four-ton cooling capacity consuming 4.03kW.

$$\text{Cooling COP} = \frac{\text{Btuh output}}{\text{kW input} \times 3{,}413}$$

$$\text{Cooling COP} = \frac{4 \times 12{,}000 \text{ Btuh per ton}}{4.03 \times 3.413} = 3.489$$

Cooling COP = 3.5 (rounded off)

When COP figures are used to compare different heat pump systems or installations, care must be taken. The COPs should be based on the same operating conditions, the same complete systems, or the same components. The COP varies with operating conditions; therefore, it cannot be used to compare systems operating over a period of time. The advantage of COP is that it uses like units. The kilowatt input is multiplied by 3,413 to convert it to Btus.

### 5.2.0 Performance Factor

The Performance Factor (PF) is used to calculate the ratio of the total heating or cooling energy delivered during a given period (usually one season). When calculated for one season, it is called the **Heating Season Performance Factor (HSPF)**. The PF is more often associated with the heating mode of the heat pump system rather than with the cooling mode.

The HSPF is found by dividing the total heating output for a season by the total energy consumed during that season. For example, to calculate the HSPF of a heat pump system that produces 102,000 Btus of heating energy while consuming 14,560 kW/hour (which includes power used to supply resistance heaters and drive all electrical components), we use the formula:

$$\text{HSPF} = \frac{\text{total seasoned heat output (Btu)}}{\text{total power input (watt-hours)}}$$

$$\text{HSPF} = \frac{\text{total Btuh output}}{\text{total kW/hour}}$$

$$\text{HSPF} = \frac{102{,}000}{14{,}560}$$

HSPF = 7.0

Calculated performance factors are affected by several variables, including: climactic conditions; sizing and application of equipment; building characteristics and their internal loads; and types of heat pump systems. The advantage of the HSPF is that it can be used to compare systems operating over a period of time. Unlike COP, however, HSPF uses mixed units (Btu output divided by kilowatt input).

Residential heat pump installations in the southern regions of the U.S. operate with higher HSPFs than similar installations in the northern regions because the average outdoor air heat source temperature is higher. Commercial and industrial installations usually operate with higher HSPFs than residential installations due to higher motor and compressor efficiency. Care must be exercised to make sure the same proportion of components are included when HSPFs are used to compare heat pump systems.

### 5.3.0 Seasonal Energy Efficiency Ratio

The Air Conditioning and Refrigeration Institute (ARI) established the **Seasonal Energy Efficiency Ratio (SEER)** as a standard by which all heat pumps are rated. Currently, a SEER of 10 is the minimum allowed; some heat pumps have SEER ratings as high as 16.

---

### Computing HSPFs
What is the HSPF of a heat pump system that produces a seasonal heat output of 200,000 Btus while consuming a total of 35,000kW/hour?

## *Heat Pump Ratings*

SEER ratings are generally used to express the cooling efficiency of air conditioning/heating systems, while COP and HSPF ratings are used to express the heating efficiency of these systems. Equipment that is certified by the ARI has an ARI label like the one shown here. For such equipment, the certified capacities, SEER, COP, and HSPF ratings, as well as other performance data, can be found by looking up the equipment by manufacturer and model number in the appropriate *ARI Performance Rating Directory*. These directories are available in printed copy and CD-ROM versions. This information is also available on the ARI web site (http://www.ari.org).

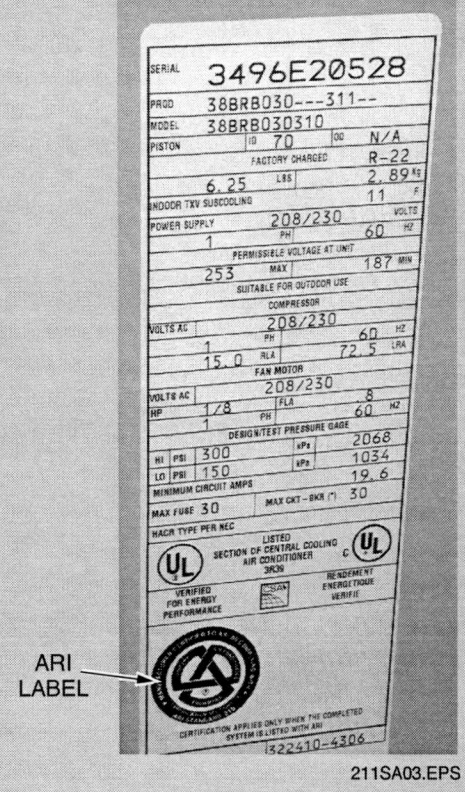

ARI LABEL

### 5.4.0 Heat Pump Efficiency Recommendations

The United States Department of Energy (DOE) Energy Star® program rates various appliances and residential heating and cooling products according to their efficiency. Citizens are encouraged to purchase Energy Star® products to help conserve energy. Many local utilities have rebate programs to encourage their customers to buy Energy Star® products. According to the DOE, the most efficient heat pump available today has a HSPF of 9.4 and a SEER of 17.65. The minimum level of efficiency required to qualify for an Energy Star® rating is a HSPF of 7.6 and a SEER of 12. The DOE publishes lists of products that qualify for Energy Star® ratings. This list is available on the Internet. Note that the higher the value for SEER, the higher the HSPF.

### 6.0.0 ♦ BALANCE POINT AND SUPPLEMENTARY HEAT

The **balance point** of a system is the outdoor temperature at which the heating capacity of the heat pump is equal to the heat loss of the building. The balance point will usually vary depending on the climate, building design, type of construction, and other factors which affect heat loss or gain. Generally speaking, the lower the balance point, the more economical a heat pump system will be. The best method of lowering the balance point is to reduce the heat loss of the structure.

*Figure 18* illustrates the effects of outdoor temperature on the operation of an air-to-air heat pump. Line A indicates that the heating capacity of the unit decreases with decreasing outdoor temperatures. At the same time, the heat loss and the heating load necessary to displace the heat

### Outdoor Thermostats

In some parts of the country, the electric utility mandates the use of outdoor thermostats to control all stages of supplemental electric heaters.

---

loss increases. This is shown by line B. The balance point of the system is the temperature at which the heat loss is equal to the heating capacity of the heat pump. The balance point, therefore, is the intersection of lines A and B; as shown in the graph, it is about 23°F for this case.

Supplementary heaters are generally used to maintain the capacity of the unit when temperatures fall below the balance point. These heaters are usually electric resistance heaters. If heat staging is used (individual banks of electric heaters are staged on as temperature decreases), some of the heaters will be controlled by outdoor thermostats.

### 7.0.0 ◆ INSTALLATION

This section covers the installation of various systems.

### 7.1.0 Split Systems

A split system (*Figure 19*) is one that has two cabinets. The outdoor unit contains the compressor, reversing valve, outdoor coil and fan, and controls. In a heat pump system, this unit would also contain one of the metering devices. The indoor unit, which is also known as the air handler or **fan coil unit**, contains the indoor fan (blower), metering device, and the indoor coil. Electric heaters, if installed, would also be located in this unit.

A variation on the split system is the triple-split system, in which the outdoor unit contains only the outdoor coil, outdoor fan, and a metering device. The compressor, reversing valve, and controls are located in a separate unit designed to be installed indoors. This arrangement is very efficient and is much easier to service in bad weather.

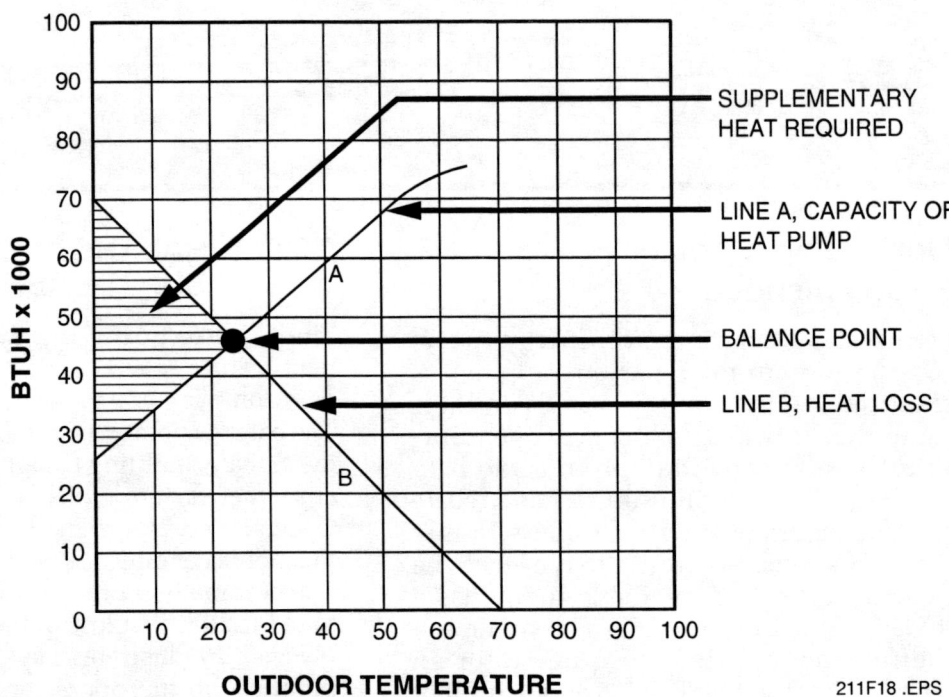

*Figure 18* ◆ Heat pump balance point.

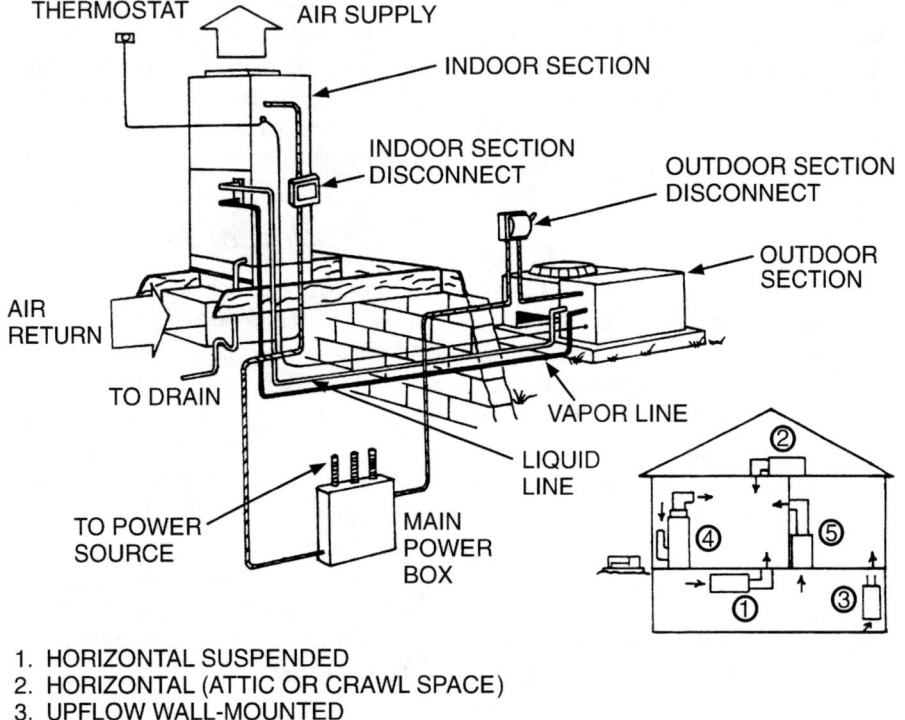

*Figure 19* ◆ Split system installation.

Another type of add-on system is the ductless split system (*Figure 20*), which is designed to serve a single room or area. The indoor coil in these applications is mounted on a wall or in a drop ceiling. These systems are especially good for add-on applications because they require no ductwork. A family room addition in a home is a good example. They may also be a good choice for historical buildings where the installation of ductwork would destroy the character of the building.

There are versions of these ductless systems in which a single outdoor unit serves two or three indoor units. These are known as *multiplexed* systems. One type of indoor unit has connections for two or three lengths of flexible duct that can distribute air to diffusers or rooms within 10' or 15' of the unit.

Several important factors must be considered before selecting the location of the indoor and outdoor units of a split-system heat pump.

### 7.1.1 Location

The location of the outdoor unit of the air-to-air heat pump should consider these factors: sound, wind direction, proximity to the structure, and the treatment of defrost and drainage. The recommended procedure is to select a site near the resi-

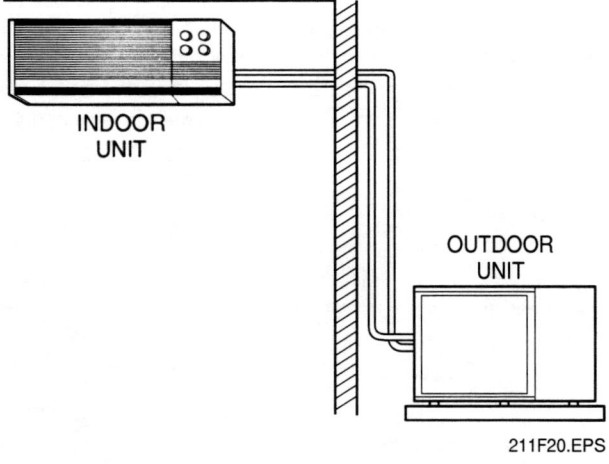

*Figure 20* ◆ Ductless split system.

dence, but away from bedroom windows or rooms where sound might be objectionable. The unit should not be located where discharge air will be directed toward windows of neighboring residences. Local ordinances often impose restrictions on the placement of outdoor units. It is very important, therefore, to consult local codes before starting the installation.

If the outdoor unit is to be placed in front of the building, it should be concealed by shrubbery so that the sound level will blend with traffic sounds. Ample clearance should be allowed around the unit for air movement and service access. Clearance dimensions are usually specified in the manufacturer's installation instructions. Local codes may also apply.

The outdoor unit should be located near enough to the building to eliminate lengthy piping runs, but it should not be placed directly below eaves or gutters. Prevailing winds can cause capacity losses due to the wind chill factor. This can be dampened by placing the unit so that wind does not blow directly across the outdoor coil. Trees, shrubs, fences, or even corners of buildings will reduce wind chill factor capacity losses.

### 7.1.2 Mounting

Once the location of the outdoor unit has been selected, the unit should be mounted in accordance with recommendations for existing climate conditions. Where there is little or no snowfall, a concrete pad is recommended. The thickness of the pad should provide about 6" between the base of the unit and the gravel bed in the ground. A recession formed in the slab under the coil will allow drainage of condensate and melting frost.

Where snowfall is heavy, it is recommended that the unit be mounted on an angle-iron frame with the supports imbedded in a concrete base. The unit should be 12" to 18" above the ground.

The unit should be level and rigidly mounted. Proper clearance for service access and airflow must be allowed. A 12" gravel bed extending out and away from the perimeter of the outdoor section should be provided to prevent mud splashing. Care must be taken to make sure that condensate does not drain directly onto areas that could become frozen and slippery and result in personal injury to pedestrian traffic.

### 7.1.3 Indoor Section

The indoor section of a split system may be placed in the basement, utility room, attic, closet, or attached garage. The indoor location should be selected to minimize the length of piping runs between the indoor and outdoor units while allowing for efficient ductwork design. Wire and conduit lengths from the main service panel to the electric heaters must also be considered. The supply and return air distribution duct runs should have canvas flex connections on either side of the plenums. Internal acoustical duct lining and vibration isolators or sound-dampening pads should also be incorporated into the unit installation procedure.

### 7.2.0 Packaged Units

A packaged unit (*Figure 21*) contains all the components in a single package, which is located outside the building. The supply and return ducts of a packaged unit penetrate the building. This is significantly different from a split system, in which only the refrigerant piping penetrates the wall or roof. For that reason, packaged units do not usually make good add-on systems. Packaged units are built in a variety of configurations, serving cooling loads ranging from a couple of tons to 75 tons or more. The larger units will have multiple compres-

---

### *Discharge Line Mufflers*

Use extreme caution when piping a split-system heat pump. The insulated vapor line becomes the compressor discharge line during the heating mode. Pressure pulsations from the compressor discharge are carried into the indoor coil, which can act like a speaker. This will transmit the pulsation noises into the ducts and throughout the structure. For this reason, many heat pump compressors are equipped with an external discharge line muffler like the one shown here. Many compressor manufacturers install an internal muffler in their compressors.

sors. Packaged units are available with gas heat for cold climates. Heat pump versions are popular in more moderate climates.

Packaged heat pump systems require the same installation considerations as the outdoor units of split systems. There are also several additional considerations. One pertains to the added weight and larger dimensions of the packaged units; since the packaged unit contains the indoor coil and blower in addition to the outdoor coil, blower and compressor, it will require sturdier mounting. This is especially true in the snow belt areas where 12" to 18" of ground clearance is recommended.

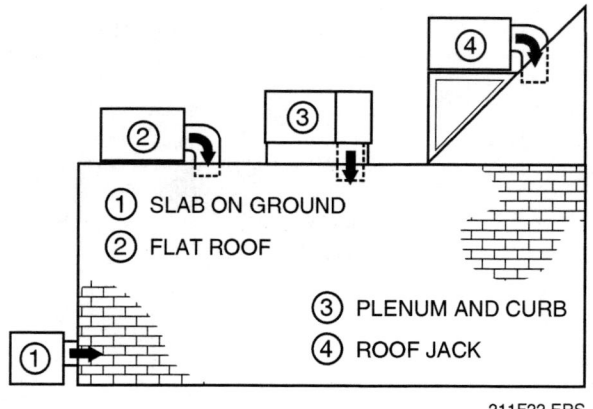

*Figure 22* ♦ Packaged unit locations.

### 7.3.0 Add-On Systems

Due to the cost of oil and gas, the add-on air-to-air heat pump (*Figure 23*) has some advantages for the homeowner. When it is more efficient or economical to heat with the heat pump (usually between 35°F and 65°F) the system automatically selects that mode. When it is more efficient or economical to heat with gas or oil, the system automatically selects that mode. Overall heating bills will be lower using this method than when heating with a single fuel. The add-on system shown in *Figure 23* features a cooling coil added to the top of the furnace and connected to a heat pump outdoor unit by refrigerant piping. This is a relatively inexpensive way to add central cooling to a building served by a furnace. The add-on heat pump uses the existing furnace as a backup heat source.

When the temperature falls below the heat pump's balance point, a **dual-fuel heating system** such as this is much more cost effective and energy efficient than using electric heaters. An outdoor thermostat is used to control a switching circuit that turns off the heat pump and turns on the furnace when the temperature falls below a selected setpoint such as 35°F.

When a heat pump is added to an existing heating system, the furnace blower must have sufficient static pressure to deliver the manufacturer's recommended cfm per ton across the evaporator coil. If the blower cannot produce this capacity, either the motor and blower assembly should be replaced with one of sufficient capacity, or the add-on unit should not be installed. Two-speed blowers are common in systems of this type.

A heat pump system generally requires 450 to 500 cfm per ton. The size of the ductwork is critical in the heating mode. The capacity of the ductwork must meet or exceed the required 450 to 500 cfm per ton.

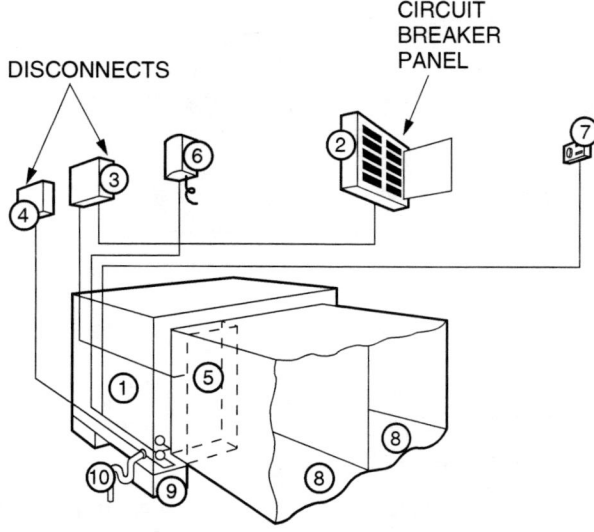

COMPONENTS OF THE PACKAGED HEAT PUMP SYSTEM

1. PACKAGED HEAT PUMP
2. MAIN SERVICE PANEL
3. OUTSIDE ELECTRIC HEAT DISCONNECT
4. HEAT PUMP OUTSIDE DISCONNECT
5. SUPPLEMENTAL HEATERS
6. OUTDOOR THERMOSTAT
7. INDOOR THERMOSTAT
8. SUPPLY AND RETURN DUCTWORK
9. MOUNTING PAD OR SUPPORTS
10. CONDENSATE PIPING

*Figure 21* ♦ Packaged unit installation.

*Figure 22* shows typical locations for packaged units; roof mounting is frequently preferred in the south and southwest. Supply and return ductwork must be insulated if it runs through an unconditioned attic or other spaces. Adequate weatherstripping and sealer must be applied where plenum or ductwork connections enter the roof or walls of the building.

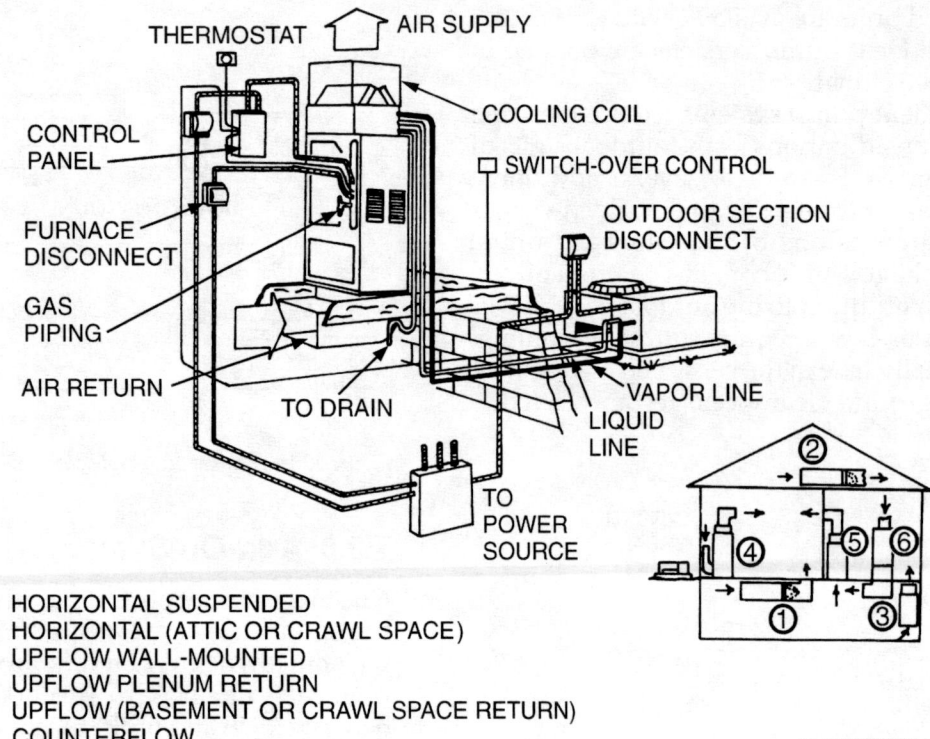

1. HORIZONTAL SUSPENDED
2. HORIZONTAL (ATTIC OR CRAWL SPACE)
3. UPFLOW WALL-MOUNTED
4. UPFLOW PLENUM RETURN
5. UPFLOW (BASEMENT OR CRAWL SPACE RETURN)
6. COUNTERFLOW

*Figure 23* ◆ Add-on heat pump installation.

**CAUTION**

The heat pump must be turned off before activating gas or oil heat. The hot air from the furnace will cause excessive high-side pressure and temperature that will cause the compressor to overheat and overload.

### 7.4.0 Installation Checklist

Before placing a heat pump in operation, check the following items, if applicable:

- Record make, model, and serial number of unit
- Check or change filter
- Check fan motor (indoor/outdoor)
- Check supply ducts for leaks
- Check return air for leaks
- Check thermostat/cycle unit
- Check all safety controls
- Clean coils
- Check crankcase heater
- Check defrost cycle operation
- Check auxiliary heat
- Check emergency heat
- Check that unit is correctly charged
- Check electrical connections and tighten, if necessary

The following data should be collected when a heat pump is placed in operation:

- Voltage and current balance on a three-phase system
- Compressor motor current draw is within manufacturer's specification
- Voltage at indoor heater at full load operation (electrical resistance heat operational)
- Measured amperage draw of each electrical resistance heat unit
- Voltage at outdoor contactor when compressor is operating
- Condensing unit pressures (psig) on both the high side and the low side
- Superheat at compressor after accumulator
- Temperature rise with heat pump only
- Outdoor temperature
- Outdoor weather conditions (cloud cover, precipitation)
- Record make, model, and serial number of indoor and outdoor units.
- Any general comments that may apply
- The date the data is taken
- The location of the site
- Record values for all the parameters just listed, and leave a copy at the site for future reference

## 8.0.0 ◆ SERVICE

A service call is recommended prior to each heating or cooling season. The check for cooling should be made when the outdoor temperature is above 70°F, and the heating check should be made when the outdoor temperature is below 55°F. A check sheet should be developed prior to servicing the system for recording information and also as a record of operational functions. When servicing a heat pump, follow these steps:

*Step 1* If a bimetal thermostat is used, check it to be sure it is level.

*Step 2* Check the thermostat for faulty wiring and loose connections.

*Step 3* Check the supply voltage and verify that it is within the allowed tolerance.

*Step 4* On a three-phase system, check the voltage and current balance.

*Step 5* Check the compressor motor current draw and verify that it is within the manufacturer's specification.

*Step 6* Change or clean the filter(s).

*Step 7* Clean the blower and the blower compartment.

*Step 8* Turn off the disconnect and check all wiring for damage and loose connections.

*Step 9* If applicable, check the blower motor drive belt for damage and proper alignment.

*Step 10* Clean the outdoor coil as necessary.

*Step 11* Apply power, then turn the thermostat down to turn on cooling. Verify that the unit runs properly in the cooling mode.

*Step 12* Set the thermostat to activate heating and verify that the heating function is working properly.

*Step 13* Use the built-in servicing feature to activate defrost and verify that the defrost function is working.

*Step 14* Check the charge and adjust as required.

*Step 15* Check the auxiliary heater(s)

## 9.0.0 ◆ HEAT PUMP CONTROLS

The control circuits of a heat pump are more complicated than those of a cooling-only unit. There are several reasons for this. Refer to *Figure 24* and locate the components or circuits as they are discussed. The numbers in parentheses in the text relate to the circled numbers on the diagram. Try to answer the questions as you progress through the discussion. They are designed to help you understand the operation of a heat pump.

The heat pump requires a defrost control circuit that changes the position of the reversing valve to switch the unit into the defrost mode. In this unit, an electronic defrost control board (1) is used. This board receives the coil temperature and outdoor air temperature from thermistors located on or near the outdoor coil.

*Question* – Is the reversing valve energized or de-energized in the defrost mode?

*Answer* – Energized. Normally open contacts of the defrost relay control the reversing valve. When the defrost board energizes the defrost relay, its contacts (2) complete the circuit to the reversing valve.

Supplementary electric heaters are usually required on a heat pump. These heaters (3) perform two functions: they augment reverse cycle heat when the temperature falls below the balance point, and they provide heat while the unit is in the defrost mode.

*Question* – How is the electric heater activated during the defrost mode?

*Answer* – A set of normally open defrost relay contacts energizes heater relay HR1 (4) when the unit goes into the defrost mode. Normally open contacts of HR1 (5) close to energize heating element #1.

A multi-stage heat-cool thermostat is needed. It should be an automatic changeover thermostat. The first stage of the heating side of the thermostat (TH1) controls reverse cycle heat. It energizes the compressor contactor when closed. The second stage, which is the first stage of electric resistance heat, is controlled by TH2. The third stage (second stage of electric heat) is controlled by an outdoor thermostat (6).

### Heat Pump Setback

For periods when buildings are unoccupied or at night, many people set back their thermostats to conserve energy. Significant savings can be realized by setting the thermostat back 5°F from the normal (occupied/daytime) position. However, heat pumps should not be set back below 60°F because it may cause them to trip on a low-pressure limit switch.

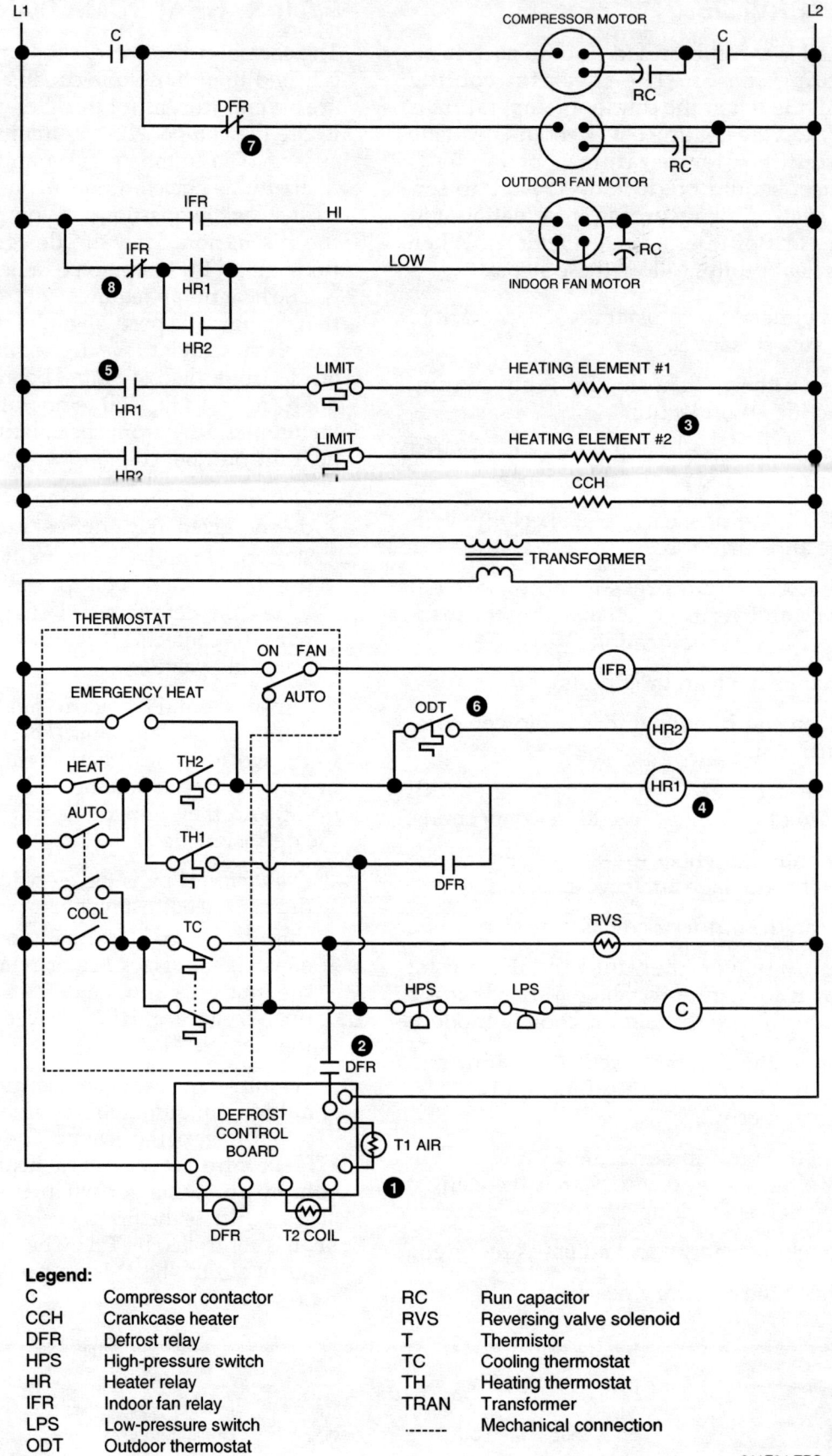

Figure 24 ♦ Heat pump ladder diagram.

*Question* – Does this circuit have any means of preventing heating element #2 from cycling on at the same time heating element #1 comes on?
*Answer* – Yes. Both heating elements are controlled by TH2. HR2 cannot be energized until TH2 and ODT are closed. Therefore, heating element #2 will never be on unless heating element #1 is also on.

Try to answer these additional questions. They will help you understand how heat pump control circuits are arranged.

*Question* – How could you modify this circuit so that the crankcase heater is on only when the compressor is off?
*Answer* – Add a crankcase heater relay with a set of normally closed contacts in series with the crankcase heater. The relay coil would be wired in parallel with the contactor coil.

*Question* – Is the outdoor fan motor on or off during defrost?
*Answer* – Off. Normally closed contacts of the defrost relay (7) are in series with the outdoor fan motor. When the defrost relay energizes, the motor is disabled. This is common in heat pump controls. If the fan is blowing cold air across the coil, it retards the defrost.

*Question* – What feature does this circuit have that makes sure the indoor blower runs whenever one of the electric resistance heaters is on?
*Answer* – Normally closed contacts of the indoor fan relay (8) will provide voltage to the low-speed side of the blower whenever either of the heater relays is energized. This feature is important when the thermostat has an emergency heat control that can be used to turn on the electric heaters if the reverse cycle heating circuits fail. Without a fan to disperse the heat created by the heating elements, the unit could overheat.

## Summary

Even on the coldest day, there is some heat in the air. Heat pumps take advantage of that heat by reversing the refrigerant cycle. The condensing coil becomes an evaporator and the evaporator becomes a condenser. Heat pumps provide much better heating efficiency than electric heat in most climates. In cold climates, they are often combined with furnaces in dual-fuel arrangements that use each type of heat to its maximum efficiency. Heat pumps are available in a wide variety of system configurations and capacities. Their greatest advantage is that one system provides both heating and cooling.

## Review Questions

1. In a water-to-air heat pump _____ in the heating mode.
   a. water is the heat source
   b. water is the heat sink
   c. air is the heat source
   d. water is both the heat source and heat sink

2. In a ground-source heat pump _____.
   a. refrigerant flows over a buried coil
   b. water flows over a buried coil
   c. nothing flows over the outdoor coil
   d. water flows from a well back to another well

3. Which of the following occurs during the heating cycle of a heat pump?
   a. Electric heaters are always energized.
   b. Air flows in the same direction as in the cooling mode, and refrigerant flow reverses.
   c. Refrigerant flows in the same direction as in the cooling mode, and airflow reverses.
   d. Refrigerant and air both flow in the opposite direction from cooling.

4. When the unit is operating in the defrost mode, what usually happens in the conditioned space?
   a. The unit continues to provide compression heat.
   b. Occupants experience discomfort.
   c. The furnace takes over.
   d. An electric heater is cycled on to provide heat.

5. A usual frequency and duration for a defrost cycle is every _____.
   a. 10 minutes for 30 seconds
   b. 90 minutes for 10 seconds
   c. 90 minutes for 10 minutes
   d. 15 minutes for 5 minutes

6. Which of the following occurs when a heat pump is in the reverse cycle heating mode?
   a. The compressor discharge is routed to the indoor coil.
   b. The compressor discharge is routed to the outdoor coil.
   c. Hot, high-pressure liquid flows through the compressor.
   d. The outdoor coil is not used.

7. Refrigerant metering in a heat pump _____.
   a. is done the same way as it is in cooling
   b. requires two metering devices
   c. requires two metering devices and two check valves
   d. is done with a single metering device and a check valve

8. Which of the following is *not* determined from the balance point?
   a. The setting of the outdoor thermostat that turns on supplementary heat.
   b. The temperature at which reverse cycle heat will no longer handle the heating load.
   c. The amount of supplementary heat needed.
   d. The SEER rating of the heat pump.

9. The term *split system* refers to _____.
   a. any system that can provide both heating and cooling
   b. a system that has two units, one indoors and the other outdoors
   c. any system that has two or more units in an outdoor unit
   d. a system in which a furnace shares the heating load with a heat pump when the outdoor temperature falls below 35°F

10. A ductless system is one in which _____.
    a. the condenser and evaporator are both indoors
    b. the ductwork is outside the building
    c. conditioned air is discharged directly into the conditioned space by the indoor unit
    d. the ductwork is both indoors and outdoors

Refer to *Figure 24* in the Trainee Module to answer the following questions.

11. Which of the following is *not* true of the outdoor fan motor?
    a. It is controlled by the compressor contactor during heating and cooling.
    b. It is on only when defrost is energized.
    c. It is off when defrost is energized.
    d. It is a single-phase motor.

12. When the unit is in the defrost mode, the _____.
    a. reversing valve solenoid is energized
    b. reversing valve solenoid is de-energized
    c. indoor fan turns off
    d. indoor fan operates at low speed

13. Which of the following is true about heating element #2?
    a. It turns on whenever TH1 is closed.
    b. It turns on whenever the outdoor temperature is below the ODT setpoint.
    c. It will not turn on unless the indoor fan is running.
    d. It will not turn on unless heater element #1 is on.

14. What would happen if the DFR contacts controlling HR1 stuck in the closed position?
    a. Heating element #1 would be on all the time.
    b. Heating element #1 would turn on whenever the thermostat called for heating or cooling.
    c. The unit would overheat.
    d. The unit would be in a continuous defrost mode.

15. The normally closed IFR contacts are designed to _____.
    a. keep the electric heaters from turning on if the indoor fan has failed
    b. turn the indoor fan on when both heaters are energized
    c. turn the fan on when either of the heaters is energized
    d. keep the indoor fan running when the unit is turned off

# GLOSSARY

## Trade Terms Introduced in this Module

*Absolute zero:* The temperature at which all molecular motion ceases. It is –460°F, –273°C, and 0°K (Kelvin).

*Balance point:* The outdoor temperature at which the heating capacity of the heat pump is equal to the heat loss of the building. The balance point varies depending on the climate, building design, type of construction, and other factors that affect heat loss or gain.

*Coefficient of Performance (COP):* The ratio of work performed in relation to energy used. A rating method for heat pumps. It is further defined as the Btuh output divided by the total electrical input (watts) required to produce this Btuh output times 3.413.

*Compression heat:* The heat produced by a heat pump when the refrigerant cycle is reversed. See *reverse cycle heat*.

*Defrost:* In a heat pump, the process of cycling hot refrigerant through the outdoor coil to melt accumulated frost due to condensation.

*Demand defrost:* A defrost cycle that is automatically turned on and off based on actual need versus elapsed time and outdoor temperature.

*Dual-fuel heating system:* A system in which a heat pump is combined with a furnace.

*Fan coil unit:* A term often applied to the indoor unit of a split system. Also known as an *air handler*.

*Four-way valve:* A heat pump reversing valve. Also known as a *switch-over* or *cross-over* valve.

*Ground-source (geothermal) heat pump:* A system in which the outdoor coil is buried in the ground and the heat exchange occurs between the earth and the refrigerant flowing through the coil.

*Heat sink:* A low-temperature surface to which heat can be transferred.

*Heating Season Performance Factor (HSPF):* A heat pump performance rating that has been adjusted for seasonal operation. It is the total heating output of a heat pump (in Btus) during its normal annual usage period for heating divided by the total electric power input in watt-hours during the same period.

*Indoor coil:* The designation given to the heat pump coil used to transfer heat to or from the conditioned space.

*Kelvin scale:* A temperature scale in which zero equals –460°F. See *absolute zero*.

*Outdoor coil:* The heat pump coil used to transfer heat to or from the outdoor air.

*Packaged unit:* A self-contained air conditioning system.

*Reverse cycle heat:* The heat produced by a heat pump when refrigerant flow is reversed. See *compression heat*.

*Reversing valve:* A valve that changes the direction of refrigerant flow in a heat pump. See *four-way valve*.

*Seasonal Energy Efficiency Ratio (SEER):* The ARI standard for measuring heat pump efficiency. It is the total cooling of a heat pump (in Btus) during its normal annual period for cooling divided by the total electric input in watt-hours during the same period.

*Split system:* An air conditioning system with an indoor coil and an outdoor coil connected with refrigerant lines.

# ANSWER KEY

## Answers to Review Questions

| Answer | Section |
|---|---|
| 1. a | 2.2.0 |
| 2. c | 2.5.0 |
| 3. b | 3.2.0 |
| 4. d | 3.3.0 |
| 5. c | 3.3.1 |
| 6. a | 4.1.0 |
| 7. c | 4.2.0 |
| 8. d | 6.0.0 |
| 9. b | 7.1.0 |
| 10. c | 7.1.0 |
| 11. b | 9.0.0 |
| 12. a | 9.0.0 |
| 13. d | 9.0.0 |
| 14. b | 9.0.0 |
| 15. c | 9.0.0 |

# REFERENCES & ACKNOWLEDGMENTS

## Additional Resources

This module is intended to present thorough resources for task training. The following reference works are suggested for further study. These are optional materials for continued education rather than for task training.

*Heat Pumps, Theory and Service*, 1993. Lee Miles. Albany, NY: Delmar Publishers, Inc.

*Inside the Heat Pump*, 1999. Syracuse, NY: Carrier Corporation

*Modern Refrigeration and Air Conditioning*, 2000. A.D. Althouse, C.H. Turnquist, A.F. Bracciano. Tinley Park, IL: The Goodheart-Willcox Company, Inc.

*Refrigeration & Air Conditioning Technology*, 2000. William C. Whitman, William M. Johnson, John A. Tomczyk. Albany, NY: Delmar Publishers, Inc.

## Figure Credits

| | |
|---|---|
| **Carrier Corporation** | 211SA01 |
| **ALCO Controls** | 211F08, 211SA02 |
| **Gerald Shannon** | 211SA03 |
| **Refrigeration Research, Inc.** | 211SA04 |

# NCCER CRAFT TRAINING USER UPDATES

The NCCER makes every effort to keep these textbooks up-to-date and free of technical errors. We appreciate your help in this process. If you have an idea for improving this textbook, or if you find an error, a typographical mistake, or an inaccuracy in the NCCER's Craft Training textbooks, please write us, using this form or a photocopy. Be sure to include the exact module number, page number, a detailed description, and the correction, if applicable. Your input will be brought to the attention of the Technical Review Committee. Thank you for your assistance.

*Instructors* – If you found that additional materials were necessary in order to teach this module effectively, please let us know so that we may include them in the Equipment and Materials list in the Instructor's Guide.

**Write:** Curriculum Revision and Development Department
National Center for Construction Education and Research
P.O. Box 141104, Gainesville, FL 32614-1104

**Fax:** 352-334-0932

**E-mail:** curriculum@nccer.org

Craft _____ Module Name _____

Copyright Date _____ Module Number _____ Page Number(s) _____

Description
_____
_____
_____
_____

(Optional) Correction
_____
_____
_____

(Optional) Your Name and Address
_____
_____
_____

Module 03212-01

*Leak Detection, Evacuation, Recovery, and Charging*

# COURSE MAP

This course map shows all of the modules in the second level of the HVAC curriculum. The suggested training order begins at the bottom and proceeds up. Skill levels increase as you advance on the course map. The local Training Program Sponsor may adjust the training order.

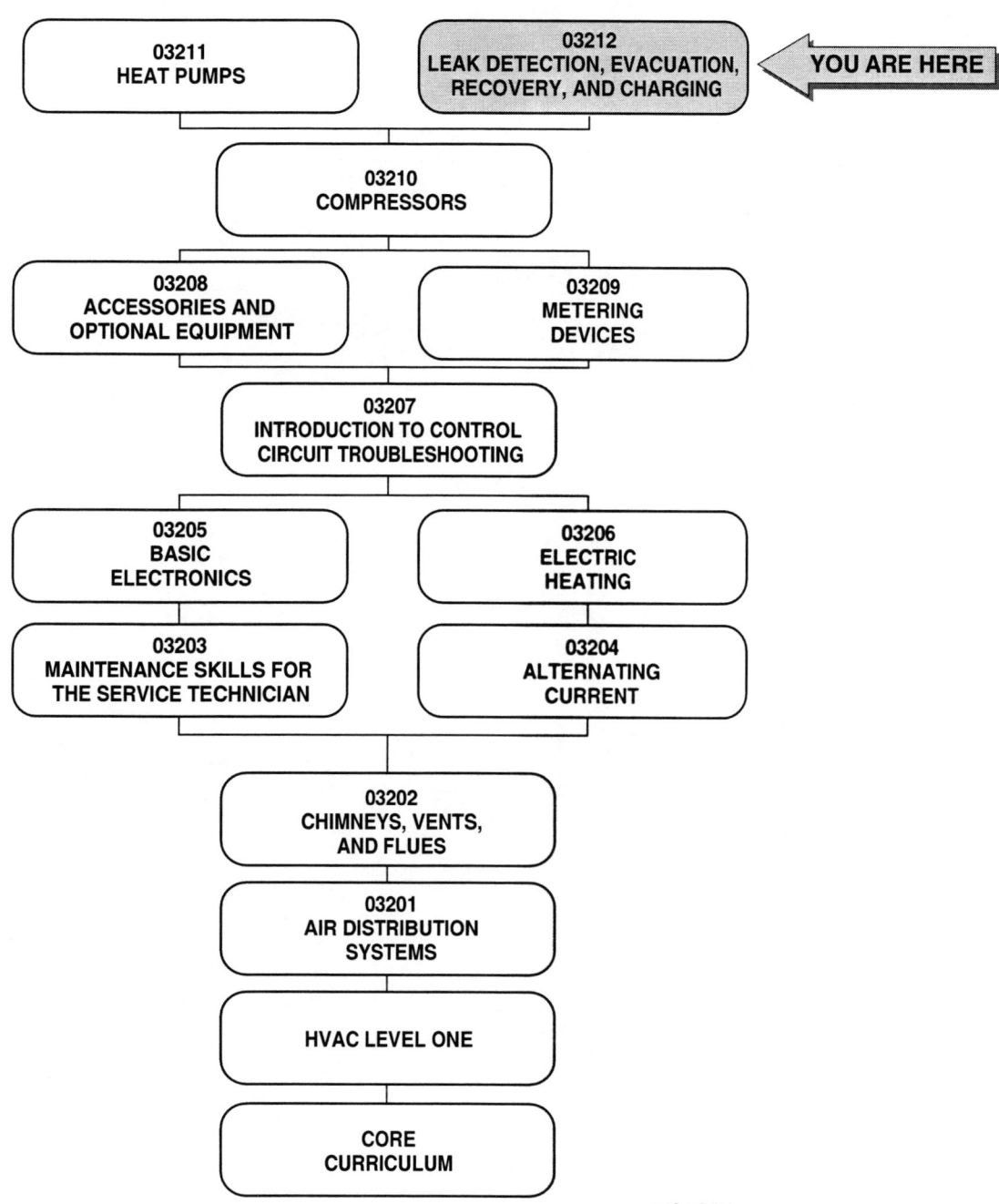

## MODULE 03212 CONTENTS

**1.0.0 INTRODUCTION** ............................................. 12.1
**2.0.0 LEAK DETECTION** ......................................... 12.1
    2.1.0 Detection Devices ...................................... 12.2
    *2.1.1 Halide Torch* ............................................. 12.2
    *2.1.2 Electronic Detectors* ................................... 12.3
    *2.1.3 Ultrasonic Leak Detectors* ......................... 12.3
    *2.1.4 Ultraviolet/Fluorescent Leak Detectors* ...... 12.4
    *2.1.5 Liquid (Bubble) Detectors* .......................... 12.5
    2.2.0 Leak Testing ............................................. 12.5
    *2.2.1 Operational System* ................................... 12.6
    *2.2.2 Systems Without a Refrigerant Charge* ....... 12.6
    *2.2.3 Systems With a Partial Charge* ................... 12.6
    *2.2.4 Disassembling Brazed Joints for Leak Repair* ............ 12.6
**3.0.0 REFRIGERANT CONTAINMENT** ....................... 12.7
    3.1.0 Refrigerant Recovery ................................. 12.7
    *3.1.1 Evacuation Requirements* ........................... 12.7
    3.2.0 Refrigerant Recovery and Recovery/Recycle Units ........ 12.8
    *3.2.1 Recovery Unit* ........................................... 12.8
    *3.2.2 Recycle Unit* ............................................. 12.9
    *3.2.3 Recovering or Recycling Different Refrigerants* ......... 12.9
**4.0.0 EVACUATION** ................................................ 12.9
    4.1.0 Service Equipment Used for Evacuation ...... 12.10
    *4.1.1 Vacuum Pump* ........................................... 12.10
    *4.1.2 Electronic Vacuum Gauge* .......................... 12.11
    *4.1.3 Gauge Manifold Set and Service Vacuum Hoses* ......... 12.12
    4.2.0 Methods of Evacuation ............................... 12.12
    *4.2.1 Deep Vacuum Evacuation Method* .............. 12.13
    *4.2.2 Triple Evacuation Method* ........................... 12.14
**5.0.0 CHARGING** .................................................. 12.16
    5.1.0 Service Equipment Used for Charging ......... 12.17
    *5.1.1 Refrigerant Cylinders* ................................. 12.17
    *5.1.2 Charging Scales* ........................................ 12.17
    *5.1.3 Charging Cylinder* ..................................... 12.18
    *5.1.4 Recovery/Recycle Unit* ............................... 12.19
    *5.1.5 System Sight Glass* ................................... 12.19
    5.2.0 Charge Determination and Accuracy ............ 12.20
    5.3.0 Charging by Weight ................................... 12.21
    *5.3.1 Liquid Charging by Weight* ......................... 12.21
    *5.3.2 Vapor Charging by Weight* ......................... 12.21

## MODULE 03212 CONTENTS (Continued)

      5.4.0    Charging by Superheat ............................. 12.24
      5.5.0    Charging by Subcooling ............................ 12.26
      5.6.0    Charging Using Pressure Charts ..................... 12.28
**6.0.0 USING ZEOTROPE REFRIGERANTS** ..................... 12.29
**SUMMARY** .................................................. 12.30
**REVIEW QUESTIONS** ........................................ 12.31
**GLOSSARY** ................................................ 12.33
**ANSWERS TO REVIEW QUESTIONS** ............................ 12.34
**REFERENCES & ACKNOWLEDGMENTS** .......................... 12.35

## Figures

| | | |
|---|---|---|
| Figure 1 | Halide torch leak detector | 12.2 |
| Figure 2 | Electronic leak detector | 12.3 |
| Figure 3 | Ultrasonic leak detector | 12.3 |
| Figure 4 | Ultraviolet/fluorescent leak detection lamp and injector | 12.4 |
| Figure 5 | Typical equipment hookup for leak detecting | 12.5 |
| Figure 6 | Typical recovery unit connected to a refrigerant system | 12.8 |
| Figure 7 | Typical combined recovery/recycle unit | 12.9 |
| Figure 8 | Vacuum pump | 12.10 |
| Figure 9 | Boiling temperatures of water at various levels of vacuum | 12.10 |
| Figure 10 | Vacuum gauge | 12.12 |
| Figure 11 | Typical evacuation gauge manifold set | 12.12 |
| Figure 12 | Service equipment connected for deep evacuation method | 12.13 |
| Figure 13 | Service equipment connected for triple evacuation method | 12.15 |
| Figure 14 | Sequence of events – triple evacuation method | 12.16 |
| Figure 15 | Selecting vapor and liquid state of refrigerant in cylinders | 12.17 |
| Figure 16 | Typical electronic refrigerant charging scale | 12.18 |
| Figure 17 | Refrigerant charging cylinder | 12.18 |
| Figure 18 | Moisture-liquid indicator (sight glass) | 12.19 |
| Figure 19 | Service equipment connected for charging liquid refrigerant by weight | 12.22 |
| Figure 20 | Service equipment connected for charging vapor refrigerant by weight | 12.23 |
| Figure 21 | Service equipment connected for charging by superheat | 12.24 |
| Figure 22 | Table used for charging for proper superheat | 12.25–12.26 |
| Figure 23 | Superheat charging calculator | 12.27 |
| Figure 24 | Charging for proper subcooling | 12.27 |
| Figure 25 | Subcooling calculator | 12.29 |
| Figure 26 | Pressure chart | 12.29 |

## Tables

| | | |
|---|---|---|
| Table 1 | Required Levels of Evacuation for Stationary Appliances Except for Small Appliances | 12.7 |
| Table 2 | Recommended Minimum Vacuum Pump Sizes for Various Applications | 12.11 |

# MODULE 03212

# Leak Detection, Evacuation, Recovery, and Charging

## Objectives

When you have completed this module, you will be able to do the following:

1. Identify the common types of leak detectors and explain how each is used.
2. Demonstrate skill in performing leak detection tests.
3. Identify the service equipment used for evacuating a system and explain why each item of equipment is used.
4. Demonstrate skill in performing system evacuation and dehydration.
5. Identify the service equipment used for recovering refrigerant from a system and for recycling the recovered refrigerant, and explain why each item of equipment is used.
6. Demonstrate skill in performing refrigerant recovery.
7. Demonstrate or explain how to use a recycle unit.
8. Identify the service equipment used for charging refrigerant into a system, and explain why each item of equipment is used.
9. Demonstrate skill in charging refrigerant into a system.

## Prerequisites

Before you begin this module, it is recommended that you successfully complete the following modules: Core Curriculum; HVAC Level One; HVAC Level Two, Modules 03201 through 03210.

## Required Training Materials

1. Pencil and Paper
2. Appropriate Personal Protective Equipment

## 1.0.0 ◆ INTRODUCTION

The following are the four basic service procedures used to troubleshoot, repair and/or maintain correct operation of the mechanical refrigeration system:

- Leak detection
- Evacuation and dehydration
- **Recovery**
- Charging

These procedures are performed when installing new systems and when servicing existing ones. Failure to perform any one of these procedures correctly can result in poor system operation, and may even cause system failure. They must also be performed in order to make sure that the venting requirements of the Clean Air Act are not violated.

## 2.0.0 ◆ LEAK DETECTION

Refrigerant leaks are one of the major causes of trouble in refrigeration systems. Leaks must be detected and repaired because they allow all or part of the system refrigerant to be lost, allow air and moisture to enter and contaminate the system, and cause our environment to become contaminated with ozone-depleting compounds. Leak testing is normally performed:

- When an operating system has low refrigerant charge
- After the assembly but before evacuating/dehydrating and charging a new system
- After making a repair on a closed refrigerant system
- When acid-moisture testing indicates moisture and/or acid in a system

## Technician Certification

As a result of the Clean Air Act that severely limits the release of refrigerants into the atmosphere, the Environmental Protection Agency (EPA) requires the certification of all technicians who service air conditioning and refrigeration equipment. This includes anyone who performs installation, maintenance, or repair on such systems. Technicians must be certified in all equipment categories that they intend to service or install. For nonmobile or nonvehicular equipment these certifications are for Small Appliance (Type I), High-Pressure Appliance (Type II), Low-Pressure Appliance (Type III), or a Universal certificate (Type IV) that covers all types.

Certification is achieved by passing the Section 608 Technician Certification Program test for each category, as given by an EPA-approved certifying agency. The list of EPA-certified agencies includes the Air Conditioning Contractors of America (ACCA), the Air Conditioning & Refrigeration Institute (ARI), and a host of other agencies. A complete list of EPA-approved certification testing agencies can be found on the EPA web site at www.epa.gov.

## 2.1.0 Detection Devices

There are many leak detection devices ranging from very simple to complex. This section describes five common leak detectors:

- Halide torch
- Electronic
- Ultrasonic
- Liquid (bubble)
- Ultraviolet/fluorescent

### 2.1.1 Halide Torch

The halide torch leak detector (*Figure 1*) depends on a flame changing color to detect the presence of halogenated refrigerant vapors. The halide torch consists of a propane or liquid propane (LP) gas tank, hose, and a special burner which contains a copper reactor plate. A sampling tube is used as a probe. The detector flame is adjusted so that the top of the flame cone is level with, or slightly above, the copper reactor plate. To check for leaks, the end of the sampling tube is placed at the point of test and the gas flame is watched closely for the slightest change in color. The sampling tube should be placed directly below the area being checked for leaks and moved up and around the area slowly when making the test. Small leaks give the flame a greenish tint, while larger leaks color the flame a vivid blue.

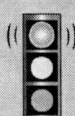

**WARNING!**
Use extreme caution when using a halide torch near combustible construction materials, including those found in attics and crawl spaces. Make sure a fire extinguisher is readily available when working with the torch.

The halide torch is not recommended for use if:

- The work area is poorly ventilated. Do not breathe the products of combustion from a halide torch when refrigerant is present. Some refrigerants break down when exposed to an open flame and produce a toxic gas.
- There is danger from explosive fumes or gases.
- There is bright sunlight.
- The work area is drafty. If possible, drafts should be reduced by shutting off fans and other devices that cause air movement.
- The leak is very small.

*Figure 1* ◆ Halide torch leak detector.

### Leak Detectors

Some electronic leak detectors are not capable of sensing some of the newer HFC refrigerants. Check with the equipment manufacturer to determine which refrigerants a particular device can detect.

Low-quality, inexpensive leak detectors may fail to detect refrigerant leaks in the field. Avoid these devices. Use only dependable, high-quality leak detectors when searching for refrigerant leaks.

#### 2.1.2 Electronic Detectors

Electronic leak detectors, commonly called *sniffers*, are accurate and easy to use (*Figure 2*). They typically detect leak rates of about ½ ounce per year. Electronic leak detectors should be operated and calibrated as directed by the manufacturer. Most of these detectors have an air filter that must be checked and replaced as needed. Usually, the leak detector sensor tip is placed next to each component or the piping in the system, and slowly moved at a rate of about 1" per second while searching for the leak. If possible, drafts should be reduced by shutting off fans and other devices that cause air movement. When a refrigerant leak is detected, the leak detector produces an audible signal, a flashing light, or both, depending on the detector in use.

#### 2.1.3 Ultrasonic Leak Detectors

Ultrasonic leak detectors are accurate and easy to use. Ultrasonic sound frequencies are sound waves beyond the range of human hearing. Ultrasonic leak detectors detect the ultrasonic sounds that a refrigerant gas makes as it leaks from a pressurized system. As the gas leaks to the atmosphere, its flow becomes turbulent. This turbulent flow has a high content of ultrasonic waves that are sensed by the detector. These devices can also detect leaks in a system under vacuum because they hear the turbulence caused by the air as it is drawn from the outside into the system. Ultrasonic leak detectors include a detector unit and a headset (*Figure 3*).

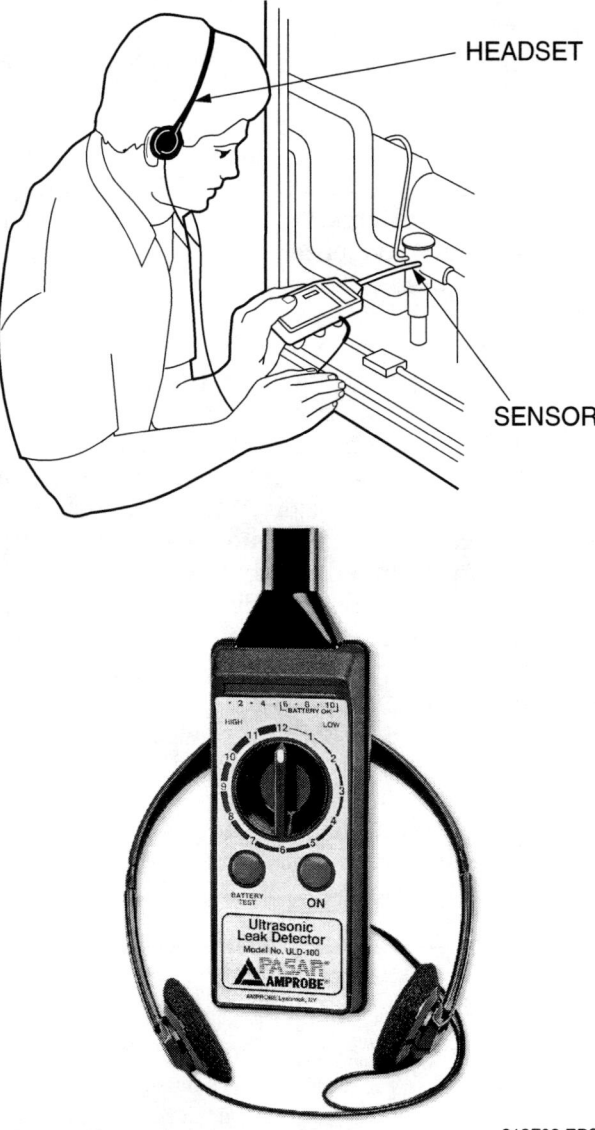

*Figure 2* ◆ Electronic leak detector.

*Figure 3* ◆ Ultrasonic leak detector.

Ultrasonic leak detectors should be operated and calibrated as directed by the manufacturer. Usually, you put on the headset, turn on the instrument, and begin listening for leaks as you slowly move the detector sensor around the system. The leak will be audible in the headphones before it is indicated by the light on the the detector.

Because the ultrasonic detector works from sound waves, it can be used to detect any type of refrigerant gas, including nitrogen. Also, it operates well in windy areas or areas with stray gases or fumes. One disadvantage of the ultrasonic detector is that sounds from other sources in the test area may affect its ability to detect a leak. Another disadvantage is that the ultrasonic detector may not be able to detect very small leaks.

### 2.1.4 Ultraviolet/Fluorescent Leak Detectors

This method involves injecting a small quantity of a fluorescent additive into the oil/refrigerant charge of an operating system. Areas of refrigerant leakage are observed as a yellow-green glow when viewed under the beam of a high-intensity ultraviolet (UV) lamp. This method of leak detection can be used to leak test CFC, HCFC, and HFC refrigerant systems as well as refrigerant blends using ester, alkylbenzene, or PAG lubricants. One manufacturer's test kit used for this purpose includes a high-intensity ultraviolet lamp (*Figure 4*), formula capsules filled with a premeasured amount of special fluorescent solution, and miscellaneous self-sealing fittings.

The fluorescent solution is injected into the system where it circulates and mixes with the refrigerant and compressor oil. The solution from the capsule can be injected into the system by: using refrigerant from a drum to move the solution into the low side while throttling refrigerant flow to produce a safe mist; connecting the capsule between the high and low sides of the system and allowing the high pressure to move the solution while throttling refrigerant flow; or by adding the solution to the oil container and then pumping it into the system. Another method commonly used to inject the solution into the system is by pouring a premeasured amount of the fluorescent solution from bottles into a special injector device (*Figure 4*).

At spots where small amounts of refrigerant are leaking from the system, the fluorescent solution also leaks out. After enough time has been allowed for the solution to circulate and mix thoroughly, the ultraviolet lamp is used to scan solder joints, fittings, couplings, and seals in the system to detect leaks. The source of any leak is shown by a bright, fluorescent yellow-green glow of the solution, which remains at the leak source.

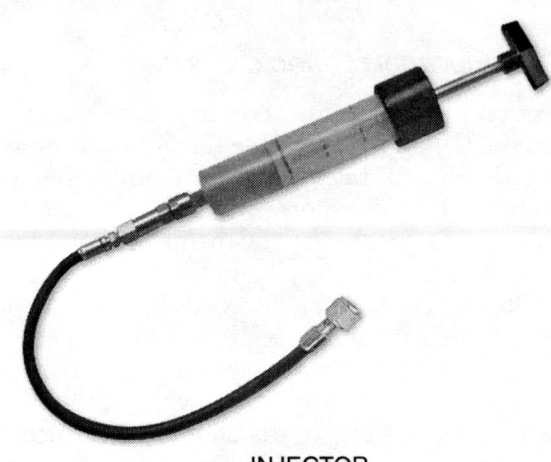

INJECTOR

LAMP

*Figure 4* ◆ Ultraviolet/fluorescent leak detection lamp and injector.

---

### Ultraviolet/Fluorescent Leak Detectors

Before injecting any leak-detecting dye, fluid, or chemical into the oil/refrigerant charge of an operating system, check with the equipment manufacturer and make sure the substance used will not harm the system. Introducing a substance to the system that is not approved by the manufacturer may void the equipment warranty.

### Tips for Using a Liquid (Bubble) Detector

- Always wipe or rinse off any residue from the liquid detector when leak detection is finished.
- Remove any hoses connected to a service valve before using a liquid detector to find leaks on the valve.
- After servicing a unit, leak check the access valves after removing the hoses.

After any leaks have been repaired, the fluorescent solution can be removed from the exterior of the leak site with any general-purpose cleaner suitable for removing oil residue. The fluorescent solution injected into the system remains in the system even after any leaks have been repaired. This allows for future leak testing of the system on a regular basis. Even if the full refrigerant charge is lost due to a leak, the solution remains mixed with the compressor oil in the system and, unless the oil is drained, it stays in the system.

#### 2.1.5 Liquid (Bubble) Detectors

Liquid detector or bubble solutions can be used to locate small leaks in a refrigerant system. Often, a liquid detector is used to find the exact source of an apparent leak that was found using one of the other types of leak detectors. The sudsy solution is brushed around the suspected area to find the leak. Escaping gas under pressure at the leak causes bubbles to form, providing a visual indication that pinpoints the leak. It may take several minutes for the system pressure to form a bubble. The main advantages of liquid detectors are their low cost and ease of use. A disadvantage is that larger leaks may blow through the solution, causing no bubbles to appear. Commercial bubble solutions are recommended because they are safer to use with metals. They also provide longer-lasting bubbles. Leak testing using solutions made from common laundry or kitchen detergents should be avoided. Many of these detergents contain chlorides which can corrode most of the metals used in refrigeration systems.

### 2.2.0 Leak Testing

This section covers leak testing. *Figure 5* shows a typical equipment hookup for a leak detection test.

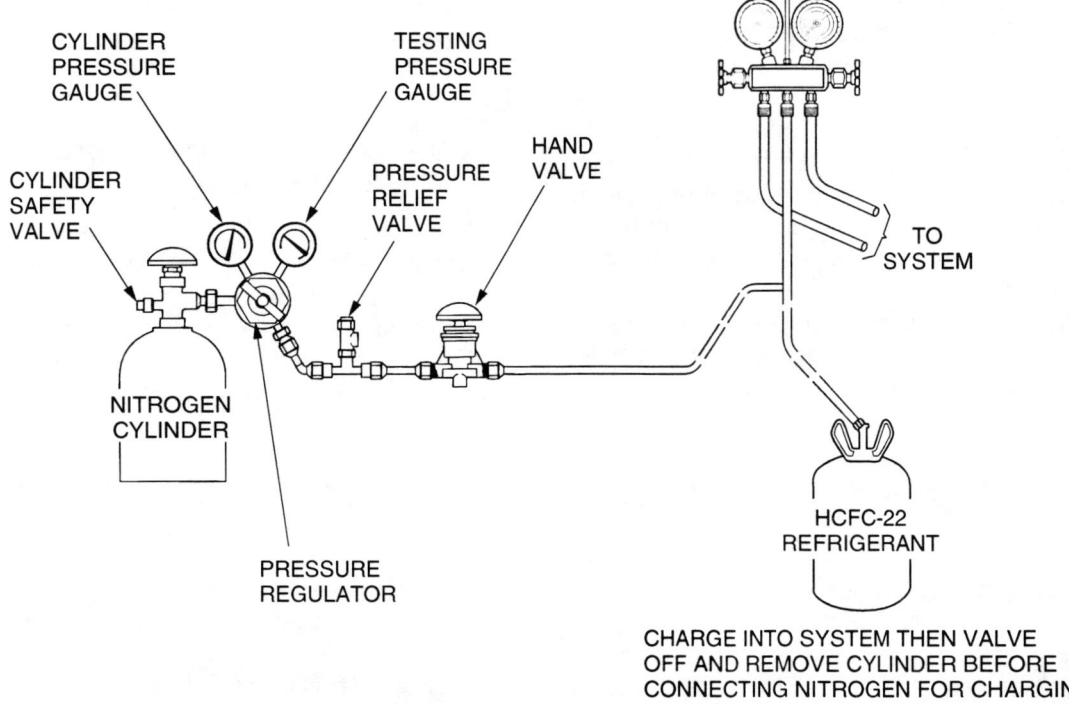

*Figure 5* ◆ Typical equipment hookup for leak detecting.

### 2.2.1 Operational System

If the system is operational, a visual check might reveal the source of a suspected leak. Since oil is mixed with refrigerant inside the system, the presence of oil around tubing joints, fittings, and on coil surfaces can indicate a leak. Look closely at all mechanical fittings, since vibration can loosen them over time. Use your eyes and ears. Large leaks can sometimes be heard. If this check leads to a component or piping suspected to be leaking, confirm that a leak exists using a leak detector. If the sight and sound method fails to pinpoint the leak, leak test the entire system with a leak detector. After finding and marking all leaks, recover the system refrigerant before attempting to make any repairs. Recover down to 0 psig to prevent contamination of the recovered refrigerant with air.

**WARNING!**
Never use oxygen, compressed air, or flammable gas to pressurize a system. An explosion will result when oil and oxygen are mixed.

Nitrogen is a high-pressure gas. At full cylinder pressure (about 2,000 psig), nitrogen can rupture a refrigerant cylinder and/or the refrigeration system. Use a pressure regulator on the nitrogen cylinder to limit the pressure. Also, use an overpressure relief valve between the cylinder pressure regulator and the system. The relief valve should be adjusted to open at about 2 psig above the system test pressure, but never more than 150 psig.

When charging the system with both a refrigerant and nitrogen, always put the refrigerant in first. Always valve off and remove the refrigerant cylinder before connecting the nitrogen cylinder.

### 2.2.2 Systems Without a Refrigerant Charge

The method approved by the EPA for leak testing a refrigeration system uses a trace amount of HCFC-22 refrigerant mixed with dry nitrogen. This method is also used to test systems that normally use a refrigerant other than HCFC-22. First, pressurize the system with a small (trace) quantity of HCFC-22 to a pressure of about 10 psig. Then, add dry nitrogen as needed to further increase the system pressure to the desired test level. When pressurizing the system with nitrogen, be sure not to exceed the maximum test pressure limits stamped on the unit's nameplate or listed in the manufacturer's service literature for the unit. A safe maximum is about 125 psig. The small amount of refrigerant used in this mixture is adequate to be detected with an electronic leak detector. The dry nitrogen provides the system pressure needed to perform the test. If you are using an ultrasonic leak detector, the system can be pressurized to the leak test level with nitrogen only. After locating and marking all leaks in the system, the mixture of trace refrigerant and nitrogen (or nitrogen alone) is vented to the atmosphere.

### 2.2.3 Systems With a Partial Charge

When a system with a partial refrigerant charge has insufficient pressure to support the leak test, the partial charge must first be recovered. The system is then pressurized with an R-22 refrigerant and nitrogen mixture the same as would be done with an empty system. It is a violation of EPA regulations to add nitrogen to an existing refrigerant charge in a system for the purposes of leak detection. The release of any such uncontrolled mixture to the atmosphere after the leak test is over is judged to be a release of pure CFCs or HCFCs.

### 2.2.4 Disassembling Brazed Joints for Leak Repair

When repairing a leak, it may be necessary to disassemble brazed piping. The procedures used to disassemble a brazed joint are basically the same as those used when assembling a brazed joint. Review the material in the HVAC Level One Module, *Soldering and Brazing*.

---

### Disassembling a Brazed Joint

When using a torch to disassemble a brazed joint, the system pressure must be 0 psig. Even the slightest pressure inside the system can cause refrigerant and oil trapped in the piping to spray out as the joint is disassembled.

To prevent pressure from building up as heat is applied, make a pressure vent line by attaching an open-ended service hose to a service port on the unit. Another option is to cut the rear of the joint with a tubing cutter.

When disassembling a brazed joint, the flux is first brushed around the joint area. It takes as much heat to take a joint apart as it does to braze it. Watch the condition of the flux to see when the filler metal has melted enough to allow the tubing to be pulled from a fitting. After the joint is taken apart, clean the disassembled parts with warm water to remove the flux.

## 3.0.0 ◆ REFRIGERANT CONTAINMENT

Since the passage of the Clean Air Act Amendments in 1990, it has been illegal to release refrigerants into the atmosphere. Most refrigerants being handled in any service procedure must be recovered, **recycled**, reclaimed, or destroyed. The EPA has specific meanings for the terms *recovery*, *recycle*, and *reclamation*:

- *Recovery* – The removal and temporary storage of refrigerant in containers approved for that purpose. Recovery does not provide for any cleaning or filtration of the refrigerant.
- *Recycle* – The circulation of recovered refrigerant through filtering devices that remove moisture, acid, and other contaminants. This does not mean that the recycled refrigerant meets the purity standards for new refrigerants.
- *Reclamation* – The remanufacture of used refrigerant to bring it up to the standards required of new refrigerant. Reclamation is not a field service procedure. It is a complicated process done only at reprocessing or manufacturing facilities.

## 3.1.0 Refrigerant Recovery

This section covers refrigerant recovery.

### 3.1.1 Evacuation Requirements

Refrigerant must be recovered from the different types of equipment (appliances) to levels established by the EPA. These levels are based on the amount of evacuation that must be achieved in a system during recovery. *Table 1* shows required evacuation levels for stationary equipment. When recovering refrigerant, except from small appliances, a recovery unit must be used that will evacuate the system to the level shown in the table. Small appliances are those in which the refrigerant is sealed within the unit at the factory and the amount of charge is five pounds or less. An example of a small appliance is a room air conditioner. For small appliances, the system is considered completely recovered when 90% of the refrigerant is removed if the appliance has a running compressor, or 80% of the refrigerant is removed if the unit has a non-operating compressor. For practical purposes, small appliances can be considered recovered to an acceptable level when evacuated to 4 in. Hg.

There are some exceptions to these EPA recovery evacuation levels. If evacuation to the specified levels is not achievable because of leaks in the system, or if recovery to these levels would contaminate the refrigerant being recovered, you must:

- Isolate the leaking components from the rest of the system, if possible.

**Table 1** Required Levels of Evacuation for Stationary Appliances Except for Small Appliances

| Type of Appliance | Inches of Mercury Vacuum (Based on Standard Atmospheric Pressure of 29.9 in. Hg) using Recovery/Recycle Unit Manufactured: | |
| --- | --- | --- |
| | Before November 15, 1993 | After November 15, 1993 |
| HCFC-22 appliance normally containing less than 200 pounds of refrigerant | 0 | 0 |
| HCFC-22 appliance normally containing more than 200 pounds of refrigerant | 4 | 10 |
| High-pressure appliance normally containing less than 200 pounds of refrigerant (includes equipment that uses CFC-12, CFC-500, CFC-502, and CFC-114, HFC-134a, or HFC-410A refrigerant) | 4 | 10 |
| High-pressure appliance normally containing more than 200 pounds of refrigerant (includes equipment that uses CFC-12, CFC-500, CFC-502, CFC-114, or HFC-410A refrigerant) | 4 | 15 |
| Very high-pressure appliance (includes equipment that uses CFC-13, CFC-503, or HFC-23 refrigerant) | 0 | 0 |
| Low-pressure appliance (CFC-11, CFC-113, or HCFC-123 refrigerant) | 25 | 25mm Hg absolute |

- Evacuate non-leaking components to the required levels, if they are to be opened.
- Evacuate leaking components to the lowest level attainable without substantially contaminating the refrigerant. This level cannot exceed 0 psig.

### 3.2.0 Refrigerant Recovery and Recovery/Recycle Units

Recovery of refrigerant from a system to the evacuation levels specified by the EPA requires the use of a certified refrigerant recovery unit or recovery/recycle unit.

### 3.2.1 Recovery Unit

Generally speaking, the greater the vacuum pulled by the recovery unit, the higher the probability that a high percentage of the refrigerant is recovered. Recovery units are not vacuum pumps and do not provide that function. If the dehydration/evacuation of a system is required, a vacuum pump must be used.

Most recovery units are capable of recovering refrigerant from a system in either the vapor or liquid state. Make sure that the unit used is capable of recovering the type of refrigerant used in the system.

Test the recovered refrigerant for the presence of acid or moisture to verify its purity before recharging it back into the system. When the refrigerant being recovered is highly contaminated, or when liquid refrigerant is being recovered, an external filter-drier should be installed in the utility hose line of the gauge manifold set connected to the recovery unit. Be sure to orient the filter-drier for correct flow direction.

Because there is a wide difference in the capabilities of the various recovery or recovery/recycle units, recovery is performed by following the manufacturer's instructions for the unit being used.

*Figure 6* shows a typical recovery unit connected to a system. Currently, recovered refrigerant can only be reused in the same system from which it was removed, or in another system belonging to the same owner.

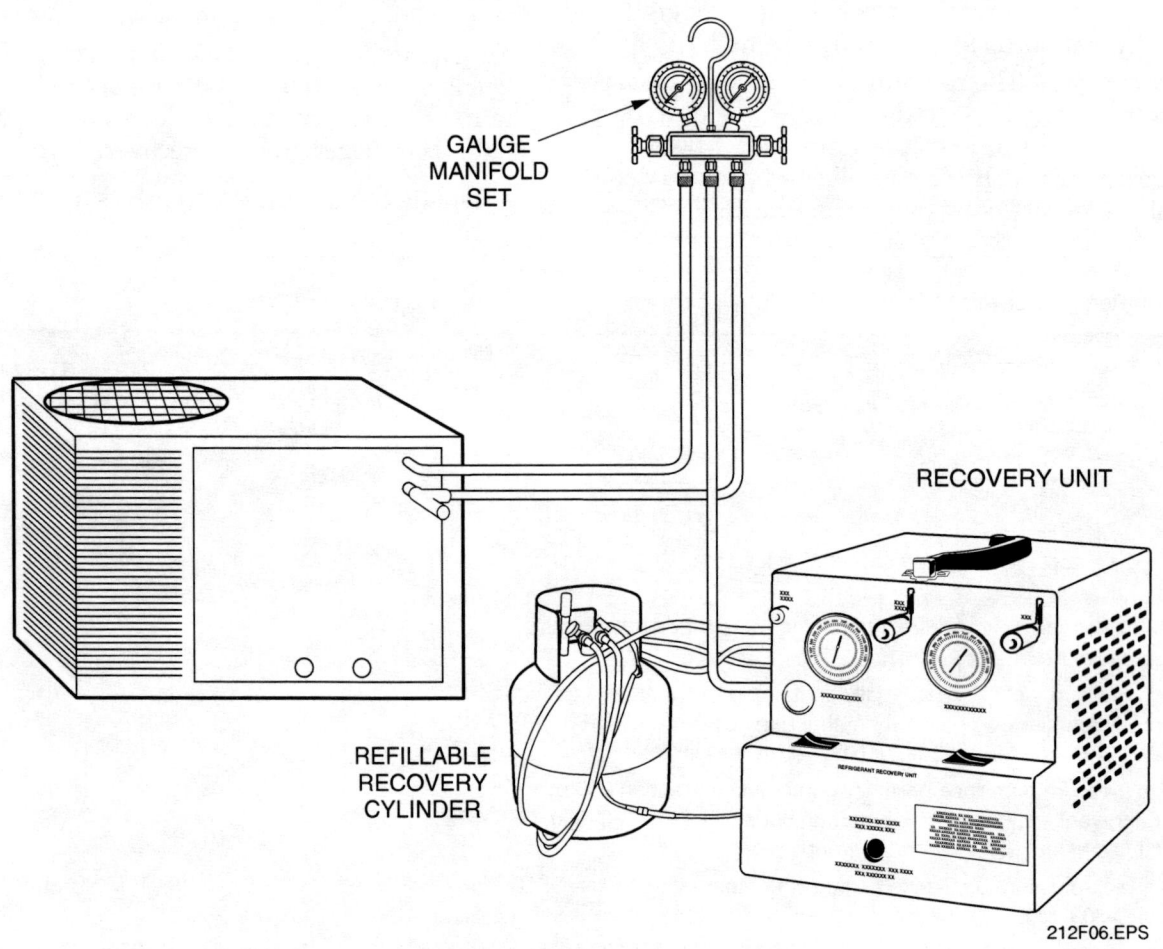

*Figure 6* ◆ Typical recovery unit connected to a refrigerant system.

Recovery must be performed:

- Before a system can be opened to make repairs
- Before pressurizing a system for leak testing
- Before disposing of any system or component
- When necessary to remove excess refrigerant from an overcharged system

### 3.2.2 Recycle Unit

Recycling of recovered refrigerant is normally done at the job site by a certified technician using a certified recycle unit (*Figure 7*). Make sure that the unit is capable of processing the type of refrigerant you plan to recycle. Follow the manufacturer's instructions for the recycle unit being used.

A recycle unit dehydrates and cleans the recovered refrigerant so that it can be returned to the system. After recycling, the refrigerant is cleaner, but its purity is not the same as that of new refrigerant. The dehydrating and purifying capability of recycle units varies from by model. Most units circulate the refrigerant through a filtration/drying process to achieve the desired quality. Acid/moisture testing of recycled refrigerant should be performed to verify the quality of the refrigerant before putting it back into the system. Currently, recycled refrigerant can only be reused in the same system from which it was recovered, or another system owned by the same customer.

### 3.2.3 Recovering or Recycling Different Refrigerants

Before using either a recovery unit or recycle unit to process a different refrigerant than was last processed, the recovery or recycle unit compressor oil should be drained and replaced with new oil. All filter-driers should be replaced and the recovery or recycle unit must be evacuated. Always make sure to use a recovery cylinder dedicated for use with the type of refrigerant to be recovered or recycled. Make sure to perform any other required service specified in the manufacturer's instructions for the unit being used. Certain refrigerants, such as R-410A, require special recovery machines.

## 4.0.0 ◆ EVACUATION

Refrigeration systems are intended to contain only refrigerant and oil. Anything else in a closed system is considered a contaminant. Contaminants such as air and moisture are major causes of compressor failure.

Evacuation is important because it is the only way that air or moisture in a system can be removed. Air is a noncondensible that can accumulate in the condenser, taking up space needed for condensing the refrigerant. This results in an increase in the condensing temperature that makes the system work harder. It also promotes the creation of acids by chemical reaction with the oil and refrigerant mixture.

Under the heat of compression, moisture will react with the refrigerant to form hydrochloric and hydrofluoric acids. These acids can cause corrosion of metals and breakdown of the insulation on the motor windings.

Moisture in the refrigerant can also cause oil sludge, which reduces the lubrication properties of the oil in the compressor. Moisture can freeze at the expansion device, slowing down or blocking the flow of refrigerant. The presence of moisture in a system can be determined using an acid/moisture test kit.

Generally, evacuation of a system is performed:

- After assembly and pressure checking, but prior to charging a system assembled in the field
- After an existing system is opened to the atmosphere as a result of parts replacement or other service procedures
- When an acid/moisture test shows the system is contaminated

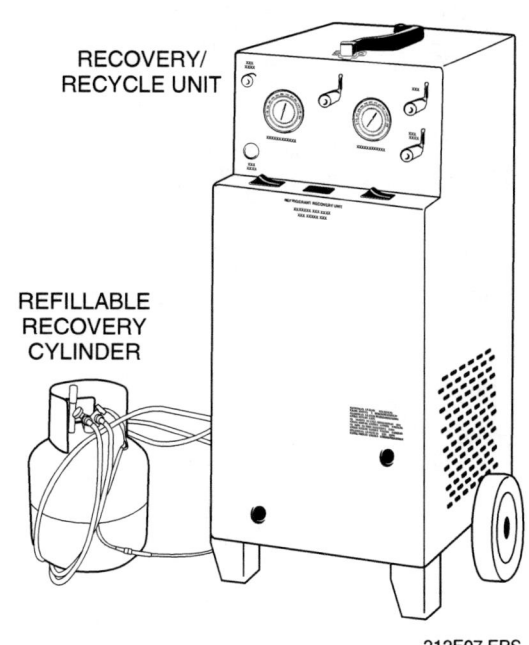

*Figure 7* ◆ Typical combined recovery/recycle unit.

## 4.1.0 Service Equipment Used for Evacuation

Evacuation of a system requires the use of a good vacuum pump and an accurate vacuum indicator. These devices are connected through a gauge manifold set to the system.

### 4.1.1 Vacuum Pump

The vacuum pump (*Figure 8*) is used to remove the air and moisture trapped in a system. The vacuum pump works to create a pressure differential between the system and the pump. This causes air and moisture vapor trapped in the system at a higher pressure to move into the lower pressure (vacuum) area created in the vacuum pump. The air and moisture vapor removed from the system are further processed through the vacuum pump and discharged into the atmosphere. When the vacuum pump lowers the pressure (vacuum) in the system enough, as determined by the ambient temperature of the system, liquid moisture trapped in the system will boil and change into vapor. Like free air, this water vapor is then pulled out of the system, processed through the pump, and exhausted to the atmosphere. *Figure 9* shows a chart of the boiling points of water at various levels of vacuum. As shown, the more the atmospheric pressure is lowered below 0.00 in. Hg vacuum, the lower the boiling point of water.

*Figure 8* ♦ Vacuum pump.

### BOILING TEMPERATURES OF WATER AT CONVERTED PRESSURES

| Temperature in °C | Temperature in °F | Inches of Vacuum | Microns* | Pounds Sq. In. (Pressure) |
|---|---|---|---|---|
| 100° | 212° | 0.00 | 759,968 | 14.696 |
| 96° | 205° | 4.92 | 535,000 | 12,279 |
| 90° | 194° | 9.23 | 525,526 | 10.162 |
| 80° | 176° | 15.94 | 355,092 | 6.866 |
| 70° | 158° | 20.72 | 233,680 | 4.519 |
| 60° | 140° | 24.04 | 149,352 | 2.888 |
| 50° | 122° | 26.28 | 92,456 | 1.788 |
| 40° | 104° | 27.75 | 55,118 | 1.066 |
| 30° | 86° | 28.67 | 31,750 | .614 |
| 26.6° | 80° | 28.92 | 25,400 | .491 |
| 24.4° | 76° | 29.02 | 22,860 | .442 |
| 22.2° | 72° | 29.12 | 20,320 | .393 |
| 20.6° | 69° | 29.22 | 17,780 | .344 |
| 17.8° | 64° | 29.32 | 15,240 | .295 |
| 15° | 59° | 29.42 | 12,700 | .246 |
| 10.7° | 53° | 29.52 | 10,160 | .196 |
| 7.2° | 45° | 29.62 | 7,620 | .147 |
| 0° | 32° | 29.74 | 4,572 | .088 |
| -6.1° | 21° | 29.82 | 2,540 | .049 |
| -14.4° | 6° | 29.87 | 1,270 | .0245 |
| -31.1° | -24° | 29.91 | 254 | .0049 |
| -37.2° | -35° | 29.915 | 127 | .00245 |
| -51.1° | -60° | 29.919 | 25.4 | .00049 |
| -56.6° | -70° | 29.9195 | 12.7 | .00024 |
| -67.8° | -90° | 29.9199 | 2.54 | .000049 |

*Remaining pressure in system in microns
1.000 inch = 25,400 microns = 2.540 cm = 25.40 mm
.100 inch = 2,540 microns = .254 cm = 2.54 mm
.039 inch = 1,000 microns = .100 cm = 1.00 mm

*Figure 9* ♦ Boiling temperatures of water at various levels of vacuum.

### Think About It: Using a Vacuum Gauge to Test for Leaks

Can a vacuum gauge be used to test for leaks? Can a vacuum gauge pinpoint the location of a leak?

---

The use of a quality, high-level vacuum pump is a must to properly evacuate and dehydrate a system. Typically, a good pump is capable of evacuating a system down to 500 microns (29.90 in. Hg vacuum). Microns are a precise measurement of pressure. One inch of mercury equals 25,400 microns. Most vacuum pumps use a direct-drive, two-stage, rotary-type vane pump driven by the attached motor. Most have an oil level sight gauge, oil port, and an oil drain. A pump with a gas ballast between the pump's first and second stages should be used. This feature permits relatively dry air from the atmosphere to enter the second stage of the pump, where it combines with the moist vapors passing through the pump from the refrigerant system. This helps to prevent the moisture from condensing into a liquid and mixing with the vacuum pump oil in the pump crankcase. The pump gas ballast valve should be opened during the early stages of a pumpdown to introduce the air into the second stage of the pump. The valve can be opened or closed at any time the pump is running. When first starting the pump, it should be closed. If it is left open, the pump oil may be discharged during the first revolution.

How quickly a vacuum pump can evacuate a system is an important service consideration. The more air the pump is capable of moving, measured in cubic feet per minute (cfm), the faster it can reach an acceptable vacuum level. As a rule of thumb, use one cfm for every ton of system capacity. The cfm requirements for a vacuum pump vary from system to system. *Table 2* lists suggested minimum amounts.

Vacuum pumps need periodic maintenance. Since the oil in the vacuum pump becomes contaminated through normal use, it should be changed after every 10 hours of pump operation. Always change the oil immediately after pumping down a wet or contaminated refrigerant system. Follow the manufacturer's instructions for servicing the vacuum pump.

#### 4.1.2 Electronic Vacuum Gauge

The electronic vacuum gauge is used to measure high vacuums. An accurate vacuum gauge must be used to measure the 500-micron range vacuum levels that must be achieved to properly evacuate and dehydrate refrigeration systems.

Even though the compound gauge on a gauge manifold set is capable of measuring a vacuum, it should not be used because the scale calibration is not accurate enough to read the specific vacuum levels needed in evacuation.

Electronic vacuum gauges are designed for use with high-vacuum pumps and can often read as low as 1.0 micron. A sensing tube is mounted at some point in the vacuum line to provide a calibrated output. This output, in microns, is displayed on an analog meter scale, a digital display, or a light-emitting diode (LED) sequence display. *Figure 10* shows a typical analog meter vacuum gauge.

**Table 2** Recommended Minimum Vacuum Pump Sizes for Various Applications

| System Size | Minimum Vacuum Pump Size |
|---|---|
| Up to 10 tons<br>Domestic refrigeration | 6 cfm |
| Up to 30 tons<br>Small residential A/C systems | 6 to 12 cfm |
| Up to 50 tons<br>Rooftop A/C systems | 10 to 15 cfm |
| Up to 70 tons<br>Large commercial systems | 12 to 20 cfm |

---

### Changing Vacuum Pump Oil

Drain the oil in the vacuum pump after every job. The drained pump will serve as a reminder to add new oil before starting the next evacuation. Frequent oil changes cost less than replacing a vacuum pump damaged by contaminated oil.

### Vacuum Gauges

A vacuum gauge can also be used to check the operation of a vacuum pump. Connect the vacuum gauge directly to the vacuum pump, then turn on the pump. The vacuum gauge should pull down to 50 microns or less for a new vacuum pump. A used pump is good if it pulls down below 200 microns. Remember to isolate the vacuum pump before turning it off. Note that a vacuum pump is not expected to hold the vacuum after it is turned off.

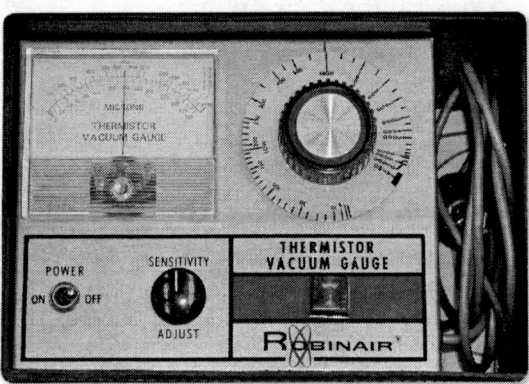

*Figure 10* ♦ Vacuum gauge.

The lower the micron reading displayed on the gauge, the closer the vacuum conditions are to a perfect vacuum (0 micron). As the vacuum levels go deeper, the reading on an analog gauge moves toward the left, indicating lower and lower micron readings. For the LED-type gauge, the LED indicators, each representing a specific vacuum level in microns, turn off sequentially from the highest micron level to the lowest as the system vacuum pressure goes deeper. The location of the sensor tube affects the vacuum reading. The closer it is located to the vacuum source, the lower the reading. When measuring the vacuum in a system, isolate the vacuum pump with a vacuum valve and allow the pressure in the system to balance before taking a reading. Always operate the vacuum gauge as directed in the manufacturer's instructions. The vacuum gauge should be hooked with a tee to the suction hose. If the micron level rises after the vacuum pump valve and the evacuation gauge manifold set are closed, there is a leak in the system or in the gauge manifold set.

#### 4.1.3 Gauge Manifold Set and Service Vacuum Hoses

Either a two-valve gauge manifold set with evacuation-quality service hoses or a high-capacity evacuation gauge manifold set can be used for evacuation. Typically, an evacuation gauge manifold set has four valves, enlarged internal passages, and ⅜" hose connections to speed up the evacuation process (*Figure 11*). The use of four valves enables the gauge manifold set to be connected to all the equipment needed to perform both the evacuation and charging of a system, without disconnecting and reconnecting the service hoses.

### 4.2.0 Methods of Evacuation

The deep vacuum and triple vacuum methods of system evacuation are the two most frequently used. However, some manufacturers may have specific requirements for evacuation. The deep vacuum method is typically used after a repair that required the system charge to be recovered and the system opened. Use of the triple evacuation method is recommended when a system has been especially wet. The amount of moisture in a system can be determined by using an acid/moisture test on the system refrigerant before or during refrigerant recovery.

*Figure 11* ♦ Typical evacuation gauge manifold set.

### Hoses for Evacuation and Charging

Most hose manufacturers produce heavy-duty black hoses with thick walls that are designed to prevent leaks when evacuating and charging a system. Hoses of this type should be dedicated to evacuating and charging. They should not be used for day-to-day maintenance activities.

#### 4.2.1 Deep Vacuum Evacuation Method

The deep vacuum method of evacuation relies on the use of the vacuum pump alone to remove moisture from the system. A deep vacuum is any vacuum of 500 microns or less (29.90 in. Hg vacuum or more). When a deep vacuum is established in a closed system, air and other noncondensibles in the system are reduced to a negligible level. As the pressure is reduced, the boiling point of water is also reduced. As long as the ambient temperature surrounding the system is higher than the boiling point of the internal moisture, the moisture will be boiled off and expelled.

If a deep vacuum is rapidly achieved with an oversized vacuum pump, water in the refrigerant lines may turn into ice. The deep vacuum evacuation method is performed as follows:

*Step 1* Connect the vacuum pump, vacuum gauge, and gauge manifold set to the service valves on the system (or component) being evacuated, as shown in *Figure 12*.

*Step 2* Turn on the vacuum pump, open the system service valves, and evacuate the system to 500 microns. During the early stages of evacuation, stop the pump at least once and monitor the vacuum gauge to see if a rapid loss of vacuum occurs. Make sure to isolate the vacuum pump from the system with a valve to prevent vacuum loss through the pump. If there is a loss of vacuum, check the system and/or vacuum pump and vacuum gauge hookup for a leak. Repair any leaks, then proceed with evacuating the system.

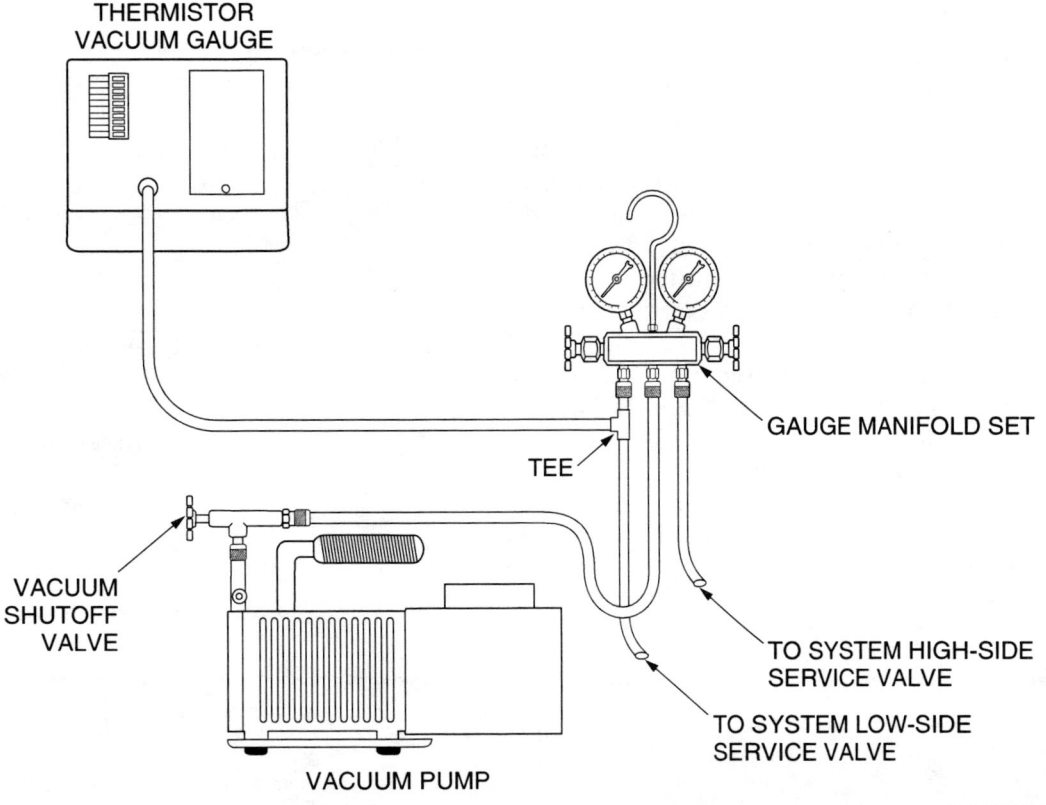

*Figure 12* ♦ Service equipment connected for deep evacuation method.

*Step 3* Once the 500-micron vacuum level is reached, close the valves to the gauge manifold set, and use the vacuum shutoff valve to isolate the vacuum pump from the system. Turn off the vacuum pump.

*Step 4* To ensure that the system is adequately evacuated, the final equilibrium pressure of the system must be checked after the system has been pumped down to 500 microns, but before it is charged with refrigerant. Vacuum leak test the system by watching the reading on the vacuum gauge to note any change in the level of vacuum in the system.

- A constant reading on the indicator of between 500 and 1,000 microns indicates a leaktight, dry system.
- If the indicator shows a pressure rise but levels off between 1,000 and 2,000 microns, this indicates that the system is leaktight, but moisture or liquid refrigerant is in the system.
- If the gauge shows a pressure rise and the pressure continues to rise without leveling off, a leak exists in the system, gauges, or the connecting tubing.

### 4.2.2 Triple Evacuation Method

Triple evacuation is recommended by most manufacturers for wet or contaminated systems. The triple evacuation method evacuates the system three times to a vacuum of about 1,000 microns (29.88 in. Hg vacuum). After the first and second evacuations, dry nitrogen is charged into the system to absorb moisture during the time period between each evacuation. The method given in this section uses dry nitrogen because it absorbs moisture well, is readily available, and compared to refrigerant, is inexpensive. Unlike refrigerant, nitrogen does not have to be recovered from the system between each evacuation. It can be released to the atmosphere.

*Figure 13* shows the hookup of the service equipment for the triple evacuation method. *Figure 14* shows the sequence of the triple evacuation process. As shown, the process takes about three hours when performed as recommended. This time can vary depending on how deep a vacuum is drawn in each step, how large the system is, and how great a capacity the vacuum pump has. The system is evacuated to 1,000 microns (29.88 in. Hg) during the three evacuations. The vacuum pump continues to run at this level or lower for at least 15 minutes during each evacuation. Evacuating to this level will cause liquid water in the system to boil anywhere with a temperature of 35°F or higher. After the first evacuation, the vacuum pump is turned off and the system is vacuum leak tested as previously described for the deep evacuation method.

Between the first and second evacuations, the system is pressurized with dry nitrogen to 10 psig or higher and allowed to sit for about an hour so that the nitrogen can absorb the moisture. A pressure of 10 psig provides more than enough nitrogen to absorb moisture in the system.

An optional triple evacuation method is to pull the system down to about 5,000 microns (29.72 in. Hg vacuum) for the first and second evacuations. For the last evacuation, a deep vacuum of 500 microns (29.90 in. Hg vacuum) is drawn and the system is vacuum leak tested in the same manner as previously described for the deep vacuum method.

Using nitrogen to pressurize systems for triple evacuation requires that certain precautions be followed.

**WARNING!**
Never use oxygen, compressed air, or flammable gas to pressurize a system. An explosion will result when oil and oxygen are mixed.

Nitrogen is a high-pressure gas. At full cylinder pressure (about 2,000 psig), nitrogen can rupture the refrigeration system. Be certain to use a pressure regulator on the nitrogen cylinder to limit the pressure. Also, make sure to use an over-pressure relief valve between the cylinder pressure regulator and the system. The relief valve should be adjusted to open at about 2 psig above the system test pressure, but never more than 150 psig. A typical nitrogen cylinder hookup used for the triple evacuation procedure is shown in *Figure 13*.

---

### Smoking Vacuum Pumps

During the initial pulldown of a vacuum, the vacuum pump can be very noisy and may emit smoke. As the vacuum deepens, the pump should operate more quietly and stop smoking. If the noise and smoke do not decrease, there may be a vacuum leak in the system.

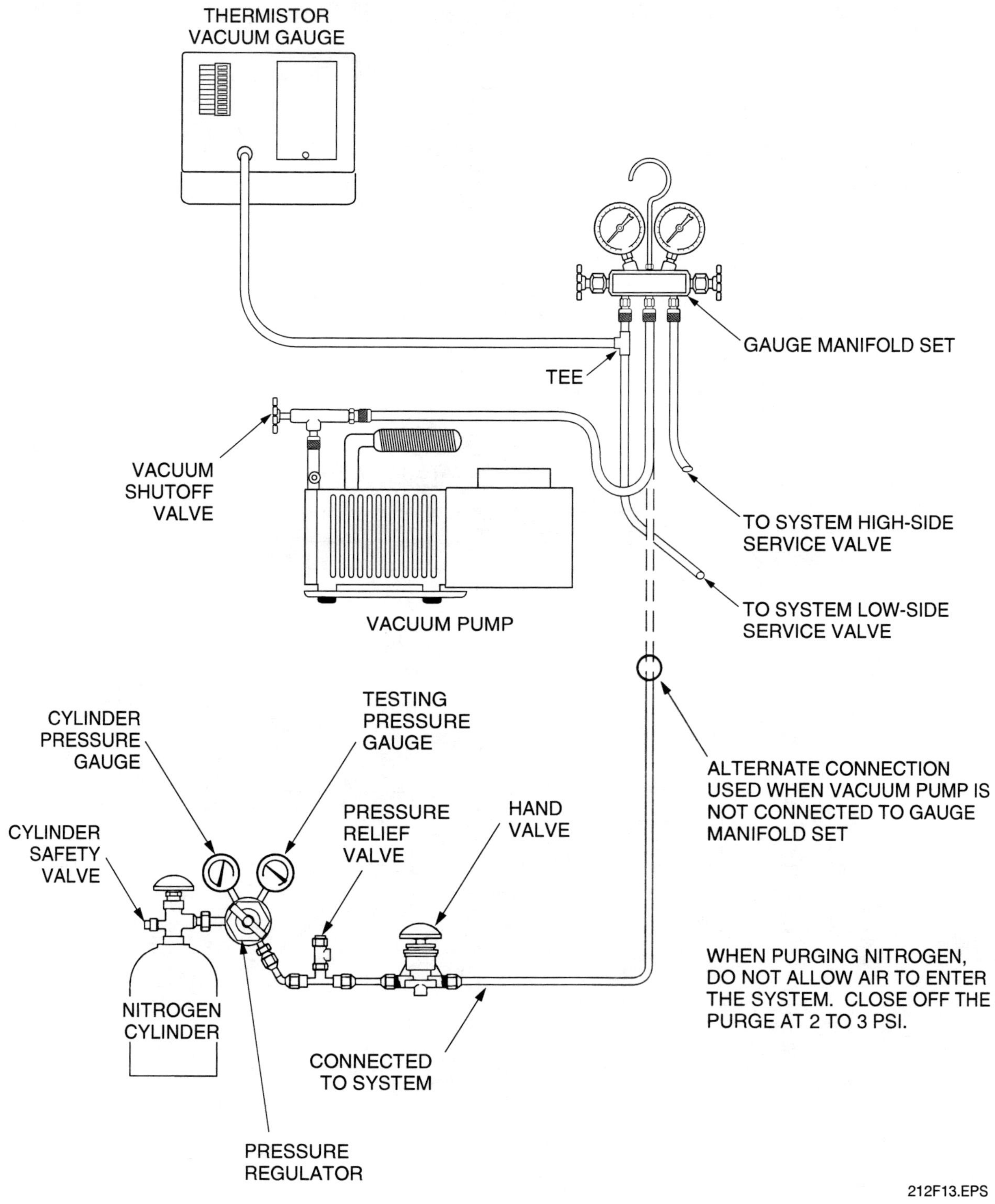

*Figure 13* ♦ Service equipment connected for triple evacuation method.

## Measuring Microns

Most vacuum microgauges do not measure accurately above 2,500 microns. When pulling a system down to levels above 2,500 microns, use an appropriate, accurate gauge capable of reading these higher values.

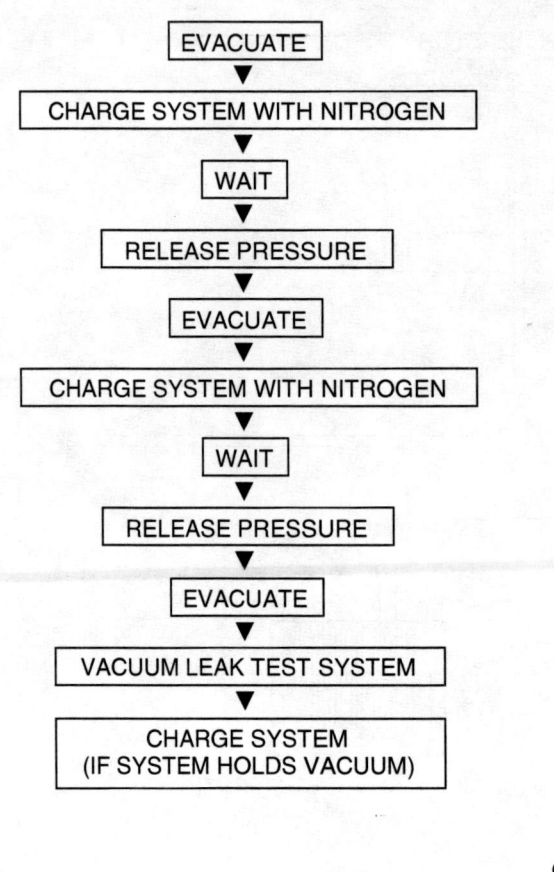

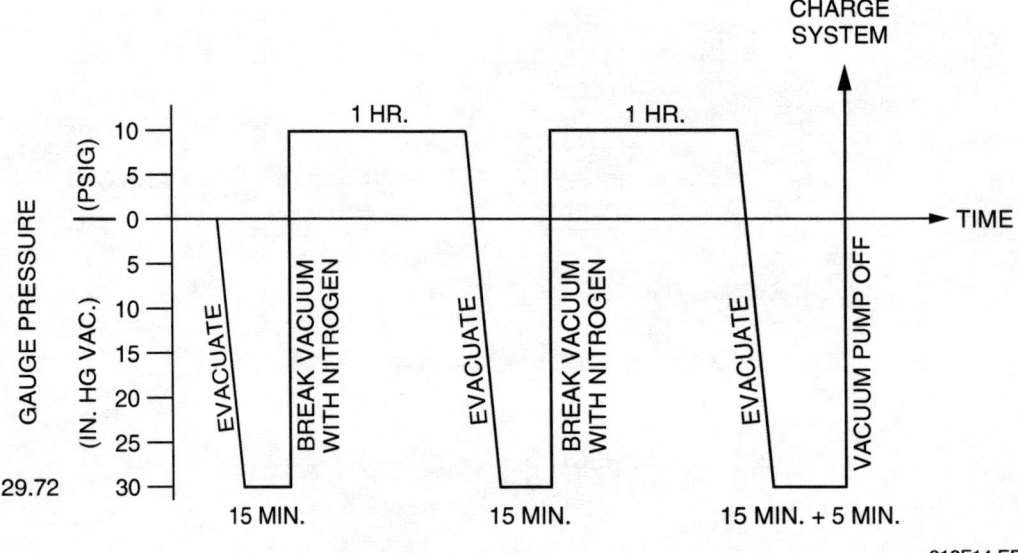

*Figure 14* ◆ Sequence of events – triple evacuation method.

## 5.0.0 ◆ CHARGING

A system must be charged with refrigerant after it has been repaired, leak tested, and evacuated, in order to return it to service. Also, new equipment may need to be charged before starting it up. Some models come factory-charged, but may need adjustment in the field before being placed into service. This section describes the service equipment and methods commonly used to charge systems, including:

- Liquid charging by weight
- Vapor charging by weight
- Charging by superheat
- Charging by subcooling
- Charging using pressure charts

### Refrigerant Cylinders

For safety and other reasons there are several specific rules that govern the handling, transportation, use, and storage of refrigerant tanks/cylinders (containers). Before handling refrigerant containers, check with your supervisor to make sure that you are aware of and understand all the rules that apply. Some of the rules are given here.

- Don't drop, dent, or abuse refrigerant containers.
- Don't tamper with container safety devices.
- Always use a proper valve wrench to open and close the valve.
- Replace the valve cap and hood cap to protect the container valve when the container is not in use or is empty.
- Secure containers in place to prevent them from becoming damaged from moving around, especially in a van or truck. Strap or chain containers in an upright position.
- Don't store containers where the temperature can exceed the container relief valve setting.
- Don't reuse disposable (nonreturnable) containers nor attempt to refill them.
- Don't fill refillable containers with more than 80% liquid. Excess liquid in a container causes hydrostatic pressure that can result in an explosion.
- Never heat a container with an open flame or place an electric resistance heater in direct contact with it.
- Make sure you are using the correct refrigerant container. Containers are color-coded and also labeled to identify their contents.

## 5.1.0 Service Equipment Used for Charging

This section will cover the service equipment used for charging.

### 5.1.1 Refrigerant Cylinders

Refrigerant may be added to a system in either a vapor (gas) or liquid state (*Figure 15*). If using refrigerant contained in a disposable cylinder to charge the system, the cylinder must be set in the upright position in order to charge with vapor. To charge with liquid, the cylinder must be turned upside down. Refillable cylinders used with recovery or recycle units remain upright for charging with either liquid or vapor refrigerant. Recovery cylinders have valves that are used to select the desired vapor or liquid state.

### 5.1.2 Charging Scales

Installing a full refrigerant charge into a system is best done by weight. This requires the use of an accurate refrigerant charging scale. Charging scales (*Figure 16*) can be used to charge both small and large systems. Selection of a charging scale is based mainly on the type and size of the system being charged. Since proper system operation depends on having the correct charge, the scale's most important feature is its accuracy. The accuracy

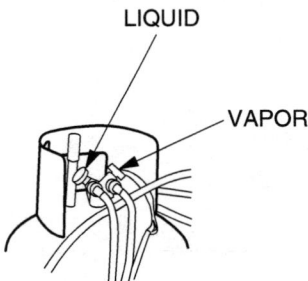

**RECOVERY (REFILLABLE) CYLINDER**

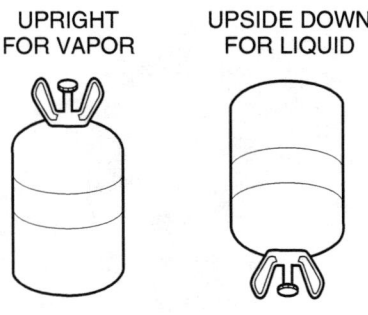

**DISPOSABLE CYLINDER**

212F15.EPS

*Figure 15* ◆ Selecting vapor and liquid state of refrigerant in cylinders.

should be matched to the size of the system being charged. A rule of thumb is that the accuracy should be within 1% of the required total system charge. To prevent damaging the scale, ensure that the scale weighing platform is strong enough to handle the largest size cylinder you intend to use.

Simple charging scales, electronic charging scales, and fully programmable automatic electronic charging scales are in common use. Depending on the model and its intended use, charging scales are calibrated to weigh refrigerant in pounds, ounces, kilograms, and/or grams. Typically, they display the cylinder/refrigerant weight using a liquid crystal display (LCD) or equivalent. Programmable models can control the flow of refrigerant and can be set to automatically dispense a preset amount of refrigerant and turn off when that amount is reached. Most stop charging if the cylinder empties before the full system charge is reached. As with other precise service instruments, the charging scale should always be operated and calibrated as directed in the manufacturer's instructions. Do not use bathroom or produce scales for refrigerant charging because they are not accurate enough. The system must be in a vacuum in order to weigh in the required charge.

### 5.1.3 Charging Cylinder

A charging cylinder is an accurate measuring device that can be used to charge either liquid or vapor refrigerant into a system (*Figure 17*). The use of a particular charging cylinder is based on the type and size of the system being charged. To maintain the accuracy of the charge, the cylinder used should be large enough to hold the total charge needed for the system. Typically, charging cylinders come in three capacities: 2½ pounds, 5 pounds, and 10 pounds.

Some charging cylinders contain an electric warming element that heats the refrigerant and builds up pressure in the cylinder. This helps push the refrigerant out of the cylinder, particularly in cold weather.

The charging cylinder contains calibrated shrouds for three or more different refrigerants. It has a sight glass to show the amount of refrigerant in the cylinder. On top of the charging cylinder is a gauge to measure the refrigerant pressure and a hand valve. On the bottom is another hand valve. The top and bottom hand valves are used to charge a system with vapor or liquid refrigerant, respectively. The bottom hand valve is also used to fill the cylinder.

*Figure 16* ◆ Typical electronic refrigerant charging scale.

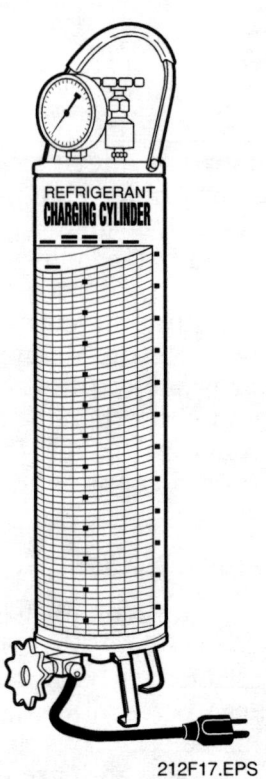

*Figure 17* ◆ Refrigerant charging cylinder.

To reduce the time needed for charging, cylinders have heaters to overcome the equalization of pressure between the system and charging cylinder. Charging cylinders normally have an automatic pressure relief valve for safety. This valve automatically resets when the safe pressure is restored.

The charging cylinder is usually evacuated and filled through the bottom hand valve with liquid refrigerant supplied from a bulk tank. A direct measurement of weight is obtained by rotating the shroud to align the sight glass with the calibration that matches the gauge pressure reading. The weight measurement is shown on the shroud at the level of refrigerant in the sight glass.

When charging the system, the top or bottom hand valve is used to control the flow of the desired vapor (top valve) or liquid refrigerant (bottom valve) into the system. At the same time, the level of the liquid refrigerant in the sight glass is watched as the charge is transferred. When the total charge is in the system, the cylinder valve is closed.

To be sure of accurate system charging, the charging cylinder should always be operated and calibrated as directed in the manufacturer's instructions.

Note that if you are filling a charging cylinder and it is necessary to remove vapor refrigerant to allow more liquid to enter, that refrigerant must be recovered.

### 5.1.4 Recovery/Recycle Unit

When charging with a recovery/recycle unit, follow the manufacturer's instructions for either vapor or liquid charging. Depending on the unit, one or more of the other methods described in this section may also have to be used in order to achieve a properly charged system.

### 5.1.5 System Sight Glass

When a system is equipped with a sight glass (*Figure 18*) in the liquid line close to the inlet of the metering device, it can be used to charge the system if the metering device is a thermostatic expansion valve (TXV, also abbreviated as *TEV*).

When the system is properly charged, only a clear stream of liquid refrigerant passes through the sight glass at a condensing temperature of 130°F.

Some sight glasses have refraction indicators to help determine flow. Bubbles or flashing usually indicate a shortage of refrigerant because they represent a mixed vapor-liquid condition.

*Figure 18* ♦ Moisture-liquid indicator (sight glass).

Only pure liquid should enter the TXV. When checking an existing charge, a clear sight glass may indicate a severe undercharge (pure gas). However, it could also indicate a proper charge or an overcharge. An overcharge looks the same as a proper charge because, in both cases, pure liquid is present at the sight glass. Because of the possibility of misinterpreting a clear sight glass, the sight glass should not be used alone for charging the system. Use other supplementary charging methods.

Once a clear sight glass is obtained, more charge may be needed. The disadvantage of using a sight glass is that it does not take into account the amount of subcooling in the liquid refrigerant.

As just noted, charging with a sight glass is only valid with systems that have a TXV. While charging to a clear sight glass, the condensing temperature should be held to 130°F.

When charging using the sight glass with a condensing temperature less than 130°F, the condenser coil should be blocked with plastic to achieve the target condensing temperature. The plastic will need to be removed from the condenser as charge is added and the condensing temperature rises above 130°F. Ultimately, the slight glass should clear while maintaining a 130°F condensing temperature.

The 130°F target condensing temperature was established to simulate a 100°F day. In many cases the technician adds 25°F to 30°F to the outdoor ambient to determine the target condensing temperature. Most days the ambient temperature will not be high enough; therefore, the artificially simulated condensing temperature of 130°F is maintained by covering the condenser coil.

This procedure should charge the TXV system correctly and provide the right level of subcooling necessary for cooling comfort. As always, you can use other charging methods to verify the correct sight glass charge.

### Electronic Sight Glass

An electronic sight glass like the one shown here can be attached to the refrigerant line for charging. It senses the condition of the refrigerant by means of ultrasonics. Two sensors are used, one for transmitting and one for receiving. These sensors are in the form of clamps that attach to the refrigeration piping. An audible beeping sound quickens as bubbles or floodback are sensed, and a row of LEDs light, simulating bubble movement in the piping.

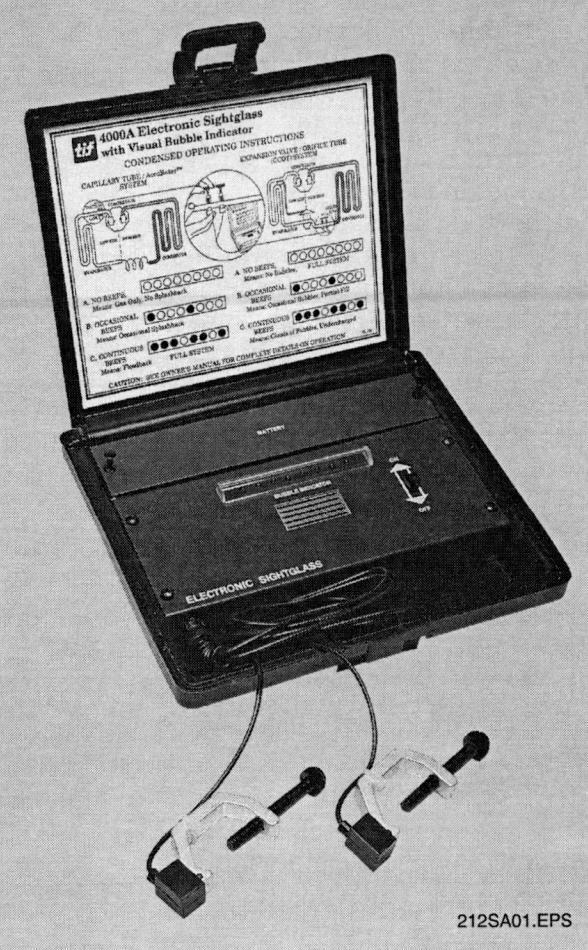

The use of a sight glass with a moisture indicator in a non-TXV system can also have value. The moisture indicator provides the technician with some insight as to the condition of the refrigerant. If the refrigerate has excess moisture it will need to be dehydrated, which may include changing the refrigerant, oil, and driers in addition to system evacuation. Do not trust a moisture indicator. Use an acid-moisture test kit to verify that moisture contamination exists in the system before beginning a lengthy refrigerant dehydration process. Occasionally, a moisture indicator will give a false moisture reading. The color of the moisture indicator varies among sight glass manufacturers.

### 5.2.0 Charge Determination and Accuracy

Regardless of system size, the operating charge of a system determines how efficiently and economically the system runs. An overcharged system can lead to high temperatures and pressures with the possibility of component failure. An undercharged system leads to insufficient cooling, and on units with hermetic compressors, may lead to compressor motor shutdown or failure. Many of the problems that occur in refrigeration and air conditioning systems are the direct result of overcharging or undercharging. Both are serious errors that must be avoided. Whether charging old or new systems, or adjusting the charge on a system, the full or partial charge must be introduced in the right way and in the right quantity.

For split systems and packaged equipment, the unit nameplate usually shows the amount of refrigerant charge required. This is also listed in the manufacturer's service literature for the equipment. The amount of charge needed for new units is determined by what charge, if any, it has when shipped from the factory. Always consult the manufacturer's literature for the unit or component to get the specific charge information. Be sure to take into account any additional refrigerant charge that needs to be added to the total system charge in order to compensate for accessories, such as filter-driers, receivers, etc.

Nameplates on split systems normally give the charge weight for the system based on the use of a standard line set length. Always be sure to consult the manufacturer's service literature to find out the length of the standard line set. If the system being serviced uses a line set that exceeds the standard length, find the amount of additional refrigerant charge that must be added to compensate for each foot of increased length.

For built-up systems, where each component is bought separately and is installed in the field, the total charge weight must be determined by adding the capacities of the individual components and accessories, plus the capacity of the connecting piping. The capacity of each component can be found in the manufacturer's literature for the component, or it may be marked on the component. One rule of thumb to estimate the charge needed for a built-up system calls for two pounds of refrigerant per each ton of system capacity. This is a ballpark estimate only. During the actual charging procedure, the charge put into the system must be precisely measured. Charging a built-up system is usually accomplished by charging the system with an adequate quantity of liquid and/or vapor refrigerant to allow the system to be started and run without tripping the safety controls. Then, a precise charge is obtained by adjusting the amount of refrigerant in the system up or down to obtain a proper superheat or subcooling temperature, or discharge pressure, depending on the type of system and the metering device.

## 5.3.0 Charging by Weight

Charging by weight is a precise method for charging a system. It is used if a complete charge is to be added and the weight of the charge is known. As previously described, the total weight can be found either from the unit nameplate or the manufacturer's literature. Be sure to include in the total charge weight any additional volume needed to compensate for added accessories or long line set lengths. Both liquid and vapor refrigerant can be charged into a system by weight.

### 5.3.1 Liquid Charging by Weight

Liquid charging goes much faster than vapor charging. This is because the density of a liquid is much greater than that of a vapor. The result is that the same size charging hoses can deliver many times more liquid charge per minute than if charging with a vapor. Because it is faster, liquid charging is the first choice to charge an empty system. When conditions are right, the entire charge may be introduced into the system in the liquid state. At other times, the flow of liquid refrigerant may slow down to a trickle or stop before the entire charge can be weighed in. When this happens, liquid charging is stopped and the remainder of the required total charge is introduced into the system using the vapor charging by weight method.

*Figure 19* shows the hookup of the service equipment used for liquid charging by weight. Liquid charging is done by adding the liquid refrigerant from the refrigerant cylinder (or charging cylinder) into the high-pressure side of the system through the high-side service valve. To prevent compressor damage, the compressor must be turned off, but the air handler should be turned on. Also, liquid refrigerant must never be added into the low-pressure side of the system or into the compressor discharge service port. If a system is equipped with a liquid charging valve between the condenser and metering device, it is better to charge the system through the liquid charging valve. Note that liquid charging valves tend to be used on systems with capacities over 20 tons.

### 5.3.2 Vapor Charging by Weight

Vapor charging by weight is normally performed in order to finish charging a system where the whole charge could not be introduced in the liquid state. Vapor charging by weight can also be used to charge an empty system, but is seldom used for this purpose. The correct method used to vapor charge a system with a partial charge depends on how much of the total charge weight is already in the system. If more than 50% is in the system, the liquid line service valve is closed, the compressor is turned on, and the refrigerant vapor is charged into the low-side of the system through the opened suction service valve (*Figure 20*). The system is vapor-charged in this manner until the required total charge by weight has entered the system.

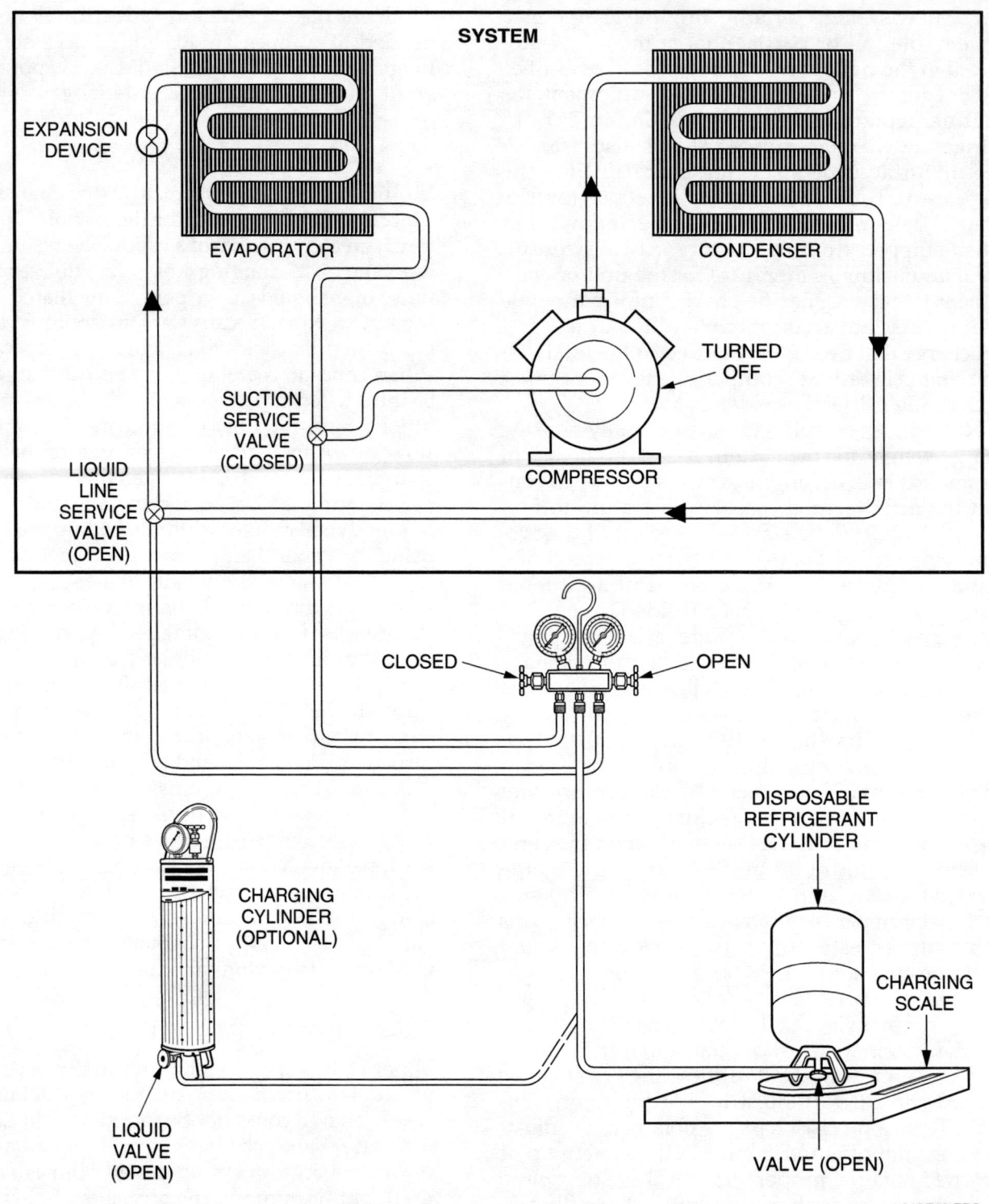

*Figure 19* ◆ Service equipment connected for charging liquid refrigerant by weight.

If less than 50% of the total charge is in the system, the compressor is turned off, the low-side and high-side service valves are opened, and the refrigerant vapor is charged into both the low and high sides of the system through the opened service valves. When more than 50% of the total charge has entered the system, the high-side service valve can be closed, allowing the remainder of the refrigerant vapor to be charged through the low-side of the system with the compressor turned on. This method would also be used to charge an empty system using the vapor charging by weight method.

As vapor is removed from the refrigerant cylinder, the system and container pressures tend to

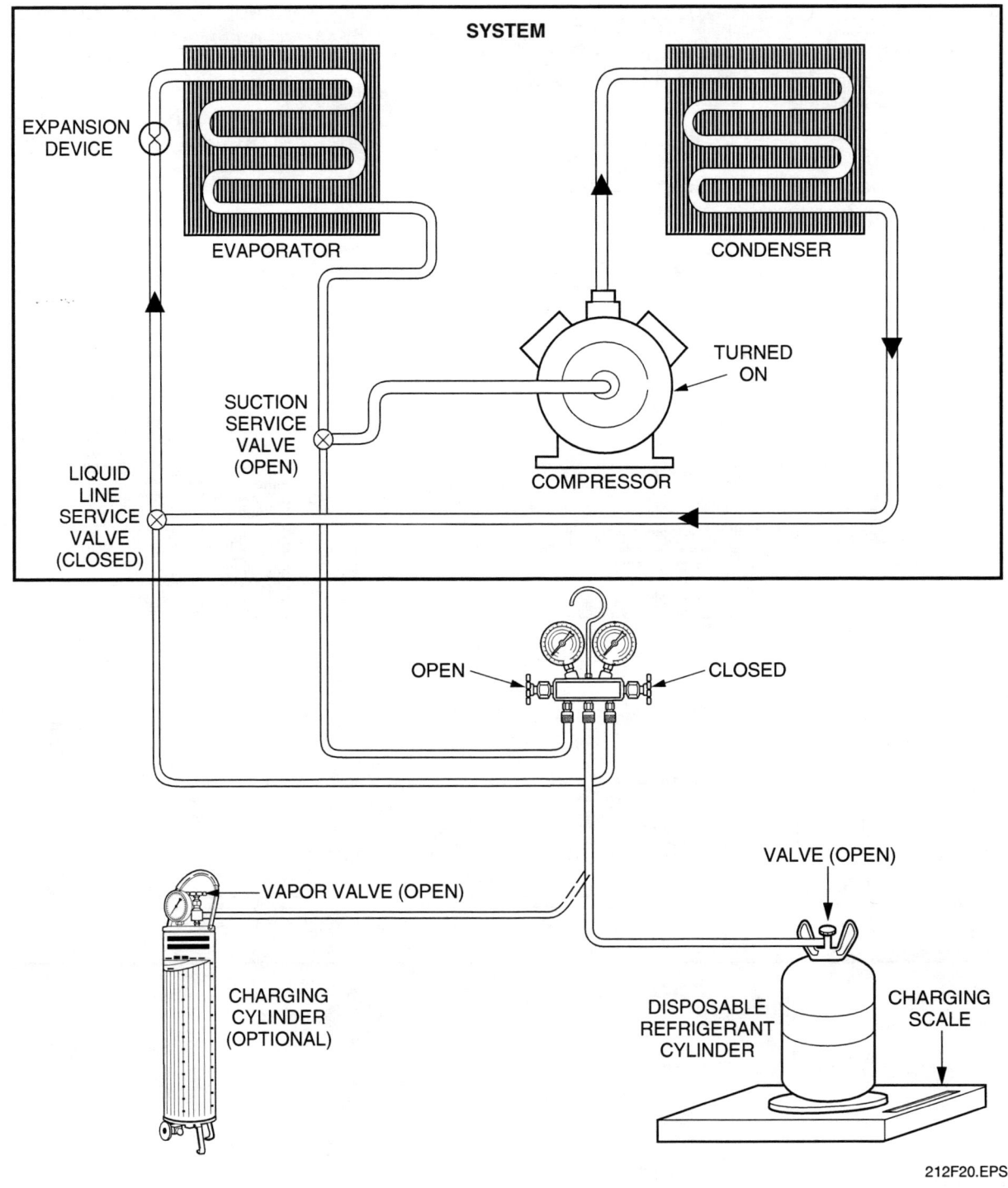

*Figure 20* ◆ Service equipment connected for charging vapor refrigerant by weight.

equalize, and the flow of vapor may slow down or even stop. When this occurs, the cylinder can be warmed by placing it in warm water. This will help to continue the flow of refrigerant by creating a pressure differential between the cylinder and the rest of the system. When warming the container, observe these precautions:

- Never heat a refrigerant cylinder above 125°F.
- Never apply a direct flame to a refrigerant cylinder.
- Never place a refrigerant cylinder on a hot plate or in direct contact with any other electric heater.

## 5.4.0 Charging by Superheat

Superheat is the heat added to a refrigerant after it has all been changed into a vapor. By knowing the amount of superheat in the suction line, you can tell if the system is properly charged.

Some manufacturers recommend the superheat method for checking or adjusting the charge of a system already in operation. Maintaining correct superheat is critical, because if liquid refrigerant returns to the compressor, it can cause damage and possible failure.

Charging for superheat is performed on systems with a fixed metering device such as a capillary tube or fixed orifice (*Figure 21*). This method uses superheat and suction line temperature tables (*Figure 22*) to first find the proper superheat level, then the required suction line temperature. These values are based on the temperatures and pressures measured in the operating system. The tables are attached to the equipment or contained in the manufacturer's service literature.

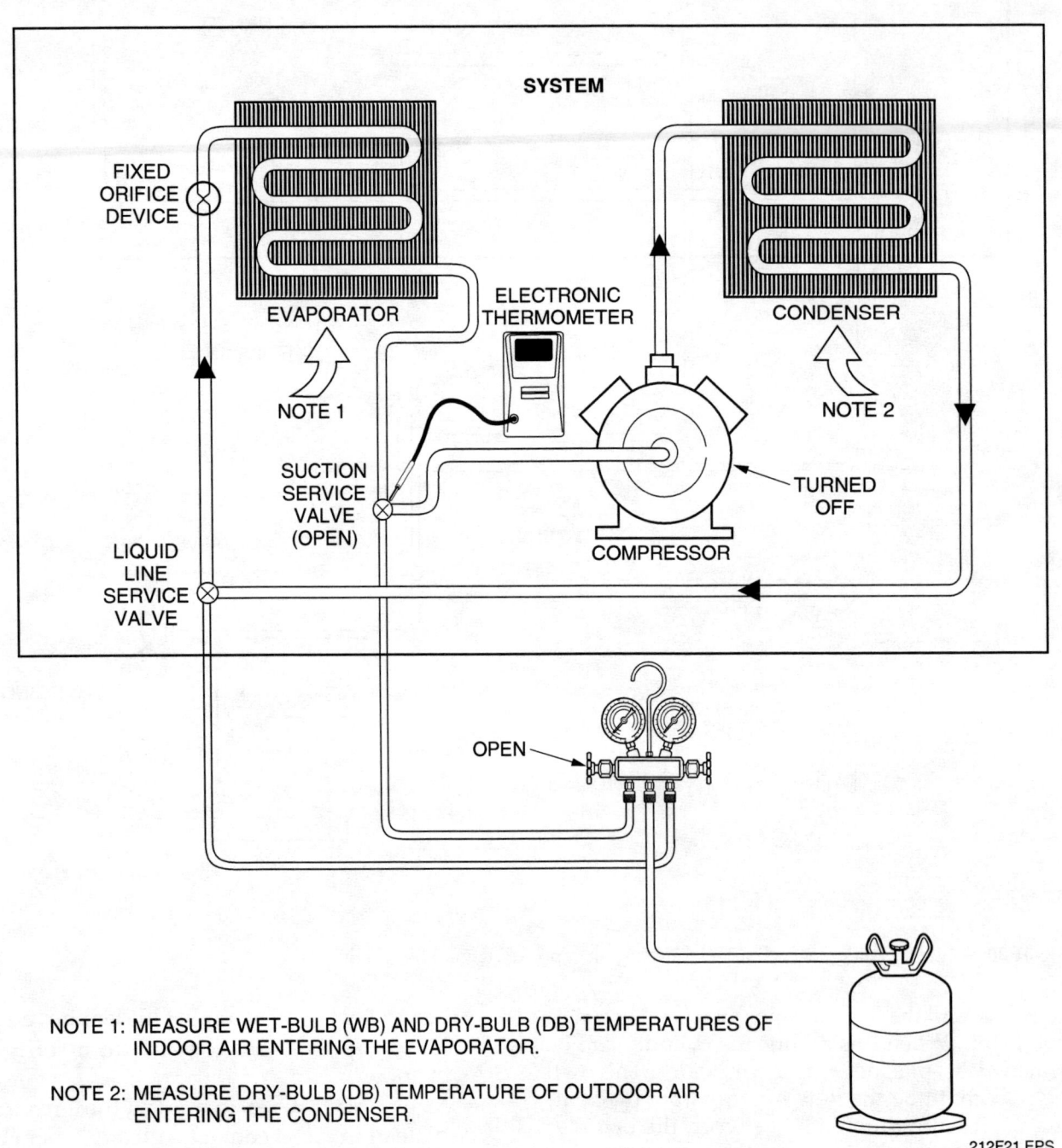

*Figure 21* ◆ Service equipment connected for charging by superheat.

## SUPERHEAT CHARGING CHART
### (SUPERHEAT ENTERING SUCTION SERVICE VALVE)

| INDOOR AIR CONDITIONS ENTERING EVAPORATOR | | | OUTDOOR AIR DRY-BULB TEMPERATURE (°F) ENTERING CONDENSER COIL | | | | | | | | | |
|---|---|---|---|---|---|---|---|---|---|---|---|---|
| WB | DB | RH % | 65 | 70 | 75 | 80 | 90 | 95 | 100 | 105 | 110 | 110 |
| 61 | 65 | 80 | 16 | 12 | 8 | 6 | Charging at these conditions can result in damage to the compressor. Charge to 5 degrees superheat and check superheat when conditions are more favorable. | | | | | |
|  | 70 | 60 | 18 | 14 | 10 | 6 | | | | | | |
|  | 75 | 45 | 20 | 16 | 12 | 8 | 6 | | | | | |
|  | 80 | 33 | 21 | 17 | 14 | 10 | 6 | | | | | |
|  | 85 | 23 | 23 | 19 | 16 | 12 | 7 | | | | | |
| 63 | 70 | 68 | 21 | 17 | 13 | 10 | 8 | 6 | | | | |
|  | 75 | 52 | 23 | 19 | 16 | 12 | 9 | 6 | | | | |
|  | 80 | 39 | 24 | 20 | 17 | 14 | 10 | 7 | | | | |
|  | 85 | 29 | 25 | 21 | 18 | 15 | 12 | 8 | 6 | | | |
|  | 90 | 20 | 26 | 22 | 19 | 16 | 13 | 9 | 7 | | | |
| 65 | 70 | 77 | 24 | 20 | 17 | 13 | 10 | 8 | 6 | | | |
|  | 75 | 59 | 25 | 22 | 19 | 15 | 12 | 10 | 7 | | | |
|  | 80 | 45 | 27 | 24 | 21 | 18 | 14 | 11 | 8 | 6 | | |
|  | 85 | 33 | 28 | 25 | 22 | 19 | 15 | 12 | 9 | 6 | | |
|  | 90 | 25 | 29 | 26 | 23 | 20 | 16 | 13 | 10 | 7 | | |
| 67 | 70 | 86 | 27 | 24 | 21 | 17 | 14 | 11 | 8 | 6 | | |
|  | 75 | 66 | 28 | 26 | 22 | 18 | 15 | 13 | 10 | 8 | 5 | |
|  | 80 | 50 | 30 | 27 | 24 | 21 | 18 | 15 | 12 | 10 | 7 | |
|  | 85 | 39 | 31 | 28 | 25 | 22 | 19 | 16 | 13 | 11 | 8 | |
|  | 90 | 30 | 32 | 29 | 26 | 23 | 20 | 17 | 14 | 12 | 9 | 6 |
| 69 | 70 | 95 | 30 | 27 | 25 | 21 | 18 | 15 | 12 | 9 | 7 | 6 |
|  | 75 | 75 | 31 | 28 | 25 | 22 | 19 | 16 | 13 | 10 | 8 | 6 |
|  | 80 | 58 | 32 | 29 | 27 | 24 | 21 | 19 | 16 | 14 | 11 | 8 |
|  | 85 | 45 | 34 | 31 | 28 | 26 | 23 | 20 | 17 | 15 | 12 | 9 |
|  | 90 | 35 | 35 | 32 | 30 | 27 | 24 | 21 | 19 | 16 | 13 | 11 |
| 71 | 75 | 82 | 34 | 31 | 29 | 26 | 23 | 21 | 18 | 15 | 12 | 9 |
|  | 80 | 65 | 36 | 32 | 30 | 28 | 24 | 22 | 20 | 18 | 15 | 13 |
|  | 85 | 51 | 37 | 34 | 31 | 29 | 26 | 24 | 21 | 19 | 16 | 14 |
|  | 90 | 39 | 38 | 35 | 32 | 30 | 28 | 25 | 22 | 20 | 17 | 16 |
|  | 95 | 30 | 39 | 35 | 33 | 30 | 29 | 25 | 23 | 21 | 18 | 17 |
| 73 | 75 | 92 | 37 | 34 | 32 | 29 | 27 | 24 | 22 | 20 | 17 | 15 |
|  | 80 | 72 | 37 | 35 | 33 | 30 | 28 | 26 | 24 | 22 | 19 | 17 |
|  | 85 | 58 | 38 | 36 | 34 | 31 | 29 | 27 | 25 | 23 | 20 | 18 |
|  | 90 | 45 | 38 | 36 | 34 | 31 | 30 | 28 | 25 | 23 | 21 | 19 |
|  | 95 | 35 | 38 | 36 | 35 | 32 | 31 | 28 | 26 | 24 | 21 | 19 |

NOTES: SUPERHEAT MEASUREMENTS SHOULD BE TAKEN AT CONDENSING UNIT SERVICE VALVES. CHARGE SYSTEM WITHIN 2°F OF SUPERHEAT INDICATED. RECOMMENDED MINIMUM SUPERHEAT IS 5°F.

WHITE AREA IN THE CHART IS THE OPTIMUM WINDOW FOR CHARGING.

DB = DRY-BULB TEMPERATURE (°F).
WB = WET-BULB TEMPERATURE (°F).
RH = APPROX. % OF INDOOR RELATIVE HUMIDITY.

FIND THE REQUIRED SUPERHEAT AT THE INTERSECTION OF THE MEASURED OUTDOOR DRY-BULB (DB) AIR TEMPERATURE AND THE INDOOR WET-BULB (WB) AND DRY-BULB (DB) AIR TEMPERATURE.

EXAMPLE: THE INTERSECTION OF A 95°F DB OUTDOOR TEMPERATURE AND A 71°F WB/90°F DB INDOOR TEMPERATURE YIELDS 22°F AS THE REQUIRED SUPERHEAT.

*Figure 22* ◆ Table used for charging for proper superheat (sheet 1 of 2).

| REQUIRED SUCTION LINE TEMPERATURE (F) (ENTERING SUCTION SERVICE VALVE) | | | | | | | | | |
|---|---|---|---|---|---|---|---|---|---|
| SUPERHEAT TEMP (F) | SUCTION PRESSURE AT SERVICE PORT (PSIG) | | | | | | | | |
|  | 61.5 | 64.2 | 67.1 | (70.0) | 73.0 | 76.0 | 79.2 | 82.4 | 85.7 |
| 0 | 35 | 37 | 39 | 41 | 43 | 45 | 47 | 49 | 51 |
| 2 | 37 | 39 | 41 | 43 | 45 | 47 | 49 | 51 | 53 |
| 4 | 39 | 41 | 43 | 45 | 47 | 49 | 51 | 53 | 55 |
| 6 | 41 | 43 | 45 | 47 | 49 | 51 | 53 | 55 | 57 |
| 8 | 43 | 45 | 47 | 49 | 51 | 53 | 55 | 57 | 59 |
| 10 | 45 | 47 | 49 | 51 | 53 | 55 | 57 | 59 | 61 |
| 12 | 47 | 49 | 51 | 53 | 55 | 57 | 59 | 61 | 63 |
| 14 | 49 | 51 | 53 | 55 | 57 | 59 | 61 | 63 | 65 |
| 16 | 51 | 53 | 55 | 57 | 59 | 61 | 63 | 65 | 67 |
| 18 | 53 | 55 | 57 | 59 | 61 | 63 | 65 | 67 | 69 |
| 20 | 55 | 57 | 59 | 61 | 63 | 65 | 67 | 69 | 71 |
| (22) | 57 | 59 | 61 | (63) | 65 | 67 | 69 | 71 | 73 |
| 24 | 59 | 61 | 63 | 65 | 67 | 69 | 71 | 73 | 75 |
| 26 | 61 | 63 | 65 | 67 | 69 | 71 | 73 | 75 | 77 |
| 28 | 63 | 65 | 67 | 69 | 71 | 73 | 75 | 77 | 79 |
| 30 | 65 | 67 | 69 | 71 | 73 | 75 | 77 | 79 | 81 |
| 32 | 67 | 69 | 71 | 73 | 75 | 77 | 79 | 81 | 83 |
| 34 | 69 | 71 | 73 | 75 | 77 | 79 | 81 | 83 | 85 |
| 36 | 71 | 73 | 75 | 77 | 79 | 81 | 83 | 85 | 87 |
| 38 | 73 | 75 | 77 | 79 | 81 | 83 | 85 | 87 | 89 |
| 40 | 75 | 77 | 79 | 81 | 83 | 85 | 87 | 89 | 91 |

FIND THE REQUIRED SUCTION LINE TEMPERATURE AT THE INTERSECTION OF THE SUPERHEAT TEMPERATURE (°F) AND THE MEASURED SUCTION LINE PRESSURE (PSIG).

EXAMPLE: THE INTERSECTION OF 22°F SUPERHEAT AND A SUCTION LINE PRESSURE OF 70 PSIG YIELDS A REQUIRED SUCTION TEMPERATURE OF 63°F.

212F22B.EPS

*Figure 22* ◆ Table used for charging for proper superheat (sheet 2 of 2).

To charge by superheat, proceed as follows:

*Step 1* Operate the system for at least 15 minutes. The filter, condenser, and evaporator coil must be clean. If applicable, make sure good airflow is established throughout the system ductwork. Measure the following:
- Suction (vapor) line pressure
- Suction line temperature
- Outdoor dry-bulb temperature entering the condenser
- Indoor wet-bulb temperature entering the evaporator
- Indoor dry-bulb temperature entering the evaporator

*Step 2* Using the Superheat Charging Chart in *Figure 22, Sheet 1*, find the required superheat at the intersection point of the measured outdoor dry-bulb air temperature and the indoor wet-bulb temperature.

*Step 3* Using the Required Suction-Line Temperature Chart in *Figure 22, Sheet 2*, find the required suction line temperature at the intersection of the superheat temperature and the measured suction line pressure.

*Step 4* Compare the actual suction line temperature measured in the system to the required suction line temperature found in the chart to determine if an adjustment in the system refrigerant charge is needed. A tolerance of ±5°F is allowed before any adjustment is required.

*Step 5* If the measured superheat level is too high, add vapor refrigerant through the low-side service port with the compressor running until the superheat drops to the correct level.

*Step 6* If the measured superheat level is too low, remove vapor refrigerant from the unit using a recovery unit and proper recovery procedures until the superheat rises to the correct level.

*Step 7* If the air temperature entering the condenser coil changes, or the suction line pressure changes, repeat the procedure and charge to the new suction line temperature indicated by the chart.

Slide rule-type superheat charging calculators used to charge systems by the superheat method are available from some of the main equipment manufacturers.

*Figure 23* shows the superheat calculator available from one such equipment manufacturer. As shown, complete instructions for use are printed on the calculator.

## 5.5.0 Charging by Subcooling

Devices like the thermostatic expansion valve (TXV) maintain a constant superheat over a wide range of load and charge conditions. Because of this, the superheat method cannot be used with systems that contain TXVs or similar devices. In this situation, some manufacturers recommend measuring subcooling in the liquid line to obtain a correct charge.

Subcooling is the temperature removed from a liquid refrigerant after all the refrigerant has condensed into a liquid. The subcooling method measures the temperature of the subcooled refrigerant in the liquid line as a means of determining if the proper amount of liquid refrigerant is being applied to the TXV metering device.

If the liquid line temperature is incorrect, it can be corrected by adjusting the amount of refrigerant charge in the system.

The subcooling method is based on the existing pressure and temperature of the liquid line in an operating system. To charge by subcooling, use the following procedure.

*Step 1* Determine the correct subcooling temperature specified by the equipment manufacturer. This can be found on the unit nameplate or in the manufacturer's service literature for the unit being serviced.

*Step 2* Operate the system until it is stabilized, then measure the system liquid line temperature and pressure.

*Step 3* Use the measured liquid line pressure and a standard pressure/temperature chart to find the saturated temperature of the refrigerant in the liquid line (*Figure 24*).

*Step 4* Calculate the subcooling in the liquid line by subtracting the actual temperature measured in the liquid line from the saturated temperature. Compare the manufacturer's value for subcooling and the actual subcooling temperature that was calculated using the measured and saturated liquid line temperatures. If the measured subcooling temperature is within ±3°F of the required temperature, no adjustment in refrigerant charge is necessary.

*Step 5* If the measured subcooling temperature is too high, use a recovery unit to remove refrigerant vapor from the system. This will lower the temperature in the liquid line to the required level.

*Step 6* If the measured temperature is too low, add refrigerant to the system.

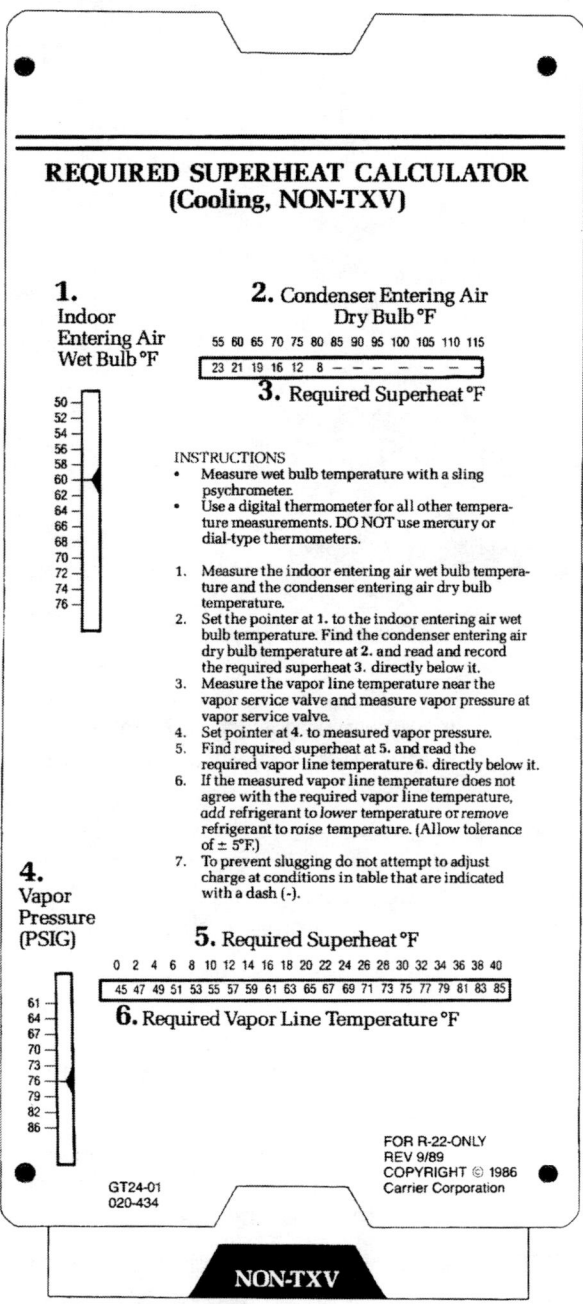

*Figure 23* ◆ Superheat charging calculator.

| °F | R-22 | R-502 | R-134A | °F | R-22 | R-502 | R-134A |
|----|------|-------|--------|----|------|-------|--------|
| 46 | 77.6 | 90.4 | 41.1 | 100 | 195.9 | 216.2 | 124.3 |
| 48 | 80.7 | 93.9 | 43.3 | 102 | 201.8 | 222.3 | 128.5 |
| 50 | 84.0 | 97.4 | 45.5 | 104 | 207.7 | 228.5 | 132.9 |
| 52 | 87.3 | 101.0 | 47.7 | 106 | 213.8 | 234.9 | 137.3 |
| 54 | 90.8 | 104.8 | 50.1 | 108 | 220.0 | 241.3 | 142.8 |
| 56 | 94.3 | 108.6 | 52.3 | 110 | 226.4 | 247.9 | 146.5 |
| 58 | 97.9 | 112.4 | 55.0 | 112 | 232.8 | 254.6 | 151.3 |
| 60 | 101.6 | 116.4 | 57.5 | 114 | 239.4 | 261.5 | 156.1 |
| 62 | 105.4 | 120.4 | 60.1 | 116 | 246.1 | 268.4 | 161.1 |
| 64 | 109.3 | 124.6 | 62.7 | 118 | 252.9 | 275.5 | 166.1 |
| 66 | 113.2 | 128.8 | 65.5 | 120 | 259.9 | 282.7 | 171.3 |
| 68 | 117.3 | 133.2 | 68.3 | 122 | 267.0 | 290.1 | 176.6 |
| 70 | 121.4 | 137.6 | 71.2 | 124 | 274.3 | 297.6 | 182.0 |
| 72 | 125.7 | 142.2 | 74.2 | 126 | 281.6 | 305.2 | 187.5 |
| 74 | 130.0 | 146.8 | 77.2 | 128 | 289.1 | 312.9 | 193.1 |
| 76 | 134.5 | 151.5 | 80.3 | 130 | 296.8 | 320.8 | 198.9 |
| 78 | 139.0 | 156.3 | 83.5 | 132 | 304.6 | 328.9 | 204.7 |
| 80 | 143.6 | 161.2 | 86.8 | 134 | 312.5 | 337.1 | 210.7 |
| 82 | 148.4 | 166.2 | 90.2 | 136 | 320.6 | 345.4 | 216.8 |
| 84 | 153.2 | 171.4 | 93.6 | 138 | 328.9 | 353.9 | 223.0 |
| 86 | 158.2 | 176.6 | 97.1 | 140 | 337.3 | 362.6 | 229.4 |
| 88 | 163.2 | 181.9 | 100.7 | 142 | 345.8 | 371.4 | 235.8 |
| 90 | 168.4 | 187.4 | 104.4 | 144 | 354.5 | 380.4 | 242.4 |
| 92 | 173.7 | 192.9 | 108.2 | 146 | 363.3 | 389.5 | 249.2 |
| 94 | 179.1 | 198.6 | 112.1 | 148 | 372.3 | 398.9 | 256.0 |
| 96 | 184.6 | 204.3 | 116.1 | 150 | 381.5 | 408.4 | 263.0 |
| 98 | 190.2 | 210.2 | 120.1 | | | | |

SATURATION TEMPERATURE (FROM TP CHART)
− LIQUID LINE TEMPERATURE (MEASURED)
———————————————————————
SUBCOOLING VALUE

EXAMPLE:
1. MEASURED LIQUID LINE PRESSURE (R-22) = 274 PSIG
2. SATURATED TEMPERATURE FROM CHART = 124°F
3. MEASURED LIQUID LINE TEMPERATURE = 114°F
4. SUBCOOLING = 124°F − 114°F = 10°F

*Figure 24* ◆ Charging for proper subcooling.

## Subcooling Value

The subcooling value for a system can be found on the unit nameplate. A typical unit nameplate is shown here.

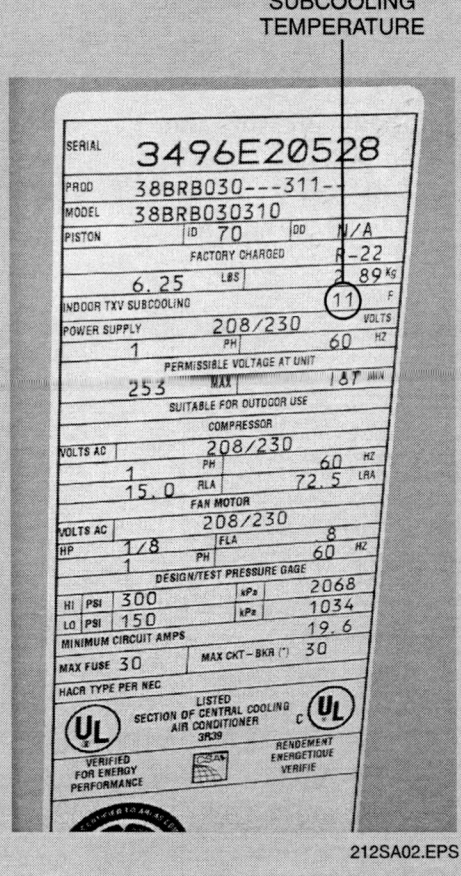

Slide rule-type subcooling charging calculators used to charge systems by the subcooling method are available from some of the main equipment manufacturers. *Figure 25* shows the subcooling calculator available from one such manufacturer. Complete instructions for use are printed on the calculator.

### 5.6.0 Charging Using Pressure Charts

Charging pressure charts are used with units where the system charge weight is unknown. Pressure charts (*Figure 26*) are normally attached to the equipment or contained in the manufacturer's service literature. To charge using pressure charts, proceed as follows:

*Step 1* Charge the system with an estimated amount of refrigerant, then operate the system until stabilized. Measure the following:

- Suction and discharge pressures
- Outdoor dry-bulb air temperature entering the condenser

*Step 2* Using the pressure chart, find the required discharge line pressure at the intersection of the measured suction line pressure and condenser entering air temperature.

*Step 3* If the measured discharge pressure matches the required discharge pressure, the refrigerant charge is considered correct and no adjustment in charge is necessary.

*Step 4* If the measured discharge pressure is too high, remove excess refrigerant using a recovery unit and proper recovery procedures to lower the discharge pressure.

*Step 5* If the measured discharge pressure is too low, add refrigerant vapor to raise the discharge pressure.

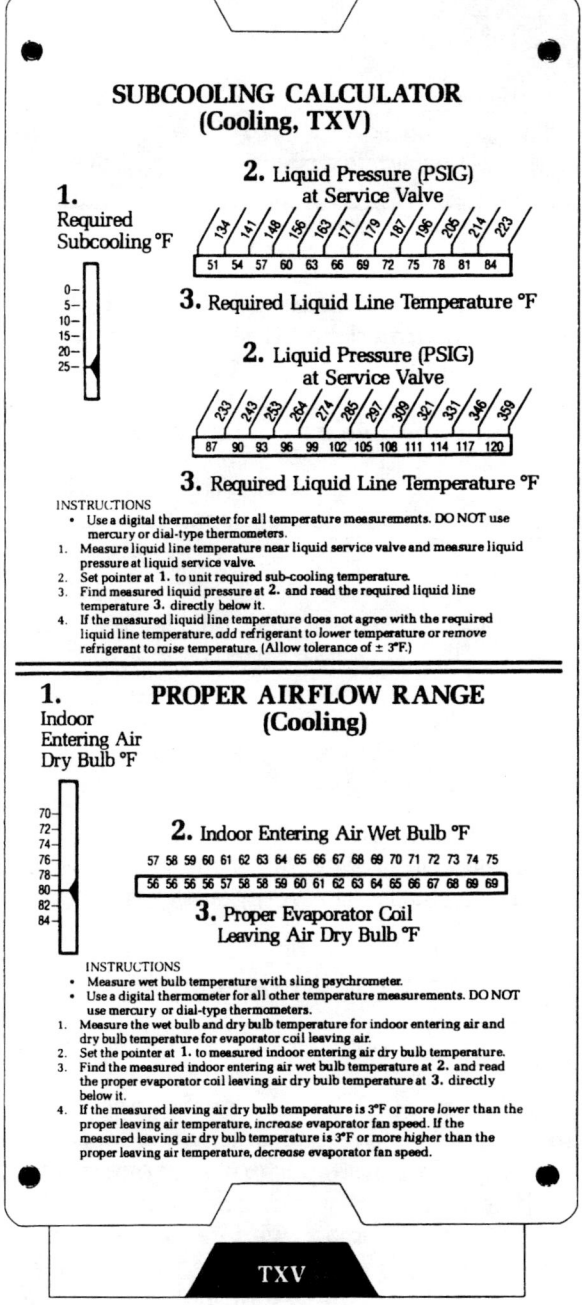

*Figure 25* ◆ Subcooling calculator.

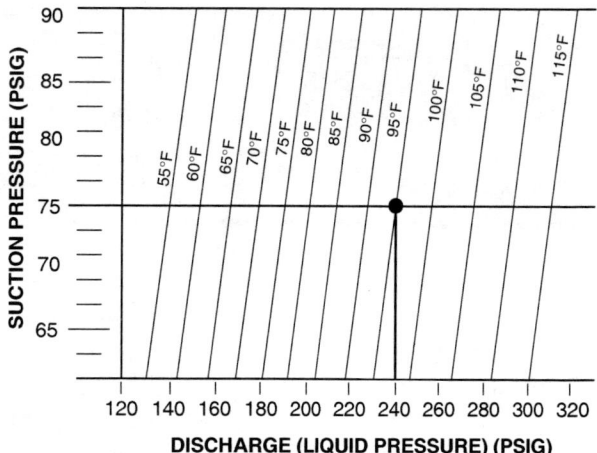

FIND THE REQUIRED DISCHARGE LINE PRESSURE AT THE INTERSECTION OF THE MEASURED SUCTION LINE PRESSURE AND CONDENSER ENTERING AIR TEMPERATURE.

EXAMPLE: THE INTERSECTION OF 75 PSIG SUCTION LINE PRESSURE AND 95°F CONDENSER ENTERING AIR YIELDS A REQUIRED DISCHARGE LINE PRESSURE OF 240 PSIG.

212F26.EPS

*Figure 26* ◆ Pressure chart.

A zeotrope is a blend of two or more refrigerants (components). Currently, most zeotropes are a blend of three refrigerants. In a zeotrope refrigerant, unlike other refrigerants, the components retain their individual characteristics and evaporate or condense over a range of temperatures at a given pressure. This causes a zeotrope to have a different liquid saturation temperature than vapor saturation temperature for a given pressure. When working with zeotropes, you must understand the meaning of two terms: **fractionation** and **temperature glide**.

- *Fractionation* – When each of the refrigerants used in a blended refrigerant retains its own chemical properties, causing each of them to leak at a different rate if released into the atmosphere. The result is that the precise proportions of the refrigerants used in the blend are altered, thereby changing the properties of the blend to something other than that specified by the manufacturer.

- *Temperature glide* – A range of temperatures in which a zeotrope refrigerant will evaporate and condense for a given pressure.

After a leak in a system using a zeotrope refrigerant is repaired, do not charge the system with the refrigerant recovered from the system and/or use a partial charge of new refrigerant to top off the system. This is because fractionation has

## 6.0.0 ◆ USING ZEOTROPE REFRIGERANTS

Refrigerants including zeotrope (near-azeotrope) blended refrigerants were discussed in detail in HVAC Level One. Some important points about zeotrope refrigerants relevant to the content of this module are briefly described here.

### Azeotropes and Zeotropes
Azeotrope refrigerants have a single boiling temperature. They mix in exact proportions and behave like pure fluids. Zeotrope refrigerants, however, are a blend of components that have different boiling points. A zeotrope refrigerant is similar to a mixture of oil and water.

altered the chemical properties of the recovered refrigerant. Recharge the system with new refrigerant after the leak is repaired and the system has been evacuated. Charging with new refrigerant guarantees that the correct system operating temperatures and pressures can be achieved. It also protects the system from possible damage resulting from the use of a refrigerant that does not have the specific properties the manufacturer intended for use in the system.

When charging a system that uses a zeotrope refrigerant, the system should be charged using liquid refrigerant. This is necessary to avoid fractionation and to be sure that the proper refrigerant blend composition is charged into the system. If it is necessary to charge an operating system with refrigerant vapor, the refrigerant must be removed from the cylinder as a liquid, then fed from the cylinder to the low side of the system through a metering device to make sure that all of the liquid refrigerant is converted to vapor before entering the system.

When using the superheat or subcooling method of charging on a system that uses a zeotrope refrigerant, the calculation of superheat or subcooling is done in a slightly different manner than with other refrigerants. This is because of the temperature glide. The pressure-temperature relationship for a zeotrope refrigerant has two temperatures for a given pressure: the saturated vapor temperature and the saturated liquid temperature. When calculating superheat, be sure to use the saturated vapor temperature in the calculation. When calculating subcooling, be sure to use the saturated liquid temperature in the subcooling calculation. Use of the wrong temperature will cause a superheat or subcooling calculation error and possible damage to the unit.

$$\begin{aligned} &\text{Actual suction line temperature} \\ -\ &\text{Suction line saturated vapor temperature} \\ =\ &\text{Superheat} \end{aligned}$$

$$\begin{aligned} &\text{Liquid line saturated liquid temperature} \\ -\ &\text{Actual liquid line temperature} \\ =\ &\text{Subcooling} \end{aligned}$$

### Summary

The four basic service procedures used to troubleshoot, repair, and/or maintain correct operation of a mechanical refrigeration system are:

- Leak detection
- Evacuation and dehydration
- Recovery
- Charging

These procedures are performed both when installing new systems and when servicing existing ones. Failure to perform any one of these procedures correctly can result in poor system operation and may even cause system failure. They must also be performed in order to make sure that the venting requirements of the Clean Air Act are not violated.

## Review Questions

1. Ultrasonic leak detection can best be used to check systems _____.
   a. under pressure or in a vacuum
   b. using liquid and vapor methods
   c. using high-side and low-side methods
   d. in highly noisy areas

2. After the application of a commercial bubble solution, a small leak may take _____ to form a bubble.
   a. several minutes
   b. 4 hours
   c. 8 hours
   d. 24 hours

3. When pressurizing a system for leak testing, always charge with the refrigerant first and remove the cylinder before adding _____.
   a. HCFC-22
   b. oxygen
   c. compressed air
   d. regulated nitrogen

4. Because of the Clean Air Act Amendments, all refrigerants must be _____.
   a. vented to the atmosphere
   b. contained
   c. charged in the liquid state
   d. charged in the vapor state

5. Recovery of a refrigerant means that the refrigerant is _____.
   a. removed from a system and stored in approved containers
   b. circulated through filtering devices to remove contaminants
   c. removed from a system and cleaned
   d. removed from a system and remanufactured to new refrigerant quality

6. Reclamation of a refrigerant means that it is _____.
   a. processed through a recycle unit
   b. processed so that it meets the standards for new refrigerant
   c. tested with a moisture/acid test kit
   d. mixed with new refrigerant

7. Recycling of a refrigerant means that it is _____.
   a. treated so that it is as dry and clean as new refrigerant
   b. run through a cleaning and decontamination filtering device
   c. mixed with new refrigerant
   d. recovered from a system more than once

8. A certified recovery unit used to recover refrigerant from a system containing 12 pounds of HCFC-22 refrigerant must be capable of evacuating the system to _____ in. Hg vacuum.
   a. 0
   b. 4
   c. 10
   d. 15

9. Contaminants that may be found in a poorly maintained system are _____.
   a. moisture
   b. air
   c. both moisture and air
   d. alcohol

10. Air in a system mainly causes _____.
    a. hydrochloric acid to form
    b. oil sludge to form
    c. corrosion of metals
    d. the system to work harder due to an increase in condensing temperature

11. When moisture and acid are present in a system, the system must be _____.
    a. destroyed
    b. evacuated
    c. steam cleaned
    d. flushed

12. The principle of evacuating a system using a vacuum pump is to lower the system pressure so any moisture can _____.
    a. form
    b. condense
    c. boil off or vaporize
    d. be drained

13. If a system is evacuated to a level of 25,400 microns, the liquid water in the system will boil at _____ and above.
    a. 32°F
    b. 76°F
    c. 80°F
    d. 86°F

14. The most accurate way to measure a vacuum is with a _____.
    a. vacuum gauge
    b. gauge manifold set
    c. compound gauge
    d. wet-bulb indicator

15. The deep vacuum method of evacuation requires that the system be evacuated to a level of _____ microns.
    a. 500
    b. 1,000
    c. 2,000
    d. 50,000

16. If a filter-drier is added, or if a long line set is used, the refrigerant charge must be _____.
    a. decreased
    b. reclaimed
    c. transferred
    d. increased

17. When charging refrigerant by weight, you can charge with _____.
    a. liquid refrigerant
    b. vapor refrigerant
    c. both liquid and vapor refrigerant
    d. nitrogen

18. When charging by superheat, the outdoor and indoor temperatures are used to find the _____.
    a. superheat in the suction line
    b. suction line pressure
    c. liquid line temperature
    d. required discharge line temperature

19. The subcooling charging method gives an indication of the amount of _____.
    a. pressure in the suction line
    b. refrigerant liquid at the expansion device
    c. refrigerant vapor just before the expansion device
    d. pressure in the liquid line

20. When charging using pressure charts, what is the required discharge line pressure if the suction pressure is 80 psig and the condenser entering air is 90°F? (Hint: Use the pressure chart in *Figure 24*.)
    a. 200 psig
    b. 220 psig
    c. 230 psig
    d. 240 psig

21. When charging a system using pressure charts, if the measured discharge pressure is higher than the required discharge pressure, you should _____ the system.
    a. remove excess refrigerant from
    b. add refrigerant vapor to
    c. make no adjustment to
    d. add dry nitrogen to

22. When each of the refrigerants used in a blended refrigerant leak at a different rate if released into the atmosphere, changing the precise proportions of the refrigerants used in the blend, the process is called _____.
    a. azeotrope reaction
    b. fractionation
    c. superheat
    d. subcooling

23. A range of temperatures in which a zeotrope refrigerant will evaporate and condense for a given pressure is called _____.
    a. fractionation
    b. subcooling
    c. temperature glide
    d. superheat

24. After a leak in a system using a zeotrope refrigerant is repaired and the system has been evacuated, the system should be charged with _____.
    a. new refrigerant
    b. recovered refrigerant
    c. a blend of recovered and new refrigerant
    d. dry nitrogen

25. When charging a system that uses a zeotrope refrigerant, the system should be charged using _____.
    a. dry nitrogen
    b. liquid refrigerant
    c. vapor refrigerant fed to the high side of the system
    d. liquid refrigerant partially converted to vapor

## GLOSSARY

# Trade Terms Introduced in This Module

*Fractionation:* A term related to refrigerant blends; refers to the process by which each of the refrigerants in the blend leaks at a different rate if released into the atmosphere.

*Reclamation:* The remanufacture of used refrigerant to bring it up to the standards required of new refrigerant. Reclamation is not a field service procedure. It is a complicated process done only at reprocessing or manufacturing facilities.

*Recovery:* The removal and temporary storage of refrigerant in containers approved for that purpose. Recovery does not provide for any cleaning or filtration of the refrigerant.

*Recycle:* To circulate recovered refrigerant through filtering devices that remove moisture, acid, and other contaminants. This does not mean that it meets the purity standards for new refrigerants.

*Temperature glide:* A range of temperatures in which a zeotrope refrigerant will evaporate and condense for a given pressure.

# ANSWER KEY

## Answers to Review Questions

| Answer | Section |
|---|---|
| 1. a | 2.1.3 |
| 2. b | 2.1.5 |
| 3. d | 2.2.2 |
| 4. b | 3.0.0 |
| 5. a | 3.0.0 |
| 6. b | 3.0.0 |
| 7. b | 3.0.0 |
| 8. c | 3.1.1 |
| 9. c | 4.0.0 |
| 10. d | 4.0.0 |
| 11. b | 4.0.0 |
| 12. c | 4.1.1 |
| 13. c | 4.1.1, Fig. 9 |
| 14. a | 4.1.2 |
| 15. a | 4.2.1 |
| 16. d | 5.2.0 |
| 17. c | 5.3.0 |
| 18. a | 5.4.0 |
| 19. b | 5.5.0 |
| 20. c | 5.5.0, 5.6.0 |
| 21. a | 5.6.0 |
| 22. b | 6.0.0 |
| 23. c | 6.0.0 |
| 24. a | 6.0.0 |
| 25. b | 6.0.0 |

# REFERENCES & ACKNOWLEDGMENTS

## Additional Resources

This module is intended to present thorough resources for task training. The following reference works are suggested for further study. These are optional materials for continued education rather than for task training.

*Modern Refrigeration and Air Conditioning*, 2000. A.D. Althouse, C.H. Turnquist, A.F. Bracciano. Tinley Park, IL: The Goodheart-Willcox Company, Inc.

*Refrigerant Service Techniques*, 1993. Syracuse, NY: Carrier Corporation.

*Refrigeration & Air Conditioning Technology*, 2000. William C. Whitman, William M. Johnson, John A. Tomczyk. Albany, NY: Delmar Publishers, Inc.

## Figure Credits

| | |
|---|---|
| **Gerald Shannon** | 212F02, 212F10, 212F16, 212SA02 |
| **Amprobe/TIF** | 212F03, 212SA01 |
| **Spectronics Corp.** | 212F04 |
| **SPX Robinair** | 212F08, 212F11 |
| **ALCO Controls** | 212F18 |
| **Carrier Corporation** | 212F23, 212F25 |

# NCCER CRAFT TRAINING USER UPDATES

The NCCER makes every effort to keep these textbooks up-to-date and free of technical errors. We appreciate your help in this process. If you have an idea for improving this textbook, or if you find an error, a typographical mistake, or an inaccuracy in the NCCER's Craft Training textbooks, please write us, using this form or a photocopy. Be sure to include the exact module number, page number, a detailed description, and the correction, if applicable. Your input will be brought to the attention of the Technical Review Committee. Thank you for your assistance.

*Instructors* – If you found that additional materials were necessary in order to teach this module effectively, please let us know so that we may include them in the Equipment and Materials list in the Instructor's Guide.

**Write:** Curriculum Revision and Development Department
National Center for Construction Education and Research
P.O. Box 141104, Gainesville, FL 32614-1104

**Fax:** 352-334-0932

**E-mail:** curriculum@nccer.org

---

Craft _____   Module Name _____

Copyright Date _____   Module Number _____   Page Number(s) _____

Description
_____
_____
_____
_____

(Optional) Correction
_____
_____
_____

(Optional) Your Name and Address
_____
_____
_____

# Index

Absolute pressure, 1.5, 1.37
Absolute zero, 11.6, 11.27
AC. *See* Alternating current
Accessibility clearances, vent systems, 2.5
Accessories, 1.2, 8.21
    carbon dioxide monitors, 8.21
    carbon monoxide monitors, 8.20–21
    comfort air conditioning, 8.3–4
    dehumidifiers, 8.6–7, 8.23
    economizers, 3.37, 8.10, 8.15–18, 8.23
    electronic air cleaners, 3.34, 3.35, 3.36, 8.12–15
    energy conservation equipment, 8.15–19
    energy recovery ventilators (ERVs), 8.10, 8.15, 8.16, 8.23
    evaporative pre-coolers, 8.18
    fire dampers, 1.25–26, 8.19–20
    heat recovery ventilators (HRVs), 8.10, 8.15, 8.16, 8.23
    humidifiers, 6.4, 8.5, 8.6–10, 8.23
    indoor air quality, 8.10–15
    mechanical air filters, 3.34, 3.35, 8.11–15
    smoke dampers, 1.26, 8.19–20
    ultraviolet light air purification systems, 8.20
    zoned control, 8.18–19
Accordion-fold packing, 3.16
Accumulators, heat pumps, 11.12–13, 11.14
Acid contamination, compressors, 10.27
Acid/moisture test kits, 10.27
Acorn nuts, 3.5, 3.6
Activated carbon disposable filters, 8.12, 8.14
Actuators, 7.55, 7.64
Add-on systems, heat pumps, 11.21–22
AI signals. *See* Analog In signals
Air conditioner branch circuits, 4.10, 4.11
Air conditioning, 8.1. *See also* Cooling system
    accessories. *See* Accessories
    circuit, 7.22
    comfort air conditioning, 8.1–4, 8.21
    defined, 8.21
    energy conservation equipment, 8.15–19
    humidity control, 8.4–10
    indoor air quality, 3.34–36, 7.26, 8.10–15
    process air conditioning, 8.1
Air contamination, compressors, 10.26–27
Air diffusers, 1.24–25
Air distribution system, 1.1–2. *See also* Airflow
    air velocity measurement, 1.33–34
    blowers, 1.2, 1.3, 1.5, 1.6–9, 1.12
    commercial and industrial system, 1.12
    duct system components, 1.18–28
    duct systems, 1.2, 1.3, 1.12–18
    electric heating, 6.15–17
    fans, 1.5–6, 1.9–12
    friction losses in, 1.3
    humidity measurement, 1.28, 1.29–30
    pressure difference to move air, 1.3
    pressure measurement, 1.4, 1.5, 1.31–32
    residential systems, 1.5–6, 1.12
    static pressure, 1.3–4, 1.10, 1.11, 1.31, 1.37
    temperature measurement, 1.28–30
    total pressure, 1.4, 1.31, 1.37
    troubleshooting, 7.26
    velocity pressure, 1.4, 1.31, 1.37
Air filters, 1.12
    cleaning, 3.35, 8.15
    disposable filters, 3.34, 8.11–12, 8.14
    electric heating, 6.4
    electronic air cleaners, 3.34, 3.35, 3.36, 8.12–15
    electrostatic filters, 3.34, 8.12
    installation, 8.14
    maintenance, 3.34–35
    mechanical air filters, 3.34, 3.35, 8.11–15
    permanent filters, 8.12
    sizing, 3.35
    types, 3.34
Air handlers, 1.2, 1.5
    blowers, 1.2, 1.3, 1.5, 1.6–9, 1.12
    defined, 1.37
    electric heating, 6.4
    furnaces, 6.1
    split system heat pump, 11.18
Air inlet, blower, 1.2
Air pressure, in air distribution system, 1.2–3
Air purification, by ultraviolet light, 8.20
Air quality. *See* Indoor air quality
Air supply, vent systems, 2.5
Air-to-air heat pumps, 11.1, 11.2, 11.3, 11.6, 11.7, 11.19, 11.21
Air-to-water heat pumps, 11.2, 11.5
Air velocity, 1.3, 1.33–34, 1.37
Air volume, 1.3
Air volume balancers, 1.34
Airflow, 1.2, 1.3
    in cold climate systems, 1.13
    cooling mode, 1.13
    duct fans, 1.10
    electric heating, 6.15–17

Airflow, *continued*
   free air delivery, 1.9
   heating mode, 1.13
   measurement of, 1.33–34
   in residential system, 1.5–6
   in warm climate systems, 1.15–19
Airflow control, 7.57
Alignment
   belt drives, 3.29–30
   couplings, 3.32–33
Alternating current (AC), 4.1
   capacitors, 4.16–17, 4.19
   circuit diagrams, 4.28–29
   circuits, 4.15–16
   history, 4.2
   induction motors, 4.17–21, 4.29
   power generation, 4.5–15, 4.6
   rectification, 5.4
   safety, 4.27–28, 6.13–16
   testing components, 4.21–27
   transformers, 4.2–5
   using, 4.16–17
Alternating current (AC) motors, variable frequency drives (VFDs), 5.15
Alternator (alternating current generator), 4.6
Aluminum, as conductor, 5.3
American National Standard thread, 3.2
Ammeters, 4.21, 4.22
   compressor checks, 10.33
   electric heating check, 6.12
   start relay checks, 7.46
   thermostat check, 6.9
Analog devices, 7.58
Analog In (AI) signals, 7.59
Analog Out (AO) signals, 7.59
Analog-to-digital converter, 7.58, 7.64
Anchor bolts, 3.7, 3.12–13
Angry customers, dealing with, 3.49
Angular ball bearings, 3.22
Angular misalignment, couplings, 3.33
Anode, 5.5, 5.27
Anti-friction bearings, 3.22–23
Anticipator, 2.9–10
AO signals. *See* Analog Out signals
Appearance, of service technician, 3.43–44
Aquastats, 7.49–50
Arbor press, 3.23, 3.25
Argumentative customers, dealing with, 3.48
Armature, generators, 4.6, 4.7
Aspect ratio, 1.19–20
Aspirating psychrometer, 1.30
Atmospheric pressure, 1.2, 1.5, 1.37
Atomic structure, 5.1–2, 5.3
Atomizing humidifiers, 8.8–9
Attic extended plenum systems, 1.17–18
Attic radial duct systems, 1.17–18
Audio card, 5.23–24
Automatic changeover thermostats, 7.4–5, 7.64
Automatic expansion valves, 9.8–9
Automatic packing, 3.15
Autotransformers, 4.3–4, 7.16, 10.24
Average building, 8.10
Axial load, bearings, 3.21, 3.53

Back electromotive force (back EMF), 4.19, 4.20
Backward-inclined centrifugal blowers, 1.8
Bag-type disposable filters, 8.12, 8.14
Balance point, 11.17–18, 11.27

Ball bearings, 3.22
Bandwidth, data transmission, 5.18–19
Base-mounted pumps, 10.36
Baseboard diffusers, 1.13, 1.24, 1.25
Baseboard heaters, 6.2, 6.18, 7.6
Battery, schematic symbol, 7.66
Battery-operated psychrometer, 1.30
Bearing heater, 3.24, 3.25
Bearing pullers, 3.23
Bearings
   anti-friction bearings, 3.22–23
   detecting failures, 3.23
   installing, 3.24–25
   lubrication, 3.33, 3.34
   over-lubrication, 3.33
   plain bearings, 3.21–22
   removing, 3.23–24
Belt-driven blowers, 1.7, 6.17
Belt drives, 3.28–30
Belts, 3.28–29
Bimetal sensor, 6.9, 6.23, 7.2
Bimetal thermostats, 7.2
BIOS, 5.18, 5.27
Bits (computers), 5.18
Bleed controls, 7.54, 7.64
Bleeder resistor, 4.25
Blind rivet, 3.9
Blower control, forced-air electric heating, 6.6–8
Blower speed, electric heating, 6.17
Blowers, 1.2, 1.3, 1.5, 1.6–9, 1.12
   belt-driven blowers, 1.7
   centrifugal blowers, 1.7–9
   control systems, 6.6–8
   defined, 1.6
   direct-drive blowers, 1.7
   electric heating, 6.4, 6.6–8, 6.17
   fan curve charts, 1.11
   motor speed controls, 7.16–17
   residential systems, 1.7
Blown fuses, 4.26, 6.15
Body comfort, 7.3
Boilers
   electric, 6.19–20
   hydronic systems, 7.49, 7.50
   low water cutoff, 7.50
Bolts, 3.2, 3.3, 3.10
Boots, 1.23
Braid-over-braid packing, 3.15, 3.16
Branch circuits, 4.10–11
Branch ducts, 1.18–22, 1.27
Brazed joint, disassembling, 12.6–7
Break-away torque, 3.11, 3.53
Bridge circuit, 5.12
Bridge rectifier, 5.7, 5.27
British thermal units (Btus), 4.16
Brushes, generator, 4.5–6
Btus. *See* British thermal units
Btuh output, 6.4, 6.13
Bubble solutions, for leak detection, 12.5
Building management systems, 7.57, 7.58
Buildings, loose, average, and tight, 8.9
Burning, CD, 5.25
Burnout, motors, 10.26, 10.35, 10.39
Burns, electrical, 4.27
Bus (computers), 5.18
Butterfly damper, 1.26
Bypass humidifiers, 8.7, 8.8
Bytes (computers), 5.18

Cache, 5.18
Cadmium sulfide flame detector, 5.13–14
Cage (bearings), 3.22
Calibration, thermostats, 7.11
Calrod-type heating element, 6.3
Campus maps, 3.38
Cap screws, 3.4
Cap tubes. *See* Capillary tubes
Capacitive circuit, 4.17
Capacitor analyzer, 4.21, 7.44
Capacitor discharging tool, 4.25, 4.26
Capacitor run motors (CSR motors), 7.42, 7.43, 7.47
Capacitor start, capacitor run motor, 4.20
Capacitor start motors, 4.19–20, 7.42
Capacitors, 4.16–17, 4.19, 7.44
   continuous-duty capacitors, 7.44
   electrical checking, 4.25–26
   PCB health hazard, 4.26
   schematic symbols, 7.66
   start capacitors, 4.19, 7.43
   testing, 4.25–26, 7.44
Capacity control, 10.13–15, 10.32, 10.39
Capillary tubes (cap tubes), 9.1, 9.3, 9.4–6, 9.22
Carbon composition resistor, 5.9
Carbon dioxide monitors, 8.21
Carbon monoxide monitors, 8.20–21
Cast-bronze bearings, 3.21
Castellated nuts, 3.5, 3.6
Cathode, 5.5, 5.27
CD-R, 5.21, 5.25
CD-ROM, 5.21, 5.25
CD-RW, 5.21, 5.25
Ceiling diffusers, 1.16, 1.24, 1.25
Central-point exhaust unit, 11.5
Centrifugal blowers, 1.7–9
Centrifugal compressors, 10.4, 10.12, 10.14
Centrifugal force, 5.27, 10.12
Centrifugal pump, 10.36
Centrifugal switch, 7.21
Charging, 12.16–29
   charge determination and accuracy, 12.20–21
   charging cylinder, 12.18–19
   charging scales, 12.17–18
   recovery/recycle unit, 12.19
   refrigerant cylinders, 12.17
   service equipment for, 12.17–19
   sight glass, 12.19–20
   by subcooling, 12.26–28, 12.29
   by superheat, 12.24–26
   system with zeotrope refrigerant, 12.30
   using pressure charts, 12.28
   by weight, 12.21–23
Charging cylinder, 12.18–19
Charging scales, 12.17–18
Chimneys, 2.6–7, 2.10–12
Chips (integrated circuits), 5.1, 5.16–17, 5.27
Circuit breakers, troubleshooting, 7.36, 7.38–39
Circuit diagrams
   AC voltage on, 4.28–29
   analyzing, 7.24
   cooling/gas heating system, 7.25
Circulating pumps, hydronic systems, 7.50–51, 10.36
Clamp couplings, 3.31
Clamp-on ammeter, 4.21, 7.9, 7.46
Clearance volume, 10.39
Clearances, vent systems, 2.5
Clevin pins, 3.8
Coal furnaces, venting, 2.3
Coefficient of performance (COP), heat pumps, 11.15–16, 11.27

Coil cleaners, 3.36
Coil energy recovery loop, 11.5, 11.6
Coils
   designations for, 7.15
   heat pumps, 11.1
   maintenance, 3.36–37
   schematic symbols, 7.66
Cold climates, duct systems, 1.13–15
Color code, resistors, 5.10–11
Combined U-inclined manometer, 1.31
Combustion. *See also* Furnaces
   carbon monoxide gas, 8.20
   flames, 2.2–3
   flue gases, 2.1, 2.3–4
   furnaces, 2.1–3
   venting, 2.1–2, 2.3, 2.4–7
Combustion air, 2.2–3
Combustion efficiency, 2.2
Comfort air conditioning, 8.1–4, 8.21
Comfort zone, 8.2, 8.3
Commercial systems
   air distribution, 1.12
   central-point exhaust unit, 11.5
   control systems, 7.11
   pneumatic control systems, 7.56
   three-phase power, 4.12–15
Commissioning job report, 3.38–39
Common, 7.43
Communicating with customers, 3.44, 3.45–50
   handling difficult customers, 3.47–50
   keeping communications positive, 3.46
   showing concern for customers, 3.46–47
   tips for, 3.46
Commutator, 4.5–6
Complete combustion, 2.2, 2.18
Compliant scroll compressor, 10.11, 10.39
Compression, 10.39
Compression couplings, 3.31
Compression heat, 11.6, 11.27
Compression rings, 10.5
Compression-type packing, 3.15, 3.17–18
Compressor. *See* Compressors
Compressor motors, 10.15–21
   alignment, 10.16
   burnout, 10.26, 10.35, 10.39
   cooling, 10.15–16
   dual-voltage rated motors, 7.36, 10.29
   electrical failures, 10.28–30, 10.35
   input power, 10.16
   operating voltage ranges, 10.28–29
   overload protection, 10.16–21
   reduced-voltage starting, 10.22, 10.24
   single-voltage rated motors, 7.35, 10.29
Compressor short-cycle timer, 7.19
Compressor valves, 10.6, 10.7, 10.33
Compressors, 10.1
   capacity control of, 10.13–15, 10.32, 10.39
   centrifugal compressors, 10.4, 10.12, 10.14
   compliant scroll compressor, 10.11, 10.39
   components, 10.1–3
   compressor short-cycle timer, 7.19
   condensers, 3.36–37, 9.1–19, 10.22
   contamination, 10.26–28
   control circuit safety switches, 7.19–20
   crankcase heaters, 10.26, 11.15
   crankshaft seals, 10.7
   cylinder unloading, 10.14
   electrical failures, 10.28–30, 10.35
   electronic head pressure control, 10.22

Compressors, *continued*
   electronically commutated motors (ECMs), 5.14
   evaporators, 9.1, 9.3, 10.21–22, 10.23
   failure, causes of, 10.16, 10.24–30, 10.35
   flooded starts, 10.25, 10.26, 10.39
   flooding, 9.2, 9.3, 9.22, 10.25, 10.39
   hermetic compressors, 7.44, 10.3
   hot gas bypass, 10.14
   inlet guide vane, 10.14
   intake slide valve, 10.14
   maximum continuous current (MCC), 10.28
   motor speed controls, 7.16–17
   motors. *See* Compressor motors
   nameplate, 10.28
   on/off cycling, 10.14
   open-drive compressor, 10.3, 10.4, 10.7
   overheating, 10.30, 10.32
   positive-displacement compressors, 10.13, 10.39
   pressure protection, 10.21, 10.23
   reciprocating compressors, 10.4–9
   refrigerant oils, 10.7–8
   replacement, 1.037, 10.33–36
   role of, 10.1, 10.3
   rotary compressors, 10.4, 10.9–10
   rotary vane compressors, 10.10, 10.11
   safety warnings, 7.44, 10.30, 10.31, 10.32, 10.34, 10.35
   schematic symbols, 7.66
   screw compressors, 10.4, 10.11–12, 10.14
   scroll compressors, 10.4, 10.10–11
   semi-hermetic compressors, 7.44, 10.3
   short-cycle time delay, 7.11
   short cycling, 7.19, 10.13, 10.22, 10.39
   slugging, 9.3, 9.22, 10.25
   speed control, 10.13–14
   stationary vane compressors, 10.9, 10.10
   subcooling, 10.2–3
   system checkout procedures, 10.30–33
   three-phase circuit, 4.13, 4.14
   time delay relay, 7.18
   troubleshooting, 10.30–33, 10.37
   types, 10.3–12, 10.37
   valve damage, 10.7
Computers, 5.18–25
   input devices, 5.21
   mainframes, 5.19
   monitors, 5.21–22
   personal computers, 5.19–24
   storage media, 5.24–25
   terms, 5.18–19
Concentric termination, vent systems, 2.13, 2.14
Concrete slabs, duct systems in, 1.14
Condensate drain, trap, 3.36
Condenser coils, 7.18
Condensers, 3.36–37, 9.1–19
Condensing furnaces, 2.1, 2.6, 2.11–14, 2.18
Condensing unit, 7.22
Conduction, body comfort and, 8.3
Conductors, 5.2, 7.65
Cone (bearings), 3.22
Constant volume HVAC system, 7.59–60
Contactor/relay contacts, checking, 7.40–41
Contactors, 7.15–16, 7.40
Contaminants
   compressor failure, 10.26
   indoor air quality, 8.10–11
Contamination, compressors, 10.26–28
Continuous-duty capacitors, 7.44
Control circuits, 7.33
   cooling systems, 7.22–24
   isolating to a faulty circuit, 7.33–35
   lockout control circuit, 7.17–18
   low-voltage circuits, 7.33
   safety switches, 7.19–20
   sequence of operation, 7.21–25
Control devices, 7.1, 7.11. *See also* Control systems
   compressor short-cycle timer, 7.19
   contactors, 7.15–16
   control circuit safety switches, 7.19–20
   defrost cycle, 11.8–10
   digital control systems, 7.57–60
   fan control, 7.20
   freezestat, 7.19
   for furnaces, 7.20–21
   heat pumps, 11.8–10, 11.23, 11.25
   for hydronic systems, 7.49–51
   inducer switch, 7.21, 7.22
   limit control, 7.20–21
   lockout relay, 7.17–18
   motor speed controls, 7.16–17
   motor starters, 7.16
   pressure switches, 7.19
   relays. *See* Relays
   sail switch, 7.57
   thermocouple, 7.21
   thermostats. *See* Thermostats
   timers, 7.18–19
Control setpoint, 10.14
Control systems. *See also* Control devices
   components, 7.12
   digital control systems, 7.57–60
   electronic furnace control, 5.17
   forced-air electric heat, 6.5–9
   furnaces, 7.20–21
   pneumatic controls, 7.51, 7.54–57
Control transformers, checking, 7.41–42
Controller, pneumatic system, 7.54
Convection, body comfort and, 8.3–4
Cooling, compressor motors, 10.15–16
Cooling compensator, 5.10, 7.3–4, 7.64
Cooling cycle, heat pumps, 11.6, 11.7
Cooling/gas heating system, circuit diagram, 7.25
Cooling mode, airflow, 1.13
Cooling-only thermostats, 7.3–4
Cooling systems. *See also* Refrigeration systems
   accessories. *See* Accessories
   airflow, 1.5
   comfort air conditioning, 8.1–4, 8.21
   compressors. *See* Compressors
   control circuit, 7.22–24
   cooling compensator, 5.10, 7.3–4
   energy conservation equipment, 8.15–19
   evaporative pre-coolers, 8.18
   heat pumps. *See* Heat pumps
   humidity control, 8.4–10
   packaged units, 6.11, 11.2, 11.20–21, 11.27
   process air conditioning, 8.1
   thermostats. *See* Thermostats
   troubleshooting, 7.42
   zoned control, 8.18–19
COP. *See* Coefficient of performance
Copper, as conductor, 5.3
Cotter pins, 3.8
Countersunk lock washers, 3.35
Couplings, 3.30–33
CPR. *See* Crankcase pressure regulator
Crankcase heaters, 10.26, 11.15
Crankcase pressure regulator (CPR), 10.28
Crankshaft seals, 10.7

Critical customers, dealing with, 3.49
Crossbreaks, 1.19
CSR motors. *See* Capacitor run motors
Cubic feet per minute (cfm), 1.10, 1.37
Cup (bearings), 3.22
Current check
    electric heating, 6.12
    thermostats, 7.8–10
Current imbalance, three-phase systems, 4.15, 10.29–30
Current monitoring devices, 10.20
Customer communication. *See* Communicating with customers
Customer interviews, for troubleshooting, 7.26
Customer relations, 3.42–45
    communicating with customers, 3.44, 3.45–50
    difficult customers, 3.47–50
    handling service calls, 3.44–50
    importance of, 3.43
    technicians' habits, behaviors, and attitudes, 3.43–44
Cylindrical bearings, 3.22

Dampers, 1.6, 1.25–26
    fire dampers, 1.25–26, 8.19–20
    inspection and cleaning, 3.37
    normally open dampers, 7.55–56
    smoke dampers, 1.26, 8.19–20
Data transmission, 5.18–19
DC. *See* Direct current
DDC. *See* Direct digital control
Deadband, 7.5, 7.64
Deep vacuum evacuation method, 12.13–14
Defrost cycle, heat pumps, 11.2, 11.6, 11.8–10, 11.27
Defrost relay (DFR), 11.9
Defrost thermostat (DFT), 11.9
Dehumidification, 8.5–6, 8.23
Delta-connected stator, 4.21
Delta transformer, 4.5
Demand defrost, 11.10, 11.27
Design load, 9.3, 9.22
Dewpoint, 1.30, 1.37
DFR. *See* Defrost relay
DFT. *See* Defrost thermostat
DI signals. *See* Digital In signals
Dielectric, 4.16
Differential, 7.64
Differential pressure gauge, 1.31–32
Difficult customers, dealing with, 3.47–50
Diffusers, 1.24–25, 7.26
Digital control systems, 7.57–60
Digital In (DI) signals, 7.59
Digital magnetic tape drives, 5.25
Digital multimeters (DMMs), 1.28, 4.8, 4.26–27, 7.37
Digital Out (DO) signals, 7.59
Digital readout megger, 4.23
Dilution air, 2.3, 2.7, 2.18
DIMMs, 5.19
Diodes, 5.4–7, 7.66
DIP switch, 5.18
Direct current (DC)
    power generation, 4.5–6
    rectification, 5.4
Direct-current (DC) motors, electronically commutated motors (ECMs), 5.14
Direct digital control (DDC), 7.57–60
Direct drive, 3.30
Direct-drive blowers, 1.7
Direct-expansion (DX) evaporators, 9.2, 9.22
Dirt contamination, compressors, 10.27–28
Discharge line mufflers, 11.20

Disconnect switches, 4.10, 4.12, 6.11
Diskette drive, 5.20
Diskettes, 5.25
Displacement, compressors, 10.13
Disposable filters, 3.34, 8.11–12, 8.14
Disposal of equipment. *See* Equipment disposal
Distributor line, 9.11, 9.22
Distributors, 9.15–16, 9.22
Diverter, 2.15
DMMs. *See* Digital multimeters
DO signals. *See* Digital Out signals
Documentation, 3.37–42
    commissioning job report, 3.38–39
    make, model, and serial number, 3.38
    roof or campus maps, 3.38
    service ticket/invoice, 3.38, 3.39
    start-up report, 3.39, 3.41, 3.42
    warranty ticket, 3.39
Double-lip seal, 3.19
Double-pole, double-throw (DPDT) relay, 7.14, 7.15
Double-wall vent connector, 2.10
Double-wall vents, 2.7
Dowel pins, 3.8
DPDT relay. *See* Double-pole, double-throw relay
Draft controls, 2.14–15
Draft diverters, 2.15
Draft hood, 2.15
Draft regulator, 2.14, 2.15
Drafts, troubleshooting, 7.26
DRAM, 5.19
Drives, variable frequency drives (VFDs), 5.15
Droop, 7.6, 7.64
Dropping point, of greases, 3.27, 3.53
Dry-bulb temperature, 1.28, 1.29, 1.37
DSL line, 5.18
Dual-fuel heating system, 11.21, 11.27
Dual-voltage rated motors, 7.36, 10.29
Duct connections, 1.12
Duct fans, 1.10
Duct heaters, 6.17, 6.18
Duct size, aspect ratio, 1.19–20
Duct system components, 1.12, 1.18–19
    air diffusers, 1.24–25
    dampers, 1.6, 1.25–26
    fittings and transitions, 1.23
    grilles, 1.24–25
    insulation and vapor barriers, 1.27–28
    main trunk and branch ducts, 1.19–22
    registers, 1.24–25
Duct systems, 1.2, 1.3, 1.12–18
    cold climate systems, 1.13–15
    components. *See* Duct system components
    in concrete slabs, 1.14
    equivalent length, 1.23
    fiberglass ducts, 1.21
    flexible duct, 1.22
    friction losses in, 1.3
    installation, 1.13
    low pressure in, 1.3, 1.4
    metal ducts, 1.19–20
    pressure difference to move air, 1.3
    warm climate systems, 1.15–19
Ductless split system, 11.19
Ducts, preinsulated, 1.27
Dust stop filters, 3.34
DVD, 5.21, 5.25
DX evaporators. *See* Direct-expansion (DX) evaporators
Dynamic-rated dampers, 1.25
Dynamic seals, 3.18, 3.53

E-P relays. *See* Electric-pneumatic relays
ECMs. *See* Electronically commutated motors
Economizers, 3.37, 8.10, 8.15–18, 8.23
Edison hookup, 4.9, 4.10
EEVs. *See* Electronic expansion valves
Effective voltage, 4.6, 4.8
Elbows, 1.23
Electric boilers, 6.19–20
Electric circuits. *See also* Circuit diagrams
   capacitive circuit, 4.17
   circuit diagrams, 4.28–29
   inductive circuits, 4.16, 4.17
   isolating to a faulty circuit, 7.33–35
   resistive circuits, 4.15–16, 4.17
   safety, 4.27–28, 6.13–16, 7.30–31
   schematic symbols, 7.65–66
   single-phase power, 4.9–11
   three-phase power, 4.12–15
Electric current. *See also* Test meters; Testing
   conductors, 5.2
   current monitoring devices, 10.20
   effects on human body, 4.27
   insulators, 5.3
   rectification, 5.4
   safety, 4.27–28, 6.13–16, 7.30–31
   static electricity, 5.15, 5.17, 7.7
Electric heating, 6.1–3
   airflow, 6.15–17
   baseboard heaters, 6.2, 6.18
   blowers, 6.4, 6.6–8, 6.17
   checking loads, 6.12
   components, 6.3–4
   determining Btuh output, 6.4, 6.13
   duct heaters, 6.17, 6.18
   electric boilers, 6.19–20
   electrical disconnects, 6.11
   fan coil with electric heat, 6.1–2, 6.3
   forced-air heating, 1.2, 6.2–3, 6.4, 6.5–11
   heating element, 6.3–5
   history, 6.2
   limit switches, 6.3, 6.4, 6.9, 6.10, 6.23
   power supply, 6.4, 6.11
   radiant heating panels, 6.19
   replacing resistance wires, 6.12
   resistive circuits, 4.15–16, 4.17
   space heaters, 6.19
   temperature rise, 6.16, 6.17, 6.23
   thermostat, 6.5–6
   troubleshooting, 6.11–15
   voltage variations, 6.11
Electric-pneumatic (E-P) relays, 7.56, 7.64
Electric power. *See also* Alternating current; Direct current; Power supply
   distribution, 4.9
   frequency, 4.7, 4.9
   generation, 4.5–15
Electric shock, 4.27, 6.14
Electrical burns, 4.27
Electrical disconnects, 6.11
Electrical system, troubleshooting, 7.26–29, 7.35–42
Electrical test meters. *See* Test meters
Electrical testing. *See* Test meters; Testing
Electroluminescence, 5.7
Electrolytic capacitor, 4.17
Electromechanical relays, 7.12–15
Electronic air cleaners, 3.34, 3.35, 3.36, 8.12–15
Electronic circuits, 5.1. *See also* Electronic components; Electronics
   integrated circuit chips, 5.1, 5.16–17
   microprocessors, 5.17
   printed circuit board, 5.15–16
Electronic components. *See also* Electronics
   cadmium sulfide flame detector, 5.13–14
   diodes, 5.4–7
   electronically commutated motors (ECMs), 5.14
   integrated circuit chips, 5.1, 5.16–17
   light-emitting diode (LED), 5.7–9
   microprocessors, 5.17
   resistors, 5.9–11
   static electricity and, 5.15, 5.17
   thermistors, 5.11–13
   variable frequency drives (VFDs), 5.15
Electronic defrost controls, 11.9–10
Electronic expansion valves (EEVs), 9.1, 9.14, 9.22
Electronic head pressure control, 10.22
Electronic hygrometers, 1.30
Electronic leak detectors, 12.3
Electronic manometers, 1.31
Electronic overloads, 10.18–20
Electronic sight glass, 12.20
Electronic solid-state relays, 7.15, 7.16
Electronic thermometers, 1.28
Electronic thermostats, 2.10, 7.2, 7.5–6, 7.10
Electronic vacuum gauge, 12.11–12
Electronically commutated motors (ECMs), 5.14, 7.42
Electronics. *See also* Electronic circuits. Electronic components
   components, 5.4–15
   defined, 5.27
   printed circuit boards, 5.15–16
   semiconductors, 5.2–4, 5.17, 5.27
   theory, 5.1–2
Electrons, atomic structure, 5.2, 5.27
Electrostatic discharge sensitivity, 5.15, 5.17
Electrostatic filters, 3.34, 8.12
Enclosure, electric heating, 6.4
Energy conservation equipment
   economizers, 3.37, 8.10, 8.15–18, 8.23
   energy recovery ventilators (ERVs), 8.10, 8.15, 8.16, 8.23
   evaporative pre-coolers, 8.18
   heat recovery ventilators (HRVs), 8.10, 8.15, 8.16, 8.23
   zoned control, 8.18–19
Energy recovery ventilators (ERVs), 8.10, 8.15, 8.16
Energy Star® program, 11.17
Enthalpy, 8.23
Enthalpy controller, 8.17
Envelope gaskets, 3.14
Equalizers, 9.11–15
Equipment disposal, capacitors and transformers, 4.26
Equivalent length, 1.23
ERVs. *See* Energy recovery ventilators
Evacuation, refrigeration systems, 12.7, 12.9–16
Evaporation, body comfort and, 8.3, 8.4, 8.23
Evaporative pre-coolers, 8.18
Evaporator coil, maintenance, 3.36
Evaporators, 9.1, 9.3, 10.22
Exhaust, blower, 1.2
Expansion boards, 5.23
Expansion-type anchor bolts, 3.7, 3.12–13
Expansion valves, 9.3, 9.7–15
   automatic expansion valves, 9.8–9
   electronic expansion valves, 9.1, 9.14, 9.22
   high-side float valves, 9.7–8
   hunting, 9.14–15, 9.22
   low-side float valves, 9.8
   manual expansion valves, 9.7
   thermal-electric expansion valves (TEEV/THEV), 9.1, 9.13, 9.14, 9.22

thermal expansion valves (TXV/TEV), 9.1, 9.9–18, 9.19, 9.22, 11.12, 11.13, 12.26
Extended plenum duct system, 1.14–15, 1.17
Extended-surface disposable filters, 3.34, 8.11–12, 8.14
Extension cords, ground fault circuit interrupters (GFCIs), 4.28, 6.14
External analog devices, 7.58
External digital devices, 7.58
External equalizer, 9.11, 9.13
External line break overloads, 10.17
External lock washers, 3.5
External static pressure loss, 1.12

Face alignment, couplings, 3.33
Fan coil unit, 11.18, 11.27
Fan coil with electric heat, 6.1–2, 6.3
Fan control, furnace, 7.20
Fan curve charts, 1.11
Fan laws, 1.10–11
Fan-powered humidifiers, 8.7, 8.8
Fan switches, checking, 7.42
Fans, 1.5–6, 1.9–12
    defined, 1.7
    duct fans, 1.10
    fan curve charts, 1.11
    fan laws, 1.10–11
    motor speed controls, 7.16–17
    propeller fans, 1.9–10
    shaded-pole motor, 4.20
Fasteners, 3.2–13
    grade designations, 3.2
    non-threaded, 3.8–10
    for sheet metal ductwork, 1.20
    thread designations, 3.2
    threaded, 3.2, 3.4–8
Fastening, fiberglass ducts, 1.21
Fault isolation, 7.28–29
Fault isolation diagrams, 7.28, 7.64
Fearful customers, dealing with, 3.47–48
Fender washers, 3.5
FHP belts. *See* Fractional horsepower
Fiberglass duct, 1.21
Filters. *See* Air filters
Fire dampers, 1.25–26, 8.19–20
Fire point, of oils, 3.26, 3.53
Fittings (in ducts), 1.23
Fixed metering devices, 9.3, 9.4–7
Fixed-orifice metering devices, 9.4, 9.6–7, 9.22
Fixed resistors, 5.10
FLA value, 10.28
Flame detector, 5.13–14
Flames, 2.2–3
Flange tightening, threaded fasteners, 3.12
Flanged couplings, 3.31
Flapper valves, 10.6
Flash gas, 9.1, 9.22
Flash point, of oils, 3.26, 3.53
Flat gaskets, 3.13, 3.14
Flat washers, 3.5
Flexible chimney liners, 2.11, 2.12
Flexible couplings, 3.31
Flexible duct, 1.22
Flooded evaporators, 9.2, 9.3, 9.22
Flooded starts, 10.25, 10.26, 10.39
Flooding, 9.2, 9.3, 9.22, 10.25, 10.39
Floor registers, 1.24, 1.25
Floor supply diffuser, 1.13
Floppy disks, 5.25

Floppy drive, 5.20
Flow control device, 9.1
Flue gases, 2.1, 2.3–4
Force-feed lubrication system, 10.8–9, 10.39
Forced-air heating, 1.2, 6.2–3, 6.5–11
    blower control, 6.6–8
    heater control, 6.6
    heating element, 6.4
    safety controls, 6.6, 6.9
    thermostat, 6.5–6, 6.9, 6.11
    transformer, 6.6
    troubleshooting, 7.26
    zoned systems, 8.19
Forms. *See* Documentation
Forward bias, diode, 5.5
Forward-curved centrifugal blowers, 1.8
Four-way valve, 11.6, 11.27
Four-wire closed delta, 4.12, 4.13
Four-wire open delta, 4.12, 4.13
Four-wire wye system, 4.13, 4.14
Fractional horsepower (FHP) belts, 3.28
Fractionation, 12.29, 12.33
Free air delivery, 1.9, 1.37
Free cooling, 8.17, 8.23
Free electrons, 5.3
Freezestat, 7.19
Frequency
    alternating current, 4.1
    generators, 4.7, 4.9
Friction losses, air distribution system, 1.3
Fuel. *See also* Combustion; Furnaces
    flue gases, 2.1, 2.3–4
    venting, 2.1–2, 2.4–7
Full-face gaskets, 3.13, 3.14
Full-wave rectifier, 5.7, 5.27
Furnace venting. *See* Venting systems
Furnaces. *See also* Gas furnaces
    air handlers, 6.1
    burner input adjustment, 2.9
    combustion, 2.1–3
    condensing furnaces, 2.1, 2.6, 2.11–14, 2.18
    control circuit, 6.8
    draft controls, 2.14–15
    electronic furnace control, 5.17
    flue gases, 2.1, 2.3–4
    high-altitude installations, 2.9
    induced-draft furnaces, 2.1, 2.3, 2.6, 2.8–11, 2.18
    inducer switch, 7.21, 7.22
    motor speed controls, 7.16–17
    natural-draft furnaces, 2.1, 2.3, 2.6, 2.7–8, 2.10, 2.18
    resistive circuits, 4.15–16, 4.17
    sizing, 2.8–9
    start-up report for, 3.42
    temperature rise adjustment, 2.9
    temperature rise range, 2.9
    thermocouple, 7.21
    thermostat heat anticipator adjustment, 2.9–10
    thermostats, 7.2–6
    venting, 2.1–2, 2.3, 2.4–7, 2.10–11
Fuses
    blown fuses, 4.26, 6.15
    checking, 4.26–27, 7.37–38
    replacing, 6.15
    schematic symbols, 7.65
    thermal fuse, 6.4, 6.9, 6.23
    troubleshooting, 7.36–38
Fusible link, 4.3

Galvanized steel ducts, 1.19–20
Game port, 5.23
Gas flame, color of, 2.2
Gas furnaces
  condensing furnaces, 2.1, 2.11–14
  draft controls, 2.14–15
  flame, 2.2
  flue gases, 2.3
  induced-draft furnaces, 2.1, 2.3, 2.6, 2.8–11
  natural-draft furnaces, 2.1, 2.3, 2.6, 2.7–8, 2.10
  venting, 2.3, 2.4–7, 2.10–11
Gas-phase air filtration, 8.13
Gaskets, 3.13–14
Gauge manifold set, 12.12
Gauge pressure, 1.5, 1.37
Generators
  alternating current, 4.6
  direct current, 4.5–6
  frequency, 4.7, 4.9
  sine wave generation, 4.6–7, 4.8
  voltage, 4.6
Geothermal heat pumps, 11.5, 11.27
GFCIs. *See* Ground fault circuit interrupters
Gib head key, 3.9
Gold, as conductor, 5.3
Grade designations, fasteners, 3.2
Graphite ribbon packing, 3.16, 3.17
Grease guns, 3.27
Greases, 3.26–27
Grilles, 1.24–25, 7.26
Ground, schematic symbol, 7.66
Ground fault circuit interrupters (GFCIs), 4.28, 6.14
Ground-source heat pumps, 11.5, 11.27
Grounded windings, 7.48
Grounding, checking motor for, 4.25
Guide bearings, 3.21

Half-wave rectifier, 5.5, 5.6, 5.27
Halide torch leak detector, 12.2
Handshake (computers), 5.18
Hard drive, 5.24, 5.25
HCAR breakers, 7.39
Head pressure, 10.36
Head pressure control, 10.22, 10.24
Heat anticipator, 2.9–10, 2.18, 7.9, 7.10
Heat exchanger, pressure drop at, 1.5
Heat load, troubleshooting, 7.26
Heat pumps, 10.2–3, 11.1–25, 11.6
  add-on systems, 11.21–22
  air-to-air heat pumps, 11.1, 11.2, 11.3, 11.6, 11.7, 11.19, 11.21
  air-to-water heat pumps, 11.2, 11.5
  balance point, 11.17–18, 11.27
  classifications, 11.1–6
  coefficient of performance (COP), 11.15–16, 11.27
  coil energy recovery loop, 11.5, 11.6
  coils, 11.1
  components, 11.10–15
  control devices, 11.8–10, 11.23, 11.25
  cooling cycle, 11.6, 11.7
  crankcase heaters, 11.15
  defrost cycle, 11.2, 11.6, 11.8–10, 11.27
  efficiency recommendations, 11.17
  electrical disconnects, 6.11
  Energy Star® program, 11.17
  geothermal heat pumps, 11.5
  ground-source heat pumps, 11.5, 11.27
  heating cycle, 11.6, 11.7
  heating season performance factor (HSPF), 11.16, 11.27
  installation, 11.18–22
  ladder diagram, 11.24
  liquid accumulators, 11.12–13, 11.14
  metering devices, 11.12
  operation, 11.6–10
  packaged units, 11.20–21
  performance factor (PF), 11.16
  pressure controls, 11.13, 11.14
  ratings, 11.15–17
  reverse cycle heat, 11.6, 11.27
  reversing valves, 11.10–12
  seasonal energy efficiency ratio (SEER), 11.16, 11.27
  service, 11.23
  setback, 11.23
  special heat sources, 11.5–6
  split systems. *See* Split systems
  start-up report for, 3.41
  subcooling, 10.2–3, 11.6
  supplementary heaters, 11.18
  thermal expansion valves (TXV/TEV), 9.1, 9.9–18, 9.19, 9.22, 11.12, 11.13, 12.26
  thermostats, 7.5, 11.13, 11.14
  uses of, 11.1, 11.2
  water-to-air heat pumps, 11.1, 11.2–4
  water-to-water heat pumps, 11.4–5
  wiring diagram, 7.52–53
Heat recovery ventilators (HRVs), 8.10, 8.15, 8.16, 8.23
Heat sink, 11.1, 11.27
Heater control, forced-air electric heating, 6.6
Heater relay (HR), 11.9
Heating-cooling thermostats, Thermostats, 7.4
Heating cycle, heat pumps, 11.6, 11.7
Heating element, 6.3–5, 6.12
Heating season performance factor (HSPF), 11.16, 11.27
Heating systems
  accessories. *See* Accessories
  airflow, 1.5, 1.13
  dual-fuel heating system, 11.21, 11.27
  electric heating. *See* Electric heating
  energy conservation equipment, 8.15–19
  forced-air heating. *See* Forced-air heating
  furnaces. *See* Furnaces
  heat pumps. *See* Heat pumps
  heating element, 6.3–5, 6.12
  humidity control, 8.4–10
  hydronic systems. *See* Hydronic systems
  thermostats. *See* Thermostats
  troubleshooting, 7.42
  zoned control, 8.18–19
Heli-coils, 3.8
Hermetic compressors, 7.44, 10.3
HFC refrigerants, 10.8
High-altitude installations, 2.9
High-pressure cutout (HPCO) switch, 7.19
High-pressure spray nozzle humidifiers, 8.8, 8.9
High-pressure switches, 7.19, 7.20
High-side float valves, 9.7–8
High sidewall registers, 1.16, 1.24, 1.25
High voltage, effects on hvac equipment, 7.35
Horsepower (hp), 1.10
Hoses, for evacuating and charging, 12.12, 12.13
Hot climates, duct systems, 1.15–16
Hot gas bypass method, 10.14
Hot water heating systems. *See* Hydronic systems
Hot wire anemometer, 1.33
Houses, loose, average, and tight, 8.9
HPCO switch. *See* High-pressure cutout switch
HR. *See* Heater relay
HRVs. *See* Heat recovery ventilators
HSPF. *See* Heating season performance factor

Humidification, 8.5
Humidifiers, 6.4, 8.5, 8.6–10
    capacity, 8.9
    defined, 8.23
    sizing chart, 8.10
    types, 8.6–9
Humidity control, 8.4–10
Humidity measurement, 1.28, 1.29–30
Hunting, 9.14–15, 9.22
Hydraulic grease fittings, 3.27, 3.28
Hydraulic packing, 3.15
Hydraulic unloading, 10.14
Hydronic systems, 7.49
    aquastats, 7.49–50
    circulating pumps, 7.50–51, 10.36
    low water cutoff, 7.50
    reset controller, 7.50
    zone valves, 7.51
Hygrometers, 1.30

IDE, 5.18
Immersion heater, compressors, 10.26
Impedance relay, 7.17, 10.22
In-line pumps, 10.36
Inches of water column, 1.4
Inclined manometer, 1.31
Inclined-vertical manometers, 1.31
Incomplete combustion, 2.2, 2.4, 2.18
Individual feeder tube metering devices, 9.6
Indoor air quality, 8.10–15
    contaminants, 8.10–11
    electronic air cleaners, 3.34, 3.35, 3.36, 8.12–15
    mechanical air filters, 3.34, 3.35, 8.11–15
    troubleshooting, 7.26
Indoor coil, 11.27
Induced-draft furnaces, 2.1, 2.3, 2.6, 2.8–11
    burner input adjustment, 2.9
    defined, 2.18
    installing, 2.8
    sizing, 2.8–9
    temperature rise adjustment, 2.9
    thermostat heat anticipator adjustment, 2.9–10
    venting, 2.10–11
Inducer switch, 7.21, 7.22
Induction motors, 4.17–21, 4.29
    single-phase motors, 4.18–20, 4.21, 4.24–25, 4.29, 7.47, 10.17, 10.18
    three-phase motors, 4.21, 4.24–25, 5.13, 7.44, 10.16, 10.20, 10.21, 10.30, 10.39
Inductive circuits, 4.16, 4.17
Inductive load, 4.24–25, 7.39–41
Industrial systems. *See* Commercial systems
Inertia, single-phase motor, 4.18
Infrared humidifiers, 8.9
Inlet guide vane method, 10.14
Input power distribution circuits, 7.32
Input voltage, checking, 7.35–36
Insulation, 1.27–28
Insulators, 5.3
Intake slide valve, 10.14
Integrated circuit, 5.1, 5.16–17, 5.27
Interference fit, bearings, 3.24
Interference fit method, couplings, 3.32
Interlocking braid packing, 3.15, 3.16
Internal equalizer, 9.11, 9.12
Internal-external lock washers, 3.5
Internal line break overloads, 10.18, 10.33
Internal lock washers, 3.5
Invar®, 7.2, 7.64

Invoices, 3.38, 3.39
ISDN, 5.18
Isolation transformers, 4.3

Jacketed gaskets, 3.14
Jam nuts, 3.5, 3.6
Journal, bearings, 3.21, 3.53

Kelvin scale, 11.6, 11.27
Keyboard (computers), 5.21
Keyboard port, 5.22
Keys, 3.8–9
Kilowatts (kW), 6.13
Klixon, 10.17

Label diagrams, 7.27–28, 7.64
Ladder diagram, 4.29, 7.64, 11.24
Laminated packing, 3.16
Lamps, schematic symbol, 7.66
LCD. *See* Liquid crystal display
Leak detection devices, 12.2–5
Leak repair, disassembling brazed joints, 12.6
Leak testing, 12.1, 12.5–7
LED. *See* Light-emitting diode
Lever-type grease gun, 3.27
Light-emitting diode (LED), 5.7–9, 5.27
Limit control, furnaces, 7.20–21
Limit switches, 6.3, 6.4, 6.9, 6.10, 6.23, 7.49–50
Line duty devices, 10.16–17, 10.39
Line marking, 3.2
Line-voltage thermostats, 6.18, 7.6
Lip seals, 3.18, 3.19
Lip-type packing, 3.15
Liquid accumulators, heat pumps, 11.12–13, 11.14
Liquid charging, by weight, 12.21, 12.22
Liquid crystal display (LCD), 5.8
Liquid detectors, 12.5
Listening skills, of service technicians, 3.46
Load changes, metering devices, 9.3
Loads
    in bearings, 3.21, 3.53
    defined, 7.33
    electric heating, 6.12
    heat load in building, 7.26
    inductive loads, 4.16, 4.24–25
Lock washers, 3.5
Locknut method, bearing installation, 3.24
Lockout control circuit, 7.17–18
Lockout relay, 7.17–18, 10.22, 10.23
Lockout/tagout, 7.30–31
Loop perimeter duct system, 1.14
Loose building, 8.9
Loss-of-charge switch, 7.19, 11.13
Low-pressure cutout (LPCO) switch, 7.19
Low-pressure switches, 7.19, 7.20
Low-side float valves, 9.8
Low sidewall registers, 1.24, 1.25
Low voltage
    effects on electric heating, 6.11
    effects on HVAC equipment, 7.35
Low-voltage circuits, 7.33
Low-voltage thermostats, 7.2
Low-voltage transformers, 4.5
Low water cutoff, 7.50
LPCO switch. *See* Low-pressure cutout switch
Lubricants
    bearing failures and, 3.23
    bearings, 3.22, 3.23
    greases, 3.26–27
    oils, 3.26, 10.7–8, 12.11

Lubrication, 3.25
    compressors, 10.8–9
    equipment for, 3.27–28
    motor lubrication, 3.33–34
    over-lubrication, 3.33
    procedure, 3.34
    sleeve bearing, 3.21
Lubrication fittings, 3.27, 3.28

Machine bolts, 3.4
Machine screws, 3.4
Mainframe computers, 5.19
Maintenance, 3.1
    air filters and screens, 3.34–36
    basic maintenance procedures, 3.33–37
    bearings, 3.21–25
    belts and belt drives, 3.28–30
    coils and condensate system, 3.36–37
    couplings and direct drives, 3.30–33
    customer communication, 3.45–50
    customer relations, 3.42–45
    damper inspection and cleaning, 3.37
    documentation, 3.37–42
    gaskets, 3.13–14
    lubrication, 3.21, 3.25, 3.27–28
    mechanical fasteners, 3.2–13
    motor lubrication, 3.33–34
    packing, 3.15–18
    seals, 3.18–21
Maintenance forms and reports. *See* Documentation
Manometers, 1.3–4, 1.31
Manual bearing puller, 3.23, 3.24
Manual coupling puller, 3.31–32
Manual expansion valves, 9.7
Masonry chimney, 2.11
MAT. *See* Mixed air thermostat
Material flexible couplings, 3.31
Maximum continuous current (MCC), compressor, 10.28
Mechanical air filters, 3.34, 3.35, 8.11–15
Mechanical cooling, 8.17, 8.23
Mechanical differential, 7.5
Mechanical fasteners, non-threaded fasteners, 3.8–10
Mechanical fasteners. *See* Fasteners
Mechanical flexible couplings, 3.31
Mechanical seals, 3.19–20
Mechanical squeeze, 3.18
Megohmmeter (megger), 4.22–24
Memory (computers), 5.19
Mercury, in thermostats, 7.2
Metal ducts, 1.19–20, 1.27
Metal packing, 3.16, 3.17
Metal vents, 2.10
Metering devices, 9.1–19, 10.3
    adapting to load changes, 9.3
    capillary tubes, 9.1, 9.3, 9.4–6, 9.22
    distributors, 9.15–16, 9.22
    expansion valves, 9.7–18
    fixed-orifice metering devices, 9.4, 9.6–7, 9.22
    function of, 9.3
    heat pumps, 11.12
    troubleshooting, 9.18–19
Metric M-profile threaded screws, 3.2
Metric MJ-profile threaded screws, 3.2
Metric screw threads, 3.2
MFD, 7.44
Microfarads, 4.16
Microminiaturization, 5.1, 5.17, 5.27
Micron, 8.23
Microns, 12.15

Microprocessor-controlled system, 5.17
Microprocessors, 5.17, 5.27, 7.47
Mineral oil, compressors, 10.7, 10.8
Misalignment
    belt drives, 3.29–30
    couplings, 3.32–33
Mixed air thermostat (MAT), 8.17
Modems, 5.24
Modulating controls, 7.57
Moisture contamination, compressors, 10.27
Monitor (computers), 5.21–22
Monitor port, 5.22
Monitoring, current monitoring devices, 10.20
Monochrome monitors, 5.21, 5.27
Motherboard, 5.23
Motor assembly, electric heating, 6.4
Motor circuits, troubleshooting, 7.42–48
Motor starters, 7.16
Motor thermostat overloads, 10.18, 10.19
Motors
    burnout, 10.26, 10.35, 10.39
    capacitor run motors (CSR motors), 7.42, 7.43, 7.47
    capacitor start motors, 4.19–20, 7.42
    checking capacitors, 4.25–26
    checking for grounded windings, 4.25
    checking inductive loads, 4.24–25
    compressor motors, 10.15–21
    current imbalance, 4.15, 10.29–30
    dual-voltage rated motors, 7.36, 10.29
    electronically commutated motors (ECMs), 5.14, 7.42
    end play, 3.23
    failure, causes of, 10.16, 10.28–30, 10.35
    high voltage, effect of on, 7.35
    induction motors, 4.17–21, 4.29
    inductive circuits, 4.16, 4.17
    inductive loads, 4.16
    low voltage, effect of on, 7.35
    lubrication, 3.33, 3.34
    megohmmeter, 4.22–24
    motor insulation tests, 4.23
    motor protection thermistors, 5.13
    motor speed controls, 7.16–17
    motor starters, 7.16
    multi-stage motors, 4.20, 7.43
    over-lubrication, 3.33
    overload protection, 10.16–21
    permanent split capacitor (PSC) motors, 4.18–19, 7.42, 7.43, 7.47
    pneumatic damper motors, 7.55
    schematic symbols, 7.66
    sequence of operation, 7.21–25
    shaded-pole motors, 4.20, 7.42
    single-phase motors, 4.18–20, 4.21, 4.24–25, 4.29, 7.47, 10.17, 10.18
    single phasing, 10.16, 10.30, 10.39
    single-voltage rated motors, 7.35, 10.29
    stepper motor, 9.14
    thermostat overloads, 10.18, 10.19
    three-phase motors, 4.21, 4.24–25, 5.13, 7.44, 10.16, 10.20, 10.21, 10.39
    time delay relay, 7.18
    troubleshooting, 7.42–48
    variable frequency drives (VFDs), 5.15
    voltage imbalance, 4.15, 10.29
    voltage phase imbalance, 7.36
    voltage tolerances, 7.35
    windings, 7.47–48
Mouse (computers), 5.21
Mouse port, 5.23

Multi-speed motors, 4.20, 7.43
Multi-stage heat control, 6.9
Multi-stage heat-cool thermostat, 11.23
Multi-stage heating-cooling thermostats, 7.5
Multimeters, 4.21
    checking capacitors, 4.25–26
    checking fuses, 4.26–27, 7.37–38
    electric heating, voltage check, 6.12
    identifying unmarked terminals, 7.47
Multiple-groove pulleys, 3.28
Multiple-vane damper, 1.26
Multiplexed system, 11.19

N-type material, 5.4
Nameplate information plate, 10.28, 12.21
Naphthenic oils, 3.26
*National Electrical Code* (NEC), 4.28, 7.30
National Environmental Balancing Bureau (NEBB), 3.39
*National Fuel Gas Code*, 2.4, 2.10
National Lubricating Grease Institute (NLGI), 3.27
Natural-draft furnaces, 2.1, 2.3, 2.6, 2.7–8, 2.10, 2.18
NEBB. *See* National Environmental Balancing Bureau
Needle bearings, 3.23
Negative ion, 5.2
Negative pressure, 1.2
Negative temperature coefficient (NTC) thermistors, 10.18
Neutral, transformer, 4.10
NIC card, 5.18
Nitrogen, pressurization, 12.14
NLGI. *See* National Lubricating Grease Institute
Non-bleed thermostat controllers, 7.54, 7.55
Non-electric thermal expansion valves (TXV/TEV), 9.14
Non-expansion anchor bolts, 3.7, 3.12
Non-threaded fasteners, 3.8–10
Nonmechanical seals, 3.18–19
Normally open dampers, 7.55–56
NTC thermistors. *See* Negative temperature coefficient thermistors
Nuts, 3.5–6

O-rings, 3.18–19
O.D. misalignment. *See* Outer diameter misalignment
Ohmmeters, 4.22
    checking cadmium sulfide flame detector, 5.13
    checking inductive loads, 4.24–25
    diode testing, 5.5, 5.6
    electric heating, current check, 6.12
Ohm's law, electric heating, 6.13
Oil burner, cad cell maintenance, 5.13
Oil-filled capacitor, 4.17
Oil furnaces, flue gases, 2.1, 2.3–4
Oil rings, 10.5
Oil seal, 3.18, 3.19, 3.20
Oils, 3.26
    changing vacuum pump oil, 12.11
    refrigerant oils, 10.7–8
Open-drive compressor, 10.3, 10.4, 10.7
Open windings, checking for, 7.47–48
Opinionated customers, dealing with, 3.48
OSHA, lockout/tagout, 7.30–31
Outdoor air dampers, maintenance, 3.37
Outdoor coil, 11.27
Outdoor thermostats, 7.20
Outer diamater (O.D.) misalignment, couplings, 3.32–33
Output voltage, transformers, 4.2
Over-lubrication, 3.33
Over-voltage, electric heating, 6.11
Overhead radial duct systems, 1.16–17
Overhead supply duct, installing, 1.18

Overhead trunk duct systems, 1.16–17
Overheating, compressors, 10.30, 10.32
Overload protection, compressor motors, 10.16–21
Oxidation, 3.26, 3.53
Oxidation resistance, of oils, 3.26
Ozone gas, electronic air cleaners, 3.34

P-E relays. *See* Pneumatic-electric relays
P-type material, 5.4
Packaged units, 11.2
    defined, 11.27
    electrical disconnects, 6.11
    installation, 11.20–21
    nameplate, 12.21
Packing, 3.15–18
    installing, 3.17–18
    removing, 3.17
    types, 3.16–17
Packing gland, 3.16, 3.53
Packing puller, 3.17
Pan-type humidifiers, 8.6
Paraffinic oils, 3.26
Parallel I/O, 5.18
Parallel misalignment, couplings, 3.32
Parallel ports, 5.22
Part-winding, 10.24
Partition, 5.19
Pascals, 1.5
PCBs, in capacitors and transformers, 4.26
Performance factor (PF), heat pumps, 11.16
Perimeter duct systems, 1.13–15
Permanent filters, 8.12
Permanent split capacitor motors (PSC motors), 4.18–19, 7.42, 7.43, 7.47
Personal computers, 5.19–24
Personal habits, of service technicians, 3.43–44
Personal hygiene, of service technicians, 3.43–44
PF. *See* Performance factor
Phases, electric circuits, 4.17
Phone jack (computers), 5.23
Photo diode, 5.8, 5.9, 5.27
Photocell, 8.19
Piezoelectric crystal, 8.8
Pilot duty devices, 10.16–17, 10.39
Pin fasteners, 3.8
Piston rings, reciprocating compressors, 10.5
Pistons, reciprocating compressors, 10.5
Pitot tubes, 1.4, 1.31, 1.32
Pixels, 5.21, 5.27
Plain bearings, 3.21–22
Plastic pipe, for vents, 2.7
Plate-type humidifiers, 8.6
Plenum duct systems, 1.14–15, 1.17
Plenums, 1.12, 1.37
Plug and play, 5.19
PN junction, 5.4
Pneumatic actuator, 7.55
Pneumatic controls, 7.51, 7.54–57
Pneumatic damper motors, 7.55
Pneumatic damper valves, 7.55
Pneumatic-electric (P-E) relays, 7.56, 7.64
POE oil. *See* Polyolester oil
Points, 7.58
Polarity, of voltage, 4.6
Polyolester (POE) oil, 10.8
Pop rivet tool, 3.10
Porous bearings, 3.21
Portable differential pressure gauge, 1.32
Positive-displacement compressors, 10.13, 10.39

Positive ion, 5.2
Positive pressure, 1.2
Positive temperature coefficient (PTC) thermistors, 7.46–47
Potentiometer, 7.16, 7.17
Pour point, of oils, 3.26, 3.53
Power circuits, 7.32
Power disconnect, 4.10, 4.12, 6.11
Power distribution, 4.9
Power formula, electric heating, 6.13
Power generation, 4.5–15
   alternating current, 4.6
   direct current, 4.5–6
   frequency, 4.7, 4.9
   sine wave generation, 4.6–7, 4.8
   single-phase power, 4.9–11
   three-phase power, 4.12–15
Power supply
   electric heating, 6.4, 6.11
   personal computers, 5.24
Power tools, ground fault circuit interrupters (GFCIs), 4.28, 6.14
Power venting, 2.16
Pratt & Whitney key, 3.9
Pre-coolers, 8.18
Preinsulated ducts, 1.27
Press fit, bearings, 3.24
Press-fitted couplings, 3.32
Press mounting method, bearing installation, 3.25
Pressure
   in air distribution system, 1.4–5
   measurement of, 1.4, 1.5, 1.31–32
   residential system, 1.5
Pressure charts, charging by, 12.28
Pressure controls, heat pumps, 11.13, 11.14
Pressure gauges, 1.5
Pressure loss
   external static pressure loss, 1.12
   outside insulation and, 1.27
   static pressure loss, 1.3
Pressure lubrication system, 10.8–9, 10.39
Pressure protection, compressors, 10.21, 10.23
Pressure switches, 7.19, 7.21
Pressurization, safety warning, 12.14
Primary air, 2.2, 2.3, 2.18
Primary reactor, 10.24
Primary resistor, 10.24
Primary winding, transformers, 4.2, 4.3
Printed circuit boards, 5.15–16
Process air conditioning, 8.1
Programmable thermostats, 7.2, 7.5–6, 7.10
Propeller fans, 1.9–10
Protons, atomic structure, 5.2
PSC motors. *See* Permanent split capacitor motors
Psychrometers, 1.29–30
Psychrometric chart, 1.29, 8.2, 8.3
Psychrometrics, 1.29, 1.37
PTC thermistors. *See* Positive temperature coefficient thermistors
Pulley misalignment, belt drives, 3.29–30
Pulse width modulation (PWM), 5.15
Pumps
   centrifugal pump, 10.36
   shaded-pole motor, 4.20
   vacuum pump, 12.10–11
PVC pipe, for vents, 2.7
PWM. *See* Pulse width modulation

R-410A refrigerants, 10.8, 10.9
R-value, for insulation, 1.27–28
Radial bearings, 3.21, 3.22
Radial blowers, 1.8–9
Radial lip seal, 3.19
Radial load, bearings, 3.21, 3.53
Radial perimeter duct system, 1.14
Radiant heating panels, 6.19
Radiation, body comfort and, 8.4
RAM, 5.19
Rational psychrometric formula, 1.29
RDRAM, 5.19
Reciprocating compressors, 10.4–9
Reclamation, 12.23
Recording instruments, 4.24
Recovery, refrigerants, 12.7–9, 12.33
Recovery unit, 12.7, 12.8–9
Rectangular duct, 1.20
Rectification, 5.4, 5.27
Rectifiers, 5.5–7, 7.66
Recycling, refrigerants, 12.9, 12.33
Redhead® drill, 3.12–13
Reduced-voltage starting, 10.22, 10.24
Reducing extended plenum duct system, 1.15
Reducing trunk, 1.6
Reed valves, 10.6
Refrigerant control, 9.1
Refrigerant cylinders, 12.17
Refrigerant distributor line, 9.11, 9.22
Refrigerant flow, air-to-air heat pump, 11.7
Refrigerant oils, 10.7–8
Refrigerant piping, slugging and, 10.25
Refrigerant recovery unit, 12.7, 12.8–9
Refrigerants
   charging, 12.16–29
   fractionation, 12.29, 12.33
   leak detection, 12.1–7
   metering devices, 9.1–19
   recovery, 12.7–9, 12.33
   recycling, 12.9, 12.33
   zeotrope refrigerants, 12.29
Refrigeration systems. *See also* Cooling systems
   charging, 12.16–29
   compressors. *See* Compressors
   evacuation, 12.7, 12.9–16
   metering devices, 9.1–19, 10.3
   with partial charge, 12.6
Registers, 1.24–25
Relative humidity (RH), 1.29, 1.30, 1.37, 8.4, 8.5
Relays, 7.12–16
   defrost relay (DFR), 11.9
   DPDT relay, 7.14, 7.15
   elctromechanical relays, 7.12–15
   electric-pneumatic (E-P) relays, 7.56
   heater relay (HR), 11.9
   impedance relay, 7.17
   lockout relay, 7.17–18, 10.22, 10.23
   pneumatic controls, 7.55
   pneumatic-electric (P-E) relays, 7.56
   principles of operation, 7.12
   schematic symbols, 7.65
   sequencer, 6.6, 6.23
   solid-state relays, 7.15, 7.16
   SPDT relay, 7.14
   SPST relay, 7.12, 7.13
   start relay, 4.19, 4.20, 7.43, 7.45–46
   time delay relay, 7.18
Removable media, for computer storage, 5.25
Reports. *See* Documentation

Reset controller, 7.50
Reset limit switch, 6.9
Residential systems
    air distribution, 1.5–6, 1.12
    blowers, 1.2, 1.3, 1.6–9, 1.7, 1.12
    in cold climates, 1.13–15
    thermostats, 7.2
    in warm climates, 1.15–19
Resistance
    measuring, 7.40
    megger, 4.22–24
Resistance check, 6.12, 7.44
Resistance wires, replacing, 6.12
Resistive circuits, 4.15–16, 4.17
Resistive loads, 7.39
Resistors, 5.9–11
    applications, 5.10
    color code, 5.10–11
    replacing, 5.11
    schematic symbols, 7.66
    variable resistor, 5.10
Retainer rings, 3.8
Return air, 2.5, 6.15
Return air grilles, 1.5, 1.12, 1.13, 1.16, 2.5
Return air plenum, 1.12
Return duct system, 1.2
Reverse bias, diode, 5.5
Reverse cycle heat, 11.6, 11.27
Reversing valves, 11.10–12, 11.27
Revolutions per minute (rpm), 1.10, 1.37
Ribbed couplings, 3.31
Rigid couplings, 3.30–31
Ring gaskets, 3.13, 3.14
Ring valves, 10.6
Rivets, 3.9–10
RLA value, 10.28
RMS voltage. *See* Root-mean-square voltage
Roller bearings, 3.22
ROM, 5.19
Roof maps, 3.38
Root-mean-square (RMS) voltage, 4.6
Rotary compressors, 10.4, 10.9–10
Rotary vane compressors, 10.10, 10.11
Rotating drum humidifier, 8.7
Rotational bearings, 3.21
Round duct, 1.20
Round fiberglass duct, 1.21
Round flexible duct, 1.22
Run capacitor, 4.18
Run capacitors, 4.19
Run-down resistance, 3.10, 3.11, 3.53
Run winding, 4.18
Runaround loop system, 11.5

SAE grade designation, 3.2
Safety, 7.30
    capacitors, 4.26
    carbon monoxide gas, 8.20
    compressor checks, 7.44, 10.30, 10.31, 10.32, 10.34, 10.35
    electrical, 4.27–28, 6.13–16
    halide torch, 12.2
    leak detection, 12.6
    lockout/tagout, 7.30–31
    mercury in thermostats, 7.2
    PCB health hazard, 4.26
    POE oil, 10.8
    power equipment, 10.30
    pressurization, 12.14
    restoring machines or equipment, 7.31
    transformers, 4.26

Safety controls, forced-air electric heating, 6.6, 6.9
Sail switch, 7.57
Saybolt Universal Seconds (SSU, SUS), 3.26
Schematic, 7.28
Screen maintenance, 3.34–35
Screw compressors, 10.4, 10.11–12, 10.14
Screws, 3.2, 3.3, 3.10
Scroll compressors, 10.4, 10.10–11
Scrolls, 10.11
SCSI, 5.19
SDRAM, 5.19
Seals, 3.18–21
Seasonal energy efficiency ratio (SEER), 11.16, 11.27
Secondary air, 2.2, 2.3, 2.18
Secondary winding, transformers, 4.2, 4.3, 4.10
SEER. *See* Seasonal energy efficiency ratio
Seizure, 3.10, 3.11, 3.53
Self-drilling sheet metal screws, 3.6
Self-locking nuts, 3.5, 3.6
Semi-hermetic compressors, 7.44, 10.3
Semiconductor devices, 5.4
Semiconductors, 5.2–4, 5.17, 5.27
Sensing bulb, 9.16–17
Sequence of operation, 7.21–25
Sequencer, 6.6, 6.23
Serial I/O, 5.19
Serial ports, 5.22
Service calls. *See also* Troubleshooting
    communicating with customers, 3.44, 3.45–50
    customer relations, 3.43–45
    handling difficult customers, 3.47–50
    heat pump, 11.23
    procedure, 3.44–45
Service entrance panel, 4.9, 4.10
Service technicians
    communicating with customer, 3.44, 3.45–50
    customer relations, 3.43–45
    documenting service work, 3.37–43
    electrical safety, 4.27–28
    handling difficult customers, 3.47–50
    handling service calls, 3.44–45
    listening skills, 3.46
    personal habits, behaviors, and attitudes, 3.43–44
Service ticket/invoice, 3.38, 3.39
Service vacuum hoses, 12.12
Set, 3.10, 3.11, 3.53
Set screws, 3.5
Setback, heat pumps, 11.23
Shaded-pole motors, 4.20, 7.42
Shading coil, 4.20
Sheet metal ducts, 1.19–20, 1.27
Sheet metal screws, 3.6
Shims, coupling misalignment, 3.32
Short-cycle time delay, 7.11
Short cycling, 7.19, 10.13, 10.22, 10.39
Shorted windings, checking for, 7.47–48
Sight glass, 12.19–20
Silicon, use as semiconductor, 5.3
Silver, as conductor, 5.3
SIMMs, 5.19
Sine wave generation, 4.6–7, 4.8
Single-operation limit switch, 6.9
Single-phase motors, 4.18–20, 4.21, 4.29
    checking inductive loads, 4.24–25
    identifying unmarked terminals, 7.47
    overload protection, 10.17, 10.18
Single-phase power, 4.9–11
Single phasing, 10.16, 10.30, 10.39
Single-pole, double-throw (SPDT) relay, 7.14

Single-pole, single-throw (SPST) relay, 7.12–14
Single-voltage rated motors, 7.35, 10.29
Single-wall vent connector, 2.10
Sizing, induced-draft furnaces, 2.8–9
Skive cut, 3.18, 3.53
Sleeve bearings, 3.21
Sleeve couplings, 3.31
Sling psychrometer, 1.29
Slip fit, bearings, 3.24
Slip rings, generator, 4.6
Sloppy customers, dealing with, 3.48, 3.49
Slotted nuts, 3.5, 3.6
Slugging, 9.3, 9.22, 10.25
Smoke dampers, 1.26, 8.19–20
Sniffers, 12.3
Soft-start couplings, 3.31
Solid-state electronics, 5.2, 7.56
Solid-state reduced-voltage starters, 10.22, 10.24
Solid-state relays, 7.15, 7.16
Sound card, 5.23–24
Space heaters, 6.19
SPDT relay. See Single-pole, double-throw relay
Speakers (computers), 5.21
Speed control, compressors, 10.13–14
Speed taps, 4.20
Spherical bearings, 3.22
Spherical roller bearings, 3.22, 3.25
Spider system, 1.18
Spillage, 2.15
Spinning disc humidifiers, 8.8, 8.9
Spiral-wound gaskets, 3.13, 3.14
Splash lubrication system, 10.8, 10.9, 10.39
Split-phase motors, 4.18
Split ring washers, 3.5
Split systems, 1.2, 7.23, 11.2
    control circuit, 7.22, 7.23
    defined, 11.27
    discharge line mufflers, 11.20
    ductless split system, 11.19
    fan coil unit, 11.18, 11.27
    installing, 10.27, 11.18–20
    nameplate, 12.21
Splitter damper, 1.26
Spray nozzle humidifiers, 8.8, 8.9
Spring pins, 3.8
SPST relay. See Single-pole, single-throw relay
Square braid packing, 3.15
Square duct, 1.20
Squeeze-bulb aspirating psychrometer, 1.30
Squirrel cage rotor, 4.18
SSU. See Saybolt Universal Seconds
Stale air, troubleshooting, 7.26
Standard frequency, alternating current, 4.1
Standard multiple belts, 3.29
Start and/or run circuits, 7.44–45
Start capacitors, 4.19, 7.43
Start relay, 4.19, 4.20, 7.43, 7.45–46
Start-up report, 3.39, 3.41, 3.42
Start winding, 4.18
Starters, schematic symbols, 7.65
Static electricity, 5.15, 5.17, 7.7
Static pressure, 1.3–4, 1.10, 1.11, 1.31, 1.37
Static pressure drop, 1.3
Static pressure loss, 1.3
Static pressure tips, 1.31, 1.32
Static-rated dampers, 1.25
Static seals, 3.18, 3.53
Stationary vane compressors, 10.9, 10.10
Stator, single-phase motor, 4.18

Stator windings, checking inductive loads, 4.24–25
Steam heating. See Hydronic systems
Steam humidifiers, 8.9
Steel-aluminum mesh filters, 8.12
Steel bolts and screws, 3.2, 3.3, 3.10–11
Step-down transformers, 4.3, 4.4, 4.9–10
Step-up transformers, 4.3, 4.9
Stepper motor, 9.14
Storage media (computers), 5.24–25
Strip chart recorders, 4.24
Stud bolts, 3.4
Stuffing box, 3.16, 3.53
Sub-base, 7.5, 7.64
Subcooling, 10.2–3, 12.26–28, 12.29
Subcooling coil, 11.6
Suction bypass unloading, 10.14, 10.15
Suction cutoff unloading, 10.14, 10.15
Superheat, 9.4, 9.22, 12.24–26
Superheat charging chart, 12.25
Supply duct system, 1.12
Surge chamber, 9.2, 9.22
SUS. See Saybolt Universal Seconds
Switches
    centrifugal switch, 7.21
    checking, 7.40–41
    control circuit safety switches, 7.19–20
    fan control, 7.20
    freezestat, 7.19
    inducer switch, 7.21, 7.22
    limit control, 7.20–21
    loss-of-charge switch, 7.19, 11.13
    pressure switches, 7.19, 7.21
    sail switch, 7.57
    schematic symbols, 7.65
Symbols, schematic symbols, 7.65–66
Synchronous speed, 4.20
System board, 5.23

T-1 line, 5.19
T-3 line, 5.19
Takeoffs, 1.23
Taper pins, 3.8
Tapered bearings, 3.22
TE expansion valve. See Thermal-electric expansion valve sensor
TEEV. See Thermal-electric expansion valves
Temperature glide, 12.29, 12.33
Temperature measurement, 1.28–30
Temperature mounting method, bearing installation, 3.24–25
Temperature rise
    electric heating, 6.16, 6.17, 6.23
    furnaces, 2.9
Test meters, 4.21–24. See also Testing
    ammeters, 4.21, 4.22, 6.12, 7.9, 7.46, 10.33
    digital multimeters (DMMs), 1.28, 4.8, 4.26–27, 7.37
    megohmmeter (megger), 4.22–24
    multimeters, 4.21, 4.25–27, 6.12, 7.37–38, 7.47
    ohmmeters, 4.22, 4.24–25, 5.5, 5.6, 5.13, 6.12
    voltmeters, 4.22, 6.12
    wattmeters, 4.21–22
Testing, 4.21–27. See also Test meters
    acid/moisture test kits, 10.27
    capacitor analyzer, 4.21, 7.44
    checking capacitors, 4.25–26, 7.44
    checking circuit breakers, 7.36, 7.38–39
    checking control transformers, 7.41–42
    checking cooling operation, 7.42
    checking current draw, 6.12, 7.8–10
    checking diodes, 5.5, 5.6

checking electric heating, 6.12
checking fan switch operation, 7.42
checking fuses, 4.26–27, 7.36–38
checking heating operations, 7.42
checking inductive loads, 4.24–25, 7.39–41
checking resistance, 6.12, 7.44
checking resistive loads, 7.39–41
checking start relays, 7.43, 7.45–46
checking start thermistors, 7.46–47
checking switch and contactor/relay contacts, 7.40–52
checking thermostats, 7.42
checking voltage, 6.12
checking windings, 7.47–48
input voltage measurements, 7.35–36
motor protection thermistors, 5.13
motors and motor circuits, 7.42–48
recording instruments, 4.24
thermal-electric expansion valve sensor, 5.12
thermostat testing, 7.8–10
TEV. *See* Thermal expansion valves
Thermal-electric expansion valve sensor, 5.12
Thermal-electric expansion valves (TEEV/THEV), 9.1, 9.13, 9.14, 9.22
Thermal expansion valves (TXV/TEV), 9.1, 9.19, 9.22
adjusting, 9.11
adjustment, 9.18
charging and, 12.26
converting to, 9.10
diagram of, 9.10
equalizers, 9.11–15
heat pumps, 11.12, 11.13
installation, 9.17
non-electric, 9.14
operating principles, 9.9–11
replacement, 9.16–18
selection, 9.16
Thermal fuse, 6.4, 6.9, 6.23
Thermal offset, 7.6
Thermistors, 5.11–13
motor protection thermistors, 5.13
positive temperature coefficient (PTC) thermistors, 7.46–47
thermal-electric expansion valve sensor, 5.12
Thermocouple, 7.21, 7.66
Thermometers, electronic, 1.28
Thermostat base, 7.5
Thermostat heat anticipator, 2.9–10, 2.18, 7.9, 7.10
Thermostats, 7.1–11
adjusting, 7.11
automatic changeover thermostats, 7.4–5, 7.64
calibration, 7.11
cooling compensator, 5.10, 7.3–4
cooling-only thermostats, 7.3–4
defrost thermostat (DFT), 11.9
diagram, 6.5
electric heating, 6.5–6
electronic thermostats, 2.10, 7.2, 7.5–6, 7.10
forced-air electric heat, 6.5–9, 6.11
freezestat, 7.19
heat anticipator, 2.9–10, 2.18, 7.9, 7.10
heat pumps, 7.5, 11.13, 11.14
heating-cooling thermostats, 7.4
heating-only thermostats, 7.2–3
installation, 7.6–11
line-voltage thermostats, 6.18, 7.6
low-voltage thermostats, 7.2
mercury safety hazard, 7.2
mixed air thermostat (MAT), 8.17
motor thermostat overloads, 10.18, 10.19
multi-stage heat-cool thermostat, 11.23

outdoor thermostats, 7.20
pneumatic control, 7.51, 7.54
principles of operation, 7.2–3
programmable thermostats, 7.2, 7.5–6, 7.10
static electricity and, 7.7
troubleshooting, 7.42
two-stage thermostats, 6.9, 6.11, 7.5
wiring, 7.8, 7.9
THEV. *See* Thermal-electric expansion valves
Thread-cutting screws, 3.7
Thread designations, fasteners, 3.2
Thread-forming screws, 3.7
Thread repair inserts, 3.8
Thread tapping, 3.12
Threaded fasteners, 3.2, 3.4–8
anchor bolts, 3.7, 3.12–13
cap screws, 3.4
flange tightening, 3.12
flat washers, 3.5
heli-coils, 3.8
installing, 3.10–13
lock washers, 3.5
machine bolts, 3.4
machine screws, 3.4
nuts, 3.5–6
self-drilling sheet metal screws, 3.6
set screws, 3.5
steel bolts and screws, 3.3, 3.10–11
stud bolts, 3.4
thread-cutting screws, 3.7
thread-forming screws, 3.7
thread repair inserts, 3.8
thread tapping, 3.12
toggle bolts, 3.7
torquing, 3.10–11
Three-phase motors, 4.21, 4.24–25, 7.44
checking inductive loads, 4.24–25
motor protection thermistors, 5.13
overloading, 10.20, 10.21
single phasing, 10.16, 10.30, 10.39
Three-phase power, 4.12–15, 10.29–30
Three-phase rectifier, 5.7
Three-phase transformers, 4.4–5
Thrust, bearings, 3.21, 3.53
Thrust ball bearings, 3.22
Thrust bearings, 3.21, 3.22
Thrust washer, 3.22
Tight building, 8.10
Time delay relay, 7.18
Timer contacts, schematic symbols, 7.65
Timers, 7.18–19
Tinner's rivet, 3.9
Toggle bolts, 3.7
Tolerances, 3.4, 3.53
Torque, 3.10–11, 3.53, 4.18
Torque wrenches, 3.10, 3.11
Total air pressure, 1.4, 1.31, 1.37
Transformers, 4.2–5
autotransformers, 4.3–4, 7.16, 10.24
components, 4.2
control transformers, 7.41–42
defined, 4.2
forced-air electric heating, 6.6
isolation transformers, 4.3
low-voltage transformers, 4.5
output voltage, 4.2
PCB health hazard, 4.26
phasing, 4.4
selection, 4.5

Transformers, *continued*
  three-phase transformers, 4.4–5
  turns ratio, 4.3
  VA rating, 4.5
  windings, 4.2, 4.3
Transitions, 1.23
Trap, condensate drain, 3.36
TRIAC, 7.17
Triple evacuation method, 12.14–16
Troubleshooting
  basic system analysis, 7.27
  checking control transformers, 7.41–42
  checking cooling operations, 7.42
  checking fan switch operation, 7.42
  checking heating operations, 7.42
  checking start and run capacitors, 7.44–45
  checking start relays, 7.43, 7.45–46
  checking start thermistors, 7.46–47
  checking switch and contactor/relay contacts, 7.40–52
  checking thermostats, 7.42
  checking windings, 7.47–48
  circuit breakers, 7.36, 7.38–39
  compressors, 10.30–33, 10.37
  customer interviews for, 7.26
  defined, 7.26, 7.64
  diagnostic equipment and tests, 7.28
  diagram, 7.28, 7.29
  electric heating, 6.11–15
  electrical system, 7.26–29, 7.35–42
  electronic furnace control, 5.17
  fault isolation, 7.28–29
  fault isolation diagrams, 7.28
  fuses, 4.26–27, 7.36–38
  HVAC, 7.60
  hvac system, 7.31, 7.32
  input voltage measurements, 7.35–36
  label diagrams, 7.27–28
  leak detection, 12.1–7
  metering devices, 9.18–19
  motors and motor circuits, 7.42–48
  physical examination of system, 7.26
  resistive and inductive loads, 4.24–25, 7.39–41
Troubleshooting diagram, 7.28, 7.29
Troubleshooting table, 7.28, 7.64
Troubleshooting trees, 7.28
Trunk ducts
  fiberglass, 1.21
  flexible duct, 1.22
  metal, 1.19–20
  overhead trunk duct systems, 1.16–17
Tube-axial fan, 1.10
Turns ratio, transformers, 4.3
Twisted braid packing, 3.15, 3.16
Two-stage thermostats, 6.9, 6.11, 7.5
Two-valve gauge manifold, 12.12
TXV. *See* Thermal expansion valves
Type B vents, 2.7
Type B-W vents, 2.7
Type L vents, 2.7

U-tube manometer, 1.31
Ultrasonic humidifiers, 8.8
Ultrasonic leak detectors, 12.3–4
Ultraviolet/fluorescent leak detectors, 12.4–5
Ultraviolet light air purification systems, 8.20
Under-voltage, electric heating, 6.11
Unified National Coarse (UNC) thread, 3.2
Unified National Extra-Fine (UNEF) thread, 3.2
Unified National Fine (UNF) thread, 3.2

USB port, 5.22
UV. *See under* Ultraviolet
V-belt drive, 3.28
V-belts, 3.28–29
VA rating, transformers, 4.5
Vacuum gauges, 12.10–11
Vacuum microgauges, 12.15
Vacuum pump, refrigerant system evacuation, 12.10–11
Vacuum pump oil, changing, 12.11
Valence electrons, 5.3, 5.27
Valves
  compressor valves, 10.6, 10.7, 10.33
  expansion valves. *See* Expansion valves
  flapper valves, 10.6
  four-way valve, 11.6, 11.27
  heat pumps, 11.10–12
  high-side float valves, 9.7–8
  intake slide valve, 10.14
  pneumatic damper valves, 7.55
  reed valves, 10.6
  reversing valves, 11.10–12, 11.27
  ring valves, 10.6
  thermal-electric expansion valves. *See* Thermal-electric expansion valves (TEEV/THEV)
  thermal expansion valves. *See* Thermal expansion valves (TXV/TEV)
  zone valves, 7.51
Vane-axial fan, 1.10
Vapor barrier, 1.27
Vapor charging, by weight, 12.21–23
Variable frequency drives (VFDs), 5.15
Variable-pitch pulleys, 3.28, 3.29
Variable resistor, 5.10
Velocity pressure, 1.4, 1.31, 1.37
Velometers, 1.33–34
Vent, defined, 2.18
Vent connectors, 2.3, 2.10, 2.18
Vent dampers, 2.14–15
Vent systems, 2.1–2, 2.3, 2.4–7
  air supply, 2.5
  chimneys, 2.6–7, 2.11
  clearances, 2.5
  codes and regulations, 2.4–5
  components, 2.6–7
  concentric termination, 2.13, 2.14
  condensing furnaces, 2.11
  draft controls, 2.14–15
  induced-draft furnaces, 2.10–11
  metal vents and vent connectors, 2.10–11
  natural-draft furnaces, 2.8
  power venting, 2.16
  requirements for, 2.4–5
Ventilation, economizers, 3.37, 8.10, 8.15–18, 8.23
Venturi, 1.9–10, 1.37
VFDs. *See* Variable frequency drives
VI. *See* Viscosity index
Video card, 5.24
Virtual memory, 5.19
Viscosity, of oil, 3.26, 3.53
Viscosity index (VI), of oils, 3.26, 3.53
Viscosity rating, of oils, 3.26
Voltage
  capacitor, 4.17
  checking circuit breakers, 7.38–39
  checing electric heating, 6.12
  compressor operating voltage ranges, 10.28–29
  effective voltage, 4.8
  of generator, 4.6
  high voltage, effect of, 7.35

input voltage measurements, 7.35–36
low voltage, effect of, 7.35
polarity, 4.6
troubleshooting, 7.35–36
Voltage imbalance, three-phase systems, 4.15, 10.29
Voltage phase imbalance, 7.36
Voltmeters, 4.22, 6.12
Volume, defined, 1.3, 1.37
VOM, 4.21

Warm climates, duct systems, 1.15–19
Warranty ticket, 3.39
Water, boiling temperature, 12.10
Water-to-air heat pumps, 11.1, 11.2–4
Water-to-water heat pumps, 11.4–5
Watt/VAR strip chart recorder, 4.24
Wattmeters, 4.21–22
Watts, 4.16
Wedge belts, 3.29
Wet-bulb temperature, 1.28, 1.29, 1.37
Wetted-element humidifiers, 8.6–7
Windings
   checking motor for grounded windings, 4.25
   grounded windings, 7.48
   open windings, 7.47–48
   preventing damage to, 10.16
   shorted windings, 7.47–48
   transformers, 4.2, 4.3
   troubleshooting, 7.47–48
Wing nuts, 3.5, 3.6
Wire-wound resistor, 5.9
Wiring, thermostats, 7.8, 7.9
Wiring diagram, 7.28
   defined, 7.64
   heat pump, 7.52–53
   schematic symbols, 7.65–66
Woodruff key, 3.9
Work order form, 3.38, 3.40
Wrapped packing, 3.16
Wye-connected stator, 4.21
Wye-delta, 10.24
Wye transformer, 4.5

Zeotrope refrigerants, 12.29
ZIP® disks, 5.25
Zone valves, 7.51
Zoned control, 8.18–19